Zohra Bellahsène   Michel Léonard (Eds.)

Advanced Information Systems Engineering

20th International Conference, CAiSE 2008
Montpellier, France, June 16-20, 2008
Proceedings

Volume Editors

Zohra Bellahsène
LIRMM–Laboratoire d'Informatique, de Robotique
et de Microélectronique de Montpellier
UMR 5506 CNRS/Université Montpellier 2
161 Rue Ada, 34392 Montpellier, France
E-mail: bella@lirmm.fr

Michel Léonard
Université de Genève
Centre Universitaire d'Informatique
Département des Systèmes d'Information
24 Rue du Général Dufour, 1211 Genève 4, Switzerland
E-mail: michel.leonard@cui.unige.ch

Library of Congress Control Number: 2008929343

CR Subject Classification (1998): H.2, H.3-5, J.1, K.4.3-4, K.6, D.2, I.2.11

LNCS Sublibrary: SL 3 – Information Systems and Application, incl. Internet/Web and HCI

ISSN 0302-9743
ISBN-10 3-540-69533-8 Springer Berlin Heidelberg New York
ISBN-13 978-3-540-69533-2 Springer Berlin Heidelberg New York

Springer is a part of Springer Science+Business Media

springer.com

Typesetting: Camera-ready by author, data conversion by Scientific Publishing Services, Chennai, India
Printed on acid-free paper SPIN: 12279258 06/3180 5 4 3 2 1 0

Preface

CAiSE 2008 was the 20th in the series of International Conferences on Advanced Information System Engineering. This edition continued the success of previous conferences, a success largely due to that fact that, since its first edition, this series has evolved in parallel with the evolution of the importance of information systems in economic development. CAiSE has been able to follow, and often to anticipate, important changes that have occurred since 1978 when the first CAiSE conference was organized by Arne Sølvberg and Janis Bubenko.

In all these years, modern businesses and IT systems have been facing an ever more complex environment characterized by openness, variety and change. Furthermore, enterprises are experiencing ever more variety in their business in many dimensions. In the same way, the explosion of information technologies is overwhelming with a multitude of languages, platforms, devices, standards and products. Thus enterprises need to manage an environment to monitor the interplay of changes in the business processes, in information technologies, and at the ontological level, in order to achieve a *sustainable* development of their information systems. Enterprises must enter the era of *sustainable information systems* to face the important developmental challenges.

During all these years, CAiSE researchers have been challenged by all these changes, and the CAiSE conferences provide a forum for presenting and debating important scientific results. In fact, CAiSE is positioned at the core of these tumultuous processes, hosting new emerging ideas, fostering innovative processes of design and evaluation, developing new information technologies adapted to information systems, creating new kinds of models, but always being subject to rigorous scientific selection.

And so, the previous CAiSE conferences have largely contributed to developing a *sustainable* conceptual platform for information systems engineering, well suited to the era of sustainable information systems. This was the main theme of this conference.

CAiSE 2008 received 273 full paper submissions from all over the world. Each submitted paper underwent a rigorous review process by at least three independent referees. The CAiSE 2008 proceedings represent a collection of 44 excellent research papers: 35 full papers and 9 short papers. The selection was very hard, due to the very high standard of the submitted papers. Several high-quality papers were selected for the CAiSE forum to stimulate open discussions of high-quality on-going research. In addition, the conference program included three keynote speeches: "Evolvable Web Services" by Mike Papazoglou, Tilburg University, "Business Entities: Unifying Process and Data Modeling" by David Cohn, IBM Watson Research Center and "Information Systems in e-Government" by Jean-Marie Leclerc, Centre des Technologies de l'Information, of Geneva. Other highlights of CAiSE 2008 were 10 top-quality pre-conference workshops and a doctoral consortium bringing together PhD students to give them the opportunity to showcase their research and providing them with feedback from senior international researchers. Contact with industry was emphasized through a special think tank on advancing innovation skills for intensive information services in Europe.

As editors of this volume, we would like to express our gratitude to the program board, the Program Committee and external reviewers for their efforts in providing very thorough evaluations of the submitted papers under significant time constraints. We would like to thank the invited speakers and authors without whom this conference would not have been possible. We would also like to thank Richard van de Stadt for his very effective support during the paper evaluation and for preparing the proceedings. Moreover, our thanks go out to the local Organizing Committee who fulfilled with a lot of patience all our wishes. Finally, many thanks to Google, Microsoft, institutional ERCIM, UM2 and CNRS, local sponsors Languedoc-Roussillon Region and the city hall of Montpellier for their sponsorship.

April 2008

Zohra Bellahsène
Michel Léonard

Organization

Advisory Committee	Janis Bubenko Jr. Royal Institute of Technology, Sweden Colette Rolland Université Paris 1 Panthéon Sorbonne, France Arne Sølvberg Norwegian University of Science and Technology, Norway
General Chair	Michel Léonard University of Geneva, Switzerland
Program Chair	Zohra Bellahsène LIRMM-CNRS/Université Montpellier 2, France
Workshop Chairs	Xavier Franch Universitat Politècnica de Catalunya, Spain Ela Hunt ETHZ, Switzerland
Tutorial and Panel Chairs	Ann Persson University of Skovde, Sweden Camille Salinesi Université Paris 1 Panthéon Sorbonne, France
Publicity Chair	Selmin Nurcan Université Paris 1 Panthéon Sorbonne, France
Sponsorship Chair	Mark Roantree Dublin City University, Ireland
Forum Chair	Carson Woo University of British Columbia, Vancouver, Canada
Doctoral Consortium Chairs	Peter McBrien Imperial College of London, UK Farouk Toumani University of Clermont-Ferrand, France
Organization Chair	Rémi Coletta LIRMM-CNRS/Université Montpellier 2, France

Local Arrangements — Céline Berger
LIRMM-CNRS/Université Montpellier 2, France

Webmaster — Fabien Durchateau
LIRMM-CNRS/Université Montpellier 2, France

Program Committee Board

Hans Akkermans, Netherlands
Sjaak Brinkkemper, Netherlands
Eric Dubois, Luxembourg
Johann Eder, Austria
Pericles Loucopoulos, UK
Andreas Opdahl, Norway
Oscar Lopez Pastor, Spain
Barbara Pernici, Italy
Anne Persson, Sweden
Klaus Pohl, Germany
Colette Rolland, France
Pnina Soffer, Israel

Program Committee

Wil van der Aalst, Netherlands
Pär Ågerfalk, Ireland
Jacky Akoka, France
Marko Bajec, Slovenia
Boualem Benatallah, Australia
Nacer Boudjlida, France
Mokrane Bouzeghoub, France
Fabio Casati, Italy
Silvana Castano, Italy
Jaelson Castro, Brazil
Corinne Cauvet, France
Dov Dori, Israel
Marlon Dumas, Australia
David Embley, USA
Joerg Evermann, New Zealand
João Falcão e Cunha, Portugal
Xavier Franch, Spain
Agnès Front, France
Paolo Giorgini, Italy
Claude Godart, France
Jaap Gordijn, Netherlands
Mohand-Said Hacid, France
Terry Halpin, USA
Manfred Hauswirth, Ireland
Patrick Heymans, Belgium
Matthias Jarke, Germany
Manfred Jeusfeld, Netherlands
Paul Johannesson, Sweden
Henk Jonkers, Netherlands
Håvard Jørgensen, Norway
Roland Kaschek, New Zealand
Marite Kirkova, Latvia
John Krogstie, Norway
Réginene Laleau, France
Marc Lankhorst, Brazil
Julio Leite, Brazil
Kalle Lyytinen, USA
Neil Maiden, UK
Peter McBrien, UK
Isabelle Mirbel, France
Michele Missikoff, Italy
Haris Mouratidis, UK
John Mylopoulos, Canada
Moira Norrie, Switzerland
Andreas Oberweis, Germany
Antoni Olivé, Spain
Jeffrey Parsons, Canada
Michaël Petit, Belgium
Yves Pigneur, Switzerland
Gert Poels, Belgium
Naveen Prakash, India
Erik Proper, Netherlands
Jolita Ralyte, Switzerland
Björn Regnell, Sweden

Additional Referees

Algirdas Laukaitis
Dejan Lavbic
Duc Minh Le
Massimiliano de Leoni
Mario Lezoche
Chen Li
Sebastian Link
Deryle Lonsdale
Giusy Di Lorenzo
Davide Lorusso
Nikos Loutas
Jenny Eriksson Lundström
Linh Thao Ly
Bernadette Farias Lóscio
Kreshnik Musaraj
Amel Mammar
Michele Mancioppi
Raimundas Matulevicius
Michele Melchiori
Jan Mendling
Harald Meyer
Wai Mok
Geert Monsieur
Stefano Montanelli
Kafui Monu
Esmiralda Moradian
Joao M. Moreira
Dominic Mueller
Lina Nemuraite
Antonio De Nicola
Moses Niwe
Michael Pantazoglou
Mike Papazoglou
Joan A. Pastor-Collado
Veronika Peralta
Thomi Pilioura
Axel Polleres
Elaheh Pourabbas
Jorge Quiane
Paul Ralph
Zornitza Rasheva
Jan Recker
Nikos Rizopoulos
Lotte De Rore
Seung Ryu
Ana Carolina Salgado
Yacine Sam
Ana Sasa
Germain Saval
Alexander Schutz
Farida Semmak
Carla Silva
Sase N. Singh
Andrew Charles Smith
Mehdi Snene
Carine Souveyet
Nikhil Srinivasan
Richard Berntsson Svensson
Francesco Taglino
Roberto Santana Tapia
Christer Thörn
Yuri Tijerino
Leonardo Tininini
Hubert Toussaint
Jean-Christophe Trigaux
Christina Tsagkani
Damjan Vavpotic
Luis Veiga
Barbara Weber
Krzysztof Wnuk
Maciej Zaremba
Jelena Zdravkovic

Gold Sponsors

Google Microsoft®

Institutional Sponsor

Local Sponsors

Laboratoire
d'Informatique
de Robotique
et de Microélectronique
de Montpellier

Table of Contents

Keynote

Duality and Process Modeling

Interoperability of IS and Enterprises

Refactoring

Information Systems in e-Government and Life-Science

Knowledge Patterns for IS Engineering

Requirements Engineering for IS

Conceptual Schema Modeling

Service Infrastructure

Service Evolution

Flexible Information Technologies

Metrics and Process Modelling

Information Systems Engineering

IS Development with Ubiquitous Technologies

The Challenges of Service Evolution*

Mike P. Papazoglou

INFOLAB, Dept. of Information Systems and Mgt., Tilburg University, The Netherlands
mikep@uvt.nl

Abstract. Services are subject to constant change and variation. Services can evolve typically due to changes in structure, e.g., attributes and operations; in behavior and policies, e.g., adding new business rules and regulations, in types of business-related events; and in business protocols. This paper introduces two types of service changes: shallow changes - where changes are confined to services or the clients - and deep changes - where cascading effects and side-effects occur. The paper introduces a theoretical approach for dealing with shallow service changes and a change-oriented service lifecycle methodology that addresses the effects of deep service changes.

Keywords: Web services, service versioning, business protocols, regulatory compliance, service contracts and policies. service contracts.

1 Introduction

Serious challenges like mergers and acquisitions, outsourcing possibilities, rapid growth, the need for regulatory compliance, and intense competitive pressures are overtaxing existing traditional business processes, slowing innovation and making it difficult for an enterprise to pursue and reach its business strategies and objectives. Such challenges require changes at the enterprise-level and thus lead to a continuous business process redesign and improvement effort.

Routine process changes usually lead to possible reorganization and realignment of many businesses processes and increase the propensity for error. To control process development one needs to know why a change was made, what are its implications and whether the change is complete. Eliminating spurious results and inconsistencies that may occur due to uncontrolled changes is therefore a necessary condition for the ability of processes to evolve gracefully, ensure stability and handle variability in their behavior. Such kind of changes must be applied in a controlled fashion so as to minimize inconsistencies and disruptions by guaranteeing seamless interoperation of business processes that may cross enterprise boundaries when they undergo changes.

* The research leading to these results has received funding from the European Community's Seventh Framework Programme under the Network of Excellence S-Cube - Grant Agreement no. 215483.

Z. Bellahsène and M. Léonard (Eds.): CAiSE 2008, LNCS 5074, pp. 1–15, 2008.

Service technologies automate business processes[1] and change as those processes respond to changing consumer, competitive, and regulatory demands. Services are thus subject to constant adaptation and variation adding new business rules and regulations, types of business-related events, operations and so forth. Services can evolve typically by accommodating a multitude of changes along the following functional trajectories:

1. *Structural changes*: These focus on changes that occur on the service types, messages, interfaces and operations.
2. *Business protocol changes*: Business protocols specify the external messaging behavior of services (viz. the rules that govern the service interaction between service providers and clients) and, in particular, the conversations in which the services can participate in. Business protocols achieve this by describing the structure and the ordering (time sequences) of the messages that a service and its clients exchange to achieve a certain business goal. Business protocols change due to changes in policies, regulations, and changes in the operational behavior of services.
3. *Policy induced changes*: These describe changes in policy assertions and constraints on the service, which prescribe, limit, or specify any aspect of a business agreement that is possible agreed to among interacting parties. Policies may describe constraints external to constraints agreed by interacting parties in a transaction and include universal legal requirements, commercial and/or international trade and contract terms, public policy (e.g., privacy/data protection, product or service labeling, consumer protection), laws and regulations that are applicable to parts of a business service. For instance, a procurement processes can codify an approval process in such a way that it can be instantly modified as corporate policies change. In most cases existing processes need to be redesigned or improved to conform with new corporate strategies and goals.
4. *Operational behavior changes:* These concentrate on analyzing the effects and side (cascading) effects of changing service operations. If, for example, we consider an order management service we might expect to see a service that lists "place order", "cancel-order," and "update order," as available operations. If now the "update-order" operation is modified in such a way that it includes available-to-promise functionality that dynamically allocates and reallocates resources to promise and fulfill customer orders, the modified operation must guarantee that if part of the order is outsourced to a manufacturing partner, the partner can fulfill its order on time to meet agreed upon shipment dates. This requires understanding of where time is consumed in the manufacturing process, what is normal with respect to an events timeliness to the deadline, and to understand standard deviations with respect to that process events on-time performance.

[1] We shall henceforth use the generic term service to refer to both services and business process. If there is a need to discriminate between simple services and fairly complex services, we shall use the terms singular service and business process, respectively.

We can classify the nature of service changes depending on the effects and side effects they cause. We may thus distinguish between two kinds of service changes:

Shallow changes: Where the change effects are localized to a service or are strictly restricted to the clients of that service.

Deep changes: These are cascading types of changes which extend beyond the clients of a service possibly to entire value-chain, i.e., clients of these service clients such as outsourcers or suppliers.

Typical shallow changes are changes on the structural level and business protocol changes, while typical deep changes include operational behavior changes and policy induced changes.

While shallow changes need an appropriate versioning strategy, deep changes are quite intricate and require the assistance of an *change-oriented service life cycle* where the objective is to allow services to predict and respond appropriately to changes as they occur. A change-oriented service life cycle provides a foundation for business process changes in an orderly fashion and allow end-to-end services to avoid the pitfalls of deploying a maze of business processes that are not appropriately (re)-configured, aligned and controlled as changes occur. The practices of this methodology are geared to accepting continuous change for business processes as the norm.

In addition to functional changes, a change-oriented service life cycle must deal with non-functional changes which are mainly concerned with end-to-end QoS (Quality of Service) issues, and SLA (Service Level Agreement) guarantees for end-to-end service networks. The objective is to achieve actual end-to-end QoS capabilities for a service network to achieve the proper levels of service required by ensuring that services are performing as desired, and that out-of-control or out-of-specification conditions are anticipated and responded to appropriately. This includes traditional QoS capabilities, e.g., security, availability, accessibility, integrity and transactionality, as well as service volumes and velocities. Service volumes are concerned with values and counts of different aspects of the service and its associated transactions, e.g., number of service events, number of items consumed, service revenue, number of tickets closed, service costs. The general performance of the service is related to service velocity i.e., the time-related aspect of business operations, such as service cycle-time, cycle-times of individual steps, round trip delays, idle times, wait-times between events, time remaining to completion, service throughput, life-time of ticket, and so on. The combination of these time-related measurements with the value-related ones provides all the information needed to understand how an enterprise is is performing in terms of its services.

The issue of service evolution and change management is a complicated one, and this paper does not attempt to cover every aspect surrounding the evolution of services. However, it introduces some key approaches and helpful practices that can be used as a springboard for any further research in service evolution. In particular, in this paper we shall concentrate only on the impact of functional services changes as they constitute a precursor to understanding non-functional service changes which are still very much an open research problem that also deserves research scrutiny.

2 Dealing with Shallow Changes

Shallow changes characterize both singular services and business processes and require a structured approach and robust versioning strategy to support multiple versions of services and business protocols. To deal with shallow changes we introduce a theoretical approach for structural service changes focusing on service compatibility, compliance, conformance, and substitutatbility. In addition, we describe versioning mechanisms developed in [1] to handle business protocol changes.

2.1 A Theory for Structural Changes

Service based applications may typically fail on the service client side due to changes carried out during the provider service upgrade. To manage changes as a whole, service clients have to be taken into consideration as well, otherwise changes that are introduced at the service provider side can create severe disruption.

In this paper, we use the term *service evolution* to refer to the continuous process of development of a service through a series of consistent and unambiguous changes. The evolution of the service is expressed through the creation and decommission of its different *versions* during its lifetime. These versions have to be aligned with each other in a way that would allow a service designer to track the various modifications and their effects on the service.

A robust versioning strategy is needed to support multiple versions of services in development. This can allow for upgrades and improvements to be made to a service, while continuously supporting previously released versions. To be able to deal with message exchanges between a service provider and a service client despite service changes that may happen to either of their service definitions (at the schema-level), we must introduce the notion of service compatibility.

Version compatibility: Is when we can introduce a new version of either a provider or client of service messages without changing the other. There are two types of changes to a service definition that can guarantee version compatibility [2]:

Backward compatibility: A guarantee that when a new version of a message client is introduced the message providers are unaffected. The client may introduce new features but should still be able to support all the old ones.

Forward compatibility: A guarantee that when a new version of a message provider is introduced the message clients who are only aware of the original version are unaffected. The provider may have new features but should not add them in a way that breaks any old clients. The assumption that underlies this definition of forward-chaining is that there is no implicit or explicit shared knowledge between the provider and the client.

Some types of changes that are both backwards- and forwards-compatible include: addition of new service operations to an existing service definition, addition of new schema elements within a service that are not contained within previously existing types. However, there are a host of other change types that are incompatible. These include: removing any existing operations, elements or attributes, renaming an operation, changing the parameters (in data type or order) of an operation and changing the structure of a complex data type.

The above definition of backward-compatibility misses the subtle possibility that a new version of a client might have a requirement to add a new feature where the client needs to reject messages that might have previously been acceptable by the previous version of the client. Consider, for instance adding a security feature for the new version of the client that rejects all messages, even if they were accepted by its previous version, unless these messages are encrypted and digitally signed. In addition, the notion of forward-compatibility is so strict that a new version of the provider is not allowed to produce any new messages that were not already produced by the old version of the provider to guarantee version consistency and type safeness!

To alleviate these problems we require an agreement between service providers and clients in the form of a shared contract.

Service Contracts

For two services to interact properly, before a service provider can provide whatever service it offers, they must come must come to an agreement or contract. A contract formalizes the details of a service (contents, price, delivery process, acceptance and quality criteria, expected protocols for interaction on behalf of the client) in a way that meets the mutual understandings and expectations of both the service provider and the service client. Introducing the notion of a service contract gives us a mechanism that can be used to achieve meaningful forward compatibility.

A service contract specifies [3]:

Functional requirements which detail the operational characteristics that define the overall behavior of the service, i.e., details how the service is invoked and what results it returns, the location where it is invoked and so on.

Non-functional requirements which detail service quality attributes, such as service metering and cost, performance metrics, security attributes, (transactional) integrity, reliability, availability, and so on.

Rules of engagement between clients and providers, known as *policies*, that govern who can access a provider, what security procedures the participants must follow, and any other rules that apply to the exchange. A point of clarity is the difference between contract and policy. A policy is a set of conditions that can apply to any number of contracts. For example, a policy can range from simple expressions informing a client about the security tokens that a service is capable of processing (such as Kerberos tickets or X509 certificates) to a set of rules evaluated in priority order that determine whether or not a client can interact with a service provider.

Given that clients may vary just as much as providers, there might be multiple contracts for a single service. In what follows and for the sake of clarity we will focus only on functional requirements. Nevertheless the provided reasoning can be generalized.

Definition 1. *Contract R, is a collection of elements that are common across the provider P and the consumer C of the service. It represents the mutually agreed upon service schema elements that are expected by the consumer and offered by the producer.*

1. Let's denote by P the set of elements (including operations) produced by the provider where $P = \{x_i, i \geq 1\}$
2. Let's denote by C the set of elements required by the client where $C = \{y_j, j \geq 1\}$
3. Let's define a partial function θ, called a *contract-binding*, that maps a set of P elements and a set of C consumed by the client, $\theta = \{\exists x_i \in P \wedge \exists y_j \in C \mid (x_i, y_j)\}$, which means the client consumes the element y_j for the operation x_i provided by the provider and $R = P_\theta(R) \cup C_\theta(R)$.

Service Compatibility

Definition 2. *Two contracts R and R' are called* backwards compatible *iff* $\forall x_i \in P_\theta(R), \exists y_k \in C_\theta(R') \mid (x_i, y_k)$.

The previous definition implies that the contract-binding is still valid despite changes in the client-side.

Definition 3. *Two contracts R and R' are called* forwards compatible *iff* $\forall y_k \in C_\theta(R), \exists x_i \in P_\theta(R') \mid (x_i, y_k)$.

The previous definition implies that the contract-binding is still valid despite changes in the provider-side.

Definition 4. *Two contracts R and R' are called* (fully) compatible *iff they are both forwards and backwards compatible, i.e. it holds that:*
$\{\forall x_i \in P_\theta(R), \exists y_k \in C_\theta(R') \mid (x_i, y_k)\} \wedge \{\forall y_k \in C_\theta(R), \exists x_i \in P_\theta(R') \mid (x_i, y_k)\}$.

Fig. 1 illustrates a contract R and its associated contract bindings.

Service Compliance

Since a version v_i^s of a service participates in a number $n, n > 0$ of relationships (either as a producer or as a consumer of other services), then it defines with them n contracts: $R_{i,k}, k = 1, \ldots, n$. Based on that we can define the notion of full compliance:

Definition 5. *Two versions v_i^s and $v_j^s, j > i$ of a service are called* compliant *iff $\forall k, k = 1, \ldots, n, R_{i,k}$ and $R_{j,k}$ are compatible.*

Compliance as we have defined it takes into account only the contracts for which the service acts as a producer. That reflects the fact the service can reconfigure itself as long as its actions do not affect its consumers.

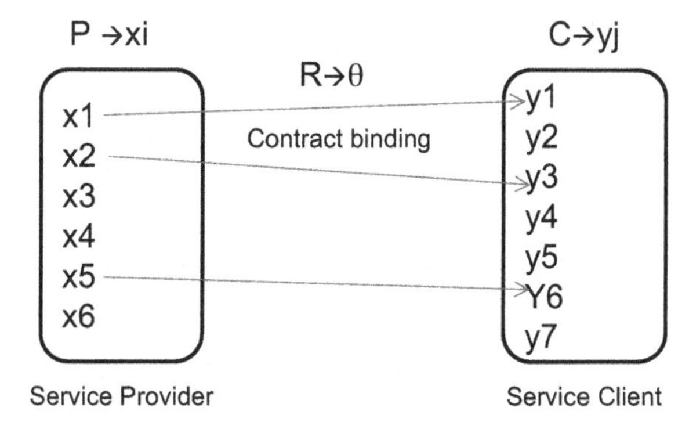

Fig. 1. Contracts and contract bindings

Service Conformance

In addition to the notion of contracts service evolution requires dealing with service arguments and return values. We need to make sure that the new version of a service can substitute an older version without causing any problems to its service clients. To guarantee service substitutatbility we rely on the notion of service conformity.

Informally, a type S conforms to a type T (written $S \triangleright T$) if an object of type S can always be substituted for one of type T, that is, the object of type S can always be used where one of type T is expected. For S to be substitutable for T requires that:

1. S provides at least the operations of T (S may have more operations).
2. For each operation in T, the corresponding operation in S has the same number of arguments and results.
3. The types of the results of the operations of S conform to the types of the results of the operations of T. The principal problem to consider is what happens to the output parameters when redefining a service. To achieve the desired effect we rely on the notion of *result-covariance* [4], [5]. Covariance states that if a method M is defined to return a result of type T, then an invocation of M is allowed to return a result of any subtype of T. A covariant rule requires that if we redefine the result of a service the new result type must always be a restriction (specialization) of the original one.
4. The types of the arguments of the operations of T conform to the types of the arguments of the operations of S. The principal problem to consider is what happens to the arguments when redefining a service. To achieve the desired effect we rely on the notion of *argument-contravariance* [4], [5]. Contravariance states that if a method M is defined with an argument of type T, then an invocation of M is allowed with an argument that is a supertype of T. A contravariant rule requires that if we redefine the argument of a service the new result type must always be an extension (generalization) of the original one.

The core of argument contravariance and result covariance concerns methods that have a functional type can be explained as follows.

Definition 6. *A subtyping relationship between functional types can be defined as follows [5], [6]: if* $T1 \leq S1$ *and* $S2 \leq T2$ *then* $S1 \rightarrow S2 \leq T1 \rightarrow T2$*, where we consider the symbol "* $\rightarrow$ *" as a type constructor.*

Assume we expect a function or method f to have type $T1 \rightarrow T2$ and therefore consider $T1$ arguments as permissible when calling f and results of type $T2$. Now if assume that f actually has type $T1' \rightarrow T2'$ with $T1 \leq T1'$. Then we can pass all the expected permissible arguments of type $T1$ without type violation; f will return results of type $T2'$ which is permissible if $T2' \leq T2$ because the results will then also be of type $T2$ and are therefore acceptable as they do not introduce any type violations.

Covariance and contravariance are not opposing views, but distinct concepts that each have their place in type theory and are both integrated in a type-safe manner in object-oriented languages [6], [5]. Argument contravariance and result covariance is required for safely substituting older service with newer service versions.

To be *fully substitutable* a service must be both compliant and conformant according the previous definitions. If we now assume without loss of generality that for $k = 1, \ldots, m, m < n$ the service participates in the contracts $R_{i,k}$ only as a producer. Then, compliance can be alternativetely defined as:

Definition 7. *Two versions* v_i^s *and* $v_j^s, j > i$ *of a service are called* fully substitutable *iff* $\forall k, k = 1, \ldots, m$*, for which the service participates as a producer,* $R_{i,k}$ *and* $R_{j,k}$ *are compatible.*

Web Service Versioning

Compatible service evolution in WSDL 2.0 limits service changes that are either backward or forward compatible, or both [7]. In accordance with the definitions in section-2.1 WSDL-conformant services are backward compatible when the receiver behaves correctly if it receives a message in an older version of the interaction language, while WSDL-conformant services are forward compatible the receiver behaves correctly if it receives a message in a newer version of the interaction language.

The types of service changes that are compatible are:

- Addition of new WSDL operations to an existing WSDL document. If existing clients are unaware of a new operation, then they will be unaffected by its introduction.
- Addition of new XML schema types within a WSDL document that are not contained within previously existing types. Again, even if a new operation requires a new set of complex data types, as long as those data types are not contained within any previously existing types (which would in turn require modification of the parsing code for those types), then this type of change will not affect an existing client.

However, there are a host of other change types that are incompatible. These include: removing an operation, renaming an operation, changing the parameters (in data type or order) of an operation, and changing the structure of a complex data type.

With a compatible change the service need only support the latest version of a service. A client may continue to use a service adjusting to new version of the interface description at a time of its choosing. With an incompatible change, the client receives a new version of the interface description and is expected to adjust to the new interface before old interface is terminated. Either the service will need to continue to support both versions of the interface during the hand over period, or the service and the clients are coordinated to change at the same time. An alternative is for the client to continue until it encounters an error, at which point it uses the new version of the interface.

2.2 Business Protocol Changes

Business protocol descriptions can be important in the context of change management as protocols also tend to evolve over time due to the development of new applications, new business strategies, changing compliance and quality of service requirements, and so on. Business protocol evolution is considered in [1] where the authors distinguish between two aspects of protocol evolution:

1. Static protocol evolution which refers to the problem of modifying the protocol definition by providing a set of change operations that allow the gradual modification of an existing protocol without the need of redefining it from scratch.
2. Dynamic protocol evolution which refers to the issue of changing a long running protocol in the midst of its execution to a new protocol. In such cases, there is a clear need for providing mechanisms for a protocol to migrate active instances running under a old protocol version to meet the new protocol requirements.

Fig. 2 illustrates the various aspects of protocol changes. In particular, it shows the notions of protocol versioning, migration, compatibility and protocol replaceability.

When evolving a protocol, it is useful to keep track of all protocol versions, revisions and variants of the protocolI. Instances of an older protocol version might still be running and used by clients, which in turn can depend on these instances. When evolving a protocol, states and transitions may be added to and removed from an active protocol. A new version of a protocol is created each time its internal structure or external behavior change. The perception that clients have of a specific protocol is called a *protocol view*. Since the client's view of a protocol is restricted only to the parts of the protocol that directly involve the client, a client might have equivalent views on different protocols. In mathematical terms, a view is a one–to–many mapping from protocols as perceived by the client, to actual protocols. Protocol views are related to many practices revolving around business protocols. For instance, clients whose views on the original and target protocols are the same, are not affected by migration practices.

Migration [1] defines the strategies adopted to implement protocol evolution, by guiding the process of changing running protocol instances. The typical options for a migration are: do not migrate any running instance, terminate all the running instances, migrate all the running instances to the new protocol, migrate instances that cannot be migrated to the new protocol, to a temporary protocol which complies to the new requirements.

Protocol compatibility (see Fig. 2) aims at assessing whether two protocols can interact, i.e. if it is possible to have a conversation between the two services despite changes to the protocol. Compatibility of two protocols can be either complete, i.e., all conversations of one protocol can be understood by the other protocol, or partial, when there is at least one conversation possible between the two protocols. Protocol revision takes place when a new protocol version is

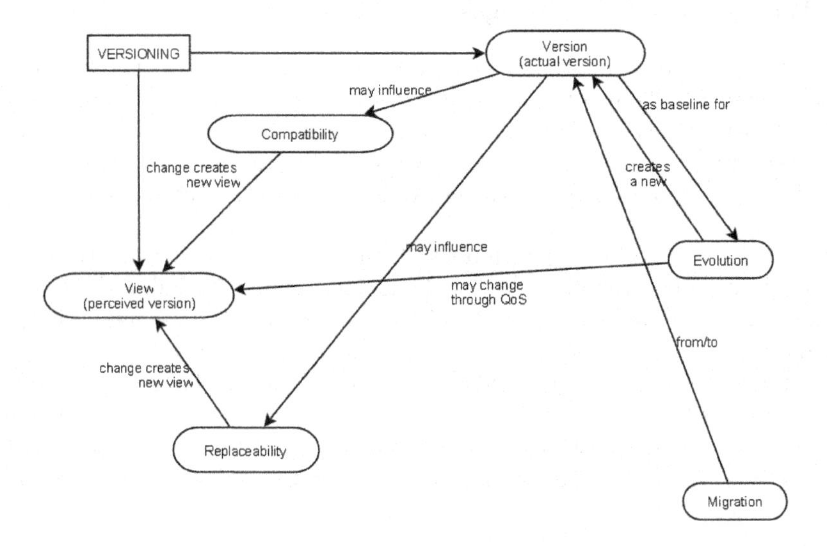

Fig. 2. Business Protocol changes

meant to supersede its predecessor. *Protocol replaceability* (see Fig. 2) deals with the problem of determining if two protocols are equivalent and which parts they have in common. The following classes of replaceability can be distinguished [1]:

- Protocol equivalence when two protocols can be used interchangeably;
- protocol subsumption when one protocol can be used to replace the other, but not vice-versa.

3 Dealing with the Effects of Deep Changes

Deep changes characterize only business processes and require that a business process be redefined and realigned within an entire business process value chain.

This may eventually lead to modification and alignment of business processes within a business process value chain associated directly or indirectly with a business-process-in-scope.This calls for methodologies to provide a sound foundation for deep service changes in an orderly fashion that allow services to be appropriately (re)-configured, aligned and controlled as changes occur. In Fig. 3 we provide an overview of the major phases in a change-oriented service life cycle. Different methodologies may subdivide the phases in a different manner, but the sequence is usually the same.

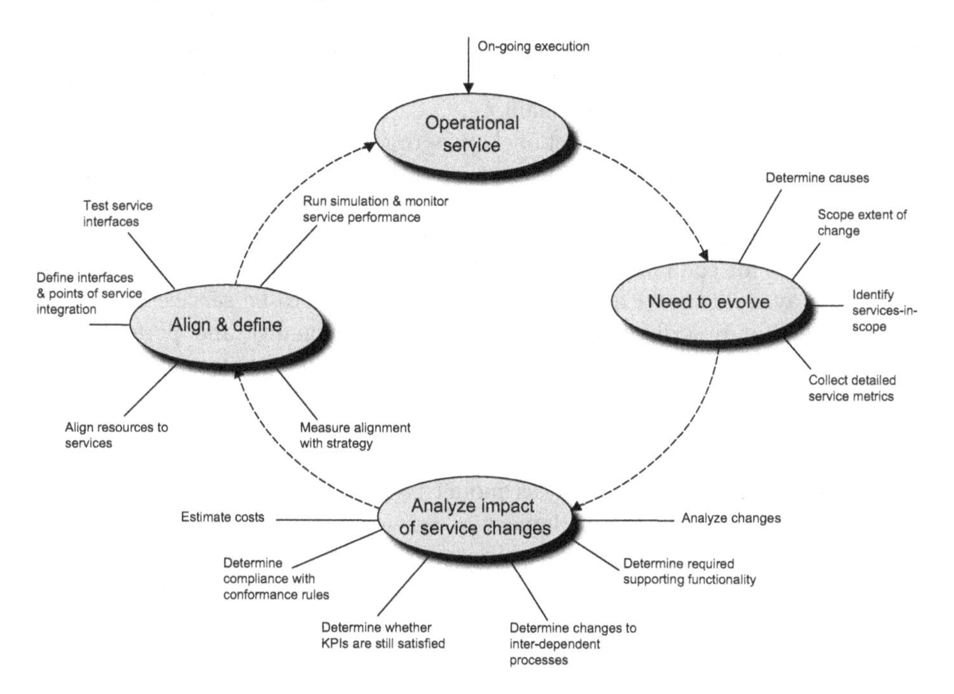

Fig. 3. Change-oriented service life cycle

The initial phase focuses on identifying the need for change and scoping its extent. One of the major elements of this phase is understanding the causes of the need for change and their potential implications. For instance, compliance to regulations is major force for change. Regulatory requirements such as HIPAA and Sarbanes-Oxley provide strict guidelines that ensure companies are in control of internal, private, public, and confidential information, and auditing standards such as SAS 70 serve as a baseline for regulatory compliance by verifying that third-party providers meet those needs. All of this may lead to the transformation of services within a business process value chain. Here, the affected services-in-scope need to be identified. In addition, service performance metrics, such as KPIs and SLAs, need to be collected. Typical KPIs include delivery performance, fill rates order fulfillment, production efficiency and flexibility, inventory days of supply, quality thresholds, velocity, transaction volumes and cost baseline. These assist in understanding the nature of services-in-scope and related services and provide a baseline for comparative purposes and determination of expected productivity, cost and service level improvements.

The second phase in Fig. 3, called *service change analysis*, focuses on the actual analysis, redesign or improvement of the existing services. The ultimate objective of service change analysis is to provide an in-depth understanding of the functionality, scope, reuse, and granularity of services that are identified for change. To achieve its objective, the analysis phase encourages a more radical view of process (re)-design and supports the re-engineering of services. Its main objective is the reuse (or repurposing) of existing service functionality in to meet the demands of change. The problem lies in determining the difference between existing and future service functionality.

To analyze and assess the impact of changes organizations rely on the existence of an "as-is" and a "to-be" service model rather than applying the changes directly on operational services. Analysts complete an as-is service model to allow understanding the portfolio of available services. The as-is service model is used as basis for conducting a thorough re-engineering analysis of the current portfolio of available services that need to evolve. The to-be services model is used as basis for describing the target service functionality and performance levels after applying the required changes. One usually begins by analyzing the "as-is" service, considering alternatives, and then settling on a "to-be" service that will replace the current service.

To determine the differences between these two models a gap analysis technique must be used. A gap analysis model is used to help set priorities and improvements and measure the impact of service changes. Gap analysis is a technique that purposes a services realization strategy by incrementally adding more implementation details to an existing service to bridge the gap between the "as-is" and "to-be" service models. Gap analysis commences with comparing the "as-is" with the "to-be" service functionality to determine differences in terms of service performance (for instance, measures of KPIS) and capabilities. Service capabilities determine whether a process is able to meet specifications, customer requirements, or product tolerances.

As service changes may spill over to other services in a supply-chain, one of the determining factors in service change analysis is being able to recognize the scope of changes and functionality that is essentially self-sufficient for the purposes of a service-in-scope (service under change). When dealing with deep service changes, problems of overlapping or conflicting functionality several types of problems need to be addressed [8] and [3]:

1. *Service flow problems*: Typical problems include problems with the logical completeness of a service upgrade, problems with sequencing and duplication of activities, decision-making problems and lack of service measures. Problems with the logical completeness of a service upgrade include disconnected activities and disconnected inputs or outputs. Problems with sequencing and duplication of activities include activities that are performed in the wrong sequence, performed more than once, and, in general the lack of rules that

prioritize flows between activities. Decision-making problems include the lack of information, such as policies and business rules, for making decisions. Lack of service measures include inadequate or no measures for the quality, quantity or timeliness of service outputs.

2. *Service control problems:* Service controls define or constrain how a service is performed. Broadly speaking there are two general types of control problems: problems with policies and business rules and problems with external services. Problems with policies and business rules include problems where a service-in-scope ignores organizational policies or specific business rules. Problems with external services include problems where external services require information that a service-in-scope cannot provide. Alternatively, they include cases where information that a service-in-scope requires cannot be provided by external services.
3. *Overlapping services functionality:* In such cases a service-in-scope may (partially) share identical business logic and rules with other related services. Here, there is a need for rationalizing services and determining the proper level of service commonality. Overlapping functionality should be identified and should be factored out. Several factors such as encapsulated functionality, business logic and rules, business activities, can serve to determine the functionality and scope of services. During this procedure, service design principles [3] such as service coupling and cohesion need to be employed to achieve the desired effects.
4. *Conflicting services functionality* (including bottlenecks / constraints in the service value stream): During this step the functionality of a service-in-scope may conflict with functionality in related services. Conflicts also include problems where a service-in-scope is not aligned to business strategy, where a service may pursue a strategy that is in conflict with is incompatible with the value chain of which it is a part, and cases where the introduction of a new policy or regulation would make it impossible for the service-in-scope to function. In addition to dealing with problems arising from overlapping and conflicting service functionality we should also unbundle functionality into separate services to the extend possible to prevent services from becoming overly complex and difficult to maintain.
5. *Service input and output problems:* These problems include problems where the quality of service input or output is low, and timeliness input or output problems where the needed inputs/outputs are not produced when they are needed.

Finally, cost estimation in the second phase involves identifying and weighing all services to be re-engineered to estimate the cost of the re-engineering project. In cases where costs are prohibitive for an in-house implementation, an outsourcing policy might be pursued.

During the service change analysis standard continuous process improvement practices such as Six Sigma DMAIC practices or Lean Kaizen [9] should be employed. These determine the services changes and define the new services and standards of performance to measure, analyze, control and systematically improve processes by eliminating potential defects.

During the third and final phase, all of the new services are aligned, integrated, simulated and tested and then, when ready, the new services are put into production and managed. To achieve this a *services integration model* [3] is created to facilitate the implementation of the service integration strategy. This strategy includes such subjects as service design models, policies, SOA governance options, and, organizational and industry best practices and conventions. All these need to be taken into account when designing integrated end-to-end services that span organizational boundaries.

A service integration model, among other things, establishes integration relationships between service consumers and providers involved in business interactions, e.g., business transactions. It determines service responsibilities, assigns duties to intermediaries who perform and facilitate message interception, message transformation, load balancing, routing, and so on. It also includes steps that determine message distribution needs, delivery-responsible parties, and provides a service delivery map. Finally, a service integration model is concerned with message and process orchestration needs. This part includes steps that establish network routes; verify network and environment support, e.g., validate network topology and environmental capacity as well as routing capabilities; and, employ integration flow patterns to facilitate the flow of messages and transactions.

The role of the services integration model ends when a new (upgraded) service architecture is completely expressed and validated against technological specifications provided by infrastructure, management/monitoring and technical utility services.

4 Summary

Services are subject to constant change and variation. Services can evolve typically due to changes in structure, e.g., attributes and operations; in operational behavior and policies, e.g., adding new business rules and regulations, in types of business-related events; and in business protocols.

We may distinguish between two kinds of service changes shallow versus deep service changes. With shallow changes the change effects are localized to a service or are strictly restricted to the clients of that service. Deep changes cause cascading types of changes which extend beyond the clients of a service possibly to entire value-chain, i.e., clients of these service clients such as outsourcers or suppliers. Typical shallow changes are changes on the structural level and business protocol changes, while typical deep changes include operational behavior changes and policy induced changes.

Shallow changes characterize both singular services and business processes and require a structured approach and robust versioning strategy to support multiple versions of services and business protocols. To deal with shallow changes we introduced a theoretical approach for structural service changes focusing on service compatibility, compliance, conformance, and substitutability. In addition, we described versioning mechanisms for handling business protocol changes. The right versioning strategy can maximize code reuse and provide a more manageable approach to the deployment and maintenance of services and protocols. It can allow for upgrades and improvements to be made to a service or protocol, while supporting previously released versions.

Deep changes characterize only business processes and require that a business process be redefined and realigned within an entire business process value chain. This may eventually lead to modification and alignment of business processes within a business process value chain associated directly or indirectly with a business-process-in-scope.To address these problems we introduced a change-oriented service life cycle methodology. A change-oriented service life cycle provides a sound foundation for deep service changes in an orderly fashion that allow services to be appropriately (re)-configured, aligned and controlled as changes occur. A change-oriented service life cycle also provides common tools to reduce cost, minimize risk exposure and improve development agility. It helps organizations ensure that the right versions of the right processes are available at all times, and that they can provide an audit trail of changes across the service lifecycle to prevent application failures and help meet increasingly stringent regulatory requirements.

Acknowledgments. I wish to thank Salima Benbernou for her help and invaluable suggestions that have considerably improved the theoretical approach for structural service changes.

References

1. Ryu, S.H., et al.: Supporting the dynamic evolution of web service protocols in service-oriented architecturesl. ACM Transactions on the Web 1(1), 1–39 (2007)
2. Orchard, D. (ed.): Extending and versioning languages. W3C Technical Architecture Group (2007)
3. Papazoglou, M.P.: Web Service: Principles and Technology. Prentice-Hall, Englewood Cliffs (2007)
4. Meyer, B.: Object-Oriented Software Construction, 2nd edn. Prentice-Hall, Englewood Cliffs (1997)
5. Castagna, G.: Covariance and contravariance: conflict without a cause. ACM Transactions on Programming Languages and Systems 17(3), 431–447 (1995)
6. Liskov, B., Wing, J.: A behavioral notion of subtyping. ACM Transactions on Programming Languages and Systems 16(6), 1811–1841 (1994)
7. Booth, D., Liu, C.K.: Web services description language (WSDL) version 2.0 part 0: Primer (2007)
8. Meyer, B.: Business Process Change. Morgan Kaufmann, San Francisco (2007)
9. Martin, J.: Lean Six Sigma for Supply Chain Management. McGraw-Hill, New York (2007)

Assigning Ontology-Based Semantics to Process Models: The Case of Petri Nets

Pnina Soffer[1], Maya Kaner[2], and Yair Wand[3]

[1] University of Haifa, Carmel Mountain 31905, Haifa, Israel
[2] Ort Braude College, Karmiel 21982, Israel
[3] Sauder School of Business, The University of British Columbia, Vancouver, Canada
spnina@is.haifa.ac.il, kmaya@braude.ac.il, yair.wand@ubc.ca

Abstract. Syntactically correct process models are not necessarily meaningful or represent processes that are feasible to execute. Specifically, when executed, the modeled processes might not be guaranteed to reach their goals. We propose that assigning ontological semantics to process modeling constructs can result in more meaningful models. Furthermore, the ontological semantics can impose constraints on the allowed process models which in turn can provide rules for developing process models. In particular, such models can be designed to be valid in the senses that the process can accomplish its goal when executed. We demonstrate this approach for Petri Net based process models.

1 Introduction

Process modeling is a complicated task and, hence, error-prone (e.g., [7][10]). Much effort has been devoted to the verification of process models leading to methods and tools for analyzing structural properties of process models and for detecting logical problems in them. These approaches are applied to already developed models, but do not provide guidance on how to develop valid models.

The syntax of process modeling languages specifies how to compose their constructs (which often have graphical notation) into process models. The semantics is believed to represent some real-world phenomena. These languages are usually defined textually or mathematically. Textual definitions are typically semi-formal or informal (e.g., "An event is something that "happens" during the course of a business process." [8]). Mathematical definitions can support precise analysis of models.

Syntactically correct process models are not necessarily meaningful or feasible to execute. This entails the need for checking completed process models for structural and behavioral properties related to whether they can be successfully executed or not.

Some research evaluated process modeling languages by mapping their constructs to ontological concepts (which are assumed to convey real-world semantics) [9]. These attempts revealed various deficiencies such as ontological incompleteness, construct overload, redundancy, and excess. In particular, no ontological meaning was identified for control flow constructs which exist in practically every process modeling language (typically manifested as splitting and merging elements).

Recently, the Generic Process Model (GPM) was used to suggest an interpretation of control flow structures [12]. GPM provides a process specification semantics based

Z. Bellahsène and M. Léonard (Eds.): CAiSE 2008, LNCS 5074, pp. 16–31, 2008.

on ontological constructs. It is intended as a framework for reasoning about process models in terms of their real-world meaning. To apply the GPM for this purpose, its constructs should be mapped to the modeling languages used, which often employ graphical notation easy for human use. Ontology-based semantics imposes modeling rules in addition to the language-based syntactical restrictions. We suggest that these rules can guide the construction of meaningful and feasible process models.

In this paper we demonstrate the use of ontological semantics for Petri net based process models, or, more precisely, Workflow nets. Petri nets are widely used, provide a high degree of formality which supports model verification, and have a graphical notation with a precisely defined mathematical semantics. An extensive body of work exists on the mathematical, structural, and behavioral properties of Petri nets and Workflow nets (e.g., [2]). Furthermore, they serve for formalizing and analyzing models in other modeling languages (e.g., EPC [1]). Petri nets employ a small set of constructs, yet possess an impressive expressive power and can be used to represent precisely the entire set of workflow patterns [3].

Petri net analysis addresses the structure of the net rather than the semantics of its elements. GPM can provide such semantics in terms of state specification and state transitions of the process domain. In this paper we show that assigning this semantics to places and transitions can lead to better-designed process models and help avoid undesired situations (which in turn can be formalized using Petri net properties).

In the following, Section 2 introduces GPM and its control flow interpretation; Section 3 maps Petri net constructs to GPM. Section 4 explores restrictions on Petri nets, and introduces additional restrictions based on GPM, and their implications for Petri net properties. Section 5 is a conclusion.

2 The Generic Process Model (GPM)

The focus of GPM analysis is a *domain*, which is a part of the world consisting of interacting *things*. We describe the behavior of the domain using concepts from Bunge's ontology [4][5] and its adaptation to information systems [13][14] and to process modeling [11][12]. A domain is represented by a set of *state variables*, each depicting a property of the domain and its value at a moment in time. A successful process is a sequence of *unstable states* of the domain, leading to a *stable state*, which is in the set of goal states (simply – *goal*). An unstable state is a state that must change due to actions in the domain (an *internal event*). A stable state is a state that only changes due to action of the environment on the domain (an *external event*). Internal events are governed by *transformation (transition) laws* that define the allowed (and sometimes necessary) state transitions (manifested as events in the domain).

We formalize these concepts as follows:

Definition 1: A domain model is a set of state functions $D=\{f_1(t)\ldots f_n(t)\}$. The value of $f_k(t)$ at a given time is termed a *state variable*, denoted x_k.

The set of state variables for domain D is denoted by $X^D=\{x_k;\ k\in I=\{1\ldots n\}\}$. The state of the domain at a given time is $s(D)=<x_1,\ldots x_n>$ (or simply s). A set of states of domain D is denoted by S(D).

Definition 2: A *transformation law* on D is a mapping $L{:}S(D)\rightarrow S(D)$

Definition 3: A domain will be said to be in a *stable state* if L(s)=s and in an *unstable state* if $L(s)\neq s$.

Definition 4: A law will be said to be *well-defined* iff it is a function.

Often, several domain states can be considered equivalent. Hence, the transformation law can be represented as a mapping between sets of states. Such a set can be specified by a predicate C(s). Specifically, the process goal is a set of stable states, specified by a predicate that manifests business objectives to be fulfilled by the process. The task of the process designer is to implement a transformation law so that the process can accomplish its goal.

To model practical situations, we consider a domain as comprising *sub-domains*, each represented by a subset of the domain state variables. Changes that occur in a sub-domain when the domain changes state, are termed the *projections* of the domain law (or domain behavior) on the sub-domain. Formally:

Definition 5: A *sub-domain* is part of the domain described by a subset of X^D.

A sub-domain D^1 of D is described in terms of $X^{D_1}\subset X^D$; $X^{D_1}=\{x_k ; k\in I^1\subset I\}$.
The state of D^1 is $s(D^1)=<x_{k_1},\ldots x_{k_{|I_1|}}>$, $k_j\in I^1$ and $k_j\neq k_l$ for $j\neq l$.

Definition 6: Let the state of D be $s=<x_1\ldots x_n>$. The *projection* of s on the sub-domain D^1 is $s/D^1=<y_1\ldots y_{|I_1|}>$ where $y_k=x_{I1(k)}$.

It is possible that several domain states will map on the same state of the sub-domain. This, in turn, can result in the same sub-domain state changing in different ways, depending on the state of the whole domain.

Definition 7: Let v be a state of D^1. The *projecting set* for v in D is the set of states of D that project v in D^1: $S(v;D^1)=\{s\in S(D) \mid s/D^1=v\}$

We now define the effect of the domain law (L) on the sub-domain D^1

Definition 8: Let $v\in S(D^1)$ and let $s(v;D^1)$ be the projecting set of v. The *law projection* of L_D on D^1 (denoted L/D^1) for v is defined by the mapping L_{D_1}: $S(D^1) \rightarrow S(D^1)$ such that $L_{D_1}(v)=\cup\{L(s)/D^1 \mid s\in s(v;D^1)\}$.

In words – the projection of the law is defined as the union of projections of the states mapped into by the law. We are interested in cases where the projected behavior of the whole domain on a sub-domain creates a well-defined function in the sub-domain. In other words, a given unstable state of the sub-domain will always map in the same way, independent of the state of the whole domain, and hence independent of the states of other sub-domains. We will then say that the sub-domain behaves *independently*. Partitioning of the domain into independently-behaving sub-domains is often a consequence of different actors acting in the domain. These actors can be people, departments, machines, computers and combinations of those.

Definition 9: A sub-domain D^1 of D will be called an *independently behaving* (in short an *independent*) sub-domain iff the law projection on D^1 is a function.

Corollary: For an independent sub-domain the law projection depends only on state variables of the sub-domain.

Note: a sub-domain might behave independently for only a subset of the state space of D. Definition 9 can be restricted to a subset of domain states.

As an independent sub-domain changes its state to a stable one, it is possible some other independent sub-domains will become unstable and will begin transforming. Thus, a sequence of transformations occurs. This sequence comprises a process.

Process models usually include split and merge points, which reflect either concurrency or choice between possible alternative paths. We now interpret these in GPM terms. First, it is possible a set of states arrived at may be partitioned so the next transformation is defined differently for each subset of states. Such partitioning might occur because the law becomes "sensitive" to a certain state variable. Consider, for example, a process where a standard product is manufactured, and then packaged according to each customer's requirements. Manufacturing does not depend on the customer (even when the customer is known). When manufacturing is completed, customer information will determine a choice between packaging actions. This situation is an exclusive choice (an XOR split). The different actions may lead to states which are equivalent for determining the next action (the law will not distinguish between different packaging options), for example, transferring the products to finished goods inventory. This is the point where the paths merge.

Definition 10: S_{sp} is an *exclusive choice* splitting point iff there exist sets of states $S_1, S_2, .. S_n$ such that $S_i \subset S_{sp}$, $S_j \cap S_k = \varnothing$, and $L(S_j) \neq L(S_k)$, $j \neq k$, $j,k=1 \ldots n$.

The corresponding form of a merge (sometimes termed simple merge) is when a single set of states is reachable by law from different sets of states.

Definition 11: Let S_1, S_2, and S_{me} be sets of states such that $S_1 \neq S_2$, $S_1, S_2 \neq S_{me}$. S_{me} is a *simple merge* iff $L(S_1)=L(S_2)=S_{me}$.

Also related to splitting and merging is *concurrency*. Since one domain cannot have concurrent transformations, concurrency should relate to transformations in different sub-domains. It means that if each sub-domain proceeds through a sequence of (projected) states, all combinations of the projected states of the different sub-domains are possible (in principle).

Lemma 1: Two sub-domains can transform concurrently only if they are independent.

Proof: Assume that two sub-domains are not independent. Then the transitions in one can depend on the state of the other. In this case, only some combinations of states of each sub-domain are possible.

It follows that a split leading to concurrency must be related to a decomposition of the domain into independently behaving sub-domains. In such a split, for the process to continue, at least one sub-domain must be unstable with respect to its (projected) law. If all these sub-domains are in unstable states for all states in the split, then this is a parallel split. Otherwise, several possibilities exist, depending on the number of the unstable sub-domains (see [12][11]). In particular, if exactly one sub-domain can be in an unstable state, then, based on Definition 10, this is an exclusive choice.

Definition 12: S_{sp} is a *parallel split* iff there exist at least two sub-domains such that at S_{sp} each sub-domain becomes independent and is in an unstable state.

For example, in the process discussed above, once products are ready, the process domain can be decomposed into two independent sub-domains: one where shipment is arranged and one where the products are transferred into the warehouse. These two sub-domains are independent and in an unstable state, thus they operate concurrently.
A decomposable domain may entail different types of merge points (see [12][11]). In particular, a simple merge - where the completion of action of any sub-domain causes the process to proceed. Here we define a synchronizing merge, where process continuation requires that all active sub-domains complete their tasks. Consider a set of states in a merge point. These states should be unstable to enable the process to continue. They should be reachable from the split, hence their projection in each sub-domain should be reachable from the split for the sub-domain. In a synchronizing merge, each sub-domain becomes stable ("waiting" for the other sub-domains). Once all the sub-domains reach the merge, the process can continue. Formally:

Definition 13: Let $D^k \subset D$, k=1...n be independent sub-domains operating concurrently following a split point S_{sp}. Let S_{me} be a set of unstable states in D, reachable from S_{sp}. S_{me} is a *synchronizing merge* iff $\forall s \in S_{me}$, $\forall k$, s/D^k is stable.

Finally, the explicit representation of process goal in GPM supports the analysis of process models for goal reachability. A process whose design ensures its goal will always be achieved under a certain set of triggering events (which are external to the domain) is termed valid [11] with respect to this set of events.

3 GPM – Petri-Net Mapping

3.1 Petri-Nets and Workflow Nets

This section provides some definitions of Petri-nets in general and Workflow-nets in particular, and their properties which are relevant for our discussion.

A Petri-net is a directed bipartite graph with two node types called places (circles) and transitions (rectangles), connected by arcs. Connections between two nodes of the same type are not allowed.

Definition 14: A Petri-net is a triple (P, T, F):

- P is a finite set of places;
- T is a finite set of transitions ($P \cap T = \emptyset$)
- $F \subseteq (PxT) \cup (TxP)$ is a set of arcs.

At any time a place contains zero or more tokens (black dots), and the state of the net is the distribution of tokens over places. The notations •t, t•, •p, p• indicate the sets of input and output places of transition t and the sets of transitions of which p is an input and output place, respectively. Given two states M_1 and M_n, M_1 M_n denotes that M_n is reachable from M_1 through a firing sequence σ.

Some Petri-nets properties, relevant for our discussion, are defined below.

Definition 15: A Petri net is *bounded* iff for each place p there is a natural number n such that for every reachable state the number of tokens in p is less than n.

Definition 16: A Petri net is a *free choice* Petri net iff, for every two transitions t_1 and t_2, $\bullet t_1 \cap \bullet t_2 \neq \emptyset$ implies $\bullet t_1 = \bullet t_2$.

A Petri net which is not free-choice usually involves a mixture of choice and parallelism, which is hard to analyze and considered inappropriate. Most of the mathematically-based properties identified and analyzed with respect to Petri nets relate to free-choice Petri nets only.

A specific form of Petri net, often used with respect to workflow modeling is Workflow net (WF net).

Definition 17: A Petri net (P, T, F) is a *Workflow net* (WF-net) iff:

(i) There is one source place $i \in P$ such that $\bullet i = \emptyset$.
(ii) There is one sink place $o \in P$ such that $o \bullet = \emptyset$.
(iii) Every node $x \in P \cup T$ is on a path from i to o.

A WF-net represents the life-cycle of a single workflow case in isolation. It has an initial state following the generation of a case, where only one token exists in place i, and a final state o after which the case is deleted. The property which is ultimately sought and verified for WF nets is soundness.

Definition 18: A procedure modeled by a WF-net PN= (P, T, F) is *sound* iff:

(i) For every state M reachable from state i, there exists a firing sequence leading from state M to state o.
(ii) o is the only state reachable from state i with at least one token in place o.
(iii) There are no dead transitions in (PN, i).

Soundness means that the modeled procedure will terminate eventually, and properly (i.e. no further transition will occur). Soundness can be verified in polynomial time for free-choice WF-nets. Another property, closely related to soundness, is well-structuredness. In a well-structured WF-net a splitting point and a merging point which correspond to each other are of the same type, namely, a split at a place (transition) corresponds to a merge at a place (transition).

3.2 Mapping Workflow-Nets to GPM

Table 1 presents the GPM interpretation of WF-net basic constructs and combinations which form control-flow basic building blocks. We focus our discussion on WF-nets, since these have a distinct termination place, which may correspond to GPM's goal concept. Nevertheless, most of the discussion is applicable to Petri nets in general. As shown in Table 1, every basic WF net construct and building block can be assigned a GPM-based interpretation. We do not attempt to do a reverse mapping, namely interpret GPM terms using WF nets, since GPM addresses issues beyond the control flow of the process, which are not in the scope of WF nets. Nevertheless, considering the proposed control flow mapping, this interpretation assumes certain domain semantics assigned to the places and transitions of a WF net. This semantics poses requirements which do not exist in the WF net syntax. If these requirements are not met, a WF net cannot be transformed into a meaningful GPM specification. It can thus be claimed that the expressive power of WF nets exceeds the expressive power of GPM with respect to control flow. Alternatively, it can be argued that WF net

Table 1. GPM interpretation of WF-Net constructs and basic building blocks

WF-net construct / building block	GPM interpretation
Place p	A set of states of a sub-domain (the projection of a set of states over a sub-domain)
Transition t	A transformation in a sub-domain
Arc	An unstable state leading to a transformation. A transformation leads to a state.
Initial place *i*	The initial set of states I
Final place *o*	The goal set G
Sequence p_1 t_1 p_2	t_1 is a transformation in a sub-domain, from a set of states p_1 into a set of states p_2
A parallel split t_1 p_1 p_2	t_1 is a transformation after which the domain becomes decomposable, where p_1 and p_2 are state projections over different sub-domains
A synchronizing merge p_1 p_2 t_1	t_1 is a transformation whose initial set of states is $p_1 \cap p_2$[1]
An exclusive choice p_1 t_1 t_2	p_1 is a set of states in which one of two sub-domain transformations is possible
A simple merge t_1 t_2 p_1	p_1 is a set of states reachable by two transformations, t_1 and t_2, separately

syntax, being anchored in mathematical semantics of graphical symbols, allows structures which are not necessarily possible in reality.

4 Mapping-Based Modeling

4.1 Modeling Requirements

Assigning a GPM meaning to WF-nets imposes additional requirements on the use of WF-nets for process modeling. In particular, we will use the goal concept of GPM to form these requirements with respect to WF-nets.

Definition 19: A WF-net is a *well-mapped domain representation* iff it can be mapped to a GPM specification.

To make this concept operational, we derive necessary conditions for a WF-net to be a well-mapped domain representation in the sense of this definition. First, every place in the net should represent a set of states of some defined sub-domain ($x_i..x_k$).

Necessary condition 1: $\forall p_k \in P$, $\exists D_k \subseteq D$ such that p_k is active iff D_k is in a given set of states $S \subseteq S(D_k)$.

We denote the sub-domain as D_{p_k}, its state variables as X^{p_k} and the set of states by S_{p_k}. S_{p_k} can be specified by a predicate $C(X^{p_k})$. p_k is active iff $C(X^{p_k})$ is TRUE.

Regarding condition 1, consider the meaning of several tokens in a place. The predicate C can be some composite expression, which may assume the value TRUE in

[1] GPM merge relates to the states, while Petri-net relates to the transition that follows the states.

more than one situation (each applicable to a subset of states of the sub-domain). In particular, if the predicate is of the form <*Expression OR Expression*>, then it may assume the value TRUE when each of the expressions becomes TRUE. In this case, each atomic expression may stand for a token. As an example, consider an order delivery process, where the customer orders several items which can be manufactured concurrently, and each item is delivered once it is ready. This can be modeled by a single place prior to delivery, whose predicate would be ((Item A=ready) OR (Item B=ready) OR…). The arrival of each item is modeled as adding a token to that place, triggering the delivery transition. The maximal number of tokens in a place is the number of atomic expressions related by an OR in the predicate that defines it. If the predicate cannot be decomposed to the form <*Expression OR Expression*>, then the maximal number of tokens allowed in that place is 1.

Lemma 2: A WF-net which is a well-mapped domain representation is bounded.

Proof: Directly from the token interpretation. It is not possible to formulate a predicate composed of an infinite number of expressions.

As a second requirement, every transition should transform the state of a defined sub-domain (based on its input places). To continue, it must place a new sub-domain in an unstable state. Hence, the domains that correspond to the set of output places of a transition should have some common state variables with the domain corresponding to each of its input places. We denote the sub-domain transformed by transition t as D_t, and its set of state variables is X^t.

Necessary condition 2: $\forall t \in T$, $p_j \in \bullet t$, $p_k \in t\bullet$, it must be that $X^t \subseteq \cup_j X^{pj}$; $X^t \cap X^{pk} \neq \emptyset$.

In other words, for every transition, the state variables of the sub-domain in which the transition occurs, should: (a) be a subset of the state variables of all sub-domains represented by the input places (leading to it), and (b) include some state variable of every sub-domain represented by places preceding it. For example: assume the transition represents manufacturing a product. The transition should "use" only state variables from its inputs (e.g., raw materials and resources), and must affect the state which triggers the next activity (e.g. packaging or shipping the ready product).

Third, for a transition to act within a sub-domain, the sub-domain law must be independent (for the set of states represented by the transition's input place).

Necessary condition 3: $\forall t \in T$, $\forall p \in P$, if $p = \bullet t$ then D_t is independent at p.

Fourth, concurrently operating threads operate over sub-domains that do not share state variables. Note that this is necessary, but not sufficient to guarantee that these sub-domains behave independently.

Necessary condition 4: Every two transitions t_1 and t_2 that may operate concurrently, satisfy $X^{t_1} \cap X^{t_2} = \emptyset$.

In particular, the following two cases can be specified:

Parallel split: Let t_s be a transition where a parallel split occurs, so $t_s\bullet = \{p_1, p_2\}$, and consider $t_1 = p_1\bullet$ and $t_2 = p_2\bullet$. Then $X^{t_1} \cap X^{t_2} = \emptyset$.

Synchronizing merge: Let t_m be a transition where a synchronizing merge occurs, so $\bullet t_m = \{p_1, p_2\}$, and consider $t_1 = \bullet p_1$ and $t_2 = \bullet p_2$. Then $X^{t_1} \cap X^{t_2} = \emptyset$.

Note that while necessary condition 4 addresses concurrency situations, we make no similar requirement with respect to choice-related splits. Since these may relate to both decomposable and non-decomposable domains, no strict rules can be formed here. Choice-related splits lead the domain in one of several possible paths, which, in turn, put the domain in one of possible (alternative) states. As opposed to the path definition in Petri nets, which relates to any sequence of connected elements, we relate to *domain paths* (D-paths), which correspond to selected sequences of states of the domain. The difference between these terms can be seen with respect to parallel splits, where the domain is decomposed into independently transforming sub-domains. In the graphical Petri net representation the sequence of elements in each sub-domain forms a different "path". However, since no choice (decision) is made, we do not address these as separate D-paths.

GPM defines a path as a set of states the domain goes through via a sequence of transitions determined by the law and by external events. Petri nets (and specifically, WF-nets) do not explicitly address external events, assuming they will occur as expected. In a WF-net, considering a sub-domain D, its state is the distribution of tokens at a given moment in all the places p_j that satisfy $X^{pj} \subseteq X^D$. We denote the state of sub-domain D by M^D.

Definition 20: Let D be a sub-domain in a WF-net. A *domain path* (D-path) of D is a sequence of states of D $<M^D_1, M^D_2, \ldots M^D_K>$, such that a sequence of transitions $<t_1, t_2, \ldots t_{k-1}>$ exists that satisfies $M^D_i \quad M^D_{i+1} \quad \ldots \quad -M^D_k$, for $1 \leq i \leq k-1$.

For clarity, we hereafter relate to domain paths as D-paths and to "ordinary" (or "traditional") Petri net paths simply as paths.

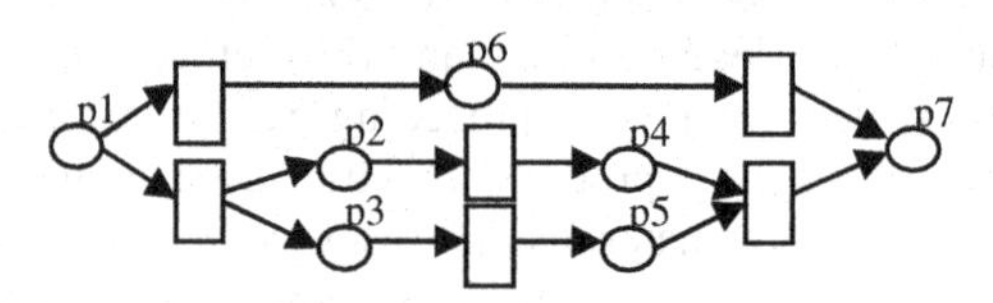

Fig. 1. D-paths example

D-paths are demonstrated with respect to the Petri net in Fig. 1, which includes four D-paths: (1) p1→ p6→ p7, (2) p1→ p2+p3→ p4+p5→ p7, (3) p1→ p2+p3→ p2+p5→ p4+p5→ p7, and (4) p1→ p2+p3→ p3+p4→ p4+p5→ p7. These are not equivalent to the three "ordinary" paths of the net. When examining the four D-paths, it is clear that three of them (D-paths 2, 3, and 4) relate to different orderings in which the concurrent transitions can be performed, while D-path 1 specifies a different way for reaching p7. This can be formalized in the following definition:

Definition 21: Let Ac be the set of places that become active in D-path C. Two D-paths C_1 and C_2 are termed *distinct* iff $Ac_1 - Ac_2 \neq \emptyset$ and $Ac_2 - Ac_1 \neq \emptyset$.

One of the basic properties of Petri nets is free-choice, which is associated with "desirable" mathematical properties of nets. Non free-choice Petri nets are associated with situations of confusion between choice and parallelism, and are considered improper (albeit syntactically possible). We do not require that Petri nets would be free choice to be well-mapped domain representations. Instead, we have a more relaxed requirement, which we term *relaxed free-choice*, specified as Necessary condition 5. In a free-choice Petri net, if two transitions share input places, then all their input places must be the same (namely, they have exactly the same triggering conditions). In contrast, the relaxed free-choice requirement is for the two transitions to share the same set of inputs (sub-domain state variables), but not necessarily at the same values (places). In other words, each of them may trigger on different combinations of values of the same set of state variables. We denote by $D_{\bullet t}$ the domain of all input places of a transition t: $X^{\bullet t} = \cup\{X^{p_k} | p_k \in \bullet t\}$.

Necessary Condition 5 (Relaxed free choice): For every two transitions t_1, t_2, if $\bullet t_1 \cap \bullet t_2 \neq \emptyset$ then $D_{\bullet t_1} = D_{\bullet t_2}$.

The relaxed free choice requirement is a result of Necessary conditions 2 and 3. Condition 2 requires D_t to be included in $D_{\bullet t}$, while condition 3 requires D_{t_1} and D_{t_2} to be independent. If D_{t_1} and D_{t_2} partly overlap (since $\bullet t_1 \cap \bullet t_2 \neq \emptyset$), then they cannot be independent of each other. In other words, at a given place a sub-domain can either transform independently or dependently of other sub-domains, both are not possible.

One of the basic concepts in GPM is the goal of a process, which we relate to the sink place *o* of a WF-net.

Necessary condition 6: The sink place in a WF net *o* marks a set of stable states of the entire process domain.

Two notes should be made. First, this set of stable states must be in the process goal. Second, a well mapped WF net is constructed so that once *o* is reached no other part of the domain can still be active. A particularly interesting case is when concurrent paths are merged by a simple merge. This structure is in violation of the well-structuredness property, yet we allow it. When concurrent paths are joined by a simple merge, the merge place may hold more than one token. The next transition may have other input places, so it may not be fired even when the merge place has tokens. For example, consider a process where two teams work concurrently to find a solution to a problem. When one team finds a solution, the process can continue. The solution found by the other team is not used. In this process model, the place representing that a solution exists may have two tokens, but only one token is used by the following transition. In the final state of the process, the entire domain is in a stable state. Nevertheless, for one sub-domain there might still be a solution "waiting" to be used, namely, to trigger additional action. For the entire domain to be stable with certainty, some action is required to "notify" the unstable sub-domain that no further transitions will take place. Technically, this can be accomplished by adding a transition from such nodes to the final place (in our example such transition may stand for archiving or discarding the "losing" solution).

Another result of Necessary condition 6 is that the net cannot include loops which action continues in parallel to the continuation of the process, namely, loops whose exit point is a parallel split, as formulated in Lemma 3.

Lemma 3: Let a WF-net be a well-mapped domain representation, and consider a transition t and three places p_1, p_2, and p_3, such that $p_1=\bullet t$, $\bullet p_2=\bullet p_3=t$. Then p_1 is not reachable from p_2 or from p_3.

Proof: Assume p_1 is reachable from p_2. When t fires, p_2 and p_3 become active, and while p_3 may lead to *o*, p_2 will lead infinitely to the sequence that activates t. This means that *o* may be reached while the loop sub-domain is unstable, in contradiction to Necessary condition 6.

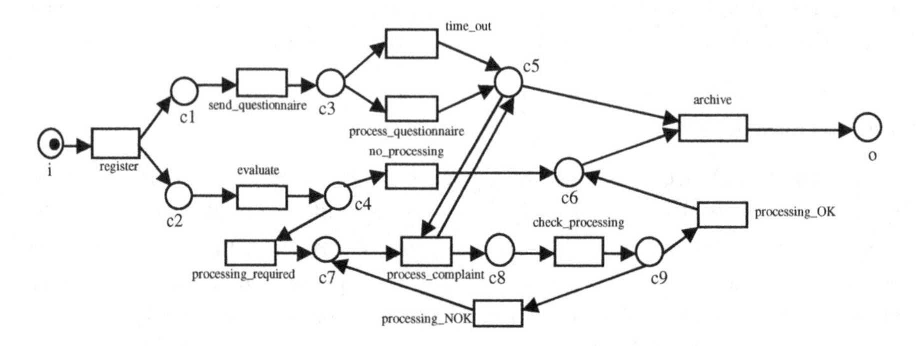

Fig. 2. A complaint processing process

To demonstrate how the necessary conditions can be used, let us examine the example given in Fig. 2 (taken from [1]). The WF-net representing complaint processing is not a well-mapped domain representation, for the following reason. The predicate that can be assigned to c5 is *(Questionnaire_status=time_out Xor Questionnaire_status=processed) Or Complaint_status=processed*. It follows that c5 is defined over a sub-domain which includes at least the Questionnaire_status and the Complaint_status state variables. c5 and c7 are input places of the transition process_complaint (a synchronizing merge), which transforms the status of a complaint. Such merge violates Necessary condition 4, which requires the sub-domains joined by a synchronizing merge to be disjoint. The modified model (Fig. 3) defines c5 over a sub-domain which does not include the complaint status, thus it is a well-mapped domain representation.

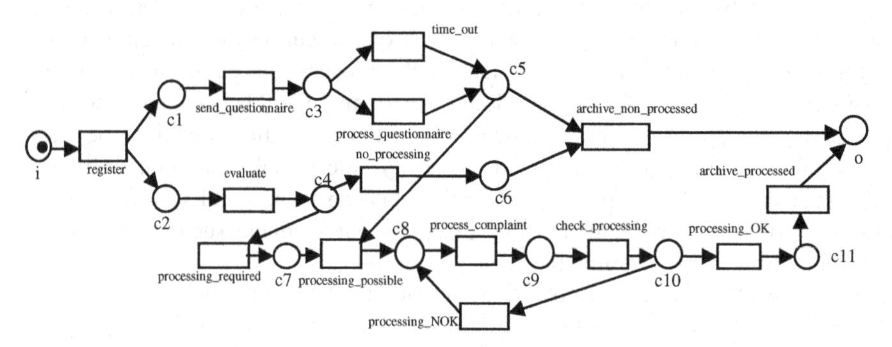

Fig. 3. A modified model of the complaint processing process

4.2 Process Validity Considerations

Validity of a process model can only be assessed with respect to a set of expected external events and to a defined goal [11]. How these are determined is outside the scope of the current analysis. We consider only the reachability of the process termination state, assuming that it represents the process goal. Assuming that all the expected external events occur, validity relates to completeness of the internal law definition (which relates to internal events) and to its consistency with the goal definition.

Incompleteness reflects potential deadlock situations. Following [6], a process instance is in deadlock iff it is not in the goal and no transition is enabled. GPM also allows for a process execution "hanging" when external events fail to occur, but we assume here no such failure happens. Thus, deadlock means that the process is in a state for which the law is not defined. In a well mapped domain representation this may occur when a transition has more than one required input place (i.e., it is a synchronizing merge which joins different sub-domains) and not all of them are enabled. Two situations are possible:

(1) Not all sub-domains have been activated at the split point.
(2) At least one of the sub-domains took a D-path which does not lead to the input place of the merge transition.

We will specify modeling rules to avoid each of these cases. Case (1) is possible if an exclusive choice is followed by a synchronizing merge. According to [12], such structure should not appear in a valid process. It also does not appear in a well-structured WF-net [1], where an exclusive choice is matched by a simple merge and a parallel split is matched by a synchronizing merge. As discussed above, we do not require well-structuredness. Instead, we only require that every synchronizing merge be preceded by a parallel split, leaving the simple merge unconstrained as to the type of split it should be preceded by. Since in Petri nets a synchronizing merge is in a transition, it should correspond to a transition in the split point (parallel split). This is formalized in Modeling rule 1.

Modeling rule (MR)1: Let x and y be two elements (i.e., places or transitions) in a well-mapped domain representation WF-net, connected by two different elementary paths leading from x to y. If x is a place then y should be a place too. If y is a transition then x should be a transition too.

To avoid the second case, the modeler needs to make sure that if two sub-domains that have alternative distinct D-paths need to synchronize, then every possible combination of these D-paths has a merging transition defined for it. To illustrate the idea, consider the examples of Fig. 4.

In Fig. 4(a), the process domain is split in t1 to two concurrently active sub-domains, and both these sub-domains have different D-paths that can be selected. The process may clearly deadlock, if one sub-domain takes a D-path leading to p7 while the other reaches p10, or if one sub-domain reaches p8 while the other takes a D-path that leads to p9. There are more combinations of D-paths that can be taken than combinations that lead to the goal of the process. The sub-domain on the left side has two distinct D-paths: (1) p2→p7, and (2) p2→p8. The sub-domain on the right

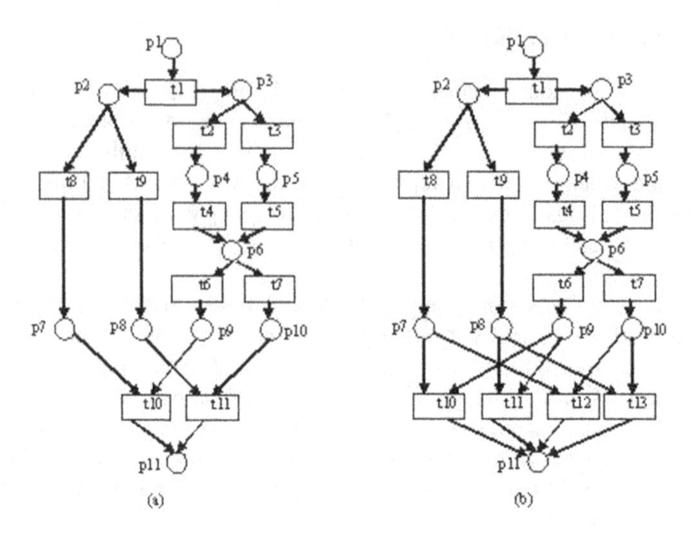

Fig. 4. D-path combinations

side has four distinct D-paths: (1)p3→p4→p6→p9, (2) p3→p4→p6→p10, (3) p3→p5→p6→p9, and (4) p3→p5→p6→p10. To eliminate the possible deadlock, we need to define action in every possible situation the process may reach. We may look for a place which is reached from all D-paths. Considering the right side sub-domain, p6 is reachable in all the distinct D-paths. Hence, it is guaranteed to be reached. Let us examine a possible correction, where p6 is connected to t11. Then t6, t7, and t11 would share p6 as an input place, while t11 is also preceded by p8 (whose domain is different). This is in contradiction to Necessary condition 5 (relaxed free choice), which proscribes a sub-domain from being both independently transforming and merging at a given place. Following this analysis, p9 and p10, which are reachable in two distinct D-paths each and together "cover" all the four D-paths, do not represent states where the sub-domain is independent (since they lead to a merge). A complete solution, addressing every possible situation, requires the net to include a transition defined for every possible combination of non-independent places in the two sub-domains, namely, (p7, p8)x(p9, p10), as shown in Fig. 4(b).

The above analysis is formalized in Modeling rule (MR) 2 that requires that if a transition depends on a combination of states of two sub-domains, and that combination is not guaranteed to happen, there must be other transitions specified for every other possible state combination.

Modeling rule (MR) 2: Let D_1, D_2 be two sub-domains, and t_1 a transition such that $\bullet t_1=\{p_1,p_2\}$, where $p_1 \in D_1$ and $p_2 \in D_2$. Let PM_1 and PM_2 be sets of places such that the domain of each place in PM_1 or in PM_2 is not independent and for every distinct D-path in D_k there is a place in PM_k, k=1,2. Then for every pair of places $p_j \in PM_1$ and $p_k \in PM_2$, there must be a transition t such that $\bullet t=\{p_j,p_k\}$.

Note: D_1 and D_2 can be identified by backtracking paths from p_1 and p_2 until the first transition which is included in both paths. For example, it is easily seen that the model in Fig. 3 is in compliance to MR2.

Inconsistency between the law and the goal definition relates to infinite loops. In WF-nets, since every element must be on a path from i to o, loops must have (at least one) exit points. These may be parallel splits or exclusive choice splits. According to Lemma 3, parallel splits cannot be exit points of loops in a well-mapped domain representation WF-net. We shall hence examine the possible structures in which loops whose exit point is an exclusive choice may become infinite. When a loop has an exclusive choice as an exit point at place p, two cases are possible:

(1) The next transition has only one input place ($\bullet t=\{p\}$). Then the exit depends on one sub-domain only. Structurally, this is not an infinite loop, and the exit from the loop depends on the decision criteria defined by the analyst.
(2) The next transition has more than one input place (i.e., it merges a number of concurrent sub-domains). In this case, if the merge deadlocks, the loop will continue infinitely. However, merging deadlocks can be eliminated by using modeling rules 1 and 2. Hence, if the modeling rules are used in a well-mapped domain representation WF-net, it does not include infinite loops.

Theorem 1: A well-mapped domain representation WF-net which satisfies Modeling rules 1 and 2 is sound.

Proof: soundness has three requirements. (ii) proper termination – follows directly from Necessary condition 6, and (iii) no dead transitions – follows from Necessary condition 2. To prove (i), namely that *o* is reachable from every state reachable from *i*, we will show that for any state M reachable from *i* there is a transition t that can be fired. Since all the elements in a WF-net are on a path from *i* to *o* and no infinite loops are possible, if any arbitrary state M transforms, *o* will be reached.

We will show that in a given state M every transition is either (a) within an independent sub-domain, or (b) a result of a merge between two (or more) sub-domains. In the first case, a transition will be fired with certainty. In the second case, by MR1, all the required sub-domains should be active, and by MR2 there is a transition defined for every possible combination of D-paths of the sub-domains. Hence a transition will be fired.

Formally: Let P(M) be the set of active places in M, and consider a place $p_1 \in P(M)$ and a transition $t_1 \in p_1\bullet$. Two cases are possible: (1) $\bullet t_1 \subseteq P(M)$, then t_1 fires at M. (2) $\bullet t_1 \cap P(M) \neq \emptyset$, $\bullet t_1 \not\subseteq P(M)$. Then t_1 cannot fire at M, but we will show that there exists a transition t_2 that can fire at M. Having more than one input place, t_1 merges two or more independent sub-domains D_i. Assume $\bullet t_1=\{p_1,p_2\}$, where p_1 is the projection of M over D_j, and $p_2 \notin P(M)$. We assume p_2 is the projection of some state M' over a sub-domain D_k ($k \neq j$). Since, by MR1, D_k is active at M, we denote the projection of M over D_k by p_3 ($p_3 \in P(M)$). According to MR2 there exist a transition t_2 and a place p_4, such that $\bullet t_2=\{p_1,p_4\}$, $Dp_4 \subseteq D_k$, and p_4 is on a D-path that includes M/D_k. Three cases should be checked: (1) $p_3 \rightarrow p_4$. Then $M \rightarrow M'$, thus t_2 can fire. (2) $p_3=p_4$. Then t_2 can fire. (3) $p_4 \rightarrow p_3$. This is impossible due to Necessary condition 5 (relaxed free choice).

In summary, following the necessary conditions and modeling rules, it is possible to construct a sound WF-net.

5 Conclusion

This paper proposed to use the ontologically-based GPM semantics for existing constructs of process modeling languages. We demonstrated how this can be done for WF-nets. We also showed how modeling guidelines, based on this semantics, can assist in avoiding process modeling problems that traditionally could only be detected by verification of the completed models. Existing verification algorithms for WF-nets can analyze in polynomial time only specific classes of models (free-choice or well-structured). The modeling rules suggested here can lead to sound WF-nets which are not necessarily free-choice or well-structured. Note that the modeling rules do not constitute a verification approach. Rather, they form a construction approach, which yields sound models when applied.

The essence of the analysis is in mapping common situations that can occur when a domain undergoes state transitions, into a WF-net representation. For a process to be guaranteed to reach its goal, its definition should fulfill three conditions: (1) no situations should arise where it "hangs", (2) completeness: all possible states should have defined transitions, and (3) no infinite loops. Process "hanging" can happen when several conditions need to be fulfilled for the process to continue – i.e. in merge situations. Merges occur because a split has occurred earlier in the process. By choosing only appropriate combinations of splits and merges, the process can be guaranteed to proceed. This was the purpose of Modeling Rule 1. Completeness requires that the process model will specify continuation for all possible states – this was the purpose of Modeling Rule 2. Both rules and the goal definition ensure the absence of infinite loops. Constructing models that conform to these two rules, therefore assures that the process, when executed, can always complete (in the sense of reaching its goal). It is important to note that the rules guide the actual construction of process model, rather than being applicable only to complete models.

As GPM concepts are generic, they can be applied to other modeling languages. We intend to do this in future research. This application would require mapping of these languages to GPM and deriving appropriate restrictions and modeling rules. In addition, we plan to empirically investigate the effectiveness of the propositions made here in contributing to the quality of models produced by modelers. Finally, we will develop a modeling tool to support the application of the modeling rules when constructing a model.

References

[1] van der Aalst, W.M.P.: Formalization and Verification of Event-Driven Process Chains. Information and Software Technology 41(10), 639–650 (1999)

[2] van der Aalst, W.M.P.: Workflow Verification: Finding Control-Flow Errors Using Petri-Net-Based Techniques. In: van der Aalst, W.M.P., Desel, J., Oberweis, A. (eds.) Business Process Management. LNCS, vol. 1806, pp. 161–183. Springer, Heidelberg (2000)

[3] van der Aalst, W.M.P., ter Hofstede, A.H.M., Kiepuszewski, B., Barros, A.P.: Workflow Patterns. Distributed and Parallel Databases 14(1), 5–51 (2003)

[4] Bunge, M.: Treatise on Basic Philosophy. In: Ontology I: The Furniture of the World, vol. 3. Reidel, Boston (1977)

[5] Bunge, M.: Treatise on Basic Philosophy. In: Ontology II: A World of Systems, vol. 4, Reidel, Boston (1979)

[6] Kiepuszewski, B., ter Hofstede, A.H.M., van der Aalst, W.M.P.: Fundamentals of control flow in workflows. Acta Informatica 39(3), 143–209 (2003)

[7] Mendling, J.: Detection and Prediction of Errors in EPC Business Process Models, PhD thesis, Vienna University of Economics and Business Administration (2007)

[8] Object Management Group (OMG), Business Process Modeling Notation Specification (2006), http://www.bpmn.org

[9] Rosemann, M., Recker, J., Indulska, M., Green, P.: A Study of the Evolution of the Representational Capabilities of Process Modeling Grammars. In: Dubois, E., Pohl, K. (eds.) CAiSE 2006. LNCS, vol. 4001, pp. 447–461. Springer, Heidelberg (2006)

[10] Sadiq, W., Orlowska, M.E.: On Correctness Issues in Conceptual Modeling of Workflows. In: Proceedings of the 5th European Conference on Information Systems, Cork, Ireland, pp. 943–964 (1997)

[11] Soffer, P., Wand, Y.: Goal-Driven Multi-Process Analysis. Journal of the Association of Information Systems 8(3), 175–203 (2007)

[12] Soffer, P., Wand, Y., Kaner, M.: Semantic Analysis of Flow Patterns in Business Process Modeling. In: Alonso, G., Dadam, P., Rosemann, M. (eds.) BPM 2007. LNCS, vol. 4714, pp. 400–407. Springer, Heidelberg (2007)

[13] Wand, Y., Weber, R.: On the Ontological Expressiveness of Information Systems Analysis and Design Grammars. Journal of Information Systems (3), 217–237 (1993)

[14] Wand, Y., Weber, R.: Towards a Theory of Deep Structure of Information Systems. Journal of Information Systems 5(3), 203–223 (1995)

On the Duality of Information-Centric and Activity-Centric Models of Business Processes

Santhosh Kumaran, Rong Liu, and Frederick Y. Wu

IBM T.J. Watson Research Center
19 Skyline Dr. Hawthorne, NY 10532, USA
{sbk,rliu,fywu}@us.ibm.com

Abstract. Most of the work in modeling business processes is activity-centric. Recently, an information-centric approach to business process modeling has emerged, where a business process is modeled as the interacting life cycles of information entities. The benefits of this approach are documented in a number of case studies. The goal of this paper is to formalize the information-centric approach and derive the relationships between the two approaches. We do this by formally defining the notion of a business entity from first principles and using this definition to derive an algorithm that generates an information-centric process model from an activity-centric model. We illustrate the two models using a real-world business process and provide an analysis of the respective strengths and weaknesses of the two modeling approaches.

1 Introduction

The role of information technology (IT) in the enterprise is to support the business operations of the enterprise. Subject Matter Experts (SMEs) document business operations using business process models which prescribe the activities that need to be performed as part of a business operation, the sequencing of these activities, and the input and output data of these activities. While business process models are good for documenting business operations, creating computer programs that support these operations has always been a challenge. Existing approaches to IT-enabling a business process take one of the following two paths:

(1) Business process models are used merely as requirement documents. From these, IT solutions are manually designed and implemented by writing new custom code, or by customizing and integrating legacy applications and packaged software; or
(2) Business process models are automatically converted into workflow definitions which are deployed on workflow engines and augmented with custom code [26].

The first approach leads to a gap between the business process models and IT solutions resulting in poor quality of the IT solutions with respect to their ability to support business processes, poor responsiveness of IT to business process changes, and inefficiency in the overall development processes. The second approach faces a number of difficulties as well, as enumerated below.

Z. Bellahsène and M. Léonard (Eds.): CAiSE 2008, LNCS 5074, pp. 32–47, 2008.

As business processes become complex and large, the workflow approach turns out to be increasingly difficult to implement, as the overall performance of the system degrades substantially and the maintenance of the resulting solution becomes extremely hard. The primary reason for this is that the workflow approach does not lend itself to componentization in a natural way [3].

Advanced features such as backward navigation, event-driven behavior, and conversational client interaction are very difficult to support in workflow models [28]. The cost of supporting these features adds more complexity, which further complicates the first issue. IT solutions often need sophisticated, user-friendly human interfaces which are not readily supported by the workflow approach. Expensive manual tweaking of the code is often needed to integrate function-rich user interfaces with workflow-based backend systems.

In response to this situation, another process modeling paradigm, which models business processes as intersecting life cycles of information entities, has been proposed. Appropriately, this approach is called *information-centric process modeling*. The information entities that are used to describe business processes in this manner have been called various names, including *adaptive documents* (ADoc) [17], *adaptive business objects* (ABO) [21], *business artifacts* [22], and lately *business entities*. In this paper, we will refer to them as business entities.

This new paradigm has been successfully tested through customer engagements. However, several problems remain. First, the concept of business entities is informal with no theoretical underpinnings. Second, in this paradigm, the key is to discover the right information entities that describe the business process. The current practice identifies these entities through intense consulting sessions. Those sessions are time consuming and demand consulting skills that typically are not common. Third, there is a lack of understanding of the relationship between this new paradigm and traditional *activity-centric process modeling* used in workflow management systems [26].

The goal of this paper is to formalize the information-centric approach and present an algorithm for transforming activity-centric process models into information-centric models. We formally define the concept of business entities and use this definition to derive an algorithm for the transformation. We use a real-world example to illustrate this transformation and analyze the respective strengths and weaknesses of the dual representations of a business process.

The remainder of the paper is organized as follows. Section 2 introduces the formal definition of business entities and gives the transformation algorithm for creating information-centric process models from activity-centric process models. A complete example is provided in Section 3 to illustrate this algorithm. Section 4 presents an analysis of the two modeling paradigms. In Section 5, we compare our work with related work. Section 6 concludes with a brief description of future work.

2 Information-Centric Process Modeling

In this section, we give formal definitions of several key notions, including *process scope, domination, business entity, and activity-centric and information-centric process models*. We start with a brief review of basic concepts of process modeling.

A *business activity* is a logical description of a piece of work that consists of human and/or automated operations and is often realized in conjunction with information processing. A *business process* links business activities by transforming inputs into outputs to achieve a business goal. A set of business processes constitutes a *business function*, provisioning a set of *business services* that defines the external view of the business function. All of the data used by the business function, including the input and output of the business services, form the information domain of the business function. The atomic elements that make up the information domain are called *information entities* (or *entities* for simplicity).

Fig. 1 shows a simple business process. This process clearly describes business activities and their execution sequence for handling a claim. In addition, activities in this process use a set of information entities, for example, *claim* and *loss event*, as the dotted lines indicate. We call Fig. 1 an activity-centric process model, as defined below.

Definition 1 (Activity-Centric Process Models). An activity-centric process model consists of business activities, connectors as control flows describing the execution sequences of these activities, and optional information entities as inputs or outputs of the activities.

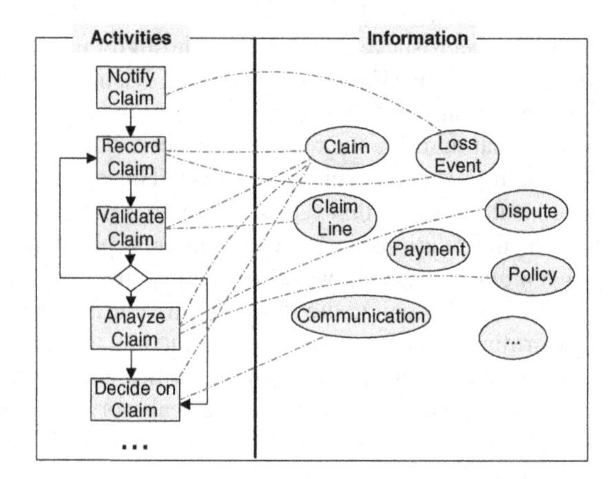

Fig. 1. Activity-Centric Business Process Model – Claim Management

Definition 2 (Process Scopes). A *process scope s* is a group of business processes, together providing a well defined end-to-end function to map input *I* to output *O*, i.e. $s: I \rightarrow O, s = \{p_1, p_2, ..., p_n\}, I = \{e_{I1}, e_{I2}, ..., e_{Ix}\}, O = \{e_{O1}, e_{O2}, ..., e_{Oy}\}$, where each *p* is a process and each *e* is an information entity.

For example, an end-to-end claim management function involves several business activities such as *record claim* and *validate claim* (see Fig. 1). The input to this process scope may be a loss event and outputs are a set of information entities including a closed claim, outgoing payments, and communication documents with the claimant.

Within a process scope, information entities are created or modified through business activities to produce the desired outputs as defined in the end-to-end function. From an information-centric point of view, information entities influence business activities in the sense that the execution of an activity is predicated on the availability of the right information entities in the right state [1]. For a set of business activities, there is a corresponding set of information entities that influence the execution of these activities. But there are differences in the degree to which a specific information entity influences the execution of the activities in the set. For example, considering the process shown in Fig. 1, *claim* information entity influences most business activities in the process. On the other hand, the influence of *claim line* is limited to only a subset of activities that are also influenced by *claim*. Therefore, we say *claim* dominates *claim line*, as formally defined below.

Definition 3 (Domination). Information entity e_1 dominates information entity e_2 in a process scope s, denoted as $e_1 \mapsto e_2$, *iff*:

(1) $\forall a \in s$, if $e_2 \in \bullet a$, then $e_1 \in \bullet a$

(2) $\forall a \in s$, if $e_2 \in a\bullet$, then $e_1 \in a\bullet$

(3) $\exists a \in s$, s.t. $e_1 \in \bullet a \cup a\bullet$, but $e_2 \notin \bullet a \cup a\bullet$, where $\bullet a$ ($a\bullet$) denotes the input (output) information entities of activity a.

In other words, e_1 dominates e_2, if (1) for every activity that uses e_2 as an input, e_1 is also used as an input, (2) for every activity that uses e_2 as an output, e_1 is used as an output, and (3) e_1 is used by at least one activity that does not use e_2.

If e_2 is only used as an input in the process scope, this domination is called *referential domination*. With *inclusive domination*, e_2 is used as an output in at least one business activity. A *dominant entity* is one that is not dominated by any other entity in a process scope. Accordingly, a *dominated entity* is an entity dominated by any other entity. The domination relationship is transitive, i.e. if $e_1 \mapsto e_2$, $e_2 \mapsto e_3$ then $e_1 \mapsto e_3$.

Fig. 2 shows sample domination relationships in an insurance process scope. *Claim* entity referentially dominates *policy* because *policy* is only a necessary input for the processing of *claim* but *policy* itself is not changed within the process scope. *Claim* inclusively dominates *claim line*, *dispute*, and *communication* entities. Some entities, for example, *loss event* and *payment*, are not involved in any domination relationship in this figure. Fig. 2 is referred to as an *entity domination graph,* which describes the domination relationships between entities within a process scope.

From a domination graph, a data model for dominant entities can be derived using containment data modeling [27]. Each dominant entity can be treated as a

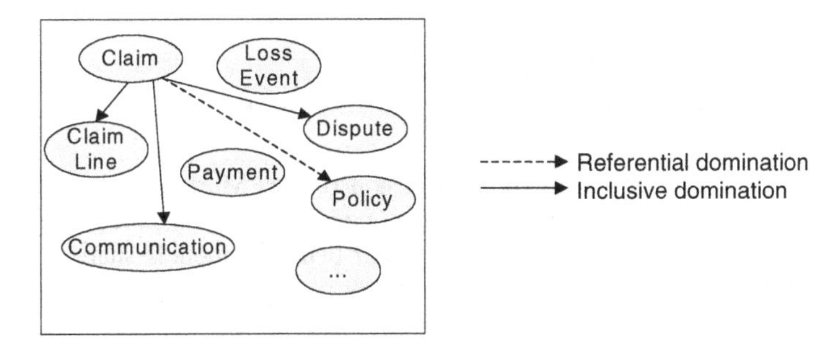

Fig. 2. Examples of Domination

"container", each inclusively dominated entity is a "contained member", and each referentially dominated entity becomes a reference member. A complete example will be provided shortly to illustrate how to derive data models for entities from domination relationships.

Definition 4 (Business Entities). A business entity is a dominant information entity with an associated data model and an associated behavior model in the context of a process scope. The data model describes the data dependencies between the dominant entity and the dominated entities as the dominant entity logically containing the dominated entities. The behavior of the business entity is modeled as a state machine where state transitions are caused by activities acting on the dominant entity.

Fig. 3 shows three business entities in the claim management process scope. Each business entity has a behavior model shown as a state machine. Business entities can be thought of as an abstraction that componentizes the information domain of a business such that the behavior models associated with these components fully capture the business process functionality. Moreover, business entities provide the information context for business activities and processes. Typically, within a process scope, the provided business functions require customer inputs. The customer inputs may initiate an instance of a business entity. The outputs of the processes may be represented as the final state of the business entity and perhaps other business entities created during the processes. Therefore, the business processes are also the process of business entities walking through their lifecycles, from their initial states to their final states.

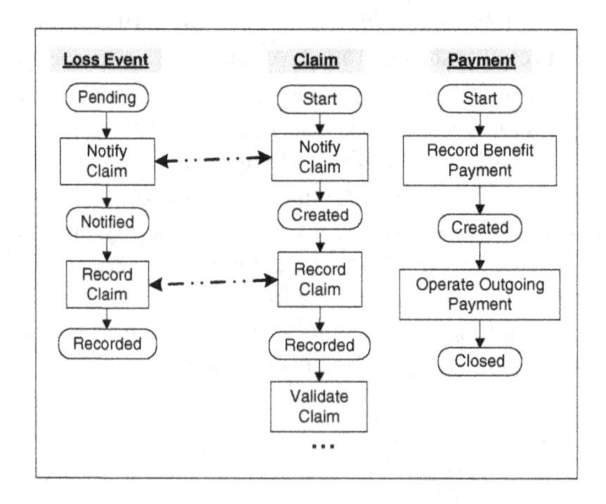

Fig. 3. Business Entities in Claim Management Process Scope

Definition 5 (Information-Centric Business Process Models). An information-centric business process model of a process scope is a set of connected business entities. Two business entities are connected if their behavior models share at least one business activity.

An information-centric process model of a process scope may contain multiple business entities. The lifecycles of these entities are linked through an instance creation pattern or a synchronization pattern [15]. In the creation pattern, an existing business entity creates a new instance of another business entity as part of a business activity. For example, in Fig. 3, when *notify claim* activity is performed on *loss event*, it creates a new instance of the *claim* entity (i.e., the state of *claim* is changed from *start* to *created*). In the synchronization pattern, two existing business entities exchange information as part of performing a business activity. The example in Fig. 3 shows *loss event* and *claim* exchanging information as part of the *record claim* activity.

Using the concept of domination, we can transform an activity-centric process model into an information-centric process model through the algorithm shown in Fig. 4. The algorithm contains four main steps: (1) Discovering business entities and constructing an entity domination graph purely based on the definitions of domination and business entities; (2) Finding input (*I*) and output (*O*) business entities of each business activity; (3) Creating an output state for each output business entity of an activity; and (4) Connecting an activity to its output business entity states and connecting each output state to the next activity touched by the business entity to construct a state machine preserving the sequence between activities as in the original activity-centric business process model. Note that an activity-centric process model may include control nodes, such as an OR-SPLIT node for alternative decisions, in addition to the activities. However, we disregard such nodes since control flow semantics is implicit in the state machine definition.

This transformation algorithm is very generic. It can be applied to any activity-centric process model that meets two conditions: (1) each activity has input and output information entities, and (2) the process model is connected such that each activity is in at least one path from the start node to the end node. There are no constraints on the format and content of an information entity. However, with respect to the complete and correct definition of an information-centric model of a business process, the information entities should exhibit the following properties.

- *Self-describing*: An information entity is self-describing if it contains metadata that describes its data content. The metadata of a business entity is composed from the metadata of its constituent information entities.
- *Non-overlapping*: Information entities partition data in the information domain of a process scope into disjoint sets.
- *Decoupled*: An entity is decoupled from others in the sense that it can evolve independently of others through business activities. Each information entity is used distinguishably by some business activities. The granularity of information entities requires that any pair of entities should not always be used by the same business activities. If so, these two entities should be merged into one. In other words, the granularity of information entities is determined by business activities.

If the information entities in a process scope possess these three properties, they are *normalized*, formally defined as follows.

Definition 6 (Normalized Information Entities). Let A be the activity set, $A=\{a_1,a_2,...,a_n\}$. Let D denote the information domain of a set of processes, $D=\{e_1,e_2,...,e_m\}$. e_i is a normalized information entity for $1 \le i \le m$, if:

(1) $e_i=(S,V)$, where V is a set of values over attribute set S (*self-describing*)

(2) For any pair of $e_i=(S_i,V_i), e_j=(S_j,V_j)$, $S_i \cap S_j = \emptyset$ (*non-overlapping*)

(3) $e_i \neq e_j$ and $\exists a \in A, s.t.\ e_i \in \bullet a \cup a \bullet, e_j \notin \bullet a \cup a \bullet$ (*decoupled*)

where $i \neq j, 1 \leq i, j \leq m$, $\bullet a\,(a\bullet)$ denotes the input (output) of activity a, and $\bullet a \subset D$, $a\bullet \subset D$

If the original set of information entities does not possess these properties, the resulting business entities may have degraded modularity as their behavior models could be connected at many shared activities. Therefore, it may be necessary to normalize the information entities by clearly defining their data schema and removing data overlap and coupling between them (By Definition 6). In addition, incomplete input and output information entities in the original activity-centric model may result in a poor-quality information-centric process model, characterized by a large number of business entities with interwoven behavior models.

```
INPUT   E        : the information entity set of process scope s, E = {e1, e2, ..., en}
        A        : the activity set in process scope s, A = {a1, a2, ..., am}
        C        : control nodes (AND - SPLIT, AND - JOIN, OR - SPLIT, OR - JOIN, START, END) in S
        Ii (Oi)  : the set of activities using ei as input (output)
        ri       : connector associated with e from node ns to nt, ri = (ns, nt, e), R = {r1, r2, ...}
OUTPUT  T        : the business entity set in scope s
        Di       : the set of dominated entities of ei
        Si       : the set of states of business entity ei
BEGIN
for every ei ∈ E                                      / * compute entity domination graph * /
      for every ej ∈ E and j ≠ i
          if ei ↦ ej then save ej to Di
      end for
end for
for every ei ∈ E and Di ≠ ∅                           / * find dominant information entities * /
     if ei ∉ Dj for any j ≠ i then save ei to T
end for
If T  ≠ ∅ then
    for every ei ∈ T
       remove activity a from Ik for any a ∈ Ik and any ek ∈ Di    / * remove dominated items * /
       remove activity a from Ok for any a ∈ Ok and any ek ∈ Di
       remove connector r = (nx, ny, ek) from R for any ek ∈ Di
       for any pair of r = (nx, ny, ei) and r' = (ny, nz, ei) where ny ∈ C / * remove control nodes * /
           remove r and r' from R, create r" = (nx, nz, ei) , add r" to R
       end for
       for any connector r = (nx, ny, ei) where nx, ny ∉ C
          create a state sij and save sij to Si         / * create states and update connectors * /
          create connectors r' = (nx, sij, ei) and r" = (sij, ny, ei)
          remove r, and add r' and r" to R
       end for
    end for
END
```

Fig. 4. Transformation Algorithm

3 Example – An Insurance Process Model

In this section, we use a real example to illustrate the transformation of business processes based on the notion of business entities. As an experiment, we examine the process models in IBM Insurance Application Architecture (IAA) [13]. IAA is a comprehensive set of best practice process models for the insurance industry. In general, these process models are used for analytical purposes, and thus are called analysis models. In practice, system analysts manually customize these analysis models case-by-case in order to implement them. This customization involves the redesign of business activities and data models. We propose to use the transformation algorithm to automatically generate information-centric process models which may then be used to implement service-oriented business process management solutions. Fig. 5 gives an example process for administering property damage claims. In this process, a claim information entity is created when a loss event is reported. This claim is validated and analyzed which could lead to one of three outcomes: rejection of the claim, acceptance of the claim, or postponement of a decision pending additional information. If the claim is accepted, the benefit in this claim is determined and then a payment is issued. If the claim is pended, arrival of additional information leads to another round of processing. In addition to activities and entities, this process contains control nodes (AND-SPLIT, AND-JOIN, OR-SPLIT, OR-JOIN, START and END) and connectors.

In general, an activity-centric process model can contain both data flows and control flows. Wang and Kumar [25] classified two types of constraints in process activities, hard constraints that arise from data dependencies, and soft constraints caused by business rules or policies. In the absence of hard constraints, control flows become necessary to sequence activities. The process model shown in Fig. 5 contains both types of flows, but some of the control flow links are redundant. For example, information entity *claim* is an output of activity *notify claim*, and it becomes an input of activity *record claim*, implying that *record claim* has to be executed after *notify claim*.

We consider this single process as a process scope. This process scope provides an end-to-end function from creating a claim after a loss event is notified to closing the claim and managing payment. Also, our investigation shows that information entities in this process model are self-describing, non-overlapping and decoupled.

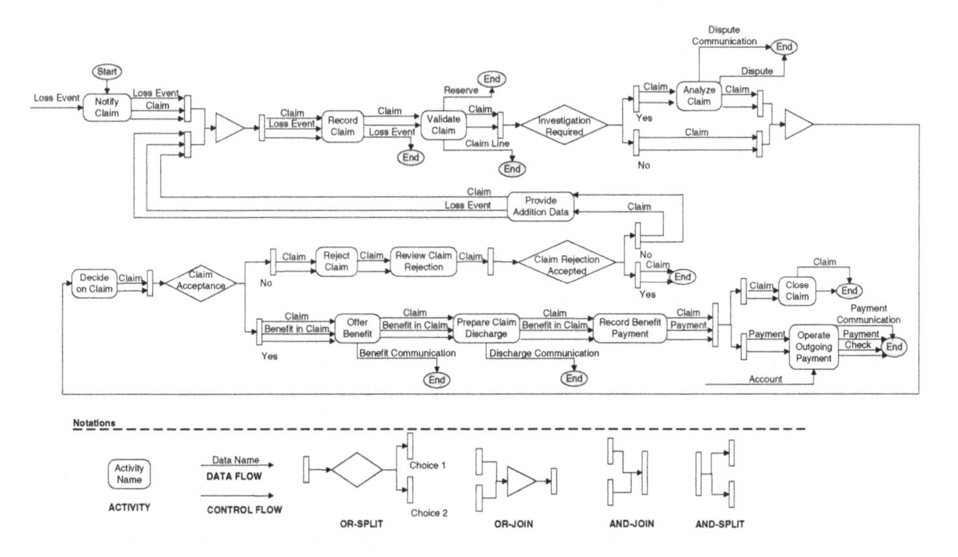

Fig. 5. Administering Property Damage Claim

Based on the concept of domination, we discover three dominant entities involved in this process: *claim*, *loss event*, and *payment*. *Claim* and *payment* business entities have several dominated entities, as shown in the entity domination graph of Fig. 6.

Next, based on the entity domination graph, we can create containment data models [27] for each business entity. Fig. 7 shows the data models. In each data model, the root container is a business entity. The *claim* business entity contains *claim line*, *dispute*, *dispute communication*, *reserve*, and *benefit in claim*, which, in turn, contains *discharge communication* and *benefit communication*. The *loss event* does not contain any entity, but it may have attributes and child items. In this paper, for simplicity, we omit the detailed attributes of each entity. In addition, there may be data relationships between business entities. For example, in Fig. 8, *claim* is created by *loss event* through activity *Record Claim*.

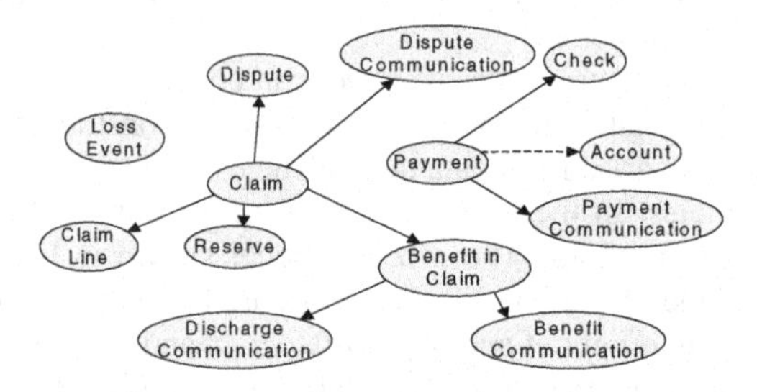

Fig. 6. Business entities – Administer Property Damage Claim

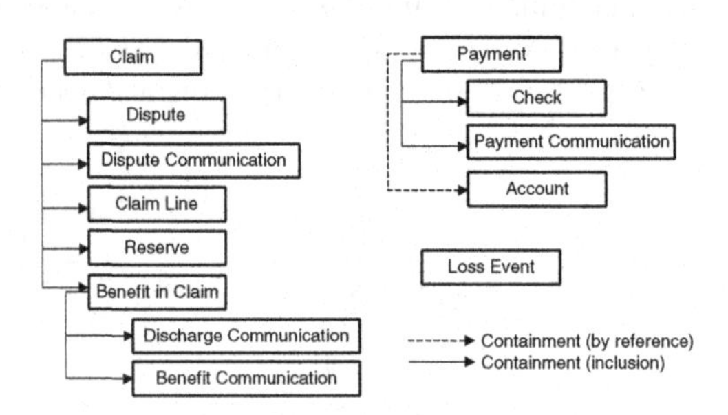

Fig. 7. Data Model of Administering Property Damage Claim

With the discovered dominant entities, we can easily construct the behavior model of each dominant entity and then convert the activity-centric model into an information-centric model using the algorithm shown in Fig. 4. Fig. 8 shows the information-centric process model consisting of three connected state machines. The state machine of *claim* describes the lifecycle of the *claim* business entity and it interacts with the other business entities, *loss event* and *payment*. For example, during *claim*'s state transition from *created* to *recorded*, the *loss event* business entity also changes its state from *notified* to the final state *recorded*. Similarly, *record benefit payment* changes the state of the *claim* business entity from *discharged* to *closed*, while creating a new instance of *payment* business entity.

The business entity behavior models provide a new perspective from which to reason about business activities. Ideally, a business activity should produce some meaningful change to a business entity, resulting in a new milestone in the lifecycle of that entity, which should be monitored and tracked for performance management. If a business activity does not bring such changes, that activity can either be removed or combined with others. Therefore, business entities actually provide guidelines for determining the right granularity of activities. For example, in Fig. 8, activity *offer benefit* is likely to notify a claimant without providing any actual changes to the *claim* business entity. This activity may be merged with the activity *prepare claim discharge*. Also, if there is a significant overlap between two behavior models, we may need to re-design the activities to decouple the business entities. Note that the algorithm in Fig. 5 does not give any state naming convention. One can name business entity states based on results produced by activities, as exemplified by Fig. 8.

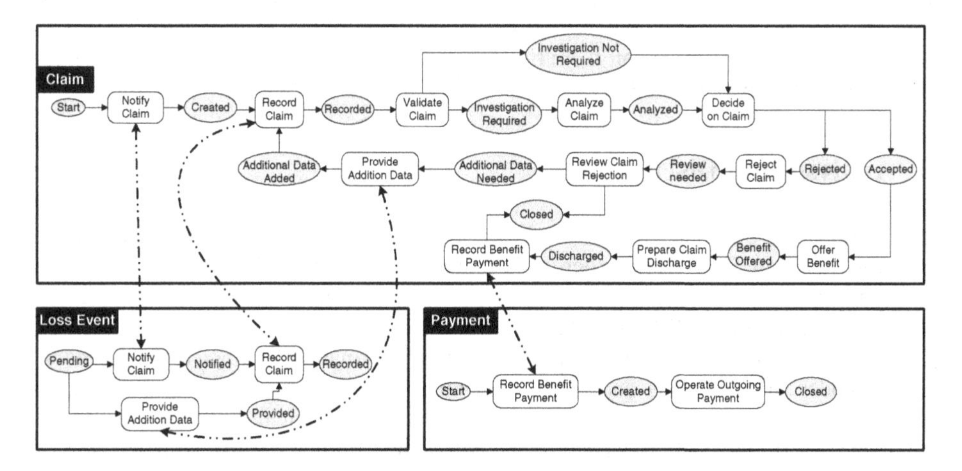

Fig. 8. Information-centric process – Administering Property Damage Claim

Compared with the original model in Fig. 5, the model in Fig. 8 has the following features. First, this information-centric process model provides better understandability because the introduced business entities highlight the focus of the process. Obviously, the process mainly deals with *claim*, tracking its behavior through its end-to-end lifecycle from creation to closure. Understandability is further improved by the decomposition of the process into three streamlined state machines each with fewer activity nodes. Empirical evidence shows that model size and average connector degree significantly affect the understandability of process models [19].

Second, the information-centric model hides IT implementation details and only describes the business entities that each business activity acts on. In Fig. 5, each business activity has detailed input and output entities. However, the information-centric model only specifies the business entities that each activity reads, updates or creates. Through the data model, an activity is able to retrieve information of the dominated entities. In practice, data access can be defined as data views by user roles and by business activities. By adding data access details for each activity, we can convert the information-centric process back to the activity-centric model. However, data access details are an IT implementation issue. We prefer to delay the definition of data access until implementation for two reasons. First, during implementation, precise data access may be defined at the attribute level instead of the entity level. Therefore, the input and output specifications in terms of entities in Fig. 5 are not sufficient. Second, in reality, data access varies with each implementation. Without data access details, an information-centric process model can be easily adapted into different process scopes. As evidence, IBM Insurance Application Architecture [13] contains seven process models, each describing a particular type of insurance claim, such as medical expense claim, life claim, and auto claim. Our analysis shows that these models can be transformed into the same information-centric process as in Fig. 8, with slightly different data graphs. For example, in the process *administering auto claim*, *benefit in claim by insurer* is one of the final output entities, instead of *payment*.

Finally, using the Model Driven Business Transformation Toolkit [17], we can generate business applications automatically from the information-centric business process models. The development and implementation time can be greatly reduced. Also, this direct transformation from business process models to IT solutions reduces the gap between business and IT.

4 Analysis and Discussion

The domination concept reveals the deep structure of the information domain of a business function. This deep structure is represented as entity domination graphs in the form of *directed acyclic graphs (DAG)*, with the dominant information entities at the source nodes of the DAGs serving as the driving force for the process flows that constitute the business function. The dominated information entities form the non-source nodes of the DAGs and play a subsidiary role in the execution of business activities. Usually, the dominated entities are created during the processing of the dominant entity and their existence depends on the dominant entity. But the dominated entities do play an important role in the lifecycle of the dominant entity. It is analogous to the growth of a tree in nature, with new branches being added as the tree progresses through its lifecycle. For example, in the insurance claim process discussed earlier, *dispute*, *claim line* and *benefit in claim* are created during the processing of the *claim* entity and their existence depends on the *claim* entity. And when a *claim* is *accepted*, one can expect to see *benefit in claim* added to the data graph associated with *claim*.

An intuitive explanation of the domination concept can be derived from the Pareto principle which states that, for many events, 80% of the effects come from 20% of the causes. When applied to business process analysis, we observe that a few information entities serve as key drivers of the flow of most activities. Using our algorithm, we are able to select the dominant entities and model their behaviors, thus leading to significant reduction in model complexity and better understanding of business operations. For example, we have observed that among 320 information entities used in IBM Insurance Application Architecture [13], only 90 qualify as dominant entities. We can also view domination as a special association rule mining [2] for discovering antecedent and consequent information entities and establishing associations between them.

The algorithm presented in this paper leverages the domination concept to transform an activity-centric business process model into an information-centric model. There are several advantages to be gained from creating such an information-centric model as discussed below.

As we have seen above, an activity-centric model of a business process enumerates all the activities in the process and defines a control flow and a data flow over these activities. As the processes grow in size and complexity, it becomes increasingly difficult to understand the business behavior using these models [3]. The traditional approach to dealing with this complexity is to resort to a hierarchical representation of business processes. It has been shown that static, hierarchical representations of business processes are not conducive to in-depth analysis and prediction of the behavior of the system under dynamic conditions [20]. The information-centric modeling approach presents an attractive alternative as it helps to analyze and predict system behavior using the lifecycle models of a few business entities in a flat structure.

Designing user-friendly human interfaces from activity-centric models to drive business process execution has been known to be a challenge [28]. Business users are knowledge workers who need contextual information to make the decisions needed in performing a business activity. This contextual information is not part of the activity-centric models of business processes and thus it becomes hard to design effective human-interfaces from such models to drive process execution. Information-centric models help in this regard since the deep structure of the information domain represented as a DAG holds this contextual information. The lack of contextual information in activity-centric process models has business-level implications as well. Since such models emphasize control flows and look at information usage only within the context of a single activity, business actors tend to focus on "what should be done instead of what can be done", hindering operational innovations [1,12].

The issue with interface design to business process execution systems gets more complex when conversational interfaces are involved. In a conversational interface, the input to a process has to be defined and refined incrementally as part of the process execution [28]. The unpredictability of the input makes it hard to precisely determine the execution order of activities during the modeling phase. For example, Zimmermann et al. [28] reported difficulties in designing conversational interfaces in a large order management process using BPEL [9]. However, an information-centric model of a process naturally supports such constraints because it models the business process as business entity lifecycles, and the set of client interactions that move a business entity though its lifecycle becomes a conversation.

Information-centric models of business processes have important implications with respect to implementations of business process management systems. Activity-centric models such as BPEL [9] and Event-driven Process Chains (EPC) [24] are executed by flow-driven workflow systems. In such systems, the process execution complexity can be classified into three types: control-flow complexity, data-flow complexity, and resource complexity [10]. Typically, the complexity increases rapidly as the number of control nodes (e.g. Pick, Flow and Switch nodes in BPEL) increases [10], thus severely impacting the scalability of this approach. In contrast, information-centric models enable the execution of the process as a set of communicating finite state machines, which significantly improves process execution efficiency [21].

The advantages of a modular design in building complex systems are well known [4], but the challenge lies in identifying the right modules. Business entities provide a natural way to modularize a business process management system. Each module implements the behavior of a business entity as a state machine and manages the information entities associated with that business entity. This approach to modularization leads to a new way to decompose business processes and implement them using service oriented architecture (SOA) [11]. With increasing industrialization of services, companies tend to decompose their business processes for selective outsourcing, selective automation, or restructuring to create decoupled composite business services which may be flexibly integrated to support end-to-end business processes [16]. Therefore, there is a need for a systematic way for companies to analyze and decouple their processes. Our algorithm does precisely this analysis and decoupling. Intuitively, each business entity along with its state machine defines a decoupled, composite business service [11]. In addition, service interfaces can be derived from the connections between business entities and the communication between these entities can be implemented as service invocations. For example, in Fig. 8, a company can define the *claim* portion as the core process which drives customer value, but outsource the *payment* portion. Both *claim* and *payment* may now be implemented using business entities as composite business services and the end-to-end claim business process can be realized via service invocations on these entities. The details about implementing information-centric business processes using SOA principles can be found in [8].

5 Related Work

Recently, information-centric modeling has become an area of growing interest. Nigam and Caswell [22] introduced the concept of business artifacts and information-centric processing of artifact lifecycles. Kumaran et al. [17] developed adaptive business documents as the programming model for information-centric business processes and this model later evolved into adaptive business objects [21]. Further studies on business artifacts and information-centric processes can be found in [6, 7, 8, 15]. [6] describes a successful business engagement which applies business artifact techniques to industrialize discovery processes in pharmaceutical research. More engagements using information-centric modeling can be found in [8]. Liu et al. [15] formulated nine commonly used patterns in information-centric business operation models and developed a computational model based on Petri Nets. [7] provides a formal model for artifact-centric business processes with complexity results concerning static analysis of the semantics of such processes. While previous work mainly focuses on completing the framework of information-centric process modeling from theoretical development to practical engagements, our work bridges the gap between activity-centric and information-centric models and shows the duality between them.

Other approaches related to information centric modeling can be found in [1, 25]. [1] provides a case-handling approach where a process is driven by the presence of data objects instead of control flows. A case is similar to the business entity concept in many respects. In [25], document-driven workflow systems are designed based on data dependencies without the need for explicit control flows. In this paper, in addition to tracking the behavior of data objects, we are interested in their deep structure.

Another related thread of work is the use of state machines to model object lifecycles. Industries often define data objects and standardize their lifecycles as state machines to facilitate interoperability between industry partners and enforce legal regulations [23]. [18] gives a technique to generate business processes which are compliant with predefined object lifecycles. Instead of assuming predefined business objects, our approach discovers business entities from process models and then defines their lifecycles as an alternative representation of process models. In addition, event-driven process modeling, for example, Event-driven Process Chains (EPC) [24], also describes object lifecycles glued by events, such as "material in stock". Our approach in this regard is also event-driven, as each business entity state can be viewed as an event. However, EPC is still an activity-centric approach as objects are added to functions as inputs or outputs and an event can be defined concerning a group of objects.

Some other notable studies that are related to our work are in the area of process decomposition and service oriented architecture. Basu and Blanning [5] presented a formal analysis of process synthesis and decomposition using a mathematical structure called metagraph. This study gives three useful criteria: full connectivity, independence, and redundancy, for examining process synthesis and decomposition. However, the metagraph approach is not applicable to process models, such as the one shown in Fig. 6, which contains many cycles when formulated as a metagraph.

6 Conclusion and Future Work

In this paper, we have presented an approach to discovering business entities from activity-centric process models and transforming such models into information-centric business process models. An algorithm was provided to achieve this transformation automatically. We illustrated this approach with a comprehensive example and tested it using reference processes from the insurance industry.

Our approach provides an alternative way to implement activity-centric process models. Instead of transforming them into BPEL processes or workflows, our approach generates information-centric models from them and implements these models using the Model-Driven Business Transformation Toolkit [21], thereby improving both the understandability of process models and their execution efficiency. Additionally, this approach provides a new way to decompose business processes into connected business entities, each of which can be implemented as a business service using SOA principles.

We are currently developing a tool based on this algorithm and applying this approach to best practice processes in the IT service delivery industry. We expect our future work to extend this algorithm to other types of process models, including BPEL and EPC models. Another research direction is to relax the concept of domination so that business entities can be discovered from models with incomplete or incorrect specifications of input or output information entities.

Acknowledgments. The authors thank Kumar Bhaskaran, David Cohn, Anil Nigam, John Vergo and other colleagues for their helpful discussion and comments.

Reference

1. Aalst, W.M.P., Weske, M., Grunbauer, D.: Case handling: a new paradigm for business process support. Data and Knowledge Engineering 53, 129–162 (2005)
2. Agrawal, R., Imielinski, T., Swami, A.: Mining Association Rules Between Sets of Items in Large Database. In: Proceedings of ACM-SIGMOD 1993, May 1993, pp. 207–216 (1993)
3. Alonso, G., Agrawal, D., El Abbadi, A., Mohan, C.: Functionalities and Limitations of Current Workflow Management Systems. IEEE Expert 12, 5 (1997)
4. Baldwin, C.Y., Clark, K.B.: Design Rules. The Power of Modularity, vol. 1. MIT Press, Cambridge (2000)
5. Basu, A., Blanning, R.W.: Synthesis and Decomposition of Processes in Organizations. Information Systems Research 14(4), 337–355 (2003)
6. Bhattacharya, K., Guttman, R., Lyman, K., Heath, I.F.F., Kumaran, S., Nandi, P., Wu, F., Athma, P., Freiberg, C., Johannsen, L., Staudt, A.: A model-driven approach to industrializing discovery processes in pharmaceutical research. IBM Systems Journal 44(1), 145–162 (2005)
7. Bhattacharya, K., Gerede, C., Hull, R., Liu, R., Su, J.: Towards Formal Analysis of Artifact-Centric Business Process Models. In: Alonso, G., Dadam, P., Rosemann, M. (eds.) BPM 2007. LNCS, vol. 4714, pp. 288–304. Springer, Heidelberg (2007)
8. Bhattacharya, K., Caswell, N., Kumaran, S., Nigam, A., Wu, F.: Artifact-centric Operatonal Modeling: Lessons learned from engagements. IBM Systems Journal 46(4) (2007)
9. BPEL, Business Process Execution Language for Web Services, version 1.1, joint specification by BEA, IBM, Microsoft, SAP and Siebel Systems (2003)
10. Cardoso, J.: Complexity Analysis of BPEL Web Processes. Software Process: Improvement and Practice Journal 12, 35–49 (2007)
11. Ferguson, D.F., Stockton, M.L.: Service-oriented architecture: programming model and product architecture. IBM Systems Journal 44(4), 753–780 (2005)
12. Hammer, M.: Deep change: How operational innovation can transform your company. Harvard Business Review, 84–93 (April 2004)
13. IBM Insurance Application Architecture (IAA), version 7.1 (2004), http://www-03.ibm.com/industries/financialservices/doc/content/solution/278918103.html
14. Kumaran, S.: Model Driven Enterprise. In: Proceedings of Global Integration Summit 2004, Banff, Canada (2004)
15. Liu, R., Bhattacharya, K., Wu, F.Y.: Modeling Business Contexture and Behavior Using Business Artifacts. In: Krogstie, J., Opdahl, A., Sindre, G. (eds.) CAiSE 2007 and WES 2007. LNCS, vol. 4495, pp. 324–339. Springer, Heidelberg (2007)
16. Karmarkar, U.: Will You Survive the Services Revolution? Harvard Business Review 82(6), 100–107 (2004)
17. Kumaran, S., Nandi, P., Heath, T., Bhaskaran, K., Das, R.: ADoc-oriented programming. In: Symposium on Applications and the Internet (SAINT), pp. 334–343 (2003)
18. Küster, J.M., Ryndina, K., Gall, H.: Generation of Business Process Models for Object Life Cycle Compliance. In: Alonso, G., Dadam, P., Rosemann, M. (eds.) BPM 2007. LNCS, vol. 4714, pp. 165–181. Springer, Heidelberg (2007)
19. Mendling, J., Reijers, H.A., Cardoso, J.: What Makes Process Models Understandable? In: Alonso, G., Dadam, P., Rosemann, M. (eds.) BPM 2007. LNCS, vol. 4714, pp. 48–63. Springer, Heidelberg (2007)
20. Modarres, M.: Predicting and Improving Complex Business Processes: Values and Limitations of Modeling and Simulation Technologies. In: Winter Simulation Conference (2006)
21. Nandi, P., Kumaran, S.: Adaptive business objects – a new component model for business integration. In: Proceedings of International Conference on Enterprise Information Systems, pp. 179–188 (2005)

22. Nigam, A., Caswell, N.S.: Business artifacts: An approach to operational specification. IBM Systems Journal 42(3), 428–445 (2003)
23. Ryndina, K., Küster, J.M., Gall, H.: Consistency of Business Process Models and Object Life Cycles. In: Kühne, T. (ed.) MoDELS 2006. LNCS, vol. 4364, pp. 80–90. Springer, Heidelberg (2007)
24. Scheer, A.W.: Business Process Engineering: Reference Models for Industrial Enterprises, 2nd edn. Springer, Heidelberg (1997)
25. Wang, J., Kumar, A.: A Framework for Document-Driven Workflow Systems. In: Proceedings of Business Process Management, pp. 285–301 (2005)
26. WfMC, The Workflow Reference Model, Issue 1.1, Document Number TC00-1003, Work-flow Management Coalition, Winchester, UK (1995)
27. Whitehead, E.J.: Uniform comparison of data models using containment modeling. In: Proceedings of the thirteenth ACM conference on Hypertext and hypermedia, Maryland, USA, pp. 182–191 (2002)
28. Zimmermann, O., Doubrovski, V., Grundler, J., Hogg, K.: Service-oriented architecture and business process choreography in an order management scenario: rationale, concepts, lessons learned. In: Proc. of the 20th SIGPLAN Conference on Object-Oriented Programming, Systems, Languages, and Applications, pp. 301–312 (2005)

A New Paradigm for the Enactment and Dynamic Adaptation of Data-Driven Process Structures*

Dominic Müller[1,2], Manfred Reichert[1,3], and Joachim Herbst[2]

[1] Institute of Databases and Information Systems, Ulm University, Germany
{dominic.mueller,manfred.reichert}@uni-ulm.de
[2] Dept. GR/EPD, Daimler AG Group Research & Advanced Engineering, Germany
joachim.j.herbst@daimler.com
[3] Information Systems Group, University of Twente, The Netherlands

Abstract. Industry is increasingly demanding IT support for large engineering processes, i.e., process structures consisting of hundreds up to thousands of processes. Developing a car, for example, requires the coordination of development processes for hundreds of components. Each of these development processes itself comprises a number of interdependent processes for designing, testing, and releasing the respective component. Typically, the resulting process structure becomes very large and is characterized by a strong relation with the assembly of the product. Such process structures are denoted as *data-driven*. On the one hand, the strong linkage between data and processes can be utilized for automatically creating process structures. On the other hand, it is useful for (dynamically) adapting process structures at a high level of abstraction. This paper presents new techniques for (dynamically) adapting data-driven process structures. We discuss fundamental correctness criteria needed for (automatically) detecting and disallowing dynamic changes which would lead to an inconsistent runtime situation. Altogether, our COREPRO approach provides a new paradigm for changing data-driven process structures at runtime reducing costs of change significantly.

Keywords: Process Coordination, Data-driven Process, Process Adaptation.

1 Introduction

In the engineering domain, the development of complex products (e.g., cars or airplanes) necessitates the coordination of thousands of processes (e.g., to design, test and release each product component). These processes and the many dependencies between them form complex and large *process structures*. While the single processes are usually implemented within different IT systems, the

* This work has been funded by *Daimler AG Group Research* and has been conducted in the *COREPRO* project (http://www.uni-ulm.de/dbis)

Z. Bellahsène and M. Léonard (Eds.): CAiSE 2008, LNCS 5074, pp. 48–63, 2008.

design, coordination, and maintenance of process structures are only rudimentarily supported by current process management technology [1]. In most cases, the different processes have to be manually composed and coordinated.

Another challenge constitutes the dynamic adaptation of process structures during runtime. When adding or removing a car component (e.g., a navigation system), for example, the process structure has to be adapted accordingly; i.e., processes for designing, testing and releasing affected product components have to be manually added or removed. Note that this also might require the insertion or removal of their synchronization dependencies with respect to other processes from the total process structure. Manually modeling and adapting large process structures requires profound process knowledge and is error-prone as well. Incorrectly specified dependencies within a process structure, however, can cause delays or deadlocks blocking the execution of the whole process structure.

Example 1. *To identify practical requirements, we investigated a variety of process structures in the automotive industry. As example consider release management (RLM) processes for electrical systems in a car. The (product) data structure (or configuration structure) for the total electrical system consists of up to 300 interconnected components which are organized by using a product data management system. The goal of RLM is to systematically test and release the different product components at a specific point in time, for example, when a certain milestone is reached. To verify the functionality of the electrical system, several processes (e.g.,* testing *and* release*) have to be executed for each electrical (sub-)component and need to be synchronized with the processes of other components. This is needed, for example, to specify that an electrical system can only be tested if all its components are individually tested before. Altogether, more than 1300 processes need to be coordinated for releasing the total electrical system [2].*

Interestingly, large process structures often show a strong linkage with the assembly of the product; i.e., the processes to be coordinated can be explicitly assigned to the different product components (cf. Example 1). Further, process synchronizations are correlated with the relations existing between the product components. In the following, we denote such process structures as *data-driven*. COREPRO utilizes the information about a product, its components and the component relations. In particular, COREPRO provides techniques for the modeling, enactment, and (dynamic) adaptation of process structures based on given (product) data structures. For example, the assembly of the (product) data structure can be used to automatically create the related process structure [2]. We have shown that COREPRO reduces modeling efforts for RLM process structures (cf. Example 1) by more than 90% [2].

When changing a (product) data structure at buildtime, the related process structure can be automatically re-created. However, long running process structures also need to be adapted during runtime. Then, it does not make sense to re-create the process structure from scratch and to restart its execution from the beginning. Instead, in-progress process structures must be adaptable on-the-fly, but without leading to faulty synchronizations like deadlocks. A particular challenge is to enable users to adapt complex process structures. When the product structure is changed, we can take benefit from the strong linkage between process structure and (product) data structure again. In COREPRO, data structure changes are automatically translated into adaptations of the corresponding process structure. Thus, changes of data-driven process structures can be introduced by users at a high level of abstraction, which reduces complexity as well as cost of change significantly. Note that this is far from being trivial considering the large number of processes to be coordinated and their complex interdependencies.

In this paper we introduce the COREPRO runtime framework, which addresses the aforementioned requirements. COREPRO enables enactment of data-driven process structures based on a precise and well-defined operational semantics. We show how to translate (dynamic) changes of a currently processed data structure into corresponding adaptations of the related process structure. Finally, we introduce consistency criteria in order to ensure that dynamic adaptations of a process structure lead to a correct process structure again. Altogether, COREPRO addresses the full life cycle of data-driven process structures including modeling, enactment, and (dynamic) adaptation phases.

Sect. 2 describes the COREPRO modeling approach for data-driven process structures and Sect. 3 specifies operational semantics for process structures. Sect. 4 shows how to adapt data-driven process structures and presents methods for detecting invalid changes that would lead to incorrect runtime situations. We discuss related work in Sect. 5 and conclude with a summary in Sect. 6.

2 COREPRO Modeling Framework

The COREPRO modeling framework allows to create data-driven process structures (cf. Fig. 1d and f) [2]. For this, we consider the sequence of states a (data) object goes through during its lifetime. A product component from Example 1 passes states like *designed*, *tested* and *released*. Generally, state transitions are triggered by executing processes modifying the respective object (e.g., *design*, *test* or *release*). An *object life cycle* (OLC) constitutes an integrated and user-friendly view on the states of a particular object and its manipulating processes (cf. Fig. 1d and f). Regarding data-driven process structures, OLC state transitions do not only depend on the processes associated with the respective object, but also on the states and state transitions of other objects. As example consider a total car, which can be only tested, if all sub-systems (e.g., engine, chassis and navigation system) are tested before. By connecting the states of different OLCs, a logical view on a data-driven process structure results (cf. Fig. 1d and f).

Example 2. *The total electrical system of a car (cf. Example 1) differs from car series to car series (e.g., series with and without navigation system). The strong relationship between data structures and process structures implies that different (product) data structures (related to the different series) lead to different RLM process structures (e.g., process structures with and without development processes for the navigation system). Each of these process structures then represents one instance of the total development process for a particular car series.*

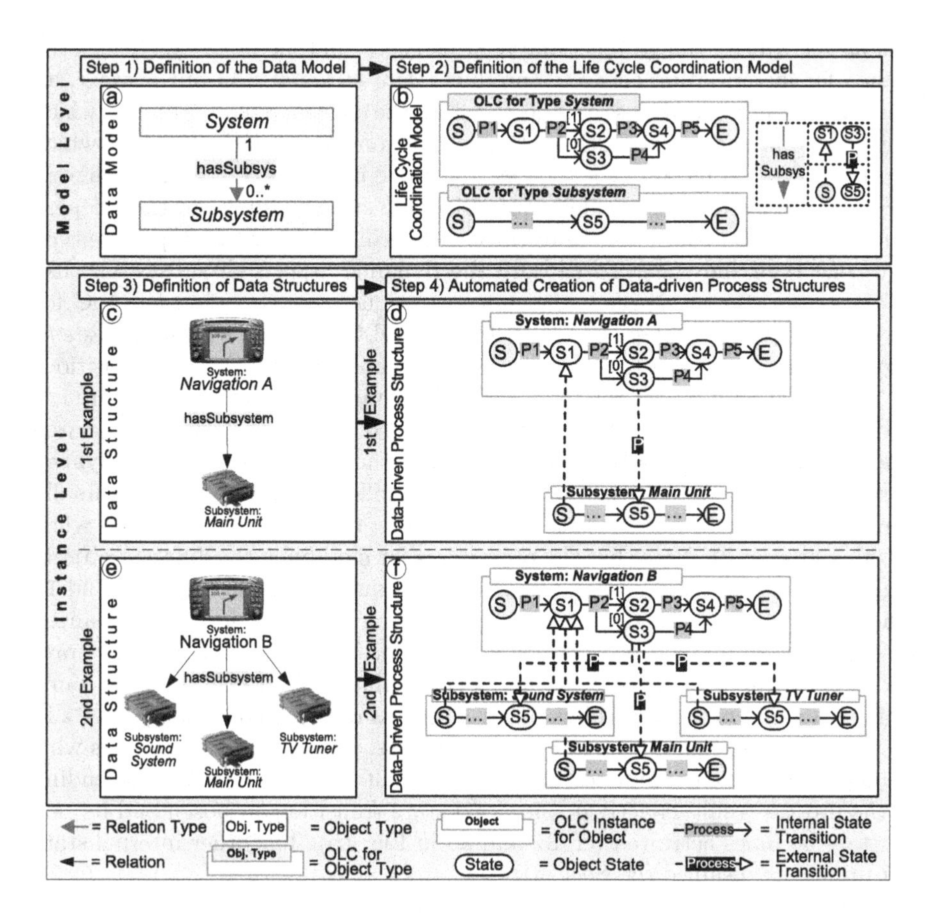

Fig. 1. Creation of Data-Driven Process Structures

Modeling OLCs for hundreds of objects (cf. Example 1) and linking them manually would be too expensive. Therefore, COREPRO follows a model-driven (respectively data-driven) approach by differing between model and instance level. First, for objects with same behavior, we model only one OLC. This OLC can be instantiated multiple times. Second, we utilize the close relationship between data and process structure when connecting different OLCs. Our case studies have shown that semantic relations between data objects can be mapped to dependencies between particular states of OLCs. Based on object (types) and relations between them, process structures can be specified at model level and then be automatically instantiated for every given data structure [2].

2.1 Defining Object Types and Object Life Cycles

Fig. 1 shows the steps necessary to model and instantiate both data and process structures. Steps 1 + 2 deal with the creation of the model level, Step 3 + 4 with the definition of the instance level. Step 1 is related to the specification of the *data model*, which defines object and relation types, and therefore constitutes the schema for instantiating concrete (product) *data structures.* In our context, an object type represents a class of objects within the data model (cf. Fig. 1a), which then can be instantiated multiples times. Fig. 1e shows a data structure which comprises one instance of type `System` and three instances of type `Subsystem`.

Step 2 is related to the modeling of object life cycles (OLC) and their dependencies. An OLC defines the states through which objects of the respective type go during their life time. To capture the dynamic aspects of subsystems like `TV tuner` and `Sound System`, the process designer models only one OLC for the object type `Subsystem`. COREPRO maps OLCs to *state transition systems* whose states correspond to object states and whose (internal) state transitions are associated with object-related processes. A state transition does not fire immediately when its source state becomes activated, but waits until its associated process has completed. In Fig. 1b, for example, the OLC for object type `System` starts with initial state `S`, then passes state `S1` followed by `S2` or `S3`, and finally completes in state `E`. The *internal state transition* from `S` to `S1` takes place when finishing process `P1`. Note that the procedures for modifying the different objects might slightly differ (e.g., different testing procedures for `Sound System` and `TV Tuner`). If the OLCs are identical for such objects, this behavior can be realized as variants within the processes to be executed. Non-deterministic state transitions are realized by associating different internal state transitions with same source state and process, and by adding a process result as condition (e.g., `P2` in Fig. 1b). This is required, for example, when associating a testing process with a component, which completes either with result *correct* or *faulty.* Depending on the process result, exactly one internal state transition is chosen and its target state becomes activated (cf. `S2` and `S3` in Fig. 1b). The other internal state transitions are disabled (cf. Sect. 3).

2.2 Modeling Object Relations and OLC Dependencies

Defining the dynamic behavior for each object type by modeling its OLC is only half of the story. We also have to deal with the state dependencies existing between the OLCs of different objects types (cf. Example 1). In COREPRO, an OLC dependency is expressed by *external state transitions* between concurrently enacted OLCs. Like an internal state transition within an OLC, an external state transition can be associated with the enactment of a process; i.e., we do not only support event-based synchronizations, but also allow for the enactment of a transition process (e.g., `send email` or `install`). Regarding the two OLCs modeled in Step 2 from Fig. 1b, for example, we associate the external state transition between state `S3` (of the OLC for type `System`) and state `S5` (of the OLC for type `Subsystem`) with process `P`. To benefit from the strong linkage between object relations and OLC dependencies, external state transitions are mapped to object relation types. Regarding our example, the aforementioned external state transition is linked to the `hasSubSys` relation connecting the two object types `System` and `Subsystem`. This information can later be utilized for automatically creating process structures out of given data structures.

The OLCs of all object types and the external state transitions of all relation types form the *Life Cycle Coordination Model* (LCM) (cf. Fig. 1b). The LCM describes the dynamic aspects of the data model and constitutes the scheme for creating data-driven process structures. This is a unique characteristic of COREPRO allowing for the data-driven configuration of process structures. In particular, different process structures (cf. Fig. 1d and Fig. 1f) can be automatically created by instantiating respective data structures (cf. Example 2).

2.3 Generating Data-Driven Process Structures

Picking up the scenario from Example 2, COREPRO allows to instantiate different data structures (cf. Fig. 1c and 1e) and to automatically create related process structures (cf. Fig. 1d and 1f). A data-driven process structure includes an OLC instance for every object from the data structure. This can be seen in Fig. 1e (data structure) and Fig. 1f (related process structure) where the numbers of objects and OLC instances correspond to each other. Likewise, as specified in the LCM, for each relation in the data structure external state transitions are inserted into the process structure; e.g., for every `hasSubsystem` relation in the data structure from Fig. 1e, associated external state transitions (with process `P`) are inserted. As result we obtain an instantiated executable process structure describing the dynamic aspects of the given data structure (cf. Fig. 1d and 1f).

To ensure a correct dynamic behavior, created process structures must be *sound*. A process structure is considered as being sound iff it always terminates properly [3]. Termination of the process structure will be guaranteed if every OLC is sound and there are no cycles caused by external state transitions. An OLC, in turn, is sound (1) if every state can be activated by firing a sequence of state transitions beginning with the start state of the OLC and (2) if the end state of the OLC can be activated by firing a sequence of state transitions beginning from every state within the OLC; further, no state transition must be processing when reaching the OLC end state (i.e., no process is running). It is important to mention that COREPRO allows for checking soundness on model level, i.e., we can ensure that each process structure derived from a sound LCM is sound as well [2]. Thereby, efforts for soundness checks do not rise with the size of the instantiated process structure but only depend on the size of the LCM.

3 Dynamic Behavior of Data-Driven Process Structures

To correctly enact data-driven process structures, a precise and formal operational semantics is needed. This is also fundamental with respect to the analysis of the correctness and consistency of process structures when dynamically changing them (cf. Sect. 4). Therefore, COREPRO uses different *markings* to reflect the runtime status of an enacted process structure. We annotate both, states and (internal as well as external) state transitions with respective markings each of them representing their current runtime status. Fig. 2 shows an example where markings describe a possible runtime status of the process structure depicted

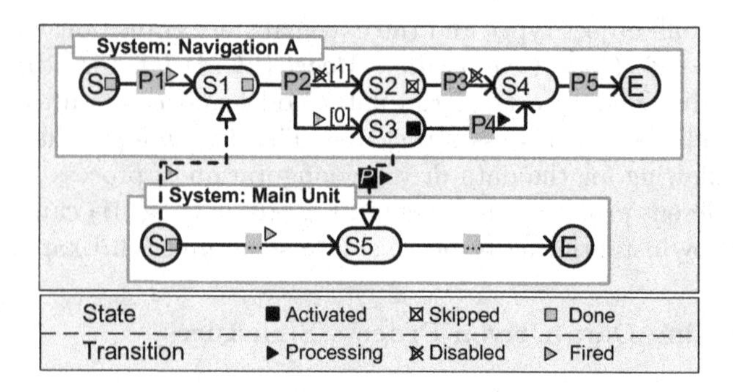

Fig. 2. Process Structure from Fig. 1d with Markings During Runtime

in Fig. 1d. By analyzing the *state markings* from Fig. 2, we can immediately figure out whether a particular state of a process structure has been already passed (e.g., state S1), is currently activated (e.g., state S3), has been skipped (e.g., state S2), or has not been reached yet (e.g., state S4). *Transition markings*, in turn, indicate whether the associated process has been started, skipped or completed. As we will see later, the use of markings eases consistency checks and status adaptations in the context of dynamic changes of process structures significantly. Therefore, markings of passed regions are preserved during runtime.

Altogether, the runtime status of a process structure is determined by the current markings of its constituent OLCs and (external) state transitions. We first introduce the operational semantics of single OLCs and then the one of the external state transitions needed to synchronize the different OLCs.

3.1 Operational Semantics of Single OLCs

Each state of an (instantiated) OLC has one of the markings `NotActivated`, `Activated`, `Skipped`, or `Done`. Fig. 3a shows the conditions for setting these markings. Initial marking of a state is `NotActivated`. An OLC state becomes activated after one incoming internal state transition has fired and all external state transitions are either fired or disabled. To realize this, marking `NotActivated` is subdivided into `IntActivated` and `ExtActivated` (cf. Fig. 3a). A state will become `Activated` if both submarkings, `IntActivated` and `ExtActivated` are reached. At the same time, the preceding state within the OLC is set to `Done`. This excludes concurrently activated states within a single OLC.

The dynamic behavior of OLCs is governed by internal state transitions. An internal state transition enters marking `Processing` and its associated process is started, when the marking of its source state switches to `Activated` (cf. Fig. 3b). When the associated process completes, the marking of the internal state transition either turns to `Fired` or `Disabled` depending on the result of the process (cf. Sect. 2.1). At the same time, the target state of the respective transition enters submarking `IntActivated` (cf. Fig. 3a).

When a transition is disabled, deadpath elimination takes place. Its target state is skipped, and succeeding internal state transitions are disabled as well.

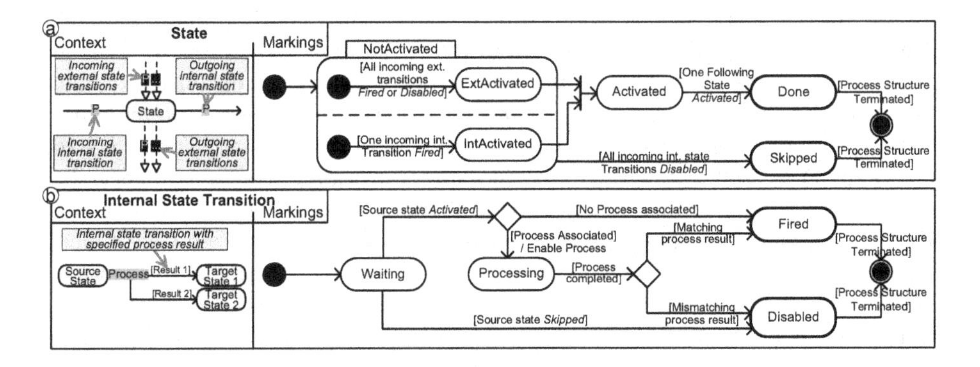

Fig. 3. Behavior of Object States and Internal State Transitions in OLCs

This is continued until the end of the dead path is reached; i.e., until a state is reached which has at least one incoming internal state transition not marked as `Disabled` (cf. Fig. 3b). Deadpath elimination contributes to avoid deadlocks caused by external state transitions which are waiting for activation (cf. Fig. 2, external state transition with process `P`). The described OLC semantics has to be extended in conjunction with loops. Due to lack of space, however, we cannot treat loop backs in this paper.

3.2 Synchronization of OLCs

As mentioned, concurrent processing of objects is a fundamental requirement in engineering processes. At the level of a single OLC, concurrency is encapsulated within the transition processes (e.g., implemented in a workflow management or a product data management system). The proper synchronization of concurrently enacted processes within different OLCs is based on external state transitions. According to internal state transitions, external state transitions are marked as `Processing` (i.e., their processes are started) when their source state becomes `Activated` (cf. Fig. 3 and Fig. 4). When the associated process is completed, in turn, the external state transition is marked as `Fired`. The target state of the fired transition can only become `Activated` if one of its incoming internal transitions has already been marked as `Fired` and all external state transitions are marked either as `Fired` or `Disabled` (cf. Fig. 3a and Fig. 4).

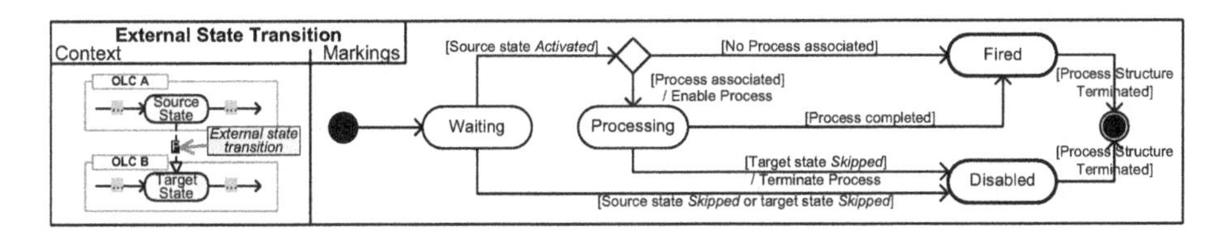

Fig. 4. Dynamic Behavior of External State Transitions Between OLCs

External state transitions whose source state is `Skipped` become `Disabled`. Otherwise the external state transition would remain marked as `Waiting` which could lead to a deadlock within its target OLC. Note that a disabled external state transition does not cause any state change within its source or target state.

An external state transitions whose target state becomes `Skipped`, has to be `Disabled` as well. First, the external state transition does not influence the target state because it has already been `Skipped`. Second, we have to ensure proper termination of the process structure. It terminates after all OLCs have reached their end states (cf. Section 2.3). In particular, soundness necessitates that no constituent process is running (i.e., no transition is processing) then (cf. Sect. 2.3). Therefore, we have to avoid external state transitions whose processes might still be enacted while their target OLC have already reached the end state.

4 Changing Data-Driven Process Structures

So far, we have investigated the modeling, instantiation and enactment of data-driven process structures. To cope with the flexibility requirements of engineering processes, we further have to look at dynamic process structure changes.

Example 3. *The creation, testing and release of the electrical system of a car takes up to several weeks. Meanwhile, changes of the electrical system, such as the addition of new objects to the electrical system, occur often (e.g., adding a new TV Tuner subsystem to the data structure from Fig. 1c). Consequently, the related process structure needs to be adapted (cf. Fig. 1d); i.e., we have to add the processes (e.g., testing) for the new component to the process structure and must synchronize them with the processes of the already existing components.*

Directly changing the process structure would not be a good solution due its size and complexity. Instead, users (e.g., engineers) must be able to perform the changes at a high level of abstraction. In COREPRO this is accomplished by adapting the respective data structure (e.g., describing the electrical system) and by translating these changes into corresponding adaptations of the process structure. Again we utilize the strong relationship between data and process structure to realize this. Furthermore, dynamic changes must not violate soundness of the process structure. To ensure this, we have to constrain changes to certain runtime states (i.e., markings) of the process structure.

First, we provide basic operations for static changes of data as well as process structures (i.e., we do not take runtime information into account). That includes mechanisms for transforming basic change operations applied to a data structure into corresponding adaptations of the process structure. Second, we define marking-based constraints for these change operations in order to ensure soundness of the modified process structure afterwards.

4.1 Static Changes of Data-Driven Process Structures

Basic operations enable the change of existing data and process structures. Concerning data structure changes, COREPRO allows to insert objects and object relations into a given data structure as well as to remove them (for an overview see the left-hand side of Fig. 5). The offered set of operations is complete; i.e., it allows the engineer to transfer a given data structure into an arbitrary other data structure, which constitutes an instance of the underlying data model (cf. Section 2.1). Regarding adaptations of a process structure, COREPRO provides basic operations for inserting and deleting OLC instances as well as OLC instance dependencies (for an overview see the right-hand side of Fig. 5). In particular, we focus on the coordination and synchronization of different OLC instances and do not consider (dynamic) changes of the internal behavior of a single OLC instance. In this paper, we describe only basic operations for adapting data and process structures and omit a discussion on high-level change operations.

As motivated, data structure changes should be automatically transformed into corresponding process structure changes in order to adapt the overall engineering process to the new (product) data structure. COREPRO accomplishes this based on the information specified at the model level; i.e., we use the life cycle coordination model (LCM) which defines the OLC for each object type and the OLC dependency for each object relation (cf. Section 2.2 and Fig. 1b).

First, we consider the addition of a new object `x` (with type `ot`) to a given data structure. To realize this change, the `addObject` operation (cf. Fig. 5a) can be used. This operation, in turn, is translated into an `addOLC` operation (cf. Fig. 5a) which inserts a corresponding OLC instance to the process structure. The OLC instance to be added is derived from the OLC linked to object type `ot` within the life cycle coordination model (LCM). When adding a new object `x` to a data structure, its relations to other objects can be specified using the `addRelation` operation (cf. Fig. 5b). The newly added relations are then automatically translated into OLC dependencies of the corresponding process structure. This is realized based on the `addOLCDependency` operation, which inserts corresponding external state transitions between the involved OLC instances. Again, information from the life cycle coordination model is used to correctly transform the newly added object relation to a corresponding OLC dependency.

To remove an object relation `r` from a data structure, we provide operation `removeRelation`. It is mapped to operation `removeOLCDependency` for the process structure, which removes all external state transitions associated with the relation `r` (cf. Fig. 5c). Removal of an isolated object (i.e., relations to other objects have been already removed) from a data structure is mapped to the removal of the associated OLC instance from the process structure (cf. Fig. 5d).

4.2 Dynamic Changes of Data-Driven Process Structures

As motivated, data-driven process structures have to be dynamically adapted when the related (product) data structure is changed. This section considers dynamic process structure changes and deals with relevant issues.

A key challenge is to preserve soundness of the overall process structure when dynamically adapting it. As long as we only consider static changes (cf. Section 4.1) soundness of an automatically changed process structure can be ensured if the underlying LCM (e.g., Fig. 1b), from which the process structure is derived,

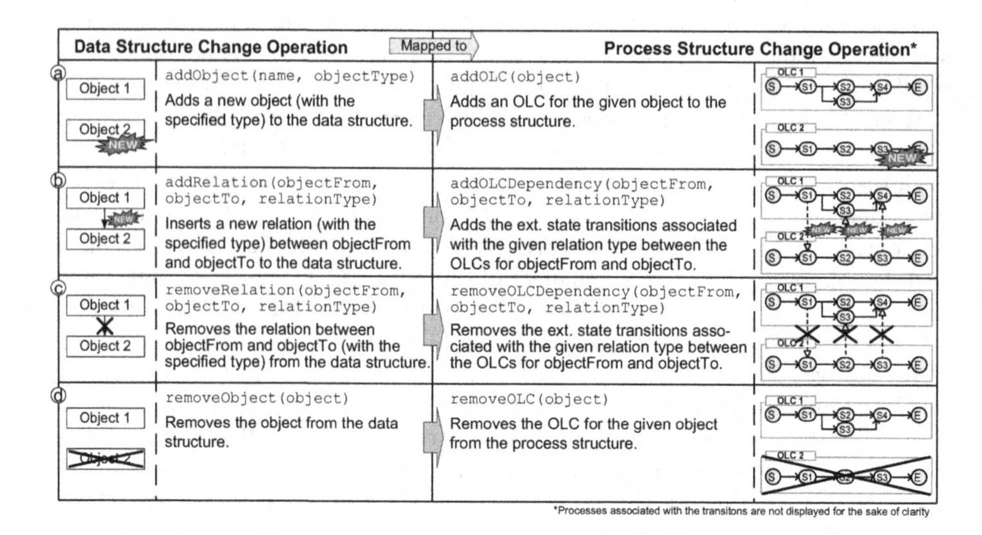

Fig. 5. Data Structure Changes and Related Process Structure Adaptations

is sound. The COREPRO modeling tool always checks this before an LCM can be released. When adapting data and process structures during runtime, we have to define additional constraints with respect to the status of the process structure in order to guarantee soundness afterwards. In particular, the addition or removal of external state transitions must not result in deadlocks or livelocks. Note that this problem is related to correctness issues discussed in the context of dynamic and adaptive workflows [4,5]. Here, one has to decide whether a structural change can be applied to a running workflow instance or not. One correctness notion used in this context is *compliance* [5]. Simply speaking, a workflow instance is compliant with a structurally modified workflow schema if its current execution history is producible on the new workflow schema as well.

In principle, the compliance criterion could be applied to process structures as well. COREPRO logs the events related to the start and completion of the processes coordinated by the process structure in an execution history. However, the compliance criterion is too restrictive for data-driven process structures. Consider, for example, the dynamic removal of an object from a data structure; e.g., an already designed component might have to be removed from the electrical system when a problem is encountered during testing. Respective runtime changes often become necessary in practice and must be supported, even if the OLC instance related to the data object has been already started or completed (i.e., corresponding entries were created in the execution log of the process structure). For this case, compliance would be violated.

COREPRO uses specific correctness constraints to adequately deal with dynamic changes. We utilize the trace-oriented markings of OLC states and the semantics of the described change operations when defining these constraints.

First, COREPRO allows to delete an already started or finished OLC instance `olc`, together with its external state transitions, as long as the execution of other OLC instances has not yet been affected by `olc`. This constraint will be satisfied if the target states of all outgoing external state transitions of `olc` have not yet been activated. In particular, this ensures that the removal of `olc` will not have any effect on other OLC instances, and therefore does also not influence soundness of the overall process structure. Note that this exactly reflects reality; i.e., as long as the removal of a data object has no effects on the status of other data objects (i.e., on the states of their related OLC instance) the object and its corresponding OLC (with its external transitions) can be removed.

Second, we constrain the addition of new OLC instances and their dependencies to other OLC instances. In particular, it must not be allowed to add an external state transition with the new OLC as source if the target state (of another OLC instance) is already marked as `ACTIVATED` or `DONE`. Otherwise soundness of the overall process structure would be violated. Apart from this, the constraint makes sense from a practical perspective; e.g., it must not be possible to insert untested components into a (product) data structure and relate them to electrical systems which have already been released.

Process structures can become very large. Therefore one goal is to efficiently check whether a dynamic change can be applied to a given process structure. It is important to mention that the operations presented in Sect. 4.1 only modify the process structure itself, but do not affect individual OLC instances (i.e., OLC states and internal state transitions are not changed by these operations). Furthermore, when analyzing the change scenarios sketched above, we can see that the applicability of a basic runtime change mainly depends on the markings of the target states of the external state transitions to be added or deleted (either explicitly or implicitly due to the deletion or addition of an OLC instance). Thus, we can reduce efforts for constraint checking to the manipulated external state transitions. More precisely, we consider the *change regions* defined by each of these transitions. Such a region comprises the transition itself as well as its source and target state. In principle, two fundamental issues emerge when dynamically adding or deleting an external state transition:

1. May an external state transition be added or deleted when considering the current marking of the states of its change region?
2. Which marking adaptations become necessary with respect to a change region when adding or removing the respective external state transition?

As example consider Fig. 6a which shows two concurrently executed OLC instances `OLC A` and `OLC B`. Assume that an external state transition (with associated process `P`) shall be dynamically added with `S1` as source and `S2` as target state. First, COREPRO checks whether this change is possible considering the current markings of the change region. Since the target state `S2` of the respective external state transition has not yet been activated, the change is allowed (cf. Fig. 6b). When applying it, in addition, the markings of the change region have to be adapted to correctly proceed with the flow of control. In the given example the newly added external state transition is automatically evaluated,

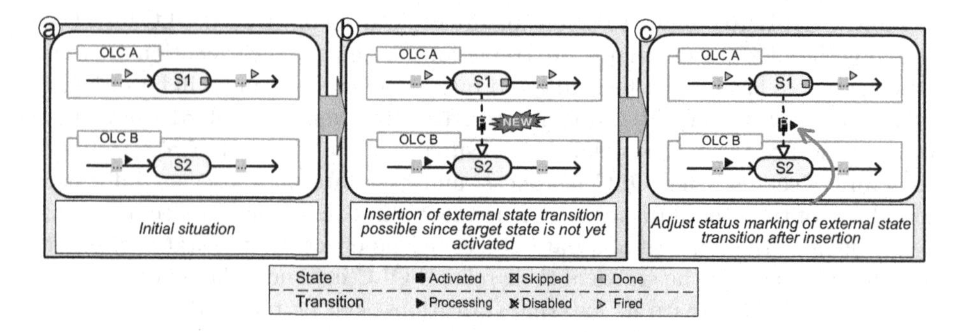

Fig. 6. Dynamic Addition of an External State Transition

which changes its marking from `WAITING` to `PROCESSING`; i.e., the process related with the transition is instantiated and started (cf. Fig. 6c). Note that a deadlock would occur in the given scenario if the newly added transition had not been marked as `PROCESSING`. As another example consider again Fig. 6a. If we tried to add an external state transition with source state `S2` and target state `S1` the change would be not allowed. Otherwise both `OLC A` and `OLC B` might be completed before the process associated with the new external state transition completes; i.e., soundness of the resulting process structure would be violated.

Generally, a structure change comprises multiple operations of which either all or none of them have to be applied to the data and process structure. To enable change atomicity and isolation, COREPRO allows to group change operations within a change transaction. Removing an object, for example, might require the removal of several relations associated with this object and finally the removal of the object itself. Temporary inconsistencies might occur, but will not become visible to other change transactions and users respectively.

4.3 Practical Impact

In the automotive domain, electrical systems comprise up to 300 components. Related process structures have more than 1300 processes and 1500 process dependencies [2]. We have already shown the need for dynamically adapting process structures in this scenario. In practice, often more than 50 dynamic changes of a process structure become necessary during its enactment (cf. Example 3). A software error within a component (e.g., the `Main Unit` of the navigation system), for example, requires the exchange of this component; i.e., a new version of the component has to be constructed (cf. Fig. 7). When exchanging the component, the related OLC has to be exchanged as well to ensure that the associated processes can be correctly re-enacted. Whether the adaptation is possible or not depends on the status of the process structure; e.g., if the `Main Unit` subsystem has been already installed and the `Testdrive` process has already started, the `Main Unit` subsystem must not be exchanged (cf. Fig. 7). COREPRO enables engineers to adapt respective process structures by changing the (product) data structure and by transforming these changes to the process structure.

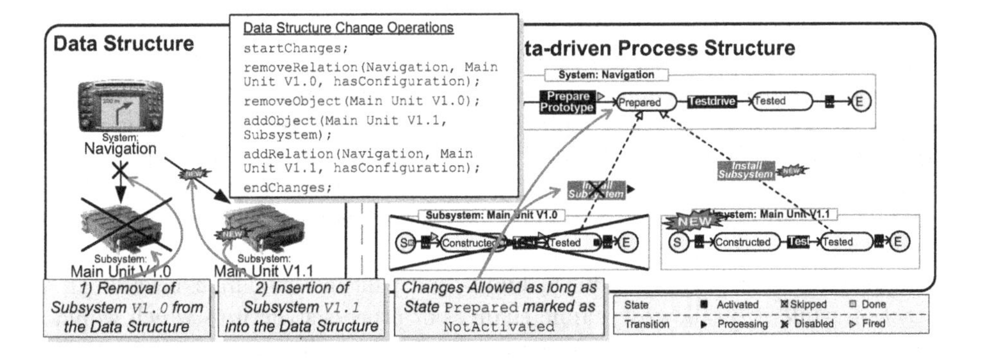

Fig. 7. Release Management Process: Creation of a System Release

5 Related Work

At first glance, conventional approaches support modeling and coordination of process structures. Choreography definition languages, for example, allow for the activity-centered specification of the dependencies between (distributed) processes [6]. Changes of process choreographies are discussed in [7,8]. The data-driven derivation and change of process structures, however, is not considered.

An approach for enacting and adapting data-driven process structures is presented in [9]. Focus is on simple data structures (e.g., lists, sets). COREPRO, by contrast, enables the definition and change of arbitrary complex data structures as well as automated creation of related process structures. Approaches for deriving process structures from bills of material are described in [10,11]. While [10] focuses on product-driven (re-)design of process structures based on design criteria like time or cost, [11] discusses how to coordinate activities based on object relations. The latter approach constitutes the basis of the *Case Handling* paradigm [12]. The idea is to model a process (structure) by relating activities to the data flow. The concrete activity execution order at runtime then depends on the availability of data. Further approaches providing support for modeling process structures consisting of OLCs are presented in [13,14]. For example, they enable the generation of activity diagrams from OLCs and vice versa. The generic definition of data-driven process structures as well as their adaptation, however, is not covered by the aforementioned approaches.

There exist approaches from the database area for describing object-oriented data structures and their behavior [15,16]. They provide a rich set of elements for modeling data structures and for mapping them to OLCs, but neglect enactment and dynamic changes. COREPRO does not focus on the data modeling part for the following reason. In the engineering domain, data modeling (incl. the expression of constraints and the avoidance of data inconsistencies) is usually done within a product data management system, which further provides techniques for versioning and variant handling. Our data model constitutes a simplified view on the PDM data model capturing the needs for coordinating and dynamically adapting process structures.

6 Summary and Outlook

IT support for enactment and consistent change of data-driven process structures is a major step towards the use of process management technology in engineering domains. COREPRO provides a new approach with respect to the automated creation and data-driven adaptation of process structures during runtime. In particular, our approach reduces modeling efforts for large process structures and ensures correct coordination of processes. Further, COREPRO enables adaptations of process structures at both, build- and runtime at a high level of abstraction while it disallows dynamic changes of process structures which would lead to an inconsistent runtime situation. Our case studies have shown that in real world scenarios it might be necessary to apply changes even if they lead to inconsistencies. Dealing with such situations requires extensive exception handling techniques (e.g., backward jumps within OLCs) which are addressed by COREPRO and will be presented in future publications.

References

1. Müller, D., Herbst, J., Hammori, M., Reichert, M.: IT Support for Release Management Processes in the Automotive Industry. In: Dustdar, S., Fiadeiro, J.L., Sheth, A.P. (eds.) BPM 2006. LNCS, vol. 4102, pp. 368–377. Springer, Heidelberg (2006)
2. Müller, D., Reichert, M., Herbst, J.: Data-driven modeling and coordination of large process structures. In: Meersman, R., Tari, Z. (eds.) OTM 2007, Part I. LNCS, vol. 4803, pp. 131–147. Springer, Heidelberg (2007)
3. Aalst, W.: Verification of workflow nets. In: Azéma, P., Balbo, G. (eds.) ICATPN 1997. LNCS, vol. 1248, pp. 407–426. Springer, Heidelberg (1997)
4. Rinderle, S., Reichert, M., Dadam, P.: Flexible support of team processes by adaptive workflow systems. Distributed & Parallel Databases 16(1), 91–116 (2004)
5. Rinderle, S., Reichert, M., Dadam, P.: Correctness Criteria For Dynamic Changes in Workflow Systems: A Survey. DKE 50(1), 9–34 (2004)
6. W3C: WS-CDL 1.0 (2005)
7. Rinderle, S., Wombacher, A., Reichert, M.: Evolution of process choreographies in DYCHOR. In: Meersman, R., Tari, Z. (eds.) OTM 2006. LNCS, vol. 4275, pp. 273–290. Springer, Heidelberg (2006)
8. Aalst, W., Basten, T.: Inheritance of workflows: an approach to tackling problems related to change. Theoretical Computer Science 270(1-2), 125–203 (2002)
9. Rinderle, S., Reichert, M.: Data–Driven Process Control and Exception Handling in Process Management Systems. In: Dubois, E., Pohl, K. (eds.) CAiSE 2006. LNCS, vol. 4001, pp. 273–287. Springer, Heidelberg (2006)
10. Reijers, H., Limam, S., Aalst, W.: Product-based workflow design. MIS 20(1), 229–262 (2003)
11. Aalst, W.: On the automatic generation of workflow processes based on product structures. Comput. Ind. 39(2), 97–111 (1999)
12. Aalst, W., Berens, P.J.S.: Beyond workflow management: Product-driven case handling. In: GROUP, pp. 42–51 (2001)

13. Liu, R., Bhattacharya, K., Wu, F.Y.: Modeling Business Contexture and Behavior Using Business Artifacts. In: Krogstie, J., Opdahl, A., Sindre, G. (eds.) CAiSE 2007 and WES 2007. LNCS, vol. 4495, pp. 324–339. Springer, Heidelberg (2007)
14. Küster, J.M., Ryndina, K., Gall, H.: Generation of Business Process Models for Object Life Cycle Compliance. In: Alonso, G., Dadam, P., Rosemann, M. (eds.) BPM 2007. LNCS, vol. 4714, pp. 165–181. Springer, Heidelberg (2007)
15. Kappel, G., Schrefl, M.: Object/behavior diagrams. In: ICDE, pp. 530–539 (1991)
16. Dori, D.: Object-process methodology as a business-process modelling tool. In: ECIS (2000)

An Aspect Oriented Approach for Context-Aware Service Domain Adapted to E-Business

Khouloud Boukadi[1], Chirine Ghedira[2], and Lucien Vincent[1]

[1] Division for Industrial Engineering and Computer Sciences, ENSM, Saint-Etienne, France
[2] LIRIS Laboratory, Claude Bernard Lyon 1 University, Lyon, France
{boukadi,Vincent}@emse.fr, cghedira@liris.cnrs.fr

Abstract. This paper proposes an architecture for a high-level structure called Service Domain which orchestrates a set a of related IT services based on BPEL specification. Service Domain was developed to enhance the Web service concept to suit e-business collaboration. Service Domains are developed to be context aware. Our approach highlights the benefits of bringing Aspect Oriented Programming to ensure context aware services. Thus, context awareness is guaranteed by enhancing BPEL execution using Aspect oriented paradigms. The proposed approach is illustrated with a running example that shows how Service Domain presents different behaviours according to the context changes.

Keywords: Web service, service adaptation, context-aware, Aspect Oriented Programming.

1 Introduction

Today, enterprises are operating in a rapidly changing market characterized by increasing customer demand for low cost and short time to market. To cope with these business conditions, enterprises have adopted a two-level solution. The first alternative is found on the inter-organizational side, in which enterprises collaborate in e-business scenarios in order to provide the best products or services. Secondly, on the organizational side, enterprises must be more dynamic, flexible, and context-aware than ever to survive.

In this endeavor, enterprise information technology (IT) systems play a crucial role. The challenge for IT infrastructures has been to help companies to respond to changes that occur in a timely, dynamic, and reliable manner without compromising organizational flexibility. This brings into focus the role of defining and implementing flexible business processes supported by corresponding flexible IT systems, which allow enterprises to collaborate with partners dynamically. Flexible IT systems are those that are malleable enough to deal with context changes in an unstable environment [1].

A contemporary approach for addressing these critical issues is the Service-Oriented Architecture and Web service technology, which offers a suitable technical foundation to ensure the flexibility required for IT systems [2]. Existing IT infrastructure can be bundled and offered as Web services with standardized and

Z. Bellahsène and M. Léonard (Eds.): CAiSE 2008, LNCS 5074, pp. 64–78, 2008.

well-defined interfaces. We call Web services arising from applying this process enterprise IT Web services, or hereafter IT services.

1.1 Limitations of the Traditional IT Service Solution

IT services are published for internal or external use. They can be combined and recombined into different solutions and scenarios, as determined by business needs. IT services promote business processes by composing individual IT services to represent complex processes, which can even span multiple organizations. However, transforming enterprise IT infrastructure into a large set of published IT services with different granularity levels has a number of drawbacks. Firstly, it may imply that an enterprise has to expose service elements, which are in isolation meaningless, to the outside world. Secondly, service consumers will undertake several low-level service combinations and this will overburden its task, thereby decreasing the added value of service provisioning. Thirdly, in this form a service consumer can compose a process which makes no sense for the service provider. To overcome these limitations, we believe that an enterprise must re-organize its IT services and presents its functionalities through a high-level service. In this work we develop a high level structure called *Service Domain* (SD), which logically represents a combination of related IT services as a single service. Service Domain orchestrates a set of IT services in order to provide a high level functionality and a comprehensible external view to the end user. Service Domain will be published as a Web service, thus hiding the complexity of publishing, selecting and combining fine grained IT services.

Furthermore, in order to satisfy enterprise adaptability to context changes, Service Domain must be more than functions provided through the Web. Indeed, it must have the capacity to adapt its own behavior by comporting appropriately to accommodate the situation in which it evolves. To meet to this objective, Service Domain has to assess its current capabilities, ongoing commitments, and surrounding environment. As a result Service Domain must be context aware.

1.2 Contribution and Paper Organization

The contribution in this paper is twofold. In the First part, the architecture of a high-level structure named Service Domain is presented. It orchestrates a set a of related IT services based on the Business Process Execution Language (BPEL) specification [3]. In the second part, we address our context aware Service Domain. Context awareness is guaranteed by enhancing BPEL execution using Aspect Oriented Programming [4].

The rest of the paper is organized as follows. The Service Domain architecture is presented in Section 2. Then, in Section 3, we present our context categorization and highlight the drawbacks of BPEL in addressing adaptability to the context changes. In section 4, the Aspect Oriented Programming and how it is used to enhance BPEL adaptability are introduced. In addition, we present a running example and implementations. Section 5 details some related work. Finally, a conclusion and possible further work is proposed.

2 Service Domain Concept (SD)

The motivation behind the Service Domain concept is to achieve manageability when dealing with a large number of IT services. The Service Domain enhances the Web service concept. In fact, its purpose is not to define new application programming interfaces (APIs) or new standards, but rather, to provide, based on existing IT services, a new higher-level structure that can mask complexities from service users, simplify deployment for service suppliers and provide self-managing capabilities. Service Domain is based on/uses Web service standards (i.e. WSDL, SOAP and UDDI).

In the future, our Service Domain will be used as major building block for implementing enterprises business processes, which will be represented as a composition of Service Domains that belong to different enterprises (see Fig.1).

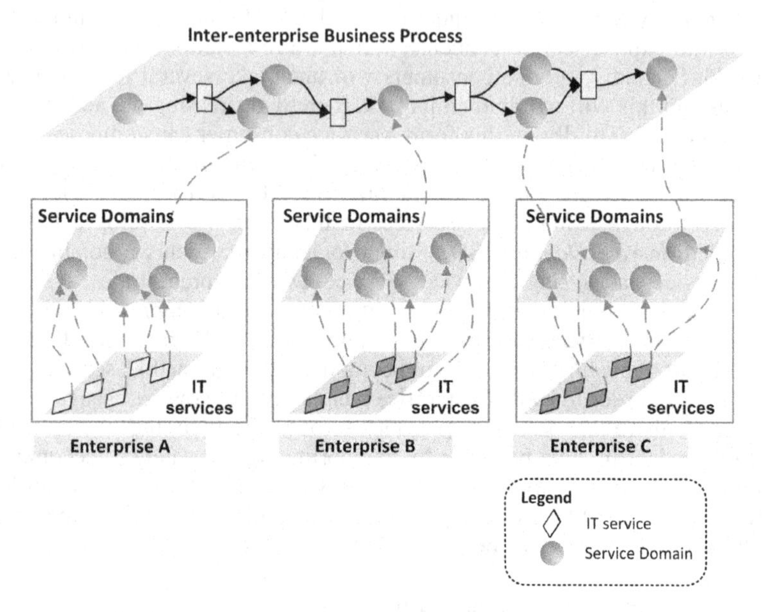

Fig. 1. Inter-enterprise collaboration based on Service Domain

The Service Domain orchestrates and manages several IT services as a single virtual service. It promotes a SOA solution which decreases the intricacy of providing business applications.

As an example of a Service Domain, consider the "logistic enterprise" that exposes a "Delivery Service Domain" (DSD), which constitutes a merchandise delivery service. DSD encapsulates five IT services: "Picking merchandise", "Verifying merchandise", "Putting merchandise in parcels", "Computing delivery price" and "Transporting merchandise". Keeping these IT services in one place facilitates manageability and avoids extra composition work on the client side as well as exposing non-significant services like "Verifying merchandise" on the enterprise side.

The Service Domain is implemented as a node consisting of an Entry Module, Context Manager Module (CMM), Service Orchestration Module (SOM) and finally an Aspect Activator Module (AAM) as presented in Fig. 2.

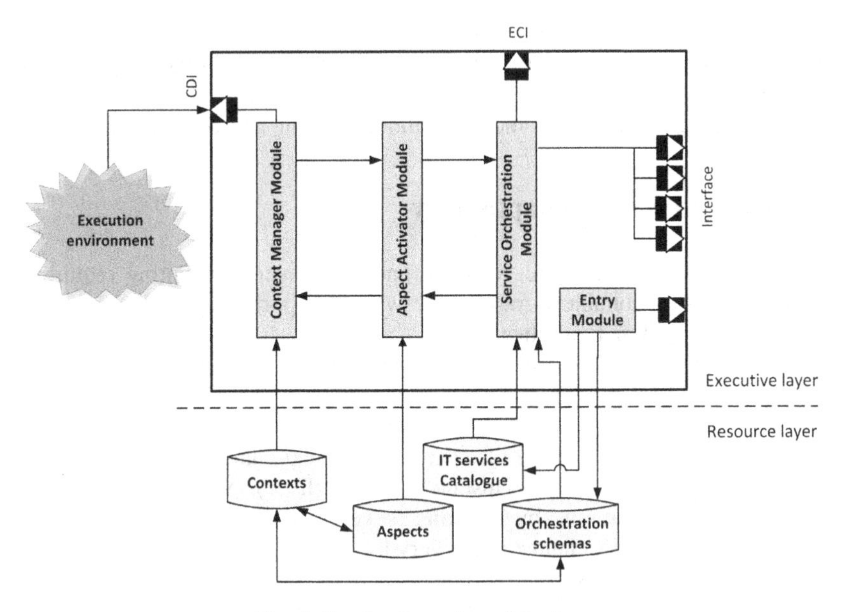

Fig. 2. Service Domain architecture

In this last Figure, three of these Modules provide external interfaces to the Service Domain node: Entry Module, Context Manager Module, and Service Orchestration Module. The Entry Module is based on Web service standard (SOAP) for receiving requests and returning responses. Aside from service requests from clients, the Entry Module also supports administrative operations for managing the Service Domain node. For example, an administrator can send a *register* command in order to add a new IT service with a given Service Domain by registering it in the corresponding IT service catalogue. The register command can also deal with a new orchestration schema which could be added in the orchestration schemas registry.

When the Entry Module receives an incoming request, it communicates with the orchestration schemas registry in order to select a suitable orchestration schema and identify the best IT service instances to fulfill the request. The selection of the orchestration schema and IT service instances, takes into account the context of the incoming request. Orchestration schemas with the set of IT service instances are delivered to the Service Orchestration Module (orchestration engine). The orchestration of different IT services belonging to one Service Domain is ensured using Web service orchestration languages like BPEL [3]. The SOM presents an external interface called Execution Control Interface (ECI) which enables a user to obtain information regarding the state of execution of the SD internal process. This interface is very useful in case of external collaboration since it insures monitoring of the internal process execution. This is the principal difference between our Service Domain and the traditional Web service. In fact, with the ECI interface, SD is based on the Glass box principles in contrast to the Web service which is based on the black box principles. Finally, the last external interface called Context Detection Interface (CDI) is used by the CMM to catch context information changes. Context detection is used to guarantee the SD adaptability. Adaptability of the SD is based on selecting and injecting the right Aspect according to the context change. To fulfill this requirement, SD uses the AAM to identify the suitable Aspect related to the context information and inject it in the BPEL process. This guarantees greater flexibility by quickly adapting the execution of the SD without stopping and redeploying it.

3 Context and BPEL Adaptability

The trend towards context-aware, adaptive and on demand computing requires that SD be equipped with suitable infrastructure which supports the delivery of adaptive services with varying functionalities.

Service Domain will be used in a context in which several factors call for dynamic execution evolution and changes (e.g., changes in the environment and/or unpredictable events).

SD must meet the requirements of customers' context changes as well as different service levels expectations. For instance, the Delivery Service Domain could advertise different behaviors by offering several delivery calculation methods depending on, for example, change in delivery location or time.

As Service Domain uses BPEL to orchestrate a set of related IT services, its adaptability to context is closely related to the BPEL support of adaptability features.

In this section, we will present the context paradigm, our context categorization and finally, discuss the shortcomings of BPEL to address the adaptability to context requirement.

3.1 Context and Context Categorization

The concept of context has been studied in various fields for quite a long time. There are a number of different definitions and uses for this term. Context appears in many disciplines as a meta-information which characterizes the specific situation of an entity, to describe a group of conceptual entities, partition a knowledge base into manageable sets, or as a logical construct to facilitate reasoning services [5]. Our definition of context follows that of Dey's [6] who says that a *context* is "*any information that can be used to characterize the situation of an entity. An entity is a person, place, or object that is considered relevant to the interaction between a user and an application, including the user and the application themselves*".

The categorization of context is important for the development of context aware applications. Context includes implicit and explicit inputs. For example, user context can be deduced in an implicit way by the service provider such as in pervasive environment using physical or software sensors. Explicit context is determined precisely by entities involved in the context. Bradely et al. depict that a variety of categorizations of context have also been proposed [7]. As a matter of fact, there are certain types of context which are, in practice, used more often than others. These major context categories are *location*, *identity*, *time,* and *activity*. Nevertheless, despite the various attempts to develop categorizations for context, there is no generic context categorization. Relevant information differs from one domain to another and depends on their effective use [8].

In this work, we propose a context categorization using an OWL ontology [9]. Fig. 3 depicts our context categorization ontology which is dynamic in the sense that new sub-categories may be added at any time. Each context definition belongs to a certain category which can be provider, customer, and collaboration related.

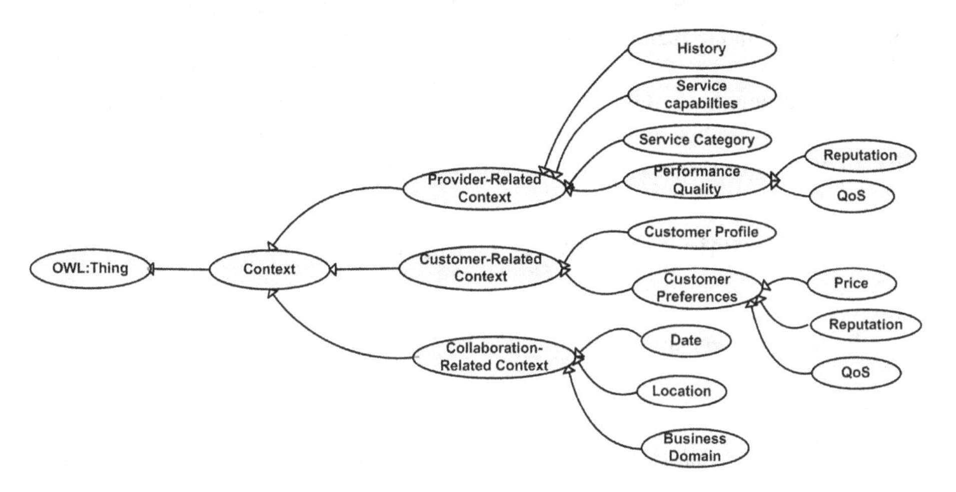

Fig.3. Ontology for categories of context

In the following, we explain the different concepts which constitute our ontology based model for context categorization:

- Provider-related context deals with the conditions under which providers can offer their Web services externally. For example, the performance qualities include some metrics which measure the service quality: time, cost, QoS, and reputation. These attributes model the competitive advantages that providers may have over each other.
- Customer-related context represents the set of available information and meta-data used by service providers to adapt their services. For example, a customer profile which represents a set of information items characterizing the customer.
- Collaboration-related context represents the context of the business opportunity. We identify three sub-categories: location, time, and business domain. The location and time represent the geographical location and the period of time within which the business opportunity should be accomplished.

3.2 Adaptability to Context in BPEL

BPEL inherited a static view of the world from workflow management systems, which did not properly support evolutionary and runtime changes [10]. Only by stopping the running process, modifying the orchestration, and restarting process execution can one simulate evolutionary and runtime changes. Obviously, this is not a viable solution, especially for long-running and collaborative processes.

Actually, BPEL is silent in regards to the specification and handling of crosscutting concerns like context information. Moreover, with BPEL it is difficult to define, modularize and manage context-sensitive behaviors. Traditionally, the implementation of adaptability extensions in BPEL gets scattered and tangled with the core functional logic. This in turn negatively impacts the system adaptability and scalability. These limitations motivate developing new principles for building such SD, and for extending BPEL capabilities with mechanisms to ease the addressing of context changes and to facilitate the development of adaptive behavior.

To overcome these shortcomings, we propose to empower BPEL with Aspect Oriented Programming (AOP) [4] to deal with Service Domain adaptation based on context. Our approach shows the straightforwardness and benefits of bringing Aspect Oriented paradigms to ensure context aware services.

4 Service Domain Adaptability Using Aspects

The proposed approach defines and implements a context adaptive Service Domain using the Aspect oriented Programming (AOP) [4].

4.1 Rationale of AOP

AOP is a paradigm that captures and modularizes concerns that crosscut a software system into modules called Aspects. Aspects can be integrated dynamically to the system thanks to dynamic weaving principle [11].

AOP introduces a unit of modularity called Aspects, containing different code fragments (*advice*), and location descriptions (*pointcuts*), to identify where to plug the code fragment. These points, which can be selected by the pointcuts, are called *join points*. The most popular Aspect language is AspectJ [12] which is based on Java. The pointcut language of AspectJ provides a set of pointcut designators such as call (for selecting method), execution (for selecting method execution), get and set (for selecting read/write field access). However, each class of application can have its own specific Aspect implementation [13]. For instance, in aspect-oriented workflow languages, the advice language should be the same as the base workflow language [14] to avoid any paradigm mismatches for the workflow designers. There are some proposals to introduce some supplemental programming language in the BPELJ [15] [16], like adding Java code snippets to the BPEL engine.

The rationale behind using AOP is based on two arguments. First, AOP enables crosscutting concerns, which is crucial for managing context information separately from the business logic implemented in the BPEL process. This separation of concerns makes the modification of context information and the related adaptability action easier. For example, in the Delivery Service Domain, we can define an Aspect related to the calculation of extra fees when there is a context change that corresponds to modifying the delivery date. This Aspect can be reused in several BPEL processes. Besides, we can attach the adaptability action (action executed as response to context change requirements) to different context information (eg. location context) without changing the orchestration logic.

Second, based on dynamic weaving principles, Aspects can be activated and deactivated at runtime. Consequently, the BPEL process can be dynamically altered at runtime.

Adding AOP to BPEL is very beneficial. However, AOP is currently used on a low level language extension [17]. In order to exploit AOP for SD adaptation, AOP techniques need to be improved to support:

- Runtime activation of Aspects in the BPEL process to enable dynamic adaptation according to context changes, and
- Aspects selection to enable customer-specific contextualization of the Service Domain.

4.2 Aspect Service Domain Specification

The core of our approach is a runtime Aspects weaving that can be injected on the existing SD BPEL process, to achieve adaptable execution based on context changes. Our key contribution consists of encapsulating context information and the corresponding adaptation actions in a set of Aspects.

A BPEL process is considered as a graph *G(V,E)* where *G* is a DAG (Directed Acyclic Graph). Each vertex $v_i \in V$ is a Web service (Web service operation). Each edge *(u, v)* represents a logical flow of messages from u to v. If there is an edge *(u, v)*, then it means that an output message produced by *u* is used to create an input message to *v*.

In this work, we use this definition of BPEL, but it is extended by adding specific constructs. Three types of vertex were identified: (i) context aware, (ii) non context aware and (iii) context manager vertexes. Theses vertexes correspond respectively to Context aware IT Services, Non-Context aware IT Service and Context Manager Services. Context manager vertexes detect context changes and usually precede the context aware vertexes.

A Context aware IT Service (CITS) may have several configurations exporting different behaviors according to the specific context. *CITS* = {*<ID-CITS, Ctx, Asp>*} where *ID-CITS* is the identifier of CITS, *Ctx* is the name of a context and *Asp* is the Aspect related to this context.

We define an Aspect as *Asp=< ID-Asp, Entry-condition, Advice, Join-points>*. Where *ID-Asp* is the identifier of the Aspect, *Entry-condition* represents the condition where the Aspect can be used, *Advice* addresses the adaptability actions related to specific context information (add, parameterize and remove IT service(s)) and *Join-points* describe the set of vertexes where possible adaptations may be required in the BPEL process.

Our adaptation approach is a three-step process (see Fig.4):

1. *Context detection* consists of checking the runtime context information, in order to detect possible context changes. These tasks are performed by the Context Manager Service which is developed as a Web service in the BPEL process.
2. *Aspect Activation* is responsible for the plug-in and the removal of pre-defined Aspects into the BPEL process using the Aspect Activator Module.

The Aspect Activator Module is conceived as an extension to the BPEL engine as was done in [18]. When running a process instance, the Aspect Activator receives the context change information from the Context Manager Service. Then it chooses and activates the appropriate Aspect that matches the values of the changed contextual information.

3. *Updating original BPEL Process* by activating the right Aspect which is executed in the BPEL process to create a contextualized process.

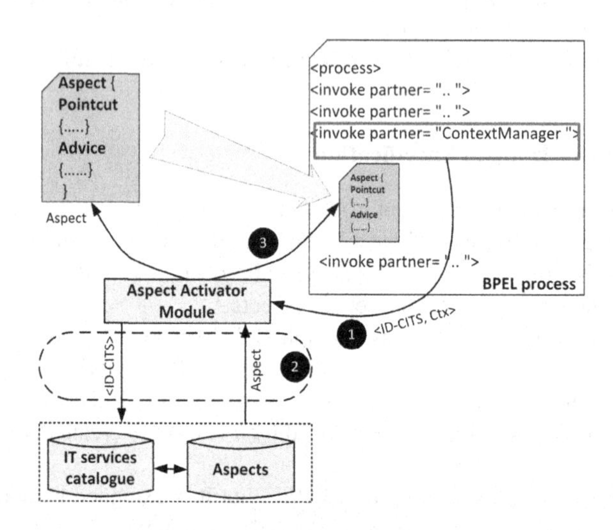

Fig.4. Aspect injection in BPEL

4.3 Running Example and Implementation

In this section, the proposed approach is applied to the case of a manufacturer of plush toys enterprise, which receives orders from its clients during the Christmas period. Once an order is received, this firm proceeds to supply the different components of plush toys. When supplied components are available, the manufacturer begins assembly operations. Finally, the manufacturer selects a logistic provider to deliver these products by the target due date. In this scenario, the focus will be only on the delivery service.

Assume that an inter-enterprise collaboration is established between the manufacturer of plush toys (service consumer) and a logistic enterprise (service provider). The logistic provider delivers parcels from the plush toys manufacturer warehouse to a specific location. The delivery service starts by picking merchandise from the customer warehouse (see Fig.5 step (i)). If there is no past interaction between the two parties involved, the delivery service verifies the shipped merchandise. Once verified, *putting merchandise in parcels service* is invoked, and followed by a *computing delivery price service*. Finally, the service transports the merchandise in the business opportunity location at the delivery due date. The delivery service is considered as a Service Domain orchestrating five IT services: *Picking merchandise*, *Verifying merchandise*, *Putting merchandise in parcels*,

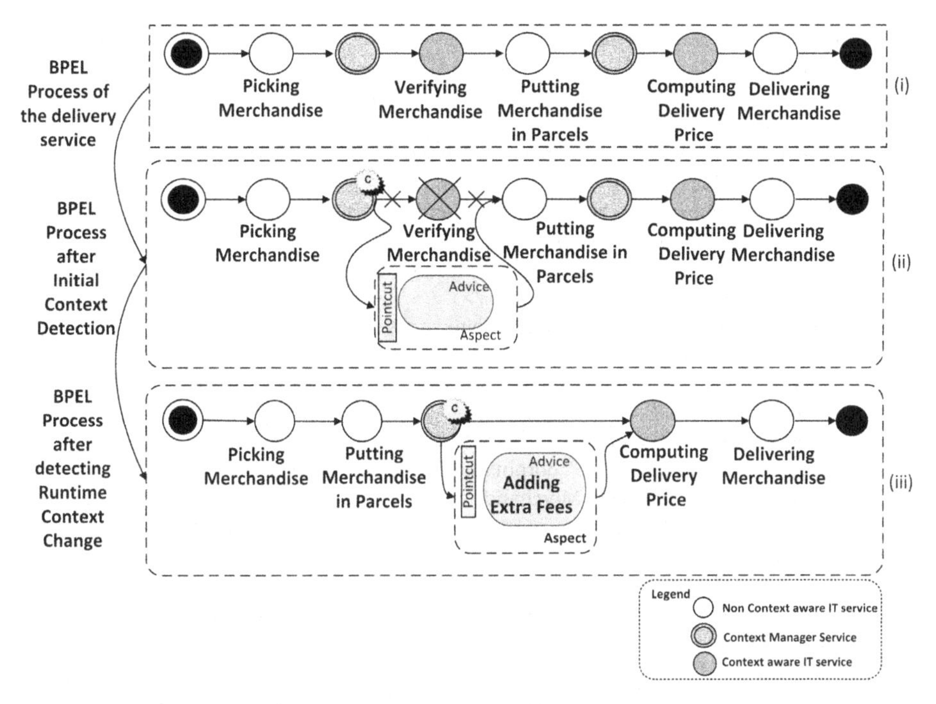

Fig. 5. The delivery service internal process

Computing delivery price and *Delivering merchandise*. Fig.5 depicts the BPEL process modeled as a graph of the delivery service and the adaptation actions according to the context changes (step ii and step iii).

Assume that *Picking merchandise* from customer warehouse, *Putting merchandise in parcels* and *Delivering merchandise* services are context independent while *Verifying merchandise* and *Computing delivery price* are context-aware (i.e., they have different behaviors according to the current customer and opportunity context). We suppose that the *Verifying merchandise* service is aware of the past interactions with customers (historical relationships). This information corresponds to *history* category defined in the context categorization. It may be either "past interaction=No" or "past interaction=Yes". In the first case, the *Verifying merchandise* service is called, but skipped in the second case. The *Computing delivery price* service is aware of runtime context changes corresponding to changes in delivery location or date. When there are changes in the date or the place, extra fees must be added to the total delivery price.

When the BPEL process (Listing 1) starts, the *Context Manager* service is invoked to collect context information (historical context category) about the plush toys enterprise (Listing 1 line 7). Assuming that the context information indicates that the plush toys enterprise is a well known customer (i.e., "past interaction=Yes"), delivery service behavior will be adapted to respond to this context information. The *Aspect Activator* will choose a suitable Aspect to be activated from the set of Aspects attached to the *Verifying merchandise* service.

```
(01) <process name = "DeliveryPackage" .../>
       <sequence>
(03)     <receive partner="client" operation="getDeliveryPackage"
          variable="request"  createInstance="yes" .../>
(05)     <invoke partner="PickingMerchandise" operation="getOkResponse"
          outputVariable="PickingResponse"/>
(07)      <invoke partner="ContextManager" operation="getContextElement"
           outputVariable="ContextResponse"/>
(09)      <invoke partner="VerifyingMerchandise" operation="VerifyMerchandises"
           outputVariable="VerifyingResponse />
(11)      <invoke partner="PuttingMerchandiseInParcels"
          operation="getOkResponse"  outputVariable="PuttingResponse"/>
(13)       <invoke partner="ContextManager" operation="getContextElement"
            outputVariable="ContextResponse"/>
(15)       <invoke partner="ComputingDeliveryPrice"
            operation="ComputePrice" outputVariable="PriceResponse"/>
(17)       <invoke partner="DeliveryMerchandise"
            operation="getOkResponse" outputVariable="DeliveryResponse"/>
(18)     </sequence>
           <assign>...</assign>
(19)       <reply partner="client" operation=" getDeliveryPackage " variable="proposition" .../>
     </sequence>
(21) </process>
```

Listing 1. The delivery process

The selected Aspect is shown in Listing 2. As mentioned before, an Aspect defines one or more pointcuts and an advice. To implement the advice code, we have chosen the BPEL specification, because the goal is to adapt the BPEL process. For the pointcuts language, we have chosen XPath [19], a language specialized for addressing parts of an XML document (a BPEL process is an XML document).

The advice part of the Aspect is expressed as a before advice activity, which is executed instead of the activity captured by the pointcut (line 3). The join point, where the advice is weaved, is the <invoke> activity that calls the Verifying merchandise service (line 5). The advice code is expressed as a <switch> activity. If *ContextResponse* ="1" (i.e., "past interaction=Yes") the advice branches to the activity *<empty>*, in order to express that it is not really necessary to perform this service. After applying the Aspect, the BPEL process of the *Delivery* service is depicted in the Fig. 5 (step (ii)).

Before invoking *Computing Delivery Price*, the *Context Manager* service checks the context information to detect possible contextual changes. Let us assume that the plush toys enterprise has decided to change the delivery date. Hence, the *Context Manager* service captures the new date (Listing 1 line 13). Then the *Aspect Activator* chooses a suitable Aspect to activate from the set of Aspects related to *Computing Delivery Price* service. The selected Aspect is shown in Listing 3. The pointcut of this Aspect (lines 3-6) selects the delivery price calculation activity in the delivery process. The context change is implemented using a *before* advice, which contains a

```
(01) <aspect name="If the enterprise is a partner, do not verify transported merchandise">
     <pointcutandadvice type= "before">
(03) <pointcut name="Verify Merchandise">
     //process[@name=" DeliveryPackage "]
(05) //invoke[@portType="VerifyPT"and
     @operation="VerifyMerchandise"]
(07) </pointcut>
     <advice>
(09)   <switch>
         <case condition="bpws:getVariableData
(11)    ('ThisProcess( ContextResponse )','PastInteraction') ==1">
          <empty/>
(13)     </case>
        </switch>
(15)  </advice>
     </pointcutandadvice>
(17) </aspect>
```

Listing 2. Context as an Aspect

switch with a case branch (lines 7-25) for calculating additional fees depending on the number of days between the initial and the new delivery date. This number will be multiplied by the daily fees already defined by the logistic enterprise.

The case branch uses an assign activity (lines 14-21) to compute the additional fees to the part *ExtraFees* of the variable *calculPrice,* which will be sent to the *Computing Delivery Price* service. The *Delivery* service BPEL process after applying the Aspect is depicted in the Fig. 5 (step (iii)).

```
(01) <aspect name=" AddExtraFees ">
     <pointcutandadvice>
(03) <pointcut name="Fees calculation">
       //process[@name=" DeliveryPackage "]
(05)   //invoke[@operation="ComputePrice"]
     </pointcut>
(07) <advice type="before">
       <switch>
(09)    <case condition=
          "getVariableData ('( ContextResponse )',' NewDate') <
(11)      getVariableData ('( clientrequest )',' DeliveryDate')">
        <!-- Here comes the action implementation of the context change -->
(13)      <sequence name="Fees calculation">
          <assign>
(15)      <copy>
          <from expression="
(17)      ((getVariableData ('( ContextResponse )',' NewDate') -
          getVariableData ('( clientrequest )',' DeliveryDate ')) *100 >
(19)      <to variable="calculPrice" part="ExtraFees"/>
          </copy>
(21)      </assign>
          </sequence>
(23)    </case>
       </switch>
(25) </advice>
     </pointcutandadvice>
```

Listing 3. Managing context change as an Aspect

5 Related Work

There are many ongoing research efforts related to the adaptation of Web services and Web service composition according to context changes [20] [21]. In the proposed work we focus specially on the adaptation of a BPEL (workflow) process.

Some other research efforts from the Workflow community address the need for adaptability. They focus on formal methods to make the workflow process able to adapt to changes in the environment conditions. For example, Casati et al. in [22] propose eFlow with several constructs to achieve adaptability. The authors use parallel execution of multiple equivalent services and the notion of generic service that can be replaced by a specific set of services at runtime. However, adaptability remains insufficient and vendor specific. Moreover, many adaptation triggers considered by workflow adaptation, like infrastructure changes, are not relevant for Web services because services hide all implementation details and only expose interfaces described in terms of types of exchanged messages and message exchange patterns.

In addition, Modafferi et al. in [23] extend existing process modeling languages to add context sensitive regions (i.e., parts of the business process that may have different behaviors depending on context). They also introduce context change patterns as a mean to identify the contextual situations (and especially context change situations) that may have an impact on the behaviour of a business process. In addition, they propose a set of transformation rules that generate a BPEL based business process from a context sensitive business process. However, context change patterns which regulate the context changes are specific to their running example with no-emphasis on proposing more generic patterns.

There are a few works using an Aspect based adaptability in BPEL. In [10, 24], the authors presented an Aspect oriented extension to BPEL: the AO4BPEL which allows dynamically adaptable BPEL orchestration. The authors combine business rules modeled as Aspects with a BPEL orchestration engine. When implementing rules, the choice of the pointcut depends only on the activities (invoke, reply or sequence). However in our approach the pointcut depends on the returned value of the *Context Manager* Web service which detects a context changes. Business rules in this work are very simple and don’t express a pragmatic adaptability constraint like context change in our case. Another work is proposed in [25] ,in which the authors propose a policy-driven adaptation and dynamic specification of Aspects to enable instance specific customization of the service composition. However, they don't mention how they can present the Aspect advices or how they will consider the pointcuts.

6 Conclusion

This paper presented an architecture of a high-level structure called Service Domain, which orchestrates a set of related IT services based on BPEL specification. Service Domain enhances the Web service concept to suit the inter-enterprise collaboration scenario. Besides, in order to address enterprise adaptability to context changes, Service Domain is developed to be context aware. Literature review has shown that BPEL, considered as the de facto standard for Web services orchestration, offers no support for dynamic adaptation of the orchestration logic according to context. To overcome these limitations, we proposed to enhance BPEL execution using the AOP. We demonstrated that it is a suitable paradigm that enables crosscutting and context-sensitive logic to be factored out of the service orchestration and modularized into Aspects. For future endeavors, we are working to improve, extend, and complete the Service Domain architecture. An empirical study to validate and test the proposed approach will be at the centre of future research. In addition, close interactions with industrial partners will be essential to validate the proposed approach.

References

1. Byrd, T.A., Turner, D.E.: An exploratory examination of the relationship between flexible IT infrastructure and competitive advantage. Information and Management 39, 41–52 (2001)
2. Papazoglou, M.P., van den Heuvel, W.-J.: Service-oriented design and development methodology. International Journal of Web Engineering and Technology (IJWET) 2(4), 412–442 (2006)
3. Andrews, T., Curbera, F.: Business Process Execution Language for Web Services (BPEL4WS) version 1.1 (2003), http://www-128.ibm.com/developerworks/library/specification/ws-bpel
4. Aspect–Oriented Software Development (2007), http://www.aosd.net
5. Benslimane, D., Arara, A., Falquet, G., Maamar, Z., Thiran, P., Gargouri, F.: Contextual Ontologies: Motivations, Challenges, and Solutions. In: Fourth Biennial International Conference on Advances in Information Systems, pp. 168–176. Springer (ED), Izmir (2006)
6. Dey, A.K., Abowd, G.D., Salber, D.: A Conceptual Framework and a Toolkit for Supporting the Rapid Prototyping of Context-Aware Applications. Human-Computer Interaction 16, 97–166 (2001)
7. Bradely, N.A., Dunlop, M.D.: Toward a Multidisciplinary Model of Context to Support Context-Aware Computing. Human-Computer Interaction 20, 403–446 (2005)
8. Mostetefaoui, S.K., Mostetefaoui, G.K.: Towards A Contextualisation of Service Discovery and Composition for Pervasive Environments. In: Workshop on Web-services and Agent-based Engineering (2003)
9. Bechhofer, S., van Harmelen, F., Hendler, J., Horrocks, I.: OWL Web Ontology Language Reference (2004), http://www.w3.org/TR/2004/REC-owl-ref-20040210
10. Charfi, A., Mezini, M.: An Aspect-oriented Extension to BPEL, World Wide Web, pp. 309–344 (2007)
11. Bockisch, C., Haupt, M., Mezini, M., Ostermann, K.: Virtual Machine Support for Dynamic Join points. In: Proceedings of the 3rd International Conference on Aspect-Oriented Software Development - AOSD 2004, Lancaster, UK, pp. 83–92 (2004)
12. The AspectJ Team, The AspectJ Programming Guide, AspectJ 1.2 edition (2007), http://dev.eclipse.org/viewcvs/indextech.cgi/~checkout~/aspectj-home/doc/progguide/index.html
13. Deursen, A.V., Klint, P., Visser, J.: Domain-Specific Languages: An Annotated Bibliography. ACM SIGPLAN Notices 35(6), 26–35 (2000)
14. Braem, M., Verlaenen, K., Joncheere, N., Vanderperren, W., Van Der Straeten, R., Truyen, E., Joosen, W., Jonckers, V.: Isolating Process-Level Concerns Using Padus. In: Dustdar, S., Fiadeiro, J.L., Sheth, A.P. (eds.) BPM 2006. LNCS, vol. 4102, pp. 113–128. Springer, Heidelberg (2006)

15. Courbis, C., Finkelstein, A.: Towards Aspect Weaving Applications. In: Proceedings of the 27th International Conference on Software Engineering, pp. 66–77. ACM Press, New York (2005)
16. BEA and IBM, BPELJ: BPEL for Java, Joint White Paper (2004), http://www-128.ibm.com/developerworks/library/specification/ws-bpelj/
17. Kiczales, G., Hilsdale, E., Hugunin, J., Kersten, M., Palm, J., Griswold, W.G.: An Overview of AspectJ. In: Knudsen, J.L. (ed.) ECOOP 2001. LNCS, vol. 2072, Springer, Heidelberg (2001)
18. Charfi, A., Mezini, M.: Aspect-oriented web service composition with A04BPEL. In: The European Conference on web Service, pp. 168–182. Springer, Germany (2004)
19. Clark, J., DeRose, S.: XML Path Language (XPath) 1.0. W3C Recommendation, November 16 (1999), http://www.w3.org/TR/xpath
20. Maamar, Z., Benslimane, D., Thiran, P., Ghedira, C., Dustdar, S., Sattanathan, S.: Towards a context-based multi-type policy approach for Web services composition. Data & Knowledge Engineering, 327–335 (2007)
21. Bettini, C., Maggiorini, D., Riboni, D.: Distributed Context Monitoring for the Adaptation of Continuous Services. World Wide Web 10(4), 503–528 (2007)
22. Casati, F., Shan, M.-C.: Dynamic and adaptive composition of e-services. Information Systems 26(3), 143–163 (2001)
23. Modafferi, S., Benatallah, B., Casati, F., Pernici, B.: A Methodology for Designing and Managing Context-Aware Workflows. Mobile Information Systems II, 91–106 (2005)
24. Charfi, A., Mezini, M.: Hybrid web service composition: business processes meet business rules. In: The 2nd international conference on Service oriented computing, pp. 30–38. ACM Press, New York (2004)
25. Erradi, A., Maheshwari, P., Padmanabhuni, S.: Towards a Policy-Driven Framework For Adaptive Web Services Composition. In: The International Conference on Next Generation Web Services Practices (NWeSP 2005), Seoul, Korea, pp. 33–38 (2005)

Modeling Service Choreographies Using BPMN and BPEL4Chor*

Gero Decker[1], Oliver Kopp[2], Frank Leymann[2],
Kerstin Pfitzner[2], and Mathias Weske[1]

[1] Hasso-Plattner-Institute, University of Potsdam, Germany
{gero.decker,weske}@hpi.uni-potsdam.de
[2] Institute of Architecture of Application Systems, University of Stuttgart, Germany
{kopp,leymann,pfitzner}@iaas.uni-stuttgart.de

Abstract. Interconnecting information systems of independent business partners requires careful specification of the interaction behavior the different partners have to adhere to. Choreographies define such interaction constraints and obligations and can be used as starting point for process implementation at the partners' sites. This paper presents how the Business Process Modeling Notation (BPMN) and the Business Process Execution Language (BPEL) can be used during choreography design. Step-wise refinement of choreographies to the level of system configuration is supported through different language extensions as well as a mapping from BPMN to BPEL4Chor. A corresponding modeling environment incorporating the language mapping is presented.

1 Introduction

Automated electronic communication between different business partners offers big optimization potential regarding the overall business process performance. However, it also comes with certain challenges that have to be tackled. Common message formats must be agreed upon and the allowed and expected interaction sequences must be clearly defined. Legal consequences of message exchanges as well as time constraints must be captured.

Choreography languages provide a means to specify the messages exchanged between different organizations along with behavioral constraints. The Business Process Modeling Notation (BPMN [2]) offers a rich set of graphical notations for control flow constructs and includes the notion of interacting processes where sequence flow (within an organization) and message flow (between organizations) are distinguished. Therefore, BPMN is a good candidate for providing a graphical notation for choreography modeling. When it comes to refining such initial choreography models, details about timing constraints and exception handling have to be added. Finally, technical configurations are introduced for reaching unambiguity "on the wire". In order to express the different levels of details in BPMN we present several extensions to this language.

* Partially funded by the German Federal Ministry of Education and Research (project Tools4BPEL, project number 01ISE08).

Z. Bellahsène and M. Léonard (Eds.): CAiSE 2008, LNCS 5074, pp. 79–93, 2008.

The Business Process Execution Language (BPEL [3]) is the de-facto standard for implementing business processes based on web services. The orchestrated web services are again exposed as services. BPEL also allows to specify ordering constraints on the messages a service accepts and produces. All in all, it only focuses specifying processes from a single organization point of view, treating the services used as opaque entities ignoring their internal structure forming separate business processes. As a consequence, choreographies cannot be described using BPEL. Therefore, we have proposed choreography extensions for BPEL in earlier work [8], adding the notion of participant topologies for gluing together different participant behavior descriptions (PBDs). PBDs are BPEL processes describing the behavior of each participant in the choreography. We propose to use BPEL4Chor as an interchange format supporting the different choreography design phases. Therefore, a transformation of BPMN choreographies to BPEL4Chor is needed.

This paper extends work from Ouyang et al. [18], where BPEL stubs are generated out of individual BPMN processes. Furthermore, this paper builds upon previous work from [10], where BPMN extensions for high-level choreography modeling were proposed, and [4], where different modeling phases and choreography viewpoints were identified. The contribution of this paper is to present the integrated usage of BPMN and BPEL4Chor during choreography design. Furthermore, we implemented a modeling environment for BPMN where BPEL4Chor choreographies are produced.

The remainder of this paper is structured as follows. The next section discusses choreography design and the use of BPMN therein. Section 3 gives an overview of BPEL4Chor. Section 4 describes the mapping of extended BPMN to BPEL4Chor and section 5 presents our modeling environment. Section 6 reports on related work in the literature, before section 7 concludes.

2 Choreography Design Using BPMN

Complex choreographies cannot be created within a single step. Whenever many different business partners and many interactions are involved, choreography design must be split up into different phases each addressing different issues of the model. As reported in [22] and [4], the following phases can be distinguished.

1. At a very high level, the interaction partners are identified. It must also be captured how many partners of a particular type are involved. E.g. in a logistics scenario a number of carriers might be involved while only one breakdown surveillance service takes part. Furthermore, business documents that are exchanged between the partners are listed and agreement on the general content of the documents must be reached. E.g. it is defined that a certain contract must carry two signatures or that a request for quote must contain the quantity of the desired product. These two first steps lead to a *high-level structural view* on the choreography.
2. Choreographies reflect what interactions are needed to fulfill a certain goal. This goal can typically be divided into sub-goals or milestones that must be

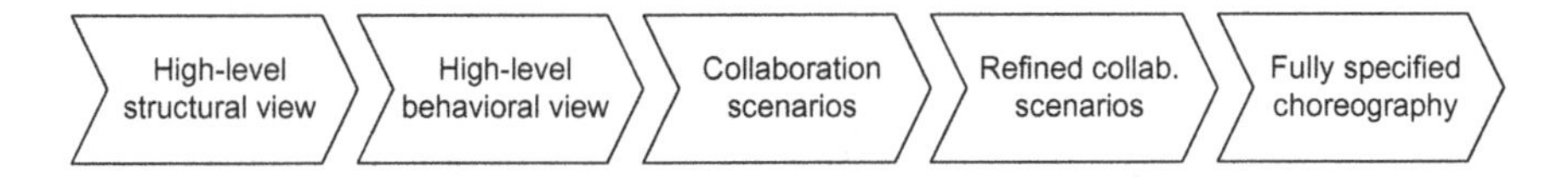

Fig. 1. Different artifacts produced in the choreography design phases

reached on the way to the overall goal. This calls for a *high-level behavioral view* on the choreography.

3. Once the milestones are defined, *collaboration scenarios* are a means to capture how to get from one milestone to another milestone. The required interaction sequences are modeled accordingly.
4. While first versions of collaboration scenario models are likely to only capture best cases, exception handling must be added subsequently. This leads to *refined collaboration scenarios* that also capture timing constraints.
5. All scenario models are aggregated into a big choreography model, including all interactions and their dependencies. Technical choices must be made, e.g. whether to use synchronous vs. asynchronous communication. This leads to the *fully specified choreography model.*

BPMN supports modeling of choreographies as collaboration diagrams. Pools model business roles or entities, while message flows represent the communication between them. High-level structural diagrams can be realized in BPMN by using empty pools and message flow between them. As it is not possible to represent that multiple participants of the same type are involved in one conversation, we added a *pool set* for this purpose.

Figure 2 is a sample structural diagram illustrating a scenario that will be used throughout this paper: A customer buys a product from a seller. The seller in turn handles payment through a payment service. Delivery is outsourced to a delivery service which in turn does not carry out the delivery by itself but rather manages the actual delivery done by a set of carriers. In some cases several carriers are involved covering a part of the overall journey by air, rail or truck.

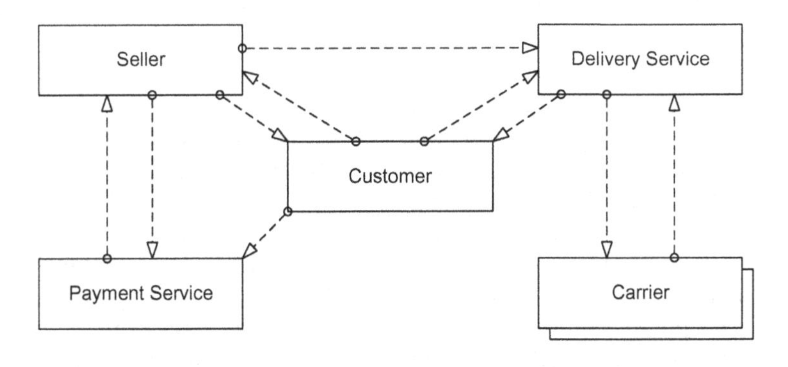

Fig. 2. High-level structural diagram in extended BPMN

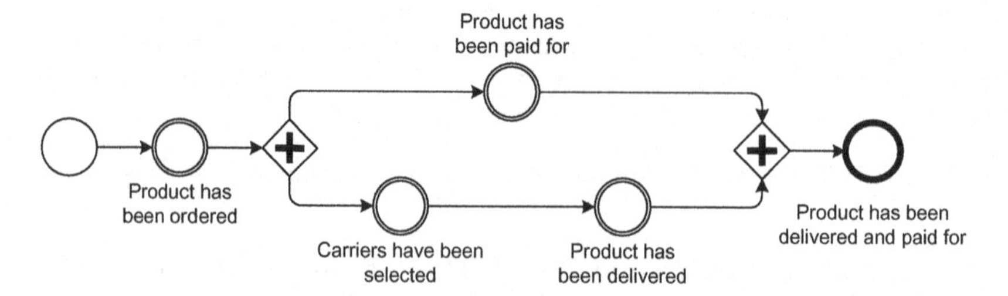

Fig. 3. High-level behavioral diagram in BPMN

The pools (rectangles) in Figure 2 represent the different participant types. Only one customer, seller, payment service and delivery service are involved in one conversation, i.e. one choreography instance. The shaded pool for type carrier represents that there might be more than one carrier involved in one conversation. The dashed arrows symbolize message flow between participants of the corresponding types, indicating who potentially sends a message to whom.

High-level behavioral diagrams can be modeled in BPMN as shown in Figure 3. Untyped events (empty circles) represent milestones which in turn are connected through control flow constructs. This example shows that the first milestone to be reached is that the customer has ordered a product. This is the precondition for the two subsequent milestones "product has been paid for" and "carriers have been selected". The first AND-gateway (diamond containing a "+") represents that the two succeeding milestones can be reached in any order. The second AND-gateway synchronizes the two branches and leads to the final milestone.

Collaboration scenarios which show how progress from one milestone to another can be achieved, are modeled as collaboration diagrams. This time, the pools are not empty but rather the ordering of the message exchanges is expressed by relating the communication activities (send and receive activities) using control flow constructs.

In Figure 4 we see how the collaboration scenario connects to other models: Two milestones from the high-level behavioral model appear again. Further connections to other models are established through the use of link events (circles containing an arrow). This is the standard BPMN way of modeling off-page connectors.

In choreographies where multiple participants of the same type might be involved, it is important to distinguish the individual participants. This is achieved by the introduction of special data object types, namely *participant references* and *participant sets*, symbolized by (shaded) artifacts with a business card icon in the upper left corner. Figure 4 illustrates how this is used in the context of different carriers that must be chosen from.

The semantics of the diagram is as follows. The seller initiates delivery by sending a delivery request to the delivery service. This service contacts all its partner carriers, asking them to check availability for the entire route or parts of the route. Upon receipt of these request, each carrier checks availability. If no capacity is available the carrier answers with a rejection message. Otherwise the carrier prepares a quote and sends it back to the delivery service. The delivery

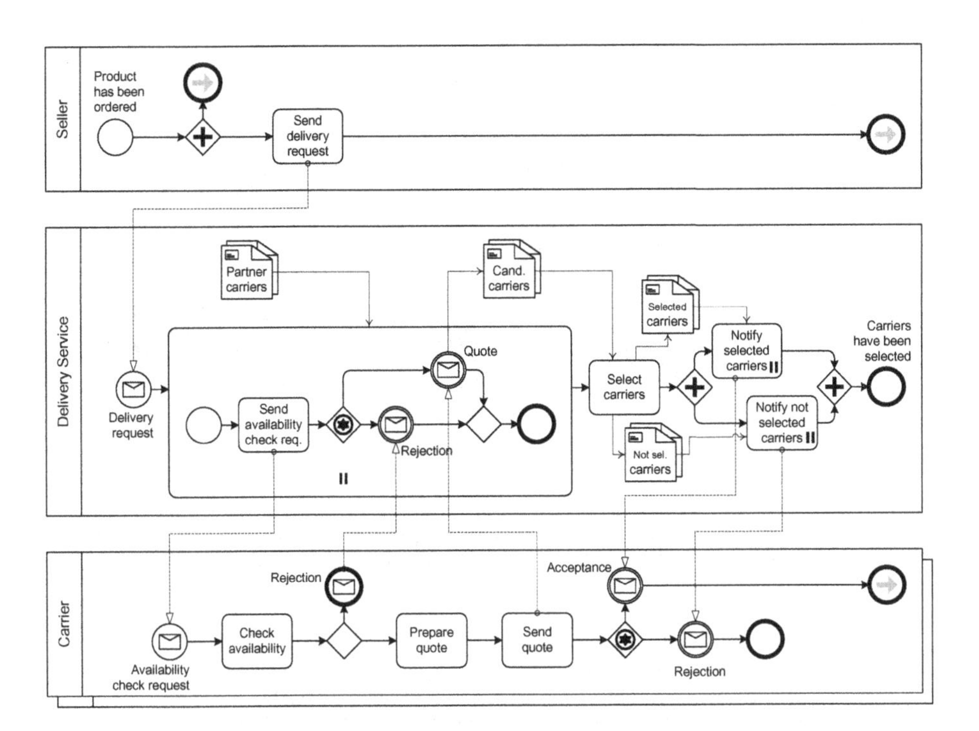

Fig. 4. Collaboration scenario in BPMN: Progressing from "product has been to ordered" to "carriers have been selected"

service collects the quotes and remembers all carriers that have sent a quote as "candidate carriers". Once all carriers have answered, the delivery service selects one or more carriers and sends notifications to the carriers telling them whether they were selected or not. After this, the scenario ends by reaching the milestone "carriers have been selected".

The diagram illustrates how participant references and participant sets affect communication activities and multi-instance activities. The set of partner carriers serves as input for the multi-instance subprocess, indicating that one instance should be spawned for each carrier in this set. Associations from participant references to send and receive activities define that the message is sent to the referenced participant and that only a message from the referenced participant will be received, respectively.

Figure 4 only covers the best case of our collaboration scenario. It is not specified yet what happens if the carriers do not respond within a given timeframe. It is also not specified what happens if no suitable carrier can be found. This might lead to notifying the customer about a delay in delivery or even completely canceling the order.

BPMN allows to model timeouts and exceptions by offering corresponding event types. Intermediate events attached to activities and subprocesses represent cancellation upon the occurrence of the event. Using these constructs

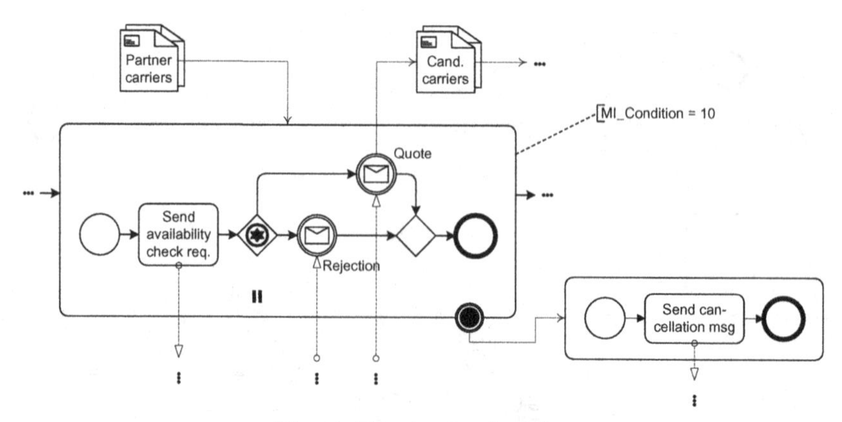

Fig. 5. Termination handlers

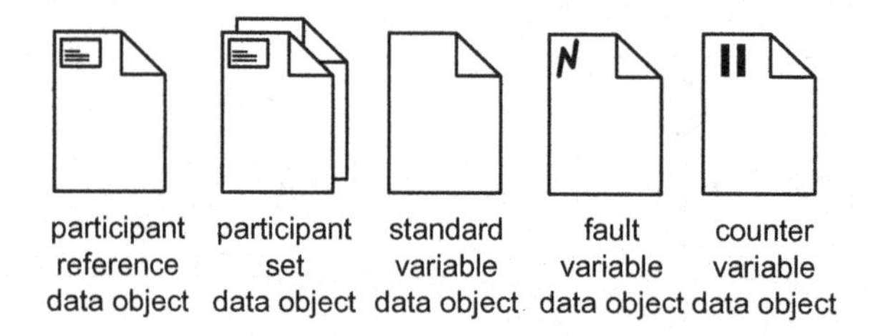

Fig. 6. Data object types

it is possible to model a wide range of exception scenarios. However, when comparing BPMN to BPEL in terms of exception handling, we find that a number of important concepts of BPEL are missing in BPMN. As we intended to allow the modeler to refine the choreography model to fully specified models, we adopted these concepts as BPMN extensions. The following list gives an overview of these extensions. A detailed discussion on these can be found in [19].

Termination handlers. The termination handler of a subprocess defines reactions to forced termination. Especially in the case of forEach constructs with completion condition, termination handlers are needed. As soon as the completion condition is fulfilled, all remaining subprocess instances are terminated. We introduce termination handlers to BPMN. We opted for a similar graphical representation as it is used for compensation handlers. Figure 5 shows a refinement of a part of the scenario from Figure 4.

Different data object types. As already mentioned we introduce participant references and participant sets as special data object types. Additionally, we distinguish between fault variable data objects, counter variable data objects and standard variable data objects. Counter variables represent the counter in a forEach activity and fault variables hold the data of a fault that was thrown or caught.

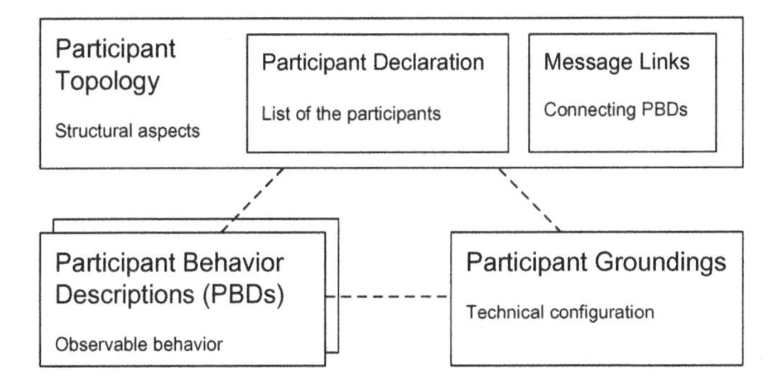

Fig. 7. BPEL4Chor artifacts

Correlation information. Correlation is the act of relating messages received to process instances and receive activities within this instance. Typically, correlation is done based on specific message content. E.g. an order id is used to route an incoming message to the corresponding instance handling the order. While there are very complex correlation mechanisms thinkable, we opted for a correlation set semantics like it is present in BPEL. Therefore, we added corresponding attributes to the invoke and receive activities.

3 BPEL4Chor Overview

BPEL4Chor is a language to describe service choreographies in BPEL. It distinguishes between three aspects: (i) the participant topology, which provides a global view on the existing participants and their interconnection using message links, (ii) participant behavior descriptions, i.e. control flow dependencies in each participant and (iii) participant groundings, i.e. concrete configurations for data formats and port types.

The high-level structural view can be captured in the participant topology. The participants are listed in the participants declarations part. Here, participants and participant sets are distinguished. Each participant carries a type, which specifies the behavior of the participant. In the example participant topology shown in Listing 1, there exists one participant for participant type DeliveryService. The delivery service knows several partner carriers, therefore the topology contains the participant set PartnerCarriers. Participant sets can be used in a forEach construct, in the sense that the forEach construct iterates over this set. The current participant for the iteration is called `currentCarrier` in the listing. The messages exchanged are modeled using message links. A message link connects two participants and states which message is sent over it. Listing 1 lists an extract of the participant topology for our scenario.

Behavioral aspects are captured in the participant behavior descriptions, expressed in BPEL. Listing 2 presents the first part of the BPEL process for the delivery service. The communication constructs are named so that they can be

Listing 1. Participant topology

```
<topology name="DeliveryTopology">
<participantTypes>
  <participantType name="Seller"
  participantBehaviorDescription="ns1:Seller" />
  <participantType name="DeliveryService" ... />
  <participantType name="Carrier" ... />
</participantTypes>
<participants>
  <participant name="Seller" type="Seller" />
  <participant name="DeliveryService" type="DeliveryService" />
  <participantSet name="PartnerCarriers" type="Carrier"
   forEach="ns2:pcarrierForEach">
    <participant forEach="ns2:pcarrierForEach" name="currentCarrier" />
  </participantSet>
  ...
</participants>
<messageLinks>
  <messageLink name="orderLink" messageName="order"
  sender="Seller" receiver="DeliveryService" />
  ...
</messageLinks>
</topology>
```

Listing 2. Participant behavior description for type delivery service

```
<process name="DeliveryService"
  <sequence>
    <receive createInstance="yes" name="ReceiveDeliveryRequest" />
    <sequence>
      <forEach name="pcarrierForEach" parallel="yes">
        <scope><sequence>
          <invoke name="SendAvailabilityCheckReq." />
          <pick>
            <onMessage wsu:id="Quote"><empty /></onMessage>
            <onMessage wsu:id="Rejection"><empty /></onMessage>
          </pick>
        </sequence></scope>
      </forEach>
      <opaqueActivity name="SelectCarriers" />
    </sequence>
    ...
  </sequence>
</process>
```

interconnected. The interconnection is formed by adding the names of the activities to message links. While the message links in the sample topology in Listing 1 have the attributes `sender` and `receiver` set, attributes `sendActivity`

Listing 3. Participant grounding

```
<grounding topology="DeliveryTopology">
  <messageLinks>
    <messageLink name="orderLink" portType="ds:deliveryService_pt"
    operation="getProduct" />
    ...
  </messageLinks>
</grounding>
```

and `receiveActivity` must also be set for referring to the communication constructs in the participant behavior descriptions.

Technical choices are reflected in the participant grounding, where concrete port types and operations come in. Each message link is assigned to a port type and operation. Listing 3 presents the grounding of one message link. The grounding can then be used to generate abstract BPEL processes which are subsequently used for executable completion.

4 Mapping BPMN to BPEL4Chor

Although our extended BPMN and BPEL4Chor have a large overlap in concepts covered, not all diagrams can be transformed to BPEL4Chor. The following BPMN elements are not allowed:

- complex gateways
- ad-hoc and transactional subprocesses
- link, rule and multiple start events
- all end events except the non-triggered ones
- cancel, rule, link, multiple or non-triggered intermediate events
- user, script, abstract, manual or reference activities

In [18] three classes of BPMN diagrams are distinguished: (i) those that can be translated using block-structured constructs only, (ii) those that require the use of control links and finally (iii) those that require event handlers, fault handlers and message passing within one process instance for realizing control flow dependencies. For instance, the occurrence of the workflow patterns arbitrary cycles and multi merge [21] make a diagram be of category (iii), as there is no direct support for these two workflow patterns in BPEL. We argue that the BPEL code resulting from (iii) is not usable as starting point for further refining it to process implementations. Therefore, we do not transform these kind of diagrams.

General Approach. We largely base our transformation on the approach presented in [18] where a subset of BPMN is transformed to BPEL. This approach is based on the identification of *patterns* in the diagram that can be mapped onto BPEL blocks. One pattern is folded into a new activity, which is associated with the generated BPEL code. We extend these patterns with the elements

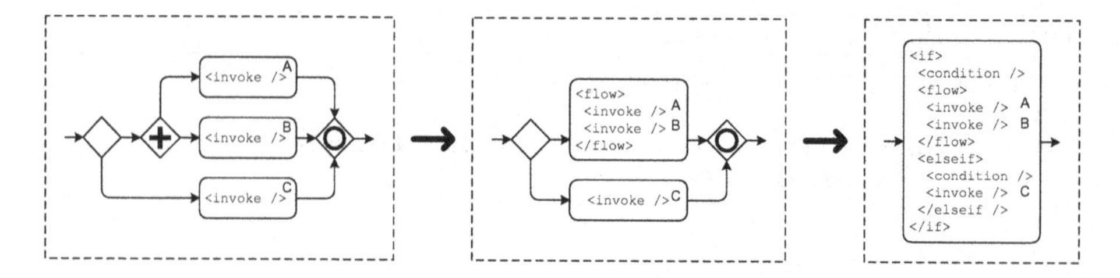

Fig. 8. Dealing with inclusive gateways

used in the extended BPMN described above. Hence, we can use that transformation for transforming processes located in a pool, pool set or subprocess to their BPEL4Chor representation. Furthermore, we loosen certain restrictions as explained in the next subsection.

Multiple start and end events. In [18] it is assumed that there is only one start event and one end event in each process. We loosen this restriction and allow certain combinations of start events as well as multiple end events. If e.g. two start events are followed by a XOR-gateway, we fold this pattern to a BPEL pick element, where the attribute createInstance is set to "yes". Also the case if they are followed by an AND-gateway can be handled and translated to BPEL4Chor. These scenarios are captured by generalized pick- and flow-patterns. While it is easy to see for these simple examples how they can be mapped, it is less obvious why some combinations are not allowed in our transformation. Imagine e.g. three start events A, B, C where A and B are merged through an AND-gateway, which in turn is merged with C through a XOR-gateway. Here, C is an alternative to the combination of A and B. Such behavior is not directly expressible in BPEL. Multiple end events are resolved by merging the different branches into an inclusive gateway.

Inclusive gateways. We allow inclusive gateways if they occur in certain combinations with other elements and can be rewritten to AND- and XOR-gateways. In order to capture these combinations, the well-structured and quasi-structured patterns from [18] are extended. This means that our transformation can handle inclusive gateways in block-structured settings only.

Figure 8 illustrates an example. It exhibits two steps to transform a BPMN diagram involving multiple invoke activities to the corresponding BPEL representation. In the first step, an AND split gateway is translated to a BPEL flow, representing concurrent invocations of A and B. In the second step, the XOR split gateway is translated to an if construct in BPEL, so that either invocations of A and B are performed concurrently or C is invoked.

Fault, compensation and termination handlers. We introduce a pattern for activities and subprocesses with attached intermediate events. This leads to the creation of a BPEL handler for each attached event. To enable direct transformation to BPEL, we only allow those fault handlers, where the outgoing control flow from the handler is directly merged with the control flow originating from the corresponding activity or subprocess.

Other constructs. The mapping of activities and events is straightforward. Variable data objects are not folded because they may be associated with flow objects in other patterns. Each pool and pool set is mapped to a participant type. For a simple pool a participant reference with its corresponding type can be generated directly. Additional references are generated from participant reference data objects. The mapping of message flows to message links depends on the connected activities, the participant reference and participant data objects associated with these activities and the message data objects associated with the message flows. As the extended transformation removes elements from the model during the folding of the patterns, the topology has to be created beforehand.

1. Generate participant types in the topology from pools and pool sets
2. Generate participant references and participant sets from the participant reference and participant set data objects
3. Generate message links from the message flow, the associated participant reference and message data objects
4. Transform the processes within the pools and pool sets
   4.1. Generate the variables from the variable data objects
   4.2. Apply the extended transformation starting with the pattern for attached events

5 Choreography Modeling Environment

We have implemented a BPMN editor and the BPMN to BPEL4Chor transformation based on the Oryx framework developed at the Hasso-Plattner-Institute[1]. Oryx is a graphical editing framework written in JavaScript that uses Scalable Vector Graphics (SVG) as rendering technology. Oryx comes with a set of stencil sets for modeling pure BPMN, the extended BPMN, workflow nets and other process modeling languages. Each stencil set defines a set of elements, including their attributes, containment relationships and connection rules. The shape definitions, i.e. the graphical appearance of elements, are defined as SVG files.

Oryx strictly follows the REST (Representational State Transfer [13]) architectural style. Each process model and each element within it are considered as resources that are uniquely identified by URIs. By addressing a process model URI in a web browser, an XHTML representation of the model is retrieved, which in turn contains all model information as embedded RDF annotations. This web page also contains links to the Oryx implementation. The browser loads these scripts which turn the web page into a graphical editor application. If models are to be imported into other applications, existing XSLT stylesheets can be applied for retrieving corresponding RDF documents. Figure 9 illustrates the system architecture using the Fundamental Modeling Concepts notation [16].

The editor provides extensibility through a plugin mechanism. We used this mechanism to integrate the BPMN to BPEL4Chor transformation functionality into the editor. The transformation plugin serializes the extended BPMN

[1] See `http://bpt.hpi.uni-potsdam.de/Oryx/`

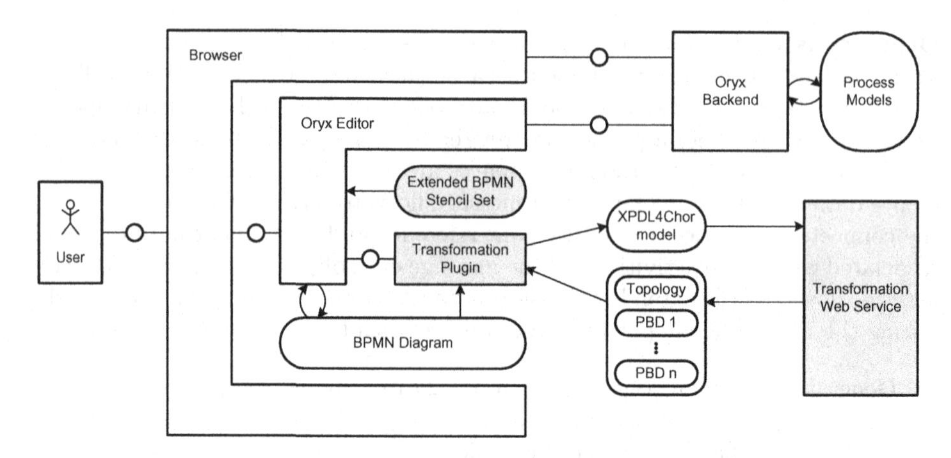

Fig. 9. Architecture of the modeling environment

diagram into an extended XPDL (XML Process Definition Language [1]) format, called XPDL4Chor. While XPDL 2.0 is a serialization format for BPMN standardized by the Workflow Management Coalition (WfMC), XPDL4Chor additionally contains the new elements and attributes we added to BPMN.

The actual transformation takes place in a separate web service. This service takes the XPDL4Chor document as input and produces the different BPEL4Chor documents. This includes the participant topology as well as the participant behavior descriptions for each participant type. The plugin offers the possibility to download these documents or to view them in the browser.

Figure 10 shows a screenshot of the Oryx editor[2]. On the left side the palette contains the different language constructs. These can be dragged onto the drawing area in the middle. Attributes of the model elements can be edited in the properties area on the right. Different editing functionality can be accessed through the buttons on the top. Output as BPEL4Chor files or output as XPDL4Chor file can be triggered through two of these buttons.

6 Related Work

There are different language proposals available for modeling choreographies. The Web Service Choreography Description Language (WS-CDL [15,5]) was released by the World Wide Web Consortium in 2005. Differences between WS-CDL and BPEL are discussed in [17]. Let's Dance [23] is another choreography language. Like BPMN, it is implementation-independent and comes with a visual notation. This language was designed to support all Service Interaction Patterns [6], a set of recurrent choreography scenarios. An assessment of WS-CDL using these patterns can be found in [9]. An earlier and less expressive choreography language is the Business Process Schema Specification (BPSS [7]). A general introduction into the different viewpoints found in inter-organizational process modeling can

[2] The editor is accessible through `http://www.bpel4chor.org/editor/`

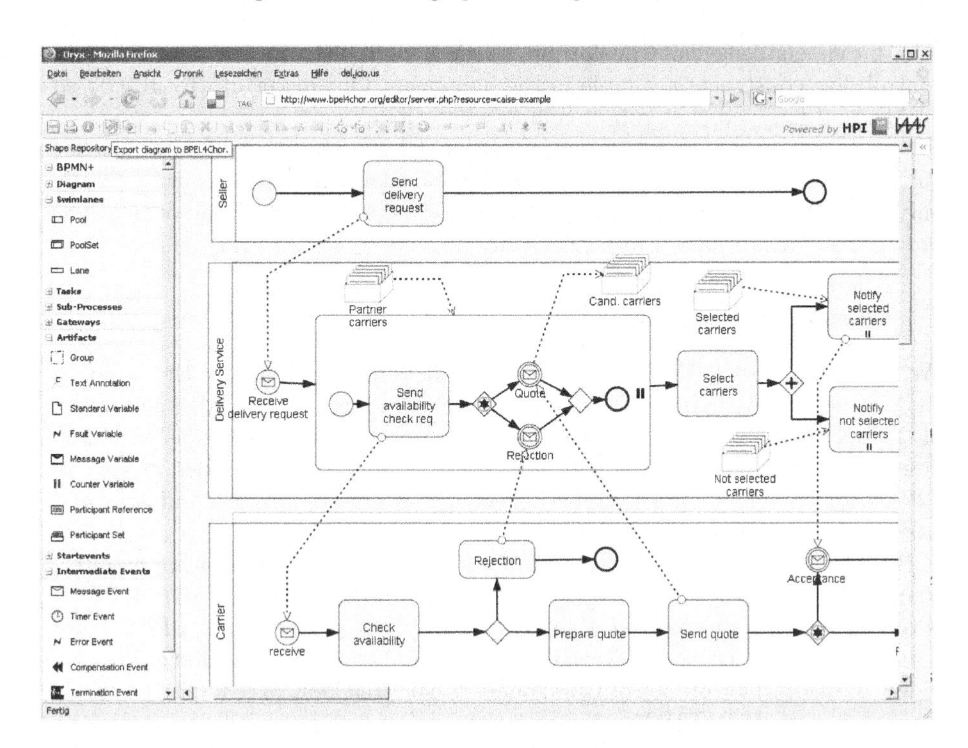

Fig. 10. Screenshot of the Oryx editor with the transformation plugin

be found in [12]. Already in [10] we have shown how the addition of the concepts pool set, participant references and participant sets leads to a significantly higher suitability of BPMN for choreography modeling. Such extended BPMN even surpasses WS-CDL in terms of Service Interaction Pattern support.

There are basically two different modeling styles manifested in choreography languages. In the case of *interconnected models*, send and receive activities are listed for each role and control and data flow dependencies are defined on a per-role-basis. In contrast to this, *interaction models* are made up of atomic interactions and control and data flow is defined globally, i.e. it is not directly assigned to any of the roles. Examples for the first group are BPMN and BPEL4Chor, but also simpler languages such as Message Sequence Charts (MSC [14]). Examples for the second group are WS-CDL, BPSS and Let's Dance. Bridging these two modeling styles is not trivial and requires for sophisticated transformation algorithms as presented in [11] for the case of interaction Petri nets and their corresponding participant behavior descriptions. This is not needed in our case, as BPMN and BPEL4Chor follow the same modeling style.

There has been some work on comparing BPMN and BPEL and carrying out transformations. Comparison was done e.g. in [20] on the general concepts covered in both languages and on the respective Workflow Pattern support: the authors' conclusion is that the expressiveness of BPMN has to be restricted if a full mapping to BPEL is desired.

A major challenge in transforming BPMN to BPEL are the differences in control flow constructs available in the languages. Ouyang et al. [18] restricted BPMN and mapped that subset completely to BPEL.

Several commercial tools allow to define BPEL-specific configurations for BPMN-models and implement transformation algorithms. However, typically only a small subset of BPMN is allowed and then translated. None of the tools provides a transformation to BPEL4Chor.

7 Conclusion and Outlook

We presented how BPMN can be used as modeling language in the different choreography design phases. By extending BPMN we reached a higher suitability of BPMN for modeling choreographies both in early design phases as well as in late phases, where exceptions and technical configurations are added. We chose BPMN for modeling choreographies, since it is widely used in the industry and has a wide tool support.

BPMN does not specify a serialization format. We use BPEL4Chor as interchange format for the choreography models at the different levels of detail. BPEL is the standard language for describing executable workflows. Since BPEL4Chor is close to BPEL, the gap between design time and runtime is narrowed. We provided a transformation of BPMN models to BPEL4Chor, extending the transformation from [18]. BPMN elements not having a corresponding notation in BPEL4Chor are not transformed. However, these details are not required by the runtime. Finally, we presented Oryx, our graphical modeling environment that runs on the web, where a transformation plugin was added. We do not provide support for round-trip engineering: only a one-way transformation from BPMN to BPEL4Chor is provided. Modifications to the BPEL4Chor artifacts are not reflected in the BPMN diagram.

Limitations of our approach are the restrictions we impose on the BPMN models that can be transformed to BPEL4Chor. As part of that, we require that BPMN models are sound and safe, i.e. deadlock-free and without multi-token flow. We currently do not check these properties prior to the transformation. Generally, the fact that only during the transformation we can detect that we cannot transform a model, can be seen as the biggest limitation of our current implementation. It is desirable to perform a check prior to starting the transformation and to give the modeler hints how to resolve the problem. This is subject to ongoing work. As part of that we are working on integrating Petri-net-based analysis functionality into the BPMN editor.

References

1. Process Definition Interface – XML Process Definition Language (October 2005), `http://www.wfmc.org/standards/docs/TC-1025_xpdl_2_2005-10-03.pdf`.
2. Business Process Modeling Notation (BPMN) Specification, Final Adopted Specification. Technical report, Object Management Group (OMG) (February 2006), `http://www.bpmn.org/`.

3. Web Services Business Process Execution Language Version 2.0 – OASIS Standard (April 2007)
4. Barros, A., Decker, G., Dumas, M.: Multi-staged and Multi-viewpoint Service Choreography Modelling. In: SEMSOA (2007)
5. Barros, A., Dumas, M., Oaks, P.: A Critical Overview of WS-CDL. BPTrends 3(3) (2005)
6. Barros, A., ter Hofstede, A.H.M., Dumas, M.: Service Interaction Patterns. In: van der Aalst, W.M.P., Benatallah, B., Casati, F., Curbera, F. (eds.) BPM 2005. LNCS, vol. 3649, pp. 302–318. Springer, Heidelberg (2005)
7. Clark, J., Casanave, C., Kanaskie, K., Harvey, B., Smith, N., Yunker, J., Riemer, K.: ebXML Business Process Specification Schema Version 1.01. Technical report, UN/CEFACT and OASIS (May 2001), http://www.ebxml.org/specs/ebBPSS.pdf
8. Decker, G., Kopp, O., Leymann, F., Weske, M.: BPEL4Chor: Extending BPEL for Modeling Choreographies. In: ICWS (2007)
9. Decker, G., Overdick, H., Zaha, J.M.: On the Suitability of WS-CDL for Choreography Modeling. In: EMISA 2006 (2006)
10. Decker, G., Puhlmann, F.: Extending BPMN for Modeling Complex Choreographies. In: CoopIS 2007 (2007)
11. Decker, G., Weske, M.: Local Enforceability in Interaction Petri Nets. In: Alonso, G., Dadam, P., Rosemann, M. (eds.) BPM 2007. LNCS, vol. 4714, pp. 305–319. Springer, Heidelberg (2007)
12. Dijkman, R., Dumas, M.: Service-oriented Design: A Multi-viewpoint Approach. International Journal of Cooperative Information Systems 13(4), 337–368 (2004)
13. Fielding, R.T.: Architectural Styles and the Design of Network-based Software Architectures. PhD thesis, University of California, Irvine (2000)
14. ITU-T. Message Sequence Chart. Recommendation Z.120, ITU-T (2000)
15. Kavantzas, N., Burdett, D., Ritzinger, G., Lafon, Y.: Web Services Choreography Description Language Version 1.0, W3C Candidate Recommendation. Technical report (2005)
16. Knopfel, A., Grone, B., Tabeling, P.: Fundamental Modeling Concepts: Effective Communication of IT Systems. Wiley, Chichester (2006)
17. Mendling, J., Hafner, M.: From Inter-Organizational Workflows to Process Execution: Generating BPEL from WS-CDL. In: OTM, Workshops (2005)
18. Ouyang, C., Dumas, M., ter Hofstede, A.H., van der Aalst, W.M.: Pattern-based translation of BPMN process models to BPEL web services. International Journal of Web Services Research (JWSR) (2007)
19. Pfitzner, K., Decker, G., Kopp, O., Leymann, F.: Web Service Choreography Configurations for BPMN. In: WESOA 2007 (2007)
20. Recker, J., Mendling, J.: On the Translation between BPMN and BPEL: Conceptual Mismatch between Process Modeling Languages. In: EMMSAD 2006 (2006)
21. van der Aalst, W.M.P., ter Hofstede, A.H.M., Kiepuszewski, B., Barros, A.P.: Workflow Patterns. Distributed and Parallel Databases 14(1), 5–51 (2003)
22. Weske, M.: Business Process Management: Concepts, Languages, Architectures. Springer, Heidelberg (2007)
23. Zaha, J.M., Barros, A., Dumas, M., ter Hofstede, A.: Let's Dance: A Language for Service Behavior Modeling. In: CoopIS 2006 (2006)

Work Distribution and Resource Management in BPEL4People: Capabilities and Opportunities*

Nick Russell and Wil M.P. van der Aalst

[1] Department of Technology Management, Eindhoven University of Technology
[2] GPO Box 513, NL5600 MB Eindhoven, The Netherlands
{n.c.russell,w.m.p.v.d.aalst}@tue.nl

Abstract. The BPEL4People and WS-HumanTask extensions to the BPEL proposal define the state of the art in resource management and work distribution in business process execution languages. In this paper, we use the workflow resource patterns as an evaluation framework to assess the capabilities of BPEL4People and WS-HumanTask and identify several areas where there is opportunity for further improvement.

Keywords: BPEL4People, WS-HumanTask, Resource Patterns.

1 Introduction

One of the major objectives of workflow systems (and process-aware information systems (or PAIS) more generally) is to facilitate the distribution and coordination of work amongst the group of human resources associated with a process. There has been explosive growth in the commercial offerings available to support this objective as organisations seek out more effective ways in which to deploy their business processes across their workforce in a predictable, reliable and controlled manner. With the rise of the internet came a consequential extension of the underpinning technologies to embrace cross-organisational processes and the concept of the web service was born together with the notion of service oriented architectures which aim to facilitate business processes on the basis of loosely coupled (and potentially widely distributed) execution capabilities.

BPEL [11] was one of the first standards initiatives that attempted to establish a common processing framework and language that distinct execution engines could adopt in order to make the notion of a distributed business process based on disparate web services a viable possibility. It met with significant commercial interest and quickly established itself as the major standards initiative in this area. Developed by an industry consortium, it is perhaps not surprising that it met with early success as many of its contributors also had specific commercial

* This research is conducted in the context of the *Patterns for Process-Aware Information Systems (P4PAIS)* project which is supported by the Netherlands Organisation for Scientific Research (NWO).

Z. Bellahsène and M. Léonard (Eds.): CAiSE 2008, LNCS 5074, pp. 94–108, 2008.

interests that were directly furthered through its publication and broad adoptance. It is ironic therefore given the level of commercial input into the overall development of the BPEL standard that it had two major omissions: (1) a lack of recognition that business processes are generally hierarchical in form (resulting in the omission of the notion of subprocesses) and (2) a lack of consideration that business processes generally have some form of human involvement. Although these may have been deliberate omissions, they limit the applicability of BPEL in real-life processes.

The WS-BPEL Extension for Sub-Processes [9] proposal resolved the first of these issues. In an attempt to address the second, the BPEL4People [4] and WS-HumanTask [3] proposals have been released. They attempt to provide a series of extensions to WS-BPEL 2.0 [11] that integrate human resources into the overall execution of business processes. As these are early stage proposals, they are still open to comment in order to ensure that they meet with general acceptance before being finalised as standards. The focus of this paper is to review the conceptual foundation of BPEL4People and WS-HumanTask using the resource patterns as an evaluation framework. Through this examination, we hope to determine where the strengths and weaknesses of these proposals lie and what opportunities there may be for further improvement.

The *resource patterns* [12] were selected as the basis for evaluating the BPEL4-People and WS-HumanTask proposals as they offer a means of examining their capabilities from a conceptual standpoint in a way that is independent of specific technological and implementation considerations. The resource patterns were developed as part of the *Workflow Patterns Initiative*, an ongoing research project that was conceived with the goal of identifying the core architectural constructs inherent in workflow technology. The original objective was to delineate the fundamental requirements that arise during business process modelling on a recurring basis and describe them in an imperative way. A patterns-based approach was taken to describing these requirements as it offered both a language-independent and technology-independent means of expressing their core characteristics in a form that was sufficiently generic to allow for its application to a wide variety of offerings. To date, 126 patterns have been identified in the control-flow [13], data [14] and resource [12] perspectives and they have been used for a wide variety of purposes including evaluation of PAIS, tool selection, process design, education and training. The workflow patterns have been enthusiastically received by both industry practitioners and academics alike. The original Workflow Patterns paper [1] has been cited by over 600 academic publications and the workflow patterns website is visited by more than 300 individuals each day. Full details can be found at `http://www.workflowpatterns.com`.

The resource patterns form part of a surprisingly small body of research into resource and organisational issues in PAIS. Relevant research in the context of this paper includes early work by Bussler and Jablonski [5] which identifies a number of shortcomings of workflow systems when modelling organisational and policy issues. Du and Shan [6] present a design for a resource manager for a workflow system which includes a high level resource model together with proposals for resource definition, query and policy languages. Similarly in [8], Huang and Shan propose a means of facilitating policy-based handling of resource assignment in a workflow context. The RBAC (Role-Based Access Control) model [7] describes a security framework for workflow that allows suitable users to be determined for a task. In [10] zur Muehlen presents a comprehensive overview of the organizational aspects of workflow technology. Several researchers [2,10] have developed meta-models describing the relationships between various workflow concepts, including aspects of work allocation, however these meta-models typically do not describe the dynamic aspects of work distribution.

The remainder of this paper proceeds as follows: Section 2 provides an overview of the BPEL4People and WS-HumanTask proposals. Section 3 presents an assessment of the two proposals using the workflow resource patterns as an evaluation framework. Section 4 discusses the results of the evaluation and identifies a number of areas where future possibilities exist for strengthening the proposals and Section 5 concludes the paper.

2 BPEL4People: Overview and Background

In this section we examine the intention and coverage provided by the BPEL4-People and WS-HumanTask proposals from various perspectives, starting with their motivation and relationship with related proposals and standards and then examining their informational and state-based characteristics on a comparative basis against those described by the workflow resource patterns.

2.1 Motivation and Related Standards

The stated intentions of the BPEL4People proposal and the closely coupled WS-HumanTask proposal are as follows [4,3]:

- BPEL4People: to support a broad range of scenarios that involve people within business processes.
- WS-HumanTask: to provide a notation, state diagram and API for human tasks as well as a coordination protocol that allows interaction with human tasks in a more service-oriented fashion, and at the same time control task autonomy.

In order to achieve these objectives, the BPEL4People proposal assumes the services of a number of related standards. Figure 1 illustrates the relationship between the various standards that are required in order to support the BPEL4People proposal. It is interesting to note that whilst BPEL4People has the most visibility, it provides minimal new capabilities from a resource perspective and essentially acts only to extend the notion of an Activity to that of a PeopleActivity hence enabling the definition of inline and local tasks carried out under the auspices of a human resource. The bulk of the new features associated with work items, work distribution and state management are actually provided

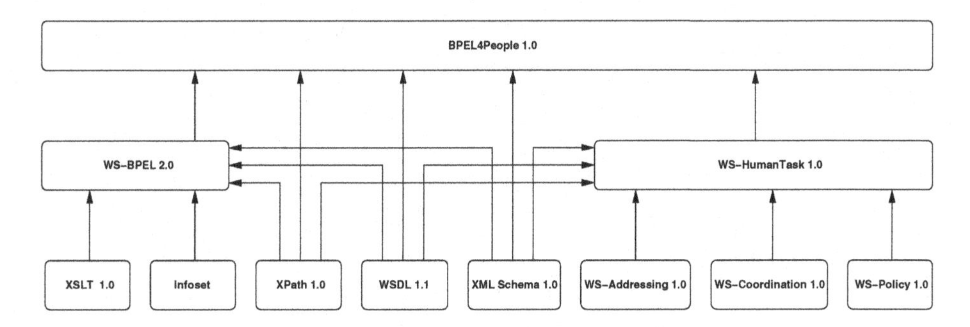

Fig. 1. Web services standards hierarchy

by the WS-HumanTask proposal which also introduces the notion of a standalone task (i.e. a task whose implementation is defined outside of the context of the BPEL process) that is undertaken by a human resource. Consequently, much of the remainder of this document will tend to focus on the capabilities defined by the WS-HumanTask proposal.

2.2 Information Coverage of the BPEL4People/WS-HumanTask Extensions

Insight into the overall capabilities of the BPEL4People/WS-HumanTask extension can be obtained by comparing the metamodels shown in Figure 2 in the form of UML class diagrams. The top model shows the main entities covered by the BPEL4People and WS-HumanTask proposals and the bottom model shows the main entities covered by the resource patterns. By comparing the two models, it is possible to analyze the differences and commonalities. Although much of the information content is common to both models, there are some noteworthy distinctions between them.

The resource patterns:

- assume a richer organisational model both to capture relationships between resources, job and organisational units, and also allow this information to be used as the basis for work distribution directives (see label (1) in Figure 2);
- include the notion of execution history (where the execution outcomes of activities in multiple concurrent cases are permanently logged) and allow this data to be used in work distribution directives (2);
- support the notion of extensible resource descriptions (via capabilities) which can be used when making decisions about distributing work items (3); and
- provides a comprehensive authorisation framework which strictly defines the work item privileges available to individual resources at runtime (4).

The BPEL4People/WS-HumanTask proposals:

- distinguish between a series of distinct task implementation strategies (local, remote, etc.) (5);

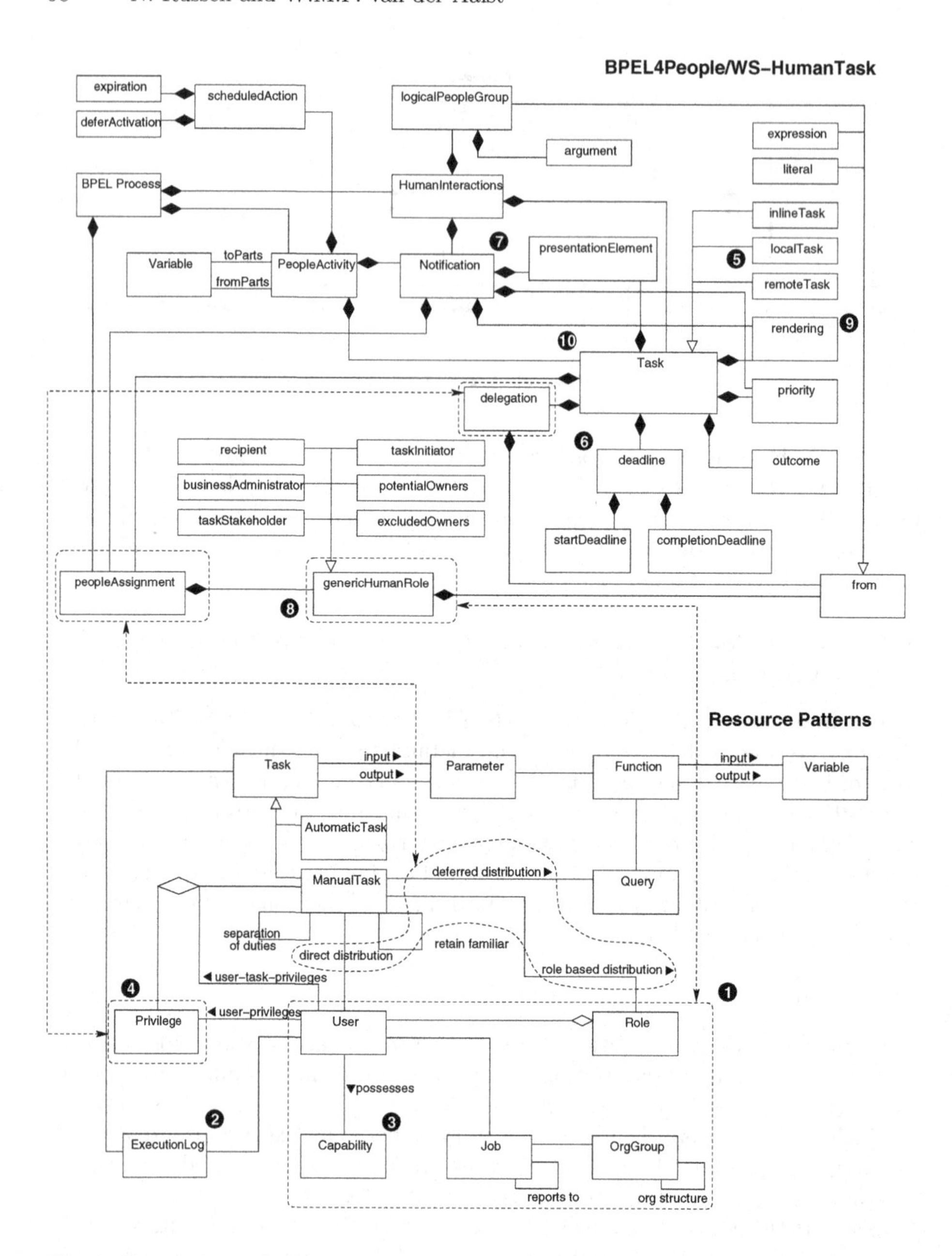

Fig. 2. Comparison of information coverage in BPEL4People [4] and WS-HumanTask [3] and the workflow resource patterns [12]

- incorporate facilities for defining commencement and completion deadlines for tasks along with the actions that should be taken when the deadline is reached. Similar capabilities exist for specifying escalations (6);

- support a series of notification capabilities to advise resources of adverse work item execution circumstances (7);
- include a series of designated roles for each task that describe specific privileges. These include task initiator and task stakeholder (8);
- incorporate the identification of rendering facilities for each task which describe the potential user interfaces that will be presented to resources undertaking the task (9); and
- include a means of representing data specific to a task instance (although interestingly, individual task data instances are only referenced by an id field and it is unclear how data elements are related to specific task instances in a specific case) (10).

Some of the distinctions outlined above are related to scope, others may indicate potential areas for improvement or enhancement and are discussed at greater length later in the paper. One observation that can be made at this point is that the two proposals consider implementation aspects for individual tasks (e.g. presentation elements, interface details and deadlines) in addition to issues associated with work distribution. In contrast, the resource patterns operate at a conceptual level and focus strictly on issues of resource management and work distribution. There is currently no consideration of functional details associated with task enactment in the workflow patterns framework and this raises the question of whether there should be further investigations into the potential for a set of *operational patterns* describing task implementation.

2.3 Dynamic Coverage of the WS-HumanTask Extension

The state models that underpin the resources patterns and the WS-HumanTask proposal[1] are analogous. Figure 3 illustrates the state transition diagrams for both of them. A major difference between them is that WS-HumanTask also includes broader consideration of error states and allows tasks that haven't yet started to be suspended (as shown by labels (1) and (2) respectively). In contrast, the resource patterns differentiate between work items offered to single and multiple resources (shown by (3)) and support a slight wider range of detour actions (as illustrated by the bold arcs).

3 Capabilities: An Assessment of Resource Pattern Support

In the following section, we provide an evaluation of the capabilities of BPEL4-People and WS-HumanTask from a resource perspective. This assessment utilises the workflow resource patterns as an evaluation framework thus providing a technologically agnostic means of examining the capabilities of the two proposals. There are seven distinct groups of resource patterns as follows:

[1] BPEL4People defers state management details to the WS-HumanTask proposal.

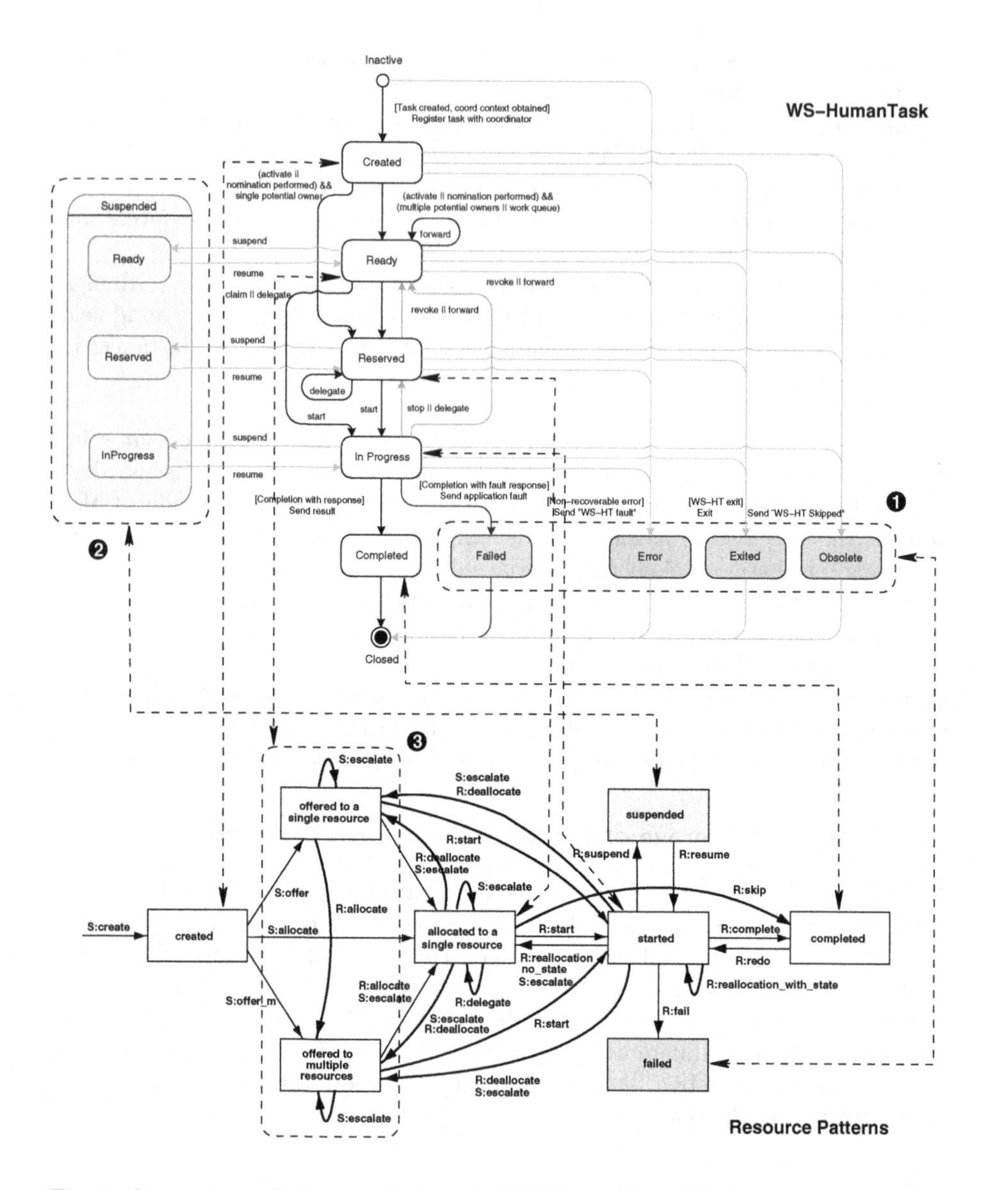

Fig. 3. Comparison of supported states in WS-HumanTask [3] and the workflow resource patterns [12]

- *creation* patterns – which correspond to limitations specified in the design time model on the manner in which a work item is executed by resources;
- *push* patterns – which characterise situations where newly created work items are proactively offered or allocated to resources by the system;
- *pull* patterns – which correspond to situations where individual resources take the initiative in committing to and undertaking available work items;

- *detour* patterns – which refer to situations where work allocations that have been made for resources are interrupted either by the system or at the instigation of individual resources;
- *auto-start* patterns – which relate to situations where the execution of work items is triggered by specific events in the lifecycle of the work item or the related process definition;
- *visibility* patterns – which describe the various scopes in which work item availability and commitment are able to be viewed by resources; and
- *multiple resource* patterns – which characterise situations where the correspondence between the resources and work items in a given allocation or execution is not 1-1.

The following sections describe the support for each of these patterns by the BPEL4People (B4P) and WS-HumanTask (HT) proposals in detail[2].

3.1 Creation Patterns

The intention of the BPEL4People and WS-HumanTask proposals – to support a broad range of scenarios that involve people within business processes – is immediately reflected by the range of creation patterns that are supported as illustrated in Table 1. As the original BPEL proposal provided no guidance in this area, the relative change is significant.

Resources are identified within the context of a BPEL process and work can be distributed directly to them by name or indirectly via role-based groupings or based on the results of queries. Through the use of these queries, separation of duties and retain familiar constraints can be specified between work items within a case. Less well-supported however is the ability to specify more precise work distribution requirements for a task in terms of organisational or history-based criteria. The organisational model supported with the BPEL4People/WS-HumanTask framework is relatively simplistic and does not explicitly identify job roles, reporting lines or relationships between organisational groupings hence these cannot be used when distributing work. Similarly, it is only possible to use the execution characteristics of work items in the same case when framing historical work distribution requirements. There is no support for adding further descriptive criteria to individual resources (i.e. capabilities) and using these when distributing work items. An additional shortcoming relates to the limited ability within BPEL4People/WS-Human-Task to impose an authorisation framework on resources and the range of actions that they are able to undertake with respect to overall process execution (other than for delegate and skip actions). Similarly, it is not possible to constrain the resources that individual tasks can be distributed to in a guaranteed way (e.g. a work item could ultimately be delegated to any resource not just one that satisfied the distribution criteria associated with the task).

[2] Details of individual pattern realisations in BPEL4People and WS-HumanTask can be found in the companion technical report BPM-07-11 at `www.BPMcenter.org`.

Table 1. Creation patterns support

Nr	Pattern	Rating	Rationale
1	Direct Distribution	+	Supported by literal assignment of potential/actual task owners (HT)
2	Role-Based Distribution	+	Supported by logical people group assignment of potential/actual task owners (HT)
3	Deferred Distribution	+	Supported by assignment of potential/actual owners based on expressions (HT)
4	Authorisation	+/–	Limited support for nominating delegation and skipping on a per task basis but no general support for user privileges (HT)
5	Separation of Duties	+	Supported via excluded owners attribute for <peopleAssignment> elements (HT)
6	Case Handling	–	No support for case handling
7	Retain Familiar	+	Supported by assigning actual owner to the same value as the actual owner of another task (HT)
8	Capability-Based Distribution	–	No support for resources to have additional capability attributes
9	History-Based Distribution	+/–	Expressions can utilise details associated with task instances for a given user via the getMyTasks function although its unclear how this can be generalised to broader history-based queries (HT)
10	Organisational Distribution	+/–	The organisational model only identifies group membership and role participation for individual resources (HT)
11	Automatic Execution	+	Directly supported by BPEL

3.2 Push and Pull Patterns

The work distribution model in WS-HumanTask is based on work being advertised to individual resources and those resources making a decision on what work they will commit to undertaking and when they will start it. The degree of support for specific push patterns is illustrated in Table 2. Work items can be offered to multiple resources or allocated to one of them, however it is not possible to offer a work item to a single resource on a non-binding basis. There is no support for randomly selecting a resource to undertake a work item or for distributing work on a round robin (i.e. an equitable) basis, however it does appear that the possibility may exist to distribute work on a shortest-queue basis where there are multiple potential resources for the same work item (although the precise means of implementing this using the provided function set is a little unclear). All work is distributed at the time the task with which it is associated is enabled. As indicated previously, under the WS-HumanTask proposal, work is advertised to resources and they commit to undertaking work items of their choice and can choose the time of commencement. The degree of support for specific pull patterns is illustrated in Table 2. There is provision for a resource to execute multiple work items simultaneously and to order and select the content of their own work queue via queries however it is not possible for the system to impose a default ordering or content for work queues.

Table 2. Push and patterns support

Nr	Pattern	Rating	Rationale
			Push patterns
12	Distribution by Offer - Single Resource	–	Not supported. If there is only one potential owner for a work item, then it is allocated to them
13	Distribution by Offer - Multiple Resources	+	Supported by setting multiple potential owners for a task instance in the Created or Ready state (HT)
14	Distribution by Allocation - Single Resource	+	Supported by setting a single potential owner for a task instance in the Created or Ready state (HT)
15	Random Allocation	–	Not supported
16	Round Robin Allocation	–	Not supported
17	Shortest Queue	+/–	It would appear that this pattern can be supported by using an expression to set the actual owner for a task instance to the potential owner with the shortest work list, however its unclear if this can be implemented with the supported functions (HT)
18	Early Distribution	–	Not supported
19	Distribution on Enablement	+	Potential owners are notified of tasks when they are created
20	Late Distribution	–	Not supported
			Pull patterns
21	Resource-Initiated Allocation	+/–	Supported via the claim function providing the work item is offered to more than one user. It is automatically started if only offered to one resource (HT)
22	Resource-Initiated Execution - Allocated Work Item	+	Supported via the start function (HT)
23	Resource-Initiated Execution - Offered Work Item	+	Supported via the start function (HT)
24	System-Determined Work Queue Content	–	No ability to limit or order the work queue for a resource
25	Resource-Determined Work Queue Content	+	The simple and advanced query functions provide the ability for resources to restrict and format the content of their worklists (HT)
26	Selection Autonomy	+	Resources can choose to start any task instance available available to them (HT)

3.3 Detour, Auto-Start, Visibility and Multiple Resource Patterns

Detour patterns provide the ability for resources (and potentially the system) to alter the normal sequence and manner in which work items are distributed for execution. A variety of distinct "detours" are supported, as illustrated in Table 3, although there is no ability to undertake work items outside of the normal execution sequence (i.e. redo/pre-do) or to rollback their execution state (i.e. stateless reallocation). Auto-start patterns correspond to mechanisms which attempt to speed up the overall throughput of work in various ways. As indicated in Table 3, BPEL4People and WS-HumanTask do not provide any capabilities in this area. Visibility patterns describe mechanisms within the workflow system for limiting the visibility of upcoming or in progress work items to selected resources. As indicated in Table 3, WS-HumanTask potentially provides support in this

Table 3. Detour, auto-start, visibility and multiple resource patterns support

Nr	Pattern	Rating	Rationale
			Detour patterns
27	Delegation	+	Supported via the delegate function (HT)
28	Escalation	+	Escalations can be specified for tasks. Both commencement and completion deadlines are supported together with logical conditions that restrict their application (HT)
29	Deallocation	+	Supported via the release function (HT)
30	Stateful Reallocation	+	Supported via the the forward function (HT)
31	Stateless Reallocation	–	Not supported
32	Suspension/Resumption	+	Supported via the suspend and resume functions (HT)
33	Skip	+	Supported via the skip function (HT)
34	Redo	–	Not supported
35	Pre-Do	–	Not supported
			Auto-start patterns
36	Commencement on Creation	–	Not supported. Task instances must be explicitly started by an owner
37	Commencement on Alloc.	–	Not supported. Task instances must be explicitly started by an owner
38	Piled Execution	–	Not supported
39	Chained Execution	–	Not supported
			Visibility patterns
40	Configurable Unallocated Work Item Visibility	+/–	The advanced query function seems to support this but its operation across process instances and also for querying work items not allocated to the requesting resource is unclear. Also it is not a mandatory part of the proposal (HT)
41	Configurable Allocated Work Item Visibility	+/–	The advanced query function seems to support this but its operation across process instances and also for querying work items not allocated to the requesting resource is unclear. Also it is not a mandatory part of the proposal (HT)
			Multiple resources patterns
42	Simultaneous Execution	+	Directly supported (HT)
43	Additional Resources	–	Not supported. There can only be one resource for a task instance

area, however it is unclear how the query function operates in the context of multiple concurrent processes. Multiple resource patterns characterise situations where the work item - resource relationship is not 1-1. As indicated in Table 3, WS-HumanTask supports the notion of simultaneous execution (i.e. one resource running multiple work items) but only allows a work item to be allocated to a single resource.

4 Opportunities

The BPEL4People and WS-HumanTask proposals provide comprehensive support for incorporating tasks undertaken by human resources within the overall process execution framework that BPEL provides. There is a broad range of ways in which human resources can be represented and grouped: individually, via roles, groups and also as a result of query execution. These strategies can also be used as the basis for work assignments. Moreover there are a number of distinct ways in which human tasks can be implemented, ranging from inline activities in which both the task definition and the associated work directives form part of the same node in the process through to standalone tasks (defined elsewhere) which are coordinated by a PeopleActivity node in a BPEL process.

Nonetheless, *the patterns evaluation undertaken in the previous section identifies a number of potential opportunities* that these two proposals could pursue to further strengthen their ability to support human resource involvement in business processes. These issues are discussed in the following sections. In order to give an indication of effort associated with addressing each of them, we have rated their complexity from * (minimal effort) to *** (significant effort).

4.1 Non-binding Offers to a Single Resource*

There is no ability in the context of WS-HumanTask to offer a work item (i.e. not allocate) to a single resource. Where a newly created work item is identified as having a single potential owner, then it is assumed to be allocated to that resource (i.e. reserved) on a binding basis. There is no option that allows the resource to decline to undertake the offered work item.

4.2 Automatic Selection of a Resource*

Where multiple potential resources are identified when seeking to distribute a work item, there is no means of selecting a single resource to whom it should be allocated. Common means of selecting a suitable resource where several are identified include round-robin (i.e. distribute work evenly), least busy user (e.g. shortest queue) and random selection.

4.3 Distinguishing Execution Instances*

There is minimal distinction made between tasks and task instances. Whilst this is inconsequential when specifying a static process model, many of the elements in the enhanced BPEL4People/WS-HumanTask proposals require specific addressing e.g. invoking a remote task requires knowledge of the remote endpoint, the process name, task name, the specific process instance and task instance being sought. Similarly, data elements are specific to a process instance (not all process instances) hence they also need to be named accordingly. Moreover there seems to be no notion of process instance or task instance identifiers in these naming schemes that facilitiate navigation to a specific instance that is currently in progress (e.g. for delivering a notification or data element).

4.4 Richer Resource Descriptions**

There is no support for more detailed definition of specific resources (e.g. via capabilities) or for the use of resource characteristics when distributing work. This limits any possibility for differentiating between specific resources on the basis of characteristics that they possess when distributing work. In effect, all resources are treated as being identical when making a decision about where to route a work item. Note that multiple processes and organisations may want to share information about resource capabilities and requirements. BPEL4People/WS-HumanTask could play a prominent role here were they able to utilise and mediate more detailed resource definitions held in distinct systems (e.g. X.500 style directory services, ERP/HR systems) for work distribution purposes.

4.5 Inclusion of an Organisational Framework***

The organisational model provided is relatively minimalistic and does not take common concepts such as jobs, reporting lines, organisational groups etc. into account nor can these characteristics be used for work distribution purposes or for identifying or grouping resources in a generic sense. As the relationships between resources cannot be described in terms of the organisational context in which they operate, it is not possible to describe a variety of common approaches to work distribution, e.g. offer the work item to a clerk reporting to the manager. Moreover, work distribution cannot be framed in terms of organisational positions or jobs. This is a common approach to describing work responsibilities in many organisations as it minimises the need to change the work distribution directives associated with a process when staffing arrangements change.

4.6 Work Distribution Based on Historical Information***

Within a process instance, there is minimal access to historical information (and at that, only that referring to preceding work items in the same case). Moreover it is not clear to what extent this can be used for work distribution purposes. This is an obvious area where further clarity can be added to these proposals. The use of historical information, particularly that based on a multitude of previously completed process instances, provides a useful means of targetting suitable resources when distributing work items and allows approaches such as "allocate to the most experienced resource" or "offer to the person who did it least recently" to be implemented.

4.7 Resource Privileges**

One notable absence is the ability to specify privileges defining what actions a resource can undertake. Ideally it should be possible to specify these on a per-task basis in order to restricting the range of actions that a resource can initiate in regard to a task (e.g. delegation, reallocation etc.).

4.8 Independent Authorisation Framework***

There is no provision for imposing an authorisation framework over the tasks in a process to limit the potential range of resources to whom they can be directed and that are able to ultimately execute them. This is particularly useful in an enterprise context in order to limit how a task can be executed, regardless of the process definition in which it appears or how it is routed.

4.9 User-Initiated Optimization**

In many situations, opportunities that may exist for optimising work throughput can best be identified by resources involved in the conduct of work associated with the actual process. There are a number of approaches to expediting the completion of a process instance, these include automatically starting tasks when they are created or allocated, automatically starting subsequent tasks and allocating to the resource that completed the preceding task in a process instance, and allocating all instances of a given task to the same resource regardless of the process instance in which they occur (i.e. chained and piled execution). None of these facilities are supported in the BPEL4People or WS-HumanTask proposals.

4.10 Provision of a Worklist Metaphor**

One of difficulties with the proposals is that there is an absence of a clearly defined user interface that describes how a resource interacts with the process engine when undertaking work items and what details associated with each work item are disclosed. Typically in workflow systems, this interface is termed a worklist handler although other metaphors are possible e.g. work queues. Access to work items is supported via user-initiated queries, however the operation of these queries is unclear when requesting work items in multiple process instances. Moreover the use of such queries also removes any potential for imposing a uniform view of work distributions to all users.

5 Conclusions

This paper has examined the support that the BPEL4People and WS-HumanTask proposals provide for extending the BPEL offering to deal with activities that are undertaken by human resources. It uses the workflow resource patterns as an evaluation framework to assess the capabilities of these proposals and in doing so identifies both their strengths and several areas where opportunities exist for further improvement.

References

1. van der Aalst, W.M.P., ter Hofstede, A.H.M., Kiepuszewski, B., Barros, A.P.: Workflow patterns. Distributed and Parallel Databases 14(3), 5–51 (2003)
2. van der Aalst, W.M.P., Kumar, A., Verbeek, H.M.W.: Organizational modeling in UML and XML in the context of workflow systems. In: Haddad, H., Papadopoulos, G. (eds.) SAC 2003, pp. 603–608. ACM Press, New York (2003)

3. Agrawal, A., Amend, M., Das, M., Ford, M., Keller, C., Kloppmann, M., König, D., Leymann, F., Müller, R., Pfau, G., Plösser, K., Rangaswamy, R., Rickayzen, A., Rowley, M., Schmidt, P., Trickovic, I., Yiu, A., Zeller, M.: Web Services Human Task (WS-HumanTask), version 1.0 (2007), http://download.boulder.ibm.com/ibmdl/pub/software/dw/specs/ws-bpel4people/WS-HumanTask_v1.pdf
4. Agrawal, A., Amend, M., Das, M., Ford, M., Keller, C., Kloppmann, M., König, D., Leymann, F., Müller, R., Pfau, G., Plösser, K., Rangaswamy, R., Rickayzen, A., Rowley, M., Schmidt, P., Trickovic, I., Yiu, A., Zeller, M.: WS-BPEL Extension for People (BPEL4People), version 1.0 (2007), http://download.boulder.ibm.com/ibmdl/pub/software/dw/specs/ws-bpel4people/BPEL4People_v1.pdf
5. Bussler, C., Jablonski, S.: Policy resolution for workflow management systems. In: Proceedings of the 28th Hawaii International Conference on System Sciences, vol. 4, pp. 831–840. IEEE Computer Society, Wailea (1995)
6. Du, W., Shan, M.C.: Enterprise workflow resource management. In: Proceedings of the Ninth International Workshop on Research Issues on Data Engineering: Information Technology for Virtual Enterprises (RIDE-VE 1999), Sydney, Australia, pp. 108–115. IEEE Computer Society Press, Los Alamitos (1999)
7. Ferraiolo, D.F., Sandhu, R., Gavrila, S., Kuhn, D.R., Chandramouli, R.: Proposed NIST standard for role-based access control. ACM Transactions on Information and System Security 4(3), 224–274 (2001)
8. Huang, Y.N., Shan, M.C.: Policies in a resource manager of workflow systems: Modeling, enforcement and management. Technical Report HPL-98-156 (1999), http://www.hpl.hp.com/techreports/98/HPL-98-156.pdf
9. Kloppman, M., Koenig, D., Leymann, F., Pfau, G., Rickayzen, A., von Riegen, C., Schmidt, P., Trickovic, I.: WS-BPEL Extension for Sub-Processes: BPEL-SPE (2005), ftp://www6.software.ibm.com/software/developer/library/ws-bpelsubproc.pdf
10. zur Muehlen, M.: Organizational management in workflow applications – issues and perspectives. Information Technology and Management 5, 271–294 (2004)
11. OASIS. Web Services Business Process Execution Language for Web Services version 2.0 (2007), http://docs.oasis-open.org/wsbpel/2.0/wsbpel-v2.0.pdf
12. Russell, N., van der Aalst, W.M.P., ter Hofstede, A.H.M., Edmond, D.: Workflow Resource Patterns: Identification, Representation and Tool Support. In: Pastor, Ó., Falcão e Cunha, J. (eds.) CAiSE 2005. LNCS, vol. 3520, pp. 216–232. Springer, Heidelberg (2005)
13. Russell, N., ter Hofstede, A.H.M., van der Aalst, W.M.P., Mulyar, N.: Workflow control-flow patterns: A revised view. Technical Report BPM-06-22 (2006), http://www.BPMcenter.org
14. Russell, N., ter Hofstede, A.H.M., Edmond, D., van der Aalst, W.M.P.: Workflow Data Patterns: Identification, Representation and Tool Support. In: Delcambre, L.M.L., Kop, C., Mayr, H.C., Mylopoulos, J., Pastor, Ó. (eds.) ER 2005. LNCS, vol. 3716, pp. 353–368. Springer, Heidelberg (2005)

Documenting Application-Specific Adaptations in Software Product Line Engineering

Günter Halmans[1,*], Klaus Pohl[2], and Ernst Sikora[2]

[1] RDS Consulting GmbH
Mörsenbroicher Weg 200, 40470 Düsseldorf, Germany
guenter.halmans@rds.de
[2] Software Systems Engineering, University of Duisburg-Essen
Schützenbahn 70, 45117 Essen, Germany
{klaus.pohl,ernst.sikora}@sse.uni-due.de

Abstract. Software product line engineering distinguishes between two types of development processes: domain engineering and application engineering. In domain engineering software artefacts are developed for reuse. In application engineering domain artefacts are reused to create specific applications.

Application engineers often face the problem that individual customer needs cannot be satisfied completely by reusing domain artefacts and thus application-specific adaptations are required. Either the domain artefacts or the application artefacts need to be modified to incorporate the application-specific adaptations. We consider the case that individual customer needs are realised by adapting the application artefacts and propose a technique for maintaining traceability between the adapted application artefacts and the domain artefacts. The traceable documentation of application-specific adaptations is facilitated by an application variability model (AVM) which records the differences between the domain artefacts and the application artefacts of a particular application. The approach is formalised using graph transformations.

Keywords: product line engineering; variability modelling; application-specific adaptations; traceability.

1 Introduction

Software-product line development differentiates between two processes [5] [12]: domain engineering and application engineering. In domain engineering, domain artefacts (domain requirements, domain architecture, domain components, domain test cases, etc.) are developed for reuse. In application engineering, domain artefacts are reused for defining and realising specific applications. Domain artefacts and application artefacts are interrelated by traceability links to support the different domain and application engineering tasks (cf., e.g. [12]). In addition, traceability links are established between requirements, architecture, components, and test artefacts, both,

* The work reported in this paper was performed while Günter Halmans was member of the Software Systems Engineering Group at the University of Duisburg-Essen.

Z. Bellahsène and M. Léonard (Eds.): CAiSE 2008, LNCS 5074, pp. 109–123, 2008.

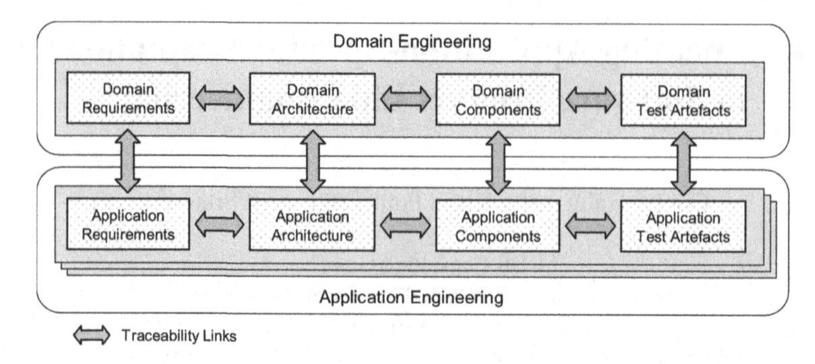

Fig. 1. Overview of domain and application artefacts including traceability links

in domain and application engineering. Fig. 1 provides a schematic overview of the two processes, the key artefacts and the traceability links.

The second, central characteristic of product line engineering is the distinction between common and variable artefacts. *Common domain artefacts* become part of each application derived from the product line. The product line variability denotes the *variable artefacts* and thus defines the possible variations among the products (applications) of a product line. Variable artefacts may (or may not) be selected for a particular application. The process of selecting the variable artefacts required for a specific application during application engineering is called the binding of variability [3]. The selection (i.e. the binding) is documented in order to ensure, for instance, the traceability of the application artefacts to the domain artefacts.

In practice, product line applications can typically not be derived 100% from the domain artefacts. For example, if a customer requirement is not part of the domain requirements, the application engineers have to create additional artefacts and/or adapt existing artefacts in order to realise the customer requirement. Thus, in addition to reusing artefacts from domain engineering, application-specific artefacts have to be defined (see e.g. [1] [4] [8] [12] [15] [16]). By *application-specific artefacts* we denote all development artefacts (including requirements, architecture, components, and test artefacts) needed specifically for a particular application. Application-specific artefacts can be realised either by extending and/or adjusting the domain artefacts and deriving the required application-artefacts from the adapted domain artefacts or by adapting the application artefacts directly.

For instance, in [10] application-specific requirements changes are realised by adapting the domain requirements. In [2], the inclusion of new components in the domain artefacts is addressed in order to realise application-specific changes. Further examples of approaches supporting the adaptation of domain artefacts to realise application-specific adaptations are [6] and [13].

In contrast, [15] and [16] address the realisation of application-specific adaptations by adapting the application artefacts. However, these approaches do not provide a systematic procedure for realising the adaptations. Thus, for instance, no traceability relationships between the adapted application artefacts and the domain artefacts are maintained. The KobrA approach foresees the adaptation of application artefacts, yet does not provide a means to document the adaptations in a traceable manner [1]. The approach in [8] realises application-specific changes by adapting application artefacts which are represented as UML diagrams. However, the changes are incorporated in each diagram, and thus, there is no central, consistent model where the adaptations are documented. In addition, the approach does not address the issue of providing traceability between the domain artefacts and the adapted application artefacts.

Summarising, none of the stated approaches supports the traceable documentation of application-specific adaptations of application artefacts throughout different artefact types (requirements, architecture etc.).

In this paper, we argue that application-specific adaptations should be incorporated at the application engineering level and that the application-specific changes must be interrelated with the domain artefacts (cf. Section 2). In Section 3, we introduce an application-specific variability model (AVM) as a means to realise and document traceable, application-specific adaptations. In Section 4, we sketch the prototypical realisation of our approach in a tool environment based on typed attributed graph transformation systems. In Section 5, we illustrate our approach using a simplified example. Section 6 summarises the paper and provides an outlook on future work.

2 Adjusting Domain Artefacts or Application Artefacts?

As mentioned in the introduction, application-specific adaptations can be realised either by adapting the domain artefacts (Section 2.1) or by adapting the application artefacts, i.e. at the application engineering level (Section 2.2).

2.1 Adaptation of Domain Artefacts

Application-specific adaptations can be realised by increasing the variability of the product line and by adding and/or adjusting domain artefacts in such a way that ***all*** required application artefacts can be derived from the domain artefacts. This approach is appealing as it simplifies the application engineering process for the specific application. However, an adaptation of domain artefacts leads to an evolution of the product line. As described, for instance, in [5] and [12], the evolution of a product line should be determined by the product line strategy of the organisation rather than by individual customer wishes. Typically, proposed changes of domain artefacts have to be approved by a change control board before being incorporated in the domain artefacts of the product line. The domain artefact change control board enforces the product line strategy and takes into account in its decisions the costs and benefits of each proposed change. For instance, the following "cost drivers" have to be considered, when deciding about the implementation of individual customer wishes through the adaptation of domain artefacts:

- *Increased complexity of domain artefacts*: Incorporating a huge number of individual customer's wishes into the domain artefacts increases the complexity of the domain artefacts as well as the variability of the product line [12]. Realising application-specific artefacts through variable domain artefacts is thus in conflict with a typical product line strategy: to facilitate testing of domain artefacts as early as possible, which requires as little variability as possible [14]. Moreover, an increase of variability leads, to a more complex product derivation process for future applications as, for instance, the application engineer has to take a higher number of decisions to derive a new application.

- *Adjustment of all applications already derived from the product line*: The applications that have already been derived from the product line or are being derived could be affected by the adaptation of the domain artefacts. One reason for integrating the changes made to the domain artefacts in all existing applications of the product line is to facilitate future maintenance of these applications (cf., e.g. [11]). However, some applications might be affected by the adaptation of the domain artefacts in an unexpected manner. Hence, the effects of the adaptations on each existing application have to be checked. For implementation artefacts the checking can be performed by an automatic regression test of the applications. Yet, requirements and architectural artefacts typically need to be checked manually which is a time and resource-intensive process, in particular, when the product line has a large number of applications.

Several researchers recommend to realise individual customer wishes by adapting application artefacts directly (cf., e.g. [4] [12] [15]) which avoids the high effort related to product line evolution. Even if satisfying the individual customer's wishes matches well the product line strategy and offers a high benefit for the entire product line, the product line organisation may decide to develop a lead product before the proposed changes are incorporated into the domain artefacts. Summarising, there are strong arguments for maintaining a clear separation between product line evolution and the realisation individual customers' wishes. Prior to adapting the domain artefacts one has perform a cost-benefit-analysis that takes into account the product line strategy and the entire set of applications.

2.2 Application-Specific Adaptation of Application Artefacts

The other option, which avoids the high effort of adapting the domain artefacts, is to realise application-specific artefacts at the application engineering level. In this case application-specific adaptations are realised by changing and extending the application artefacts derived from the domain artefacts. Thus, the adaptations only affect the application under development instead of all existing or future applications.

When changing the derived application artefacts to incorporate application-specific requirements, obviously, the resulting application artefacts differ from the domain artefacts. If the product line organisation does not establish a systematic approach for dealing with those differences, severe problems may occur that undermine the advantages of product line engineering (cf. [11]). There is thus a need to record the interrelations between the domain artefacts and the application-specific adaptations of application artefacts. We illustrate this by the following two cases:

- ***Case A:*** Assume that a particular domain requirement is selected for the application and that this requirement is slightly modified to satisfy the customer need. If there is no documentation of this adaptation and no relation to the original domain artefact, the application architect may not be aware of the fact that the application-specific requirement is based on a domain requirement. Thus, the architect might not know that there are domain components which realise the domain requirement (which the application-specific requirement is based on) and which could be modified to fulfil

the application-specific requirement. When designing the application architecture, the architect would assume that the components required for implementing the application-specific requirement have to be developed anew. Likewise, the application tester is not able to identify the domain test artefacts that, with some modifications, could be reused for testing the application-specific requirement. Summarising, if there is no adequate traceability information from the specific application requirement to the reused domain requirement, it is close to impossible to identify, later on in the development process, that the application-specific requirement is based on a particular domain requirement (i.e. that there was a reuse with modifications).
- ***Case B:*** Assume that two variable domain requirements exclude each other, for instance, because each requirement alone causes about 60% system load. One of the two domain requirements is 100% reused for an application and the other is adapted for the application. If the application-specific requirement is not linked with the original domain requirement, it is very difficult for an application architect to recognise the conflict. Consequently there is a risk that an application is developed that does not fulfil its performance requirements since the application developers are not aware of the existing conflict.

We conclude that application-specific adaptations of application artefacts should be traceable and the application artefacts should be interrelated with the domain artefacts. In general, a particular application-specific adaptation most likely influences all or most types of application artefacts (requirements, architecture, components, test cases, etc.). For instance, a home security product line might include two authentication mechanisms for unlocking the front door: to enter a personal identification number in a key pad or to swipe a personal access card through a card reader. Incorporating a new authentication mechanism, for instance, biometric authentication, for a specific home security system would change the requirements, possibly the architecture, the implementation as well as the test cases. In the following, we discuss two approaches to tracing adaptations in application artefacts.

- ***Documenting Adaptations within Application Artefacts:*** Keeping track of the differences between application artefacts and domain artefacts can be achieved by documenting the differences individually for each type of application artefacts. In this case, application-specific changes in application requirements are interlinked with the corresponding domain requirements, changes in the application architecture artefacts are interlinked with the domain architecture artefacts and so forth. Using this approach, the traceability information for a particular change is split across the various artefact types. The splitting entails that the impact of a particular application-specific adaptation on the different application artefacts is difficult to recognise.
- ***Documenting Adaptations in a Dedicated Model*:** In order to avoid the splitting of traceability information across the various artefact types, we suggest to use a dedicated model to document the application-specific adaptations. The key idea behind this approach is to interpret application-specific changes as variation of the application artefacts with respect to the domain artefacts, i.e. to interpret the adaptation as application-specific variability. For instance, the change of the authentication mechanism of a specific home security system can be regarded as a variation

between the specific system and the home security product line, regardless which artefacts have to be adapted in order to realise this change.

We call the dedicated model used to record application-specific variations *application variability model.* The idea of the application variability model is similar to the orthogonal modelling of the variability at the domain level (cf. [12]). Documenting the application-specific variations in an application variability model has, essentially, the same advantages as documenting product line variability in a dedicated model and not in the domain artefacts themselves (cf. [3]), for instance:

- The application variability model can be used to identify, communicate, and reason about application-specific changes as opposed to having to consider the adaptations made in different types of artefacts where each adaptation only provides a partial view on an application-specific change.
- The application variability model serves as an entry point for navigating to all application artefacts (requirements, architecture, test cases etc.) affected by an application-specific change. This makes it easier to keep the adaptations consistent across the different artefact types and development phases.

3 Application-Specific Variability Model

This section sketches our solution for documenting application-specific adaptations using an application variability model (AVM). In Section 3.1, we briefly recapitulate how product line variability is documented using an orthogonal variability model (OVM). In Section 3.2, we introduce the application variability model for documenting application-specific adaptations.

3.1 Orthogonal Variability Modelling

Product line variability is typically represented by variants, variation points and dependencies between variants and variation points. We use our orthogonal variability model (OVM) to document product line variability. A detailed description of the variability modelling language is given in [12]. Fig. 2 depicts a simplified example of a variability model for a home security system. A variation point is represented as a triangle. The variation point named *door lock mechanism* in Fig. 2 is related to the two variants *key card* and *key pad* which are represented as rectangles. The dashed lines between the variants and the variation point represent optional variability dependencies. A variant with an optional variability dependency can, but does not have to be selected during the binding of variability. In contrast, a mandatory variability dependency (represented by a solid line) means that the corresponding variant has to be selected. The model further indicates by the arc annotated with the cardinality *[1..1]* that exactly one of the two variants of the variation point *door lock mechanism* must be selected. Selecting the variant *key pad* means that the resulting home security system requires the user to enter a personal identification number using a key pad for authentication. Selecting the variant *key card* means that the home security system performs the authentication by checking a key card and the identification number saved on it. The variants of the variability model are connected with domain artefacts by so called *artefact dependencies*, which are represented by dotted lines with arrow heads. The artefact dependencies depicted in Fig. 2 express that, e.g. the requirements *R 12* and *R 13* specify the variant *key pad*. So called constraint dependencies are used to model that two variants, two variation points or a variant and a variation point either exclude each other or that one requires the other.

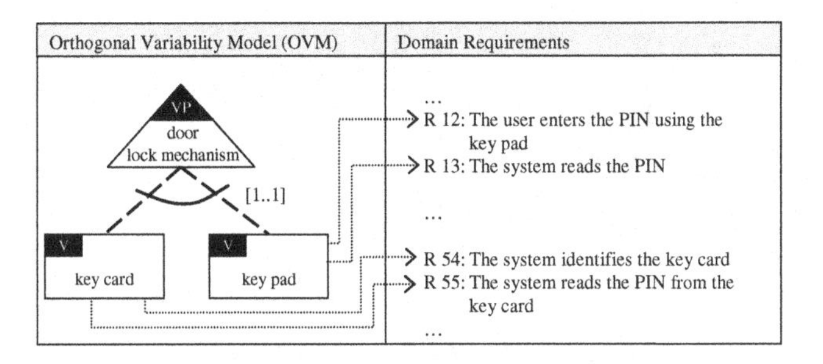

Fig. 2. Example of an orthogonal variability model

3.2 Application Variability Model

As outlined in Section 2, we propose to document application-specific changes (e.g. additional variants or additional variation points) in an *application variability model* (AVM). As indicated in Fig. 3, an AVM is defined for each application to document the application-specific adaptations with respect to the domain variability model (DVM). The AVM captures the application-specific variability of all types of application artefacts such as requirements, architecture, etc. As the application engineering process proceeds, requirements variability is mapped to architecture variability and so forth (cf. [12]). The traceability between the AVM and the different types of application artefacts is established by means of artefact dependency links.

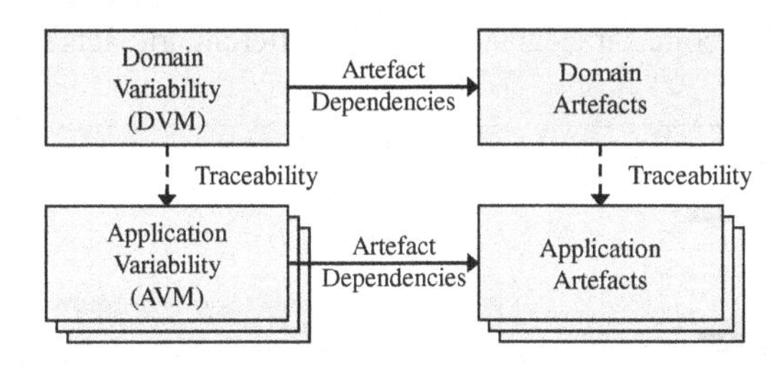

Fig. 3. DVM and AVM

In the following, we outline four main cases of how the AVM supports the traceable documentation of application-specific adaptations. The four cases apply to all types of artefacts. Yet, for simplicity, we use requirements to illustrate the cases.

Case 1 – 100% reuse: The application requirement can be defined completely by reusing domain requirements. In this case, no application-specific adaptation is required. Nevertheless, the binding of the selected variant has to be documented. By selecting a variant, the domain requirements related to this variant are completely reused. If, e.g., the variant *key card* (cf. Fig. 2) is selected, the requirements R54 and R55 become part of the application requirements. The result of this variability binding is illustrated in Fig. 4: The selected variant, the related variation point *door lock mechanism*, and the dependencies to the application requirements *R54* and *R55* become part of the bound AVM.

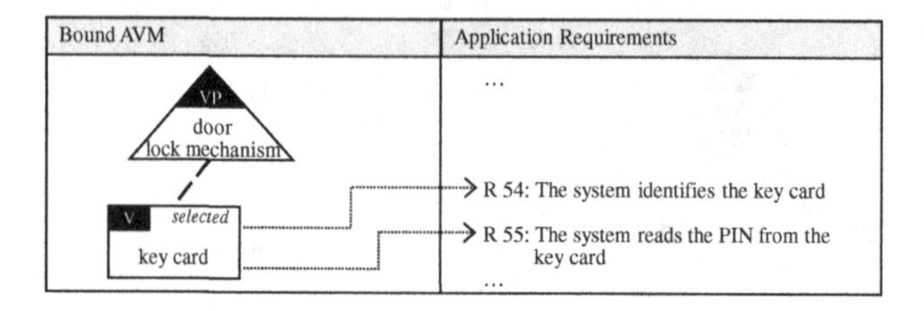

Fig. 4. Bound AVM with complete reuse

Case 2 – No reuse: In this case, an application requirement cannot be fulfilled by reusing domain requirements and must therefore be developed from scratch. Assume that an application requirement demands a door locking mechanism with a finger print scanner, which is not offered by the domain requirements. In this case, a new variant *finger print* is introduced in the AVM and related to the corresponding requirements (cf. Fig. 5). As the variant is required for the application it is also selected for this application. When realising the application-specific variant *finger print*, the application engineers may either relate the application-specific adaptations of system components, test cases etc. to the variant finger print, or define additional variants and variation points that represent the application-specific adaptations in the architecture, test cases etc. The addition of a variation point for representing architectural adaptations is illustrated in Section 5. Since the architectural artefacts, test cases etc. required for the new functionality are related to the elements of the AVM, the impact of the application-specific adaptations becomes traceable across the different artefacts types.

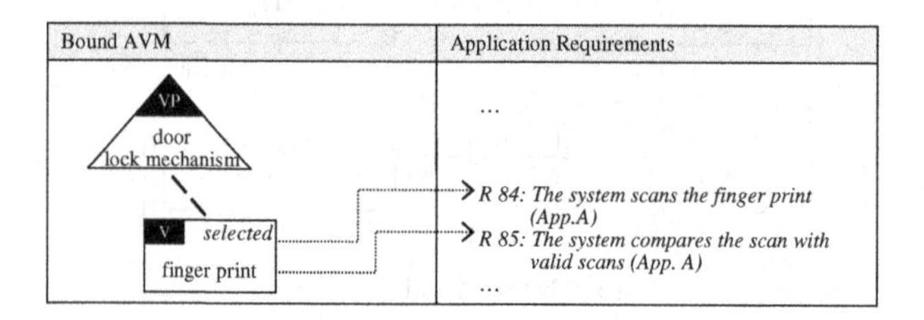

Fig. 5. New variant with related requirements

Case 3 – Partial reuse: In this case, the application requirement can be partially defined by reusing domain requirements. In other words, domain requirements can be reused but some application-specific adaptations are needed. Assume, for example, that the application-specific requirement *R12a* is defined by partially reusing the domain requirement *R12*. *R12a* contains a restriction of the PIN length whereas *R12* has no such restriction (cf. Fig. 6). To record this adaptation, the artefact dependency from the variant *key pad* to *R12* is marked as deleted and labelled with the string *App.A* to denote that this element is application-specific for the application *A*. Moreover, the new application requirement *R12a* is included, and a new artefact dependency is introduced from the variant *key pad* to *R12a*. This artefact dependency is also labelled with *App.A*, and a reference is added to requirement *R12a* to record that this requirement is the partially reused domain requirement *R12* (cf. *App.A, former R12*) thus providing traceability.

Note that the described solution for the documentation of the specific application requirement *R12a* is one of many possible solutions. Alternatively, a new variant could have been introduced and new artefact dependencies from the new variant to the application requirements *R12a* and *R13* could have been documented. Since different modelling choices exist to document the same kind of adaptation, the documentation of adaptations cannot be automated. Rather, the engineer has to decide, which modelling choice is appropriate for the particular case. However, once the engineer has chosen a particular modelling option, the validity of the resulting AVM with respect to the DVM can be checked automatically.

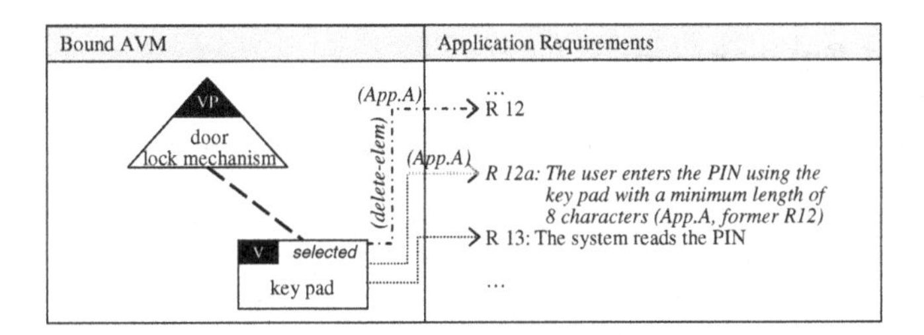

Fig. 6. Partial reuse

Case 4 – Conflict with the product line variability: In this case, the application-specific extension is in conflict with the variability defined in the DVM. This conflict must be resolved in order to realise the application-specific adaptation. For example, assume an application requirement for a specific application requires both door locking mechanism depicted in the DVM of Fig. 6. In this case, a conflict exists between the domain variability model and the specific application requirement. In the DVM, the cardinalities of the alternative group, which encompasses the two variants *key card* and *key pad*, are defined as *[1..1]* and thus one (and only one) of the two variants can be selected for an application. To resolve this conflict, the restriction of the variant selection with respect to the variation point *door lock mechanism* has to be suspended for the application. Thus, in the AVM the upper bound of the cardinalities is changed to 2 (cf. ❶ in Fig. 7) and thereby the selection of both variants for this application is enabled (in terms of the variability model). The label *App.A* on the alternative group represented by the arc of a circle makes the adaptation traceable.

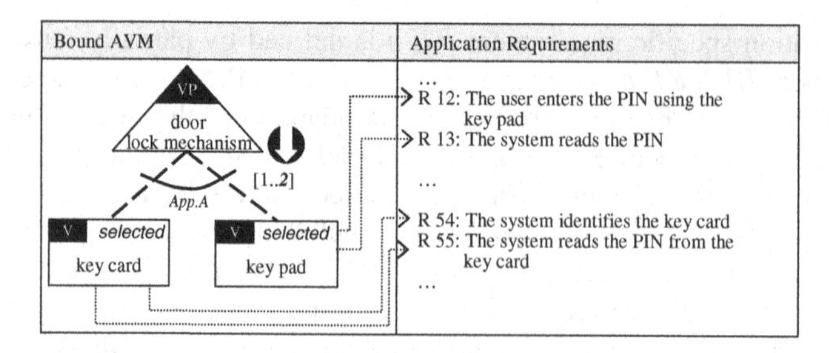

Fig. 7. Conflict with DVM

To summarise, we document application-specific adaptations in the AVM (application variability model) and relate all application artefacts affected by an application-specific adaptation to the corresponding elements in the AVM. Moreover, if an application-specific artefact is based on a domain artefact, the application artefact is additionally related to the domain artefact it is based on thus ensuring traceability. Deleted elements are marked by appropriate labels. Thus, the AVM facilitates the documentation of all application-specific adaptations in one central model and thereby facilitates consistency checks as well as an easier comprehension of the adaptations.

4 The Tool Environment

To support the documentation of the application-specific changes and to ensure the consistency between the AVM and the DVM we have formalised our approach using graph transformations. We decided to use the integrated development environment for graph-transformation AGG (Attributed Graph Grammar System) because, among others, AGG is well-founded by the theory of categories, has been successfully used in various settings and provides a comprehensive set of techniques for validating graphs and graph transformation systems, e.g. parsing, critical-pair-analysis, and consistency checks (cf. [7] for more details).

4.1 Graph Transformation Systems as a Basis for Adapting and Binding Variability Models

We realised the following two graph transformation systems in AGG as a basis for defining variability, recording adaptations of variability models and for binding variability:

- ***VM-GTS*:** The typed attributed graph transformation system *VM-GTS* provides the formalism to model and adapt variability models including the needed traceability. The VM-GTS contains a type graph, which represents the variability meta model. The type graph defines node types (i.e. variation points, variants, application requirements, etc.) and edge types (i.e. variability dependency, constraint dependency, artefact dependency). Moreover, in the type graph, attributes of nodes and edges are defined. These attributes are used for the unique identification of nodes and edges as well as for holding traceability information. In addition to the type graph, the VM-GTS contains a set of graph transformation rules. These rules represent the permissible operations on a variability model such as the insertion of new variants for a given variation point. The rules are generic in the sense that they contain variables for the different attributes that are instantiated during the application of a specific rule (Section 4.2). Finally, the VM-GTS contains pre- and post-conditions, which ensure that the variability model is well-formed. The pre- and post-conditions prohibit, for instance, that two variants are connected by both, an *exclude* and a *requires* dependency.

- ***BV-GTS*:** The BV-GTS (Binding Variability Graph Transformation System) supports the binding of the variability for the application. The BV-GTS uses an AVM to generate a bound AVM. The input for the BV-GTS is a labelled AVM, i.e. an AVM where the selected variants are indicated by labels. The derivation of the BV-GTS results in a bound AVM that includes only selected variants. Derivation of the BV-GTS means that all graph transformation rules defined in the BV-GTS are applied until no graph transformation can be applied any more. The BV-GTS ensures that only valid bindings are performed, i.e., for instance, all variation points, variability dependencies, and artefact dependencies of the selected variants as well as variants that are required by the selected variants are bound. The resulting AVM is thus a valid variability model.

4.2 Generating an AVM

Based on the VM-GTS we are able to define a graph transformation system specifically for an application, which documents the adaptations, makes the adaptations visible in the AVM, and labels the selected variants. The so called **A**pplication **V**ariability **M**odel **G**raph **T**ransformation **S**ystem (AVM-GTS) includes the type graph of the VM-GTS and a set of instantiated rules from the VM-GTS rule-set including the corresponding pre- and post-conditions. To define an AVM-GTS, firstly, the needed graph transformation rules from the rule-set of the VM-GTS have to be selected. The selection is determined by the modeller, since, typically, more than one solution is available to achieve a required adaptation. Secondly, the selected rules are instantiated. During instantiation, for instance, the generic rule for inserting a variant (from the VM-GTS) is concretised with values such as the short name for the variant. These concrete values for specific attributes are also used to find a particular element in the DVM, e.g. the variation point to which the new variant is related. Third, the AVM-GTS is derived based on the DVM. Derivation of the AVM-GTS means that the selected and instantiated graph transformation rules of the AVM-GTS are applied until no rule is applicable anymore. This activity is performed automatically. The output is an AVM that has been generated from the DVM by applying the transformation rules defined in the AVM-GTS.

4.3 Evaluation

Our experience in defining the VM-GTS and the BV-GTS in AGG indicates that the definition of the different graph transformation rules, the type graph as well as the conditions is fairly simple. We applied our approach to several examples of varying complexity (in terms of the size of the variability model and the number of adaptations) for generating and binding AVMs. The performance results obtained for generating adapted AVMs from two different DVMs are shown in Table 1. The full set of examples including the performance results has been published in a thesis (cf. [9]). The experimentation with our prototypical implementation based on AGG supports our claim that our approach for the traceable documentation of application-specific adaptations is feasible. In particular, by using the tool prototype, we gained the following insights about our approach:

- Our orthogonal variability model in combination with graph transformations provides a sound basis for the precise documentation of application-specific adaptations in product line engineering.
- The recording and management of inter-model trace links for product line artefacts can be greatly eased by the use of graph transformation rules.
- The generation of application artefacts from domain artefacts including application-specific adaptations and the binding of variability can be accomplished quite easily by using a graph transformation environment.

The observed increase in runtime between the first and the second example depicted in Table 1 can be explained by the complexity of graph matching. The reduction of the time for generating the AVM is a topic of further investigations.

Table 1. Performance results for the prototypical implementation with AGG (excerpt)

#Variation points in DVM	#Variants in DVM	#Artefact dependencies in DVM	#Adaptations in AVM-GTS	Time for generating the AVM [seconds]
10	20	60	5	5
20	40	120	15	45

5 Example

In this section, we illustrate the application of our approach on a simplified example from the e-commerce domain. The example consists of a domain variability model, a domain use case and an excerpt of a domain components model including the relevant artefact dependencies between the domain variability model and the other domain artefacts (see Fig. 8). We omit technical details as these have already been described in Section 4. The domain variability model depicted in Fig. 8 allows to choose between two payment methods allowing the application engineer to derive applications either supporting credit card payment or direct debit payment. The choice affects the example use case (either step 2 [V1] or step 2 [V2] is selected) and the example components model (either the credit card payment plug-in or the direct debit payment plug-in is selected).

In the application engineering process, the customer expresses the following requirements that demand application-specific adaptations:

- The application shall offer both payment methods, credit card and direct debit.
- The user shall be able to choose the payment method before entering the required payment details.

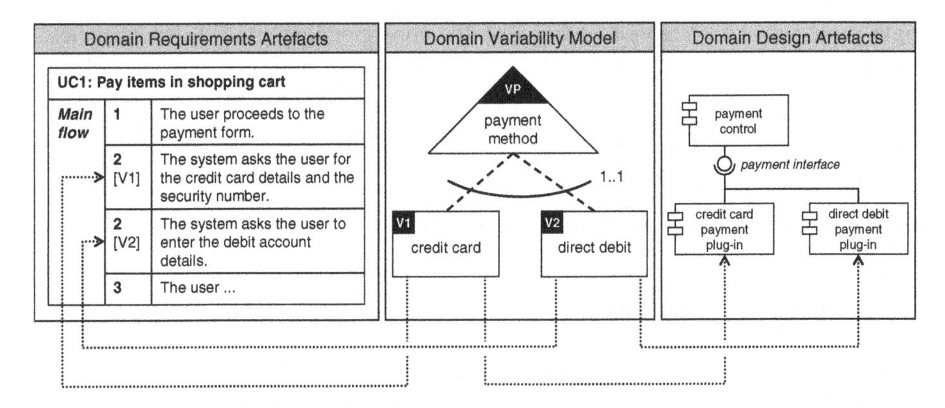

Fig. 8. Example of a domain variability model with associated domain requirements and domain design artefacts (excerpt)

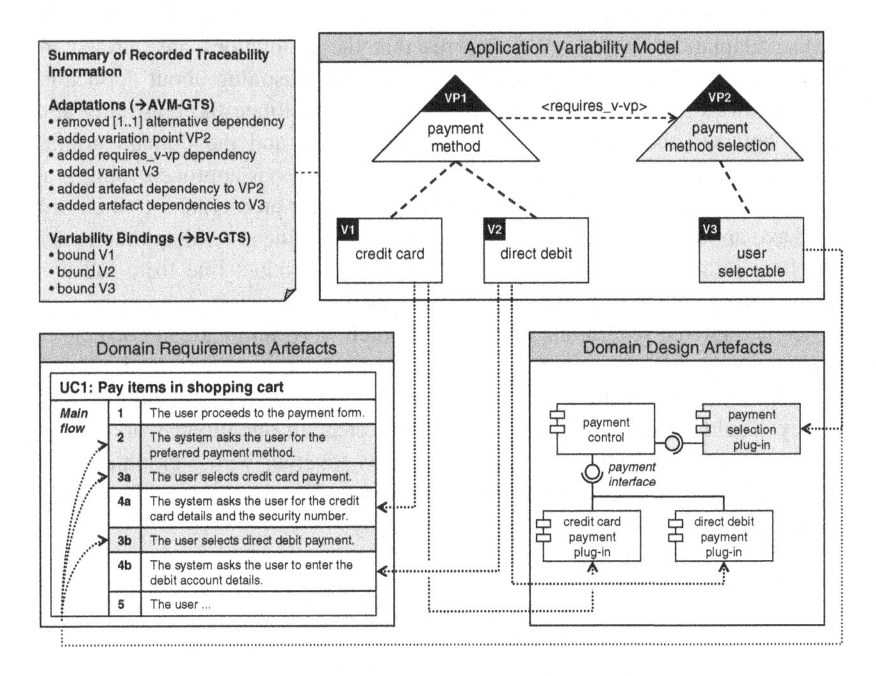

Fig. 9. Example of an application variability model with recorded traceability information (as natural language text) and application artefacts including modifications

Fig. 9 shows the resulting application variability model and the associated application artefacts. In the variability model, variants *V1* and *V2* have been bound. To enable the binding of both variants for an application, the alternative dependency between *V1* and *V2* had to be deleted. Furthermore, the required adaptations in the application use case and the application design have been documented in the application variability model by inserting the variation point *payment method selection* and a single variant *user selectable* together with the corresponding artefact dependencies. In addition, variants *V*1, *V*2 and *V*3 have been bound. An informal summary of the traceability information recorded by our approach is shown on the upper left of Fig. 9. In our tool, the stated adaptations and bindings are recorded by selecting and instantiating the needed graph transformation rules from the VM-GTS and the BV-GTS respectively as described in Section 4. With the recorded information, the AVM for the specific application can be generated from the DVM.

6 Summary and Outlook

When developing product line applications, typically not all application requirements can be fulfilled by reusing domain artefacts designed and developed in domain engineering. Thus, application-specific adaptations of the domain artefacts (including extensions) are required. We proposed, in this paper, to record application-specific adaptations in a dedicated model, the application variability model (AVM). Documenting the adaptations in an AVM ensures that the adaptations are defined in a central location (the AVM) and provides the basis for reasoning about application-specific adaptations across different artefact types. To facilitate the reasoning about consistency between the application-specific adaptations and the domain artefacts (including the product line variability) we have formalised our approach using graph transformation systems which have been implemented in a prototypical tool environment. We have applied our prototypical environment to define several applications for a product line with different complexities, including a product line from the home automation domain. When defining application-specific adaptations for several applications, we encountered several inconsistencies which were automatically detected by our tool environment. Those inconsistencies, especially with respect to the domain variability model, were in many case not easy to detect, i.e. without an automated tool support they would have most likely not been detected. In our future work, we will develop a better graphical visualisation of the AVM together with a graphical editor which eases the task of documenting application-specific adaptations. We will then use the extended tool environment in a real case study.

Acknowledgments. This paper was partially funded by the DFG projects PRIME (Po 607/1-1) and IST-SPL (Po 607/2-1). We would like to thank Prof. Dr. Michael Goedicke for the fruitful discussions during the elaboration of our approach.

References

1. Atkinson, C., Bayer, J., Bunse, C., Kamsties, E., Laitenberger, O., Laqua, R., Muthig, D., Paech, B., Wüst, J., Zettel, J.: Component-Based Product-Line Engineering with UML. Addison-Wesley, UK (2002)
2. Baum, L., Becker, M., Geyer, L., Molter, G.: Mapping Requirements to Reusable Components using Design Spaces. In: Proc. of the 4th Intl. Conference on Requirements Engineering (ICRE 2000), pp. 159–167. IEEE Computer Society, Los Alamitos (2000)

3. Bühne, S., Lauenroth, K., Pohl, K.: Modelling Requirements Variability across Product Lines. In: Atlee, J.M. (ed.) 13th IEEE Intl. Conference on Requirements Engineering, pp. 41–50. IEEE Computer Society, Los Alamitos (2005)
4. Bosch, J., Ran, A.: Evolution of Software Product Families. In: Van der Linden, F. (ed.) Software Architectures for Product Families, International Workshop IW SAPF 3, Las Palmas de Gran Canaria, Spain, pp. 169–183. Springer, Heidelberg (2000)
5. Clements, P., Northrop, L.: Software Product Lines – Practices and Patterns. Addison-Wesley, Boston (2001)
6. Eriksson, M., Börstler, J., Borg, K.: The PLUSS Approach - Domain Modeling with Features, Use Cases and Use Case Realizations. In: Obbink, H., Pohl, K. (eds.) SPLC 2005. LNCS, vol. 3714, pp. 33–44. Springer, Heidelberg (2005)
7. Ehrig, H., Ehrig, K., Prange, U., Taentzer, G.: Fundamentals of Algebraic Graph Transformation. Springer, Heidelberg (2006)
8. Gomaa, H.: Designing Software Product Lines with UML. Addison-Wesley, Boston (2004)
9. Halmans, G.: Ein Ansatz zur Unterstützung der Ableitung einer Applikationsanforderungsspezifikation mit Integration spezifischer Applikationsanforderungen (in German). Doctoral Dissertation, Logos Verlag, Berlin (2007)
10. Mannion, M., Kaindl, H., Wheadon, J.: Reusing Single System Requirements from Application Family Requirements. In: Proc. of the 21th Intl. Conference on Software Engineering (ICSE 1999), pp. 453–462. ACM Press, New York (1999)
11. Mohan, K., Ramesh, B.: Change Management Patterns in Software Product Lines. Communications of the ACM 49(12), 68–72 (2006)
12. Pohl, K., Böckle, G., van der Linden, F.: Software Product Line Engineering – Foundations, Principles, and Techniques. Springer, Heidelberg (2005)
13. Padmanabhan, P., Lutz, R.R.: Tool-Supported Verification of Product Line Requirements. In: Automated Software Engineering, vol. 12(4), pp. 447–465. Springer, Heidelberg (2005)
14. Reuys, A., Kamsties, E., Pohl, K., Reis, S.: Model-Based System Testing of Software Product Families. In: Pastor, Ó., Falcão e Cunha, J. (eds.) CAiSE 2005. LNCS, vol. 3520, pp. 519–534. Springer, Heidelberg (2005)
15. Raatikainen, M., Soininen, T., Männistö, T., Mattila, A.: Characterizing Configurable Software Product Families and their Derivation. In: Software Process Improvement and Practice, vol. 10(1), pp. 41–60. Wiley, Chichester (2005)
16. Weiss, D.M., Lai, C.T.R.: Software Product-Line Engineering, A Family-Based Software Development Process. Addison-Wesley, Boston (1999)

Refactoring Process Models in Large Process Repositories

Barbara Weber[1] and Manfred Reichert[2]

[1] Quality Engineering Research Group, University of Innsbruck, Austria
Barbara.Weber@uibk.ac.at
[2] Institute of Databases and Inf. Systems, Ulm University, Germany
manfred.reichert@uni-ulm.de

Abstract. With the increasing adoption of process-aware information systems (PAIS), large process model repositories have emerged. Over time respective models have to be re-aligned to the real-world business processes through customization or adaptation. This bears the risk that model redundancies are introduced and complexity is increased. If no continuous investment is made in keeping models simple, changes are becoming increasingly costly and error-prone. Though refactoring techniques are widely used in software engineering to address related problems, this does not yet constitute state-of-the art in business process management. Process designers either have to refactor process models by hand or cannot apply respective techniques at all. This paper proposes a set of behaviour-preserving techniques for refactoring large process repositories. This enables process designers to effectively deal with model complexity by making process models better understandable and easier to maintain.

1 Introduction

Process-aware Information Systems (PAIS) offer promising perspectives for enterprise computing and are increasingly used to support business processes at an operational level [1]. In contrast to data- or function-oriented information systems (IS), PAIS strictly separate process logic from application code, relying on explicit *process models* which provide the schemes for process execution. This allows for a separation of concerns, which is a well established principle in computer science to increase maintainability and to reduce cost of change [2].

With the increasing adoption of PAIS large process repositories have emerged. Over time corresponding process models have to be adapted at different levels to meet new business, customer and regulatory needs, and to ensure that PAIS remain aligned with the processes as executed in real world. Typical adaptations include the customization of (reference) process models to specific needs of a customer [3,4] or – at the operational level – the adaptation of running process instances to cope with exceptional situations [5]. Like software programs degenerate when adding more and more code or introducing changes by different devlopers [6], process adaptations bear the risk that model repositories are becoming increasingly complex and difficult to maintain over time.

Z. Bellahsène and M. Léonard (Eds.): CAiSE 2008, LNCS 5074, pp. 124–139, 2008.

In software engineering (SE), refactoring techniques have been widely used to address related problems and to ensure that code bases remain maintainable over time [7,8]. Refactoring allows programmers to restructure a software system without altering its behaviour. Refactoring is typically used to improve code quality by removing duplication, improving readability, simplifying software design, or adding flexibility [9]. Examples of SE refactoring techniques include the renaming of a class to foster understandability or the extraction of a method from an existing code block to reduce redundant code fragments.

Process modeling is often referred to as *programming in the large* [10,11]. Thereby, a process schema is comparable to a software program specifying the inputs and outputs of activities as well as the control and data flow between them. Despite these similarities refactoring is not yet established in the field of business process management (BPM) and existing process modeling tools only provide limited refactoring support. Consequently, process designers either have to refactor process models by hand or cannot apply respective techniques at all.

This paper adapts SE refactoring techniques to the needs of process modeling and complements them with additional refactorings specific to BPM. In particular, we describe techniques suitable for refactoring large process repositories, where we can find both collections of inter-related process models and process variants derived from generic models (e.g., reference process models). The former consist of a set of models, which may refer to each other (e.g., a parent process refers to a child process) resulting in *model trees*. In contrast, *process variants* are part of a *process model family*, and are derived from a *generic process model* through a sequence of adaptations. This approach is often referred to as *model customization* or *configuration* [3,4]. Like in SE, tool support is essential as a refactoring applied to one model might require changes in other models as well. To avoid the introduction of inconsistencies and errors through refactorings, their application must be behaviour-preserving and should be accomplished automatically. The final decision whether to apply a refactoring or not, however, is left to the process designer.

In this paper we focus on refactoring techniques for the control flow aspect of executable process models. For each proposed refactoring we describe its intent, give examples for its applicability and use (similar to *code smells* in SE [8]), and discuss its effects in respect to process model quality metrics (e.g., measuring control flow complexity) [12,11].

Section 2 provides background information. Section 3 gives an introduction into refactorings for process model trees. Section 4 suggests a refactoring to effectively deal with process variants and Section 5 introduces advanced refactorings for model evolution considering process history data. Related work is discussed in Section 6. Finally, Section 7 concludes with a summary and an outlook.

2 Background Information

This section describes basic concepts, notions and metrics used in this paper.

2.1 Basic Concepts and Notions

A PAIS is a specific type of information system which provides process support functions and allows for the separation of process logic and application code. At *build-time* process logic has to be explicitly defined in a *process schema*, while at *run-time* the PAIS orchestrates processes according to their defined logic.

For each business process to be supported, a *process type* represented by a *process schema S* has to be defined. In the following, a process schema corresponds to a directed graph, which comprises a set of *nodes* – representing *activities* or *control connectors* (i.e., XOR/AND-Split, XOR/AND-Join) – and a set of *control edges* between them. The latter specify precedence relations. Further, activities can be atomic or complex. While an *atomic activity* is associated with an invokable application service, a *complex activity* contains a sub process or, more precisely, a reference to a (sub) process schema S'. This allows for the hierarchical decomposition of schemes resulting in a *process model tree* (cf. Fig. 1a). Generally, different schemes $S_1 \ldots S_n$ may refer to a (sub) process schema S'. Fig. 1a shows a schema S modeled in BPMN notation consisting of seven nodes. Thereby, `A`, `B` and `D` are atomic activities, `C` and `E` are complex activities referring to (sub) process schemes $S1$ and $S2$ respectively, and `XOR-split` and `XOR-Join` are control connectors. $S2$ itself refers to schema $S3$ resulting in a process model tree with *depth* three.

Process schemes can either be created from scratch or through configuration, i.e., customization of a *generic process model* (e.g., a reference model). From

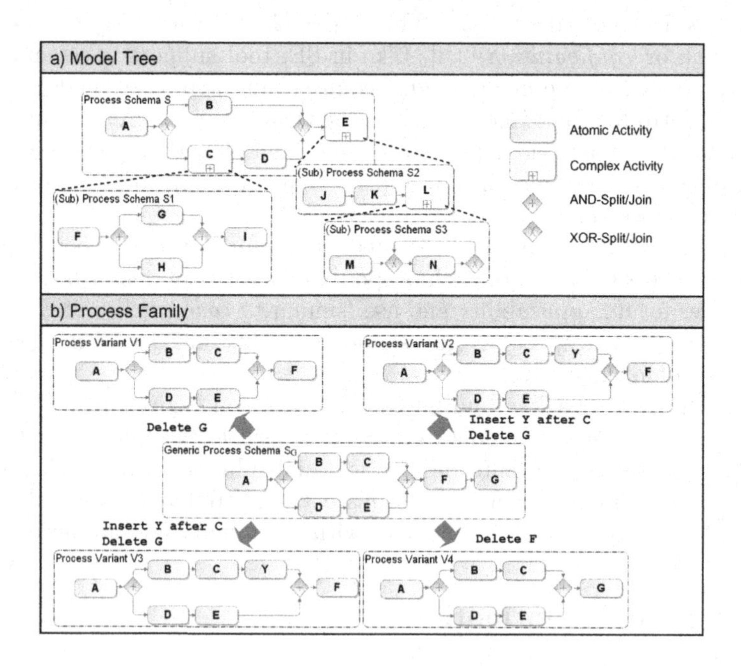

Fig. 1. Core Concepts

such a generic model several *process variants* (each with own schema) can be derived based on a restricted set of change operations [5,13]. Thereby, for a given variant we denote the set of change operations needed to transform the generic model into the variant as *bias*. Usually, the aim is to minimize the number of operations required in this context. The total set of all variants derived from a generic process model is called *process model family*. Fig. 1b shows a generic process schema S_G and four variants $V_1, \ldots, V_4$ derived from it. For example, the transformation of S_G to V_1 requires deletion of Activity G.

Most refactoring techniques are not only applicable to activities, but also to sub process graphs with single entry and exit nodes (also denoted as *hammocks*). We use the term *process fragment* as generalizing concept for all these granularities; e.g., in Fig. 1a the sub-graph of schema S containing Activities B, C, and D and the two control connectors constitutes a hammock. Based on schema S, at run-time new *process instances* can be created and executed. The latter is reflected by the execution *traces* of these instances.

Definition 1 (Execution Trace). *Let $\mathcal{PS}$ be the set of all process schemes and let $\mathcal{A}$ be the total set of activities (or more precisely activity labels) based on which schemes $S \in \mathcal{PS}$ are specified (without loss of generality we assume unique labelling of activities). Let further $\mathcal{Q}_S$ denote the set of all possible execution traces producible on schema $S \in \mathcal{PS}$. A trace $\sigma \in \mathcal{Q}_S$ is then given by $\sigma = < a_1, \ldots, a_k >$ (with $a_i \in \mathcal{A}$) where the temporal order of a_i in σ reflects the order in which activities a_i were completed over S.*

For example, $\sigma_1 = < A, B, D, C, E, F >$ and $\sigma_2 = < A, B, C, D, E, F >$ both constitute traces producible by process variant V_1 in Fig. 1b.

Schemes S and S' are called *trace equivalent* if and only if the same set of execution traces can be produced based on S as well as on S'.

Definition 2 (Trace Equivalence). *Two process schemes S and S' are trace equivalent iff $\mathcal{Q}_S = \mathcal{Q}_{S'}$.*

To determine whether two (hierarchically) composed process schemes S and S' are trace equivalent, the respective process model trees need to be expanded. For this, each complex activity needs to be replaced by the (sub) process schema it refers to. Consequently, the trace of an activity does not contain the complex activity directly, but the trace of the associated sub process. A possible execution trace for schema S in Fig. 1a is $\sigma_1 = < A, B, J, K, M, N >$.

Finally, to decide whether a process instance I can be executed according to a process schema S we use the notion of *compliance*.

Definition 3 (Compliance). *Let I be a process instance with execution trace σ. Let further S be a process schema. Then: I is compliant with S iff σ is producible on S.*

2.2 Quality Metrics for Business Process Models

In SE, metrics have been used since the 60s to measure software quality. Main purpose is to improve software design, resulting in better understandable and maintainable code [14,15]. BPM research has recently started to adopt quality metrics to specific needs of process modeling [10,11,16,17] and to empirically validate these metrics [10,12]. Similar to SE our goal is to use refactoring techniques to obtain more comprehensive and better maintainable process models. In the following we apply popular *metrics* for measuring process model quality with the design goal of comprehensive and maintainable models in mind. We use these metrics to illustrate the effects of the proposed refactorings (cf. Fig. 2). Note that the latter have effects on many other metrics, which cannot be all discussed in this paper due to lack of space.

Quality Metrics for Business Processes		
Let S = (N, E) be a process model with N denoting the set of nodes and E the set of edges.		
Metric	Description	Metrics calculated for Fig. 1
Size [11, 18]	$Size(S) = \|N\|$ measures the number of nodes in process schema S	Size(S) = 7
Process Depth [18]	$Levels(S)$ = number of process levels of the model tree with S as root	Levels(S) = 3
Control-Flow Complexity [10]	Let ANDSplits, XORSplits and ORSplits denote the node sets of S comprising respective split nodes. Let further $\delta^{+}(n)$ denote the number of direct successors of node n (number of control edges outgoing from n). Then: $CFC(S) = \|ANDSplits\| + \sum_{c \in XORSplits} \delta^{+}(c) + \sum_{c \in ORSplits} (2^{\delta^{+}(c)} - 1)$ is the sum over all connectors weighted by their potential combinations of states after the split	CFC(S) = 2
Change Distance [20]	$Dist(S1, S2)$: Minimal number of high-level change operations (e.g., MOVE activity) needed to transform schema S1 to schema S2	$Dist(S_G, V_1) = 1$

Fig. 2. Selected Quality Metrics for Process Models

Quality metrics can help process designers to identify quality problems and potential refactoring options, and to measure effects on model quality. However, what a high or low value for a particular quality metric is cannot be answered in general, but highly depends on the concrete process model(s). Therefore, like in SE it is up to the process designer to decide whether applying a particular refactoring is worthwhile. As the application of a particular refactoring may affect several schemes it is not sufficient to look only at the quality metrics of a single schema in isolation, but to apply metrics to the entire collection of schemes as well. For this purpose we introduce functions *sum* and *avg*, which we use later on for comparing process models before and after refactorings.

$$sum : 2^{\mathcal{PS}} \times Metrics \times Params \mapsto \mathbb{N}_0 \text{ with } sum(mset, m, p) := \sum_{S \in mset} m(S, p)$$

$$avg : 2^{\mathcal{PS}} \times Metrics \times Params \mapsto \mathbb{R}_0^+ \text{ with } avg(mset, m, p) := \frac{sum(mset, m, p)}{|mset|}$$

For example, the total change distance for the process family depicted in Fig. 1b is $sum(\{V_1, \ldots, V_4\}, Dist, S_G) = 6$, while the average change distance is 1.5.

3 Refactorings for Process Model Trees

This paper describes 11 refactoring techniques which allow process designers to improve the quality of process models (cf. Fig. 3). In our context refactorings constitute model transformations which are behaviour-preserving if certain pre- and postconditions are met. Implementation of these refactorings can be based on the restricted use of change patterns as presented in [13,18]. We use *trace equivalence* (cf. Def. 2) as formal notion for most refactorings to ensure that process model behaviour is not changed due to their application. If for a model tree with root S_i the same trace sets can be produced before and after the respective refactoring, process behaviour will be preserved.

We divide our refactorings into *basic* ones, which can be applied to a single schema, and *composed refactorings* applicable to a collection of inter-related process schemes. Basic refactorings transform a schema S into a new schema S' by applying a refactoring operation op. This transformation might also imply changes of a model tree, e.g., when a fragment is extracted from a process model and replaced by a reference to a sub process. Composed refactorings, in turn, will refer to a collection of process schemes $S_1 \ldots S_n$ and apply basic refactorings to them if they meet the respective pre-conditions.

For each of the proposed refactorings we describe its intent, give examples, provide a description of the refactoring operation (with pre- and postconditions) and its implementation, and describe their effects on selected quality metrics. We organize our refactorings into three groups. The first one is introduced in this section and contains *refactorings for process model trees*. The second one suggests a *refactoring for process model variants* (cf. Section 4). The third group describes model refactorings, which support *model evolution* considering process history data (cf. Section 5).

First, we describe 8 **refactorings for process model trees**. Refactoring *RF1 (Rename Activity)* can be applied when the name of an activity is not intention revealing and *RF2 (Rename Process Schema)* allows altering the name of a schema. Using *RF3 (Substitute Process Fragment)* process designers can substitute a fragment within a schema by another one which is simpler in structure, but has the same behaviour. *RF4 (Extract Process Fragment)* allows extracting a process fragment into a sub process to remove model redundancies, to foster reuse, and to reduce the size of a schema. Applying *RF5 (Replace Process Fragment by Reference)* a process fragment can be replaced by a complex activity referring to a (sub) process schema containing the respective fragment. *RF6 (Inline Process Fragment)* can be applied to collapse the hierarchy by inlining a fragment. *RF7 (Re-Label Collection)* is a composed refactoring, which supports re-labelling of certain activities within an entire process collection. Finally, *RF8 (Remove Redundancies)* allows for combined use of RF4 and RF5 to remove redundant fragments from multiple schemes in a model collection at once.

RF1 (Rename Activity). RF1 allows altering the name of an activity x to y if x is not intention revealing. RF1 is comparable to the *Rename Method* refactoring in SE [8]. Renaming an activity does not alter the behaviour of the schema S as only labels are changed. However, the notion of trace equivalence is not

Refactoring Catalogue		
Name	Refactoring Operation	Short description of refactoring
Refactorings for Process Model Trees		
RF1: Rename Activity	`renameActivity(S,x,y)`	Changes the name of an activity from x to y in schema S *Pre-Condition:* No activity from S is labelled with y
RF2: Rename Process Schema	`renameSchema(S,S')`	Renames schema from S to S' and updates all references to S *Pre-condition:* There exists no schema with label S' in the repository
RF3: Substitute Process Fragment	`substituteFragment(S,G,G')`	Substitutes sub-graph G in S by sub-graph G' *Pre-condition:* G and G' constitute hammocks and are trace equivalent
RF4: Extract Process Fragment	`extractFragment(S,G,x,S')`	Extracts sub-graph G in S and substitutes it with complex activity x referring to S' *Pre-condition:* There is no activity with label x in S; G is a hammock
RF5: Replace Process Fragment by Reference	`replaceFragment(S,G,x,S')`	Substitutes sub-graph G in S by complex activity x referring to schema S' *Pre-condition:* No activity from S is labelled with x; G is a hammock, and G and S' are trace equivalent
RF6: Inline Process Fragment	`inlineFragment(S,x)`	Inlines the sub process schema activity x refers to in S and deletes the respective sub process schema, if it is unused after the refactoring *Pre-condition:* Activity x is a complex activity
RF7: Re-label Collection	`relabelCollection(C,x,y)`	Applies RF1 to every schema $S_1,\ldots, S_n$ in model collection C where $x \in S_i$
RF8: Remove Redundancies	`removeRedundancies(C,G,x,S')`	Applies RF4 to the first schema S_i in model collection C meeting the pre-conditions and RF5 to all other schemes
Refactoring for Process Variants		
RF9: Generalize Variant Changes	`generalizeVariantChanges(S_G, VariantSet,ChangeSet)`	Generalizes variant changes by applying changes from ChangeSet to generic model S_G and by re-linking all variants from VariantSet to the new generic model S_G' (i.e., their biases are re-calculated with respect to S_G')
Refactorings for Model Evolution		
RF10: Remove Unused Branches	`removeUnusedBranch(S,G)`	Removes an unused branch G from schema S. *Pre-condition:* G constitutes a branch within a conditional branching, which was not entered when executing instances of S.
RF11: Pull Up Instance Change	`pullUpInstChange(S,InstSet, ChangeSet)`	Pulls frequent changes that happened at the process instance level up to the type level schema S. Change are described in terms of a set of applied change operations.

Fig. 3. Refactoring Catalogue

suitable in this context. Instead, we use a correctness notion based on the *Replace Process Fragment* change patterns [13,18]. For each trace σ produced on schema S with an entry of x there exists a respective trace μ on S' which is identical to σ, except that for every x in σ a y in μ can be found at the same position. Applying RF1 does not have effects on the quality metrics described in Fig. 2. However, names which reveal the intention of process designers more clearly improve understandability of the model and consequently result in decreased costs of change and reduced errors [19].

RF2 (Rename Process Schema). RF2 allows designers to rename a schema S to S'. A similar refactoring in SE is *Rename Class* [20]. To guarantee that RF2 does not alter process behaviour, all references to S are updated. Obviously, trace equivalence can be used as formal notion for RF2 ensuring that the behaviour of the model collection remains unchanged. Like RF1 this refactoring does not affect quality metrics, but improves model clarity.

RF3 (Substitute Process Fragment). RF3 allows substituting a fragment by another one with simpler structure, but same behaviour. Applying RF3 requires both fragments to contain activities with same labelling. The *Substitute Algorithm* refactoring [8] known from SE is comparable to RF3. Scenarios in which RF3 is useful include unnecessarily complex parallel branchings (cf. Fig. 4a) or unneeded control edges due to transitive relations. RF3 can be implemented based on change pattern *Replace Process Fragment* [13,18]. As formal criterion trace equivalence can be used (cf. Def. 2). Substituting a fragment by a simpler one allows designers to improve model quality along several dimensions: by removing unnecessary parallel branchings and edges not only model clarity is increased, but also size and control-flow complexity (CFC) are decreased.

RF4 (Extract Process Fragment). RF4 can be used to extract a process fragment from schema S, e.g., to eliminate redundant fragments or to reduce size of S. The fragment to be extracted must constitute a hammock. The intent of RF4 is similar to *Extract Method* as known from SE [8]. It results in the creation of a new (sub) process schema $S1$ containing the respective fragment. In addition, the fragment is replaced by a complex activity referring to $S1$. As formal notion for reasoning about behaviour preservation, trace equivalence is used. RF4 can be implemented based on change pattern *Extract Process Fragment* [13,18]. Extracting parts of a schema often results in a reduced CFC (cf. Fig. 5). Similarly, in SE the *Extract Method* refactoring is suggested as remedy for high cyclomatic complexity [21]. RF4 can also be used to reduce size of large schemes and the overall number of nodes in the process repository by removing redundancies. Further, removing redundancies reduces cost of future process changes as same changes do not have to be performed at multiple places.

RF5 (Replace Process Fragment by Reference). RF5 is used to replace a process fragment by a complex activity referring to a trace equivalent (sub) process schema. RF5 is often used in combination with RF4. It can be implemented based on change pattern *Replace Process Fragment* [13,18]. Regarding qualitiy metrics similar considerations hold than for RF4.

RF6 (Inline Process Fragment). RF6 can be used to collapse the hierarchy of a model by inlining the process fragment, e.g., if it is not justifying its induced overhead. Similarly, in SE *Inline Method* [8] allows programmers to inline the body of a method. By inlining a fragment $S1$ into S the complex activity referring to $S1$ is substituted by the fragment corresponding to $S1$. Again trace equivalence can be used as formal notion. RF6 can be implemented based on the *Inline Process Fragment* change pattern [13,18]. In particular, RF6 allows designers to collapse the hierarchy of a process model tree resulting in a decrease of levels. Note that metrics Size and CFC might increase when applying RF6.

RF7 (Re-Label Collection). RF7 is a composed refactoring for re-labelling a particular activity in all schemes of a model collection. For this, RF1 is applied to all schemes containing the activities to be re-labelled.

RF8 (Remove Redundancies). RF8 is a composed refactoring based on RF4 and RF5. It can be applied to a collection of schemes $S_1 \ldots S_n$ to remove redundancies. For this, RF4 is applied to one of these schemes to extract the redundant fragment. To all other schemes, RF5 is applied for replacing the respective fragment by a reference to the (sub) process schema created before.

Example. *Fig. 4 shows the combined usage of the basic refactorings described so far. For schema S Activity* `A` *is renamed to* `A'` *using RF1. RF2 is used to rename schema $S3$ to $S3'$. As process schemes S and $S1$ contain complex Activity* `M` *referring to $S3$ the references in* `M` *need to be updated to $S3'$. A further refactoring* *option is given by schemes S, S1 and S2, all containing a process fragment with same behaviour. However, fragment G in schema S has a more complex structure than G1 in schemes S1 and S2. First, RF3 is used to replace the fragment in S with the one of S1 or S2. Next, RF4 is applied to either S, S1 or S2 to extract the redundant process fragment to a (sub) process schema S5. Finally, RF5 is applied to the two other schemes to replace the respective fragment by a reference to S5. Instead of RF4 and RF5 the composed refactoring RF8 could be used alternatively. Schema S4 only consists of a single activity and is therefore inlined in schema S2 using RF6.*

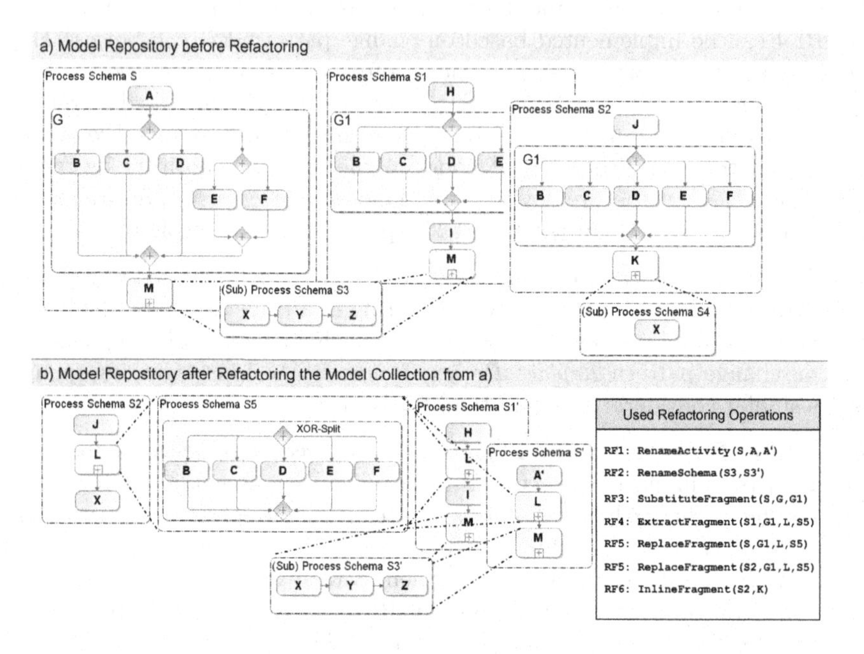

Fig. 4. Refactorings for Process Model Trees (Toy Example)

Effects on Quality Metrics. In the following we show for the refactorings in Fig. 4 how metrics can be used to measure their effects. Note that Fig. 4 constitutes a toy example, whose purpose is to show the application of the proposed patterns and its effects on quality metrics. Usually, refactorings are not applied in isolation, but in combination with other refactorings and to a collection of models. Consequently, refactoring has an impact on the collection of process models. In Fig. 4 the combined use of refactorings RF3, RF4, RF5 and RF6 reduces the total number of nodes in the given model collection from 34 to 20 and decreases average CFC of the schemes by factor 1:4 (cf. Fig. 5). In all cases no changes of model behaviour have been performed. In particular, application of RF3 allows for the removal of two unnecessary connector nodes, reducing size by two and CFC by one; RF4 and RF5 remove existing redundancies leading to an additional saving of 11 nodes. Finally, RF6 reduces size by one.

Before Refactoring (Fig. 4a)				After Refactoring (Fig. 4b)			
	Size	CFC	Levels		Size	CFC	Levels
S	11	2	2	S	3	0	2
S1	10	1	2	S1	4	0	2
S2	9	1	2	S2	3	0	2
S3	3	0	-	S3	3	0	-
S4	1	0	-	S4			
S5				S5	7	1	-
Sum	34	4		Sum	20	1	
Avg.	6.8	0.8		Avg.	4	0.2	

Fig. 5. Effects on Quality Metrics (with respect to Fig. 4)

As illustrated in Fig. 5 the proposed refactorings do not only result in smaller and less complex models, but also decreases costs of future changes by removing redundancies. For example, assume that Activity `D` in Fig. 4 shall be replaced by a sequence consisting of Activities `D1` and `D2`. Without the described refactoring this change would require to modify schemes S, $S1$ and $S2$ by applying three change operations to each of these schemes resulting in a total change distance of 9. In contrast, considering the refactoring only schema $S5$ needs to be modified (`Delete(S5,G)`, `SerialInsert(S5,D1,XOR-Split)` and `SerialInsert(S5,D2,D1)`) reducing the total change distance by 66,67 % to 3. Removing redundancies does not only result in smaller change distance, but also reduces the risk of introducing inconsistencies or errors. Finally, the exact change distance depends on the intended change and the used meta-model.

Due to the very simple nature of Fig. 4a it can be discussed whether much is gained from applying refactorings. However, for more realistic models refactorings can significantly improve understandability and maintainability as our case studies in the healthcare and automotive domains revealed. When elaborating 30 process models of a Women's hospital, for example, we detected redundancies in more than 60% of them [22]. Particularly, larger models with more than 20 activities often contained redundant process fragments (e.g., for making appointments with medical units or for exchanging medical reports). As we learned, these redundancies can be abolished using the proposed refactorings.

4 Refactoring for Process Variants

Another challenge is to manage the process variants belonging to a process family (cf. Fig. 1b). Usually, respective variants are derived from a generic schema S_G by applying a set of change operations to it. In general, configuration of new variants and adaptation of existing variants can be done most effectively when the average change distance (cf. Section 2) between generic schema S_G and its variants $V_1, \ldots, V_n$ is minimal (i.e., the average number of change operations needed to transform S_G to V_i is minimal). However, to keep the average change distance small, continuous efforts have to be made to evolve the generic model over time. Otherwise, more and more redundant changes have to be performed to different variants to keep them aligned with the real-world processes. Though respective variants are often similar, slight differences make refactorings RF4 and RF5 inapplicable in many situations. Therefore, an additional refactoring technique is needed, which supports designers in maintaining generic models.

RF9 (Generalize Variant Changes). RF 9 allows designers to pull changes, which are common to several variants, up to the generic model (similar to *Pull Up Method* and *Push Down Method* in SE [8]). This allows removing redundancies and decreasings costs of future changes. As example consider Fig. 1b, which shows a generic model S_G and variants $V_1, \ldots, V_4$ derived from it. Analysis of S_G and its variants shows that Activity `G` has been deleted for 3 of the 4 variants. Refactoring `GeneralizeVariantChanges(`S_G`,`$\{V_1, \ldots, V_4\}$`,`$\{$`Delete(G)`$\}$`)` can be applied to generalize the respective change by pulling the deletion of `G` up to the generic model S_G (not shown in Fig. 1b). As Activity `G` is deleted from the generic model, `G` needs to be inserted in variant V_4 to keep the behaviour of variant V_4 unchanged. This results in a reduction of the total change distance from 6 to 4 and a decrease of the average change distance from 1.5 to 1.0.

In a case study we did in the healthcare domain we identified 10 variants for medical order handling with similar behaviour [22]. Though respective variants were similar, slight differences existed and redundant fragments could not be extracted to (sub) processes. However, by applying RF9 we are able to reduce redundancies resulting in easier to configure and better maintainable variants.

Implementing RF9 necessitates a framework for coping with generic schemes and variants derived from them. First, advanced techniques for analyzing process variants and for identifying variant changes to be pulled up to the generic model are needed. In MinADEPT [23], for example, a generic model S'_G can be derived from a set of variants `VariantSet` such that the change distance between S'_G and the variants becomes minimal. Second, when applying RF9 the change operations in `ChangeSet` (cf. RF9 in Fig. 1b) are applied to S_G resulting in a new version S'_G of the generic model. All variants in `VariantSet` need to be re-linked from S_G to S'_G and for each variant $V_i \in$ `VariantSet` its bias is re-calculated in respect to S'_G [24]. Third, effective techniques are needed for internally representing generic models, its variants and related biases. Note that RF9 does not alter variant behaviour. Applying the updated bias of a variant V_i to S'_G results in the same variant-specific schema as applying the old bias to S_G. Thus trace equivalence can be used as formal notion. RF9 bears a high potential for full automation.

5 Refactorings for Model Evolution

This section describes refactoring techniques, which become applicable when process models are executed by PAIS and historic data on process instances is available in execution or change logs [25,26]. These logs can be analyzed and mined to discover potential refactoring options. In this context *RF10 (Remove Unused Branches)* allows process designers to remove unused paths from a process model and *RF11 (Pull Up Instance Change)* enables generalization of frequent instance changes by pulling them up to the process type level. Several mining methods for discovering such situations already exist [25,23]. We therefore do not look at mining techniques, but use them for realizing refactorings based on historical data.

RF10 (Remove Unused Branches). RF10 allows designers to remove unexecuted process fragments from a schema S. It can be implemented based on change pattern *Delete Process Fragment* [13,18] and on standard process mining techniques. Note that trace equivalence is not suitable as formal basis since the behaviour *producible* on the respective process schema is altered by RF10. Therefore we use the notion of *compliance* (cf. Def. 3). RF10 can be applied to schema S if the traces of all instances on S are re-producible on the new schema; i.e., *observed* behaviour remains unchanged. Obviously, compliance can be guaranteed when removing unused execution paths. While unused branches can be automatically deteced, RF10 is not automatically applied, but the designer has to ensure that the misalignment between model and log was not caused by design errors or an execution log not covering all relevant traces. Depending on the concrete application scenario the time window for which events from the log are considered can be narrowed. Applying RF10 decreases both model size and control flow complexity. Fig. 6a shows a schema S with its execution log comprising the traces of completed instances. Mining this log reveals that the path with activities E and F was never executed. RF10 could be applied to remove the unused fragment. This reduces size of S from 9 to 7 and CFC from 3 to 2. After removing E and F all instances in the log are compliant with schema S'.

RF11 (Pull Up Instance Change). RF11 can be used to generalize frequently occurring instance changes by pulling them up to the process type level (similar to RF9 where variant changes are generalized). Like for RF9 the overall goal is to reduce average and total change distance between type schema and instance-specific schemes; e.g., to learn from instance changes and to reduce the need for adapting future instances [24]. The implementation of RF11 is similar to RF9.

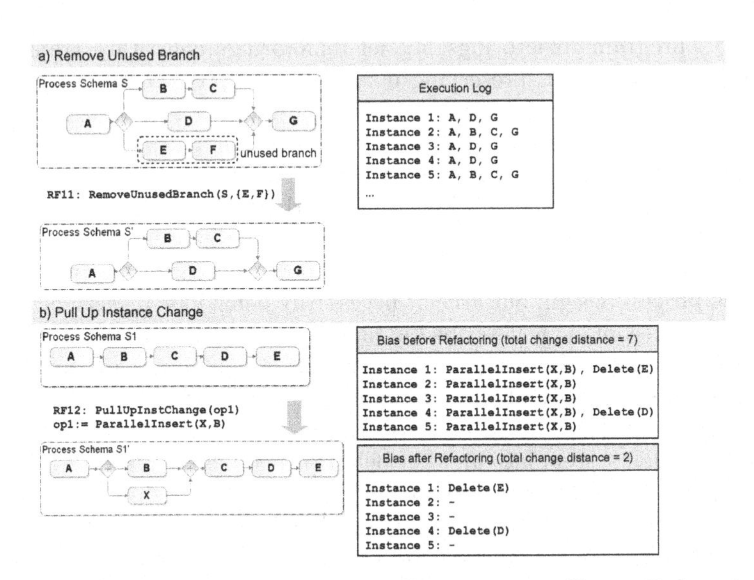

Fig. 6. Remove Unused Branch and Pull Up Instance Change Refactorings

In contrast to RF9, however, trace equivalence cannot be used to ensure that no errors are introduced when applying RF11. By pulling changes from the instance level to the type level behaviour producible on the respective schema is always altered. Therefore, compliance is used as formal notion like in RF10. Like RF9, RF11 has the potential for full automation.

Fig. 6b shows a process schema $S1$ and for each process instance $I_1, \ldots I_5$ its deviation from $S1$. Activity X was inserted parallel to B for each of these instances. For I_1, Activity E was additionally deleted and for I_4 Activity D was deleted. To pull up the insertion of Activity X (which is common to all instances) to the type level and to reduce the need for future instance adaptations, RF11 could be applied. Using RF11 reduces the total change distance from $sum(\{I_1, \ldots, I_5\}, Dist, S1) = 7$ to $sum(\{I_1, \ldots, I_5\}, Dist, S1') = 2$.

6 Related Work

Refactoring techniques for improving software design were first proposed by Opdyke [7]. He suggested a set of refactorings for C++ which are semantic preserving when certain preconditions are met. The first notable refactoring tool has been the Refactoring Browser [20] for Smalltalk, which automatically performs the refactorings proposed by Opdyke plus some additionally techniques [27]. As all refactorings provided by this tool constitute behaviour-preserving transformations it is ensured that no errors or information losses are introduced. Tool support for languages like C++ and Java recently emerged. The provided refactorings are usually not provably behaviour-preserving. Therefore, refactorings need to be backed up by automated regression tests to detect behavioural changes in the software and to avoid errors [8].

Similar to program refactorings, model refactorings constitute transformations, which are behaviour-preserving if certain pre-/post-conditions are met. Existing approaches focus on UML model transformations [28], while refactoring has not been elaborated in detail for business process models. There exist a few approaches which provide specific refactorings in a narrow context (e.g., a particular process modeling formalism). In [29] refactoring techniques for event-driven process chains (EPCs) are described. Refactoring techniques have been also discussed in connection with model merging [30]. The proposed transformations aim at improved process design, but are not necessarily behaviour-preserving. A specific refactoring technique is described in [31] where algorithms for transforming unstructured processs models into block-structured models are proposed. Synthesis of Petri Nets, in turn, offers techniques which take a transition system and generate a Petri net from it [32]. This approach can be used to transform a Petri Net via a transition system into another behaviour-equivalent Petri net. Respective techniques allow to elimate unnecessary net elements (e.g., silent activities, unnecessary places) [32] or to discard OR-joins from process models [33].

This paper complements existing work dealing with *process redesign* [34] or *process adaptation* [5]. Both refactoring and *process redesign* [34] may require model transformations. However, scope of process redesign is much broader and goes beyond structural adaptations. Redesign is primarily business driven and aims to improve one or more performance dimensions of a process (e.g., time, quality, costs or flexibility) [34]. Therefore, process redesign often affects external quality of a PAIS and its results are visible to the customer. In contrast, refactoring techniques primarily impact the internal quality of the PAIS, ensure conceptual integrity, and foster maintainability. Similar to refactorings, *process adaptations* [5] refer to structural changes of a process schema (e.g., using change patterns) [13,18,5]. In contrast to refactorings, process adaptations are usually affecting the behaviour of a process model. We build upon existing research in this area and extend it to be applicable for process model refactorings.

Existing BPM tools only provide limited refactoring support. Renaming of activities and process schemes is supported by most tools (e.g., ARIS). However, more advanced refactoring support is missing.

7 Summary and Outlook

We proposed 11 refactorings specifically suited for large process repositories. These techniques allow process designers to better deal with model complexity and to make process models easier to change and better understandable. With the increasing adoption of PAIS and the emergence of large process repositories systematic support for model management is getting increasingly important. We are currently working on a reference implementation of a tool for refactoring process models to support users in both identifying refactoring options and applying behaviour-preserving or compliance-ensuring refactorings. We further plan to integrate this with our previous work on change patterns [13,18], model evolution [35], and process change mining [23] to provide integrated support for the management of process models throughout the entire process life cycle.

References

1. Weske, M.: Business Process Management: Concepts, Methods, Technology. Springer, Heidelberg (2007)
2. Dijkstra, E.W.: A Discipline of Programming. Prentice-Hall, Englewood Cliffs (1976)
3. Rosemann, M., van der Aalst, W.: A Configurable Reference Modelling Language. Information Systems (2005)
4. Rosa, M.L., Lux, J., Seidel, S., Dumas, M., ter Hofstede, A.: Questionnaire-driven Configuration of Reference Process Models. In: Krogstie, J., Opdahl, A., Sindre, G. (eds.) CAiSE 2007 and WES 2007. LNCS, vol. 4495, pp. 424–438. Springer, Heidelberg (2007)
5. Reichert, M., Dadam, P.: ADEPT$_{flex}$ – Supporting Dynamic Changes of Workflows Without Losing Control. JIIS 10, 93–129 (1998)
6. Parnas, D.L.: Software Aging. In: Proc: ICSE 1994, pp. 279–287 (1994)
7. Opdyke, W.F.: Refactoring Object-Oriented Frameworks. PhD thesis, Univ. of Illinois (1992)
8. Fowler, M.: Refactoring - Improving the Design of Existing Code. Addison-Wesley, Reading (2000)
9. Beck, K.: Extreme Programming Explained. Addison-Wesley, Reading (2000)

10. Cardoso, J.: Process Control-Flow Complexity Metrics: An Empirical Validation. In: Proc. IEEE SCC 2006, pp. 167–173 (2006)
11. Vanderfeesten, I., Cardoso, J., Mendling, J., Reijers, H., van der Aalst, W.: Quality Metrics for Business Process Models. In: 2007 BPM & Workflow Handbook (2007)
12. Mendling, J.: Detection and Prediction of Errors in EPC Business Process Models. PhD thesis, Vienna Univ. of Economics and Business Administration (2007)
13. Weber, B., Rinderle, S., Reichert, M.: Change Patterns and Change Support Features in Process-Aware Information Systems. In: Krogstie, J., Opdahl, A., Sindre, G. (eds.) CAiSE 2007 and WES 2007. LNCS, vol. 4495, pp. 574–588. Springer, Heidelberg (2007)
14. McCabe, T.: A Complexity Measure. IEEE ToSE 2, 308–320 (1976)
15. Yourdon, E., Constantine, L.: Structured Design: Fundamentals of a Discipline of Computer Program and Systems Design. Prentice Hall, Yourdon Press (1979)
16. Nissen, M.E.: Redesigning Reengineering through Measurement-Driven Inference. MIS Quarterly 22, 509–534 (1998)
17. Reijers, H., Vanderfeesten, I.: Cohesion and Coupling Metrics for Workflow Process Design. In: Desel, J., Pernici, B., Weske, M. (eds.) BPM 2004. LNCS, vol. 3080, pp. 290–305. Springer, Heidelberg (2004)
18. Weber, B., Rinderle, S., Reichert, M.: Change Support in Process-Aware Information Systems - A Pattern-Based Analysis. Technical Report TR-CTIT-07-76, University of Twente (2007)
19. Becker, J., Rosemann, M., Uthemann, C.v.: Guidelines of Business Process Modeling. In: BPM 2000, pp. 30–49 (2000)
20. Brant, J., Roberts, D.: Refactoring Browser: st-www.cs.uiuc.edu/users/brant/refactoringbrowser/
21. Glover, A.: Refactoring with Code Metrics (2006), `www.ibm.com/developerworks/java/library/j-cq05306/`
22. Reichert, M., Dadam, P., Schultheiss, B., Konyen, I.: Modeling and analysis of healthcare processes in a woman's hospital. project reports no. dbis-27, dbis-28, dbis-29, dbis-16, dbis-15, dbis-14, dbis-7, dbis-6, dbis-5 (1996-1997)
23. Li, C., Reichert, M., Wombacher, A.: Issues in process variants mining. Technical Report TR-CTIT-08-10, CTIT, University of Twente, Enschede (2008)
24. Rinderle, S., Weber, B., Reichert, M., Wild, W.: Integrating Process Learning and Process Evolution – A Semantics Based Approach. In: van der Aalst, W.M.P., Benatallah, B., Casati, F., Curbera, F. (eds.) BPM 2005. LNCS, vol. 3649, pp. 252–267. Springer, Heidelberg (2005)
25. Van der Aalst, W., van Dongen, B., Herbst, J.: Workflow Mining: a Survey of Issues and Approaches. Data and Knowledge Engineering, 237–267 (2003)
26. Rinderle, S., Reichert, M., Jurisch, M., Kreher, U.: On Representing, Purging, and Utilizing Change Logs in Process Management Systems. In: Dustdar, S., Fiadeiro, J.L., Sheth, A.P. (eds.) BPM 2006. LNCS, vol. 4102, pp. 241–256. Springer, Heidelberg (2006)
27. Roberts, D., Brant, J., Johnson, R.: A Refactoring Tool for Smalltalk. Theory and Practice of Object Systems, 253–263 (1997)
28. Sunye, G., Pollet, D., Traon, Y.L., Jezequel, J.: Refactoring UML Models. In: Gogolla, M., Kobryn, C. (eds.) UML 2001. LNCS, vol. 2185, pp. 134–148. Springer, Heidelberg (2001)
29. Fettke, P., Loos, P.: Refactoring von Ereignisgesteuerten Prozessketten. In: EPK 2002, pp. 37–49 (2002)
30. Küster, J., Koehler, J., Ryndina, K.: Improving Business Process Models with Reference Models in Business-Driven Development. In: BPM 2006 Workshops (2006)

31. Liu, R., Kumar, A.: An Analysis and Taxonomy of Unstructured Workflows. In: van der Aalst, W.M.P., Benatallah, B., Casati, F., Curbera, F. (eds.) BPM 2005. LNCS, vol. 3649, pp. 268–284. Springer, Heidelberg (2005)
32. Cortadella, J., Kishinevsky, M., Lavagno, L., Yakovlev, A.: Deriving petri nets from finite transition systems. IEEE Transactions on Computers 47(8), 859–882 (1998)
33. Mendling, J., van Dongen, B., van der Aalst, W.: Getting rid of the OR-Join in business process models. In: EDOC 2007, pp. 3–14 (2007)
34. Reijers, H.A.: Design and Control of Workflow Processes: Business Process Management for the Service Industry. Springer, Heidelberg (2003)
35. Rinderle, S., Reichert, M., Dadam, P.: Correctness Criteria for Dynamic Changes in Workflow Systems – A Survey. DKE 50, 9–34 (2004)

Service-Oriented Information Systems Engineering: A Situation-Driven Approach for Service Integration*

Nicolas Arni-Bloch and Jolita Ralyté

University of Geneva, CUI, 24 rue General Dufour
CH-1205 Geneva, Switzerland
{Nicolas.Arni-Bloch,Jolita.Ralyte}@cui.unige.ch

Abstract. In this work we propose a Metamodel of Information System Service (MISS) and introduce a situation-driven approach for ISS integration. This approach is based on situational method engineering principals and is defined as a collection of inter-related method chunks.

1 Applying SOA to Information System Engineering

The community of IS developers increasingly adopts service-oriented approach to IS engineering. In this work we consider a particular type of services – IS Services (ISS) – that have to be integrated into enterprise legacy IS in order to avoid IS fragmentation. Following the traditional SOA [1, 5], integration of an ISS into an IS would be limited to the exchange of messages between services. That means that only needs for services and their capabilities to response would be considered but not the information overlap that different services could have. This allows to resolve "point-to-point" integration and represents a step towards integrated IS but this not allows to resolve the problem of information overlap management. The overlap supported by data redundancy maintained between services will continue to generate extra cost and bad quality of data and processes.

Before publishing a service in the registry of the enterprise, it is necessary to guaranty the integrity of the data but also the alignment of the rules and processes on the IS policies. Besides, the impact of ISS integration has to be evaluated taking into account technical, informational and business aspects. At the technical level it means to consider the cost of the interoperability between language, framework or hardware components. The evaluation at the business level consists in analysing the capability of the enterprise to support new processes and to provide the necessary data. Finally, at the informational level links between data and the compatibility between processes and rules has to be considered. Therefore, we argue that the ISS integration cannot be limited to the exchange of messages between services but has to deal with information overlap. To enable the integration of ISS, it is necessary to extend service description with its informational knowledge: (1) the definition of service data structure and semantics, (2) the definition of service behaviour in terms of actions that can be

* This work was partially supported by the Swiss National Science Foundation. N° 200021-103826.

Z. Bellahsène and M. Léonard (Eds.): CAiSE 2008, LNCS 5074, pp. 140–143, 2008.

executed by the service and effects that these actions provoke, and (3) the definition of rules (data integrity and process rules) to be respected when realising the service. Taking into account the complexity of the integration process is critical and requires an advanced methodological support. In particular, in order to deal with the diversity of integration situations, the method for service integration has to be flexible, modular and configurable.

2 Metamodel of Information System Service (MISS)

We define an Information System Service (ISS) as an autonomous coherent and interoperable component of an information system that offers capabilities and owns resources to realize these capabilities. These resources can be technical (hardware or software), informational (information, behaviour and rules) and organizational (organizational role, actors). At the informational level, an ISS is considered through three spaces: static, dynamic and rules. Fig. 1 represents the Metamodel of Information System Service (MISS).

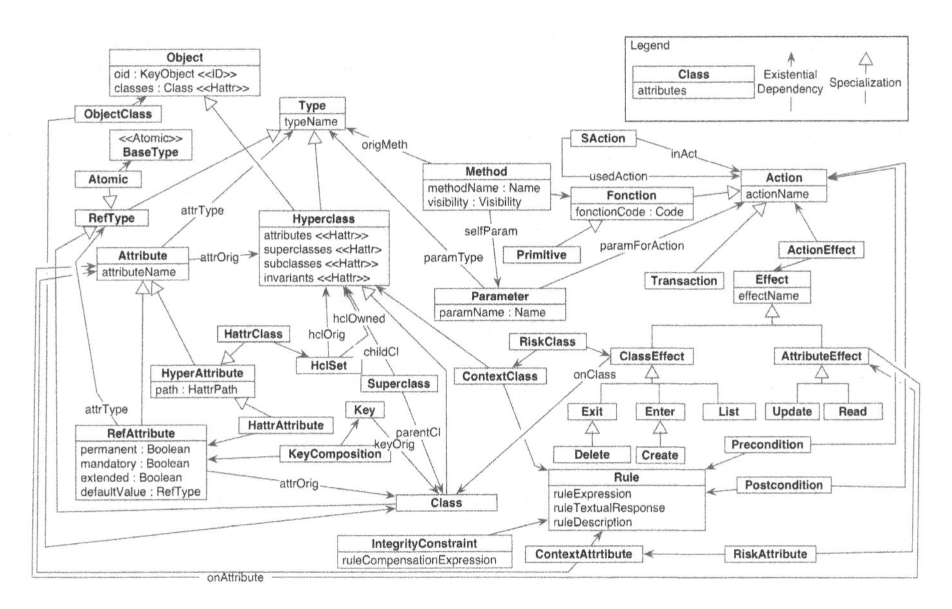

Fig. 1. Metamodel of Information System Service: the informational part of an ISS

Static Space. The static space of the MISS represents the data structure of an ISS by using the following concepts: *class*, *key*, *attribute*, *method* and *object*. Besides, we use the concept of *hyperclass* [4] in order to define the set of classes required by the service to realize its capabilities and to guaranty the completeness and coherence of its data structure.

Dynamic Space. The dynamic space of an ISS represents the behaviour of the service capabilities. The main concepts of the dynamic space are: *action* and *effect*. An action is an object that defines a behavior having effects on other objects. An action accepts *parameters* that denote objects to be given to the action for its execution. An action is described by a *process* to be executed and produces one or more *effects* that specify the type of its result. Finally, the execution/enactment of an action is constrained by a set of *preconditions* and a set of *postconditions*. The effects of an action are defined by using a set of primitives: create, enter, exit, update, read, list and call.

Rule Space. The objective of this space is to preserve the coherence, correctness and consistency of the ISS during its exploitation. A *rule* is an expression/algorithm that returns a boolean value when evaluated. The classes and attributes that participate in the validation of the rule define its validation context. Rules are used as a basis for the specification of the *integrity constraints* and *conditions*. An integrity constraint is a rule that has to be verified in each state of the services or at each modification of it. A *condition* is a rule that has to be valid at some point of process execution.

3 A Situational Approach for ISS Integration

The process of ISS integration cannot be limited to a simple prescribed set of activities because of the multitude of integration situations that has to be considered. To be able to deal with this diversity of integration situations we construct our method following the principles of Situational Method Engineering (SME) and particularly the chunk-driven SME approach [2]. We define our approach as a collection of interrelated method chunks each of them addressing some specific activity in the ISS integration process and organised into a map-based [3] process model.

The ISS integration mainly consists in resolving the information overlap situations between the selected ISS and the legacy IS. We identify four main intentions that have to be considered in the ISS integration process:

1. To identify and characterise the information overlap between the ISS and the IS,
2. To adapt the overlap part in the ISS specifications in order to allow their integration with the IS specification,
3. To integrate the IS specifications and the ISS specifications, and
4. To consolidate the integrated specifications.

Therefore, our process model (Fig. 2) is based on these four intentions and identifies several strategies that have to be considered in order to achieve these intentions. The first step consists in identifying overlap situations in the specifications of the ISS and the IS, analysing these situations, identifying elements to be integrated (e.g. classes to be merged) and detecting elements that need to be unified before their integration. The *semantic analysis* strategy helps to identify naming inconsistencies in the static space of the IS and the ISS, while the *functional analysis* strategy focus on the identification of similar capabilities provided by both IS and ISS and specified in their dynamic space. The *structural*, *action-driven* and *rule-driven* strategies continue the identification and characterisation of the overlap in the static, dynamic and rule spaces respectively.

Next, the overlap situations requiring some unification have to be settled by applying the appropriate *semantic unification* and/or *transformation* operators. The third step consists in selecting the appropriate integration operator for each overlap situation and applying it. The strategy "*with integration operator*" in fact represents a bundle of exclusive strategies each of them dealing with specific static, dynamic or rules space integration operator.

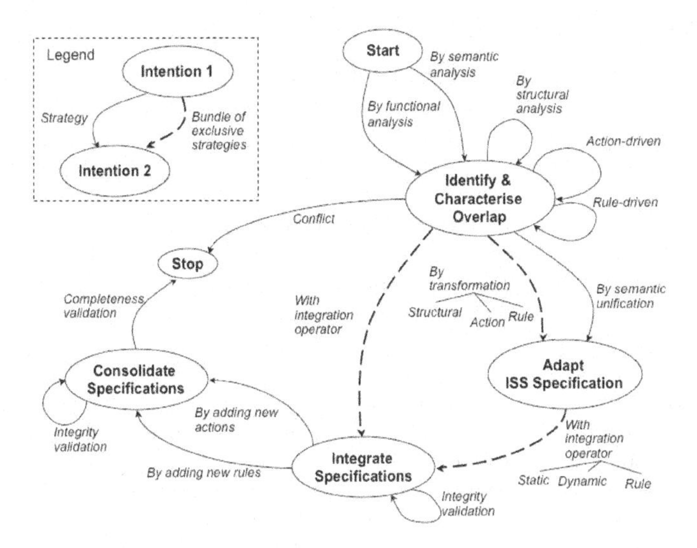

Fig. 2. Process model for ISS integration (Map formalism [3])

Finally, the integrated specification has to be consolidated. The integration of an ISS into an IS can create new situations which didn't exist neither in the initial IS nor in the ISS. Therefore, it can be necessary to *add new integrity rules* or new *actions* in order to guarantee the completeness and the coherence of the integrated specification.

Our approach supporting ISS integration aims to provide one or more method chunks for each section (a triplet <*source intention, target intention, strategy*>) of this map. Currently, we focus our effort on identifying and evaluating different situations that can occur in the ISS integration process and defining method chunks satisfying these situations. A tool support is also under development to store and enact the method chunk knowledge.

References

1. OASIS Reference model for service oriented architecture 1.0. Technical report (2006), http://docs.oasis-open.org/soa-rm/v1.0/soa-rm.pdf
2. Ralyté, J., Rolland, C.: An Approach for Method Reengineering. In: Kunii, H.S., Jajodia, S., Sølvberg, A. (eds.) ER 2001. LNCS, vol. 2224, pp. 471–484. Springer, Heidelberg (2001)
3. Rolland, C., Prakash, N., Benjamen, A.: A Multi-Model View of Process Modelling. Requirements Engineering 4(4), 169–187 (1999)
4. Turki, S., Léonard, M.: Hyperclasses: towards a new kind of independence of the methods from the schema. In: Proc. of ICEIS 2002, pp. 788–794 (2002) ISBN: 972-98050-6-7
5. Erl, T.: Service-Oriented Architecture (SOA): Concepts, Technology, and Design. Prentice Hall PTR, Englewood Cliffs (2005)

When Interaction Choices Trigger Business Evolutions

Guillaume Godet-Bar, Sophie Dupuy-Chessa, and Dominique Rieu

Laboratoire LIG, University of Grenoble
681, rue de la Passerelle – BP 72
38402 St Martin d'Hères, France
firstname.surname@imag.fr

Abstract. In the context of development methods, early collaborations between specialists (SE, HCI, business, usability experts ...) allows having a broader view of the development possibilities, notably in terms of user-system interaction. Consequently, in-depth transformations of the processes and concepts can be considered with minimum financial or temporal impact. We discuss in this paper the opportunities for business process evolutions emerging from the application of our collaborative method, based on the choices made for designing the interaction.

Keywords: Information Systems, Human-Computer Interaction, Collaborative Design, Evolution

1 Introduction

The evolution of computer technologies, in terms of communication (wireless networking) and interaction device (visualization headsets, tactile gloves) deeply alter the classical, implicit perception of Human-Computer Interaction (HCI). The user can now evolve in environments blending real and virtual entities. We shall use the concept of "Augmented Reality system" (AR system) to designate any interactive system that superimposes virtual data onto the real world. A classical example of AR system can be found in [1].

These new opportunities sometimes lead HCI specialists to design the future application from a totally different, usability-oriented, point of view, and therefore explore aspects of the future system that are sometimes unforeseen from a purely "functional core side". In order to address these issues, we have proposed an adaptation of the Symphony development method [2]. While this method is originally Software Engineering oriented and focuses on the design of the functional core and business aspects of the system, we have integrated HCI practices and models for designing classical as well as complex interfaces. This method is also a medium for showing the impact of HCI choices on business evolution, from the bottom-most levels of refinement of the specifications, up to the business definition level.

We summarize in Section 2 the essential notions on which the Symphony method is based. Then, Section 3 describes the principles for envisaging business evolution. We conclude this paper by giving some perspectives on future works.

Z. Bellahsène and M. Léonard (Eds.): CAiSE 2008, LNCS 5074, pp. 144–147, 2008.

2 The Symphony Method

We use the Symphony method as a medium for merging SE and HCI activities. It is a user-oriented, business component-based development process originally proposed by the UMANIS company. It has already been extended by [2] and [3], mainly in order to improve reusability of components, and lately to integrate the design of complex interfaces such as Augmented Reality systems.

Symphony is organized into three design branches, similarly to 2TUP, into a Y-lifecycle: functional aspects are treated in the left branch, technical concerns in the right branch, and a central branch merges both developments. As this development process has already been addressed previously, we shall focus on the aspects pertinent for business evolution, which occurs in the left branch of the development cycle:

- The functional branch features two essential phases: Specification of requirements and Analysis of requirements,
- All phases aim at refining models and scenarios outlined in the previous phase,
- SE and HCI-oriented activities are realized in parallel, by design actors specialized either in usability, business expertise, Software Engineering or Human-Computer Interaction, setting out from common scenarios elaborated during the Inception phase. Therefore it is likely that two different approaches to the final system will be developed, both valuable in terms of the aspects they are expected to concentrate on (for instance functionalities for SE specialists, usability for HCI experts),
- Collaboration points are envisaged at specific stages of development, in order to ensure consistency of adopted design options, synchronize points of view, establish conceptual links between models or take collegial decisions on design choices.

The functional branch ultimately produces a set of dynamic and static models –mostly UML models– for describing the application. We have described in [3] a specific model type for describing business concepts (Business Objects) and interactional concepts (Interactional Objects).

The following section details the different stages of business evolution, starting from the end of the Analysis phase and using Business Objects and Interactional Objects, up to the Specification of requirements.

3 Different Stages of Business Evolution

We explore in this section how, starting from a coordination activity for modifying the Business Objects model and the Interactional Objects model, an evolution of the business space may be triggered at different levels of refinement of the Information System.

- **Business Objects model evolution**
  Even though Interactional Objects are generally more complex than most Business Objects, due to the intrinsic complexity of interactions and interfaces, we have set up a systematic analysis of the structural imbalance between Interactional Objects and Business Objects.
  In particular, we evaluate whether the Interactional Objects model aspects of the system which in fact correspond to business concerns. Our analysis aims at deciding whether the Information System would benefit from transferring and adapting this data to the business space.
  For instance, Augmented Reality systems often feature topological data, usually used for positioning the user or artifacts in tridimensional virtual environments superimposed on the physical world. A possible transfer of competences would consist in augmenting the Business Objects managing location concepts with architectural plans. Subsequently, Interactional Objects would remain responsible for managing the visual representation of the architectural plans.
- **Activity evolution**
  In the context of the Symphony method, an essential aspect of the specification of the business space at the Activity level is the identification of computerized tasks realized by internal actors, using Use Cases. The latter are organized into logical packages corresponding to Business Process Components (i.e., units representing uninterrupted exchanges between external actors and the Business Process).
  While focus at the Business Objects model level was on transfers of competences, we concentrate at the Activity level on the way the resulting new data (in our example, topological data) may be collected, organized and used.
  From a development process point of view, Use Cases for carrying these new activities consequently appear.
- **Business Process evolution**
  At the Business Process level, the Symphony method focuses on describing the interaction between external actors and the system.
  One of the essential activities at this level of description is the identification of Business Process Components, that is, uninterrupted exchanges between actors and the Business Process. In order to achieve this, these interactions are represented using UML sequence diagrams.
  Driving business evolution to this level means capitalizing on the new Use Cases introduced at the Activity level. In particular, use of the new data by different (or new) Business Processes Components and the intervention of new external actors, within the Information System, may be envisaged. This may enable automatization of manual tasks, facilitate certain processes...
  Concerning our example, topological data could be used for estimates, providing access to location-related data *in situ* (using head-mounted display and positioning technologies), providing location-dependent services etc.
- **Business definition evolution**
  Beyond reorganizing Business Processes, business evolution may be taken to the point that new services are handled by the Information System, thus

changing the definition of business. New Business Processes may be added, that will need their own entire iteration of the development process.

Consequently, new interaction choices will be made for these new processes and new actors that may also eventually affect the business space.

In our example, integrating topological data into the business space could be used for organizing virtual tours for potential clients.

4 Conclusion and Perspectives

We have detailed in this paper how describing the interactional aspects of a system may affect the design of the business space, when an effort for rationalizing the balance of competences between the two conceptual domains is made. The design of complex interactive systems puts even more strains on this problematic, because of their intrinsic requirements for intuitiveness, continuity of interaction, the numerous inputs they need to integrate...

In the context of the Symphony development method, we have proposed a process for capitalizing on the evolutions induced by interaction choices, from concepts reorganization to the redefinition of the business.

In the current instance of the Symphony method, we tackle the specifications evolution, which is closely linked to that of data reorganization, by relying on the concept of Iteration Plan, from the Rational Unified Process [4]. Thus, we do not need to identify cluttering "iteration pathways" for each activity. Instead, programmed evolutions are progressively added to the coming Iteration Plan. Final decision on the application of features of the iteration can thus be regularly discussed with stakeholders.

As is often the case with new constructions in methodology, we need to confirm our propositions with further experiments. This may be delicate, given that this method applies to complex interactive systems integrated into larger Information Systems. Consequently, we shall focus future works on the adequate description of Iteration Plans centered on business evolution, both in terms of models and process.

References

1. Ishii, H., Ullmer, B.: Tangible bits: towards seamless interfaces between people, bits and atoms. In: CHI 1997: Proceedings of the SIGCHI conference on Human factors in computing systems, pp. 234–241. ACM, New York (1997)
2. Hassine, I., Rieu, D., Bounaas, F., Seghrouchni, O.: Symphony: a conceptual model based on business components. In: SMC 2002, IEEE International Conference on Systems, Man, and Cybernetics, vol. 2 (2002)
3. Godet-Bar, G., Rieu, D., Dupuy-Chessa, S., Juras, D.: Interactional objects: Hci concerns in the analysis phase of the symphony method. In: 9th International Conference on Enterprise Information Systems ICEIS 2007, Funchal, Madeira, June 2007, pp. 37–44 (2007)
4. Jacobson, I., Booch, G., Rumbaugh, J.: The Unified Software Development Process. Addison-Wesley, Reading (1999)

GATiB-CSCW, Medical Research Supported by a Service-Oriented Collaborative System

Konrad Stark[1], Jonas Schulte[2], Thorsten Hampel[2], Erich Schikuta[1], Kurt Zatloukal[4], and Johann Eder[1,3]

[1] University of Vienna, Dept. of Knowledge and Business Engineering, Austria
[2] University of Paderborn, Heinz Nixdorf Institute, Germany
[3] University of Klagenfurt, Dept. of Informatics Systems, Austria
[4] Medical University Graz, Institute of Pathology, Austria

Abstract. Medical research is a collaborative process in an interdisciplinary environment that may be effectively supported by a Computer Supported Cooperative Work (CSCW) system. Such a system imposes specific requirements in order to allow flexible integration of data, analysis services and communication mechanisms. Persons with different expertise and access rights cooperate in mutually influencing contexts (e.g. clinical studies, research cooperations). Thus, appropriate virtual environments are needed to facilitate context-aware communication, deployment of biomedical tools as well as data and knowledge sharing. We systematically elaborate the main requirements of a medical CSCW system and present a conceptual model, as well as an architectural proposal satisfying the demands. We design a prototypical virtual workbench to support research and routine activities in the context of the GATiB (Genome Austria Tissue Bank) initiative.

Keywords: Medical Research, CSCW, Service-Oriented Architecture.

1 Introduction

Medical care and medical research are cross-fertilising areas. Courses of disease may be monitored and analysed from a scientific perspective, whereas treatment development and medication design benefit from the the results of research. Vast amounts of patient records containing information about diagnosis, laboratory tests, radiology images and medications are created continuously. For example, a medium-sized hospital in Austria provides medical treatment for 512,000 patients within one year (*http://www.klinikum-graz.at*). Although most of the data is recorded in medical information systems, they do not support collaborative work. Thus, there is a strong need to share, contextualise and annotate data allowing inter-organisational and interdisciplinary collaboration. A wide range of biomedical applications have been developed for very specific purposes, e.g. image processing, gene expression analysis. However, they also lack support for collaborative processes. In this paper we designed a *Computer Supported Cooperative Work (CSCW)* system that meets the requirements of the medical research

Z. Bellahsène and M. Léonard (Eds.): CAiSE 2008, LNCS 5074, pp. 148–162, 2008.

domain, especially out of the field of flexible service integration (SOA) and object oriented data integration (integration of persistency layers). Standard "out of the box" CSCW systems do not meet these requirements. Therefore, some new architectural elements had to be developed. Collaboration in the medical research field is characterised by high complexity and high variation of the collaborative situations. Data is distributed over several institutes and underlays various restrictions in accessibility for different persons (roles). Access to patient data is for example highly limited by the context or role of application. Data is collected, restructured, analysed and shared with each other in different settings. This paper addresses such questions by defining the requirements for a CSCW system. The resultant architecture of our CSCW system is also presented to highlight how the requirements can be attained in the field of medical research. However, in its characteristic as a collaborative knowledge management system our GATiB-CSCW system can be easily used as an e-learning platform. Our means of flexible configuration of access rights and the architectures ability to define contextual views on the presented data allows it to be used as a powerful collaborative learning platform. Much work has been done in developing IT infrastructure for the biomedical research. Some grid-based platforms support effective co-working of researchers on distributed data sets [1], expose various data sources on the grid and allow access by web services [2] or enhance interoperability between distributed medical models in a grid portal [5]. Other approaches provide an infrastructure for collaboratively using distributed computational services and data resources [8,6,12]. Though, few approaches [9,4] exist to explicitly support collaboration processes in the biomedical context, we encourage to close the gap by implementing a collaboration-aware biomedical infrastructure based on a service-oriented architecture. This work was developed in the context of the biobank initiative GATiB (Genome Austria Tissue Bank) which is part of the Austrian Genome Program (*http://www.gen-au.at*). GATiB aims at the establishment of a tissue bank which builds on a collection of diseased and corresponding normal tissues representing all diseases at the natural frequency of occurrence from a non-selected central European population of more than 700.000 patients. Major emphasis is placed on annotation of archival tissue with comprehensive clinical data including follow-up. A more detailed description of the biobank initiative is given in [3]. The paper is structured as follows: To describe the background of the requirements for a CSCW system useable for medical research, Section 2 presents example collaboration scenarios and summarises the main requirements for such a system. Some practical examples and a prototypical virtual workbench is presented in Section 3. The conceptual components for the GATiB-CSCW system are described in Section 4. The architecture of the CSCW system is specified in Section 5. Our paper ends with a description of the current status and an overview of ongoing and future work in Section 6.

2 Scenarios and Requirements

Collaboration in the GATiB project focuses on both the medical research and routine activities of medical scientists and supporting staff. Data that is locally distributed over institutes and research groups is accessed in manifold ways and for various purposes. The following three example scenarios describe different types of collaboration in medical research.

Scenario 1: A patient diagnosed with mamma carcinoma is assumed to suffer from a rare subtype of breast cancer. A group of expert pathologists are concerned with the correct cancer classification. Each of the experts needs detailed access to anamnesis data of the patient and her family. Additionally, virtual slides made from the instantaneous section are used to cooperatively mark and annotate sections of the image. A set of similar cases as well as selected state-of-the-art publications are provided for comparative analysis.

Scenario 2: For an extensive evaluation of the course of disease of liver cancer over the past 20 years information about diagnoses, family anamnesis and follow-up data (medication, therapy, resection) of relevant cases is needed. Data from the institutes pathology, oncology and surgery is accessed as well as survival data in order to support statistical analyses. The distribution of liver cancer subtypes is calculated and an inventory of the biobank lists all associated tissue samples. The results are used both in revision reports as well as in a publication.

Scenario 3: Due to a cooperation between the hospital and a pharmaceutical company a group of suitable human-tissue donors is to be identified to support a drug discovery study. Therefore, pathological diagnoses, survival data and tissue images of patients that have signed an informed consent are required. After searching and structuring the information confidential patient data has to be protected since an external organisation is involved in the study. Hence, identifying attributes (name, day of birth) are eliminated and quasi-identifying attributes are k-anonymised [10,14]. Life style data is included by filling out questionnaires. Further, a tissue microarray of the relevant cases is made in order to test candidate tumour markers. The results should be made available to other research groups.

These scenarios demonstrate the diversity of collaboration types that may occur. A cooperative system in a biomedical research environment requires a high degree of flexibility and extensibility. Distributed data is accessed in different levels of grading considering data privacy issues, it is annotated and analysed. We use the above-mentioned scenarios to deduce general requirements a CSCW system has to comply with.

Requirements

- $R(1)$ **User and role management.** The CSCW has to be able to cope with the organisational structure of the institutes and research groups of the hospital. Data protection directives do have to fit in the access right model of the system. Though, the model has to be flexible to allow the creation of new research teams and information sharing across organisational borders.
- $R(2)$ **Transparency of physical storage.** Although data may be stored in distributed locations, data retrieval and data storage should be solely

dependant on access rights, irrespective of the physical location. That is, the complexity of data structures is hidden from the end user. The CSCW system has to offer appropriate search, join and transformation mechanisms.
- $R(3)$ **Flexible data presentation.** Since data is accessed by persons having different scientific background (biological, medical, technical expertise) in order to support a variety of research and routine activities, flexible capabilities to contextualise data are required. Collaborative groups should be able to create on-demand views and perspectives, annotate and change data in their contexts without interfering with other contexts.
- $R(4)$ **Flexible integration and composition of services.** A multitude of data processing and data analysis tools exists in the biomedical context. Some tools act as complementary parts in a chain of processing steps. For example, to detect genes correlated with a disease, gene expression profiles are created by measuring and quantifying gene activities. The resulting gene expression ratios are normalised and candidate genes are preselected. Finally, significance analysis is applied to identify relevant genes [16]. Each function may be provided by a separate tool; for example by Genespring®and Genesis® [7,15]. In some cases, tools provide equal functionality and may be chosen as alternatives. Through flexible integration of tools as services with standardised input and output interfaces a dynamic composition of tools may be accomplished. From the systems perspective services are technology neutral, loosely coupled and support location transparency [11]. That is, the execution of services is not limited to proprietary operation systems and any service caller does not know the internal structure of a service. Further, services may be physically distributed over departments and institutes, e.g. image scanning and processing is executed in an own laboratory where the gene expression slides reside.
- $R(5)$ **Support of cooperative functions.** In order to support collaborative work suitable mechanisms have to be supplied. One of the main aspects is the common data annotation. Thus, data is augmented and shared within a group and new content is created cooperatively. Therefore, Web 2.0 technologies like wikis and blogs procure a flexible framework for facilitating intra- and inter-group activities.
- $R(6)$ **Data-coupled communication mechanisms.** Cooperative working is tightly coupled with excessive information exchange. Appropriate communication mechanisms are useful to coordinate project activities, organise meetings and enable topic-related discussions. On the one hand a seamless integration of email exchange, instant messaging and voip tools facilitates communication activities. We propose to reuse the organisational data defined in $R(1)$ within the communication tools. On the other hand, persons should be able to include data objects in their communication acts. E.g., images of diseased tissues may be diagnosed cooperatively, whereas marking and annotating of image sections supports the decision making process.
- $R(7)$ **Knowledge creation and knowledge processing.** Cooperative medical activities frequently comprise the creation of new knowledge. Data sources are linked with each other, similarities and differences are detected,

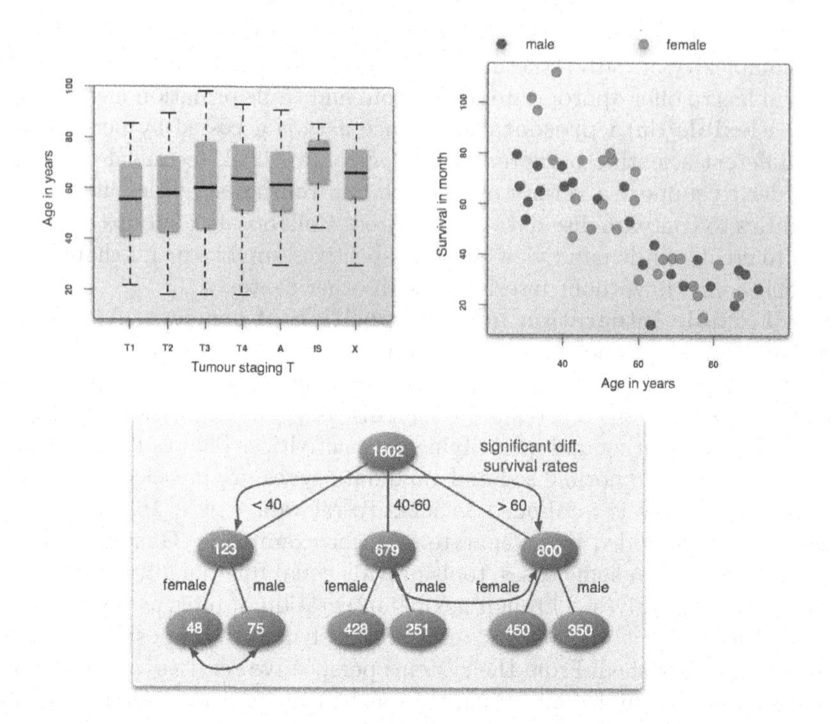

Fig. 1. TUMour Staging, Age, Survival Plots

and involved factors are identified. Consider a set of genes that is assumed to be strongly correlated with the genesis of a specific cancer subtype. If the hypothesis is verified the information may be reused in subsequent research. Thus, methods to formalise knowledge, share it in arbitrary contexts and deduce new knowledge are required.

3 Practical Examples and Virtual Workbench

In order to illustrate different types of data and knowledge we have to cope with in the GATiB project we give some practical examples. A key aspect of the biomedical collaboration is analysing the data collected about tumours in multiple ways. In Figure 1 three examples of graphical presentation of patient records are given. In the top left part of the figure the age distribution of patient groups is visualised by boxplots. Patients are grouped by tumour staging T (TNM classification). The top right part illustrates survival periods and age of female and male patients by a scatterplot. These visualisations may be a helpful presentation for scientists to formulate new hypotheses. The stratification tree at the bottom gives a detailed characterisation of the whole data set. This presentation is useful to survey frequencies of attribute value combinations. The original stratification tree was augmented with results of statistical tests. Hence, significant differences among subgroups of the data set are displayed by red arrows. In our example there is a significant difference of survival periods between male and female patients under 40. The dependency is obvious when looking at the scatterplot of Figure 1. Further, the survival period of patients older than 60 is significantly different from both younger age groups. Various types of data and knowledge presentation may serve different purposes. A tabular presentation of knowledge may be useful for final documentation, either to be exported into publication drafts or into presentations, while the visualisation by stratification trees may support generation of hypotheses in ongoing research. We propose to introduce formal representations of detected relationships to capture knowledge structures that may be reused in similar contexts and constitute the basis for reasoning in medical decision support systems. The details of knowledge representation are topics of ongoing research and beyond the scope of the paper. In Figure 2 a prototypical view of the scientific workbench is given. A group of persons cooperate in a virtual environment in the context of an annotation project. All project members currently working in the *context* of the project are marked as online. Online persons may communicate immediately by instant messaging. Additionally, further contact information (phone number or mail address) may be retrieved by clicking on the second icon. A set of records of liver cancer patients represents the shared data objects. Individual objects may be accessed, altered and marked as 'completely annotated'. A set of relevant functions is displayed to visualise and analyse data. Further, related publications are presented and discussion forums (data and knowledge topics) are available.

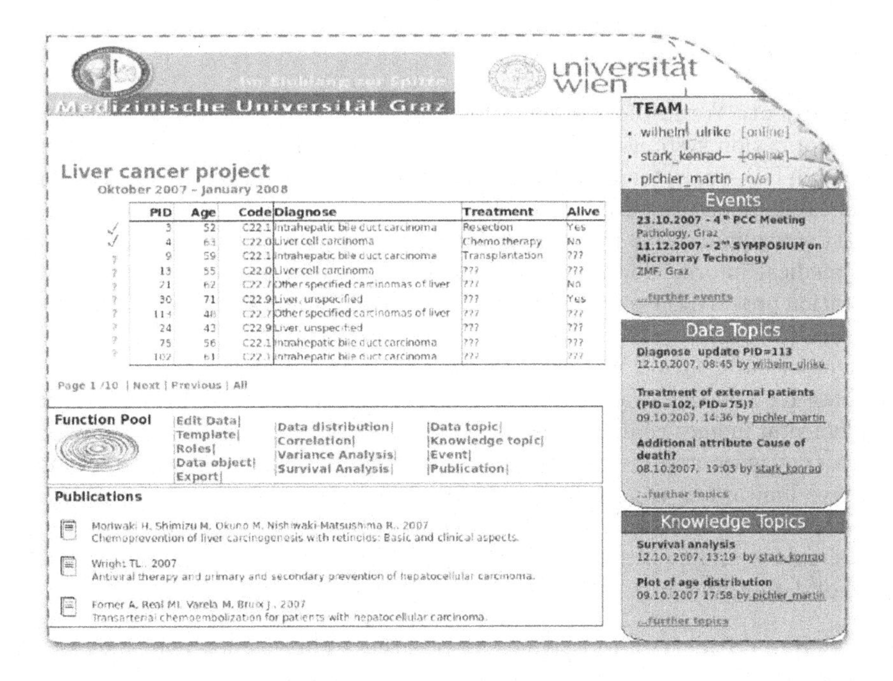

Fig. 2. Scientific Workbench

4 Conceptual Model

The example applications show that data used for collaboration has to be differentiated. Furthermore, the environment for a collaboration has to be specified and the users have to be represented in the CSCW system in a suitable way. We elaborated a conceptual model in which we specify how the requirements of the previous section could be met.

4.1 Resource

The example applications described in Section 2 highlight that the same data can be analysed and visualised in various ways. Our GATiB-CSCW system has to distinguish between the basic data on which the analyses are performed and the results of analyses. A resource is either a data or knowledge object that is accessed in a cooperative process. While data objects consist of restructured or transformed data, knowledge objects are built by deducing relationships and conclusions from data objects. Thus, a knowledge object always has at least one associated data object. Data and knowledge objects have some common characteristics: the content of both data and knowledge objects may be personal. Access to data objects can be restricted due to legislative regulations to guarantee data privacy. Further, the content of knowledge objects is to be protected in cases of ongoing research work. Moreover, both data and knowledge objects have to be protected from unauthorised modifications. The CSCW system needs a user management handling authentication and authorisation before accessing data. To protect the content of knowledge objects in ongoing research, a version management is required.

Data Object. We consider a data object as an information that was extracted from information systems or documents or was entered manually. In the context of the GATiB project emphasis is put on biomedical data. Biomedical data consists of data records from medical hospitals and research facilities that are of interest for collaborative research. It includes patient-specific records extracted from medical information systems or data manually entered as a result of an annotation process. Data may originate from clinical studies, anamnesis, lifestyle, and survival data as well as gene expression profiles. We do not put limits on the type of data nor on the size. Hence, a data object may consist of a set of database records that was returned as a result of a specific query linked with supplemental images as illustrated in Figure 3. Further, a data object may be a text document containing fragments of a publication. The CSCW system must not have limitations concerning the data stored and accessed. Since arbitrary data types have to be supported, applications used to visualise or modify the data should be made available within the CSCW system. We propose flexible levels of data granularity to satisfy requirement $R(2)$.

Knowledge Object. A knowledge object is the result of a non-empty sequence of functions applied to at least one data object. Generally, each knowledge object enhances the information that is currently available in data objects. We

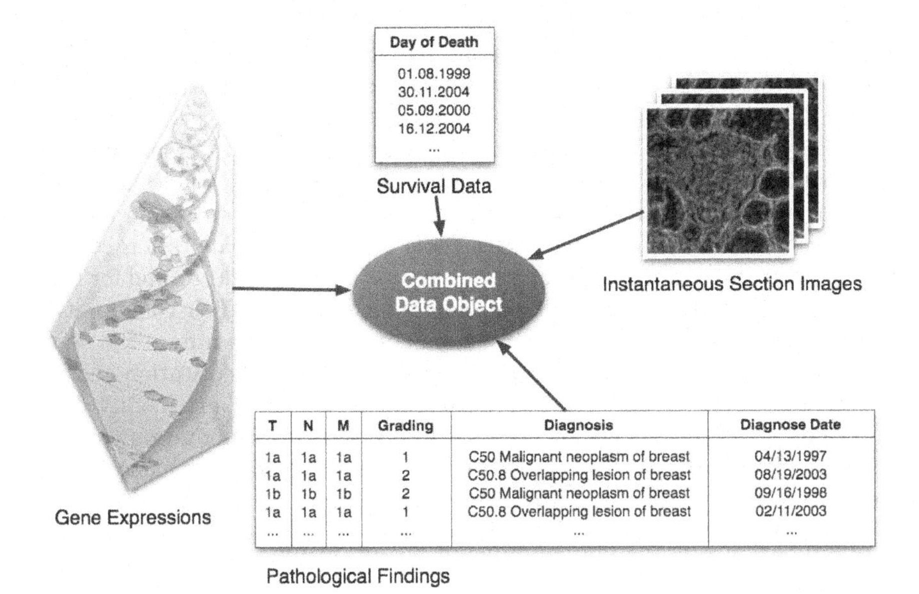

Fig. 3. Combined Data Object

distinguish various types of knowledge objects: graphical objects (plots, diagrams, etc.), tabular objects presenting summarised information (e.g. the results of aggregation operations) and knowledge structures (e.g. annotations). Semantic structures may be used to formalise knowledge. Relationships between objects or groups of objects are stored in a processible way in order to build a knowledge repository. The repository integrates conclusions from various contexts allowing the user to explore consolidated knowledge and deduce new knowledge. Making knowledge persistent and processable allows us to fulfil requirement $R(7)$. The post-processing of accessible data objects is a specific but also important criterion for a medical CSCW system research. Since there should be no limitations concerning the data stored and accessed, the CSCW system has to support all required post-processing steps and must be flexible to integrate new functions.

4.2 Collaboration Context

A collaboration context specifies the type of collaboration. It provides the general framework of relationships between individuals, data and knowledge. It may be considered as a basic template defining how data and knowledge is accessed and modified by individuals in a certain type of collaboration. In our collaborative medical research system we use three basic types of contexts: **patient-centred**, **project-centred** and **disease-centred** context. These patterns are applicable to the scenarios described in Section 2 which correspond to common use cases of medical routine work as well as medical research activities.

Generally, the patient-centred and project-centred contexts access more sensitive data. A patient-centred context starts with the patients data and will be used for medics working with the patient directly. Hence, involved individuals have to have access rights to original patient-related data. Whereas in the disease-related context access is given to anonymised and/or summarised data of the same disease type. As a consequence, instead of focusing on specific patients, a particular disease type, its related therapies and medications are of main interest for the collaboration act. The project-centred context provides a specific environment to coordinate collaborative actions (e.g. data sharing and annotation, discussion forums, communication tools etc.). The collaboration context conforms to an environment which present users a common background for their work. It presents data needed as background information for their collaboration since all data available in the context can be assumed as known from the users. It is obvious that a collaboration context must consider authentication and authorisation in order to limit the user's capabilities to access and modify data. Further important is the view on the data available in a collaboration context. Section 4.5 details the need for different views on the same data. The CSCW system has to be able to present these three context types. Though, to develop a flexible system, it has to support the definition and usage of arbitrary contexts.

4.3 Knowledge Spaces

Knowledge spaces are our representation and structure of presenting the different context types of collaboration situations to the users/actors in the collaboration process. Users need a knowledge space as a virtual environment for their collaboration activities in which they can meet and work. We define a knowledge space as an *actual use case* in a *collaboration context.* In scenario 3, a knowledge space is set up for those persons participating from the hospital and the pharmaceutical company, which are therefore a member of this specific collaboration group. Although knowledge spaces are separate virtual concepts, data and knowledge exchange between knowledge spaces is encouraged. An important capability is to upload data or link to remote data which is already available. To suffice the individual users' needs for their collaboration, they should organise and structure the knowledge space completely on their own. Consequently, users are responsible for the organisation of their knowledge space which implies highest flexibility. However, reorganisation of one specific knowledge space might be restricted to a limited group of users in order to avoid unauthorised modifications. Such self-organisation conforms to the self-organisation forms of knowledge in the Web 2.0 (tagging). Further concepts of the Web 2.0 like annotating available information are important for a useful CSCW system. Knowledge spaces are extendable in order to invite other users of the CSCW to join the collaboration.

Communication is the most important criterion to perform a successful collaboration: In particular, supporting discussions is crucial for the usability of a CSCW system since it is an important tool in the context of collaboration between users located at different places or contributing at different times. To offer a useable platform for collaborative work, different communication ways have to be supported. A classical and popular form of synchronous communication is to use instant messaging. Initially, the chat protocol is only privately accessible for the participants, however, they can make it accessible for all users of the knowledge space. In the context of a platform for medical research, discussion forums for knowledge and data topics are helpful. Furthermore, Web 2.0 concepts like wikis and blogs should be supported by the CSCW system. Wikis can be generated for a dedicated collaboration context or be accessible for a subset or all individuals of the CSCW system. To add comments to knowledge objects or data objects, the CSCW system offers annotation capabilities. Thereby, data and knowledge objects can be annotated in the same way and the CSCW system is responsible for granting access to annotations according to the users' access rights. Hence, we may fulfil requirements $R(5)$ and $R(6)$. Realising user-awareness in the CSCW system is crucial to support communication. User-awareness implies that users are represented in the virtual world by avatars, photos, or an icon. This enables other users to see which individuals are registered, are author of a wiki entry, annotation etc., are currently logged in, and can be contacted.

4.4 Individual

An individual is a person participating in collaborative acts. Individuals may be for instance researchers from the medical or biological domain, medical students or project managers. Individuals are categorised into internal and external persons in order to differentiate between access to sensitive patient-related and research-related data and access to anonymised and summarised data. Internal individuals are for example employees of the Medical University Graz and external individuals may be members of companies or research institutes cooperating with the university. The CSCW system has to ensure that individuals are distinguished in order to reflect the activities of one individual in the real world as a one-to-one mapping in the virtual world. This is important in the context of collaboration (discussions, annotations, ...) as well as for the data access (authentication and authorisation). The organisational mapping accomplishes the user and role management specified in $R(1)$.

4.5 Roles

Individuals having the same access rights and using the same set of functions may be integrated to groups and certain roles may be assigned to users/groups as illustrated in Figure 4. Roles are used to adjust the access of individuals and groups to resources. That is, a role is used to bundle access parameters that are classified into three different categories: **view parameters**, **modification parameters** and **extension parameters**. Flexible presentation of data and data annotation capabilities satisfy requirement $R(3)$. Individuals can be part of several groups and a group can consist of subgroups and roles may be assigned to both individuals and groups. Further, roles are coupled with knowledge spaces allowing to access a resource in various contexts.

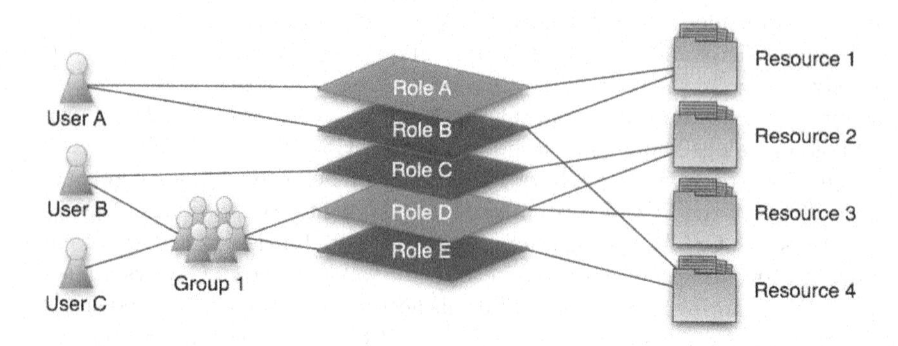

Fig. 4. Role Assignment

4.6 View Parameters

Different knowledge spaces have a different focus on the same data. This implies that the data is visualised differently according to the selected knowledge space. View parameters are utilised to specify how the data is presented. Although the same data is shared among a group of individuals, sometimes part of the data is to be hidden or presented in an alternative way. For instance, consider sensitive information like patient name and day of birth. In a clinical study the identifying attributes of a patient record have to be presented to assign further attributes like radiology parameters correctly. In order to guarantee data privacy, these attributes should not be presented to supporting staff like medical students. Hence, the identifying attributes are suppressed for supporting staff and displayed for scientific staff by setting the view properties in the corresponding roles appropriately. The CSCW system has to support a fine-grained definition of view parameters. Supporting the self-organisation of the knowledge spaces, view parameters must be able to be defined by the users themselves. Here, the definition might be available for a specific group of users that are responsible for the organisation of the knowledge space.

4.7 Modification Parameters

The content of a resource may be changed during a review, discussion, modification or annotation process. Parts of the content may be limited by read-only access while other parts may be edited to correct or to supplement data. Thus, we allow to assign read/write properties at the finest granularity level offered by a resource. In case of a data table structured by attributes, modification properties may be set for each attribute separately as well as for each table entry. Consider a clinical data annotation project where a group of medical scientists creates a data table collaboratively. In order to protect table entries that are assumed to be complete from inadvertent modifications, those entries may be set to read-only. In addition, some data that was accumulated from a reliable source (e.g. date of surgery from a clinical information system) is set to read-only to preserve data integrity. The CSCW system has to offer users the functionality to set modification parameters. Sensible standard parameters are essential for the usability. Internally, the system can realise the modification parameters by disabling all users' authorisation to write the data. Consequently, the CSCW system needs to guarantee an exact authorisation method to prevent data from unauthenticated read-access and unauthorised modifications.

4.8 Extension Parameters

Although well-defined data structures ease and standardise the collaborative data entry process, they lack flexibility and extensibility. As collaborative work is an evolutionary process, data may be restructured in order to fit to upcoming demands. We allow two types of data structure modifications: property addition at the conceptual level and property addition at the instance level. A shared data resource may be extended by defining a property template including a set of (property name, property type) pairs. The original view on the resource is augmented with these properties and supplementary information may be entered for each data table entry. Moreover, additional information may be assigned to a single table entry. Consider the follow-up data of one patient when recorded in a different hospital. We want to mark the patient with a 'Follow-up data outside hospital' information. Thus, we require object extension at the instance level.

5 Architecture Overview

The Wasabi framework enabling collaborative work as described in the previous sections is a service-oriented architecture [13]. It bases on the JBoss Application Server (AS) in order to fulfil requirements for enterprise solutions and to provide the scalability and performance needed for a CSCW system which is used in a distributed manner like in the GATiB project. Service orientation is an important characteristic in the context of flexibility, adaptability and maintainability, as already mentioned in requirement $R(4)$. Note that the service orientation of Wasabi can be realised since the underlying JBoss AS is also service oriented. Service orientation is also an essential characteristic for a CSCW system since the data stored might be of arbitrary formats and located in arbitrary repositories. Supporting storing and handling content in a flexible way opens CSCW systems to a wide field of collaboration activities and is no longer limited to specific collaborative environments or applications. The Wasabi's functionality enables on the one hand data to be stored in databases or in file systems reachable via the network. On the other hand it can link already existing databases and data sources to provide users access to those data under consideration of their rights. This is particularly beneficial for the GATiB project since the data collected in different hospitals and institutes is thereby available for the whole community without transferring the data for collaboration to a specific repository. As depicted in Figure 5 the server consists of four main components. The core of the Wasabi enterprise server architecture implements the framework for a CSCW system and aggregates all services of Wasabi (marked with (4) in Figure 5).

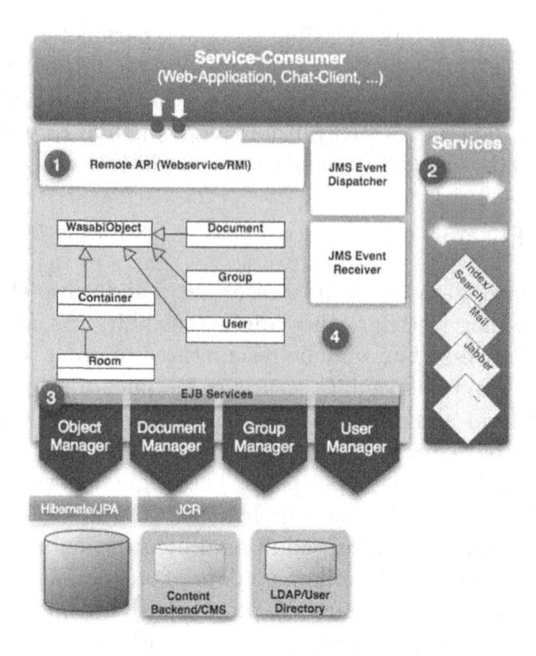

Fig. 5. Main Architecture

The event dispatcher and receiver are responsible for handling internal events, including the arrival of messages, by calling appropriate methods of the internal EJB services (marked with (3)). These services are responsible for handling the objects of the Wasabi core, e.g. the **UserManager** is the service responsible for handling, modifying, and extracting information of **user** objects. These services classify the EJB services as basic services since they implement basic functionalities on the objects of the Wasabi core. Since their tasks focus on the modification and provisioning of data stored in the Wasabi core objects, they can be further classified as data-centric basic services. EJB services are also used to realise flexible user authentication mechanisms as well as to be adaptable to various content backends/repositories.

Third, the Remote API provides an interface for client-server communication (marked with (1)). Therewith a common interface can be used to send requests to different adapted services, in the case the services expect the same input data. This simplifies the enhancement of Wasabi Beans by adapting new services with little effort. The fourth component is responsible for the message exchange with adapted services (marked with (2)). It generates the outgoing messages according to the defined interfaces of the remote webservices and processes and extracts information from incoming messages. After receiving an incoming message, appropriate methods of the Wasabi services are called. Internally individuals, groups, and rooms are objects persistently stored in the central database. Since also the related classes are subclasses of `WasabiObject`, it is necessary to assign to each object an unique identifier (UUID). Storing all `WasabiObjects` in a central database, supports to generate a unique identifier for a `WasabiObject`, i.e. no room or individual can have the same UUID. To provide best possible adaptability, entities of the Wasabi architecture provide a remote API. Main classes are entities like `WasabiObject`, `Document`, `Group`, `User`, `Container`, and `Room`, since these classes realise the data organisation and implement our concept of virtual rooms. To make objects persistent, it is necessary to implement the EJB services (`DocumentManagerService`, `ObjectManagerService`, and `UserManagerService`) as data-centred basic services. The EJB services are data-centred, since they are responsible for data storage and data access. We support different persistency layers and encapsulate this functionality through the EJB services. This encapsulation of functionalities into basic services allows modification on the data access and data storage functionalities without entailing any changes on the server core.

6 Conclusion

In this paper, we presented the main requirements for a CSCW system supporting collaboration in medical environments and outlined a conceptual model and a system architecture fulfilling the specific demands. We found the Wasabi framework capable of supporting the cooperation processes in our project, as it allows the creation and management of knowledge spaces based on a flexible object model and a service-oriented architecture. Though, there is still adjusting work left. We have to integrate data resources into persistency layers, map organisational data, and wrap biomedical applications into services appropriately. We implemented a workbench to access and anonymise distributed data sources. We are working on the extension of the workbench in order to implement the project-centred CSCW client as presented in Figure 2. In our future work we will focus more intensely on service composition in the biomedical context and on formalisation of medical knowledge. That is, the functionality of our GATiB-CSCW will be considerably enhanced and it may be utilised as a virtual platform for collaborative research.

References

1. Armendolia, S.R., Estrella, F., McClathey, R., et al.: Managing pan-european mammography images and data using a service oriented architecture. In: IDEAS-Workshop on Medical Information Systems (2004)
2. Assel, M., Krammer, B., Loehden, A.: Management and access of biomedical data in a grid environment. In: Cracow Grid Workshop (2006)
3. Asslaber, M., Abuja, P., Stark, K., Eder, J.: The genome austria tissue bank (gatib). Pathobiology 74, 251–258 (2007)
4. Bouillon, Y., Wendling, F., Bartolomei, F.: Computer-supported collaborative work (cscw) in biomedical signal visualization and processing. Trans. on Information Technology in Biomedicine 3 (1999)
5. Chu, X., Lonie, A., Harris, P., Thomas, S., Buyya, R.: A service-oriented grid environment for integration of distributed kidney models and resources. In: Concurrency and Computation: Practice and Experience (CCPE) (2007)

6. Arbona, A., et al.: A service-oriented grid infrastructure for biomedical data and compute services. IEEE Trans. on Nanobioscience 6 (2007)
7. GeneSpring. Cutting-edge tools for expression analysis, `www.silicongenetics.com`
8. Li, W., Krishnan, S., Mueller, K., Ichikawa, K., et al.: Building cyberinfrastructure for bioinformatics using service oriented architecture. In: Sixth IEEE Int. Symposium on Cluster Computing and the Grid Workshops (2006)
9. Makris, L., Kamilatos, I., Kopsacheilis, E.V., Strintzis, M.G.: Teleworks: A cscw application for remote medical diagnosis support and teleconsultation. Trans. on Information Technology in Biomedicine 2 (1998)
10. Sweeney, L., Samarati, P.: Protecting privacy when disclosing information: k-anonymity and its enforcement through generalization and suppression. In: Proc. of the IEEE Symposium on Research in Security and Privacy (1998)
11. Papazoglou, M.P.: Service -oriented computing: Concepts, characteristics and directions. In: wise, vol. 00, p. 3. IEEE Computer Society Press, Los Alamitos (2003)
12. Sartipi, K., Yarmand, M., Down, D.: Mined-knowledge and decision support services in electronic health. In: Int. Workshop on Systems Development in SOA Environments (2007)
13. Schulte, J., Hampel, T., Bopp, T., Hinn, R.: Wasabi framework an open service infrastructure for collaborative work. In: The third Int. Conference on Semantics, Knowledge and Grid (2007)
14. Stark, K., Eder, J., Zatloukal, K.: Priority-Based k-Anonymity Accomplished by Weighted Generalisation Structures. In: Tjoa, A.M., Trujillo, J. (eds.) DaWaK 2006. LNCS, vol. 4081, pp. 394–404. Springer, Heidelberg (2006)
15. Sturn, A., Quackenbush, J., Trajanoski, Z.: Genesis: Cluster analysis of microarray data. Bioinformatics 18(1), 207–208 (2002)
16. Tusher, V.G., Tibshirani, R., Chu, G.: Significance analysis of microarrays applied to the ionizing radiation response. In: Proc. Natl. Acad. Sci., vol. 98 (2001)

Strategic Alignment in the Context of e-Services – An Empirical Investigation of the INSTAL Approach Using the Italian eGovernment Initiative Case Study

Gianluigi Viscusi[1], Laure-Hélène Thevenet[2,3], and Camille Salinesi[2]

[1] Department of Informatics, Systems and Communication (DISCo)
Università degli Studi di Milano-Bicocca - Italy
[2] Centre de Recherche en Informatique, Université Paris 1 Panthéon-Sorbonne
90 rue de Tolbiac 75013 Paris
[3] BNP Paribas, Banque de Détail France – Informatique
41 rue de Valmy 93100 Montreuil Sous Bois
viscusi@disco.unimib.it, laure-helene.thevenet@malix.univ-paris1.fr, camille.salinesi@univ-paris1.fr

Abstract. Strategic alignment is an issue that is not just met in companies, but also in governments, governmental agencies and public administrations. This paper investigates the issues raised by strategic alignment in the context of eGovernment initiatives focused on e-services. The analysis reports a case study that explores the strategic alignment issue with the INSTAL method. Two facets are tackled: (a) the modeling of strategic alignment using a formal approach, and (b) the elicitation of evolution requirements based on the analysis of strategic alignment models.

Keywords: alignment model, strategic alignment, e-government, e-services.

1 Introduction

Strategic alignment (SA) of information systems (IS) is a primary concern arising from the eGovernment objectives in the context of the Italian health service. This paper investigates the INSTAL approach, which aims to model SA, analyse it, and elicit evolution requirements. The research question is how SA can be tackled in the eGovernment context and how the INSTAL approach can support the planning of e Government initiatives focused on design and development of e-services. Quality of services for the citizens is a major issue in the eGovernment context. Furthermore, quality is rarely considered in the SA perspective. Taking this into account, we explored the following hypotheses:

<h1: The integration of strategic alignment and quality issues provides insights for the choice of suitable eGovernment initiatives>;

<h2: Modeling strategic alignment helps eliciting evolution requirements for eGovernment initiatives>

The paper is organized as follows. Section 2 introduces the case study. Section 3 discusses the case study report. Section 4 concludes and gives perspectives.

Z. Bellahsène and M. Léonard (Eds.): CAiSE 2008, LNCS 5074, pp. 163–166, 2008.

2 Case Study Presentation

In the initial situation, the Italian public administration (PA) is composed of central (e.g. ministries) and local agencies (about 8000). The organization is agency-centric processes, with little inter-agency, inter-organizational, and cross sector relationships. In particular, in the context of public health, information is rarely shared between different agencies, front offices and back offices. Citizens must make contact with local agencies to ask for agency-specific services. We focus in the rest of this paper on the health services and in particular on the process of changing family doctors.

In the initial situation, health care services faced difficulty in fulfilling users' demand, due to the bureaucratic procedures and PAs organization. In Italy, any citizen must have a family doctor. To choose one, citizens have to (i) make the request to the health PA of residency (at opening time), (ii) exhibit their health card, and (iii) proceed to the choice of a doctor (in the area). Several certificates are asked. For citizens who do not live in the area of the health center, a specific committee has to decide. For foreign citizens with no residency, the choice of doctor can be valid from three months to a year and renewable; in the case of residents the choice is valid for a year and automatically renewed. In 2002, a plan was initiated to change the initial PA organization for service provision, by having a user oriented perspective and developing online services.

3 Case Study Report

This section reports how the INSTAL (Intentional Strategic Alignment) method [1] was applied on the case study of the Italian eGovernment.

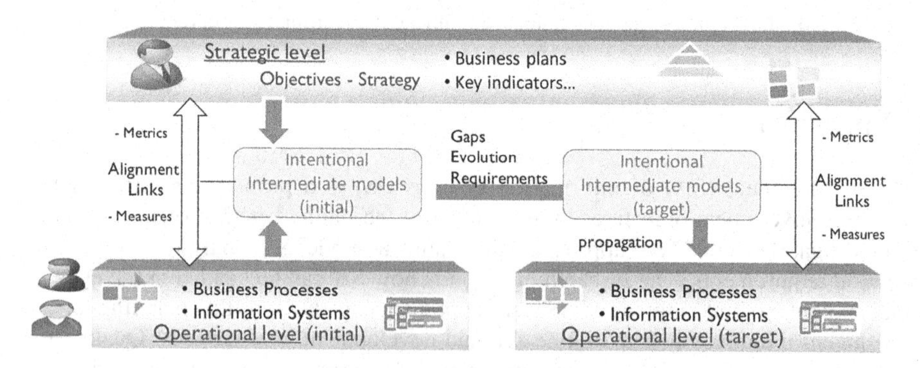

Fig. 1. INSTAL overview

As shown in Fig. 1, the main characteristics of the INSTAL method are: (1) to model strategic alignment intentions at an intermediate level (SA maps), and (2) to define strategic alignment links between strategic and operational elements. In INSTAL, only elements that share a same intention can be aligned and thus a SA link modeled. Metrics (at strategic level) and measures (at operational level) can also be attached to SA links to provide a quantitative assessment of SA. The map formalism was used at the intermediate level because (1) it is intentional and (2) it allows tackling the variability within and between the two levels.

As shown in [1], to define SA maps, we searched for public administration *issues* (e.g. increasing reliability, decreasing cost; attaining the most users, avoiding redundant activities), *qualities* (that should be increase, e.g.: availability, accessibility, ease of use, transparency) and *resources* (that should be controlled, e.g. dematerialized data, time, complexity control, competencies) in strategic and operational elements. The SA map presented in Fig. 5 was first specified, it represents the Italian eGovernment strategy under the perspective of its implementation at the operational level. A SA map is drawn according to the MAP formalism [3]. Two main goals are presented in this map: (b) *Increase the access quality of service* and (c) *Maintain data.* These goals are ambivalent. Indeed, they can represent the organization's strategy but also tackle the operational level. The goal (b) *Increase the access quality* to service is an important goal that justifies the strategy undertaken. Access quality covers availability, accessibility, ease of use etc. The other important goal is (c) *Maintain data,* since every service is based on data. Data is the main resource specifically provided by eGovernment to citizens.

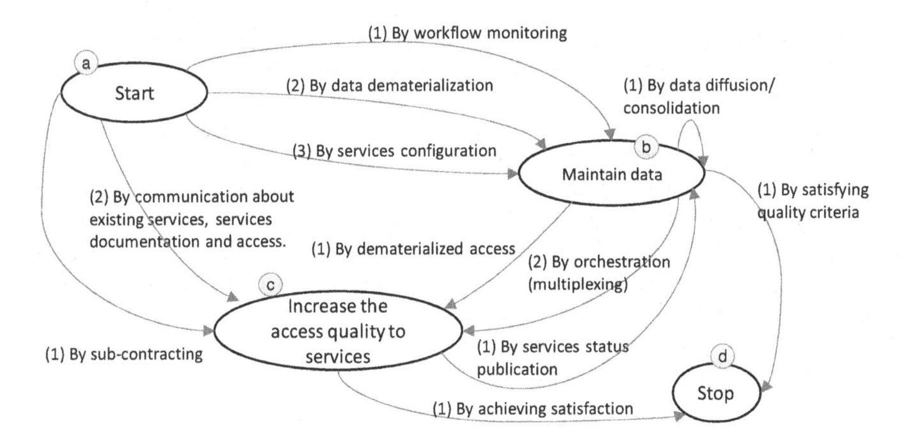

Fig. 2. SA map "Improve accessibility to Italian Public Administration Services"

For the sake of space, we do not present the eleven SA links attached to the sections. Each one is named by the section it relates to (i.e. the target goal and strategy behind the link), and is defined between the eGovernment strategy and the operational elements in order to detect the cases of good alignment and misalignment. For example, elements of the initial business process for changing doctors are involved in some SA links with a *contradictory* role. In fact, the SA link relating to the section ab2 (maintain data by data dematerialization) involves the business process with a *contradictory* role since any exchange of data is paper-based. The analysis of elements roles as well as the new SA requirements makes emerge evolution requirements. Moreover, we used three quality level criteria (defined in [2]): the temporal, economic, and procedural efficiency, to uncover metrics completing the initial SA link models. To each metric is associated a current and a target value that characterize how the SA should be reached in the future. For example, the service time available is currently 30 hours a week, whereas the required value is 72 hours a week. Based on SA models and SA links analysis (with attached metrics), evolution requirements can be found, such as:

- *A health-website at a regional level* should provide most important health services to citizens and businesses (no more restricted to the services available in local agencies), even if requests can be made on different channels such as Internet and local agencies.
- *A multichannel strategy* oriented to users should be adopted (e-mail center, call center, authoring system). For the change of doctor request, citizens can make requests on the Internet, download certificates if needed, and follow the status of their request (by status notification, by email) etc.
- *Information sharing* between all health organizations. For the change of doctor request, citizens do not have to provide documents. Health system (in local agencies or by internet) can search information (automatically and transparently to user) by IT services, etc.

4 Conclusion

This paper explored the issue of developing services by analyzing their alignment with strategic Italian eGovernment objectives. This case study has shown: (i) SA models enriched with quality issues provide a purposeful view on alignment (confirming *h1*), (ii) SA models can be used as a central point to draw and organize complex SA links between organization elements (confirming *h1*), (ii) roles of SA links can be used to analyze initial SA and identify evolution requirements (confirming *h2*). However, a more complete evaluation should be carried out, one that more specifically considers other eGovernment initiatives (in the same and other countries), and also focuses on the interactions with businesses in the private sector. Future work is to consider: (i) studying the difference between public and private sector and if needed adapt the INSTAL method; (ii) integrating a social facet in the SA model to understand the social impact of the new services, and to facilitate their adoptions.

Acknowledgements

The research work has been partially supported by the Italian FIRB project NeP4B.

References

1. Thevenet, L.-H., Salinesi, C.: Aligning IS to Organization's Strategy: The INSTAL Method. In: Krogstie, J., Opdahl, A., Sindre, G. (eds.) CAiSE 2007 and WES 2007. LNCS, vol. 4495, pp. 203–217. Springer, Heidelberg (2007)
2. Viscusi, G., Batini, C., Cherubini, D., Maurino, A.: A Quality Driven Methodology for eGovernment Project Planning. RCIS:97-106 (2007)
3. Rolland, C.: Capturing System Intentionality with Maps. In: Krogstie, J., Opdahl, A.L., Brinkkemper, S. (eds.) Conceptual Modelling in Information Systems Engineering. Springer, Heidelberg (2007)

Understanding and Improving Collective Attention Economy for Expertise Sharing

Yunwen Ye[1,3], Kumiyo Nakakoji[1,2], and Yasuhiro Yamamoto[2]

[1] SRA Key Technology Laboratory, Inc., 3-12 Yotsuya, Shinjuku, Tokyo 160-0004, Japan
[2] RCAST, University of Tokyo, 4-6-1 Komaba, Meguro, Tokyo, 153-8904, Japan
[3] L3D Center, University of Colorado, Boulder, CO80309-0430, USA
yunwen@colorado.edu, kumiyo@kid.rcast.u-tokyo.ac.jp, yxy@kid.rcast.u-tokyo.ac.jp

Abstract. The importance and benefits of expertise sharing for organizations in knowledge economy are well recognized. However, the potential cost of expertise sharing is less well understood. This paper proposes a conceptual framework called collective attention economy to identify the costs associated with expertise sharing and provide the basis for analyzing and understanding the cost-benefit structure of different communication mechanisms. To demonstrate the analytical power of the conceptual framework, the paper describes a new communication mechanism—Dynamic Mailing List (DML)—that is developed by adjusting certain cost factors.

Keywords: collective attention economy, expertise sharing, socially aware.

1 Introduction

Despite the advance of information systems that makes it easier to store and access knowledge, knowledge workers who are faced with complex knowledge-intensive work still routinely rely on their peers for knowledge and expertise [1]. Knowledge held by its members constitutes one of the most important assets of an organization. To fully utilize such valuable assets, one of the key challenges in the design of information infrastructure for organizations is to facilitate the easy transfer, sharing and integration of knowledge held by members with computer-mediated communication mechanisms that connect an expertise seeker, who is looking for specific knowledge to solve his or her own problem, with an expertise provider, who holds the sought after knowledge. Much research on expertise sharing has been conducted on providing support for expertise seekers, through the design of different mechanisms of finding who have the needed knowledge to become expertise providers and who are available to help [3].

Despite its great importance and benefits, expertise sharing, however, comes with a cost. When an expertise seeker posts a question, a *seeking cost* occurs because the expertise seeker has to spend some time to formulate the question and to decide whom to ask. When an expertise provider offers an answer, an *answering cost* occurs because the expertise provider has to pay attention to read the question and formulate an answer. The real cost, however, could be far greater than the sum of the *seeking*

Z. Bellahsène and M. Léonard (Eds.): CAiSE 2008, LNCS 5074, pp. 167–181, 2008.

cost and the *answering cost*, depending on the communication mechanisms used. For example, if a community-wide communication mechanism such as mailing list is used to post the question, all members of the community who pay any form of attention—from receiving the question, scanning the subject, skimming the contents, to answering the request—have paid a cost of attention, which is a scarce resource. Collectively, the cost of expertise sharing could be very high, and even outweighs its benefits and lower the group productivity [17].

To understand the cost of expertise sharing better, we introduce the notion of *collective attention* to denote all the attention that is consumed by all parties involved in a transaction of expertise sharing; and present a conceptual framework called the *economy of collective attention* to analyze the cost-benefit structure of communication mechanisms used for expertise sharing. The framework guides us to articulate design requirements for information systems in support of *situated expertise sharing* to improve the utilization of collective attention. We use the term *situated expertise sharing* to refer to a particular kind of expertise sharing in which a knowledge worker asks peers questions for the purpose of obtaining the specific knowledge that is needed in his or her own work. A new communication mechanism called Dynamic Mailing List (DML) is introduced and its implementation in a system is briefly described.

2 Problem Context: Situated Expertise Sharing

Expertise sharing takes place in many different situations and for different purposes. Despite many shared aspects, the practice of sharing expertise faces different challenges when the types of knowledge and the external constraints vary. Our research focuses on expertise sharing situations in which a group of knowledge workers engage, under strong pressures of productivity and quality, in the collaborative construction of a common knowledge artifact that is decomposed into parts that have complicated inter-dependency. The common artifact is made possible through the integration of distributed work and knowledge of each knowledge worker. Each worker is responsible for constructing some parts by bringing their unique set of knowledge. Due to decomposition, each knowledge worker only has partial knowledge of the artifact and of the process; and due to the inter-dependency, each knowledge worker often needs to seek knowledge from peer workers of the same group to carry out his or her work efficiently and effectively.

This kind of *situated expertise sharing* is not for the general purpose of learning or creating awareness in which knowledge is not immediately coupled with the task at hand. Rather, it is a clearly purposed act that serves the goal of the accomplishment of an individual worker's current task, and it arises on an as-needed basis and requires quick resolution. Software development is one typical example. Programming requires undivided attention, and in general programmers prefer to work in solitary with long periods of uninterrupted time during which they can concentrate. It is this kind of solitary work that gets the code written. However, due to the interdependency of their work, programmers also have to engage in situated expertise sharing with peers to seek knowledge necessary to accomplish their individual task. A study has shown that this type of ad hoc and situated expertise sharing takes up to 41% of a programmer's time [19].

The above settings entail the following constraining factors that have to be balanced by information systems that support situated expertise sharing:

(1) *The collective attention has a limited capacity due to the fixed size of the project group.* This is different from volunteer-based community projects such as Wikipedia where the number of contributors can be increased through strategies of turning passive users into active contributors.
(2) *Situated expertise sharing is not a one-time affair; it has to be sustainable* because its continuous enactment is required throughout the lifecycle of the project. A member's engagement in one sharing act should not result in his or her reluctance to participate in further sharing acts down the road. No absolute experts exist. Depending on the context, a knowledge worker often assumes the role of seeker or provider of expertise at different times.
(3) *The costs and benefits of situated expertise sharing have to be considered together with the group productivity.* On the one hand, if a knowledge worker is unable to obtain the knowledge in the head of peers timely, he or she cannot carry out his or her task effectively, and thus lowers his or her own productivity, which in turn lowers the productivity of the project group. On the other hand, if a knowledge worker is frequently interrupted for offering help, the expert's productivity is significantly reduced, resulting in lower group productivity. It has been observed that such costs could even outweigh its benefits and lower the group productivity [17].

3 Collective Attention Economy

Attention is directed, involved awareness, and is the intellectual energy invested toward some purpose. Attention is an intrinsically scare resource because everyone has only a certain stock of supply. In this Internet era full of overwhelming information, many of us feel short of enough attention. Goldhaber and others have eloquently argued that we are entering a world where our lives are guided more by the laws of the economics of attention because attention is quickly becoming the scarcest resource in our society [9].

Attention economy is concerned with the use or the patterns of allocation of attention for the best possible benefits. Collective attention economy is concerned with the effective use of the sum of attentions of members in a group. To improve collective attention economy within a group is to improve the patterns of attention allocation of its members for the purpose of achieving better expertise sharing results and increase the collective attention capacity of the group.

3.1 The Cost of Collective Attention in Situated Expertise Sharing

In an act of situated expertise sharing, both the *asker* (expertise seeker) and the *recipients* (those who receive the question asked by the asker) consume attention. Those recipients who provide answers to the question are *helpers*. Those recipients who do not provide answers are *onlookers*. Helpers consume more attention than onlookers.

An asker needs to find out who has the expertise. Previous research has shown that such awareness of who knows what takes extensive time to develop, and its utilization consumes intensive attention [13]. We denote the attention cost as C_{Find}.

The question needs to be formulated and articulated, and we denote this attention cost as C_{Ask}. Rhetorical strategies, linguistic complexity and word choice of the question all influence the likelihood of receiving replies to a question [4, 5]. The asker also needs to make a decision based on social cues whether the potential experts could be interrupted, and to determine opportune times to interrupt [6].

When the recipients receive the question asked by an asker, all of them are interrupted and distracted from their current work. The cost of attention, denoted as $C_{Interrupt}$, includes not only the attention spent on attending to the interrupting event but the disruption of flow and the accompanied work resumption efforts [10, 23].

Recipients need to make a conscientious decision to respond to it or not. A number of factors affect this decision-making process: whether they have sufficient expertise or interest on the topic [24]; how they value their contributions by answering the question [16]; how many efforts does it take to post a reply [11]; how they perceive their relationship with the asker [5]. To make this decision, they at least need to skim the question by finding out who is the asker and what is the topic [10]. We denote this attention cost as C_{Skim}, and this cost applies to all recipients.

If a recipient decides to respond to the help request, he or she needs to spend time and attention in thinking and composing the response. The cost of attention for answering the question is denoted as C_{Answer}, and this cost is incurred only on helpers.

Upon receiving an answer, the asker needs to evaluate its quality and interpret its meaning in terms of his or her task. Not all responses are of equal value, and some of them may not be very helpful. We denote this cost of attention as $C_{Evaluate}$.

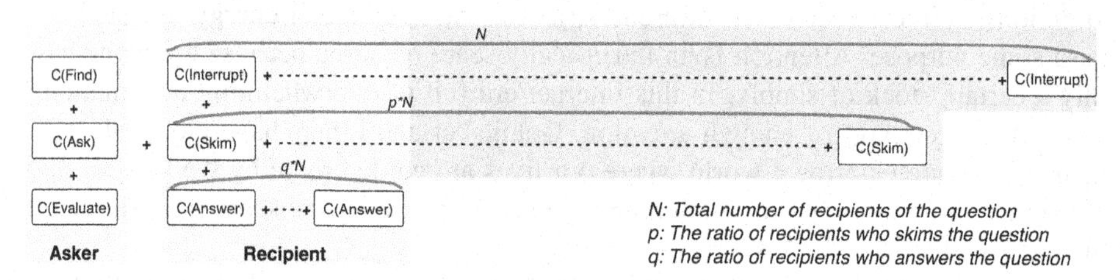

Fig. 1. Cost of Collective Attention (CoCA) in Situated Expertise Sharing

Fig. 1 shows the total *cost of collective attention* (CoCA for short) for an act of situated expertise sharing. Suppose the question posted by the asker is sent to N recipients, then all the recipients will have to shoulder a cost of $C_{Interrupt}$. Among the N recipients, some of them (we denote the ratio as p) will read the question, and some of them (we denote the ratio as q) will answer the question.

3.2 Approaches to Improving Collective Attention Economy

Two major variables affect the effectiveness of the total collective attention consumed for situated expertise sharing. The fist variable is the total number of recipients (N). If only the experts who can provide helpful answers receives the question, then the attention is well spent. The second variable is the success rate (denoted as r) of expertise-seeking attempts because only successful expertise sharing acts return benefits. The two variables are not independent: r tends to increase together with N because when N increases, the possibility of someone who is able and willing to answer increases. However, when N increases, the *CoCA* for each act of expertise sharing also increases.

Improving collective attention economy can be approached from two directions. First, one can try to raise the success rate r by increasing the ratio of posted questions that receive timely and high quality replies over the total number of questions that are asked. Second, one can try to reduce *CoCA* for each act of expertise sharing. The most significant waste of attention is that consumed by onlookers: $(1\text{-}q)*N*C_{Interrupt}+(p\text{-}q)*N*C_{Skim}$: the attention spent by those who are interrupted or read the question but are not interested in providing answers (the darker area in Fig. 1).

One might argue that these attentions are not wasted because by reading only albeit not actively participating, the onlookers also might learn something useful for the future. Moreover, such seemingly wasted communication often contributes to increasing shared context and awareness. However, shared awareness comes at the cost of collective attention cost. In general, knowledge workers "are 'not interested' in the enormous contingencies and infinitely faceted practices of colleagues unless they may impact our own work " and workers "routinely expect not to be exposed to the myriad detailed activities" of others [21].

Our goal is to pursue the right balance among the degrees of awareness, connectedness, and the needs for solitary concentration craved by knowledge workers. To reduce *CoCA*, our approach is to find a way to reduce N without reducing r. At the same time, we try to complement the plausible side effect of decreasing shared context and awareness through other means, such as the accumulation of discussion archives which can be used as a means to provide contexts and as learning materials by those not directly involved.

4 The Dynamic Mailing List Approach

This section describes the Dynamic Mailing List (DML) mechanism that we has designed to improve the collective attention economy for situated expertise sharing.

4.1 Basic Strategies

Our approach is based on the following strategies:

- Keep C_{Find} to be zero; namely, expertise seekers can ask questions without knowing or thinking of to whom it should be sent.
- Reduce the number of those who receive the question with minimal effects on r. We use heuristic strategies to route questions to the people who are most likely and willingly to answer the question, and reduce the wasted attention of onlookers.
- Raise the ratio r by reducing the number of unsuccessful expertise sharing acts through the improvement of the quality of each act of expertise sharing.

Due to the wide diversity and fine specialization of knowledge, not all members in a group have expertise or are interested in all topics discussed in all acts of situated expertise sharing. The attention ($C_{Interrupt}$ + C_{Skim}) consumed by those members who receive the question but have no expertise on the specific topic of the question is underutilized. Even if an answer is offered, the answer is often not helpful due to the lack of matching expertise, and it also increases the cost of attention ($C_{Evaluate}$) consumed by the asker in evaluating the answers because he or she has to sift through all answers to find the useful ones. This leads to the first design principle for the DML mechanism.

Principle 1: *Members who do not have matching expertise shall not receive the question.*

Having expertise is only a necessary not a sufficient condition for a peer to share his or her expertise: the peer has to be willing to share the sought after expertise. Some knowledge workers might not want to provide expertise on certain topics for various reasons. Some members might get bored by answering repeatedly questions that they deem too simple to worth their time and expertise; and some might want to guard their certain expertise to retain their "market value" in the organization [18]. When workers are forced into sharing expertise they are not willing to share, they often use "verbal and intellectual skills as a defense to keep a person with a problem from consuming too much of their time," and their answers are often "impressive-sounding" but not helpful [5]. As a result, the attention consumed by both the expertise seeker and the unwilling expertise provider is wasted. These empirical observations lead to the second design principle.

Principle 2: *Members who are not willing to share the matching expertise shall not receive the question.*

Another factor that affects a recipient's willingness of sharing expertise is his or her perceived social relationships with the asker and with the group at large. Favorable inter-personal relationship facilitates expertise sharing due to preexisting trust and mutual understanding [2]. When an expert chooses to pay attention to an asker, the social relationship between the two people gets changed. A recipient decides to help the asker based on his or her perceived existing relationship or desired future relationship with the asker. Given the varieties of human relationships in a group, it is natural that a member may want to be related only with a certain subset of the whole group. Because an arduous relationship between an expertise seeker and an expertise provider often leads to the failure of expertise sharing [5], a recipient who does not like to work with the asker is not likely to provide help. People have very nuanced preferences concerning how and with whom they like to share expertise and like to maintain control of their social interaction [1]. Hence we have the following design principle.

Principle 3: *Members shall be able to decide with whom to share expertise.*

4.2 Creating Dynamic Mailing Lists for Situated Expertise Sharing

A dynamic mailing list (DML) differs from traditional mailing list in that a new mailing list is created every time when an asker posts a question, with the recipients decided dynamically based on the three principles identified in section 4.1. The bottom half of Fig. 2 shows the general architecture of a server that supports the DML approach. The top half shows an illustrative use scenario. Suppose *Harry* has a question about a Java API method named *exec*. He first searches the *Discussion Archives* that store previous discussion emails exchanged through the DML server. If Harry does not find answers in archives, he can post a question, and the DML server uses two steps to decide the recipients of the question: *expert identification* and *expert selection.*

Based on the *Human-Topic Relationship* database, which stores the *expertise profiles* for members, the expert identification step creates a list of *candidate experts* who have expertise on the topic of the question (*exec*). The expertise profile of a member represents what topics on which the member has expertise. A member can manually specify his or her expertise, but this is usually too costly. In practice, this expertise profile should be automatically generated using data mining techniques by analyzing preexisting documents, artifacts and expertise-sharing activities. Following Principle 2, this expertise profile should also include topics on which a member is not willing to share his or her expertise.

From the list of *candidate experts*, the expert selection step chooses, based on the *Human-Human Relationship* database that represents the social relationships among members, those who have the highest possibility of helpping the asker (*Harry*). A member's social relationships with other members and the group are represented by his or her *social profile* that consists of the following four kinds of basic relationships:

- *help<A, B>*: This represents how many times that *A* has provided answers to questions posted by B.
- *include<A, B>*: *A* states that *A* is willing to help *B* if he or she has any level of expertise on a topic that *B* asks.
- *exclude<A, B>*: *A* states that *A* is not wiling to share expertise with *B* on whatever topics.
- *email<A, B>*: The represents the number of regular emails that *A* has sent to *B* outside of the DML server. This is extracted by analyzing the mailbox of each member, and represents the existing social relationships that *A* and *B* had before they started using the DML approach.

With the relationships defined above, the expert selection step in DML uses the following process to choose from the list of *candidate experts* a predefined number of members who are most likely to answer *Harry*'s question on *exec*. The selection process consists of 5 passes, and stops processing the next pass when the predefined number of experts has been selected.

Pass 1: Based on the preference of unfavorable relation. For each person *X* in the list of *candidate experts*, if *exclude<X, Harry>* exists, *X* is removed from the list of *candidate experts* because *X* has indicated no intent to help *Harry* on whatever topics.

Pass 2: Based on the preference of friendly relation. For each *X* in the remaining list of *candidate experts*, if *include<X, Harry>* exists, *X* is selected because *X* declares his or her willingness to help *Harry*.

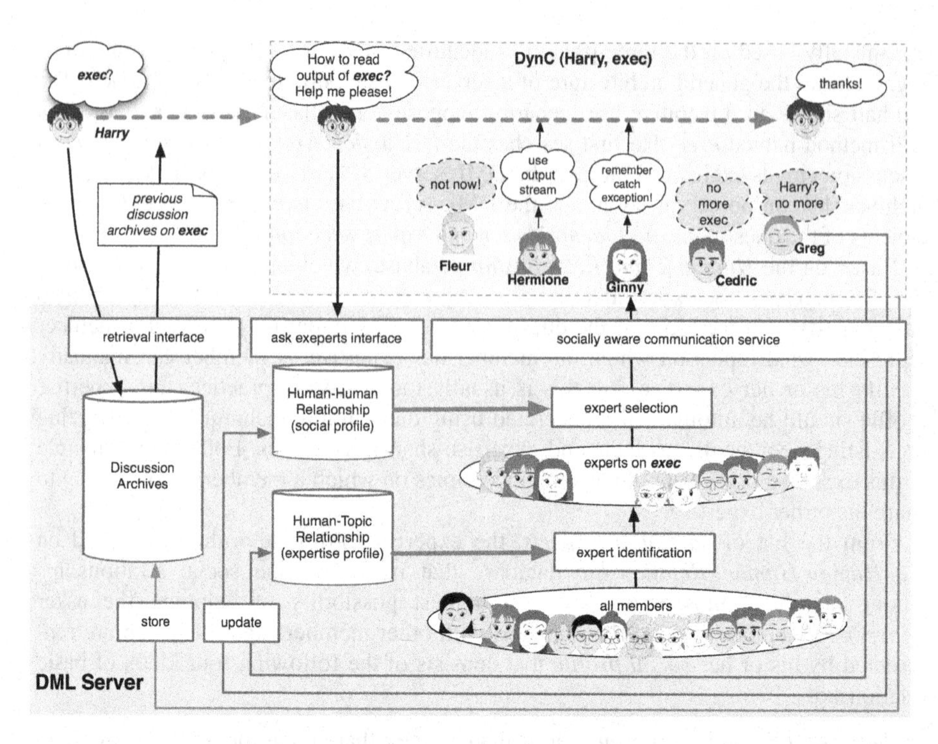

Fig. 2. The general architecture of Dynamic Mailing List and an illustrative scenario

Pass 3: Based on the rule of direct reciprocity. For each *X* in the remaining list of *candidate experts*, if *Harry* has helped *X* more times than *X* has helped *Harry* (i.e. *help<Harry, X>* >= *help<X, Harry>*), *X* is selected because social norms in general requires that *X* reciprocate the favor that he or she has received from *Harry* in the past.

Pass 4: Based on the rule of generalized reciprocity. From the remaining list of *candidate experts,* this pass chooses those who have been helped more in the whole group than they have helped others, regardless their direct relations with *Harry.* Although they may not bear direct social relations with *Harry*, they should offer help due to the principle of generalized reciprocity in the whole group.

Pass 5: Based on outside communication. If no sufficient number of experts has been selected till this pass, the DML mechanism resorts to preexisting social relationships that members have before they started using DML. From the remaining list of *candidate experts,* this pass chooses the experts who have sent most emails to *Harry* based on *email<X, Harry>*, because *X* must know *Harry* to a certain degree if he or she has sent emails to *Harry*, and might be willing to help *Harry.* Unlike the initial 4 passes that use the relationships resulted from the social interactions within the DML mechanism, this pass uses the social interaction history outside of the DML mechanism, and this pass is heavily used at the initial stage of the DML mechanism when there is no sufficient history of interactions within the DML mechanism.

The chosen experts become the participants of the dynamically created mailing list *DynC(Harry, exec)*, and will receive the question asked by *Harry* on *exec* through emails. If a recipient (*X*) replies to the question, his or her reply is sent to the same members of the *DynC(Harry, exec)*, and the DML server automatically increases *help(X, Harry)* by 1 to update the social profile of X.

The dynamic mailing list is disposed either manually by the asker when he or she decides there is no more need for expertise sharing on the topic, or automatically by the DML server when there is no more email exchanges for a predefined period of time. The messages are stored in the Discussion Archives.

As a result of this two-step processing by the DML server, only those members who have both high level of expertise on the topic and high possibility of helping the asker are chosen to receive the question. By so doing, the success rate *r* of expertise exchange can be improved because the intensity of engagement by each expertise provider is directly related to his or her willingness to actively inquire into and understand the asker's problem and then shape their answer to the problem in generating help [5]. This would also translate into a lower cost of $C_{Evaluate}$ on the asker's side. Because the list of expert recipients is automatically generated, the C_{Find} remains to be minimal. *CoCA* is reduced because those members who do not have the relevant expertise on the particular question or who are not likely to help the asker do not receive the question and do not need to consume their attention in receiving and reading the question. The potential learning benefits of the onlookers can be similarly achieved through the browsing and searching of the accumulated Discussion Archives.

4.3 Socially Aware Communication

When expertise providers specify their preferences on the kinds of expertise sharing activities they do not want to participate, they are saying "no" explicitly to their peers. In a community where social norm expects active collaboration, their decision of no participation may cause damages to relationships with other members and risk disruption of group cohesion. Many people who feel guilty when they have to say no loudly and publicly might choose not to participate in the community all together or be forced to endure the waste of their attention repeatedly. Both deteriorate the economy of the collective attention, with the former decreasing the total available resources of expertise and the latter wasting the attention of onlookers.

Improving the economy of collective attention, therefore, needs to be complemented with socially aware communication mechanisms. Socially aware communication refers to the transmission of information or signals that does not violate social norms, and therefore is not punishable by the iron hand of social pressure that establishes group cohesion through enforcing the required individual behaviors [15]. In face-to-face communications, many non-linguistic social signals are used to encourage or discourage the further occupation of attention in a socially acceptable way, without causing unwanted damages to social relationships.

Although information and communication technology has advanced greatly in easing the transmission of non-linguistic social signals to approximating the effects of fact-to-face communication, it also offers opportunities of devising alternative ways of conducting socially aware communication other than those used by people in face-to-face communications. The DML approach adopts the following principle of *asymmetric disclosure of information* to achieve socially awareness: To make the decision of no participation socially acceptable, DML ensures that an expertise provider's refusal of further allocation of attention on unwanted expertise sharing be not known by other members, and therefore makes the refusal socially plausible.

First, all choices made by an expertise provider regarding his or her preference of participation in the kinds of expertise sharing must be strictly limited to the eyes of the expertise provider only. No other members should know the decision. This can be realized with login names and passwords. Second, when an asker posts a question through DML, the recipients of the question are not made public. Only the recipient knows that he or she gets the question. No other members, including the asker and other members who receive the same question, know who else receives it. On the other hand, if a recipient replies, his or her name is revealed to all participants of the DML, and is revealed to all other members in the Discussion Archives. This asymmetric disclosure of information removes the undesired social implications of no participation but highlights and encourages cooperative behaviors with explicit acknowledgement of participation. With the existence of such socially aware communication mechanisms, members can freely say no—actually they do not say it aloud, they only choose no secretly—if they are unable to help others at any given time.

Each question sent through a DML is associated with two links that allows its recipients to update his or her expertise profile and social profile. As shown in the scenario of Fig. 2, if a member finds that he receives a question on a topic that he or she does not want to answer any more, he or she can click one link to update his or her expertise profile so that he or she will no longer receive questions on the same topic in the future. If a member finds that he or she is receiving questions from a person he or she does not like to work with, he can click on the link to update his or her social profile to add an *exclude* relationship so that he or she will not receive any question from the asker in the future. Both links reduces the burden of maintaining updated expertise profile and social profile. Updating profiles costs a little bit more attention than just ignoring the question, but this one-time extra attention cost will lead to the reduction of future attention cost of dealing with questions that the member does not want.

5 Implementing DML in the STeP_IN System

As a way to illustrate the DML mechanism, we have implemented it in the STeP_IN system (standing for Socio-Technical Platform for in situ Networking) [25, 26]. A huge reusable class library is one of the major benefits brought by Java, but it also poses great challenges for programmers to learn to use those library methods. Most programmers only know a portion of them, and the expertise of the library is asymmetrically distributed among programmers. This gives rise to many needs of situated expertise sharing that often take place in the middle of programming when a programmer needs to use an unknown API method. An effective way of learning is to ask those who are experts on the given method. The STeP_IN system implements the DML mechanism to help programmers to learn from their peers by asking questions about Java API methods. The system also provides other technical support, but this paper will focus on the issues related to the DML mechanism, for more details on other aspects of the system, please see [26].

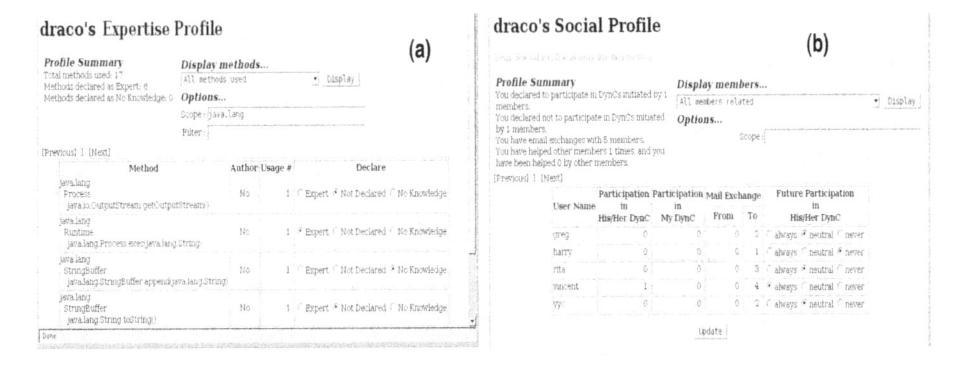

Fig. 3. Expertise and Social Profiles in STeP_IN

At the core of the DML mechanism is the creation and use of *expertise profiles* and *social profiles*. Because social profiles represents social relationships resulted from social interactions among members that are domain independent, the social profile in STeP_IN is defined in the same way as described in Section 4.2. It has four kinds of relationships: *help*, *include*, *exclude*, *email*. Expertise profiles are domain dependent. In STeP_IN, a programmer's expertise profile has two sets of Java API methods. The first set is *known methods*, which include those methods of which the programmer has expertise. The second set is *uninterested methods*, which include those methods on which the programmer does not to share expertise with others.

To use STeP_IN, a user has to register first. Upon a user's registering to the system, an initial expertise profile for the user is automatically created by analyzing all the Java programs that he or she has written. The number of API method usage is extracted and stored in his or her initial expertise profile. A user can edit his or her expertise profile through the expertise profile management interface in STeP_IN (Fig. 3a). A user can select *Expert* in the *Declare* column to add the method to his or her *known methods* set no matter whether he or she has ever used it or not, or a user can select *No Knowledge* to add the method to the *uninterested methods* set in his or her expertise profile so that he or she will not receive questions about the method. As we have discussed in Section 4.3, such a selection can also be made once he or she receives a question on the method through DML emails, to ease the task of maintaining updated profiles.

At the registration time, an initial social profile is automatically created by analyzing the user's mailbox. It contains the number of emails that the user has received from other members. The other key element in the social profile is the *help* history the user has with other members. The number in the *Participation in His/Her DynC* column (Fig. 3b) indicates the number of help that the user has given to the member, whose name is shown at the first column; and the number in the *Participation in My DynC* column indicates the number of help the user has received from the same member. Through the profile management interface, the user can declare *include* and *exclude* relation with the member by choosing *always* or *never* in the *Future Participation in*

His/Her DynC. Similarly to expertise profile, this personal preference can be made when he or she receives a question form the member through DML.

We will now continue to use the scenario in Fig. 2 to illustrate how a user interacts with the STeP_IN system for situated expertise sharing. When Harry posts a question on *exec* in STeP_IN, the system first creates a list of members whose *known methods* include *exec* and then removes from the list those whose *uninterested methods* contain *exec*. The resulted list is the list of *candidate experts* on *exec*.

From the list of *candidate experts*, STeP_IN chooses 5 members who have established social relationship with *Harry* based on each person's social profile. It first excludes those who declared *never* to participate in *Harry*'s DynC. For example *Draco* would be removed because he have chosen *never* in his relationship with Harry (Fig. 3b) although he declared himself an expert on *exec* in his expertise profile (Fig. 3a). From the remaining list of *candidate experts*, the DML server in STeP_IN first chooses those whose social profile includes an *always* declaration regarding *Harry*; and then chooses according to the numbers appeared in the columns 2 and 3 in the social profile (Fig. 3b), i.e. those who owe *Harry* because they were helped more often by *Harry* than they helped *Harry*. If the system cannot choose 5 members from those who directly interacted with *Harry,* it chooses those who have received more help from the group. If the above process fails to reach the number 5, it uses the *email* relationship and chooses members who have sent most emails to *Harry*.

The question posted by *Harry* is then sent to the selected experts (*Fleur, Hermione, Ginny, Cedric and Greg*). However, *Harry* does not know who receives the question. Each recipient does not know who else receives the question. *Greg* finds the question is from *Harry* whom he is not fond of, so he clicks one of the embedded links that takes him to his social profile management interface and chooses *never* regarding *Harry*. *Cedric*, who finds he gets yet another question on *exec*, decides to change his expertise on *exec* to *No Knowledge* to avoid getting further questions on *exec*. *Hermione* and *Ginny* replied to the DML. *Fleur*, who is preoccupied with her own work, stays silent. Due to the asymmetric disclosure of information, all members know that *Hermione* and *Ginny* have helped *Harry*, but no one knows that *Fleur, Cedric* and *Greg* have received the question and *Cedric* and *Greg* have changed their preferences.

Harry is satisfied with the help he gets from *Hermione* and *Ginny* and goes to STeP_IN to evaluate the DynC as helpful. The DynC is then discontinued and the emails exchanged are archived and linked to the method *exec*.

An evaluation of the STeP_IN system [26] shows that the DML mechanism may miss some experts who are eager to help others regardless of their social obligations. This problem can be solved if the eager helpers set their participation preferences to *always* for all members so that they will be included in all expertise sharing acts concerning topics on which they have expertise. They can even choose *Expert* on all methods so that they will be included in all acts of expertise sharing. The DML mechanism will act like a traditional mailing list for members with the above generous settings, but other less eager helpers will still have their choices and options to control their allocation of attentions.

6 Concluding Remarks

Most of the research on expertise sharing has focused on helping users find the right expert [2, 7, 12, 13]. Such systems aim to reduce the cost incurred on the asker, mainly C_{Find}. Reder points out, however, that automated attempts to "pin people down" may not bring about better communication or enhanced productivity because successful expertise sharing requires intensive engagement of expertise providers [17]. Most existing research focuses only on the benefits an asker receives, and ignores the cost that helpers shoulder as well as the potential adverse impacts on group productivity. However many empirical studies have concluded that the cost of interruption [10, 14] and the overload of communication brought by ubiquitous connectivity [6] create the "dearth of attention"[22]. We do not have a systematic way to address the attention cost resulted from communication technologies.

In this paper we attempted to balance the needs of askers and the burdens of helpers in expertise sharing. The necessity to balance attention and communication are recognized in [8], which suggests two strategies to conserve attention resources in communication by providing information asynchronously and by reducing the frequency of interruption through the aggregation of information. These strategies can be subsumed in reducing the cost of $C_{Interrupt}$ in the conceptual framework of the collective attention economy. However, as we can see from our analysis, this cost is only a portion of the cost of collective attention in collaboration.

We are fully aware that to model concepts as complicated and subjective as attention should not be taken lightly. The proposed notion of *CoCA* is not meant to compute the absolute value of attention cost. The main goal is to use this relatively simple framework to analyze the factors that affect the economic utilization of the collective attention of all parties involved, either actively or passively, in expertise sharing, and to use it as a guidance to design alternative communication mechanisms that have different cost-benefit structure by manipulating some variables in *CoCA* for different expertise sharing situations. By trying to change the number *N*, we devised the DML mechanism that is neither direct email nor mailing list nor BBS, but something in between email and mailing list with the feature of persistent storage of discussions. The comparison is not meant to rank the absolute superiority of communication mechanisms, but gives a clear understanding of each mechanism so that users and organizations can choose the most appropriate communication channel for their varied and nuanced communication needs in their specific socio-technical environment.

Among many systems that support expertise sharing [2, 12, 13], Answer Garden 2 [2] is most similar to the DML approach. Both approaches go through the expert identification and expert selection steps. They differ in the strategies of defining social relationships. DML defines social relationships based on inter-personal interaction histories while Answer Garden 2 uses organizational and physical proximities. The more important difference is the DML approach gives high priority to the individual preferences of experts, granting experts the full control of allocating their attentions with the introduction of socially aware communication mechanisms. The availability of choices and options helps the development of favorable attitudes toward expertise sharing [20], and this favorable attitude is critical for expertise sharing to become sustainable in an organization.

References

1. Ackerman, M.S., Halverson, C.: Sharing Expertise: The Next Step for Knowledge Management. In: Social Capital and Information Technology, pp. 333–354. MIT Press, Cambridge (2004)
2. Ackerman, M.S., McDonald, D.W.: Answer Garden 2: Merging Organizational Memory with Collaborative Help. In: Proceedings of CSC 1996, pp. 97–105 (1996)
3. Ackerman, M.S., Piipek, V., Wulf, V.: Sharing Expertise: Beyond Knowledge Management. MIT Press, Cambridge (2002)
4. Arguello, J., Butler, B.S., Joyce, E., Kraut, R., Ling, K.S., Rosé, C., Wang, X.: Talk to Me: Foundations for Successful Individual-Group Interactions in Online Communities. In: Proceedings of CHI 2006, Montréal, Canada, pp. 959–968 (2006)
5. Cross, R., Borgatti, S.P.: The Ties That Share: Relational Characteristics That Facilitate Information Seeking. In: Huysman, M., Wulf, V. (eds.) Social Capital and Information Technology, pp. 137–161. The MIT Press, Cambridge (2004)
6. Dabbish, L.A., Kraut, R.: Controlling Interruptions: Awareness Displays and Social Motivation for Coordination. In: Proceedings of CSCW 2004, pp. 182–191 (2004)
7. Dieberger, A., Dourish, P., Höök, K., Resnick, P., Wexelblat, A.: Social Navigation: Techniques for Building More Usable Systems. Interactions 7, 36–45 (2000)
8. Fussell, S.R., Kraut, R.E., Lerch, F.J., Scherlis, W.L., McNally, M.M., Cadiz, J.J.: Coordination, Overload and Team Performance: Effects of Team Communication Strategies. In: Proceedings of CSCW 1998, Seattle WA, pp. 275–284 (1998)
9. Goldhaber, M.H.: The Attention Economy. First Monday 2 (1997)
10. Jackson, T., Dawson, R., Wilson, D.: The Cost of Email Interruption. Journal of Systems and Information Technology 5, 81–92 (2001)
11. Lakhani, K.R., von Hippel, E.: How Open Source Software Works: Free User to User Assistance. Research Policy 32, 923–943 (2003)
12. McDonald, D.W., Ackerman, M.S.: Expertise Recommender: A Flexible Recommendation System Architecture. In: Proceedings of CSCW 2000, pp. 101–120 (2000)
13. Mockus, A., Herbsleb, J.: Expertise Browser: A Quantitative Approach to Identifying Expertise. In: Proceedings of 2002 International Conference on Software Engineering, pp. 503–512 (2002)
14. O'Conaill, B., Frohlich, D.: Timespace in the Workplace: Dealing with Interruptions. In: Proceedings of CHI 1995 Conference Companion, pp. 262–263 (1995)
15. Pentland, A.: Socially Aware Computation and Communication. Computer 38, 33–40 (2005)
16. Rashid, A.M., Ling, K., Tassone, R.D., Resnick, P., Kraut, R.E., Reidl, J.: Motivating Participation by Displaying the Values of Contribution. In: Proceedings of CHI 2006 (2006)
17. Reder, S.: The Communication Economy of the Workgroup: Multi-Channel Genres of Communication. In: Proceedings of CSCW 1988, pp. 354–368. ACM Press, New York (1988)
18. Reichling, T., Veith, M.: Expertise Sharing in a Heterogeneous Organizational Environment. In: Proceedings of 9th European Conference on Computer-Supported Cooperative Network, pp. 325–345 (2005)
19. Robillard, P.N.: The Role of Knowledge in Software Development. CACM 42, 87–92 (1999)
20. Salancik, G.R., Pfeffer, J.: A Social Information Processing Approach to Job Attitudes and Task Design. Administrative Science Quarterly 23, 224–253 (1978)

21. Schmidt, K.: The Critical Role of Workplace Studies in CSCW. In: Luff, P., Hindmarsh, J., Heath, C. (eds.) Workplace Studies: Recovering Work Practice and Informing System Design, pp. 141–149. Cambridge University Press, Cambridge (2000)
22. Simon, H.A.: The Sciences of the Artificial, 3rd edn. The MIT Press, Cambridge (1996)
23. Szoestek, A.M., Markopoulos, P.: Factors Defining Face-To-Face Interruptions in the Office Environment. In: Proceedings of CHI 2006, pp. 1379–1384 (2006)
24. von Krogh, G., Spaeth, S., Lakhani, K.R.: Community, Joining, and Specialization in Open Source Software Innovation: A Case Study. Research Policy 32, 1217–1241 (2003)
25. Ye, Y., Yamamoto, Y., Nakakoji, K.: A Socio-Technical Framework for Supporting Programmers. In: Proceedings of 2007 ACM Symposium on Foundations of Software Engineering (FSE 2007), pp. 351–360 (2007)
26. Ye, Y., Yamamoto, Y., Nakakoji, K., Nishinaka, Y., Asada, M.: Searching the Library and Asking the Peers: Learning to Use Java APIs on Demand. In: Amaral, V., Veiga, L., Marcelino, L., Cunningham, H.C. (eds.) Proceedings of 2007 International Conference on Principles and Practices of Programming in Java, pp. 41–50. ACM Press, Lisbon (2007)

Exploring the Effectiveness of Normative *i** Modelling: Results from a Case Study on Food Chain Traceability

Alberto Siena[1], Neil Maiden[2], James Lockerbie[2], Kristine Karlsen[2], Anna Perini[1], and Angelo Susi[1]

[1] Fondazione Bruno Kessler - Irst, Trento, Italy
{siena,perini,susi}@fbk.eu
[2] Centre for HCI Design, City University, London
{N.A.M.Maiden@,J.Lockerbie@soi.,
Kristine.Karlsen@soi.}city.ac.uk

Abstract. This paper evaluates the effectiveness of an extension to *i** modelling – normative *i** modelling – during the requirements analysis for new socio-technical systems for food traceability. The *i** focus on modelling systems as networks of heterogeneous, inter-dependent actors provides limited support for modelling system-wide properties and norms, such as laws and regulations, that also influence the specification of socio-technical systems. In this paper we introduce an extension to *i** to model and analyse norms, then apply it to model laws and regulations applicable to European food traceability systems. We report an analysis of the relative strengths and weaknesses of this extended form of *i** with its traditional forms, and use results to answer two research questions about the usefulness and usability of the *i** modelling extension.

1 Introduction

Analysts are increasingly using *i**, the strategic goal modelling approach [21], to model and analyse requirements. *i** has been applied successfully to model requirements for air traffic management tools [8, 9] and decision support aids in agriculture [11] as well as to support individuals and groups in the work of charitable organisations [17]. Reported benefits to our projects have included automatic requirements generation from *i** models [9] and detection of omissions from UML requirements specifications [8]. However, the focus on modelling systems as networks of heterogeneous but inter-dependent actors provides limited capabilities for addressing the broader, system-wide properties and norms that also influence the specification and design of socio-technical systems. Examples of such norms include laws and regulations, which constrain and influence how actors in these systems shall operate. In this paper we report an extension to the *i** modelling approach to model norms in socio-technical systems, then investigate the effectiveness of this extension through its application to a large-scale case study – introducing new traceability technologies into two European food chains.

Whilst *i** has many strengths that have contributed to its increasing adoption, the representation of laws and regulations, as well as actors' adherence to these laws and

Z. Bellahsène and M. Léonard (Eds.): CAiSE 2008, LNCS 5074, pp. 182–196, 2008.

regulations, is recognized as problematic because laws and regulations are difficult to represent using the standard actor-goals-dependencies metaphor found in the basic modelling approach – referred to as "basic" *i** in this paper. For example, whilst actors in socio-technical systems might seek to achieve compliance with reported regulations such as for food hygiene, regulation compliance on its own is often not a strategic goal or softgoal of actor. Furthermore, inclusion of new actors who generate and impose laws and regulations detract from the main analytic purpose of *i**. As a consequence, representing laws and regulations is often overlooked in early requirements work, with consequences for downstream analysis and design of socio-technical systems.

Previous work has introduced new *i** modelling concepts to represent and analyse norms to represent laws and regulations [15]. However, like all extensions to basic *i**, such as SecureTropos [5], the addition of new modelling semantics and syntax can increase the complexity and reduce the usability and adoption of the *i** modelling approach. Therefore, studies were needed to explore the coverage, effectiveness and usability of *i** modelling extensions prior to their widespread adoption.

In this paper, we report the extension of *i** modelling with norms to investigate whether such extensions deliver advantages such as the induction of new actor goals from the adoption of a norm, explanation of existing goals due to the imposition of norms, and the discovery of new roles/actors due to the imposition of a norm. We investigated TRACEBACK, a EU-funded Integrated Project seeking to introduce new technologies to improve traceability in European food chains. We used results to seek answers to two research questions:

Q1: Were analysts using the extended *i** semantics and notation with the norm concept able to infer new properties of the system being modelled?

Q2: Were analysts using the extended *i** modelling approach able to represent concepts related to norms, such as legislation, rules, etc. in an efficient manner (compared to basic *i**)?

The paper is structured as follows: section 2 reports the basic *i** modelling framework; section 3 introduces the normative *i** framework; section 4 reports the models obtained for the food traceability information system with both basic *i** and normative *i**; section 5 evaluates empirically the new framework by comparing the results of its application and the basic *i** models; section 6 discusses the results and tries to answer the raised questions; finally, section 7 concludes the paper.

2 *i** and Redepend

In TRACEBACK we applied the RESCUE requirements process with the *i** modelling approach and REDEPEND tool. RESCUE [6] supports a concurrent engineering process in which different modelling and analysis processes take place in parallel. Each stream has a unique and specific purpose in the specification of a socio-technical system:

1. Human activity modelling provides an understanding of how people work, in order to baseline possible changes to it [14]. In TRACEBACK we observed and documented the work activities and behaviour of actors in, for example, food plants producing milk-based products;

2. System modelling enables the team to model the future system boundaries, actor dependencies and most important goals of actors in the dairy food chain using the *i** approach [21] and REDEPEND tool [7];
3. Use case modelling and scenario-driven walkthroughs enable the team to acquire complete, precise and testable requirements from stakeholders [18]. For example, we specified the behaviour of how food chain actors would work with new micro-devices and service-based information systems in improved traceability practices, then walked through scenarios to discover more complete requirements on these devices and information systems;
4. Managing requirements enables the team to handle the outcomes of the other 3 streams effectively as well as impose quality checks on all aspects of the requirements document [13].

In this paper we focus on the second stream using the *i** approach to food chains in terms of actor dependencies and goals.

*i** is an approach originally developed to model information systems composed of heterogeneous actors with different, often-competing goals that depend on each other to undertake their tasks and achieve these goals. *i** can be applied effectively to model food and food-related information chains, as we will demonstrate. Due to the physical characteristics of food production, actors and resources generated and consumed later in the food chain depend on actors and resources earlier in the food chain, which can be represented using dependency relationships central to the *i** approach.

Basic *i** modelling supports 2 basic types of model. The first *i** model produced was the Strategic Dependency (SD) model, which describes a network of dependency relationships among actors. The opportunities available to these actors can be explored by matching the depender who is the actor who "wants" and the dependee who has the "ability". Since the dependee's abilities can match the depender's requests, the system-wide strategic model is developed. For example, in TRACEBACK, the actor *Primary Dairy Producer* depends on a second actor *Farm* to attain the goal *hygiene standards met*, achieve the softgoal *quality milk received*, and obtain the resource *fresh milk*. More details of these models are reported in Section 4.

The second type of *i** model is the Strategic Rationale (SR) model, which provides an intentional description of how each actor achieves its goals and softgoals. An element is included in the SR model only if it is considered important enough to affect the achievement of some goal. The SR model includes the SD model, so it describes which actors may be able to accomplish something by themselves, or by depending on other actors. It specifies goals, tasks, resources and softgoals linked by dependency links from the SD model, task decomposition links, means-end links, and the contributes-to-softgoal links [21]. For example in TRACEBACK, the actor *Primary Dairy Producer* performs the task *undertake contamination recall procedures*, which achieves the softgoal *target recall undertaken successfully*, which in turn contributes positively to the actor *regulator* achieving the softgoal *contaminated products recalled efficiently*. Again more details of these models are reported in Section 4.

RESCUE is supported by REDEPEND [7], a tool based on Microsoft Visio and designed to provide systems engineers with *i** modelling and analysis functions. It provides drag-and-drop capabilities to visually develop *i** Strategic Dependency (SD) and Rationale (SR) models. REDEPEND also provides systems engineers with simple model verification functions for large-scale SD and SR models. In RESCUE we applied basic *i** to model essential actors, dependencies, goals and tasks in the dairy food chain. The next section reports methodological extensions to *i** to model normative contexts in socio-technical systems also applied to the dairy food chain.

3 Normative *i**

What distinguishes socio-technical organisations from simple groups of interacting people are norms [10]. Various types of norms exist in the real world, but, as pointed out in [16], the one that gains relevance at requirements time is the behavioural norm – essentially, behavioural norms impose actions to perform, goals to be achieved, resources to be used or principles to be respected.

Recent studies in requirements engineering address the problem of modelling regulations for requirements compliance. A survey on current approaches is given in [12]. Worthy of mention here is a proposal that relies on the analogy between regulations and requirements documents to model the objectives stated in the regulations [3]. However, the adopted goal-oriented framework – Kaos [2] – misses the capability of supporting agency in the models. In [1] the focus is on automatic extraction of obligations and rights from legal texts, so usefully supporting the analysts in parsing the law documents, but not in representing them. In [4], traceability links are used to map *i** models of the regulations into the *i** models for the stakeholders. We took these approaches into consideration before introducing a new modelling framework. However, our need attains principally to supporting the analyst in the *discovery* and *integration* of legal requirements, so none of the approaches were satisfactory for us in a domain like the food chain.

As with [3], we propose to use a goal-oriented approach, based on *i**, for modelling norms, but in contrast to the above mentioned work we focus on the interaction of norms, actors and goals during the requirements elicitation process. More specifically, as introduced in a previous work [15], our idea is to model contextually and homogeneously, but separately, the normative context of a domain and its stakeholders with their intentionality. We adopt the definition of "norm" as a means for communicating standards of behaviour [19], and which acts as an abstraction for any kind of deontic prescription (such as laws, regulations and so on). On the basis of this definition, in the present work we derive three properties of norms that are relevant for the requirements acquisition: i) the normative commitment relation; ii) the schema of the norm; iii) the compliance intentions.

The normative commitment relation. Intuitively, when we think of laws or regulations, we think of artefacts, i.e., text documents, that contain prescriptions. It is interesting to notice that, when a law commits something (the prescription) to someone, the commitment establishes a relation. The relation involves two subjects: the one who created the norm – the source of the norm; and the one who is addressed by the norm – the addressee [19]. So, as depicted in Figure 1, in *i** diagrams we represent laws in a ternary relation that links the source, the addressee and the legal artefact that contains the prescription. The double arrow represents the commitment direction, whilst the triangle represents the norm. The link between *EU*, *EC178/2002*

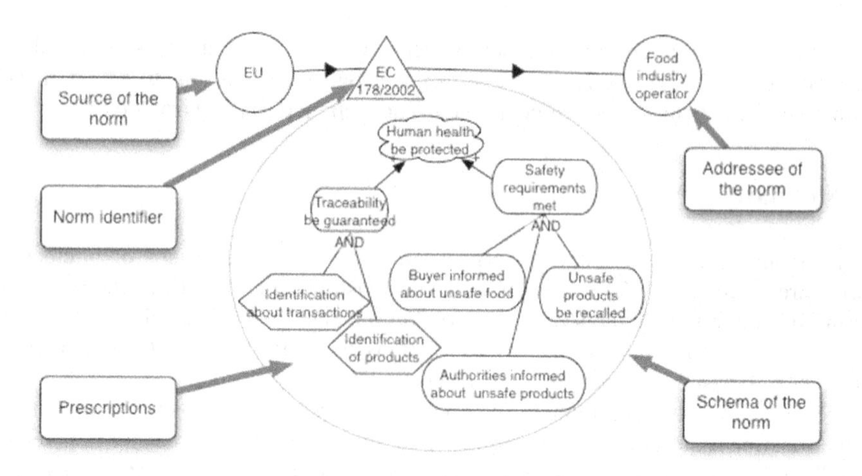

Fig. 1. Normative *i**: the normative commitment relation between two stakeholders and the schema of the norm

and *Food industry operator* can be read as follows: the European Union has laid down the EC178/2002 law, which addresses all the operators that work in the food industry.

The schema of the norm. A norm artefact (e.g. a law's text) typically imposes some prescriptions. With the term *Schema of the norm* we refer to the behavioural pattern that the norm imposes to the addressee, that is, ways of acting, goals and principles to be adopted. In Figure 1 the schema of the norm is depicted as a balloon collecting a set of *i** intentions – goals, softgoals and tasks – and, possible relations among them – like decomposition or means-end. The depicted norm's schema can be read as follows: *Food industry operators* must ensure that the minimum requirements for food safety are met (hardgoal *Safety requirements met*), i.e., they must ensure that, in case of known, unsafe food, the products are recalled, and both the buyers and the authorities are informed (hardgoals *Buyer informed about unsafe food*, *Authorities informed about unsafe products*, and *Unsafe products be recalled*). They must label their products with an identification code (task *Identification of products*), and register any transaction (task *Identification about transactions*); but they must ensure the traceability of the food products they process, whatever other actions they do (hardgoal *Traceability be guaranteed*). While performing these tasks or fulfilling these goals, the leading principle that should inspire their conduct should always be the protection of the health of the consumers (softgoal *Human health be protected*); i.e., the accomplishment of the hardgoals has to be evaluated with regard to the root softgoal, and no other interpretations should be accepted.

The compliance intentions. We want to understand the actual impact of the law on the involved stakeholders, namely what do they put into action – if they do – to accomplish to the law imposition. In Figure 2, the interleaving between the actor's intentions and the law's schema shows how the *Food industry operator* intends to comply with the law. In the example, the law lays down for the actor the responsibility of recalling products if they are known to be unsafe. However, the lack of safety of products is not known *a priori* by the operator. So it will need to keep its products monitored, and so is how the goal *Products safety be monitored* is generated inside the actor's rationale. Such a goal is then further decomposed into two specific tasks (*Monitor unsafety of milk* and *monitor unsafety of dairy products*). In the intention of the food chain operator, the two tasks should ultimately contribute to the compliancy with the norm. So now we know that the monitoring activities have been undertaken by the operator for the specific need of complying with a prescription of the EC178/2002 law.

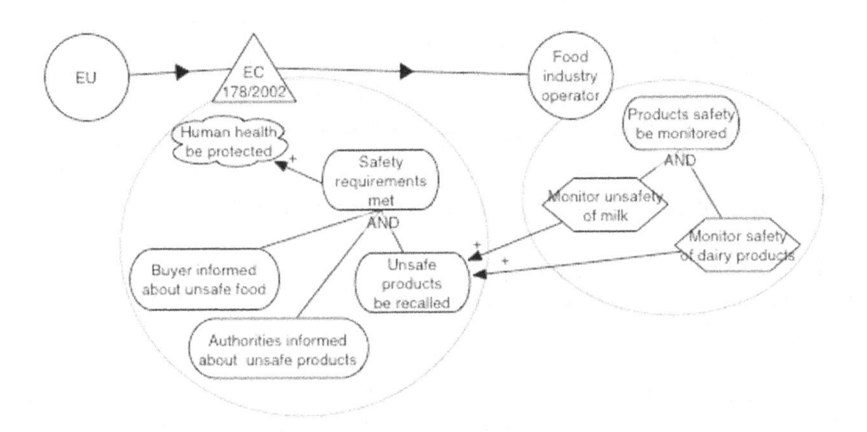

Fig. 2. Schema of the norm EC178/2002 with the intentional entities inside the balloon representing the responsibilities established by the law

When performing requirements elicitation, we interleave *i** modelling of domain stakeholders and normative modelling, as described below with the help of Figure 3. Figure 3(a) depicts a typical scenario that occurs while exploring a regulated domain. Let us suppose that we observe only Actor1, while Actor2, Actor3 and Actor4 are hidden. Here *hidden* means that the interviewed stakeholder(s) did not explicitly mention any norms, or if they did, they did this without highlighting their role. This is a typical problem of *tacit knowledge*. Returning to the example, we know that Actor1 is called to comply with two laws, Norm1 and Norm3, and so we proceed with the analysis of such laws (Figure 3(b), step 1). If the laws address other actors, they are added to the domain model (step 2). At this point Actor3 is still *hidden*. However, by analysing the source of Norm3 (step 3), we are able to find Norm2 (step 4), which in turn leads us to Actor3 (step 5). We store all this information in a norm diagram such as the one in Figure 3(b) for further analysis of the model. So we have discovered Actor2 and Actor3; but are those actors actually part of the domain? For sure we only want to model those actors that are relevant for the requirements specification. For this purpose, the analysis of the norm's schema allows us to discard those actors that are irrelevant for the problem under study. For instance, having discovered Norm1, we could observe that it lays down prescriptions attaining different topics, not in our interest. So, Actor4 will not enter in the description of the domain.

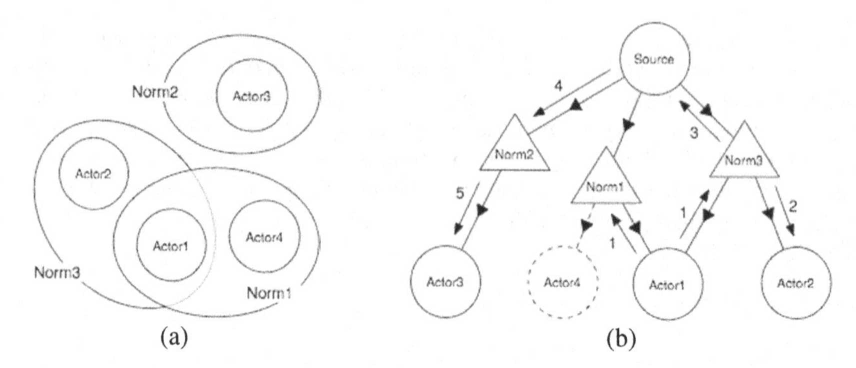

Fig. 3. (a) the scope of three norms in a domain with four actors; (b) the process of norms/actors discovery and representation in the same domain

4 The Traceback Food Traceability Case Study

As Section 2 reports we applied the RESCUE process, *i** modelling approach and REDEPEND tool to the EU-funded TRACEBACK Integrated Project. Assuring the total traceability of food and feed along the whole chain from production to consumption is a cornerstone of EU policy on the quality and safety of food. This is a complex procedure involving identification, detection and processing of a vast amount of information. Profit margins of food producers and processors are already very tight, so they require a tracking mechanism that is not only reliable and easy to use, but does not entail a major cost burden. With a concerted effort and input from expert institutions, modern technology could provide such a system. TRACEBACK is developing innovative solutions based on micro-devices and innovative service-based architectures to provide innovative new information services to actors from primary food producers to consumers and health authorities. Solutions, which will include new micro-devices and a service-oriented reference architecture for traceability information systems (RATIS), are to be trialled on two major product chains – feed/dairy and tomatoes. In this paper we focus on models developed for one of the selected food chains – dairy products such as milk-based products.

During application of the RESCUE process a team of 3 analysts, produced *i** SD and SR models describing actors in the dairy food chain. The models were developed using information from descriptions of current processes and workflows in the dairy food chains in Europe, one-on-one interviews with stakeholders who fulfil modelled actor roles in these food chains, *i** modelling workshops at project partner sites, and electronic distribution of SD and SR models to stakeholders for comment and feedback. Overall the process lasted 6 months. Key results are reported in 4 basic *i** models – 1 SD and 1 SR model each for the 2 TRACEBACK-enhanced food chains in the European dairy and tomato food chains.

In addition to the RESCUE work, normative *i** models were developed by another analyst following the process sketched in Section 3. The analyst independently explored the domain with the purpose of both discovering the applicable norms and finding related stakeholders. Using documentation and information gathered from a one-on-one stakeholder interview, 7 models were developed with a drawing tool that can export to Visio/REDEPEND.

The SD and SR models for the dairy food chain and an excerpt from the normative *i** models are reported in the next section.

4.1 The Basic *i** SD and SR Models

The basic *i** SD model of actors in the dairy food chain is depicted in Figure. 4, and the inset shows part of the model in a readable form. The model expresses 79 strategic dependencies between 13 actors from *feed suppliers* to *transportation* and even the *media* in a dairy food chain. The inset shows dependencies between the *Feed supplier* and *Farm* actors. For example, the *Farms* depend on the *Feed supplier* to achieve the softgoal *feed contamination detected early*.

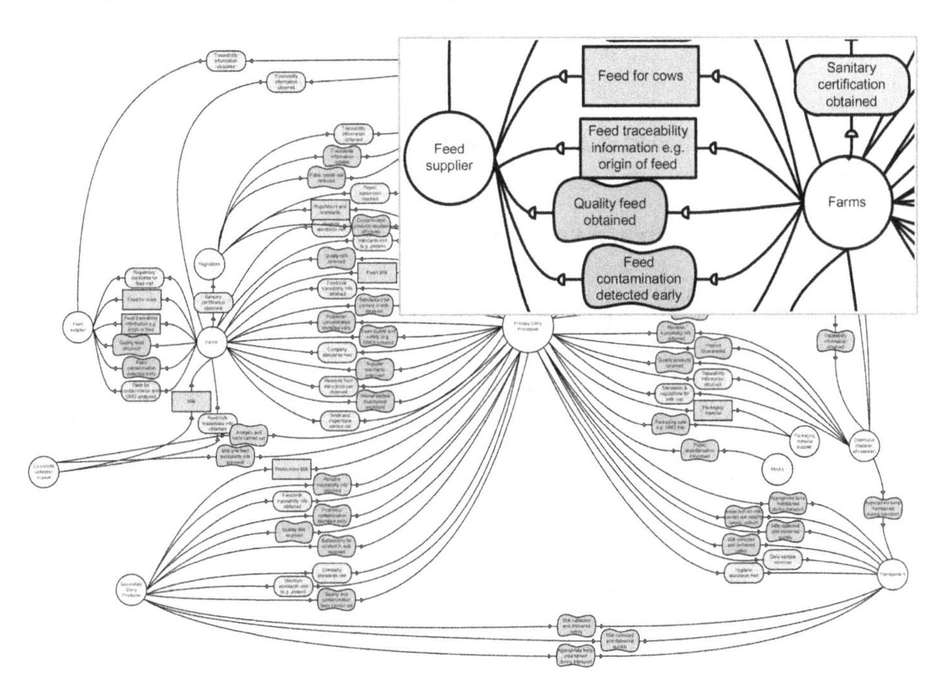

Fig. 4. The basic *i** SD model of actors in the dairy food chain, with an inset showing dependencies between the *feed supplier* and *farm* actors

The basic *i** SR Model for the same dairy food chain actors is depicted in Figure 5. The model specifies 251 different process elements and 257 different associations between these elements. The inset demonstrates part of the SR model, the *feed supplier* actor, in a readable form. The *feed supplier* undertakes the task *supply feed to farms*. To do this the *feed supplier* provides *feed traceability data* and uses the resource *feed for cows*, and seeks to achieve the softgoal *quality product stocked*.

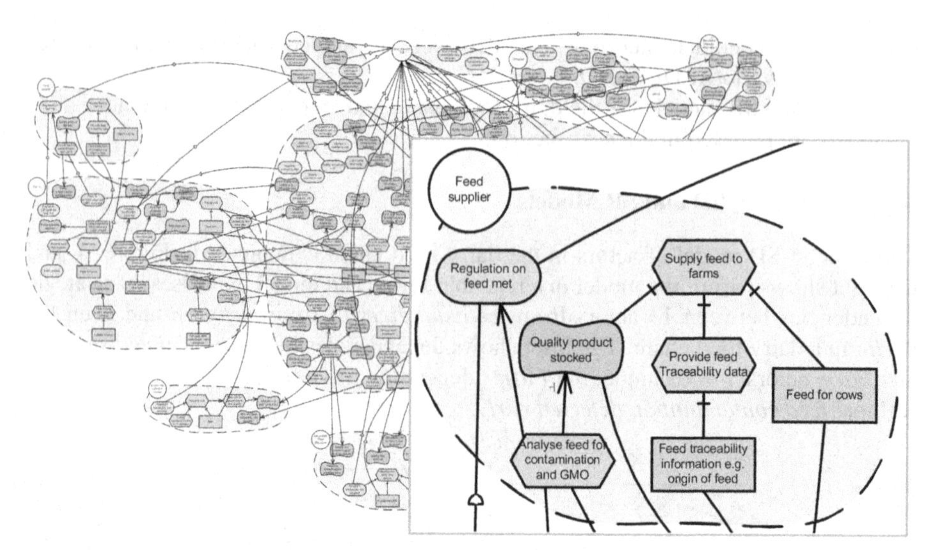

Fig. 5. The basic *i** SR model of actors in the dairy food chain, with an inset showing the expanded *food supplier* actor

4.2 The Normative *i** SD and SR Models

An excerpt of the normative *i** models is depicted in Figure 6. The actor *Food Safety Authority* has been instituted by the EC178/2002 for monitoring the entire food market, whilst the *Rapid Alert System*, which is comprised by the national governments and the EU bodies, is in charge of receiving and dispatching alerts on food-related events. In Figure 6 are also depicted the results of the norm's schema analysis, based on the same EU178/2002. In the following we discuss the four elements that are pointed out by the dashed arrows labelled 1, 2, 3 and 4:

1. The *Rapid Alert System* is devoted to the collection and forwarding of recalls across Europe, so the *Food industry operators* depend on it for dispatching alerts. At the same time, the *Rapid Alert System* depends on the food operators for having detailed traceability information to dispatch.
2. Some goals that had emerged as *Food industry operator* goals did actually come from EU laws. Recalling unsafe products or warning customer is not a free choice of producers, but are needed to comply with the law.
3. To minimise the impact that the recalling policy has on the budget, food industry operators try to discover potential unsafeties as early as possible (softgoal *unsafe products early detected*), so they monitor the quality of the raw materials and, when possible, the production processes of their suppliers. For example, in the picture we show how operators that work in the dairy production prescribe to the farmers a sort of non-legislative regulation, the Good Manufacturing Practices (GMP), to ensure the achievement of internal goals.
4. *The farmers*, in turn, put into action tasks and generate goals to be able to comply with the GMP.

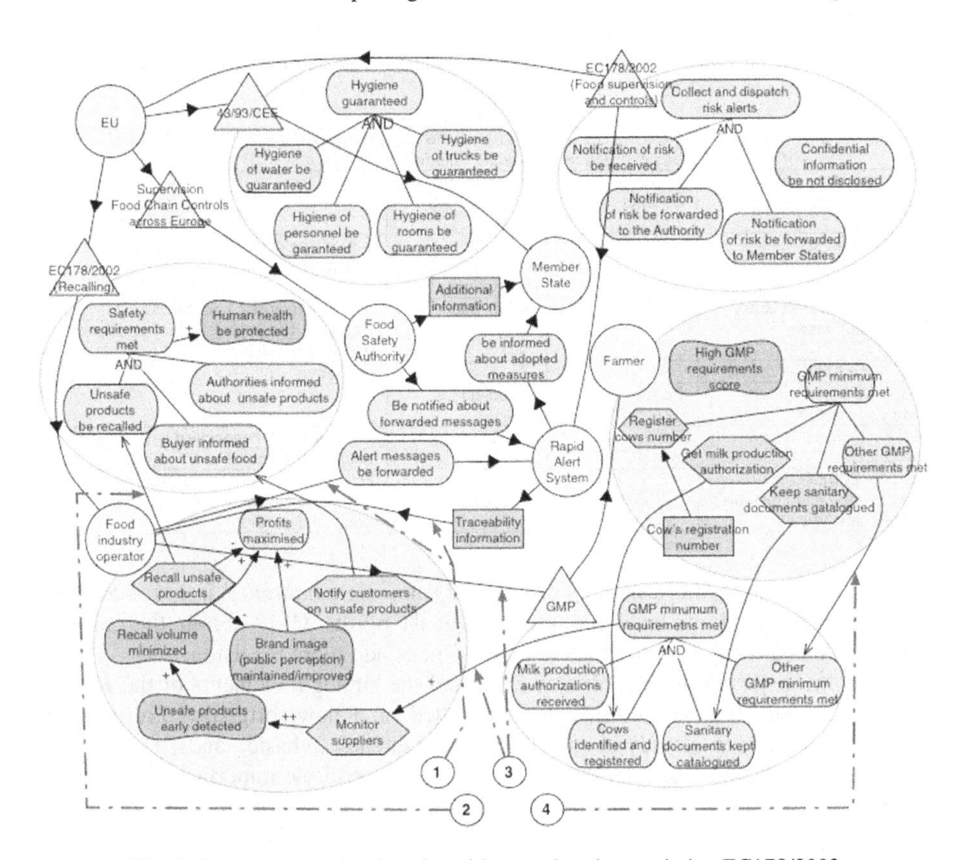

Fig. 6. A snapshot on the domain, with regard to the regulation EC178/2002

5 Empirical Analysis of the *i** Basic and Normative Traceback Models

Whilst undertaking our RESCUE goal modelling process using basic *i**, we were aware of the existence and importance of laws and standards of behaviour but did not model these explicitly. Instead, we modelled these implicitly through the goals of the actors – for example, *Farms* seek to attain the goal *sanitary certification obtained.* However, a comparative analysis of the basic *i** and normative *i** models for the feed/dairy chain (Figure 5 and Figure 6) revealed that the identification and subsequent modelling of norms added some useful detail that was either overlooked or not clearly expressed in the basic *i** model. This analysis is described further below and summarised in Table 1.

Taking the GMP norm for *Farms* as an example, we can see that the norms approach leads to 2 new goals being introduced – *GMP minimum standards met* and *other GMP requirements met.* As mentioned before, we touched upon some of the GMP areas using the basic *i** approach such as *sanitary certification obtained* (goal),

Table 1. Summary of the comparative analysis undertaken between the feed/dairy basic *i** SR model and its normative *i** equivalent

Normative / *i** Actor	Matches to basic *i** model	Additions to basic *i** model	Amendments to basic *i** model
Food industry operator / Primary dairy producer	3 softgoals 2 tasks 1 goal	1 task	
Farmer/Farms		3 tasks 1 softgoal 1 resource	5 existing goals to be reconciled with 2 new goals
EC178/2002 / Primary dairy producer	1 goal	3 sub-goals	Must review the goal boundary between the primary dairy producer and the regulator
EC178/2002 / Regulator	1 softgoal		As above
43/93/CEE / Primary dairy producer	1 goal	4 sub-goals	
GMP / Primary dairy producer		3 goals	7 goals to be reconciled with 2 new goals and 2 norms
EC178/2002 (Food supervision and controls) / none specific		8 goals 3 actors	Primary diary producer and regulator actors to be reconciled with new goals and actors

whereas undertaking the task *keep sanitary documents catalogued* is introduced in order to meet the high level requirement (goal) of the GMP. From this simple example we can see that the normative approach can add more precision.

Whilst we believed we had already modelled the strategic elements of the *Farms* actor, the introduction of the GMP norm resulted in: 1 new softgoal (10 already in basic), 2 new goals (6 already in basic), 3 tasks (8 already in basic), and 1 resource (6 already in basic). The resource, softgoal and 3 tasks constitute important additions to the model, whereas the 2 new goals encompass 5 of the original 6 goals related to standards, certification, analysis and inspections. Through further analysis of the GMP these goals could be aligned to the norm or modified accordingly.

Looking at the *Primary Dairy Processor/Food Industry Operator* actor boundary, we can see that the GMP norm contributed one additional task – *monitor suppliers* – whilst it is apparent that the other elements featured in the norms model were derived from the original *i** model. Under the basic approach, the goals contained within the GMP boundary should, in theory, feature within the actor boundary of the norm creator – in this instance the *Primary Dairy Processor.* A review of this actor boundary reveals 7 goals related to standards, regulations and requirements. However, none of these goals explicitly refers to the GMP goals of *milk production authorisations received, cows identified and registered* or *sanitary documents kept catalogued*, therefore we could argue for their inclusion within the basic *i** model.

Returning to the 7 goals relating to standards, regulations and requirements mentioned above, it is interesting to note that the high level goals *hygiene standards met* and *safety standards met* are elaborated upon in the norms model. The norm 43/93/CEE provides us with the additional detail of 4 hygiene-related sub-goals, whilst the EC178/2002 norm details 3 additional safety-related sub-goals. EC178/2002 also provides us with the softgoal *human health be protected* that is touched upon within the Regulator actor boundary in the basic *i** model by the softgoal *public health risk reduced.* This brings us back a limitation of basic *i** mentioned earlier, the issue of whose actor domain the goal belongs to – is it the goal of the regulator, the diary processor or both? Normative *i** provides us with the opportunity to treat the normative layer of the domain as a separate concern in domain modelling, hence removing this issue and supporting more effective analysis.

Another area completely overlooked by the basic *i** model was that of collecting and dispatching risk alerts, as addressed by the norm EC178/2002 (food supervision and controls). The normative model draws our attention to 3 new actors – *Food Safety Authority*, *Rapid Alert System* and *Member State* – which provide us with 8 additional goals. It is possible that overlooking these actors in the basic *i** approach may have had consequences further down the line for the analysis and design of the TRACEBACK socio-technical systems.

As mentioned earlier, we originally applied our standard RESCUE goal modelling process to TRACEBACK and did not explicitly model laws and regulations using basic *i**. Therefore, there is clearly an overhead associated with using the normative *i** approach that needs to be analysed with respect to the additional benefits it provides. We can divide our analysis into four main activities: interaction with stakeholders, inspection of documents, analysis of norm scope, and building the models. Such activities were mostly interleaved, but approximately we can estimate a 1-day interview with stakeholders; 3 days for deepening the knowledge on the norms; 7 days for exploring the norms scope and to identify the relevant ones; and finally 5 days to synthesize them and build the actual models. So, in total we can estimate that 16 person-days were spent applying the normative *i** approach to TRACEBACK.

6 Discussion

We used results reported in Section 5 to answer the 2 research questions about the *i** normative modelling extension. The answer to Q1, were analysts using the extended *i** semantics and notation, able to infer new properties of the system related to norms and legislation, is a tentative yes. In purely quantitative terms, 3 new actors and 24 new process elements, including 18 new goals and 1 new softgoal, were expressed and analysed in models developed for 3 separate pieces of legislature that impacted on two existing actors – the *primary dairy producer* and *farm* actors.

The comparative analysis we undertook showed that applying the normative approach generally added more detail to the standards-related goals already present in the basic *i** model – such added detail included cow registration and sanitary documentation cataloguing. In essence, we were able to disambiguate a number of high-level goals and derive more precise properties of the system being modelled. Furthermore, explicitly modelling the laws and standards adds richness to the models that can provide benefits later on in the software development process. As TRACEBACK is developing a service reference architecture that will provide multiple instantiations of traceability information systems, knowledge of each individual domain including GMP and EU laws is important. The normative *i** can be used as a reference model from which analysts explore the finer details to discover important system properties and final specifications.

Another point to note in support of the normative *i** approach is its usefulness and effectiveness where stakeholder access is limited. For example, we did not have the means to access the farms directly, so we obtained documentation from the dairy producers about the GMP and used normative *i** models to infer, from scratch, the missing knowledge. In this case the norms approach was a useful and effective way to better understand the domain and capture more detailed requirements.

Results from applying normative *i** to TRACEBACK also provided qualitative evidence to support our initial assertions. The basic *i** goal/actor metaphor cannot support a sufficiently complete representation and exploration of normative contexts in complex domains such as food traceability. Several problems identified and addressed subsequently in the project were a further exploration of important goal boundaries between the *primary dairy producer* and *regulator* actors or between the *primary dairy producer* and the *farmers*. Evidence from TRACEBACK indicated that stakeholders often did not venture knowledge and model feedback beyond the boundaries of the actors representing them on the *i** models, and the modelling of norms helped us to overcome this limitation of the goal/actor metaphor.

The modelling process applied in TRACEBACK also provides an interesting insight with which to interpret our answer to Q1. Draft basic *i** models were already available when the normative modelling began. Clearly the basic *i** models did not explicitly model the norms. Instead, with hindsight, stakeholders' perceptions of norms can be inferred from the basic models. So for instance, the goal *Feed regulation met* in the SR model of the actor *Feed Supplier* depicted in Figure 6 represents the actor perception of the law EU178/2002.

In contrast to Q1, we were unable to answer Q2 conclusively and determine whether analysts using the extended *i** modelling approach were able to represent concepts related to norms, such as legislation, rules, etc. in an efficient manner (compared to basic *i**). We estimated the time in TRACEBACK to produce and analyse the normative models against the advantages reported previously. A crude quantitative analysis of the number of modelled elements per day revealed a productivity measure of 1.7 elements/day (27 new model elements divided by 16 person-days). Although this modelling rate is low we also need to take into account the qualitative benefits of the normative *i** approach. Also, further analysis of the data in Table 1 suggests little overlap between the modelled elements in the two models, with 9 matches to the basic *i** version compared with 27 additions. This result implies that normative *i** complements its basic equivalent giving us benefits that appear cost-effective.

Overall, our subjective opinion is that our application of normative *i** to TRACEBACK was cost-effective, but further research and a detailed cost benefit analysis would need to be undertaken to provide a more objective and definitive answer to this question.

Interestingly, the laws we considered were generally quite clear and readable. It was apparent that the well-organised structure and unambiguous nature of the legislature supported the cost-effectiveness of the normative *i** approach. In contrast, scope analysis resulted in being the most time-expensive activity, due to the large number of laws, several of them cross-referring each other and mostly out of scope. Building models of the legal documents is also quite time-consuming, but less than scope analysis, since norms are expressed in natural language, and to reduce ambiguity they tend to be extremely analytic. In order to get useful information from them to represent their intentional characteristics, we need to synthesize them.

7 Conclusions

In this paper we evaluated the effectiveness and efficiency of the normative *i** modelling, an extension to *i**, which aims at supporting requirements elicitation in domains articulated by norms. The analysis was performed on a case study based on a real project, TRACEBACK, devoted to the improvement of the traceability in European food chains. We used the normative *i** notation for modelling laws and regulations of the European food supply chain, and the resulting models have been compared with corresponding models, built previously with the basic *i** approach (basic *i**). Along with the comparison we addressed specific questions aimed at finding evidence of the effectiveness and the efficiency of normative *i**. Concerning effectiveness, from this experience it turned out that using normative *i** we were able to infer about the existence of several new goals and actors strictly related to the normative context, which were otherwise probably ignored. As for the efficiency of using normative *i**, we tried to characterize it in terms of extra time costs for this further analysis of the domain, resulting in about 5% of the overall time spent in modelling-related activities. An extra cost to be contrasted with the gain in modelling effectiveness. As a concluding remark, we consider our experience significant towards proving the effectiveness and efficiency of normative *i** modelling. Large-scale applicability could be evaluated through an empirical study, asking two groups of analysts to perform basic *i** and normative *i** modelling in parallel [20], but to be feasible, this type of analysis will require a lab-size case-study.

From this experience we derived some interesting work directions for the future.

- The normative *i** framework needs to be supported by a formal semantics. A conceptual meta-model will complete the framework and make it comparable to other approaches. Work is currently ongoing in this direction.
- The normative and basic *i** could be integrated into one single interleaved methodology, also in order to minimize the possible model reconciliation effort.
- As pointed out in [12], a major problem in se is the traceability of normative prescriptions. Being able to separate normative from strategic requirements is the first step towards supporting traceability along the different phases of the software development.

Acknowledgements. This work was funded by the EU-funded FP6 TRACEBACK project FP6-2005-FOOD-036300.

References

1. Breaux, T.D., Vail, M.W., Anton, A.I.: Towards Regulatory Compliance: Extracting Rights and Obligations to Align Requirements with Regulations. In: Proceedings of the 14th IEEE International Requirements Engineering Conference (RE 2006), pp. 49–58. IEEE Society Press, Los Alamitos (2006)
2. Dardenne, A., van Lamsweerde, A., Fickas, S.: Goal-Directed Requirements Acquisition. Science of Computer Programming 20, 3–50 (1993)
3. Darimont, R., Lemoine, M.: Goal-oriented analysis of regulations. In: International Workshop on Regulations Modelling and their Verification & Validation (2006)

4. Ghanavati, S., Amyot, D., Peyton, L.: A Requirements Management Framework for Privacy Compliance. In: The 10th Workshop on Requirements Engineering (WER 2007), pp. 149–159 (2007)
5. Giorgini, P., Massacci, F., Mylopoulos, J., Zannone, N.: Requirements Engineering meets Trust Management: Model, Methodology, and Reasoning. In: Proc. of the 2nd International Conference on Trust Management (iTrust 2004) (2004)
6. Jones, S.V., Maiden, N.A.M.: RESCUE: An Integrated Method for Specifying Requirements for Complex Socio-Technical Systems. In: Mate, J.L., Silva, A. (eds.) Requirements Engineering for Socio-Technical Systems, pp. 245–265. Ideas Group (2005)
7. Lockerbie, J.A., Maiden, N.A.M.: REDEPEND: Extending i* Modelling into Requirements Processes. In: Proceedings 14th IEEE International Conference on Requirements Engineering, pp. 361–362. IEEE Computer Society Press, Los Alamitos (2006)
8. Maiden, N.A.M., Jones, S.V., Manning, S., Greenwood, J., Renou, L.: Model-Driven Requirements Engineering: Synchronising Models in an Air Traffic Management Case Study. In: Persson, A., Stirna, J. (eds.) CAiSE 2004. LNCS, vol. 3084, pp. 368–383. Springer, Heidelberg (2004)
9. Maiden, N.A.M., Manning, S., Jones, S., Greenwood, J.: Generating Requirements from Systems Models using Patterns: A Case Study. Requirements Engineering Journal 10(4), 276–288 (2005)
10. North, D.C.: Institutions, Institutional Change, and Economic Performance. Cambridge University Press, Cambridge (1990)
11. Perini, A., Susi, A.: Designing a Decision Support System or Integrated Production in Agriculture. An Agent-Oriented approach. Environmental Modelling and Software Journal 19(9) (September 2004)
12. Otto, P.N., Antón, A.I.: Addressing Legal Requirements in Requirements Engineering. In: 15th IEEE Inter. Requirements Engineering Conference, pp. 5–13 (2007)
13. Robertson, S., Robertson, J.: Mastering the Requirements Process. Addison-Wesley, Reading (1999)
14. Vicente, K.: Cognitive work analysis. Lawrence Erlbaum Associates, Mahwah (1999)
15. Siena, A.: Engineering Normative Requirements. In: 1st International Conference on Research Challenges in Information Science (RCIS 2007) (2007)
16. Stamper, R., Liu, K., Hafkamp, M., Ades, Y.: Understanding the Role of Signs and Norms in Organisations - a semiotic approach to information systems design. Journal of Behaviour and Information Technology (2000)
17. Sutcliffe, A.G.: Analysing the Effectiveness of Socio-technical Systems with i*, in Requirements Projects: Some Experiences and Lessons. In: Giorgini, M., Mylopoulos, Y. (eds.) Social Modeling for Requirements Engineering. MIT Press, Cambridge (2007)
18. Sutcliffe, A.G., Maiden, N.A.M., Minocha, S., Manuel, D.: Supporting Scenario-Based Requirements Engineering. IEEE Transactions on Software Engineering 24(12), 1072–1088 (1998)
19. Van Kralingen, R.: A Conceptual Frame-based Ontology for the Law. In: First International Workshop on Legal Ontologies (1997)
20. Wohlin, C., Runeson, P., Hoest, M., Ohlsson, M., Regnell, B., Wesseln, A.: Experimentation in Software Engineering - An Introduction. Kluwer Academic Publishers, Dordrecht (2000)
21. Yu, E., Mylopoulos, J.M.: Understanding "Why" in Software Process Modelling, Analysis and Design. In: Proceedings, 16th International Conference on Software Engineering, pp. 159–168. IEEE Computer Society Press, Los Alamitos (1994)

Towards a Catalogue of Patterns for Defining Metrics over *i** Models

Xavier Franch and Gemma Grau

Universitat Politècnica de Catalunya (UPC)
UPC – Campus Nord, Omega building, 08034 Barcelona, Spain
franch@lsi.upc.edu, ggrau@lsi.upc.edu

Abstract. Metrics applied at the early stages of the Information Systems development process are useful for assessing further decisions. Agent-oriented models provide descriptions of processes as a network of relationships among actors and their analysis allows discerning whether a model fulfils some required properties, or comparing models according to some criteria. In this paper, we adopt metrics to drive this analysis and we propose the use of patterns to design these metrics, with emphasis in their definition over *i** models. Patterns are organized in the form of a catalogue structured along several dimensions, and expressed using a template. The patterns and the metrics are written using OCL expressions defined over a UML conceptual data model for *i**. As a result, we promote reusability improving the metrics definition process in terms of accuracy and efficiency of the process.

1 Introduction

Measuring is a central task in the Information Systems (IS) development process. Some measures are used to evaluate an already built IS, for instance, by establishing its size according to the number of classes or lines of code, or by checking that the resulting system accomplishes its non-functional requirements fit criteria. However, measures can also be taken at the early stages of the IS development process, where they allow predicting some of the quality factors of the system-to-be, and planning corrective actions if needed. Therefore, there are many approaches that propose metrics over specification artefacts such as Statechart Diagrams [14], Use Cases [28] or OCL expressions [27]. Other metrics focus on the business process using Workflow Diagrams [26] or propose a mixed approach using Maps for representing the business process and UML class and transition diagrams for representing the system [8]. In this paper we are interested in the definition of metrics over a particular type of specification artefacts, namely goal-oriented models [24] expressed using the *i** framework.

The *i** framework [33] is currently one of the most widespread goal-oriented modelling and reasoning frameworks in IS development. One of its strengths is that it proposes different kinds of constructs that allow representing in a single model both the strategic needs of the business process and the operational specification of the IS. Because of that, it is currently used in different disciplines such as requirements

Z. Bellahsène and M. Léonard (Eds.): CAiSE 2008, LNCS 5074, pp. 197–212, 2008.

engineering [12], [25], business process modelling and reengineering [32], organizational modelling [23], or architecture representation [11], among others. As a consequence of this intensive work, the *i** framework has gained a solid position in the community and several research results can be considered as consolidated such as ontological issues [20], [21], metamodelling [30], development methods [2], [19], or reasoning algorithms [15], [29]. But on the other hand, this continuous research issues many new challenges to overcome, being one of them the analysis of IS represented with *i** models. One way to conduct such an analysis is by using metrics. Having good suites of metrics allows not only analysing the quality of an individual model, but also comparing different alternative models with respect to some properties in order to select the most appropriate alternative. Therefore, we claim that having a comprehensive catalogue of *i** metrics would be a significant contribution to the state of the art in *i** and, therefore, for IS development.

Aligning with this belief, in our group, we are using metrics on *i** models from some time ago [10], [9], [19] and we have formulated several metrics in different disciplines (see Section 2). One of our current lines of research is constructing a comprehensive framework of metrics over *i** models. During this research, we are facing two particular problems: (1) for some concepts to be measured, the expression of the metrics is cumbersome; (2) even in metrics with quite different purpose, there are some elements in the process that appear over and over, mainly the type and combination tactic of *i** elements to conform the metrics. The purpose of this paper is to formulate the basis of a catalogue of metrics over *i** models that helps to overcome the two problems mentioned above. With this aim, we are presenting a catalogue of patterns aimed at supporting the definition of metrics over *i** models.

The remainder of the paper is organized as follows. In section 2 we present an overview of the antecedents on metrics over *i** models and in section 3 we present a metamodel for the *i** framework (basic knowledge on *i** is assumed). Our final purpose is to define a catalogue of patterns to be used when defining metrics over *i** models and, for doing this, in section 4 we propose a template for documenting the patterns. In order to guide their use, patterns can be classified into several categories by forming a structured catalogue, which we present in section 5. In section 6 we show an example of application. Finally, we end with the assessment, the conclusions and future work in section 7.

2 Antecedents

Roughly speaking, we may classify existing proposals for analysing *i** models into quantitative and qualitative depending on the dominant dimension. In the quantitative-dominant side, we mention the AGORA method [22] that provides techniques for estimating the quality of requirements specifications with emphasis in the AND/OR decomposition of goals. Sutcliffe and Minocha [31] propose the analysis of dependency coupling for detecting excessive interaction among users and systems; they combine quantitative formulae based in the form of the model with some expert judgment for classifying dependencies into a qualitative scale. Bryl *et al.* propose structural metrics for measuring the Overall Plan Cost of an agent-based system [5]. For qualitative-predominant techniques, we mention Yu's seminal contribution [33] concerning ability, workability and commitment, and also [15], [29], which combine qualitative assessment with some rules for forward and backward propagation over AND/OR decomposition graphs in the context goal satisfaction analysis.

Concerning our own work, our first proposal appeared in [11]. In this paper, we used *i** to model component-based software architectures and defined six different metrics aimed at informing the selection of the most appropriate architecture with respect to some software requirements: diversity, vulnerability, packaging, data self-containment, uniformity and connectivity. Then, in [10] we defined a general framework. We focused on the following metrics: data privacy, data accuracy, process agility and responsibility dissolution. This framework was formalised in [9], defining a UML metamodel and using OCL as metrics definition language. We explored as metrics: predictability of models, and segregation of duties (in the extended version of [9] as technical report). Finally, in [17] and [18] we applied our approach in reengineering processes obtained by customization of ReeF, a generic Reengineering Framework [16]. In [17], we explored reengineering of software architectures over a documented case study and for this purpose we defined over the *i** framework two classical metrics, coupling and cohesion. In [18], we targeted reengineering of software processes and we focused mainly on defining the functional size of a software system in the COSMIC-FFP framework (using then cfsu, COSMIC functional size unit, as metric); we also included some results about process agility and ease of communication in the considered organizational alternatives. Other metrics are not currently available in the form of publications.

From ours and others' previous research, we observed the following facts:

- *i** is a versatile framework to represent concepts at different levels of abstraction (organizations, processes, architectures, etc.) and allows stating pertinent metrics over the models built. However, for some metrics, some additional information is required, as the type of actor, the criticality of a resource or task, etc.
- Purely qualitative or quantitative approaches do not exist, therefore we must be able to combine techniques of both kinds in the framework to make it usable.
- The process of definition of metrics in *i** may be cumbersome. An example is actors' predictability as defined in [9], which required six OCL predicates along with some auxiliary let-expressions for its definition.
- However, even in metrics with quite different purpose, there are some elements in the process that appear over and over, making thus both feasible and convenient to define a catalogue of patterns for defining *i** metrics.

3 A Metamodel for the *i** Framework

The *i** framework is both a goal- and agent-oriented framework with the aim of modelling and reasoning about organizational environments and their information systems. For doing so, it offers a formal representation of the involved actors and their behaviour. *Actors* can be specialized into *agents*, *roles* and *positions*. A position *covers* roles. The agents represent instances of actors within the organization and they *occupy* positions (consequently, they *play* the roles covered by positions).

The *i** framework proposes two types of models for modelling systems each one corresponding to a different level of abstraction: the Strategic Dependency (SD) and the Strategic Rationale (SR). An SD model consists of a set of nodes that represent actors and a set of dependencies that represent the relationships among them. Dependencies express that an actor (*depender*) depends on some other (*dependee*) in order to obtain some objective (*dependum*). Depending on the *dependum* kind, the *depender* depends on the *dependee* to bring about a certain state in the world (goal dependency), to attain a goal in a particular way (task dependency), for the availability of a physical or informational entity (resource dependency) or to meet some non-functional requirement (softgoal dependency).

An SR model allows visualizing the intentional elements into the *boundary* of an actor in order to refine the SD model with reasoning capabilities. The dependencies of the SD model can be linked to the appropriate intentional elements (also classified as goals, softgoals, tasks and resources) inside the actor boundary. The elements inside the SR model are decomposed accordingly to three types of links. *Means-end links* establish that one or more intentional elements are the *means* that contribute to the achievement of an *end*. The "end" can be a goal, task, resource, or softgoal, whereas the "means" is usually a task. In *Means-endContribution* links, with a *softgoal* as end, it is possible to specify if the contribution of the means towards the end is negative or positive. *Task-decomposition links* state the decomposition of a task into different intentional elements. We refer to [1] for details about usage of links. *Scenario paths* (also called routines in [33]) are composed of tasks and goals. For more details on the *i** framework, we refer to [33].

In Fig. 1 we present a metamodel for the *i** framework that represents the explained concepts that will be used for defining metrics. The metamodel is essentially the same as in [9] (which in its turn is similar to other existing proposals) but including a

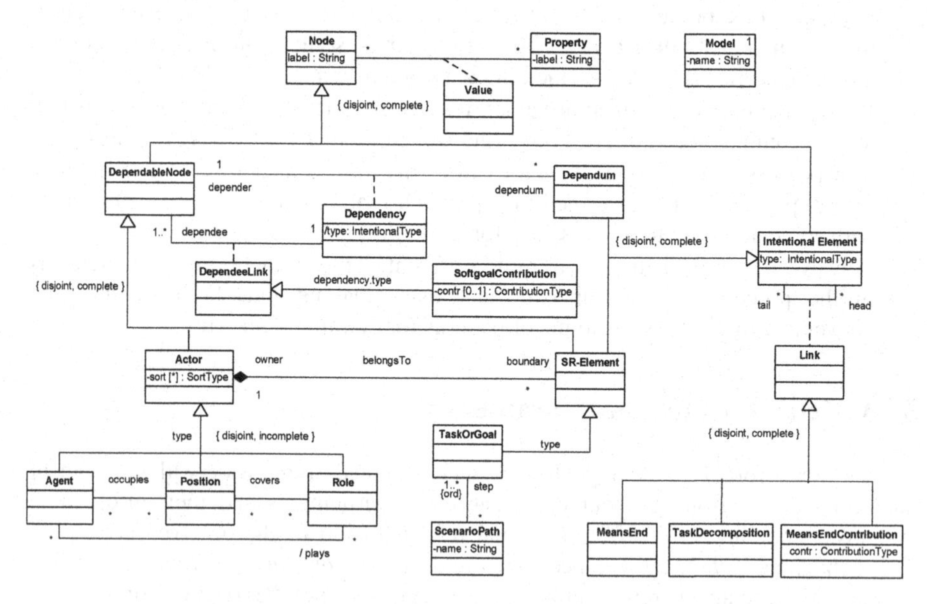

Fig. 1. A UML class diagram for *i** (some specializations are omitted due to the lack of space)

Table 1. Example of pattern: the Sum pattern

Name	**Sum**
Context	A metric is defined over two types of different model elements such that elements of one type (aggregated) contain elements of the other (aggregee)
Problem	There is a need of computing the aggregated metric in terms of aggregee's
Solution	Define aggregate's metric, Aggregated::metric, as the sum of aggregee's
Involved Classes and Types	Aggregated: <<Node>> -- aggregated's class in the *i** metamodel Aggregee: <<Node>> -- aggregee's class in the *i** metamodel Type: <<DataType>> -- the type of the metric
Assumptions	– The metric ranges onto a numerical data type – The Aggregated class is an aggregation (either direct or transitive) of Aggregee – There is a definition of the metric over the Aggregee, Aggregee::metrics
Required Knowledge	– The relationship of aggregation from Aggregated to Aggregee, Aggregated::aggregees(): Set(Agregee)
Form	**context** Aggregated::metric(): Type **post**: result = self.aggregees().metric()->sum()
Related Patterns	– Numerical patterns (e.g., Normalization) to manipulate the result – Navigational patterns (e.g., All Elements of a Kind) to define aggregees – Discrimination patterns (e.g., Discrimination By Type) to filter the aggregees
Example of Use	In the context of summing the size of the resources managed by an actor: **context** Actor::size(): Integer **post** result = self.allResources().size()->sum()

singleton class for models, types of actors, and the possibility to attach properties that may influence metrics to model elements, improving thus expressiveness. OCL integrity constraints are not included for the sake of brevity.

4 The Pattern Template

As mentioned in section 1, when analysing *i**-based metrics, we observed some patterns that appeared over and over. Their identification and classification would help in metrics definition and reuse. In this section we define the information enclosed in patterns definition, which yields to a proposal of template.

The structure of the pattern is as follows:

- **Name**, **context**, **problem**, **solution**, **related patterns**, and **example of use.** These are usual components of every pattern-based proposal such as [13] and have proven their usefulness for documentation purposes.
- **Involved classes** and **types.** Elements from the *i** metamodel that appear in the definition of the pattern. Also, the data types required appear here.
- **Assumptions.** The concept of assumption as proposed in [4] is included in order to embody the intuitive knowledge about the pattern. Assumptions provide the basis over which the metrics are defined.
- **Required knowledge.** Domain-related information that has to be provided in order to effectively use the pattern.
- **Form.** An OCL expression that defines the pattern in terms of the involved classes and types, the required knowledge and under the stated assumptions.

Table 1 shows an example of the documentation template for the Sum pattern. This pattern computes the value of a metric applied on a model element (*aggregated*) in terms of the same metric applied on some other model element (*aggregee*) which is related with the former by an aggregation relationship (*aggregees*). This is one of the most used patterns when a metric is decomposed top-down, e.g. a model metric as the sum of the metrics applied to its actors, or an actor metric as the sum of the metrics applied to its intentional elements. Often, the result is modified with some Numerical pattern (e.g., the result may be normalized into a given interval). Also, the aggregee's metric is sometimes applied just to those aggregees that satisfy some property (e.g., software actors, or resource intentional elements) using some Discrimination pattern. Last, some Navigational templates may be used to generate the *aggregates* function, as it would happen for the allResources that appears in Table 1. These types of related patterns are presented in the next section.

A concept that is fundamental when using the patterns is that of pattern instantiation. When a pattern is fully instantiated, it becomes an OCL expression completely determined, ready to be evaluated. A complete and correct instantiation requires specifying which actual model elements, types and knowledge play the parts identified in the pattern, which of course must fulfil the stated assumptions. A complete and correct instantiation of the Sum pattern presented in Table 1, to obtain the metric defined in the Exemple of Use part of the template, would be declared as:

```
size ::= Sum[Aggregated ::= Actor, Aggregee ::= SR-element, Type ::= Integer,
              aggregees() ::= allResources()]
```

5 The Catalogue of Patterns

In order to facilitate the definition and reuse of the proposed patterns, we adhere to the catalogue definition stated in [13], which includes classification criteria as part of the pattern. From our experience, we have identified a four-dimension classification of the patterns, organized into categories and subcategories.

Table 2 presents these main categories and subcategories as well as s representative sample of patterns (we have near one hundred of such patterns and therefore it is not feasible to present all of them here). In the process of conforming metrics, several of these patterns can be instantiated either sequentially or nested. The definitions are provided below (together with some examples; we skip the Examples of Use clause since all these patterns are used in the example of section 6):

- **Metrics declaration.** The first decision to take when conforming metrics is to decide their concrete form. We have two different criteria to identify patterns:
  - The subject of measure. The metrics may apply to the whole model, an individual model element (e.g., an actor; see Table 3) or in the middle, a set of model elements (e.g., metrics to analyse pairs of actors). Depending on this granularity, it is possible to determine the context of the OCL expression used for the metric.
  - The objective of the metrics. Its effect is to determine the OCL expression return value's type: enumeration or string; numerical; Boolean; or aggregation of model elements. It may be a classification instrument (i.e., a nominal metric), a measuring instrument (from an ordinal, absolute or ratio scale), a condition-checker (checking if a given domain property is attained) or a

locator (searching for a model element, or aggregate of model elements, that satisfy a condition). A measuring instrument may be used as the basis to obtain instruments or the other type. For instance, a Boolean metric may be defined as a numerical metric compared to a certain threshold value; or a sequence of model elements may be ordered with respect to the numerical metric value.

Table 2. Overview of the proposed catalogue of patterns

Category	Subcategories			Pattern
Metrics Declaration	Subject			Model
				Set of model elements
				Individual Element
	Result			Classification Instrument
				Measuring Instrument
				Condition-Checker
				Locator
Metrics Definition	Qualitative			By Criterion
				Individual
				Global Information
	Quantitative	Structural	Aggregation	Sum
				Count
			Discrimination	By Type
				By Type and Value
			Element-Based	Actor-Based
				Dependency-Based
		Property-Based		
Metrics Transformation	Numerical	Inverse		
		Average		
		Normalization		
Metrics Auxiliary Elements	Navigational	All Elements of a Kind		
		Superclass		
		Transitive Clousure		

- **Metrics core definition.** To define the metrics (i.e., the body of the OCL expression), the discussion about qualitative and quantitative predominance issued in section 2 drives our further classification:
  - Qualitative-predominant definition. Their use appears when the metric is strongly domain-dependent, or manages concepts that do not appear explicitly in the model (e.g., actors' tacit knowledge). In these cases, a domain expert must provide the needed knowledge. We have mainly three subcategories:
    - ◊ By criterion. Each model element that satisfies some criterion (e.g., being a resource dependency, or a human actor) has a given value assigned.
    - ◊ Individual. Each model element has a value assigned (see Table 4).
    - ◊ Global information. A particular item of information which represents global knowledge that may affect many metrics. This information is considered as owned by the Model single instance (i.e., it may be considered as an attribute of the class Model).
  - Quantitative-predominant definition. The model encloses all the necessary information for computing the metrics. We distinguish two big subcategories:

◊ Structural. The metric is computed from the form of the model, using its structure: actors, dependencies, etc. There are quite a lot of patterns belonging to this category, see Table 2 for some of them. This subcategory is divided into three. Aggregative patterns generalize the idea explained in section 3 for the Sum pattern, which in fact is a particular case, as the Count pattern, which are the two most used aggregative patterns. Discrimination patterns use some criteria on model elements to compute the metric (see Table 5). Element-based patterns define a metric on a basic element of the model, typically dependencies or actors (see Table 6).

◊ Property-based. The metric uses domain properties that are embodied in the model using the *property* operation from the Node class.

- **Metrics transformation.** They modify the value by applying some transformation, usually numerical. Typical examples are the inverse function and the average and normalization within a range (see Table 7).
- **Metrics auxiliary elements.** They capture repetitive situations when defining metrics. The most significant subcategory is navigational patterns, which provide OCL expression types to navigate and obtain (an aggregate of) a model element (e.g., all the tasks that form a routine, or a dependency's depender).

Categories may have a template bound, which makes possible to implement the catalogue hierarchy by using the concept of pattern specialization: a subcategory

Table 3. Example of pattern: the Subject->Individual Element pattern

Name	Individual element (Metrics Declaration -> Subject)
Context	The process of conforming a metric has just started
Problem	The metric has no sense for most types of model elements except one
Solution	Define the metric with this type of element as context
Involved ...	Elem: <<Node>> -- class corresponding to this type of element in the *i** metamodel
Assumptions	N/A
Required ...	N/A
Form	**context** Elem::metric(): STILL TO KNOW **post**: STILL TO KNOW
Related patterns	– Result patterns (e.g., Locator) to declare the result – Metrics definition patterns to complete the definition

Table 4. Example of pattern: the Qualitative->Individual pattern

Name	Individual (Metrics Definition -> Qualitative)
Context	In the process of defining a metric, a particular type of model element must be assessed
Problem	– Just the SD model is available (not the SR) and this implies lack of information, or – Quantitative analysis may be unacceptably costly or not feasible
Solution	Provide individual qualitative assessment for each model element
Involved ...	Type: <<DataType>> -- the type of the metric
Assumptions	N/A
Required knowledge	function to represent expert judgement on each individual model element of the type, judgement: String → Type such that domain(judgement) = Node.allInstances().label
Form	**context** Node::ExpertJudgement(): Type **post**: result = judgement(self.label)

Table 5. Example of pattern: the Discrimination by Type pattern

Name	**By Type (Metrics Definition -> Quantitative -> Structural -> Discrimination)**
Context	Some metrics that are defined over one particular type of node (e.g., actor, dependency, SR-element) have a value that depends on the subtypes of that type (e.g., for actor: agent, position and role).
Solution	Apply polymorphism on the model, applying e.g. hook strategy: – Declare an operation for that metric over the type of node which returns that value (hook version of the pattern) – Redefine the operation on those subtypes that assign a different value
Involved Clas-ses and Types	Parent: <<Node>> -- class whose metric' value is being computed Type: <<DataType>> -- the type of the metric
Assumptions	– Parent is root of a hierarchy, being $Heir_1$, …, $Heir_k$ its subclasses
Required knowledge	A total function assign which maps each subtype to the appropriate value, ass: <<Node>> → Type, such that domain(assign) = Parent's subclasses
Assumptions	$\forall j$: $1 \leq j \leq k$: $Heir_j$::metric is not defined explicitly $\Rightarrow$ ass($Heir_j$) = Value
Form	– **context** Parent::metric(): Type **post**: result = Value – $\exists i$: $1 \leq i \leq k$: [$\forall j$: $1 \leq j \leq i$: **context** $Heir_j$::metric(): Type **post**: result = *ass*($Heir_j$)]

Table 6. Example of pattern: the Dependency-Based pattern

Name	**Dependency-Based (Metrics Definition -> Quantitative -> Structural)**
Context	Some metrics have sense when applied to dependency links
Problem	The metrics will depend not just on the characteristics of the dependency link itself, but also on the two actors that act as depender and dependee
Solution	Identify three different factors that influence the metrics: one bound to the dependency link itself (probably related with the type of its dependum), and the others to the two actors, depender and dependee
Involved …	N/A
Assumptions	N/A
Required knowledge	– The effect of the depender, the dependee and the dependum in the metric, represented by three functions: filter: Dependum → Float correctionFactorDepender: Actor → Float, correctionFactorDependee: Actor → Float
Form	**context** DependencyLink::metric(): Type **let** ownerActor(x: DependableNode): Actor = **if** x.oclIsTypeOf(Actor) **then** x **else** x.owner **in**: **post**: result = self.dependency.dependum.filter() * ownerActor(self.dependency.depender).correctionFactorDepender() * ownerActor(self.dependee).correctionFactorDependee()

Table 7. Example of pattern: the Normalization pattern

Name	**Normalization (Metrics Transformation -> Numerical)**
Context	Some metrics may have a value that depend on the number of elements of a certain type that have influenced the metric
Problem	Often, the value of the metric should not depend on the number of elements processed
Solution	Normalize the value into some interval
Involved …	N/A
Assumptions	– The metric ranges onto a numerical data type – The normalization interval is [0.0, 1.0]
Required knowledge	– The value to be normalized, Value – The number of elements used to compute Value, Size
Form	**context** Element::metric(): Type **post**: Size = 0 **implies** result = 1.0 **post**: Size > 0 **implies** result = Value / Size

Table 8. Template for the Aggregation category

Name	**Aggregation (Metrics Definition -> Quantitative -> Structural)**
Context	A metric is defined over two types of different model elements such that elements of one type (aggregate) contain elements of the other (aggregee)
Problem	The value of a metric applied over the aggregate depends on the value applied over the aggregee
Solution	Define aggregate's metric as the combination of aggregee's
Involved ...	Identical to Table 1
Assumptions	– Similar to Table 1
Required knowledge	– The relationship of aggregation from Aggregate to Aggregee, Aggregate::aggregees(): Set(Aggregee) – The combination function of aggregee's values into aggregate's, *aggregationFunction*

Table 9. Example of pattern specialization: from the Aggregation category into the Sum pattern

Name	**Sum**
Subtype-of	Aggregation[aggregationFunction ::= sum]
... (rest of the Sum pattern, see Table 1)	

specializes its supercategory by adding detail to some of the parts of the template. In this process, a partial instantiation is possible to bind some of the parameters of the supercategory pattern. As an example, in Table 8 we show an outline of the template for the Aggregation category and in Table 9 its specialization into the Sum pattern presented in Table 1. To make specialization explicit, we add a specialization clause in patterns. Eventually, we may get rid of some parts of the specialized template if there is redundancy.

6 Example of Application

In this section we show the applicability of the metrics pattern catalogue to one particular case, namely predictability as defined in [9].

Predictability is used in [23] as one of the properties of interest when analysing organizational styles. To obtain the metric as done in [9], we follow the following process (row's references in the text refer to Table 10):

- **Row 1.** From [23], we concluded that predictability is a measuring instrument for actors, therefore we declare the metric using the Individual Element and Measuring Instrument patterns. In [9], we defined actor predictability as the sum of the predictability of its stemming dependencies, including both the dependencies stemming from the actor itself and the dependencies stemming from the actor's intentional elements; to obtain these dependencies we apply a Navigation pattern. To obtain a normalized measure, between 0 and 1, we apply the normalization pattern to the value obtained from the sum described above. The process is described in detail in Fig. 2.

Table 10. Application of the metric definition process for Predicatibility

Row 1	See fig. 2, where this step is presented in detail and graphically

Row 2	Dependency::predictability = DiscriminationByType[Parent ::= Dependency; Type ::= Float; ass ::= dependencyTypePredictability]
	context Dependency::predictability(): Float
	  post: result = 1.0
	context GoalDependency::predictability(): Float
	  post: result = self.goalPredictability()
	context SoftgoalDependency::predictability(): Float
	  post: result = self.softgoalPredictability()

Row 3	SoftgoalDependency::softgoalPredictability =
	Dependency-Based[filter ::= dependency.knowHow; correctionFactorDepender ::= dependerExpertise; correctionFactorDependee ::= 1] -- does not affect the result
	context SoftGoal::softgoalPredictability(): Float
	  let ownerActor(x: DependableNode): Actor =
	    if x.oclIsTypeOf(Actor) **then** x **else** x.owner **in**
	  post: result = ownerActor(self.dependency.depender).dependerExpertise() * self.dependency.dependum.knowHow()

Row 4	Actor::dependerExpertise = ByCriterion[type ::= Float; criterion ::= true; judgement ::= Actor::expertise] **context** Actor::dependerExpertise(): Float **post**: result = expertise(self.label)

Row 5	**context** Dependency::knowHow(): Float -- no pattern applied
	  pre: self.type = Softgoal
	  let theModel = Model.allInstances()->any() **in**:
	  post: result = 1 - theModel.slope / contributionsToSoftgoalDep()+1
	Dependency::contributionsToSoftgoalDep = Count[Aggregate ::= Dependency; Aggregee ::= DependeeLink; Type ::= Integer; aggregee ::= self.dependeeLink]
	context Dependency::contributionsToSoftgoalDep(): Integer
	  post: result = self.dependeeLink.contributionsToSoftgoalDep()->size()
	DependeeLink::contributionsToSoftGoalDep = DiscriminationByTypeAndValue [Node ::= DependeeLink; type ::= Boolean; function ::= hasContributionLabel (as defined below)]
	context DependeeLink::hasContributionLabel(): Boolean **post**: result = false
	context SoftgoalContribution::hasContributionLabel(): Boolean
	  post: result = self.contr->notEmpty()

Row 6	GoalDependency::goalPredicatibility = Inverse[Type ::= Integer; Value ::= self.nbTaskCombinations(); ResultIfZero ::= 0]
	GoalDependency::allTaskCombinations = TransitiveClousure [Elem :: = GoalDependency; Result ::= Set(Task); Expr ::= --not shown for the space reasons]
	context GoalDependency::goalPredictability(): Float
	  let nbTaskCombinations(x: GoalDependency): Integer =
	    x.allTaskCombinations()->size() **in**
	  post: self.nbTaskCombinations() = 0 **implies** result = 0
	  post: self.nbTaskCombinations() > 0 **implies** result = 1 / self.nbTaskCombinations()

- **Row 2.** Yu provides some rationale about the degree of freedom bound to dependencies [33, p. 15]. Analysing this rationale, we concluded that task and resource dependencies are totally predictable whilst goal and softgoal ones are not. Therefore, we define predictability of dependencies in terms of their type applying the Discrimination By Type pattern.
- **Row 3.** For softgoal dependencies, since softgoal satisfaction involves a compromise among depender and dependee, we use the Dependency-Based structural pattern, in which the depender side expresses the expertise of the depender actor to take informed decisions, whilst the dependee side measures the available know-how about that dependency (measured in terms of contributions to softgoals inside the dependee's SR). Since we need to refer to the owner actor of a dependency, that may be established in terms of the actor itself or some intentional element therein, we use a Navigational pattern to locate that actor.

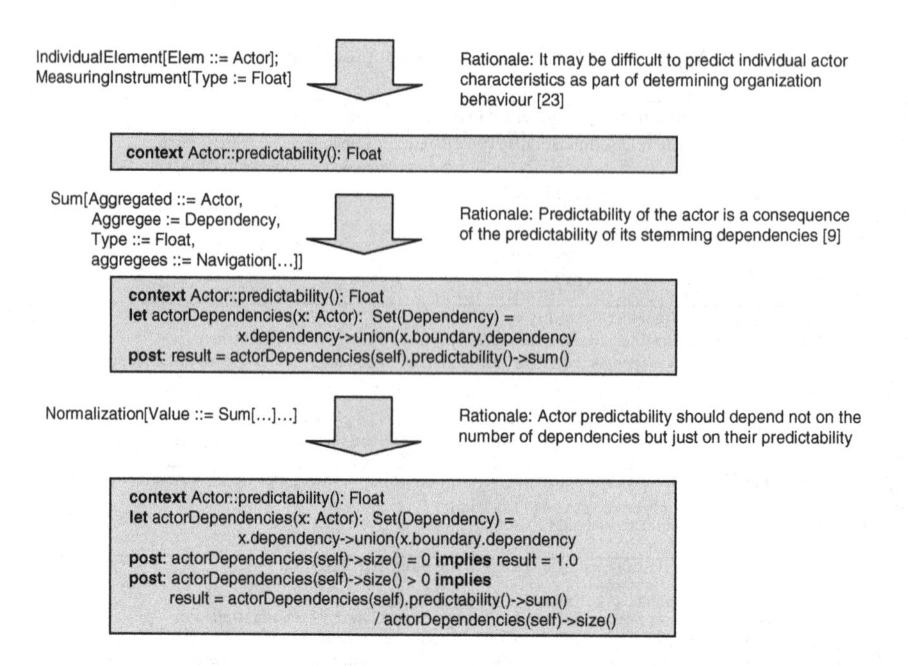
IndividualElement[Elem ::= Actor];
MeasuringInstrument[Type := Float]
Rationale: It may be difficult to predict individual actor characteristics as part of determining organization behaviour [23]
context Actor::predictability(): Float
Sum[Aggregated ::= Actor,
Aggregee := Dependency,
Type ::= Float,
aggregees ::= Navigation[...]]
Rationale: Predictability of the actor is a consequence of the predictability of its stemming dependencies [9]
context Actor::predictability(): Float
let actorDependencies(x: Actor): Set(Dependency) =
x.dependency->union(x.boundary.dependency
post: result = actorDependencies(self).predictability()->sum()
Normalization[Value ::= Sum[...]...]
Rationale: Actor predictability should depend not on the number of dependencies but just on their predictability
context Actor::predictability(): Float
let actorDependencies(x: Actor): Set(Dependency) =
x.dependency->union(x.boundary.dependency
post: actorDependencies(self)->size() = 0 implies result = 1.0
post: actorDependencies(self)->size() > 0 implies
result = actorDependencies(self).predictability()->sum()
/ actorDependencies(self)->size()

Fig. 2. Application of patterns in the first step of predictability definition

- **Row 4.** Concerning depender expertise, we considered that this knowledge cannot be computed from the model and then we apply a Qualitative pattern, namely By Criterion (of actor). Since the metric is applied to all actors, the criterion predicate must evaluate always to true.
- **Row 5.** Concerning know-how, we apply both the Count and Discrimination by Type and Value patterns to compute the number of dependees that state a contribution value to the dependum. In this case, however, we need to manipulate the result in order to create an inverse function as done in [9].
- **Row 6.** For goal dependencies, predictability was measured as the different ways of fulfilling the goal, generating all feasible task combinations using a Navigational pattern and then counting them (row 6). Then the Inverse Numerical pattern is applied to obtain the final result.

7 Conclusions and Future Work

In this paper we have motivated the need for having a catalogue of patterns for defining metrics over *i** models, proposed the general structure of such a catalogue, presented some patterns therein and illustrated their use with an example. Our framework facilitates the objective of analysing and comparing *i** models with respect some giving criteria in different contexts: business process reengineering, requirements validation, architecture assessment, etc.

We have validated this proposal using both our work and the related work. We comment here our own work. We have applied the framework retrospectively (as done in section 6) to 16 other metrics and obtained the following data: (i) 32 patterns of metrics definition (16 for Subject and 16 for Result); (ii) 44 for metrics definition, being Structural Discrimination patterns the most used by far (24 applications); (iii) 18 Numerical patterns applied, most of the times one for pattern as last manipulation to normalize the result; (iv) 11 Navigational patterns. It must be said that the metrics analyzed are not so complex as the Predictability studied in this paper. Also, we have checked with related work the applicability of the metrics with success. For instance, the metric proposed in [31] is a typical example also of applicability of Structural Discrimination patterns with qualitative assessment. This also happens with [22]. As an example, they define completeness as $\#\{i \in InitialGoal \mid \exists f \in FinalGoal \cdot AllPositive(i,f)\} / \#InitialGoal$. We may apply the sequence of patterns: Model+Classification Instrument for definition; Discrimination By Type and Value to apply the metric on goal satisfying the condition that are initial; Count for counting; Normalization to divide; Navigation to generate AllPositive. Similar for the others metrics in [22].

As a summary of this assessment, we may remark as fundamental characteristics of our approach the following:

- **Efficiency.** The process of defining metrics is greatly improved since the engineer needs just to identify the prototypical traits of the metrics and choose the appropriate patterns. The classification schema supports this selection. The existence of patterns aimed at capturing some time-consuming and cumbersome behaviour (numerical manipulation, navigational patterns) also supports efficiency.
- **Executability.** The use of OCL as metrics language allows using different types of tools, from OCL editors to validators and execution tools.
- **Expressivity.** Since the framework operates around the *i** conceptual data model, all the model elements may be considered. The addition of the concept of property allows defining not just structural metrics but also others more domain-oriented.
- **Robustness.** Using patterns, errors when defining the metrics are reduced once the patterns are validated. Numerical and navigational patterns are good examples of that. The use of assumptions in the pattern definition helps to establish explicitly which are the correctness conditions of the pattern.
- **Understandability and uniformity.** Using patterns, similar situations in different metrics are treated the same way, and the resulting metrics look similar, making easier their understanding.
- **Versatility.** The catalogue allows designing the metrics according to different concepts (the classification criteria): type of knowledge available, effort to invest, predominant model element, etc. Some of the patterns recognize the fact that it may be necessary to evaluate model elements in an individual basis, with some kind of qualitative judgement.

In relation to [7], we may say that our proposal helps in overcoming most of the drawbacks identified therein (at some degree, all, except modularity): refinement, because metrics can be defined at different levels of abstraction assessing the development process and helping in choosing between refinement alternatives; repeatability, because model similarity can be assessed by comparing values on appropriate metrics; complexity management, using the appropriate metrics (project-oriented metrics) to drive model management; traceability, considered in terms of what parts of the model correspond to which domain concepts; reusability, because metrics can be used to decide if one model can be used in some context; scalability, because metrics help to analyse large models from several points of view; domain applicability, since not just the plain concepts of a domain may be represented but also metrics already defined for this domain.

As future work, we mention:

- To complete the catalogue with new, validated patterns and metrics constructed with them. As part of this goal, we aim at identifying other domains that may benefit from the existence of such catalogue, e.g. configuration management. Also, we plan to complement the catalogue with some classification schema to allow browsing the catalogue in a systematic way.
- To incorporate the catalogue into our current *i** modelling tools, REDEPEND-REACT for modelling component-based system architectures [17] and J-PR*i*M for driving business process reengineering processes [19]. Both tools currently allow defining structural metrics using some forms.
- To generalize the framework from *i** to a more general context. Since most of the concepts presented here are not particular of *i**, this is a feasible and logical goal to abstract the patterns into its metamodel level, in order to obtain a more generic form to be customized into a particular language and model. We think that this approach may be applied to the family of modelling languages structurally similar to *i** models, with a graph-oriented form, which includes several goal-oriented and other proposals. In this line of research, we plan to pay special attention to the analysis of completeness of the catalogues.
- To improve the definition of patterns by using metamodeling approaches to software metrics definition as those proposed in [3], [6]. We think that both approaches are complementary and would benefit from each other: patterns are methodologically-oriented whilst metamodeling is more foundational-oriented.

References

1. Ayala, C.P., Cares, C., Carvallo, J.P., Grau, G., Haya, M., Salazar, G., Franch, X., Mayol, E., Quer, C.: A Comparative Analysis of i*-Based Goal-Oriented Modeling Languages. In: Proceedings 17th SEKE International Conference (2005)
2. Bresciani, P., Perini, A., Giorgini, P., Giunchiglia, F., Mylopoulos, J.: Tropos: An Agent-Oriented Software Development Methodology. Journal of Autonomous Agents and Multi-Agent Systems 8(3) (2004)
3. Franch, X., Burgués, X., Ribó, J.M.: A MOF-Compliant Approach to Software Quality Modeling. In: Delcambre, L.M.L., Kop, C., Mayr, H.C., Mylopoulos, J., Pastor, Ó. (eds.) ER 2005. LNCS, vol. 3716, pp. 176–191. Springer, Heidelberg (2005)
4. Briand, L., Morasca, S., Basili, V.R.: An Operational Process for Goal-Driven Definition of Measures. IEEE Transactions on Software Engineering 28(12) (2002)
5. Bryl, V., Giorgini, P., Mylopoulos, J.: Designing Cooperative IS: Exploring and Evaluating Alternatives. In: Meersman, R., Tari, Z. (eds.) OTM 2006. LNCS, vol. 4275, pp. 533–550. Springer, Heidelberg (2006)

6. Cachero, C., Calero, C., Poels, G.: Metamodeling the Quality of the Web Development Process' Intermediate Artifacts. In: Baresi, L., Fraternali, P., Houben, G.-J. (eds.) ICWE 2007. LNCS, vol. 4607, pp. 74–89. Springer, Heidelberg (2007)
7. Estrada, H., Martínez, A., Rebollar, O., Pastor, J.: An Empirical Evaluation of the i* in a Model-Based Software Generation Environment. In: Dubois, E., Pohl, K. (eds.) CAiSE 2006. LNCS, vol. 4001, Springer, Heidelberg (2006)
8. Etien, A., Rolland, C., Salinesi, C.: Measuring the Business / System Alignment. In: Proceedings 1st REBNITA International Workshop (2005)
9. Franch, X.: On the Quantitative Analysis of Agent-Oriented Models. In: Dubois, E., Pohl, K. (eds.) CAiSE 2006. LNCS, vol. 4001, pp. 495–509. Springer, Heidelberg (2006)
10. Franch, X., Grau, G., Quer., C.: A Framework for the Definition of Metrics for Actor-Dependency Models. In: Proceedings 12th IEEE RE International Conference (2004)
11. Franch, X., Maiden, N.A.M.: Modeling Component Dependencies to Inform their Selection. In: Erdogmus, H., Weng, T. (eds.) ICCBSS 2003. LNCS, vol. 2580, pp. 81–91. Springer, Heidelberg (2003)
12. Fuxman, A., Liu, L., Mylopoulos, J., Pistore, M., Roveri, M., Traverso, P.: Specifying and analizing early requirements in Tropos. Requirements Engineering Journal (REJ) 9(2) (2004)
13. Gamma, E., Helm, R., Johnson, R., Vlissides, J.M.: Design Patterns: Elements of Reusable Object-Oriented Software. Addison-Wesley, Reading (1995)
14. Genero, M., Miranda, D., Piattini, M.: Defining and Validating Metrics for UML Statechart Diagrams. In: Proceedings 5th ICEIS International Conference (2003)
15. Giorgini, P., Mylopoulos, J., Nicciarelli, E., Sebastiani, R.: Formal Reasoning Techniques for Goal Models. In: Spaccapietra, S., March, S.T., Kambayashi, Y. (eds.) ER 2002. LNCS, vol. 2503, Springer, Heidelberg (2002)
16. Grau, G., Franch, X.: ReeF: Defining a Customizable Reengineering Framework. In: Krogstie, J., Opdahl, A., Sindre, G. (eds.) CAiSE 2007 and WES 2007. LNCS, vol. 4495, pp. 485–500. Springer, Heidelberg (2007)
17. Grau, G., Franch, X.: A Goal-Oriented Approach for the Generation and Evaluation of Alternative Architectures. In: Oquendo, F. (ed.) ECSA 2007. LNCS, vol. 4758, pp. 139–155. Springer, Heidelberg (2007)
18. Grau, G., Franch, X.: Using the PRiM method to Evaluate Requirements Models with COSMIC-FFP. In: Proceedings MENSURA International Conference (2007)
19. Grau, G., Franch, X., Maiden, N.A.M.: PRiM: an i*-based process reengineering method for information systems specification. In: Information and Systems Technology (IST), vol. 50(1-2), Elsevier, Amsterdam (2008)
20. Guizzardi, R., Guizzardi, G., Perini, A., Mylopoulos, J.: Towards an Ontological Account of Agent-Oriented Goals. In: Choren, R., Garcia, A., Giese, H., Leung, H.-f., Lucena, C., Romanovsky, A. (eds.) SELMAS. LNCS, vol. 4408, pp. 148–164. Springer, Heidelberg (2007)
21. Jureta, I., Faulkner, S.: Tracing the Rationale Behind UML Model Change Through Argumentation. In: Parent, C., Schewe, K.-D., Storey, V.C., Thalheim, B. (eds.) ER 2007. LNCS, vol. 4801, Springer, Heidelberg (2007)
22. Kaiya, H., Horai, H., Saeki, M.: AGORA: Attributed Goal-Oriented Requirements Analysis Method. In: Proceedings 10th IEEE RE International Conference (2002)
23. Kolp, M., Castro, J., Mylopoulos, J.: Organizational Patterns for Early Requirements Analysis. In: Eder, J., Missikoff, M. (eds.) CAiSE 2003. LNCS, vol. 2681, Springer, Heidelberg (2003)
24. van Lamsweerde, A.: Goal-Oriented Requirements Engineering: A Guided Tour. In: Proceedings 5th ISRE International Symposium (2001)

25. Maiden, N.A.M., Robertson, S.: Integrating Creativity into Requirements Processes: Experiences with an Air Traffic Management System. In: Proceedings 13th IEEE RE International Conference (2005)
26. Reijers, H.A., Vanderfeesten, I.T.P.: Cohesion and Coupling Metrics for Workflow Process Design. In: Desel, J., Pernici, B., Weske, M. (eds.) BPM 2004. LNCS, vol. 3080, pp. 290–305. Springer, Heidelberg (2004)
27. Reynoso, L., Genero, M., Piattini, M., Manso, E.: Assessing the impact of Coupling on the Understandability and Modificaiblity of OCL expressions within UML/OCL combined models. In: Proceedings 11th METRICS International Symposium (2005)
28. Saeki, M.: Embedding Metrics into Information Systems Development Methods: An Application of Method Engineering Technique. In: Eder, J., Missikoff, M. (eds.) CAiSE 2003. LNCS, vol. 2681, Springer, Heidelberg (2003)
29. Sebastiani, R., Giorgini, P., Mylopoulos, J.: Simple and Minimum-Cost Satisfiability for Goal Models. In: Persson, A., Stirna, J. (eds.) CAiSE 2004. LNCS, vol. 3084, pp. 20–35. Springer, Heidelberg (2004)
30. Susi, A., Perini, A., Mylopoulos, J., Giorgini, P.: The Tropos Metamodel and its Use. Informatica 29(4) (2005)
31. Sutcliffe, A., Minocha, S.: Linking Business Modelling to Socio-technical System Design. In: Jarke, M., Oberweis, A. (eds.) CAiSE 1999. LNCS, vol. 1626. Springer, Heidelberg (1999)
32. Yu, E., Mylopoulos, J.: Understanding Why in Software Process Modelling, Analysis, and Design. In: Proceedings 16th IEEE ICSE International Conference (1994)
33. Yu, E.: Modelling Strategic Relationships for Process Reengineering. PhD. thesis, University of Toronto (1995)

Business Process Modelling and Purpose Analysis for Requirements Analysis of Information Systems*

Jose Luis de la Vara, Juan Sánchez, and Óscar Pastor

Department of Information Systems and Computation, Technical University of Valencia, Camino de Vera s/n, 46022, Valencia, Spain
{jdelavara,jsanchez,opastor}@dsic.upv.es

Abstract. Although requirements analysis is acknowledged as a critical success factor of information system development for organizations, problems related to the requirements stage are frequent. Some of these problems are lack of understanding of the business by system analysts, lack of focus on the purpose of the system, and miscommunication between business people and system analysts. As a result, an information system may not fulfil organizational needs. To try to prevent these problems, this paper describes an approach based on business process modelling and purpose analysis through BPMN and the goal/strategy Map approach. The business environment is modelled in the form of business process diagrams. The diagrams are validated by end-users, and the purpose of the system is then analyzed in order to agree on the effect that the information system should have on the business processes. Finally, requirements are specified by means of the description of the business process tasks to be supported by the system.

Keywords: Business process modelling, system purpose, BPMN, Map, task description.

1 Introduction

Requirements analysis has been widely acknowledged as a critical success factor of software projects [31]. If not properly addressed, requirements can cause a project to fail. Nevertheless, practical experience proves that problems can easily arise from the requirements stage of information system (IS) development for organizations. Some of these problems are lack of understanding of the business by system analysts, lack of focus on the purpose of the system, and miscommunication between business people and system analysts. Since these problems can hinder business/IT alignment [20][25], the IS does not fulfil organizational needs.

Requirements must be defined in terms of phenomena that occur in the business environment [36]. However, it is common for requirements documentation to be solution-oriented, to not reflect the business environment, or to only consist of a data model in the

* This work has been developed with the support of the Ministry of Education and Science of Spain under the project SESAMO TIN2007-62894 and the program FPU, and cofinanced by FEDER.

Z. Bellahsène and M. Léonard (Eds.): CAiSE 2008, LNCS 5074, pp. 213–227, 2008.

form of a class or entity-relationship diagram. As a solution, the requirements engineering community has acknowledged the importance of the use of business concerns to drive requirements elicitation [29]. More specifically, the importance of organizational modelling during requirements analysis [18] and the role that system analysts must play as business analysts [16] have both been acknowledged. Organizational models depict the structure and behaviour of an organization and are very useful in helping system analysts to properly understand the business environment and the system requirements. Among other approaches, business process modelling has been declared as a good approach for organizational modelling and also as a must for IS development [2][13].

Nevertheless, business process models are not always enough to analyze the business context. Organizations often decide to introduce or modify an IS to solve a specific problem or need. These problems or needs correspond to the goal that the system must fulfil, i.e., the system purpose. Therefore, it is important for system analysts to explore the goals of different stakeholders and the activities that they carry out so that they can define purposeful requirements [26]. When the system purpose is not very complex, it can be directly analyzed on business processes, but sometimes it requires a deeper analysis. In the latter case, the use of a goal-driven approach that facilitates purpose analysis helps to better understand the system purpose and, consequently, to better respond to end-user needs.

Furthermore, good communication between business people and system analysts is essential at the requirements stage [17]. However, it can be difficult to achieve because of the existence of a gap between business and computing domains [32] that can cause mismatches between what end-users say and what system analysts understand. One reason for miscommunication is that the requirements models used to interact can be hard to understand and validate by end-users because of their lack of computing background. Therefore, models that facilitate communication during requirements analysis should be used. According to the software community [8] and to cognitive experiments [1], the use of process models when developing an IS can make human understanding and communication easier, and can help interaction between system analysts and end-users.

This paper presents a requirements analysis approach based on business process modelling and purpose analysis in order to try to prevent the problems described above. It is the result of a project between the Technical University of Valencia and the company CARE Technologies (http://www.care-t.com). The approach is characterized by the use of business processes for organizational modelling and as starting point of requirements analysis, the focus on system purpose, the use of BPMN [23] for business process modelling, the use the goal/strategy Map approach [27] for system purpose analysis, the detailed specification of functional requirements, and the involvement of end-users. This involvement is very positive when modelling an organization [30] and is a success factor in software development [31].

Organizations are modelled in the form of business process diagrams. The diagrams are validated by end-users, and the system purpose is then analyzed in order to come on agreement on the effect that the IS should have on the business processes. Finally, requirements are specified by means of the description of the business process tasks to be supported by the IS. These well-defined requirements are the input of the subsequent development stages.

The paper is organized as follows: section 2 describes the notations of the approach; section 3 describes the case study; section 4 presents the description of the approach; subsections 4.1 and 4.2 describe purpose analysis and functional requirements specification in detail; section 5 describes the practical experience using the approach; section 6 revises related work; finally, section 7 presents our conclusions and future work.

2 Notations of the Approach

The approach uses two notations: BPMN and Map. This section describes them briefly and justifies their selection for business process modelling and purpose analysis. Their joint use is also explained and justified. For further details about BPMN and Map, see [23] and [27], respectively.

2.1 BPMN

We use BPMN for business process modelling. Its creators argue that it offers a notation that is understandable by all business process users (process analysts, IS developers, process managers…). Therefore, BPMN provides a standard that fills the gap between business models and their implementation.

The notation consists of a diagram, called Business Process Diagram (BPD), whose aim is to provide a means for the development of graphical models of business process operations. A BPD is designed from a set of graphical elements that make diagrams simple to develop and easy-to-understand. The graphical elements are flow objects (activities, gateways and events), connecting objects (sequence flows, message flows, and associations), swimlanes (pools and lanes), and artefacts (data objects, annotations, and groups).

With regard to the choice of BPMN, several surveys have evaluated its adequacy for business process modelling and have compared it with other notations. These surveys are based on different criteria, such as workflow patterns (e.g. [34]), quality principles (e.g. [22]), or the BWW representation model (e.g. [28]). From the result of these studies and our own experience, BPMN has three main advantages: it is probably the most expressive notation, it is easy to use and understand, and it is receiving strong support from practitioners and vendors. As a result, BPMN is considered the de facto standard for business process modelling.

2.2 Map

Map is a goal-driven approach whose aim is to capture the intentions (goals) of an enterprise or system and determine the strategies that can contribute to the fulfilment of these intentions. The emphasis on the concept of strategies as ways to achieve goals distinguishes Map from other goal-based approaches. This emphasis is motivated by the fact that stakeholders do not naturally make the distinction between goals and strategies, and, as a consequence, pitfalls can arise [27]. The size of a goal model can unnecessarily increase when strategies are expressed as goals, alternative ways to make the business can be more difficult to discover, and recognizing stable elements in a business (intentions) versus more versatile ones (strategies) can be more difficult. In addition, Map promotes variability analysis at the requirements stage.

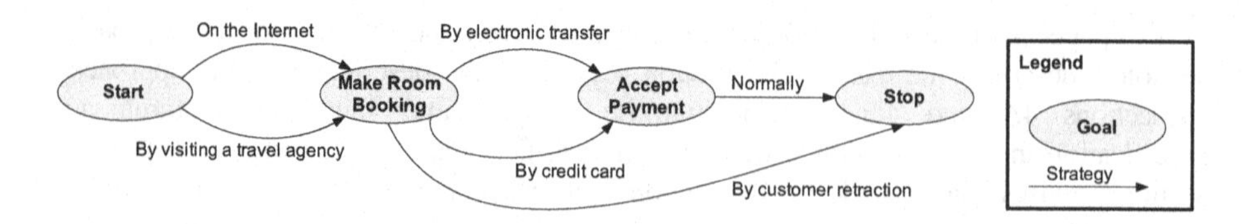

Fig. 1. Example of a map for a booking [27]

Map diagrams consist of a graph (called map) whose nodes are intentions and whose edges are strategies. An example is shown in Fig. 1. An edge entering a node identifies a strategy that can be used to achieve the intention of the node, so a map shows which intentions can be achieved by which strategies. Each map has two special intentions, Start and Stop, associated with the initial and final state, respectively. The aggregation of a source intention, a target intention, and a strategy is called section. Sections can be refined in another map.

Of all the goal-based approaches available within requirements engineering and business process reengineering, we use Map for the following reasons: it focuses on strategies to achieve goals; it has only two main concepts (goal and strategy), thus facilitating its use and understanding; and it does not deal with tasks, which we prefer to consider on a BPD instead of on a goal diagram.

Map has been used in several projects and in different areas, including business process modelling (for detailed references, see [27]). An important advantage of Map is that it encourages customer participation and helps to solve communication problems between business people and system analysts. However, we think that other notation should be used for business process modelling. We base this opinion on the criteria that are usually analyzed when evaluating business process notations, as described above for BPMN. The use of just two main concepts (goal, strategy) is useful for purpose analysis, but it is a drawback when addressing business process modelling because the models are neither detailed nor expressive enough to have a deep understanding and knowledge of business processes.

In summary, we think that BPMN and Map can complement each other. BPMN is better suited for business process modelling, but it does not provide any mechanism for system purpose analysis so that a new business process that fulfils organizational needs is designed. Map focuses on system purpose, and the use of strategies allows its graphs to be simpler than the goal diagrams of other approaches. As a result, the approach benefits from the features of both BPMN and Map.

3 Case Study

As a case study, we will use the business processes for the product development of a software company (Fig. 2).

The organization develops a software product that is provided to several customers. The product is standard, so no customer has a personalized product. However, customers can request improvements in the product, and the requests are included in future versions of the product.

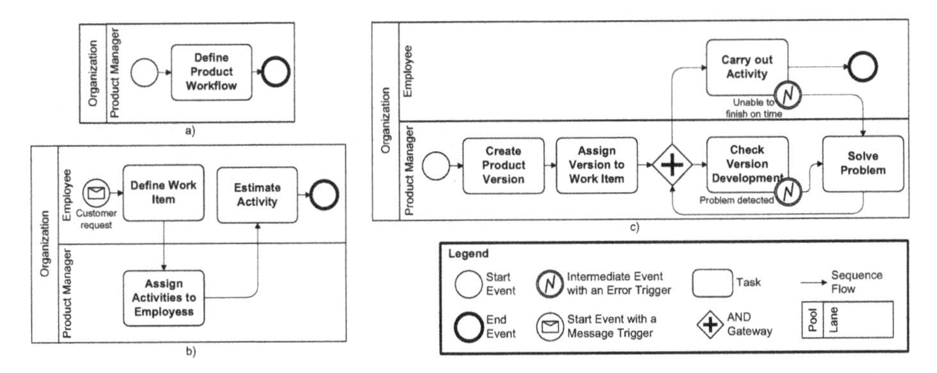

Fig. 2. BPDs for product development: a) definition of product workflow; b) request management; c) version development

The product manager defines the set of activities that has to be carried out to develop the product through product workflow. When a customer requests a new improvement, an employee defines the work item that is necessary to provide the customer with the request. Next, employees are assigned the activities that are necessary to develop the work item, and employees have to estimate how long the activities will take. The product manager is also responsible for the periodical creation of product versions, which have a strict deadline, and must decide the version in which a work item will be developed. Afterwards, employees carry out the activities in order to finish the work item and deliver the requested improvement, and the product manager checks that version development is correct. However, problems may arise while developing versions. Employees may not be able to finish the activities they are responsible for due to time constraints. If a problem arises, the product manager has to try to solve it.

4 Approach Description

The approach (Fig. 3) consists of three stages: organizational modelling, purpose analysis, and functional requirements specification. The first one depicts the current business environment (As-Is), which has a problem or need that could be fulfilled by an IS. The organization will change to solve the problem (To-Be), and the change will have an effect on business processes.

In the first stage, the organization for which the IS is going to be developed is modelled. For this purpose, the information gathered is a glossary, the business events, the business rules, and a role model. BPDs of the organization are created from this information. The diagrams must be validated by the end-users in order to guarantee that the organization has been properly modelled and understood. Several iterations are usually needed to get the final version.

The organizational problem or need is analyzed during purpose analysis stage. The aim is to find system strategies that can solve the organization problem, determine how to operationalize the strategies, and agree on the effect that the development of the IS may have on the business processes. This effect is the support or control that

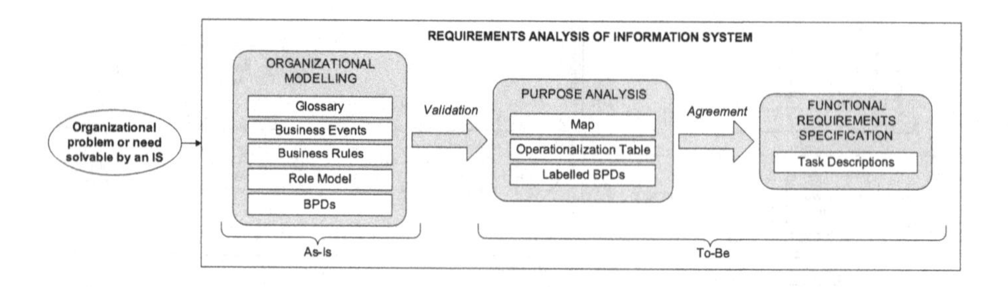

Fig. 3. Approach overview

the IS will have on them. As a result, business process elements are labelled according to the effect that the operationalization of the strategies will have on them, and changes in the business process can occur.

Finally, functional requirements are specified by means of the description of the business process tasks to be supported by the IS. Every task will have a textual template that describes it. The set of templates will be the starting point of the rest of the development stages.

For brevity, the organizational modelling stage is not described in the following sections, so only the purpose analysis and functional requirements specification stages are detailed.

4.1 Purpose Analysis

After organizational modelling, the system analyst has enough knowledge about the business processes to properly understand the organization activity. Nevertheless, the analyst also needs to understand the organizational problem or need. Consequently, purpose analysis is carried out from the business processes and the organizational problem or need to be fulfilled.

The introduction or modification of an IS in an organization can cause business process reengineering. An IS can initiates and supports reengineering before, during or after a process is designed [3]. In our approach, IS is a facilitator of the changes in the business processes because its effect is taken into account while designing the new processes. The new business processes (To-Be) are designed from the original ones (As-Is), the organizational needs, and the solutions that the IS can provide. IS will support business processes, and business processes are designed in terms of the IS capabilities.

Purpose analysis consists of three stages: map construction, map operationalization, and BPDs creation and labelling.

4.1.1 Map Construction

The organizational problem or need is modelled in a map where the solutions that the IS can provide are analyzed. The map is created in a participative manner between the system analyst and end-users, so they can agree on the solution. First, a map is created to analyze the problem or need. Second, the goals that the end-users want to achieve in order to solve the problem through the use of the IS are modelled as intentions (nodes). Third, system features that can fulfil the end-user goals are modelled as strategies (edges), which link the nodes. Finally, sections are refined if needed. These guidelines are based on the map construction process [27], but we have adapted them to the specific use of defining system strategies to fulfil end-user intentions.

The map that corresponds to the case study is shown in Fig. 4. The organization has been experiencing problems with delivery requests. Lack of knowledge about version development has caused requests to be delivered later than expected by customers. As a result, the customers have complained, and the strategic goal “Keep customer satisfaction” is not met. The main reason for the delay is that activity development is not always performed as planned because of the great amount of work that employees have to do. The product manager needs to be able to better project, for example, if an employee will miss working days, or if an employee has spent more time than planned on an activity. The product manager needs to foresee problems and find solutions quickly. In addition, employees need to be able to determine more accurately the time they have at their disposal to finish the activities that are in their charge of and how long these activities will take.

To solve these problems, employees wanted the IS to facilitate the work item development and to improve the knowledge they have about the status of the activities that they have to carry out. The product manager wanted the IS to improve the knowledge about the status of the versions and to minimize the time that takes a request to be delivered. The system analyst proposed system features that could fulfil these intentions and modelled them in the map in accordance with end-users.

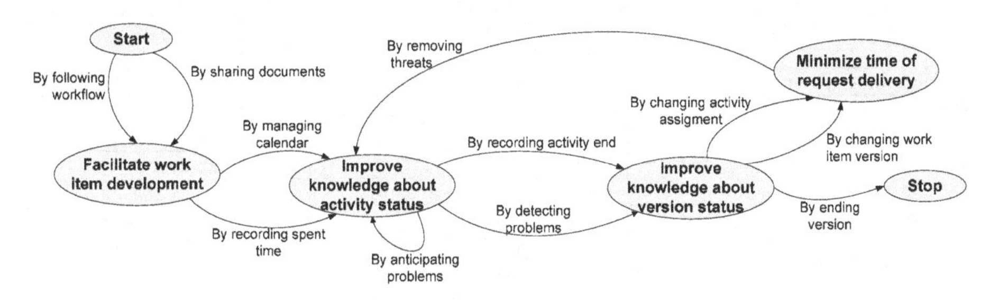

Fig. 4. Map for product development process

4.1.2 Map Operationalization

When the map is finished, the system analyst has to determine how to operationalize the map strategies, and come to an agreement on the effect that the operationalization will have on the old business process. Existing BPD elements can be removed or maintained, and new elements may be introduced. To facilitate this analysis, a table with three columns is created: a column to list the strategies; a column to list the BPD elements that will operationalize each map strategy and specify if the element has been removed (R), maintained (M), or it is new (N); and a column to specify the participant that will be in charge of the element.

Table 1 shows the BPD elements that operationalize each map strategy for the case study. There are several new elements: “Start activity” refers to the task in which an employee begins the performance of an activity and has to receive the necessary documents to carry out the activity; “Finish Activity” refers to the task in which an employee finishes an activity and has to share the documents related to the activity performance; "Manage Calendar" refers to the task in which an employee divides the time that can spent in a working day; "Need to start activity" refers to the condition in which an employee must be notified that an activity has to be started in order to finish the work item before version deadline; "Change Activity Assignment" refers to the task in which a product manager changes the employee that is responsible for an activity in order to finish a work item before version deadline; "Change Work Item Version" refers to the task in which the product manager changes the version of a work item due to some problem; "Version deadline" refers to the moment in which the date of version release is reached; "Release Version" refers to the task in which the product manager releases a finished version.

Table 1. BPD elements that operationalize the map strategies of the case study

Map strategy	BPD element	Participant
By following workflow	Define product workflow (M)	Product Manager
	Assign Activities to Employees (M)	
	Start Activity (N)	Employee
	Carry out Activity (M)	
	Finish Activity (N)	
By sharing documents	Start Activity (N)	Employee
	Finish Activity (N)	
By managing calendar	Manage Calendar (N)	Employee
By recording spent time	Carry out Activity (M)	Employee
	Finish Activity (N)	
By anticipating problems	Estimate Activity (M)	Employee
	Need to start activity (N)	
By recording activity end	Finish Activity (N)	Employee
By detecting problems	Check Version Development (M)	Product Manager
	Problem detected (M)	
	Carry out Activity (M)	Employee
	Unable to finish on time (M)	
By changing activity assignment	Change Activity Assignment (N)	Product Manager
By changing work item version	Change Work Item Version (N)	Product Manager
By ending version	Version deadline (N)	Product Manager
	Release Version (N)	
By removing threats	Carry out Activity (M)	Employee
	Notify changes (N)	Product Manager

With regard to how the BPD elements operationalize the map strategies, we will use the map strategy "by following workflow" as example. The BPD elements that operationalize the map strategy are "Define Product Workflow" because it is the task in which the activities and documents of the product workflow are defined, and "Assign Activity to Employee", "Start Activity", "Finish Activity" and "Carry out Activity" because they refer to activities of the product workflow. Therefore, the workflow is followed because of the execution of these tasks.

4.1.3 BPDs Creation and Labelling

Finally, the system analyst and the end-users agree upon the design of the new business process. First, changes in the BPDs are modelled, i.e., elements can be removed or introduced according to the operationalization of the map strategies. Next,

BPD elements are labelled according to the IS support on them. Tasks, events with triggers, and gateways that depict decisions are labelled as: "O" (out of the system), if the element will not be part of the IS; "IS" (controlled by the system), if the IS will be in charge of its control and execution with no human participation; or "U" (executed by a user), if the element will be executed by a person that interacts with the IS.

For the case study, Fig. 5 shows the new business processes. The business processes have changed as a result of the introduction of the IS. Several new tasks and events have been defined, and there is a new business process (calendar management) and a sub-process (problem resolution). An interesting fact is that, as acknowledged by other authors [4], the new BPDs (software development-oriented) are more detailed than the original BPDs (organizational documentation-oriented).

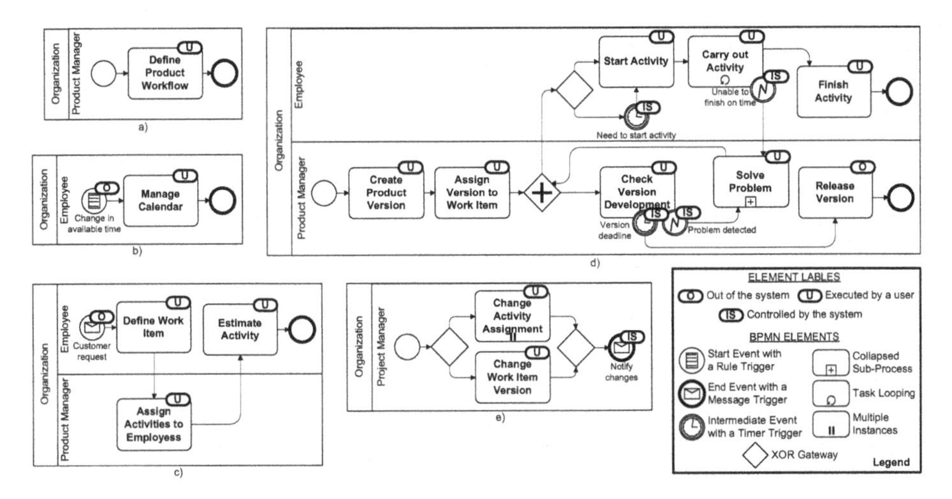

Fig. 5. Labelled BPDs for: a) definition of product workflow; b) calendar management; c) request management; d) version development; e) problem resolution

After labelling, the data objects that are input and output of every task to be supported by the system are defined. The state of the data objects before and after the execution of the task is specified. In order to keep Fig. 5 as simple as possible, the data objects are not shown for this case study.

4.2 Requirements Specification

In the last stage of the approach, functional requirements are specified from the labelled BPDs. For this purpose, business process tasks to be supported by the system are described through a textual template.

The content of the template is based on essential use cases [7] and task & support descriptions [19]. An essential use case is a simplified form of use case that depicts an abstract scenario for a complete and intrinsically useful interaction with a system from the perspective of users. A task & support description is a way to express what the system actors want to perform, including domain-level information and how a new system could support an activity to solve a problem. Both essential use cases and task & support descriptions contain the fewest presuppositions about technology.

An example of task template is shown in Table 2. It corresponds to the task "Carry out Activity" of the case study. A task template includes the business process to which the task belongs, the name of the task, the role responsible for its execution, the triggers, preconditions and postconditions of the task, the input and output data and their states, a specification of the interaction between a user and the IS through user intention and system responsibility, and the business rules that affect the task.

Table 2. Template for the task "Carry out activity"

Business Process: Version development			
Task: **Carry out Activity**		Role: Employee	
Triggers			
-			
Preconditions			
-			
Postconditions			
-			
Input		Output	
Data Object	State	Data Object	State
Activity	In progress	Activity	In progress
		Time register	
User intention		System responsibility	
		1. Show the activities in progress assigned to the employee	
2. Select an activity			
3. Carry out activity			
4. Indicate performance begin and end			
		5. Record time register	
		6. Update employee time	
Business Rules			
• An employee can only carry out an activity at the same time			
• An employee can only carry out activities that have been assigned to him			

All the information of a task template comes from its BPD. The name of the business process and the name of the task are the same as in the BPD. The role is the participant in the business process that is in charge of the task. The triggers correspond to the events with a trigger that precede the task in the business process and are part of the IS. The preconditions correspond to the gateways that precede the task, represent decisions, and are part of the IS. The postconditions correspond either to the gateways that follow the task, represent decisions that can make the business process iterate and are part of the IS, or to the end events with a trigger that follow the task and are part of the IS. The input and output of the task are its data objects in the business process. User intention and system responsibility are defined from the behaviour of the participant in charge of the task when executing it and how the system will support it. User intention may include actions that do not represent interactions with the system, and the system responsibility must be agreed upon with the end-user. Finally, the business rules specified in the template correspond to business rules that define or constrain the task, could not be modelled graphically, and have to be supported by the system.

After every business process task has been described, the subsequent development stages will be based on the task templates in order to provide the organization with an IS that fits its needs, its structure, and its behaviour.

5 Practical Experience

As explained above, the approach is the result of a project with a company, CARE Technologies. The purpose of the project is to solve problems related to the requirements stage by trying to link business and software domains.

After analyzing the requirements practices of the company, we identified problems related to business understanding, purpose analysis, and communication with end-users. The company uses OO-Method [24], a methodology for automatic software generation based on conceptual modelling. The data conceptual schemas consist mainly of a class diagram that is enriched with functional information about the result of class service execution. Analysts just provide some textual description about the requirements and validate them on the class diagram or on the generated application.

Although analysts feel comfortable with this technique, we think it could be improved. Some authors have stated that class diagrams alone might not be appropriate for communicating and verifying requirements, that there are few studies addressing the ability of end-users to understand class models, and that they can be complex for people that have not been trained in object-oriented modelling [12]. In addition, objects might not be a good way of thinking about a problem domain [33].

The approach has been used in four real projects in order to evaluate it and try to find improvements. It has been refined gradually based on the comments of both customers and analysts. The organizations for which the systems were developed were an apartment rental company, a car rental company, the organization of a golf tournament, and the organization of this case study. They were small/medium size projects. CARE had developed software for the organizations previously, so both the technique they usually use and our approach could be compared.

The system purpose was analyzed on the BPDs before the case study project. The reason for this is that the purpose analysis of the other projects was less complex, and thus easy to analyze. The introduction of the new IS did not change the business processes significantly, and the result was automation rather than reengineering. However, the system analysts said that some technique for purpose analysis would be helpful for the product development project.

We held between 2 and 5 meetings with end-users to obtain the business process models of the organizations, 1 meeting for purpose analysis, 1 or 2 meetings to define the task templates, and 1 meeting to validate the entire requirements specification. Another meeting was held in order to talk about the experience with end-users and analysts. Each meeting took approximately 2 hours.

As expected, the end-users stated that they could understand and validate the requirements models of the approach more easily than the class diagrams, thus facilitating communication and interaction. They also felt more involved in system development and claimed that they had a more participative attitude.

When asking analysts about the usefulness of the approach, we obtained different opinions. Although all the analysts stated that the approach allowed them to better understand the organizations, the system purpose, and, consequently, the requirements, there were some analysts who did not think that the approach could improve their job significantly and would probably not use it.

We do not find these comments about the approach discouraging. The analysts who did not think that the approach was very useful were senior analysts that are already very skilled in using OO-Method and interacting with customers. They usually model the systems while the customer describes what the system should do, so they can quickly generate it, validate it, and fix it if needed. However, most of the junior analysts, who have less experience in dealing with customers and, therefore, in understanding what is needed, considered that the approach could really help them.

We think these results are a reflection of common practices in IS development. Models are only used when they are believed to be useful [10]. In our case, some senior analysts do not think that the approach can accelerate their job, whereas junior analysts think that it can improve their performance.

Another interesting subject that arose while discussing the approach was the viewpoint to use when defining strategies in the map. We recommended the system analysts to include both system and enterprise strategies in order to better analyze the problem. However, they argued that the company mission was software development, not business consultancy. System analysts believed that BPDs and end-users goals were enough to properly understand the organization. Therefore, they did not consider business strategies that would not be supported by the IS to be useful for their work.

Finally, this practical experience has some limitations. First, the approach has to be used in more projects to draw definite conclusions. Second, the opinions of end-users and the analysts were obtained by discussing with them informally, so we are now designing a form to survey the next projects. Last but not least, we want to asses the approach by means of experiments with students and analysts from other companies.

6 Related Work

The need of organizational modelling and system goal analysis has been widely acknowledged within requirements engineering. Many approaches consider them to be the first step in software development, and some approaches use business process modelling. Nevertheless, the approaches usually focus on only one of the issues, and the use of models and techniques that facilitate the communication is not common.

Goal-oriented approaches have played an important role within requirements engineering. They try to solve the problem of systems that do not properly respond to organizational and user needs. Despite the acknowledged contributions of goal-oriented approaches, they have some weaknesses [26]. In our opinion, the main weaknesses are that goals might not provide a good starting point for requirements analysis, goal reduction is not straightforward, and most of the approaches do not pay enough attention to business concerns and business process reengineering. Apart from Map, two well-known approaches are i* [35] and KAOS [9].

i* is focused on the modelling of dependencies among the organizational actors in order to achieve organizational goals. It has been used in several software development methods, such as Tropos [6] or RESCUE [21]. However, several weaknesses have been identified by practitioners (e.g. [15]). i* diagrams might be too complex, and they should better support granularity and refinement.

KAOS requirements models are built from organizational goals. These goals are systematically refined to operational requirements through refinements patterns. KAOS also uses an object model, an agent responsibility model, and an operation model. When comparing the use of KAOS with Map, we have obtained simpler goal models with Map because of its focus on strategies. In addition, KAOS does not provide any mechanism to model and analyze business processes.

Among organizational modelling-based approaches, EKD [5] and ARIS [11] provide ways of analyzing an enterprise by using enterprise modelling. EKD is composed of a goal model, a business rules model, a concepts model, a business process model, an actor and resources model, and a technical components and requirements model. ARIS consists of five views: organizational view, data view, control view, function view, and product/service view. In our opinion, when using EKD and ARIS for tailored software development, their requirements specifications should be more detailed, and they should be more focused on goal analysis. EKD also lacks tool support that facilitates the development and maintenance of all its models. In addition, as we have stated above for Map, BPMN is better suited for business process modelling than the notations that EKD and ARIS propose.

Some approaches use UML for organizational and business process modelling (e.g. [14]). These approaches use elements that are close to those elements used in the software development area. This fact is a drawback, because the models are easy to use and understand by system analysts but might be too complex to be validated by end-users. In addition, the UML-based approaches do not focus on goal analysis.

7 Conclusions and Future Work

Requirements analysis is still a stage of software development where mistakes are common. Therefore, it can be the source of problems in subsequent development stages and can cause an IS not to fulfil the real needs of the organization where the IS has to be modified or introduced. Some of the mistakes detected in practice are the lack of understanding of the business by system analysts, the lack of focus on system purpose, and miscommunication between business people and system analysts.

This paper has described an approach to try to prevent these problems based on the modelling of an organization by means of BPMN and the goal/strategy Map approach. The approach allows system analysts to properly understand and analyze the organization, its needs, and the system goals in a participative way with end-users. Business people and system analysts share a common language that is understandable to both of them thanks to BPMN, Map, and task templates. BPDs are the basis for the end-user to validate that the organization structure and behaviour have been properly understood so that the system analyst can propose solutions based on the system purpose. Furthermore, the approach tries to mitigate the weaknesses of a separate use of BPMN and Map, and benefit from the advantages of their joint use.

Apart from the improvement in the surveys, the next steps in the project are the development of a tool that supports the approach, the introduction of a technique for the analysis of non-functional requirements, and the linking of the approach with

OO-Method. In addition, we want to extend the approach by introducing information about the user interface in the task template in order to derive an abstract description of the interaction between the users and the IS.

References

1. Agarwal, R., Prabuddha, D., Sinha, A.P.: Comprehending Object and Process Models: An Empirical Study. IEEE Transactions on Software Engineering 25(4), 541–556 (1999)
2. Alexander, I., Bider, I., Regev, G.: REBPS 2003: Motivation, Objectives and Overview. Message from the Workshop Organizers. In: CAiSE Workshops (2003)
3. Attaran, M.: Exploring the relationship between information technology and business process reengineering. Information & Management 41, 585–596 (2003)
4. Becker, J., Kugeler, M., Rosemann, M. (eds.): Process Management. Springer, Heidelberg (2003)
5. Bubenko, J., Persson, A., Stirna, J.: EKD User Guide (2001), http://www.dsv.su.se/~js
6. Castro, J., Kolp, M., Mylopoulos, J.: Towards requirements-driven information systems engineering: the Tropos Project. Information Systems 27, 365–389 (2002)
7. Constantine, L., Lockwood, L.: Software for Use. Addison-Wesley, Reading (2002)
8. Curtis, B., Kellner, M., Over, J.: Process Modelling. Communications of the ACM 35(9), 75–90 (1992)
9. Dardenne, A., Lamsweerde, A., van Fickas, S.: Goal-directed Requirements Acquisition. Science of Computer Programming 20, 3–50 (1993)
10. Davis, I., et al.: How do practitioners use conceptual modelling in practice? Data & Knowledge Engineering 58, 359–380 (2006)
11. Davis, R., Brabänder, E.: ARIS Design Platform. Springer, Heidelberg (2007)
12. Dobing, B., Parsons, J.: Understanding the role of use cases in UML: a review and research agenda. Journal of Database Management 11(4), 28–36 (2000)
13. Dumas, M., van der Aalst, W., ter Hofstede, A. (eds.): Process-Aware Information Systems. Wiley, Chichester (2005)
14. Eriksson, H., Penker, M.: Business Modelling with UML. John Wiley and Sons, Chichester (2000)
15. Estrada, H., Rebollar, A.M., Pastor, Ó., Mylopoulos, J.: An Empirical Evaluation if the i* Framework in a Model-Based Software Generation Environment. In: Dubois, E., Pohl, K. (eds.) CAiSE 2006. LNCS, vol. 4001, pp. 513–527. Springer, Heidelberg (2006)
16. IIBA. Business Analysis Body of Knowledge (2006), http://www.iiba.com
17. Holtzblatt, K., Beyer, H.: Requirements gathering: the human factor. Communications of the ACM 38(5), 31–32 (1995)
18. Kirikova, M., Bubenko, J.: Enterprise Modelling: Improving the Quality of Requirements Specification. Information Systems Research Seminar in Scandinavia, IRIS-17 (1994)
19. Lauesen, S.: Task Descriptions as Functional Requirements. IEEE Software 20(2), 58–65 (2003)
20. Luftman, J., Raymond, R., Brier, T.: Enablers and Inhibitors of Business-IT Alignment. Communications of AIS 1(11), 1–33 (1999)
21. Maiden, N., Jones, S.: An Integrated User-Centered Requirements Engineering Process, Version 4.1 (2004), http://hcid.soi.city.ac.uk/research/Rescue.html
22. Nysetvold, A., Krogstie, J.: Assessing Business Process Modelling Languages Using a Generic Quality Framework. In: EMMSAD 2005, CAiSE Workshops (2005)

23. OMG: Business Process Modelling Notation (BPMN) Specification (online) (2006), http://www.bpmn.org
24. Pastor, O., Molina, J.C.: Model-Driven Architecture in Practice. Springer, Heidelberg (2007)
25. Reich, B., Benbasat, I.: Factors That Influence the Social Dimension of Alignment Between Business and Information Technology. MIS Quarterly 24(1), 81–113 (2000)
26. Rolland, C., Salinesi, C.: Modeling Goals and Reasoning with Them. In: Engineering and Managing Software Requirements, pp. 189–217. Springer, Heidelberg (2005)
27. Rolland, C.: Capturing System Intentionality with Maps. In: Conceptual Modelling in Information Systems Engineering, pp. 141–158. Springer, Heidelberg (2007)
28. Rosemann, M., et al.: A Study of the Evolution of the Representational Capabilities of Process Modeling Grammars. In: Dubois, E., Pohl, K. (eds.) CAiSE 2006. LNCS, vol. 4001, pp. 447–461. Springer, Heidelberg (2006)
29. Sommerville, I., Sawyer, P.: Requirements Engineering: A Good Practice Guide. John Wiley and Sons, Chichester (1997)
30. Stirna, J., Persson, A., Sandkuhl, K.: Participative Enterprise Modeling: Experiences and Recommendations. In: Krogstie, J., Opdahl, A., Sindre, G. (eds.) CAiSE 2007. LNCS, vol. 4495, pp. 546–560. Springer, Heidelberg (2007)
31. The Standish Group. Chaos Reports, http://www.standishgroup.com
32. Taylor-Cummings, A.: Bridging the user-IS gap: a study of major information systems projects. Journal of Information Technology 13, 29–54 (1998)
33. Vessey, I., Coner, S.: Requirements Specification: Learning Object, Process, and Data Methodologies. Communications of the ACM 37(5), 102–113 (1994)
34. Wohed, P., et al.: On the Suitability of BPMN for Business Process Modelling. In: Dustdar, S., Fiadeiro, J.L., Sheth, A.P. (eds.) BPM 2006. LNCS, vol. 4102, pp. 161–176. Springer, Heidelberg (2006)
35. Yu, E.: Modelling Strategic Relationships for Process Reengineering. PhD Thesis, University of Toronto (1995)
36. Zave, P., Jackson, M.: Four Dark Corners of Requirements Engineering. ACM Transactions on Software Engineering and Methodology 6(1), 1–30 (1997)

Supporting the Elicitation of Requirements Compliant with Regulations

Motoshi Saeki[1] and Haruhiko Kaiya[2]

[1] Dept. of Computer Science, Tokyo Institute of Technology
Ookayama 2-12-1, Meguro-ku, Tokyo 152, Japan
[2] Dept. of Computer Science, Shinshu University
Wakasato 4-17-1, Nagano 380-8553, Japan
saeki@se.cs.titech.ac.jp, kaiya@cs.shinshu-u.ac.jp

Abstract. This paper presents a technique to check the compliance of requirements with regulations while eliciting requirements. In our technique, we semantically represent a regulation with combinations of case frames resulting from Case Grammar technique. We match a newly elicited requirement sentence with the case frames of regulation sentences and then check if the requirements include the obligation acts specified by the matched regulation sentences and if they do not have prohibited acts. If we find that a requirement sentence does not follow the regulation, the addition or removal of the illegal acts included in the requirements are suggested.

1 Introduction

Eliciting requirements from customers and users is a crucial step to develop information and software intensive systems of high quality. If we fail in eliciting mandatory requirements and/or elicit inconsistent requirements, we have developed the systems that are unsatisfactory to the customers and users, and we should spend much more costs and labor efforts on developing the systems because of re-developing them.

Recently, more laws and regulations related to information technology (simply, regulations) are being made in order to implement properly business processes and to avoid the dishonest usage of information systems by malicious users. A typical example is Japanese Act on the Protection of Personal Information [1] that specifies the proper handling of personal information such as names, addresses and telephone numbers in order to prevent from making misuse of this information. In this situation, we have to develop the systems compliant with these regulations. In particular, eliciting the requirements incompliant with the regulations has large harmful influences on later development activities and system release. Suppose that we develop an Internet auction system such as eBay. In this system, an individual that would like to put her goods up for an auction needs to register her membership at first. On registering her, the system collects her personal information such as her name, age, address, telephone number etc. Japanese Act on the Protection of Personal Information provides that, except special cases, an entity must notify the person of the purpose of the usage of acquired personal information or publicly announce it, when acquiring the personal information. If a requirements analyst fails to elicit the functional requirements on notifying or announce the purpose

Z. Bellahsène and M. Léonard (Eds.): CAiSE 2008, LNCS 5074, pp. 228–242, 2008.

of the usage of personal information and the system is developed without this function, we have the system incompliant with Act on the Protection of Personal Information and should re-develop the system later.

This paper presents a technique to check requirements if a newly elicited requirement is legal or not while eliciting requirements, in order to avoid the development of an information system incompliant with regulations. It is the technique to check semantically a newly elicited requirement against regulations and to give an analyst the suggestions on modifying the requirements if a compliance failure is detected, during her elicitation tasks. Usually, regulations and requirements specifications are written in natural language, and thus we have to develop a kind of semantic processing technique for natural languages. In our technique, we represent regulations with combinations of case frames resulting from Case Grammar technique [5]. When a modification on the current version of requirements occurs to maintain regulatory compliance, it should be recorded as a modification rationale. Suppose that an analyst modified a requirement A to B in order to maintain the compliance with a regulation. If she tries to modify B further, the rationale on the modification of A to B can suggest to her that this further modification can violate the compliance with the regulations again. To support the usage of rationale, we adopt linking mechanism among requirements, regulations and modification records. The rest of the paper is organized as follows. The next section presents the technique on the semantic representation of regulations with case frames in the next section. We show a requirements elicitation process following our technique and an example in sections 3 and 4 respectively. Section 5 is for listing related work.

2 Representing Regulations

2.1 Structure of Regulation Sentences

According to [4], a regulation sentence consists of 1) the descriptions of a situation where the sentence should be applied and 2) the descriptions of obligation, prohibition, permission and exemption of an entity's acts under the specified situation. For example, the Article 18, No. 1 of Act on the Protection of Personal Information provides that

> When having acquired personal information, an entity handling personal information must, except in cases in which the Purpose of Use has already been publicly announced, promptly notify the person of the Purpose of Use or publicly announce the Purpose of Use.

We can consider that "when having acquired personal information, except in cases in which the Purpose of Use has already been publicly announced" is a situation where this act should be applied, while "notify" and "announce" represent the acts of "the entity". Note that these acts are obligations that the entity should do. Furthermore, we can consider the situation as a logical conjunction ($\wedge$) of more atomic situation "having acquired personal information" and the negation ($\neg$) of "the Purpose of Use has already been publicly announced". The obligation acts are a logical disjunction ($\vee$) of the two acts "notify" and "announce". That is to say, we can represent more complicated regulation sentences by combining atomic descriptions with logical connectives such as $\wedge$, $\vee$ and $\neg$.

2.2 Case Frame

Requirements sentences and regulation ones are described in natural language. We need a technique of their semantic representation We adopt a technique of case frames originated from Fillmore's Case Grammar and this technique has been popularly used for developing specification languages of pseudo natural language [15], modeling conceptually information systems [11], and eliciting requirements [14], etc. A case frame consists of a verb and semantic roles of the words that frequently co-occur with the verb. These semantic roles are specific to a verb and are called *case*. For example, the case frame of the verb "get", having the cases "actor", "object" and "source", can be described as "get(actor, object, source)", where "get" denotes the acquisition of the thing specified by the object case. The actor case represents the entity that performs the action of "get" and that will own the thing as the result of the "get" action. The source case denotes the entity from where the actor acquires the object. By filling these case slots with the words actually appearing in a sentence, we can obtain its semantic representation. In the example of the sentence "an entity handling personal information acquires from a member her personal information", we can use the case frame of "get" and have "get(entity handling personal information, personal information, member)" as its semantic representation.

Used case frames depend on regulations. In our approach, we use a text-mining tool [9] to extract words that frequently occur in a regulation document and to find their relationships. After extracting the words and their relationships, an analyst selects relevant verbs and their relationships to other words, and composes case frames from them by manual. Figure 1 illustrates case frames included in Article 18, which is obtained as the result of applying our text-mining approach to the document of Act on the Protection of Personal Information. We use class diagram notation in the figure, and a class and an association correspond to an extracted word and the relationship among words. In the figure, a verb is depicted as an association and nouns participating in the association can be considered as case slots of the verb. For example, the verb "get" is an association among "Entity", "Personal Information" and "Person", and their cases are Actor, Object and Source in turn.

2.3 Representation of Regulations

Figure 2 depicts the structure of regulations, i.e. a meta model of regulations. A regulation document comprises articles and an article can have a tree structure. An instance of Base Article expresses a leaf in a tree structure and is a regulation sentence in the lowest level. Since an article may often refer to the other articles, we has an association "refer" between Article Component. As mentioned in section 2.1, each regulation sentence consists of a situation and an act with modality. In the figure, a situation and an act are instances of Situation class and Act respectively. They can be represented with logical combinations of case frames, and a case frame comprises a verb and case slots relevant to the verb. The case slots are filled with words appearing in a regulation sentence to represent the meaning of the regulation sentence.

Figure 3 shows an example of Article 18, No.1, which is a part of duties of entities handling personal information (Articles 15 - 36 of Act on the Protection of Personal Information).

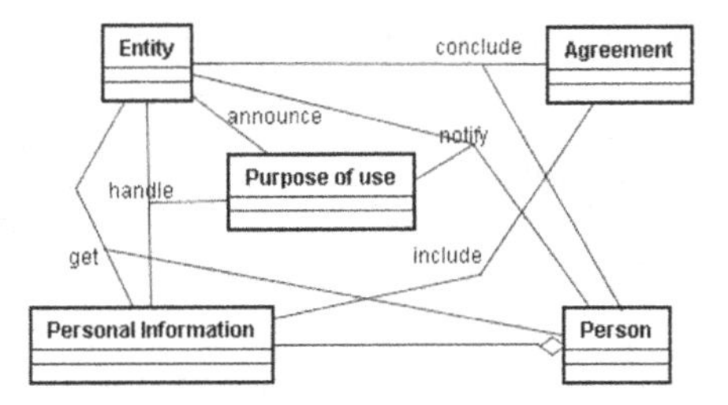

Case Frame:
handle(actor:Entity, object:Personal Information, purpose:Purpose of use)
announce(actor:Entity, object:Purpose of use)
notify(actor:Entity, object:Purpose of use, destination: Person)
get(actor:Entity, object:Personal Information, source:Person)
conclude(actor:Entity, object:Agreement, destination:Person)
include(actor:Agreement, object: Personal Information)
aggregate(owner:Person, owned: Personal Information)

Fig. 1. Case Frames in Article 18

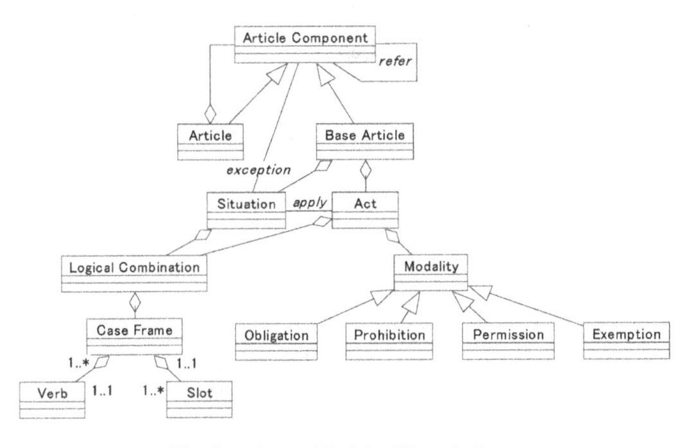

Fig. 2. A Meta Model of Regulations

3 Requirements Elicitation Process

We assume that requirements are described in natural language and an analyst elicits a new requirement using a certain method such as goal-oriented analysis and scenario analysis method, etc. During this elicitation process, our approach monitors if a newly elicited requirement together with already existing requirements are compliant with regulations or not, and suggests how to modify the requirements if they can be incompliant with the regulations.

3.1 Basic Idea

Figure 4 depicts the essential point of our approach. In the figure, the case structure of the requirement sentence bbb is analyzed and its semantic representation is generated

Article 18 (Notice of the Purpose of Use at the Time of Acquisition, etc.)

1. When having acquired personal information, an entity handling personal information must, except in cases in which the Purpose of Use has already been publicly announced, promptly notify the person of the Purpose of Use or publicly announce the Purpose of Use.

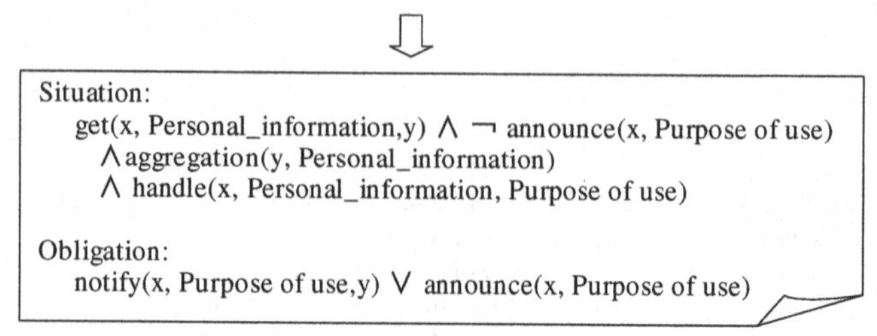

Situation:
get(x, Personal_information,y) ∧ ¬ announce(x, Purpose of use)
∧aggregation(y, Personal_information)
∧ handle(x, Personal_information, Purpose of use)
Obligation:
notify(x, Purpose of use,y) ∨ announce(x, Purpose of use)

Fig. 3. An Example of Semantic Representation of a Regulation

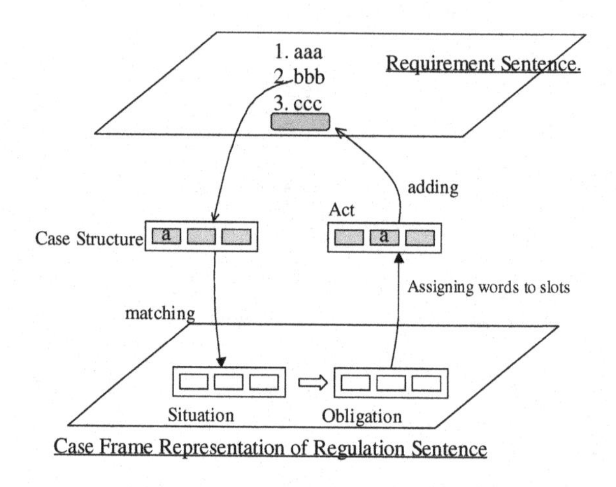

Fig. 4. Mapping to Semantic Representation

by filling the case slots with words appearing in bbb. We look for a regulation sentence whose situation part is matched with bbb in their semantic representation level, i.e. the level of case frames. If we can find a matching, there is a possibility that the requirement bbb denotes the situation where the matched regulation should be applied, and thus we pay attention to the modality and the act parts of the matched regulation sentence. In this example, since the modality is Obligation, we check if the act is included in the requirement sentences including bbb. If it is not included yet, we are suggested that the act should be added to the requirements. The information on the words assigned to the case slots can be used for us to compose the newly added requirement sentence. The matching process can be automated so as to list up the candidates of the matched regulations, and it uses a thesaurus such as WordNet in order to deal with synonyms, near-synonyms, hypernyms and hyponyms.

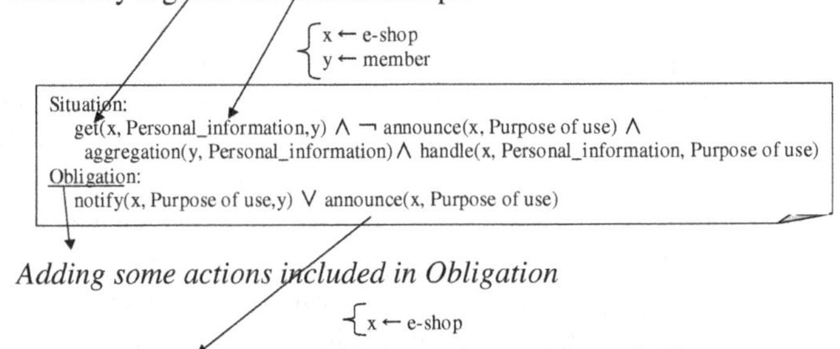
The e-shop obtains personal information of members when they register their memberships.
x ← e-shop
y ← member
Situation:
get(x, Personal_information,y) ∧ ¬ announce(x, Purpose of use) ∧
aggregation(y, Personal_information) ∧ handle(x, Personal_information, Purpose of use)
Obligation:
notify(x, Purpose of use,y) ∨ announce(x, Purpose of use)
Adding some actions included in Obligation
x ← e-shop
The e-shop announces personal information and then obtains personal information of members when they register their memberships.

Fig. 5. An Example of Article 18

However, we should decide whether in the requirements we truly have the situation that the regulation should be applied, and whether we finally add a new requirement denoting the obligation act. This decision is recorded to support the future additions and changes of the requirements.

Figure 5 illustrates the application of Article 18, No.1 of Act on the Protection of Personal Information. This is an example of an Internet shop (simply, e-shop) and the shop collects personal information of customers when it registers them as shop members. Article 18 provides that the shop has an obligation to announce publicly or notify the person of the purpose of usage of the collected personal information, when it gets her personal information. The requirement sentence "The e-shop obtains personal information of members when they apply to register their memberships", which our analyst elicited, is not compliant with the Article 18, No.1. In our approach, the automated tool extracts its case structure, and then tries to find a regulation having a case frame of the verb that is a synonym of "obtain" appearing in the sentence as a main verb. Since the situation part of the semantic representation of Article 18 is the case frame get(x, Personal_information, y) and it is matched with the requirement, an analyst checks if the article should be truly applied to the requirement or not. For example, she checks if ¬ announce(x, Purpose of use), which is included in the situation part of the article, holds or not. By the results of checking, she decides that the article should be applied and adds the obligation act to the current version of the requirement. In the figure, two suggestions, adding "notify" or "announce" act, can be considered and she selects the case frame announce(x, Purpose of use) from a list of the obligation acts, fills words with its case slots using the matching information and revise the requirement. As a result, she assigns "e-shop" to actor slot "x" and adds the description of e-shop's announcing the purpose of usage of personal information. In the next sub section, we explain how to suggest the modification candidates in our approach.

3.2 Generating Suggestions

We assume that an analyst elicits a requirement sentence by sentence. Suppose that the analyst has got the requirements $R_1, \ldots, R_{i-1}$ in order, and she is eliciting now R_i.

The semantic representation of a regulation sentence L_j consists of a situation SL_j, a modality ML_j and an act AL_j, and the situation and the act are represented with logical combination of case frames. For simplicity, let a situation be described in conjunctive form where each conjunct $SL_{i,j}$ ($1 \leq i \leq m$) is a case frame or a case frame with negation $\neg$. That is to say, we have $SL_j = SL_{1,j} \wedge \cdots \wedge SL_{m,j}$. On the other hand, an act is represented in disjunctive normal form, i.e. $AL_j = AL_{1,j} \vee \cdots \vee AL_{n,j}$, where $AL_{k,j} = AL_{1,k,j} \wedge \cdots \wedge AL_{n_k,k,j}$ ($1 \leq k \leq n$) and $AL_{s,k,j}$ ($1 \leq s \leq n_k$) is a case frame or a case frame with $\neg$. Although L_j can have more complicated logical connective structure, we can decompose and rewrite it into a set of the expressions having the above form. We can have the predicate $match$ for investigating whether a requirement sentence can match to a case frame existing in the regulation sentences, as follows;

$match(R_i, L) \triangleq \exists l\, (CR_{i,l} = L \circ \theta)$,
where
1) $CR_{i,l}$ is a case structure included in the semantic representation of R_i,
2) L is $SL_{k,j}$ or $AL_{s,k,j}$,
3) θ is an assignment of words to case slots, and
4) $L \circ \theta$ is a result of applying the assignment θ to L.

Roughly speaking, $match(R_i, SL_{k,j})$ presents that R_i can be matched with the situation part of L_j, while $match(R_i, AL_{s,k,j})$ is used to show the matching of R_i with the act part of L_j.

The suggestions provided for an analyst depend on which part of a regulation is matched and its modality. Table 1 shows the suggestions to the analyst. Suppose the case that the newly elicited requirement R_i can satisfy the situation part of L_j, i.e. $match(R_i, SL_{k,j})$. If L_j presents an obligation (ML_j = Obligation), its act $AL_{u,j}$ ($= AL_{1,u,j} \wedge \cdots \wedge AL_{n_u,u,j}$) should be included in some of the requirements. The second column of the table suggests the addition of $AL_{1,u,j}, \cdots, AL_{n_u,u,j}$ as one of the alternatives in order to keep the regulatory compliance, if they are not included. If L_j presents a prohibition and a conjunct of $AL_{u,j}$ for each u is included in some of the requirements, it should be excluded or the requirements satisfying the situation of L_j should be changed so that the situation cannot hold.

When the requirements sentences are changed by reason of regulatory compliance, additional information such as rationale is attached in the form of links. The suggestions "Create a link" in the table are for recording the requirements changes. Its detail will be mentioned in the next sub section.

3.3 Linking to Regulation Sentences

Requirements changes frequently occur by various reasons. For future maintenance activities, it is important to record the requirements changes that regulatory compliance resulted in. Figure 6 depicts the structure to record requirements changes by regulatory compliance. Basically, it has links that hold information on the contents of the changes and the regulations that resulted in the changes. A link from the requirement to the situation part of a regulation sentence specifies the information on whether their

Table 1. Suggestions to Analysts

	ML_j = Obligation	ML_j = Prohibition
$match(R_i, SL_{k,j})$	If $\neg\exists u, \forall s, \exists l\ match(R_l, AL_{s,u,j})$, 1) Select u and add all of the acts denoted by $AL_{1,u,j}, \cdots, A_{n_u,u,j}$ as new requirements R_{i+1} or 2) Re-consider $R_1, \cdots, R_{i-1}$ so that $SL_{1,j}, \cdots, SL_{k-1,j}, SL_{k+1,j}, \cdots,$ or $SL_{m,j}$ cannot hold, or 3) Re-consider R_i. Create links according to a selected suggestion.	If $\forall u, \exists s, \exists l\ match(R_l, AL_{s,u,j})$, 1) Select u and remove all of the acts denoted by $AL_{1,u,j}, \cdots, A_{n_u,u,j}$ from $R_1, \cdots, R_i$, or 2) Re-consider $R_1, \cdots, R_{i-1}$ so that $SL_{1,j}, \cdots, SL_{k-1,j}, SL_{k+1,j}, \cdots,$ or $SL_{m,k}$ cannot hold, or 3) Re-consider R_i. Create links according to a selected suggestion.
$match(R_i, AL_{s,k,j})$	4) If $\forall u, \exists l\ match(R_l, SL_{u,j})$, create a link.	If $\forall u, \exists l\ match(R_l, SL_{u,j})$, 4) Remove all of the acts denoted by $AL_{1,k,j}, \cdots, AL_{s_k,k,j}$ from R_i, or 5) Re-consider $R_1, \cdots, R_i$ so that $SL_{1,j}, \cdots, SL_{k-1,j}, SL_{k+1,j}, \cdots,$ or $SL_{m,k}$ cannot hold. Create links according to a selected suggestion.
	ML_j = Permission	ML_j = Exemption
$match(R_i, SL_{k,j})$	Create a link annotated with "Possible to add" to $AL_{s,u,j}$ for any s.	If $\exists l, \exists u, \exists s\ \ match(R_l, AL_{s,u,j})$, create a link annotated with "Possible to remove" to $AL_{s,u,j}$.
$match(R_i, AL_{s,k,j})$	If $\forall u, \exists l\ match(R_l, SL_{u,j})$, create a link annotated with "Possible to add" to $AL_{s,k,j}$.	If $\forall u, \exists l\ match(R_l, SL_{u,j})$, create a link annotated with "Possible to remove" to $AL_{s,k,j}$.

situations are satisfied or not (Satisfy or Deny), while a link to the act part presents that the act is included in the requirement. The requirement changes can be categorized into addition (Add), removal (Remove) and replacement (Modify) and we can attach this categorization to a link as well as annotations described with free texts. Since the requirements changes cause version changes, we have another type of link denoting the correspondence of a requirement sentence of the older version to the newer one. The class Version and their association in the figure is for recoding this information, and the roles +previousVersion and +newVersion express the older version and the newer one respectively.

Figure 7 illustrates the link structure when the new requirement is added as shown in Figure 5.

4 Example

The aim of this section is to discuss the feasibility and the limitation of our technique through the example analysis. We use Japanese Act on the Protection of Personal Information [1] and the following example system that our technique is applied to.

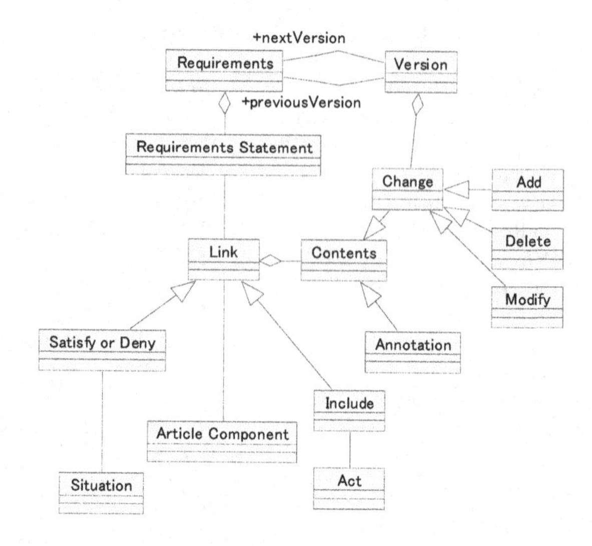

Fig. 6. Links from Requirements to Regulations

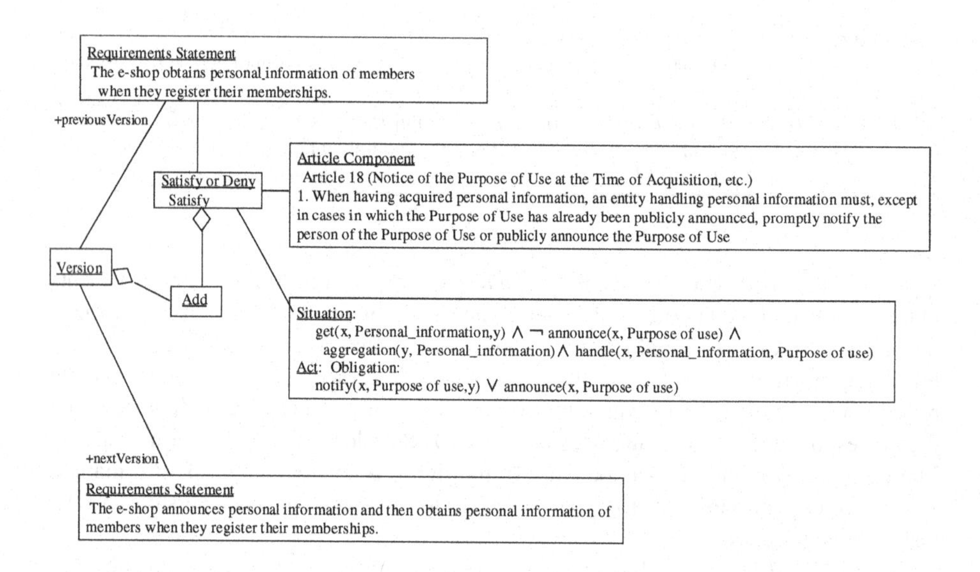

Fig. 7. Recording a Requirements Change

- The system to be developed is a customer management system in a satellite television company.
- Each subscriber can receive TV programs by using a television receiving set provided by the company.
- Each TV receiving set has its own product ID, and the ID is put in front of the set.

- The TV receiving set has a communication interface for Internet or for a telephone circuit, and it enables the company to maintain the set remotely.

Our requirements analyst starts with them by eliciting requirements stepwise, as shown in the successive sub sections. Note that this example scenario was written in Japanese and we directly translate the scenario to English.

Requirements for the system version 0

The current system, say version 0, supports workers in the company to ship a receiving set to subscribers and to invoice them by postal mail (not by email nor via Internet). The system includes both customer database and a formatter for postal tags, thus the requirements for the system are elicited as follows.

- (Req.1) The system gets the information on subscriber's name and address, when the worker of the company receives the name and address by postal mail and inputs them into the system.
- (Req.2) The system shall print a shipping tag for a receiving set.
- (Req.3) The system shall print an invoice for a monthly license fee.

In the same way as shown in Figure 5, our method suggests to a requirements analyst that she adds the following requirement based on Req.1 and Article 18 in [1], because the verb "get" in Req.1 is matched with the situation part of the article.

- (Req.4) The system shall print a letter to notify subscribers of the purpose of the usage of their names and addresses.

In Article 25, an entity, e.g. a company, can be requested to disclose retained personal data. This article can be written in the following form.

Article 25

Situation: retain(x, Personal Information) $\wedge$ request(y, Disclose, x)
Obligation: disclose(x, Personal Information, y) $\vee$ disclose(x, no information, y)

We have the Req.1 having "get" as a main verb. Although "get" is not a near-synonym of the verb "retain", the situation part of Article 25 can be matched with the Req.1 if we can have the thesaurus such as [6]. In this thesaurus, we have several specific relationships among ontological concepts for software development, and one of the examples is "cause" relationship which expresses behavioral causality. In the case of Article 25 and Req.1, since it can be assumed that we have the information on "get(x,y) causes retain(x,y)" in the thesaurus and the inference rule "(verb1 *cause* v2 $\wedge$ match(R,v1)) $\rightarrow$ match(R,v2)", our analyst can get the matching of Article 25 with Req.1 and then add the following requirement by using the suggestion 1) for obligation in Table 1.

- (Req.5) The system can print a letter to disclose a name and an address of a subscriber when the subscriber requests the company to disclose them.

This example of the requirement addition means that we need additional thesauruses or ontologies in addition to WordNet, e.g. domain specific ontologies such as [6].

In Article 19, the company must maintain personal information accurate and up to date. This article can be written in the following form.

Article 19

Situation: retain(x, Personal Information) $\wedge$ change(y, Personal Information)
Obligation: maintain(x, Personal Information)

Similar to the above example, Article 19 can be matched with Req.1 and the analyst finds that a name and an address can be changed from the verb "change" in the article. Thus the system has to maintain such changes continuously, and more concretely, the information on subscribers' names and addresses stored in the system can be updated when some subscribers request to update their names or addresses. According to Article 19, Req.1 and a suggestion 1) for obligation in Table 1, the analyst adds the following requirement.

- (Req.6) The system shall update a name and/or an address of a subscriber whenever she submits a letter to request the change of her name and/or address.

Note that the analyst refines the word "maintain" to "update" during the addition of this requirement.

Security control is requested by Article 20 if there can be leakage, loss or damage of personal data.

Article 20

Situation: (handle(x, Personal Information, u) $\wedge$ leak(Personal Information))
 $\vee$ (handle(x, Personal Information, u) $\wedge$ loss(Personal Information))
 $\vee$ (handle(x, Personal Information, u) $\wedge$ damage(Personal Information))
Obligation: takeSecurityMeasure(x, Personal Information)

Note that its situation part is not a form of logical conjunctive formula, because it includes $\vee$. However, we can decompose it to several expressions having $\wedge$ only, and for simplicity, we keep it as above.

Since Req.1 is matched with the situation part of this article through the verb "handle", our analyst should consider its application. In Req.6, there is no security measure, especially an authentication measure, for handling personal information, thus a name and an address can be maliciously changed. This is a kind of damage or loss of the name and the address. Therefore, the analyst takes some authentication mechanism into account, and modifies Req.6 according to Article 20 and a suggestion 3) for obligation in Table 1 as follows.

- (Req.6') The system shall update a name and/or an address of a subscriber whenever she submits a letter to request its change, and the letter should include her current registered name, address and a product ID of her receiving set as a kind of PIN (Personal Identification Number) for authentication.

Requirements for the system version 1: introducing Internet

The company decides to use Internet for managing information of subscribers in addition to the services in the system version 0. The system version 1 is proposed according to this decision. The initial requirement for the system version 1 is as follows.

- (Req.7) The system shall get subscriber's name and address via Internet by the system.

According to the thesaurus, the verb "get" is a hyponym of "handle". It is well known that there is no confidential mechanisms for standard Internet communication, thus the analyst has to take threats such as leakage, loss or damage into account according to the Article 20. After the analyst considers the usage of SSL that is a kind of security measures, she updates Req.7 as follows according to a suggestion 1) for obligation.

- (Req.7') The system shall get subscriber's name and address via Internet under SSL protocol by the system.

In the same way as Req.4 and 6', the following requirements are added according to Articles 18, 19 and 20.

- (Req.8) The system shall show the purpose of the usage of their names and addresses just after receiving them via Internet.
- (Req.9) A subscriber can update her name or address when she submits its change via Internet using her current registered name, address and a product ID of her receiving set as a kind of PIN (Personal Identification Number) for authentication.

The analyst has to add a requirement corresponding to Req.5, a requirement for disclosure of current information according to Article 20 and a suggestion 1) for obligation.

- (Req.10) The system shall show a name and an address of a scriber when the subscriber requests the company to disclose them via Internet with her name, address and product ID of her receiving set.

Requirements for the system version 2: reducing costs

The company becomes worrying about the postal costs related to Req.4 and 5 for the system version 0, and tries to charge postal fee to the subscribers who do not still use Internet but postal service. The analyst changes Req.4 and Req.5 as follows, i.e. adding the function of making an invoice for sending letters.

- (Req.4') The system shall print a letter to notify subscribers of the purpose of the usage of their names and addresses and print an invoice.
- (Req.5') The system can print a letter to disclose a name and an address of a subscriber when the subscriber requests the company to disclose them and print an invoice.

The words "notify" and "disclose" are matched with the situation part of Article 30, which specifies the permission of entity's collecting charges for taking her acts, and the analysts decides the above two requirements are regulatory compliant.

Article 30

Situation: notify(x, Purpose of use, y) $\vee$ disclose(x, Personal Information, y)
Permission: charge(x, notify, y) $\vee$ charge(x, disclose, y)

Note that Reqs.4 and 5 are linked to Articles 18 and 25 respectively, because these articles force the analyst to add the requirements. The analyst checks the compliance of these changes with the Article 18 and 25 and finds their compliance.

Requirements for the system version 3: regional services

The company decides to start regional services based on the address of subscribers. For example, subscribers living in Texas can receive local TV programs such as local news in Texas and also advertisement promotion. To achieve this decision, the analyst adds the following requirement.

- (Req.11) The system can retrieve the address of a receiving set by using its product ID as a retrieval key.

Note that the company can control all TV receiving sets via Internet or telephone by specifying each product ID.

In Req.11, the product ID is used to identify the region where the TV receiving set with the ID is located, in other words, the system handles the product ID for region query. If our thesaurus has the information that "retrieve" is a hyponym of "handle", the analyst can get the matching of Req.11 with the following Article 16.1, providing that using personal information without any consent is restricted.

Article 16.1

Situation:
$\neg$ consent(y, Purpose of use, x) $\wedge$ handle(x, Personal Information, Purpose of use)
Prohibition: handle(x, Personal Information, Purpose of use)

However, no subscribers have had consent to this usage of product ID yet. According to a suggestion 5) for prohibition in Table 1, the following requirement for getting a consent from a subscriber is added to satisfy "consent(y, Purpose of use, x)" in Article 16.1.

- (Req.12) The system shall print a letter to get subscriber's consent about regional services.

5 Related Work

The techniques to develop information systems and software compliant with regulations are being actively studied. Many of them are the approaches to represent regulations with formal expressions and verify their consistency to formal specifications by using theorem provers or model checkers. The state of the art of this research direction and some achievements can be found in [10]. However, it stays in the status where some regulations were described using formal methods yet. Although formal modeling methods for regulations can be established, we should solve a scalability problem of consistency checking for formal specifications and regulations of practical size. On the other hand, our approach is applicable to cases of practical size because 1) it is an incremental checking of compliance whenever a new requirement is elicited and 2) it is a lightweight method of semantic analysis using case frames.

In [8], regulations are represented as XML documents and a tool to retrieve the relevant regulation sentences using XML tags has been developed. Although the retrieval process itself can be automated, the appropriateness of tags for semantic retrieval and the costs of human efforts for tagging regulations may be problematic. In our approach, since we use a synonym dictionary and case-frame matching technique to retrieve the relevant regulation sentences, we can automate the retrieval process in a certain degree using natural language processing techniques.

Katayama proposed a new discipline called Legal Engineering in [7], where processes to compose laws are clarified using software engineering technique and computerized supporting tools to enact laws are developed. Its aim is different from ours, but since it includes the automated analysis techniques of law sentences, it is very useful to automate the translation of regulation sentences to case frames.

Anton et. al. developed the semantic representation technique called KTL and proposed a methodology to detect ambiguity included in regulation using KTL [2]. They constructed by hand the KTL semantic representation of rights and obligation parts of privacy rules included in the regulation U.S. Health Insurance Portability and Accountability Act (HIPAA). Since KTL is essentially the same as hierarchical case frame, we can use their methodology to compose the case frames of regulations as their semantic representation. Differently from their methodology, our approach uses an automated text-mining technique to reduce the human efforts.

6 Conclusion and Future Work

This paper proposes a technique to elicit regulation-compliant requirements. In this technique, we represent the meaning of regulations with case frames, and check semantically the regulations against requirements sentences to detect the missing obligation acts and the prohibition acts in the requirements. The detection results are shown to an analyst and how to modify incompliant requirements is suggested.

Our technique for matching requirements sentences to regulation ones uses just word matching with a synonym dictionary and thus more sophisticated approach may be necessary. For example, we can adopt these techniques to detect the regulation sentences relevant to the requirements, such as structural similarity predicates [12], similarity measure based on word occurrences [3] and the algorithm to detect overlaps between specifications [13].

The future work can be listed up as follows.

1. Elaborating the supporting tool and its assessment by case studies,
2. Considering semantic interdependency relationships among the requirements in a requirements document,
3. Handling with the integration of multiple regulations,

4. Combining tightly our approach to requirements elicitation methods such as goal-oriented analysis and scenario analysis,
5. Managing the requirements that have the potentials for being incompliant with regulations and developing metrics of measuring compliance,
6. Managing evolution processes of requirements together with version control.

Acknowledgements

The authors are very grateful to anonymous reviewers for their valuable comments in order to improve the earlier version of this paper.

References

1. Act on the protection of personal information (2003), `http://www5.cao.go.jp/seikatsu/kojin/foreign/act.pdf`
2. Breaux, T., Vail, M., Anton, A.: Towards Regulatory Compliance: Extracting Rights and Obligations to Align Requirements with Regulations. In: Proc. of 14th IEEE International Requirements Engineering Conference, pp. 49–58 (2006)
3. och Dag, J., Regnell, B., Carlshamre, P., Andersson, M., Karlsson, J.: Evaluating Automated Support for Requirements Similarity Analysis in Market-Driven Development. In: Proc. of REFSQ 2001 (2001)
4. Eckoff, T., Sundby, N.: RECHTSSYSTEME (1997)
5. Fillmore, C.: The Case for Case. Rinhehart and Winston, Holt (1968)
6. Kaiya, H., Saeki, M.: Using domain ontology as domain knowledge for requirements elicitation. In: Proc. of 14th IEEE International Requirements Engineering Conference (RE 2006), pp. 189–198 (2006)
7. Katayama, T.: Legal Engineering – An Engineering Approach to Laws in E-Society Age. In: Proc. of 1st Workshop on JURISIN (2007)
8. Kerrigan, S., Lawa, K.H.: Logic-based Regulation Compliance-Assistance. In: Proc. of 9th International Conference on AI and Law, pp. 126–135 (2003)
9. Kitamura, M., Hasegawa, R., Kaiya, H., Saeki, M.: An Integrated Tool for Supporting Ontology Driven Requirements Elicitation. In: Proc. of 2nd International Conference on Software and Data Technologies (ICSOFT 2007), pp. 73–80 (2007)
10. Laleau, R., Lemoine, M. (eds.): International Workshop on Regulations Modelling and Their Validation and Verification (REMO2V), CAiSE2006 Workshop (2006)
11. Rolland, C., Proix, C.: A Natural Language Approach for Requirements Engineering. In: Loucopoulos, P. (ed.) CAiSE 1992. LNCS, vol. 593, pp. 257–277. Springer, Heidelberg (1992)
12. Salinesi, C., Etien, A., Zoukar, I.: A Systematic Approach to Express IS Evolution Requirements Using Gap Modelling and Similarity Modelling Techniques. In: Persson, A., Stirna, J. (eds.) CAiSE 2004. LNCS, vol. 3084, pp. 338–352. Springer, Heidelberg (2004)
13. Spanoudakis, G., Finkelstein, A., Till, D.: Overlaps in Requirements Engineering. Automated Software Engineering 6(2), 171–198 (1999)
14. Watahiki, K., Saeki, M.: Scenario Patterns based on Case Grammar Approach. In: Proc. of 5th IEEE International Symposium on Requirements Engieenring (RE 2001), pp. 300–301 (2001)
15. Zhang, H.H., Ohnishi, A.: A Transformation Method of Scenarios from Different Viewpoints. In: Proc. of 11th Asia-Pacific Software Engineering Conference (APSEC 2004), pp. 492–501 (2004)

On the Impact of Evolving Requirements-Architecture Dependencies: An Exploratory Study

Safoora Shakil Khan, Phil Greenwood, Alessandro Garcia, and Awais Rashid

Lancaster University, UK
{shakilkh,greenwop,garciaa,marash}@comp.lancs.ac.uk

Abstract. Architecture design plays a significant role in the evolution of software systems, as it provides the prime realization of the driving requirements and their inter-dependencies. With the increasing volatility of software requirements nowadays, it is necessary to understand the correlation between evolving classical requirements dependencies and their impact on the architectural decomposition. In the context of this analysis, two questions arise: (i) what are the conventional categories of requirements dependencies that are more architecturally significant in terms of change impact? and (ii) to what extent those evolving dependencies tend to generate ripple effects through architectural modules and interfaces. In order to address these two questions, this paper first presents an analysis model that categorizes requirements dependencies. Second, we have performed an exploratory study, based on the change history analysis of a real-life Web-based information system, in order to gather the most architecturally-significant requirements dependencies from our model. We have systematically analyzed ten system releases, based on some qualitative and quantitative indicators, with respect to how the requirements-architecture dependencies and compositions evolved.

Keywords: Dependency analysis, traceability, software architecture, change impact analysis.

1 Introduction

Software architecture of information systems is the pivotal realization of requirements and their inter-dependencies as it encompasses the driving design decisions in order to satisfy stakeholders' needs [19]. Software architectures are often decomposed into a set of modules (components) and interfaces, typically designed for future evolution in order to avoid that the system succumbs in the presence of changes [19][20]. Thus, architectural design forms the backbone of the target system, and frequent non-systematic modifications to requirements dependencies can make the architecture fragile and cause short-term design degeneration [15][16][17]. As a result, software architects often need to understand, predict and trace the impact of evolving requirements dependencies on architectural designs [1][2][3].

However, there is not much empirical knowledge on the influence of different categories of changing requirements on architecture elements [1][2][3] so that designers can forecast change impact. As a result, they cannot thereafter effectively

Z. Bellahsène and M. Léonard (Eds.): CAiSE 2008, LNCS 5074, pp. 243–257, 2008.

trace key problem-solution dependencies. Not surprisingly, requirements-architecture traceability and change impact analysis techniques are still in its infancy and limited support is provided to software architects. There is a growing body of traceability techniques emerging in the literature [15][16][17][22], but they do not provide an adequate end-to-end tracing between categories of requirements dependencies and architectural elements. Most of these approaches are based on mapping requirements to architecture using trace matrix or trace graph [24][25]. There are a few approaches [2][5][6][8][9] that focus on characterizing requirements dependencies but these approaches have not been extended to cope with evolving requirements-architecture relations. Most empirical studies in the literature focus on characterizing architectural changes [18], keeping architecture and implementation in synchronization [15][17] [22], or supporting change impact analysis on implementation artifacts [23].

This paper presents a first exploratory study in order to identify the potential factors associated with evolving requirements dependencies and their corresponding effects or architectural changes. In order to be able to perform the investigation, we have defined a requirements dependency model (Section 2); we also discuss how each category of requirements dependency is likely to affect the architectural decomposition in the presence of change to it. The dependency model is based on a systematic analysis of classical requirements engineering techniques [1][2][3][5]. In a second step, we have undergone a set of experimental procedures (Section 3) in order to analyze the impact of evolving requirements dependencies and architecture changes through the releases of a real-life Web-based information system called Health Watcher (HW). The goal of the analysis was to characterize (Section 4):

(i) how the nature of requirements dependencies can lead to tight or loose interconnections with architectural elements;
(ii) the most architecturally-significant requirements dependencies; and
(iii) how the requirements dependencies tend to evolve in a typical Web-based information system.

Our analysis was based on qualitative and quantitative indicators. Classical change impact metrics have also been used to quantify the requirements and architecture co-relations. We also contrast our findings with related work (Section 5), and provide some concluding remarks (Section 6).

2 From Requirements to Architecture: A Dependency Model

This section presents a dependency model that provides support to software analysts to understand how the requirements are being realized to the architecture components and their compositions. This model assists a number of software evolution tasks, such as: (i) the identification of dependencies that lead to tight or loose interconnections among requirements elements and architectural decompositions, and (ii) basic support for understanding significant architecture implications from the perspective of requirements changes. We have defined a dependency taxonomy based on a systematic analysis of conventional requirements engineering approaches [7][14]. The analyst must keep in mind that more than one dependency can hold between requirements to architecture and it may be rare that individual dependency exist due to requirements characteristics. We have discussed six types of dependencies from our dependency model:

Goal Dependency. Goal dependency relates system's quality attributes at problem domain to their realization in solution domain (architecture and implementation). This dependency has been adapted from the conventional goal-oriented requirements engineering approaches [1][2][3] that define characteristics of systems. Dependency relates to requirements specifying quality of service (security, availability, performance, etc) and development (compatibility, adaptability, interoperability, etc) to component at architecture level. Consider the requirement 'system must be flexible in terms of the storage format [4]', i.e., to enhance variability and provide the user with the multiple options of storing data, such as arrays or different databases. This requirement will be linked to component providing persistence at the architecture level as it defines the objective of the system under construction.

Service Dependency. Service dependency relates the requirements expressing behavioral or functional characteristics of the system to corresponding operations and functions at architecture. The trace connection among requirements-level operations to architecture will usually be intricate or fine grained, as it will relate to classes, operations, or interfaces at the architecture level. A service dependency may coexist with goal, conditional, temporal, etc., dependency as it provides operation to perform when certain goal or condition is met. Consider the requirement '...system provides user with the queried data ...' [4], forms a service dependency with architecture. Operation for searching and retrieving the requested data of a particular query type is invoked at architecture level.

Conditional Dependency. Conditional dependency defines events that trigger services, processes, and tasks based on certain conditions, constraints, or decisions taken at the requirements level to their realization at the architectural level. The triggering can be autonomous or non-autonomous reaction to a condition, constraints, or decisions, for example, when smoke is detected the autonomous reaction of the system is to open the doors. This dependency has been inspired from programming and Meyer [26] 'Design by Contract'. Consider the requirement 'employee can make changes to when authenticated by the system as an employee [4]'. This requirement has conditional and service dependency with components of the architecture. The conditional dependency exists as the employee will not be granted access to perform restricted operations unless they have been verified as a valid employee.

Temporal Dependency. Temporal dependency relates requirements specifying time frame of an event to occur, processes to complete, or condition to hold true, to their realization at architecture. Temporal dependencies manifest often in requirements associated with real-time systems and distributed systems. Temporal dependency is closely related to conditional dependency and may usually co-exist. Consider the requirement 'terminate user's request if system does not respond within 5 seconds [4]', has temporal and conditional dependency with the architecture. The dependencies hold as condition needs to be true within the specified time frame for the system to proceed further.

Task Dependency. Task dependency traces the connection between artifacts which require response, input, or feedback from user for their completion. Task dependency forms a medium between user and system, allowing user to request for systems services. Consider the requirement 'employee chooses one of the given options (review, update, delete)' has task dependency as the system needs users input to invoke the corresponding service based on users request. The common notion between conditional and task dependency is that systems halts for response. The distinguishing notion between the two dependencies is that conditional dependency depends on input, response or feedback from other operations/services in the system. Retrospectively, a task dependency depends on input, response or feedback from user.

Infrastructure Dependency. Infrastructure defines the hardware and software of system. Infrastructure dependency relates the resources, infrastructures (networks, telecommunications, mobile, etc.), technical standard/details, and compatibility issues specified in stakeholder's requirement to the architecture conception/construction. This dependency has been adapted from Ramesh and Jarke [8] *resource dependency* and Grady [9] *implementation requirements*. Consider the requirement, 'application should be accessible via internet' , it has infrastructure dependency with architecture components as .the Web service is implemented using servlets.

3 Case Study and Evaluation Procedures

This section describes in detail the requirements and architectural characteristics of the application used in our exploratory study (Sections 3.1), and the evaluation and analysis procedures (Sections 4.1 to 4.2). Health Watcher (HW) [10][11][12] is a typical Web information system and a real-life application that was chosen to support the empirical analysis in our exploratory study. The reason for selecting the HW case study is threefold. First, a rich set of HW artifacts and their releases were made available. For instance, for the analysis we have requirements specification (available from [13]), both use case descriptions and goal models. The architecture design is specified according to two fundamental architectural views, the module and component-connector views. Also, deployment-related decisions are embedded in these views. As a result, these requirements and architecture artifacts (Sections 3.1) capture respectively a rich, complementary set of requirements-architecture dependencies according to our model (Section 2).

Second, the original HW implementation and its ten releases [4] are available in three programming languages, Java, AspectJ, and CaesarJ. As a consequence, HW architecture is basically realized according to two architectural designs: a layered OO version and a layered AO version. None of these artifacts have been specially prepared or modified for our exploratory study, which in turn makes the analyzed base of changes more representative from realistic software maintenance scenarios. Third, this application has been used for implementation-level maintenance analyses [10][11][12] allowing us to correlate our findings with their results. Finally, in the investigation reported in [10], additional changes were applied to the HW application leading to ten implementation releases.

The study is divided as follows: (i) definition of change metrics (section 3.3), (ii) identification of requirements to architecture dependencies (section 4.1), (iii) application of change scenarios to assess change impact (architectural models and code are visited) and measure change propagation to identify the architecture-level changes in terms of classes, interfaces and compositions (section 4.2), and (iv) analysis of the assessments in order to identify the architecturally-significant dependencies (section 4.3) in the presence of changes.

3.1 Health Watcher Requirements and Architecture

The Health Watcher application is a Web-based information system which allows online access to register complaints, read health notices, and query regarding health issues. Employees can record, update, delete, print, search, change the records stored in the HW repository (in form of tables) after being authenticated as HW employee, i.e., by providing correct login name and password. Citizen can register complaints that system registers in the repository and generates a complaint code. The initial version of the HW system lacked flexibility and incapability to support generic Web applications as it was bound to specialized Web services. Also, the initial version provided limited functionality for a limited set of data, for example the system only allowed to query and update health units and complaints.

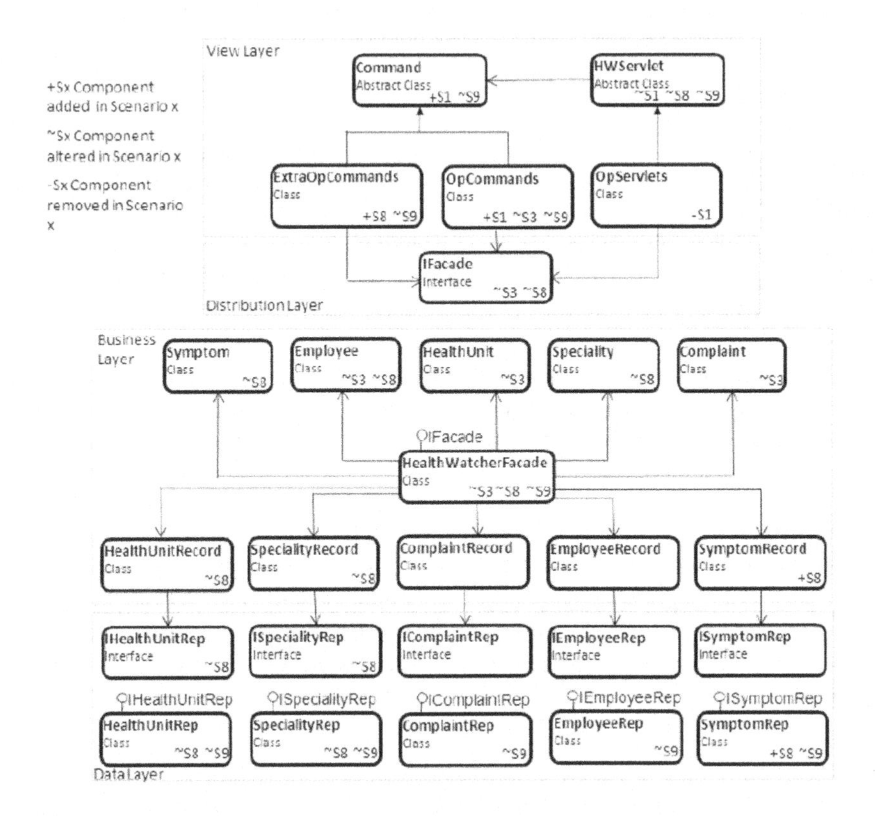

Fig.1. Module view of the Health Watcher system [10]

We have analyzed both OO and AO architectures of the HW system. The AO architecture modularizes concurrency, distribution, and persistence as aspects. Due to space limitation, our description focuses on the OO architectural design; detailed discussion about the AO architecture version is available at multiple sources [4][10][14][21]. Figure 1 shows the module view of the OO version for the HW system [10] that realizes the layered architectural style. It comprises of four layers: view, distribution, business, and data. Citizens access Web pages to query and/or register complaints. Multiple users can access the HW system simultaneously through Java Servlets, captured by the view layer (Figure 1), which decomposes into two main modules `HWServlet` and `OpServlets`. The Distribution module provides the interface `IFacade` to enable the access the HW services implemented in Business layer. The latter comprises of a number of modules, such as: `HealthUnitRecord`, `SpecialtyRecord`, `ComplaintRecord`, `EmployeeRecord`, and `Symptom Record`. Each of these modules is invoked for specific operations requested by citizen/employee. For example, if the citizen has requested to query a health unit then `IFacade` provides access to business layer module `HealthUnitRecord` which accesses the HW database `HealthUnitRep` using interface `IHealthUnitRep`. A complete description of the HW architecture and implementation is available elsewhere [4][10][11][12]. Figure 1 also represents the implemented change scenarios. In particular, it points out the impacted modules for each scenario, which are sub-scripted by change type and scenario number.

3.2 Change Metrics

Our quantitative assessment is based on a metrics suite to identify propagation of requirements change on the elements of the OO and AO architecture design. The architectural views, the module and component-connector views consists of module/components that have class(es), operation(s), and port(s)/interface(s). The metrics will give quantitative values to analyze the dependency from perspective of change and architectural significance. The quantitative metrics to access change at architecture level are:

Concentration (C): It measures requirements dependency to the architecture components and their composition. Set of requirements may trace to a layer or more than one layer comprising of components. In the AO architecture version an aspect will be treated as a layer. The lowest value of *C* is one, which is also an indicator that the change will be occurring at intra-level, i.e., only affecting a layer. The highest value of *C* is the total number of layers in the architecture design, which is also an indicator that the change may cause ripple effect in the architecture.

Dispersion (D): It measures the percentage of components impacted by change in a layer or multiple layers. In equation 1, *Ec:* is number of effected classes, operations, and interfaces in architecture due to change and *Tc:* is the total number of components in a layer.

$$D = \left(\frac{Ec}{Tc}\right) \times 100 \quad (1)$$

We will calculate *D* for each layer and then take an average. For example, if change is concentrated (*C*) in two layers, then *D* will individually calculated for each layer and then an average will be taken to calculate the final value of *D*. If dispersion percentage is lower than 25% (1/4) it may be considered as a mild dispersion. If inter-component dispersion percentage is greater than 33% (1/3), even if change is concentrated in a layer it is considered severe.

Inclusion (I): It measures the number of components added when change is incorporated. It is not mandatory that each change introduces a new component, therefore, *I* can be 0 or any number of components added.

4 Empirical Results and Constraints

This section reports the evaluation outcomes and discussion, based on a systematic analysis of the nine change scenarios implemented in Java and AspectJ, which has led to ten releases for the HW architecture (Section 3.1).

4.1 Requirements-Architecture Dependency Analysis

This section provides a summary of requirements-architecture dependency analysis for HW using the model described in Section 2. This analysis involved the trace of the requirements to their module and composition counterparts in the architectural models. Figure 2 shows a representative set of examples on dependencies that are likely to exist from the HW requirements elements to the architecture components and their compositions. A few of requirements have been shortened due to space limitations:

R1 to R5 form *goal* and *infrastructure dependencies* with architecture layers of HW system. For example, R2 have *goal* and *infrastructure dependencies* as one of a few goals of health watcher is ease of access (i.e., available online) and usability, which are satisfied by implementing health watcher as a Web-based online application and servlet for GUI. R6, R7, and R9 have *task dependency* with view layer as user interacts with the system providing input or feedback to system to proceed further.

R8 forms *service dependency* with business and data layer. *Service dependency* holds as health watcher performs operations to store information entered by the user, parse the data entered by the user, creates a new instance of the appropriate complaint type, generates a unique identifier and assigns this to the new complaint, complainers address is parsed and saved.

R13 forms *service dependency* with business and data layer as `searchComplaint (int code)`and `search(String login)` classes are invoked to list the complaints and employees to be updated of HW system. R11 forms *conditional and service dependencies* because user can not access the restricted operations: update and register unless system verifies (user's login and password) them as valid user.

R15 and R16 are representative example of a few errors occurrence during the operation of the HW system. Error handling has *service and conditional dependencies* with view, distribution, business, and data layer, depending where the error occurred. Distribution, persistence, and concurrency are modularized as aspects in the AO architecture. Therefore, the requirements which did not have explicit trace dependency for OO version form dependencies in AO architecture design. R20 forms *service and conditional dependency* with concurrency component `(HWManagedSync` and `HWTimeStamp)`. Timestamp provides functionality to avoid data inconsistency by applying timestamp field on the most recently modified data, storing it in the persistence mechanism.

Requirements	Dependency	Architecture
R1: System should be an online Web-based service	Goal & Infrast.	View
R2: System should have an easy to use GUI	Goal & Infrast.	View
R3: System must provide flexible storage mechanism	Goal & Infrast.	Business & Data
R4: System should be capable of running on separate machines	Goal & Infrast.	Distribution
R5: System must be able to handle 20 simultaneous users	Goal & Infrast.	Distribution
R6: ... to register complaint citizen choose a complaint type: animal, food, or special	Task	View
R7: ...user provides complaint details, place, and date/time	Task	View
R8: ...complaint is stored on the server assigning an identification number to each stored complaint	Service	Business & Data
R9: ... to query any information user selects query type: healthunit, specialty, or complaint	Task	View
R10: ... based on selection of query type system retrieves the list	Service	Business & Data
R11: ... to access restricted operations employee verifies themselves	Conditional & Service	View, Business & Data
R12: ... verified employee selects healthunit, specialty, or complaint to update	Conditional & Task	View
R13: For particular selection: healthunit, specialty, complaint data is retrieved	Service	Business & Data
R14: updates for healthunit, specialty, or complaint are stored at server	Service	Business & Data
R15: ... raise error message if invalid data is entered	Cond. & Service	View
R16: ... if the system does not respond in 5sec raise an exception	Temp., Cond., & Service	View & Distribution

Fig. 2. Functional and non-functional requirements of Health Watcher system [13][14]

4.2 Evolving Dependencies' Impact on the Architecture

Our findings report how changing requirements dependencies (Section 4.1) tend to entail four categories of architecture-level changes, namely: adaptive restructuring, perfective modifications, incremental changes, and behavioral modifications. The analysis was guided by using an existing categorization of architectural modifications in the OO and AO module view [18]. This reference model systematically characterizes both the level of impact and severity of each type of architectural change. However, it provides an investigation on the correlation of requirements dependencies and architecture change categories, which is the key aim of our analysis.

Adaptive Restructuring of Layers. There were two change scenarios that implied adaptive restructurings. First, change scenario 1, involved restructuring of HW software to provide extensibility by separating servlets and promote GUI decoupling. This scenario evolved *goal and infrastructure dependencies* of R2 to service dependency with view layer. The changes only spanned over the inner architectural elements of the view layer. On the other hand, there was an evidence of high change dispersion (Table 1) as there were changes in `HWServlet` and the removal of

OpServlets in view layer. This also led to the inclusion (I) of two components Command and OpCommand in view layer of the OO architecture.

Second, change scenario 5 led to deployment restructuring of the persistence mechanism in order to allow data storage in memory or database repository. *Goal and infrastructure dependencies* of R3 with business and data layer evolved to *goal, conditional, and infrastructure dependencies* in OO architecture, whereas in AO architecture *goal, conditional, and infrastructure dependencies* are formed with persistence (an aspectual component) and data layer. The reason for *conditional dependency* to emerge is because the change scenario has introduced a strategic design choice that altered the architecture design based on storage mechanism chosen at deployment time.

Table 1. Change scenarios and change assessment

	Description	**Type of Change**	**Evolving Req. dependencies**	**Assessment of Impacted Architecture**			
				OO/AO	**C**	**D**	**I**
1	Restructure Web based health watcher system to improve extensibility	Adaptive	Goal and Infrastructure	OO	1	50%	2
				AO	2	50%	1
2	Disable multiple updates once complaint state is CLOSED	Corrective	Services	OO	1	25%	1
			Conditional and Service	AO	1	25%	1
3	Improve maintainability by disassociating Update functionality from HW functions: health unit, specialty, and complaint	Perfective	Service	OO	1	28%	-
			Conditional and Service	AO	1	33%	1
4	HW system should support use of different distribution configurations	Perfective	Goal and Infrastructure	OO	2	39%	-
				AO	-	-	-
5	System must flexible in term of data storage	Adaptive	Goal and Infrastructure	OO	2	22%	1
				AO	2	33%	1
6	Ease the process of adding GUI	Perfective	Service	OO	1	33%	-
				AO	1	33%	-
7	Generalize distribution mechanism	Perfective	Goal and Infrastructure	OO	1	22%	1
				AO	-	-	-
8	Provide functionality to query more data types: symptoms and disease	Perfective	Service	OO	3	83%	3
			Conditional and Service	AO	6	50%	-
9	Modularize error handling strategies and provide better error recovery mechanisms	Perfective	Conditional and Service	OO	3	54%	-
				AO	4	57%	1

Perfective Modifications of Pivotal Architectural Services. Change scenario 3 is a perfective change modularizing update function, decoupling it from the rest of HW system, such as the disassociation of the health unit, specialty, and complaint entities in business layer. The set of requirement had *service dependency* with business layer and data layer. Change is concentrated in business layer with a dispersion of 28%, decoupling update functionality from HealthWatcherFacade, Employee, HealthUnit, and Complaint components in business layer. For the AO architecture, update function formed *service and conditional dependencies* with concurrency (aspect) and business layer. The reason to form dependencies with concurrency (aspect) is to achieve synchronization and maintain data consistency, which is achieved by applying timestamp field to the retrieved/updated complaint, health unit, and specialty data.

Change scenario 9, modularizes error and exception handling, decoupling it from the rest of the health watcher system. Similarly, change scenario 4 performs a perfective change as it modularizes the distribution mechanism, decoupling it from the rest of the health watcher system for OO architecture in order to facilitate deployment of different distribution configurations. But change scenario 4 is not applicable as distribution mechanism was modularized as an aspect in the initial AO architecture design.

Incremental Change of Component Interfaces. Change scenario 6 is an increment of release 2 that implements change scenario 1. It generalizes the request and response servlet parameters to enable inclusion of new operations and GUI. R2 entails *service dependency* to the view layer. *Dependency* remains unchanged due to the perfective nature of the modifications. Change is concentrated on the view layer with change dispersion lower than of change scenario 1 as it impacts a few operations in component interfaces within the view layer.

Change Scenario 7 is an increment of release 5 which implements change scenario 4. Incorporated change allows number of different distribution mechanisms to be configured for multiple servlets. Change is concentrated on the business layer of OO architecture with change dispersion lower than of release 5 (change scenario 4). Change scenario 7 has no effect/applicable as distribution mechanism was modularized as an aspect in the initial design of AO architecture.

Behavioral Modifications. Change scenarios 2, 8, and 9 modified/changed the intended behavior of system functions. Change scenario 2 refined the update complaint functionality, by allowing the complaint status to be set to close once complaint had been modified by employee. R14 implied on a *service dependency* with business and data layers for OO architecture, in AO architecture it formed *service and conditional dependencies* with concurrency (aspect) and business layer. For AO architecture the dependencies remained the same, whereas for OO architecture the dependency evolved to *conditional and service dependencies*, as every time a complaint is requested for update its status was checked. From the architectural perspective, a few operations in classes of `Complaint` component underwent some impact. In fact, the dispersion percentage was relatively low, i.e. 25%. The change has also encompassed the introduction of `State` component in business layer of OO and AO architecture.

The initial version of HW system only provided the option to query the health unit, complaint, and specialty. New functionality and querying options are provided with respect to the symptoms and diseases entities in scenario 8. Querying functionality had *service dependency* with business layer in OO architecture. Interestingly, the change has propagated to 3 layers and rippled through the system. The change added `SymptomRecord` class which formed *service* dependency with business layer and `SymptomRep` class which formed *goal and infrastructure dependencies* with data layer of OO architecture. While, querying functionality forms *service and conditional dependencies* with concurrency (aspect), persistence (aspect), and business layer of AO architecture. Change to query behaviour impact 6 layers and added observer pattern. `SymptomRecord` and `SymptomRepositoryRDB` classes were added in AO architecture.

Scenario 9 comprises of two architecture-level changes: a) perfective and b) behavioral. The changes in scenario 9 had been incorporated in parallel therefore they were not considered under the incremental change category. The change introduced new exceptions that were not in the initial intended OO and AO architecture design: `CommunicationException`, `SQLPersistenceMechanismException`, and `RepositoryException`.

4.3 On the Architecturally-Significant Dependencies

This section discusses some findings on which types of changing requirements dependencies (Section 2) are likely to have widely-scoped, moderate or localized impact at the architecture design. The previous quantitative and qualitative analysis (Section 4.1) are used as the basis.

Architectural Pull. An analysis of Table 1, externalizes the fact that dependencies are orthogonal in nature (Section 2) and it is impossible to have strict separation between the dependencies due to nature of requirements. These dependencies may form weak or strong interconnection with the architecture elements, which we refer to as architecture pull. The dependencies pull the architecture in various directions which may provide an insight of dominant dependencies at architectural level from perspective of change, as discussed below.

Dominant Dependency for a Particular Type of Change. The orthogonal nature of dependencies raises some questions: how coexisting dependencies evolve? which dependency has dominant impact during change? does the impact degree of a dependency category vary based on heterogeneous change types?

Table 1, shows *goal and infrastructure dependencies* co-existed, they were impacted by adaptive change (scenario 1 and 5). The adaptive modifications led to high dispersion of architecture-level change. It is well understood that a goal dependency is significant at architecture level [1] but for the specific change *infrastructure dependency* was dominant as the changes were focused on the software architecture being adapted to new standards/techniques in order to improve both server and client performance and this change did not cause the system to deviate from its original design and objectives and, as a result, the *goal dependency* was of limited impact. Change in *goal dependency* may lead to significant changes as the system's objectives are changed which will lead to degeneration of architecture design.

According to Table 1, *usability and service dependencies* (scenario 6) were affected by perfective change. For the specific change scenario *usability dependency* played a minimum role in the architectural modifications, the dominant dependency was *service dependency* as it refined the operations provided by system. Based on the assumption of corrective change in scenario 2, dependencies might have equal importance and may not be dominant over the other from perspective of change.

Many requirements formed *conditional and service dependencies* with OO and AO architecture, as seen in Table 1. In our analysis, we observed that *service dependency* led to high dispersion when it involved changes in system functionality/behavior. We have noticed the similar trend for the *service dependency* co-existing with *conditional dependency*. *Conditional dependency* captures the behavior and structure in form of architectural design choices, decisions, and constraints, which form core of the architecture, which are implied on number of architectural elements. When architectural decisions and/or choices change it may lead to addition of new structure or behavior in the architecture causing a break down, degradation, or enhancement in the architecture. Therefore a *conditional dependency* plays a dominant role and qualifies as a significant dependency when it co-exists with *service dependency* as all the dependant process or services are checked to see if they satisfy the new condition, decision, or constraint.

Independent Dependencies. Up to this point, we have discussed the orthogonal or overlapping dependencies. Now the question arises: if it is possible for dependencies to exist independently. A careful analysis of the outcomes in section 4.1and 4.2 makes it evident that *goal dependency* is unlikely to exist independently. *Goal dependency* defines the system objectives or quality attributes that are achieved by operations, services, software, and technical infrastructure, which have clear realization at architecture.

Dependencies with Minimum Architectural Impact. From analysis of the change scenarios we identified dependencies that are least likely to impact the architecture. The least significant dependency is *task dependency*. As task *dependency* facilitates user's interaction with the system through a medium (e.g., Web-browser, command prompt, etc.). Even if the backend software is enhanced/ modified (as in scenario 1) or service is modified/added (scenario 6) the front end remains the same, i.e., a Web browser. Based on scenario 1, if the GUI is changed, i.e., addition of radio buttons, drop down list, or check boxes, it will not introduce any change at architecture level, but at code level.

4.4 Study Constraints

Even though this study fully satisfies our initial goal of providing a first empirical investigation on the impact of evolving requirements dependencies on architectural changes, our procedures have some limitations. These limitations will contribute to further explorations using other experimental procedures and systems as targets.

Our analysis concentrates on the change history of one software system providing a single point of observation. Our target case study is a representative choice of Web-based information systems for several reasons: the HW realizes n-tier architecture that is one of the most common design alternatives implemented by deployed Web systems [10][21], HW functional and non-functional requirements are consistently part of applications in this specific domain, changes are heterogeneous and representative of real change requests within software projects from this nature. More importantly, the design of the HW system realizes the best design practices [4] and have being systematically being enhanced through the years [10]. It provides evidence that the requirements-architecture co-changes are not merely a matter of lack of systematic design choices.

Of course, strong conclusion can not be drawn by analyzing a single system as dependencies may (or may not) vary depending based on web-based information system's quality attributes. For example, an online banking system has security and privacy as its key quality attributes. Similarly an online comparison system has completeness and correctness as its quality attributes, b) we analyze a single architectural view, therefore, we are unsure of the dependencies that may exist for other views, how the architecture will be affected, on incorporating change and will similar set of dependencies be architecturally significant, and c) we only focused on analyzing the dependencies that evolved due to incorporated change scenarios. This does not give us a clear idea on change impact on other dependencies.

5 Related Work

A few conventional requirements modeling languages and methodologies [1][2][3] explicitly attempt to support a more straightforward derivation of architecture designs. They rely on the provision of means to enable the requirements engineers to reason about the system goals and how they are operationalized in terms of architectural elements. However, the sole use of these approaches is not sufficient to estimate the impact of requirements changes on the derived architectures. Requirements change is inevitable and incorporating these changes involve: huge cost and effort, risk of architecture degeneration, and undesirable system deviation from its original design.

One way to assess effect of change is to apply impact analysis techniques. There are many code-level change impact analysis approaches, but a scarcity of impact analysis support targeted at architectural evolution. Only a few authors [15][16][17] define impact analysis techniques for architecture designs, but their focus is restricted to the scope of architecture-level changes. They are oblivious to the way changes to the requirements and their inter-dependencies impact architectural decompositions. As a consequence, it makes it difficult to architects to estimate how specific change requests on requirements will affect the architecture elements. For understanding impact of requirements evolution on architecture it is fundamental to understand the trace dependencies that may hold between entities of the two artifacts.

Ramesh and Jarke [8] defined requirements trace dependencies: *goal, task, resource, and temporal* for trace managing requirements on the basis of nature of dependency. Grady [9] took a step further to define architectural requirements by defining *design, implementation, interface, and physical* requirements, which only provide a mechanism to analyze how these requirements are realized at architecture. William and Carver [18] characterized architectural change and specified their affect on logical and runtime architectural structure. Inspiring from works of Ramesh and Jarke [8], Chung et. al. [2], William and Carver [18], and Grady [9] we have defined requirements to architecture dependency model to predict impact of requirements change based on dependencies and change type (corrective, adaptive, and perfective) on architecture. This has help predict change impact of dependencies and identify architecturally significant dependencies.

6 Conclusion and Future Work

This paper presents the outcomes of an exploratory study aimed at assessing the impact of typical changes in requirements dependencies on architecture design. From our exploratory study we have externalized the fact that requirement dependencies are often orthogonal in nature and, in many cases, there is no strict separation between the dependencies. We have analyzed how co-existing dependencies evolve, which dependencies have dominant impact during change and led to different "architectural pulls". Our study has shown that evolution of co-existing *conditional and services dependencies* is architecturally significant as for evolving *service dependency* it needs to be checked if the corresponding condition, constraints, and decision are satisfied or for evolving *conditional dependency* it needs to be checked if the process or services correspond to the condition. Similarly, co-existing *goal and infrastructure dependencies* are architecturally significant as change to system objectives may lead to architectural degeneration. Strong conclusion can not be drawn by analyzing a single system as dependencies may (or may not) vary depending based on Web-based information system's quality attributes. In order to validate our dependency model and identify other architecturally-significant dependencies, other studies should further analyze applications from different domains.

Acknowledgements. This work was partially supported by the European Commission grant IST-33710 - Aspect-Oriented, Model-Driven Product Line Engineering (AMPLE), and the TAO project, funded by Lancaster University Research Committee.

References

1. van Lamsweerde, A.: From System Goals to Software Architecture. In: Bernardo, M., Inverardi, P. (eds.) SFM 2003. LNCS, vol. 2804, pp. 25–43. Springer, Heidelberg (2003)
2. Chung, L., Nixon, A.B., Yu, E., Mylopoulous, J.: Non-functional Requirements in Software Engineering. Kluwer Academic Publishing, Dordrecht (1999)
3. Herold, S., et al.: Towards Bridging the Gap between Goal-Oriented Requirements Engineering and Compositional Architecture Development. In: SHARK-ADI 2007 (2007)
4. Greenwood, P., et al.: Aspect Interaction and Design Stability: An Empirical Study (2007), http://www.comp.lancs.ac.uk/computing/users/greenwop/ecoop07
5. Jacobson, I., Chirsterson, M., Jonsson, P., Overgaard, G.: Object-Oriented Software Engineering: A Use Case Driven Approach, 4th edn. Addison-Wesley, Reading (1992)
6. Chitchyan, R., et al.: Semantics-based Composition for Aspect-Oriented Requirements Engineering. In: AOSD 2007, Vancouver, Canada, pp. 36–48 (2007)
7. Chitchyan, R., et al.: Survey of Aspect-Oriented Analysis and Design. AOSD-Europe Project Deliverable No: AOSD-Europe-ULANC-9
8. Ramesh, B., Jarke, M.: Towards Reference Models for Requirements Traceability. IEEE Transactions on Software Engineering 27(1) (January 2001)
9. Grady, R.: Practical Software Metrics for Project Management and Process Improvement. Prentice-Hall, Englewood Cliffs (1992)
10. Greenwood, P., et al.: On the Impact of Aspectual Decompositions on Design Stability: An Empirical Study. In: Ernst, E. (ed.) ECOOP 2007. LNCS, vol. 4609, pp. 176–200. Springer, Heidelberg (2007)
11. Sant'Anna, C., et al.: On the Modularity of Software Architectures: A Concern-Driven Measurement Framework. In: Oquendo, F. (ed.) ECSA 2007. LNCS, vol. 4758, pp. 207–224. Springer, Heidelberg (2007)

12. Cacho, N., et al.: Composing Design Patterns: A Scalability Study of Aspect-Oriented Programming. In: AOSD 2006, pp. 109–121 (2006)
13. TAO: A testbed for Aspect Oriented Software Development (2007), http://www.comp.lancs.ac.uk/~greenwop/tao/
14. Khan, S.S., et al.: Material for On the Impact of Evolving Requirements-Architecture Dependencies: An Exploratory Study (2007), http://www.comp.lancs.ac.uk/~shakilkh/caise08
15. Feng, T., Maletic, I.J.: Applying Dynamic Change Impact Analysis in Component-based Architecture Design. In: 7th ACIS International Conference on Software Engineering, Artificial Intelligence, Networking, and Parallel/Distributed Computing (SNPD 2006) (2006)
16. Riebisch, M., Wohlfarth, S.: Introducing Impact Analysis for Architectural Decisions. In: ECBS 2007, pp. 381–392 (2007)
17. Zhao, J., et al.: Change Impact Analysis to Support Architectural Evolution. Software Maintenance: Research and Practice 14(5), 317–333 (2002)
18. Williams, J.B., Carver, C.J.: Characterizing Software Architecture Changes: An Initial Study. In: 1st Intl. Symposium on Empirical Software Engineering and Measurement (ESEM 2007), pp. 410–419 (2007)
19. Clement, P., et al.: Documenting Software Architectures: Views and Beyond. SEI Series in Software Engineering. Addison-Wesley, Reading (2002)
20. Shaw, M., Garlan, D.: Software Architecture: Perspectives on an Emerging Discipline. Prentice-Hall, Inc., Englewood Cliffs (1996)
21. Soares, S., et al.: Distribution and Persistence as Aspects. Software Practice and Experience (2006)
22. Murta, G.P.L., et al.: ArchTrace: Policy-Based Support for Managing Evolving Architecture-to-Implementation Traceability Links. In: ASE 2006, pp. 135–144 (2006)
23. Lee, M., Offutt, J.: Change Impact Analysis of Object-Oriented Software, 193 pages, George Mason University (1998)
24. Browning, T.R., et al.: Applying the Design Structure Matrix to System Decomposition and Integration Problems: A Review and New Directions. IEEE Transaction on Engineering Management 48(3), 292–306 (2001)
25. Sangal, N., et al.: Using Dependency Models to Manage Complex Software Architecture. In: OOPSLA 2005, San Diego, California, USA (2005)
26. Meyer, B.: Applying Design by Contract. Computer 25(10), 40–51 (1992)

The IT Organization Modeling and Assessment Tool for IT Governance Decision Support

Mårten Simonsson, Pontus Johnson, and Mathias Ekstedt

Department of Industrial Information and Control Systems, KTH, Royal Institute of Technology. Osquldas väg 10, SE 100 44 Stockholm, Sweden
{ms101,pj101,mek101}@ics.kth.se

Abstract. This short paper describes the IT Organization Modeling and Assessment Tool (ITOMAT) and how it can be used for IT governance decision making support. ITOMAT consists of an enterprise architecture metamodel that describes IT organizations. ITOMAT further contains a Bayesian network for making predictions on how changes to IT organization models will affect the IT governance performance as perceived by business stakeholders. In order to make such predictions accurately, the network learns from data on previous experience. Case studies at 35 companies have been utilized for calibration of the network.

Keywords: IT governance, IT organization, Enterprise Architecture, modeling, metamodel.

1 Introduction

One of the most common approaches to IT management today is enterprise architecture. Enterprise architecture (EA) is a model-based approach to IT management that encompasses architecture descriptions of business, information, applications and infrastructure of an organization [1], [9], [11]. The basic idea is that instead of changing information systems and their surrounding business using trial and error, models are used to analyze the future behavior of potential scenarios.

IT governance denotes how the IT organization is managed and structured and provides mechanisms that enable the development of integrated business and IT plans, allocation of responsibilities within the IT organization, and prioritization of IT initiatives [3], [6], [7]. It is important to ensure that the IT governance of an organization is not only designed to achieve internal efficiency in the IT organization, such as if rights and responsibilities are distributed over the appropriate people in an explicit manner, if processes formalized for important tasks are implemented, and if documentation exists. We call this internal IT organization efficiency *IT governance maturity*. The final goal of good IT governance is rather to provide the business with the support needed in order to conduct business in a good manner, which is here called *IT governance performance*. Current IT governance frameworks, including the Weill & Ross framework, ITIL and COBIT, are not explicit with how IT governance maturity affects IT governance performance [4], [8], [10].

Z. Bellahsène and M. Léonard (Eds.): CAiSE 2008, LNCS 5074, pp. 258–261, 2008.

2 The IT Organization Modeling and Assessment Tool

The IT Organization Modeling and Assessment Tool (ITOMAT) is an IT governance decision support tool based on enterprise architecture models. It relies partly on the COBIT framework for IT governance maturity assessments of the IT organization. ITOMAT is further enhanced with information about statistical correlations to external IT governance performance [4]. ITOMAT is therefore able to statistically predict IT governance performance from IT organization scenarios. This section presents the different steps involved in such IT governance decision making and how it is supported by the ITOMAT, cf Fig. 1.

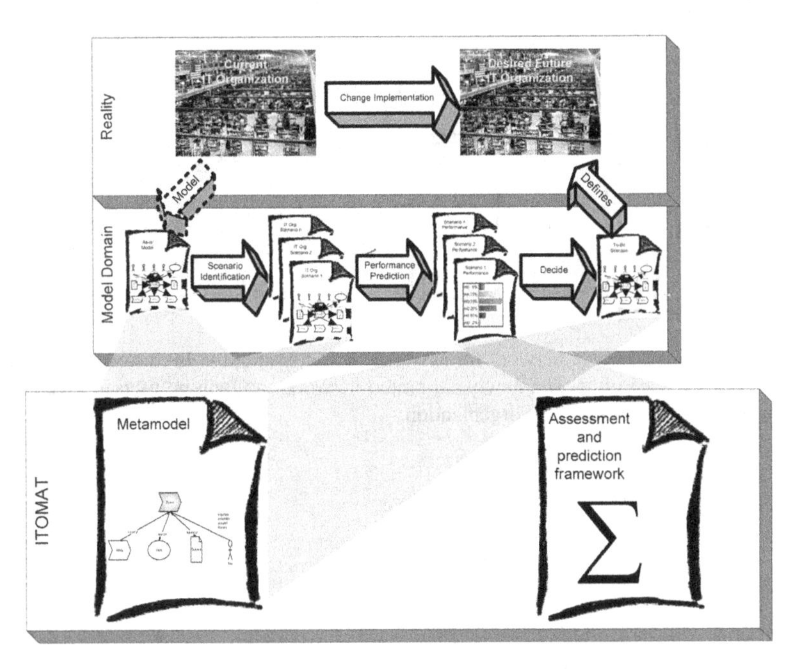

Fig. 1. Workflow for IT governance decision making and how the ITOMAT's supports this process with enterprise architecture scenario development and assessment

1. **Create As-is Model of Current IT Organization**
   This first step concerns the development of a model of the current IT organization. The result is an as-is model that follows a predefined metamodel, c.f. Fig. 1. The metamodel prescribes what entities and relations that are allowed for IT governance maturity modeling, including processes, activities, documents, roles & responsibilities, and metrics. It also includes entities for IT governance performance modeling, including a number of objectives taken from Weill & Ross [10].
2. **Identify Change Scenarios**
   As a second step, a number of possible future scenarios of the IT organization are modeled.

3. **Predict IT Governance Performance**
   So far, only descriptive models have been developed. As a third step a normative evaluation is conducted. The ITOMAT's assessment and prediction framework analyses the different change scenarios' outcome in terms of overall IT governance performance, c.f. Fig. 1. The predictive ability of the ITOMAT is made through the use of a Bayesian network [5] taught with the experience from previous case studies on currently 35 organizations, cf Fig. 2. Fig. 3 exemplifies results from the ITOMAT's Bayesian network representation of IT governance performance for an as-is model, and predictions for two change scenarios that involve ITIL implementation and enhanced project management respectively.

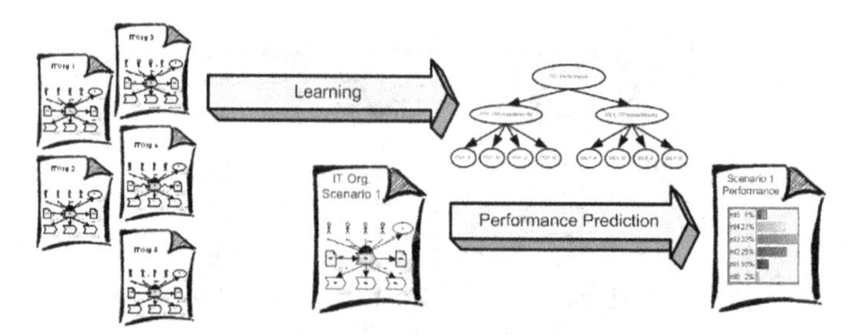

Fig. 2. The learning Bayesian network for assessment and prediction. In this example, five case studies are used for learning. The newly calibrated network can then be used to predict the IT governance performance in a sixth organization.

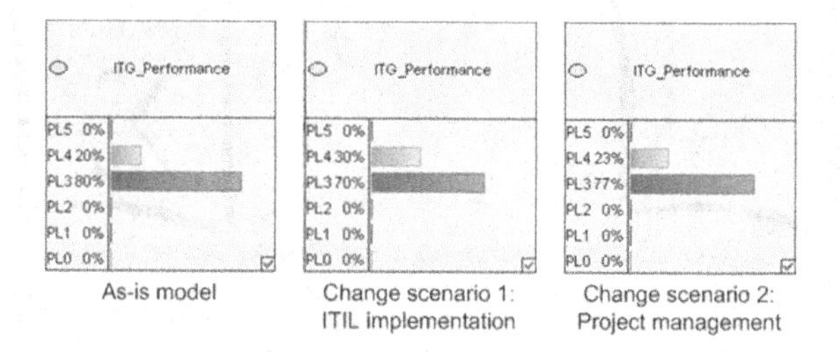

Fig. 3. IT governance performance predictions for as-is model and two change scenarios. The performance graphs are created in the GeNIe tool [2] and show probability of each performance level to occur.

4. **Decide on To-Be Scenario**
   The previous step results in a prediction value of the IT governance performance of each developed scenario. This thus provides the decision maker with an estimate of the overall "goodness" of a scenario. Ceteris paribus the scenario with the highest goodness value is the rational choice.
5. **Implement To-Be Scenario**
   The four steps above describe a strategic planning exercise. In order to achieve the benefits of the new organization, the to-be scenario of course also has to be implemented in reality. This is however out of the scope for the ITOMAT.

3 Discussion and Conclusion

This paper describes an ongoing research project for predicting IT governance performance, ITOMAT. In essence, the ITOMAT is contributing to academia and practitioners in two ways. Firstly it formalizes how analysis and assessments of the quality of IT organizations is to be performed on enterprise architecture models. Secondly, it makes use of knowledge from previous case studies to make predications about the IT governance performance, i.e. the impact of the IT on the business, from enterprise architecture models. It combines model-based decision-making with best practice from the field of IT governance, and the predictive capabilities of Bayesian networks. An important factor for achieving credible results from the predictions of ITOMAT is to have much empirical material so that the underlying Bayesian network can be further refined. As of today, the predictions are based on data from 35 case studies ranging from municipalities, large banks, industrial companies and small consulting firms. Data has been collected through 158 interviews and 60 surveys in 35 organizations. The case study findings have been validated in meetings with 15 IT governance experts. Altogether, the ITOMAT provides useful support for IT management decision making on IT organizational changes and it clearly bridges the void between IT governance and Enterprise architecture.

References

1. DoDArchitecture Framework Working Group: DoD Architecture Framework Version 1.0. Department of Defense (2003)
2. The GeNIE tool for decision theoretic models, `http://genie.sis.pitt.edu/`
3. Henderson, J.C., Venkatraman, N.: Strategic Alignment - Leveraging Information Technology for Transforming Organizations. IBM Systems Journal 32(1), 472–485 (1993)
4. IT Governance Institute: Control Objectives for Information and Related Technology, 4.1th Edition (2007)
5. Jensen, F.V.: Bayesian Networks and decision graphs. Springer, New York (2001)
6. Korac-Kakabadse, N., Kakabadse, A.: IS/IT Governance - Need for an Integrated Model. Corporate Governance 4(1), 9–11 (2001)
7. Loh, L., Venkatraman, N.: Diffusion of Information Technology Outsourcing: Influence Sources and the Kodak Effect. Information Systems Research 3(4), 334–359 (1993)
8. Office of Government Commerce: Service Strategy Book. The Stationery Office (2007)
9. TOG, The Open Group: The Open Group Architecture Framework Version 8.1.1, Enterprise Edition. (2006)
10. Weill, P., Ross, J.W.: IT governance – How Top Performers Manage IT Decision Rights for Superior Results. Harvard Business School Press, USA (2004)
11. Zachman, J.: A Framework for Information Systems Architecture. IBM Systems Journal 26(3) (1987)

Ensuring Transactional Reliability by E-Contracting*

Ting Wang, Paul Grefen, and Jochem Vonk

Information Systems Subdepartment, Department of Technology Management
Eindhoven University of Technology, The Netherlands
{t.wang,p.w.p.j.grefen,j.vonk}@tue.nl

1 Introduction

The increasing complexity of business processes makes reliable execution of such processes more and more complicated. Thus, execution reliability has long been a challenge and attracted a lot of attention from both academia and industry. On the one hand, the demand for reliability from the business world has pushed IT to advance in transaction management. On the other hand, the progress in IT has provided more opportunities for the business world to enhance reliability of process execution. Motivated by this observation, we have studied an e-contracting case from a transactional perspective and discovered a gap between the business and IT world in their awareness of the transactional reliability. This gap leads to the inadequacy of transactional agreements between collaborating parties and exposes the process execution to many potential threats, thus decreasing reliability. To bridge the gap, we proposed the TxQoS (Transactional Quality of Service) approach in [1], by specifying and ensuring transactional performance of services/processes in contracts. In this paper, we continue this line of research by illustrating with an overview scenario. The focus is the mechanism of specification, which is essential to realize the contractual approach for transaction management. The unique feature is the interpretation of transactional reliability from a business perspective, which use contracts for specification and ensuring purposes to bridge the gap. Due to space limitations, we present briefly the TxQoS scenario and the specification attributes. Interested readers are refer to [2] for more details.

2 TxQoS Scenario

The TxQoS approach is illustrated in Figure 1. The scenario describes the non-functional aspects of services that address the QoS and SLA issues. First, a service provider designs its 'TxQoS Template' which can be differentiated into multiple 'TxQoS Offers' based on its transactional reliability performance. Meanwhile, a potential user may also design a 'Requirement Template' for each service

* The research reported in this paper has been conducted as part of the eXecution of Transactional Contracted Electronic Services (XTC) project (No. 612.063.305) funded by the Dutch Organization for Scientific Research (NWO).

Z. Bellahsène and M. Léonard (Eds.): CAiSE 2008, LNCS 5074, pp. 262–265, 2008.

it invokes based on its reliability requirement. Each template is instantiated into one 'TxQoS Requirement' document that may or may not be met by existing offers. Second, the user looks up in the 'Performance Repository' to decide which TxQoS offer is the best match. Negotiations between the provider and user may take place if no ready-to-agree 'TxQoS Offer' is suitable. Third, after an offer and a requirement are matched, a TxSLA (Transactional SLA) is established and the agreed TxQoS specification is enclosed into the service contract. Monitoring of the running transaction performance is then realized by checking the compliance of the runtime statistics with the specification.

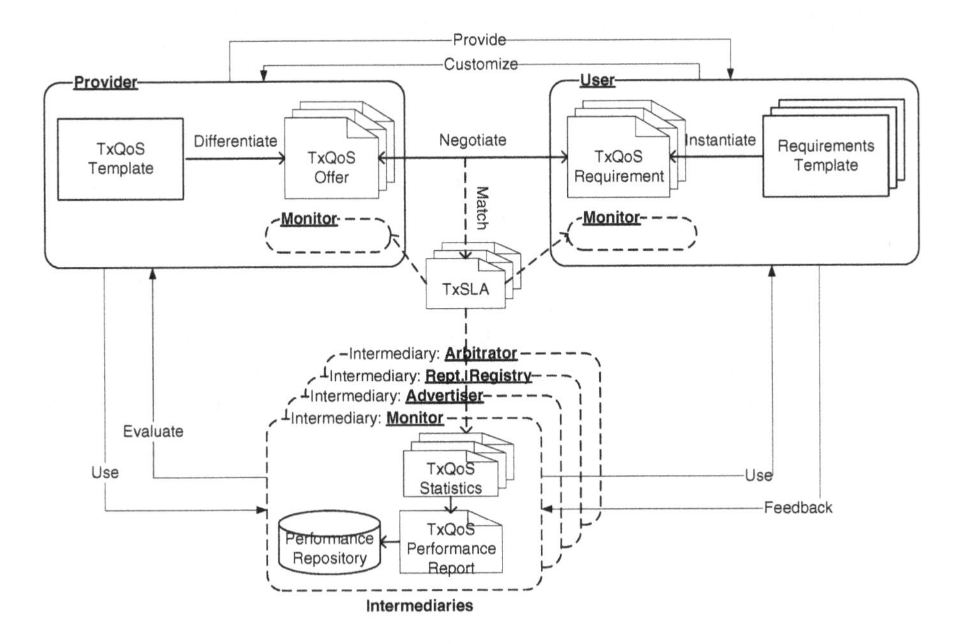

Fig. 1. TxQoS Scenario

The monitoring module is indispensable thus should be provided by at least one party. If a provider and a user do not trust each other, an intermediary (i.e. the 'Monitor') is delegated. Otherwise the provider or user can host such a module, which appeared at each party in Figure 1 to show the possible existence. The other intermediaries (i.e. advertiser, reputation checker, arbitrator) can also be omitted from the scenario in case a provider and a user are tightly coupled and sufficient trust between them has been established. In a word, all four types of intermediaries can either be omitted, or be (partially) used depending on the trust level between providers and their users. Their functions are realized by accessing the 'Performance Repository' for 'TxQoS Performance Reports' generated on the basis of runtime 'TxQoS Statistics'. In the next section, we introduce a method to specify transactional reliability performance, which is the key to realize the TxQoS approach.

3 TxQoS Attributes: FIAT

A TxSLA states metrics and measurement of the TxQoS specification. A TxQoS specifications specifies the transactional reliability of a service. We have designed the FIAT attributes for this purpose: *Fluency, Interferability, Alternation, Transparency.* The design is based on our knowledge gained of the related areas (e.g. transaction management, e-contracting, QoS management) and a real-life case analysis. The FIAT design meets the following criteria: 1) A TxQoS attribute should reflect reliability at service level guaranteed by transaction management mechanisms; 2) A TxQoS attribute should be understandable by the business world; 3) A TxQoS attribute should be precisely specifiable and monitorable like other functional attributes (e.g. time, cost, capacity etc.); 4) The use of a TxQoS attribute should benefit both the service provider and the user. Below we outlined the definition of the four attributes and use 'Fluency' as an example to show how to use it for contracting purpose.

Fluency indicates the smoothness of service execution and is monitored by counting the number of unexpected breakdowns, which suspend execution and demand for a fix to continue. If canceled by a user, it is counted as an expected exception instead of a breakdown, and therefore belongs to 'Interferability'. While monitoring of 'Fluency' is easy (by counting), the specification requires complicated calculations based on the past performance statistics to predict future performances. We use the NonHomogeneous Poisson Process (NHPP) model, which is widely adopted in software reliability engineering [3]. We assume breakdowns happen stochastically that meet the following two conditions: (1)No simultaneous breakdowns can happen at any time; (2)The causes of the past breakdowns are fixed and do not affect future execution. The probability of exact n breakdowns occurring in the time interval $(a, b]$ is given by

$$P(n) = \frac{\left[\int_a^b \lambda(t)d(t)\right]^n e^{-\int_a^b \lambda(t)d(t)}}{n!}, \quad \text{for } n = 0, 1, \ldots \tag{1}$$

where the past statistics determines $\lambda(t)$ (breakdown happening rate), and prediction of future 'Fluency' can be calculated. For example, suppose there is a service with its execution time $T(T = max(t))$. Any execution after T is viewed as a failure and is excluded in the fluency statistics. During the test, it shows the rate $\lambda(t)$ is a constant so that the GO NHPP model [4] is adopted. Then a fluency function $f(n)$ can be defined as the probability of having no more than n breakdowns during execution (i.e. within the time interval $(0, T]$):

$$f(n) = \sum \frac{\left[m(T + \frac{e^{-rT}-1}{r})\right]^n e^{-m(T+\frac{e^{-rT}-1}{r})}}{n!} \tag{2}$$

Based on the calculation of $f(n)$, a TxQoS offer can be specified using statement such as '*we guarantee no more than n breakdowns during the execution*', or in quantitative values such as '*Min(Fluency)=n*'.

Interferability describes the control of users upon a service being invoked. This property is especially suitable in outsourcing scenarios where different levels of control are necessary. Note that a user only has interferability to the activities that are specified as transparent (see 'Transparency' below). Interferability can be interpreted as the set of commands from the users to intervene an activity (viewed as a node in execution path), plus the allowed timings to issue these commands. At runtime, each user command is checked for its validity and unspecified commands are blocked.

Alternation describes the choices that are pre-defined in case of (expected) exceptions/errors during execution. At the design phase, 'Alternation' is specified as a set of allowed execution graphs and these predefined graphs that are grouped as 'the preferred path' and its 'alternatives'. Runtime monitoring is enabled by comparing the ongoing execution path with the paths specified under the 'Alternation' attribute.

Transparency describes the visibility of a service and is specified as the set of activities that are visible to the users at the external level of a process. This is the only attribute that needs to be specified but does not need runtime monitoring. Here we assume a log of each instance execution status and parameters is kept to settle potential disputes.

Besides the FIAT attributes for design-time specification and runtime monitoring, a framework consisting of an architecture, a contracting model, and the monitoring mechanism has also been developed with the details in [2].

4 Conclusions and Future Work

The TxQoS approach ensures transactional reliability by e-contracting for contract-driven, service-oriented processes, which we believe points out a new research direction in the related areas (e.g. transaction management, e-contracting, QoS). Our future work falls in two categories. First, we are performing another case study in the healthcare domain, where exceptions (e.g. not enough ward capacity, unexpected symptoms) are very likely to occur and reliability of medical processes is of top priority. Therefore, a methodology to validate and apply the TxQoS approach will be developed. Second, we are going to extend our approach by a XML-based specification language for TxQoS and a refined framework.

References

1. Wang, T., Vonk, J., Grefen, P.: TxQoS: A contractual approach for transaction management. In: Proc. 11th IEEE Int. Conf. Enterprise Computing (EDOC 2007), pp. 327–338. IEEE Computer Society Press, Los Alamitos (2007)
2. Wang, T., Grefen, P., Vonk, J.: TxQoS: concept, scenario, and framework. Technical report, Eindhoven University of Technology (2007)
3. Pham, H.: Software Reliability (1999)
4. Goel, A., Okumoto, K.: Time-dependent error-detection rate model for software reliability and other performance measures. IEEE Trans. Reliability R-28, 206–211 (1979)

Drawing Preconditions of Operation Contracts from Conceptual Schemas

Dolors Costal, Cristina Gómez, Anna Queralt, and Ernest Teniente

Departament de Llenguatges i Sistemes Informàtics, Universitat Politècnica de Catalunya
{dolors,cristina,aqueralt,teniente}@lsi.upc.edu

Abstract. Conceptual schemas include the definition of integrity constraints which must be satisfied in each state of the Information Base. Integrity constraints have a considerable impact on the specification of operations since operations should preserve the Information Base consistency. In this paper, we present an approach that automatically generates the preconditions that basic operations must include to ensure that a set of predefined integrity constraints is satisfied after their execution. Our approach is independent of the conceptual modelling language used. We also describe a prototype tool that implements our proposal for UML conceptual schemas.

Keywords: conceptual modelling, operation contracts, integrity constraints.

1 Introduction

An information system must include a representation of the *knowledge* of the domain, i.e. the Conceptual Schema (CS), and of the *state* of that domain, i.e. the Information Base (IB), to perform its functions.

The goal of automating information systems building was already stated in the late sixties [1]. However, and thanks to the definition and standardization of the MDA [2], this goal has revived and seems now more feasible than ever. For this reason, there has recently been a significant amount of work aimed at providing an automatic generation of (parts of) the software system from its specification.

In this context, we may find several proposals that provide an automatic definition of the basic operations (such as entity insertion or deletion, attribute modification, etc.) from a conceptual schema which allow updating the contents of the IB [3, 4, 5, 6]. Their main drawback is that either they do not take into account the integrity constraints to be preserved during the automatic generation of the operations or they consider them only up to a limited extent. Nevertheless, the automatic generation of the software elements required to ensure that the IB always satisfies the constraints of the CS is a crucial issue in software automation [7].

Our approach in this paper represents a step forward in this direction. Given a set of basic operations that update the contents of the IB (which may be either manually or automatically generated), a conceptual schema and a set of predefined integrity constraints, we are able to automatically determine the weakest precondition that must be considered for each basic operation so that integrity constraints are never violated when the operation is executed. Since we only consider adding preconditions,

Z. Bellahsène and M. Léonard (Eds.): CAiSE 2008, LNCS 5074, pp. 266–280, 2008.

integrity enforcement is achieved by avoiding the operation execution when its precondition is not satisfied. Our approach is independent of the conceptual modelling language used, although we will use UML and OCL in our examples.

In this way, our approach facilitates the automatic model-driven development of the information system from its initial specification since it simplifies the manual computation of the operation preconditions during software development. We have also developed an implementation of our approach which is integrated in a CASE tool.

As an example, consider the conceptual schema of Figure 1 which contains information about the employees assigned to projects and their supervisors. The schema contains three textual and two graphical constraints.

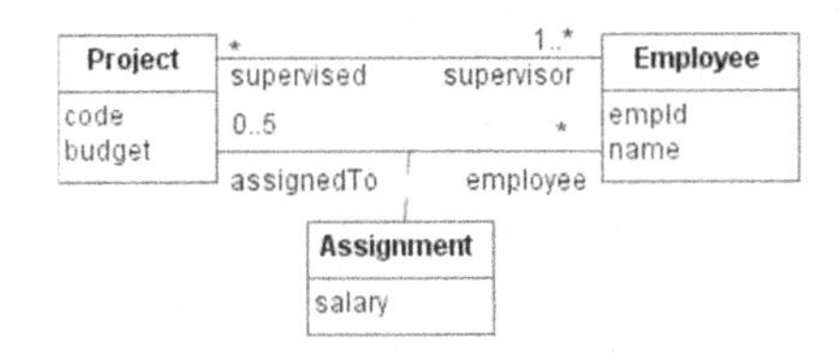

1. Projects are identified by code and employees by empId
   context Project inv: Project.allInstances()->isUnique(code)
   context Employee inv: Employee.allInstances()->isUnique(empId)

2. The salary of an assignment must be greater than 100€.
   context Assignment inv: self.salary>100

3. The supervisor of a project must not be one of its employees.
   context Project inv: self.employee->excludes(self.supervisor)

Fig. 1. Conceptual schema of our example application

Figure 2 shows a natural specification of the operation that assigns employees to projects. We assume that the parameters are provided as objects but their identifiers could be used as well.

Operation: newAssignment(e: Employee, p: Project, sal: Float)
Pre: *--the employee is not assigned to the project*
e.assignedTo -> excludes(p)
Post: *--a new instance of Assignment is created*
Assignment.allInstances()->exists(a | a.oclIsNew() and
a.salary = sal and a.project=p and a.employee=e)

Fig. 2. A sample partial contract for the operation *newAssignment*

It can be easily seen that the previous contract does not take integrity constraints into account since its precondition does not ensure that all constraints are satisfied. For instance, it allows assigning an employee to a project even if he is its supervisor. Therefore, this precondition must be extended to guarantee that the operation execution always leads the IB to a consistent state. Doing this by hand is time-consuming and error prone since it is not easy to identify the integrity constraints that may be violated by the operation execution and the additional required preconditions.

The contract of *newAssignment* that incorporates all the knowledge provided by the integrity constraints is shown in Figure 3 and it can be automatically obtained with our approach.

An automatic computation of the preconditions required to ensure that the operation contracts do not violate any integrity constraint provides two important contributions. First, it improves the quality of the specified operations since human

Operation: newAssignment(e: Employee, p: Project, sal: Float)
Pre:
--the employee is not assigned to the project e.assignedTo -> excludes(p)
--the salary is greater than 100 sal > 100
--the employee does not supervise the project p.supervisor -> excludes(e)
--the employee is not assigned to five projects e.assignedTo -> size()<5
Post:
--a new instance of Assignment is created
Assignment.allInstances()->exists(a | a.oclIsNew() and
a.salary = sal and a.project=p and a.employee=e)

Fig. 3. The full contract for the operation *newAssignment*

mistakes can be completely avoided. Second, software development is accelerated since integrity-preserving contracts can be automatically obtained.

The rest of the paper is organized as follows. The next section reviews some preliminary concepts. Section 3 describes a set of basic predefined operations. In section 4, we describe the conflicts that arise between integrity constraints and operations and we present our proposal for the automatic generation of preconditions. Section 5 describes a tool that implements our proposal. Related work is reviewed in section 6 and, finally, section 7 presents some conclusions and points out future work.

2 Preliminary Concepts

A CS consists of a taxonomy of entity types together with their attributes, a set of relationship types, and a set of integrity constraints [8]. A relationship type has several participants, i.e. entity types that play a certain role in the relationship type. In this paper, we deal with relationship types that have two participants (i.e. binary). Some relationship types are reified and, thus, they may have attributes and participate in other relationship types.

An information system maintains a representation of the state of a domain in its IB [9]. The state of the IB is the set of instances of the entity types and relationship types defined in the CS. The integrity constraints of the CS define conditions that each state of the IB must satisfy. Those constraints can have a graphical representation or can be defined through a particular language.

Additionally, a CS includes a set of operations, and the content of the IB changes as a result of their execution. The effect of each operation on the IB is specified by an operation contract. An operation contract is defined by a precondition, which expresses a condition that must be satisfied when the call to the operation is made, and a postcondition, which expresses a condition that the new state of the IB must satisfy [10].

Integrity constraints are closely related to operations, since the former must hold in every state of the IB, and the latter are the ones that change its content. Then, an operation contract must guarantee that the integrity constraints defined in the schema hold after its execution. We consider the following predefined integrity constraints (a more detailed description can be found in [11]).

An *identifier constraint* specifies a set of properties that uniquely identifies each instance of an entity type. Let E be an entity type and $\{p_1,...,p_n\}$ a set of properties, which can be attributes or roles. An identifier constraint specifies that a subset $\{p_i,...,p_j\}$ of these properties uniquely identifies the instances of E.

Recursive relationship type constraints, referred to as ring constraints in [12], are constraints that apply over recursive binary relationship types to guarantee that the relationship type fulfils a certain property. We consider five such constraints: *symmetric, asymmetric, antisymmetric, irreflexive* and *acyclic* constraints.

Let E be an entity type and R a recursive relationship type over E. A symmetric constraint over R guarantees that if a and b are instances of E and a is R-related to b, then b is R-related to a. An asymmetric constraint guarantees that if a and b are instances of E and a is R-related to b, then b is not R-related to a. An antisymmetric constraint over R guarantees that if a and b are instances of E, a is R-related to b and b is R-related to a, then a and b are the same instance. An irreflexive constraint over R guarantees that if a is an instance of E then a is never R-related to itself. An acyclic constraint guarantees that if a and b are instances of E and a is R-related to b, then b or instances R-related directly or indirectly to b are not R-related to a.

Path comparison constraints restrict how to relate the population of one role or role sequence (i.e. a path) to the population of another [12]. *Path inclusion, path exclusion* and *path equality* are all examples of this type of constraint and apply to an entity type A related to an entity type B via two different paths $r_1 ... r_i$, and $r_j ... r_n$. A path inclusion constraint guarantees that if a is an instance of A, the set of instances of B related to a via $r_1 ... r_i$ includes the set of instances of B related to a via $r_j ... r_n$. In a similar way, a path exclusion constraint ensures that the intersection between the populations of both paths is empty while a path equality constraint guarantees that both populations contain exactly the same instances.

Value comparison constraints restrict the values of an attribute by comparing it with a constant or with another attribute value [13]. Let E be an entity type, let a_i be an attribute of E, let v be either a constant or the value of an attribute accessible from E, and let op be an operator of type $<$, $>$, $=$, $<>$, $\leq$, or $\geq$. A value comparison constraint restricts the values of a_i with respect to the value of v according to op.

Cardinality constraints for binary relationship types restrict the number of instances that can be related to another instance through the relationship type. Let R be a binary relationship type such that entity type E_1 plays role p_1 and entity type E_2 plays role p_2 in it. A cardinality constraint from p_1 to p_2 in R indicates the minimum and maximum number of instances of type E_2 that may be related with any instance of type E_1 through R [14]. Cardinality constraint from p_2 to p_1 in R is defined similarly.

Disjointness and *covering* constraints impose restrictions on the population of a set of entity types. A disjointness constraint for entity types $E_1,...,E_n$ indicates that a particular entity can be instance of at most one E_i [15]. A covering constraint between an entity type E and a set of entity types $E_1,...,E_n$ indicates that every instance of E is instance of at least one E_i [15].

3 Basic Operations

A CS must be complemented with a set of operations that define how the users may modify the contents of the IB. In this paper, we deal with basic operations. We describe our basic operations in terms of their postconditions because our approach only depends on them to generate operation preconditions. We consider the following set of basic operations which correspond to the categories identified in [16] to describe operation postconditions. For the sake of generality, we use external identifiers instead of objects in the operation signatures. Therefore, each instance to be modified is identified by a set of attribute values and not by its object reference.

InstanceCreation. The operation `createE(v1,…,vn: Set(String))` creates an instance of entity type E and gives values $v_1,\ldots,v_n$ to attributes $a_1,\ldots,a_n$ of E. The postcondition of this operation can be specified in OCL [17] as follows:

```
post: E.allInstances()-> exists(e |e.oclIsNew()and e.a1=v1 and
      … and e.an=vn)
```

As a result of this operation, the new instance belongs to E and all its supertypes.

InstanceDeletion. The operation `deleteE(id1,…,idn: Set(String))` deletes an instance of entity type E identified by parameters $id_1,\ldots,id_n$. Its postcondition is:

```
post: not(E.allInstances()->exists(e|e.p1=id1and …and e.pn=idn))
```

where $p_1,\ldots,p_n$ are the paths that identify the instances of E. We assume that all the relationships in which the instance participates are deleted, and that the instance is deleted from E and all its supertypes.

AttributeValueModification. The operation `modifyAfromE(id1,…,idn,nv: Set(String))` modifies attribute a of an instance of the entity type E. The instance to modify is identified by parameters $id_1,\ldots,id_n$ of the operation. The new value for the attribute is nv. Its postcondition is:

```
post: E.allInstances()-> select(e |e.p1=id1 and … and
      e.pn=idn).a = nv
```

where $p_1,\ldots,p_n$ are the paths that identify the instances of E.

RelationshipCreation. The operation `createR(id11,...,id1n,id21,...,id2m: Set(String))` creates an instance of the relationship type R between two instances $i1$and $i2$ playing roles $r1$ and $r2$ in R. The instances to relate are identified, respectively, by the parameters $id1_1,\ldots,id1_n$ and $id2_1,\ldots,id2_m$, and can be obtained as follows from them:

```
let i1: E1 = E1.allInstances()-> select(e| e.p11=id11 and …
             and e.p1n=id1n)
let i2: E2 = E2.allInstances()-> select(e| e.p21=id21 and …
             and e.p2m=id2m)
```

The postcondition of this operation is: `post: i1.r2->includes(i2)`

RelationshipDeletion. The operation `deleteR(id11,...,id1n,id21,...,id2m: Set(String))` deletes the instance of the relationship type R between two instances $i1$and $i2$ playing roles $r1$ and $r2$ in R. These instances are identified, respectively, by the parameters $id1_1,\ldots,id1_n$ and $id2_1,\ldots,id2_m$, and can be obtained as in the previous operation. The postcondition of this operation is: `post: i1.r2->excludes(i2)`

InstanceGeneralization. The operation `generalizeEitoE(v1,…,vm: Set(String))` establishes that an instance of E which is identified by values $v_1,\ldots,v_m$ for paths $p_1,\ldots,p_m$, respectively, is not an instance of E_i after its execution (although it has not been deleted from the IB and it is still an instance of E). The OCL postcondition of this operation is:

```
let i1: E = E.allInstances()-> select(e |e.p1=v1 and ...
      and e.pm=vm)
post: not (i1.oclIsTypeOf(Ei))
```

We assume that all the relationships in which *i1* participates are deleted.

InstanceSpecialization. The operation `specializeEtoEi(v1,…,vm: Set(String), nv1,…,nvk: Set(String))` establishes that an instance *i1* of *E* is also an instance of E_i after its execution. Additionally, it takes values $nv_1,\ldots,nv_k$ for the attributes $a_1,\ldots,a_k$ of E_i. The instance is identified by values $v_1,\ldots,v_m$ for paths $p_1,\ldots,p_m$, respectively. Its postcondition is:

```
let i1:E=E.allInstances()-> select(e |e.p1=v1 and ... and e.pm=vm)
post: i1.oclIsTypeOf(Ei) and i1.a1=nv1 and ... and i1.ak=nvk
```

We consider also other basic operations whose postcondition can be stated as a combination of those of the basic operations specified so far. They are the following:

WeakInstanceCreation. The operation `createW(id1,…,idn: Set(String), v1,…,vm: Set(String))` creates an instance of entity type *W*, gives values $v_1,\ldots,v_m$ to attributes $a_1,\ldots,a_m$ of *W* and relates it through a relationship type *R* to an instance *i* of entity type *S* playing role *rs* in *R*. The instance *i* is identified by the parameters $id_1,\ldots,id_n$ and can be obtained as follows from them:

```
let i:S=S.allInstances()->select(e|e.p1=id1 and ... and e.pn=idn)
```

The postcondition of this operation is:

```
post: W.allInstances()-> exists(e |e.oclIsNew()and e.a1=v1 and
      ... and e.am=vm and e.rs=i)
```

ReifiedRelationshipCreation. The operation `createRR(id11,...,id1n, id21,...,id2m: Set(String), v1,…,vk: Set(String))` creates an instance of the reified relationship type *R* that relates instances *i1*and *i2* playing roles *r1* and *r2* in *R*. Additionally, it takes values $v_1,\ldots,v_k$ for the attributes $a_1,\ldots,a_k$ of *R*. The instances to relate are identified, respectively, by $id1_1,\ldots,id1_n$ and $id2_1,\ldots,id2_m$, and can be obtained as described in the RelationshipCreation operation. Its postcondition is:

```
post: R.allInstances()-> exists(e |e.oclIsNew()and e.r1=i1
    and e.r2=i2 and e.a1=v1 and ... and e.ak=vk)
```

We define *WeakInstanceDeletion* and *ReifiedRelationshipDeletion* in a similar way; as well as *InstanceChangeOfSubclass* which mixes *InstanceGeneralization* and *InstanceSpecialization*. We omit their formal definition due to space limitations.

4 Automatic Generation of Operation Preconditions

We describe in this section the approach we propose to automatically generate the weakest preconditions required by our set of basic operations in order to guarantee that their execution does not violate any of the predefined integrity constraints. By weakest we mean the necessary and sufficient conditions that allow ensuring that the constraints will not be violated after applying just the minimum changes specified by the postcondition when the operation precondition is satisfied (i.e. without requiring compensatory actions to restore the IB consistency).

It may happen that the execution of a basic operation postcondition always leads to an integrity constraint violation. Then, no weakest precondition exists. Our approach is able to identify these situations and it discards the definition of such operations.

We identify in section 4.1 the conflicts that arise between predefined constraints and basic operations. Then, in section 4.2, we describe how the weakest preconditions can be automatically obtained.

4.1 Conflicts between Constraints and Operations

The following table summarizes the conflicts that exist between integrity constraints and operations. Columns correspond to the predefined integrity constraints, and rows to the basic operations. A cross in a cell represents that there is a conflict between the corresponding constraint and operation, meaning that the constraint may be violated when the postcondition of the operation is satisfied. Thus, some preconditions must be added to the operation to prevent the violation in these cases.

Table 1. Conflicts between predefined constraints and basic operations

	Identifie	**Irreflexi**	**Symmetr**	**Asymm**	**Antisym**	**Acyclic**	**PathIncl**	**PathExcl**	**PathEq**	**ValueCo**	**Min.**	**Max.**	**Disjoint**	**Covering**
InstanceCreation	×									×	×			×
InstanceDeletion							×		×		×			
AttributeValueModif.	×									×				
RelationshipCreation		×	×	×	×	×	×	×	×			×		
RelationshipDeletion			×				×		×		×			
InstanceGeneralization							×		×		×			×
InstanceSpecialization	×									×	×		×	×
WeakInstanceCreation	×						×	×	×	×	×	×		×
WeakInstanceDeletion							×		×		×			
ReifiedRelationshipCre	×	×	×	×	×	×	×	×	×	×	×	×		×
ReifiedRelationshipDel			×				×		×		×			
InstanceChangeOfSubty	×						×		×	×	×			

The explanation of all marks in the table will be provided in the next section while identifying the preconditions required by the operations in each case.

4.2 Drawing Preconditions

The preconditions that are generated for each basic operation are the following.

InstanceCreation

The operation `createE(v1,…,vn: Set(String))` may violate identifier, value comparison, minimum cardinality and/or covering constraints.

Identifier. The violation of an identifier constraint for the entity type *E* or one of its supertypes occurs when the values for the identifying properties of the created instance are equal to those values for an already existing instance. To prevent it, the following precondition must be added to the operation:

```
pre: not(E.allInstances()->exists(e| e.a_i=v_i and … and e.a_j=v_j))
```

where $a_i,\ldots,a_j$ are the identifier attributes and $v_i,\ldots,v_j$ are the new values of the created instance for them.

For example, an instance creation operation `createEmployee(ei:String, nm:String)` for the conceptual schema shown in Figure 1 requires the following precondition since there is an identifier constraint which states that employees are identified by their *empid*:

```
pre: not(Employee.allInstances()-> exists(e |e.empid=ei))
```

Value Comparison. A value comparison constraint a_i *op* v for any attribute a_i of *E* or one of its supertypes that is initialized by the operation is violated if the specified comparison is not satisfied by the new instance. Thus, the following precondition is needed for each such a_i attribute:

```
pre: v_i op v
```

where v_i is the value of the created instance for a_i.

Minimum Cardinality. Let *R* be a binary relationship type such that entity type *E* plays role *p* and an entity type E_1 plays role p_1 in it. A minimum cardinality constraint from *p* to p_1 in *R* is always violated by the operation, since it creates an unrelated instance. The violation cannot be prevented by means of a precondition and, consequently, the operation cannot be executed in any case. Therefore, the InstanceCreation operation is discarded in this case.

Covering. Any covering constraint between entity type *E* and a set of entity types $E_1,\ldots,E_n$ is violated since the operation creates an instance in a single entity type. Again, the violation occurs in any case and the operation cannot be executed.

InstanceDeletion

An instance deletion operation, `deleteE(id_1,…,id_n: Set(String))`, may induce the violation of path inclusion, path equality and/or minimum cardinality constraints.

Path Inclusion. A path inclusion constraint which states that a first path includes a second path can be violated if the operation deletes an instance of one of the entity types that is traversed by the first path. The violation occurs when, after the deletion, the set of instances related to an instance i via the first path does not include the set of instances related to i via the second one. The following precondition is then required:

```
pre: Start.allInstances()-> forAll(s |newPath1(s)->
        includesAll(newPath2(s)))
```

Start is the origin entity type of both paths. *NewPath1(s)* and *newPath2(s)* define the set of instances that are reached from instance *s* by the first and second paths, respectively, assuming that the postcondition of the operation holds.

Path Equality. A path equality constraint between two paths can be violated by an instance deletion if the operation deletes an instance of an entity type in any of the two paths. The violation occurs when, after the deletion, the set of instances related to an instance i via the first path is not equal to the set of instances related to i via the second one. Therefore, the following precondition must be added to the operation:

```
pre: Start.allInstances()->forAll(s |newPath1(s)=newPath2(s))
```

where *Start, newpath1(s)* and *newpath2(s)* are defined as in the path inclusion case.

Minimum Cardinality. Let E be an entity type such that one of its instances is deleted by the operation. Let R be a relationship type such that an entity type E_1 plays role p_1 and entity type E plays role p in it. A minimum cardinality constraint from p_1 to p in R is violated if there is an instance i belonging to E_1 that was related to the deleted instance and that, after the deletion, does not satisfy the minimum cardinality any more. The violation can be prevented by the precondition:

```
pre: delInst.p1->forAll(e1 |e1.p->size()>min)
```

where *delInst* defines the deleted instance.

Note that the previous precondition will always evaluate to false if R has a maximum and a minimum cardinality constraints restricted by the same value. Therefore, the operation should be discarded in this case.

For instance, our running example of Figure 1 depicts a minimum cardinality constraint to ensure that all projects have at least one supervisor employee. Therefore, the following precondition must be generated for `deleteEmployee(ei:String)`, aimed at deleting an employee with code *ei*.

```
pre: delInst.supervises->forAll(pr|pr.supervisor->size()>1)
```

where *delInst* defines the deleted instance:

```
let delInst : Employee = Employee.allInstances()->
    select(e |e.empid=ei)
```

Assuming that we had a subtype *JuniorEmployee* of *Employee* in our example, the basic operation `deleteJuniorEmployee(ei:String)` would also require the previous precondition.

AttributeValueModification

An attribute value modification operation, `modifyAfromE(id1,…,idm,nv:Set(String)`, may violate identifier and/or value comparison constraints.

Identifier. This operation violates identifier constraints if the values of the updated instance for the identifying properties are equal to those values for another instance, after the modification. To avoid it, the following precondition is needed:

```
pre: not(E.allInstances()-> exists(e |e.pi=k.pi and … and
    e.a=nv and … and e.pj=k.pj))
```

where $p_i,\ldots,a,\ldots,p_j$ are the E identifier properties specified by the constraint and k is defined as the instance updated by the operation.

Value Comparison. A value comparison constraint for the updated attribute, *a op v*, is violated if the specified comparison is not satisfied by the new value. The following precondition must be added to the operation:

```
pre: nv op v
```

RelationshipCreation

The operation `createR(id1₁,…,id1ₙ,id2₁,…,id2ₘ:Set(String))` creates an instance of a relationship type *R* between instances *i1* and *i2* of entity types E_1 and E_2 playing roles r_1 and r_2 in *R*. As can be seen in table 1, this operation may violate several constraints, many of them when *R* is recursive.

Irreflexive. If *R* has an irreflexive constraint, the violation happens when *i1*=*i2*. The precondition to be added is:

```
pre: i1 <> i2
```

Symmetric. If *R* has a symmetric constraint, the violation happens if *i2* is not R-related to *i1*, i.e. when an instance that is symmetric to the new one does not exist. Since the IB must be consistent before the execution of any operation, the symmetric instance needed will never exist. Thus, the violation cannot be prevented by means of a precondition and the operation should be discarded.

Antisymmetric. When the relationship has an antisymmetric constraint, the violation happens when *i2* is R-related to *i1*, unless *i1* and *i2* are the same instance. In this case, the following precondition must be added to the operation:

```
pre: i2.r1->includes(i1) implies i2=i1
```

Asymmetric. On the contrary, an asymmetric constraint in a recursive relationship type is violated when *i2* is already R-related to *i1*. The following precondition has to be added to prevent the previous violation:

```
pre: i2.r1->excludes(i1)
```

Acyclic. If the relationship type has an acyclic constraint, it is violated when *i2* is R-related (directly or indirectly) to *i1*, both of them instances of E_1.

```
pre: i2.successors()->excludes(i1)
```

where `successors()` recursively obtains all the instances that are R-related to an instance of E_1. It is defined as follows:

```
context E₁ def:
successors():Set(E₁) = self.r₁->union(self.r₁.successors())
```

For relationship types that are not necessarily recursive, the constraints that may be violated are path constraints and maximum cardinality constraints.

Path Inclusion, Equality and Exclusion. A path inclusion constraint that traverses *R* is violated when, after the creation, the set of instances related to an instance *i* via the first path does not include the set of instances related to *i* via the second one. Violations of path exclusion and path equality constraints can be explained analogously. The preconditions to be added are the same than in the instance deletion operation.

Maximum Cardinality. A maximum cardinality from r_1 to r_2 is violated when *i2* is already related to *max* instances of *E1*. The violation can be prevented by adding the following precondition:

```
pre: i1.r2->size() < max
```

If the maximum cardinality constraint is from r_2 to r_1, the precondition needed is:

```
pre: i2.r1->size() < max
```

As before, the operation is discarded if there is a maximum and a minimum cardinality constraint restricted by the same value.

RelationshipDeletion

When `deleteR(id1`$_1$`,…,id1`$_n$`,id2`$_1$`,…,id2`$_m$`:Set(String))` operation deletes the instance of the relationship type *R* between two instances *i1* and *i2* of entity types E_1 and E_2, playing roles *r1* and *r2* in *R*, the constraints that may be violated are the symmetric, path inclusion, path exclusion and minimum cardinality constraints.

Symmetric. If *R* is recursive and symmetric, this constraint is violated when, after the deletion, *i2* is R-related to *i1*. This will always happen, since the operation `deleteR` deletes a single instance. Thus, this violation cannot be prevented in any case.

Path Inclusion and Equality. A path inclusion constraint that traverses *R* is violated when, after the deletion, the set of instances related to an instance *i* via the first path does not include the set of instances related to *i* via the second one. The reason for the violation of path equality is analogous. The preconditions to be added for these cases are the same than in the previous operation.

Minimum Cardinality. A minimum cardinality constraint from r_1 to r_2 is violated when, after the deletion, *i2* is related to less than *min* instances of E_1. The following precondition must be added:

```
pre: i1.r2->size() > min
```

If the minimum cardinality constraint is from r_2 to r_1, the precondition needed is:

```
pre: i2.r1->size() > min
```

Again, the operation is discarded if there is a maximum and a minimum cardinality constraint restricted by the same value.

InstanceGeneralization

An instance generalization may violate path inclusion, path equality, minimum cardinality and/or covering constraints.

In some respects, an instance generalization is similar to an instance deletion since, in both cases, the particular entity affected by the operation is no longer an instance of an entity type after its execution. Thus, violations of path inclusion, path equality or minimum cardinality constraints are like those described above for instance deletions and can be prevented by similar preconditions.

Additionally, an operation `generalizeE`$_i$`toE(v`$_1$`,…,v`$_m$`: Set(String))`, may violate a covering constraint between entity type *E* and a set of entity types $E_1,\ldots,E_n$ that include E_i. The violation occurs if the involved entity is not an instance of any $E_1,\ldots,E_n$ after the execution of the operation. We need the following precondition:

```
let i1: E = E.allInstances()-> select(e |e.p1=v1 and …
    and e.pm=vm)
pre: i1.oclIsTypeOf(E1) or … or i1.oclIsTypeOf(Ei-1) or
    i1.oclIsTypeOf(Ei+1) or … or i1.oclIsTypeOf(En)
```

InstanceSpecialization
It may violate identifier, value comparison, minimum cardinality, disjoint and/or covering constraints.

An instance specialization is similar to an instance insertion because an entity starts to be an instance of a certain entity type after the execution of both operations. Thus, violations of identifier, value comparison, minimum cardinality or covering constraints are like those described above for instance insertions.

The operation `specializeEtoE`$_i$`(v`$_1$`,…,v`$_m$`: Set(String), nv`$_1$`,…,nv`$_k$`: Set(String))`, may also violate a disjointness constraint for a set of entity types $E_1,\ldots,E_n$ that include E_i. This happens if the involved entity is an instance of more than one $E_1,\ldots,E_n$ after the execution. The precondition that avoids the violation is:

```
let i1: E = E.allInstances()-> select(e |e.p1=v1 and …
    and e.pm=vm)
pre: not(i1.oclIsTypeOf(E1) or … or i1.oclIsTypeOf(Ei-1) or
    i1.oclIsTypeOf(Ei+1) or … or i1.oclIsTypeOf(En))
```

We omit the description of the preconditions that are generated for the rest of basic operations because those preconditions can be seen as combinations of the cases that have already been described.

For instance, the operation *newAssignment* in our example of the introduction is a ReifiedRelationshipCreation operation whose effect is defined by combining an InstanceCreation and a RelationshipCreation operations. Then, the preconditions added to the contract in Figure 3 correspond to a violation of a *Value Comparison* constraint of the InstanceCreation and a violation of a *Path Exclusion* and a *Maximum Multiplicity* constraints of the RelationshipCreation.

5 Prototype Tool

We have developed a prototype tool that allows the automatic computation of the preconditions of the operation contracts, along the ideas developed in this paper, on top of Poseidon® 4.1 since this CASE tool provides an extension mechanism by means of Java plug-ins.

The designer may specify an operation as provided by Poseidon®. Then, with our plug-in, he may make use of the basic operations to state its postcondition. In Figure 4, we show the specification of an instance creation operation *newPerson* aimed at creating instances of the class *Persona.* Once this is done, he can press the button *Normalize* to automatically obtain the preconditions required for the operation contract. As can be seen in Figure 5, the resulting contract includes a precondition to prevent the violation of the specified identifier constraint.

Our prototype allows the definition and treatment of most of the basic operations considered in this paper. In particular, it is able to handle InstanceCreation, InstanceDeletion, AttributeValueModification, RelationshipCreation, Relationship Deletion, WeakInstanceCreation and WeakInstanceDeletion.

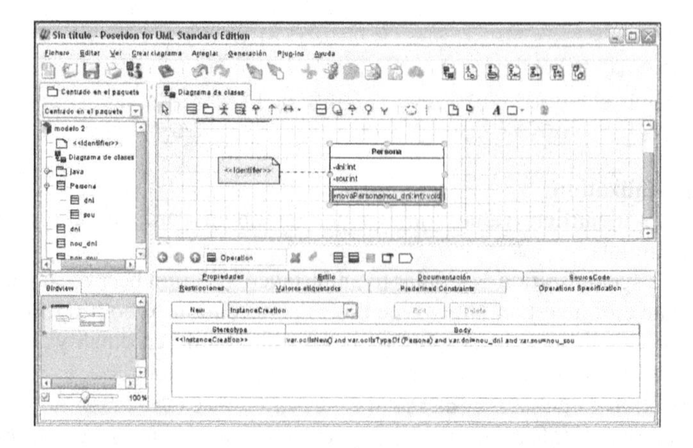

Fig. 4. Specification of an operation contract for *newPerson*

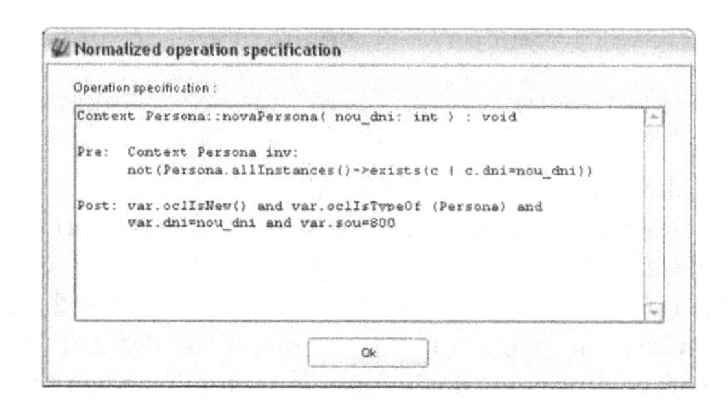

Fig. 5. Automatic generation of the precondition of *newPerson*

6 Related Work

The problem of identifying preconditions of an operation is not new. It has been addressed in the database field and in conceptual modelling of information systems as part of the checking and integrity maintenance problem (see, among others, [4] and [18]). [4] automatically generates elementary operations from an extended ER model of a database application. These operations contain additional manipulations, known as update propagations, to maintain some integrity constraints defined in the conceptual model. Preconditions to guarantee cardinality constraints and other general constraints have to be added to the specification of complex operations (sequence of elementary ones) by the designer. [18] draws automatically a transaction specification from a conceptual model and identifies conditions (preconditions) and repair actions to preserve integrity constraints. This method does not deal with cardinality constraints.

Ackermann and Turowski [13] propose a set of OCL specification patterns that facilitate the definition of some preconditions (as class instance existence, value specification of input parameter and so on). The use of these patterns simplifies the specification of operations although preconditions for each operation must be identified manually by the designer.

In [19] an identification process of preconditions for operations to modify instances of a data model (only a subset of the OMT object model is considered with classes and relations) is defined. This process is not systematic and requires interaction with the designer. An initial precondition for an operation must be provided by the designer and then the *Z-EVES* theorem prover is used to verify whether this precondition is needed for the operation.

A goal similar to ours is addressed in [20], that proposes an approach to identify the weakest preconditions to be added to the operations such that their execution does not violate any integrity constraint. They consider UML class diagrams but they assume that constraints and operation contracts are specified in the B language. Our approach, however, is independent of the conceptual modelling language used. We have shown how to apply it in OCL, which is the language most frequently used. Another difference is that their approach is based on performing general reasoning on the relevant B expressions while we provide an ad-hoc treatment endowed to the particular semantics of each basic operation and predefined constraint.

7 Conclusions and Future Work

Conceptual schemas usually include an important amount of integrity constraints, which must be satisfied in each state of the IB. These constraints may have a graphical representation or can be defined by means of a particular language. The content of the IB changes due to the execution of operations. The effect of an operation is defined by means of a postcondition, which expresses a condition that the IB must satisfy after applying it. Preconditions, which must be satisfied before the execution of the operation, must guarantee that it leaves the IB in a state satisfying all the constraints.

Due to the great amount of constraints that a schema may include, the task of manually determining which preconditions are needed by each operation is time consuming and error prone. To overcome this limitation, we have presented an approach to automatically generate the preconditions needed to guarantee that an operation satisfies the integrity constraints defined in the schema after being executed. Our approach is able to deal with a set of predefined integrity constraints and basic operations and allows to determine the weakest precondition which ensures that the postcondition can be safely applied. As an additional result of this automation, software development will also be performed faster. We have implemented our approach and integrated it in a CASE tool.

Future research may involve drawing preconditions from complex non-basic operations, i.e., operations defined as combinations of the basic ones studied in this work. Additionally, we plan to deal with other types of frequent general constraints. It may also be worth studying how the violation of an integrity constraint may be solved by including some corrective action instead of forbidding the operation execution.

Acknowledgements. We would like to thank Quim Vilà for developing the prototype and the GMC group for helpful discussions on this paper. We are also grateful to the anonymous referees for their useful comments. This work has been partially supported by the Ministerio de Ciencia y Tecnología under project TIN2005-06053.

References

1. Teichroew, D.: Methodology for the Design of Information Processing Systems. In: Proc. Fourth Australian Computer Conference, pp. 629–634 (1969)
2. OMG: MDA Guide Version 1.0.1. (2003)
3. Costal, D., Sancho, M.-R., Olivé, A., Roselló, A.: The Role of Structural Events in Behaviour Specification. In: Tjoa, A.M. (ed.) DEXA 1997. LNCS, vol. 1308, pp. 673–686. Springer, Heidelberg (1997)
4. Engels, G., Gogolla, M., Hohenstein, U., Hüllmann, K., Löhr-Richter, P., Saake, G., Ehrich, H.-D.: Conceptual Modelling of Database Applications Using an Extended ER Model. Data & Knowledge Engineering 9, 157–204 (1992)
5. Laleau, R., Polack, F.: Specification of Integrity-Preserving Operations in Information Systems by Using a Formal UML-based Language. Information and Software Technology 43, 693–704 (2001)
6. Cabot, J., Gómez, C.: Deriving Operation Contracts from UML Class Diagrams. In: Engels, G., Opdyke, B., Schmidt, D.C., Weil, F. (eds.) MODELS 2007. LNCS, vol. 4735, pp. 196–207. Springer, Heidelberg (2007)
7. Olivé, À.: Conceptual Schema-Centric Development: A Grand Challenge for Information Systems Research. In: Pastor, Ó., Falcão e Cunha, J. (eds.) CAiSE 2005. LNCS, vol. 3520, pp. 1–15. Springer, Heidelberg (2005)
8. Olivé, A.: Conceptual Modeling of Information Systems. Springer, Heidelberg (2007)
9. ISO/TC97/SC5/WG3: Concepts and Terminology for the Conceptual Schema and Information Base. ISO (1982)
10. Meyer, B.: Object-Oriented Software Construction, 2nd edn. Prentice-Hall, Englewood Cliffs (1997)
11. Costal, D., Gómez, C., Queralt, A., Raventós, R., Teniente, E.: Improving the Definition of General Constraints in UML. Software and Systems Modeling (2008) DOI: 10.1007/s10270-007-0078-4
12. Halpin, T.: Information Modeling and Relational Databases: From Conceptual Analysis to Logical Design. Morgan Kaufmann, San Francisco (2001)
13. Ackermann, J., Turowski, K.: A Library of OCL Specification Patterns for Behavioral Specification of Software Components. In: Dubois, E., Pohl, K. (eds.) CAiSE 2006. LNCS, vol. 4001, pp. 255–269. Springer, Heidelberg (2006)
14. Liddle, S.W., Embley, D.W., Woodfield, S.N.: Cardinality Constraints in Semantic Data Models. Data and Knowledge Engineering 11, 235–270 (1993)
15. Lenzerini, M.: Covering and Disjointness Constraints in Type Networks. In: Proc. ICDE 1987, pp. 386–393. IEEE Computer Society Press, Los Alamitos (1987)
16. Larman, C.: Applying UML and Patterns, 3rd edn. Prentice-Hall, Englewood Cliffs (2004)
17. OMG: UML2.0 OCL Specification, OMG Adopted Specification (2005)
18. Pastor, J.A., Olivé, A.: Supporting Transaction Designs in Conceptual Modeling of Information Systems. In: Iivari, J., Rossi, M., Lyytinen, K. (eds.) CAiSE 1995. LNCS, vol. 932, pp. 40–53. Springer, Heidelberg (1995)
19. Ledru, Y.: Idenitfying pre-conditions with the Z/EVES theorem prover. In: Proc. 13th International Conf. on Automated Software Engineering. IEEE Computer Society Press, Los Alamitos (1998)
20. Mammar, A., Gervais, F., Laleau, R.: Systematic Identification of Preconditions from Set-Based Integrity Constraints. In: INFORSID, pp. 595–610 (2006)

Decidable Reasoning in UML Schemas with Constraints

Anna Queralt and Ernest Teniente

Universitat Politècnica de Catalunya
{aqueralt,teniente}@lsi.upc.edu

Abstract. In this paper we propose an approach to reason on UML schemas with OCL constraints. We provide a set of theorems to determine that a schema does not have any infinite model and then provide a decidable method that, given a schema of this kind, efficiently checks whether it satisfies a set of desirable properties such as schema satisfiability and class or association liveliness.

Keywords: Conceptual modeling, Reasoning, Decidability.

1 Introduction

A conceptual schema consists of a taxonomy of classes together with their attributes, a taxonomy of associations among classes, and a set of integrity constraints over the state of the domain, which define conditions that each instance of the schema must satisfy. These constraints may have a graphical representation or can be defined by means of a particular general-purpose language.

The Unified Modeling Language (UML) has become a de facto standard in conceptual modeling of information systems. In UML, a conceptual schema is represented by means of a class diagram, with its graphical constraints, together with a set of user-defined constraints, which are usually specified in OCL.

Due to the high expressiveness of the combination of both languages, checking the correctness of a UML conceptual schema manually becomes a very difficult task, specially when the set of textual constraints is large. For this reason, it is desirable to support the designer in reasoning on a conceptual schema.

There are several reasoning tasks that can be performed to determine the correctness of a schema, such as satisfiability of the schema, liveliness of a class or association or reachability of certain states of the information base. Several efforts have already been devoted to reasoning on conceptual schemas. There are automatic procedures for the verification of some properties of schemas in Description Logics [1,2] or to reason about different kinds of cardinality constraints [3,4,5], but they do not deal with general integrity constraints.

Since the problem of reasoning with integrity constraints in its full generality is undecidable, two different approaches can be followed, either based on decidable procedures for certain restricted kinds of constraints [6,7], or based on semidecidable procedures for highly expressive constraints [8].

Z. Bellahsène and M. Léonard (Eds.): CAiSE 2008, LNCS 5074, pp. 281–295, 2008.

In this work we take a mixture of both directions. In particular, our approach consists in translating a UML schema, with OCL constraints satisfying mild syntactical restrictions, into a logic representation to reason on it. Then we provide a set of conditions that guarantee that the schema does not have any infinite model, so that the problem becomes decidable. In this way, any theorem prover or reasoning method can be used knowing that any reasoning task performed on the schema will terminate.

Additionally, once we have determined that all the models of a schema are finite, and since the logic representation of our UML and OCL schemas follows a specific syntactical structure, we provide a reasoning procedure that always terminates and works more efficiently than in the general case.

2 Base Concepts

Throughout the paper, a, b, c, a_1, b_1,... are constants. The symbols X, Y, Z, X_1, Y_1,... denote variables. Sets of constants are denoted by $\bar{a}$, $\bar{b}$, $\bar{c}$, $\bar{a_1}$, $\bar{b_1}$,... and $\bar{X}$, $\bar{Y}$, $\bar{Z}$, $\bar{X_1}$, $\bar{Y_1}$,... denote sets of variables. Predicate symbols are p, q, r, p_1, q_1,... A term is either a variable or a constant. If p is a n-ary predicate and T_1, ..., T_n are terms, then $p(T_1, ..., T_n)$ is an atom, which can also be written as $p(\bar{T})$ when n is known from the context. An atom is ground if every T_i is a constant. An ordinary literal is defined as either an atom or a negated atom, i.e. $\neg p(\bar{T})$. A built-in literal has the form of $A_1 \omega A_2$, where A_1 and A_2 are terms, and the operator ω is either $<, \leq, >, \geq, =$ *or* $\neq$.

A normal clause has the form

$$A \leftarrow L_1 \wedge ... \wedge L_m \text{ with } m \geq 0$$

where A is an atom and each L_i is a literal, either ordinary or built-in. All the variables occurring in A, as well as in each L_i, are assumed to be universally quantified over the whole formula. A is often called the head and $L_1 \wedge ... \wedge L_m$ is the body of the clause. A set of normal clauses is called a normal program. The definition of a predicate symbol p in a normal program P is the set of all clauses in P that have p in their head. A normal program P is hierarchical if there is a partition $P = P_1 \cup ... \cup P_n$ such that the following condition holds for i = 1, 2, ..., n: if an atom $r(\bar{T})$ occurs positively or negatively in the body of a clause in P_i then the definition of r is contained within P_j with $j < i$.

Terms, literals and the syntactic structures made of them are expressions. If E is an expression, then *constants(E)* and *variables(E)* are the sets containing the constants and variables, respectively, occurring in E.

A substitution θ is a set of the form $\{X_1/T_1, ..., X_n/T_n\}$, where each variable X_i is unique and each term T_i is different from X_i. The term T_i is called a binding for X_i. θ is called a ground substitution if each T_i is a constant.

Let E be an expression and $\theta = \{X_1/T_1, ..., X_n/T_n\}$ a substitution. Then $E\theta$ is the expression obtained from E by simultaneously replacing each occurrence of the variable X_i in E by the term T_i.

A fact is a normal clause of the form: $p(\bar{a}) \leftarrow$, where $p(\bar{a})$ is a ground atom.

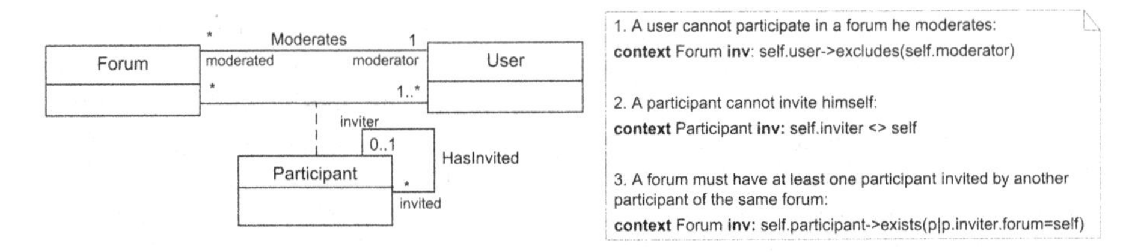

Fig. 1. Conceptual schema for an internet forum

A deductive rule is a normal clause of the form: $p(\bar{T}) \leftarrow L_1 \wedge ... \wedge L_m$ with $m \geq 1$ where p is the derived predicate defined by the deductive rule.

A condition is a formula of the (denial) form: $\leftarrow L_1 \wedge ... \wedge L_m$ with $m \geq 1$.

A database schema S is a tuple *(DR, IC)* where *DR* is a finite set of deductive rules and *IC* is a finite set of conditions. Literals occurring in the body of deductive rules and conditions in S are either ordinary or built-in. The predicate symbols in ordinary literals range over the extensional database (EDB) predicates, which are the relations that will be stored directly in the database, and the intensional database (IDB) predicates, which are the relations defined by the deductive rules in *DR*. EDB predicates cannot be derived. Conditions in *IC* define the integrity constraints of the schema S.

Deductive rules as well as conditions are required to be safe, that is, every variable occurring in the head or in negated or built-in atoms of their body must also occur in an ordinary positive literal of the same body.

Our logic representation of a schema has a specific structure, since it is obtained from the translation of a UML conceptual schema, with its constraints specified using a subset of the OCL. In particular, the only IDB predicates that appear are those needed to make the conditions safe. Thus, the conditions of the resulting schema are such that their literals correspond to either (positive or negative) EDB predicate symbols or negative IDB predicate symbols. Predicate symbols occurring in ordinary literals in the bodies of deductive rules are positive EDB predicate symbols, i.e. derivation rules do not include derived predicates. To satisfy this restriction, although OCL constraints can include the operations `includes`, `includesAll`, `notEmpty`, `exists` and `one`, these specific operations cannot be recursively combined in an OCL expression.

For a schema S = *(DR, IC)*, a state, instance or model D is a tuple *(E, S)* where E is an EDB, that is, a set of ground facts about EDB predicates. *DR(E)* denotes the whole set of ground facts about EDB and IDB predicates that are inferred from a database state D = *(E, S)*. *DR(E)* corresponds to the fixpoint model of $DR \cup E$.

An instance D violates a condition $\leftarrow L_1 \wedge ... \wedge L_n$ if there exists a ground substitution θ such that $D \models (L_1 \wedge ... \wedge L_n)\theta$. In other words, when $L_1\theta, ..., L_n\theta \subseteq DR(E)$. D is consistent when it violates no condition in *IC*.

We present in this section a simple UML conceptual schema with a set of OCL constraints. We will use this example to illustrate our approach to determine the absence of infinite models. For the sake of simplicity we do not specify any attributes in the classes.

As can be seen in Fig. 1, each *Forum* is related to a *User* that moderates it, and to a set of users that are its *Participants*, which can be invited by other participants. The textual constraints impose that the moderator of a forum cannot participate in it, and that each forum must have at least one invited participant, which must be invited by some other participant in the same forum.

The translation of the schema into logic results in the following rules, which specify the implicit, graphical and textual constraints that appear in the UML schema. According to our previous work [8], classes are translated into unary predicates, whereas n-ary predicates represent the associations (or association classes). Conditions *1* to *6* correspond to the referential constraints of associations. In this case, since all associations are binary, two such constraints are needed for each of them (one constraint for each association end). Condition *7* specifies that there cannot be two instances of *Participant* with the same *Forum* and *User*, which is an implicit constraint of association classes. Conditions *8* to *11* are the cardinality constraints of associations (the upper and lower bounds of *moderator* in *Moderates*, the lower bound of *User* in the association class *Participant*, and the upper bound of *inviter* in *HasInvited*). Finally, conditions *12* to *14* are the textual constraints of the schema.

```
 1. ← Moderates(F,U) ∧ ¬Forum(F)
 2. ← Moderates(F,U) ∧ ¬User(U)
 3. ← Participant(P,F,U) ∧ ¬Forum(F)
 4. ← Participant(P,F,U) ∧ ¬User(U)
 5. ← HasInvited(P,I) ∧ ¬IsParticipant(P)
    IsParticipant(P) ← Participant(P,F,U)
 6. ← HasInvited(P,I) ∧ ¬IsParticipant(I)
 7. ← Participant(P,F,U) ∧ Participant(P2,F,U) ∧ P≠P2
 8. ← Forum(F) ∧ ¬OneModerator(F)
    OneModerator(F) ← Moderates(F,U)
 9. ← Moderates(F,U) ∧ Moderates(F,U2) ∧ U≠U2
10. ← Forum(F) ∧ ¬OneParticipant(F)
    OneParticipant(F) ← Participant(P,F,U)
11. ← HasInvited(P,I) ∧ HasInvited(P2,I) ∧ P≠P2
12. ← Moderates(F,U) ∧ Participates(P,F,U)
13. ← HasInvited(P,P)
14. ← Forum(F) ∧ ¬OneInvited(F)
    OneInvited(F) ← Participant(P,F,U) ∧ HasInvited(P,I)
    ∧ Participant(I,F,U2)
```

3 Determining the Decidability of Reasoning on a Schema

In this section we present our method to identify whether a schema is such that any reasoning task performed on it will terminate, which happens when all its models are finite.

To determine the absence of infinite models, we obtain a graph from the set of constraints of the schema that shows the existing dependencies between them. We formalize the construction of the dependency graph for a given schema in section 3.1, and in section 3.2. we explain how to analyze the graph in order to determine whether all the models of a schema are finite.

3.1 The Dependency Graph

As seen in section 2, a condition consists of a set of positive literals, a set of negative literals and a set of built-in literals. Positive literals are the ones that may violate the constraint, whereas negative literals repair the constraint in case it is violated or, in other words, avoid violation in case that all the positive literals hold in the EDB.

Definition 1. *A literal* $p(\bar{X})$ *is a* potential violation *of a condition* ic *if it appears positively in its body. We denote by* V(ic) *the set of potential violations of condition* ic.

For example, *V(ic1)* = *{Moderates(F, U)}* since the existence of a fact *Moderates(a,b)* in the EDB causes the violation of *ic1* if the EDB does not contain the fact *Forum(a)*.

Definition 2. *Given a condition* ic, *there is a* repair $R_i(ic)$ *for each negative literal* $\neg L_i$ *in the body of* ic. *If* L_i *is base then* $R_i(ic) = \{L_i\}$, *otherwise* $R_i(ic) = \{p_1(\bar{X}_1),...,p_n(\bar{X}_n)\}$, *where each* $p_j(\bar{X}_j)$ *is a literal that appears positively in the body of the derivation rule that defines* L_i.

Each repair of a condition gives an alternative way to avoid its violation. In our example, all conditions have a single repair. For instance, the repair of condition 1 is *R(ic1)* = *{Forum(F)}*, and the repair of condition 14 is *R(ic14)* = *{Participant(P,F,U), HasInvited(P,I), Participant(I,F,U2)}*. There are some constraints, such as *ic7*, that cannot be repaired once they are violated, unless their potential violations are removed from the EDB.

In the proposed approach, the set *IC* of constraints of a schema is associated with a directed graph *G*, that we call *dependency graph.* This graph shows, for each condition ic_i of the schema, which conditions may be violated as a result of each possible repair of ic_i.

Definition 3. *A dependency graph* G *is a graph such that each vertex corresponds to a condition* ic_i *of the schema. There is an arc from* ic_i *to* ic_j, *labeled* $R_k(ic_i)$, *if there exists a predicate* p *such that* $p(\bar{X}) \in R_k(ic_i)$ *and* $p(\bar{Y}) \in V(ic_j)$.

Note that *G* is sometimes a multigraph, since two different repairs of a condition ic_i may lead to the violation of a same other condition ic_j.

Figure 2 depicts the dependency graph built from the conditions of our example. For the sake of clarity, conditions 5 and 6 have been collapsed in a single vertex, since the predicates belonging to their sets of potential violations and repairs coincide. For instance, it can be seen that the repair of conditions *ic5* and *ic6* can violate *ic4*, *ic7*, *ic3* and *ic12*, since the predicate *Participant*, which is a repair for conditions *ic5* and *ic6*, belongs to their sets of potential violations.

However, not all the arcs that appear in the graph represent the violation of the condition in the terminal vertex when repairing the initial one. Sometimes, the existence of an arc means the opposite: that the repair of the condition in the initial vertex guarantees the non-violation of the condition in the terminal

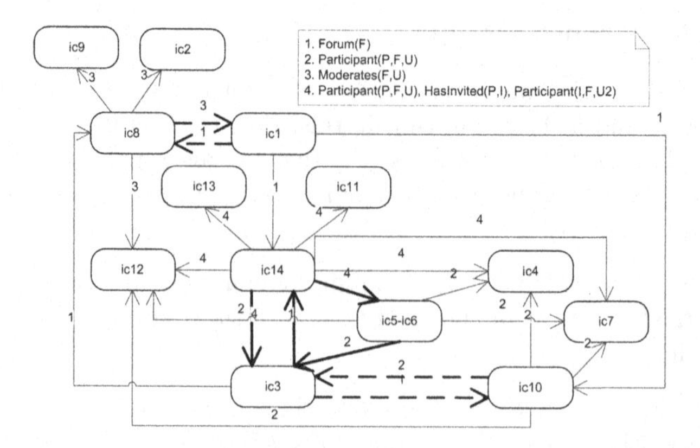

Fig. 2. Dependency graph. Superfluous arcs are dashed, and cycles are highlighted.

vertex. We say this kind of arcs are *superfluous*. Examples of superfluous arcs, which are depicted with dashed lines in Fig. 2, are the ones between conditions *ic1* and *ic8*. When *ic1* is violated due to the insertion of a fact *Moderates(a,b)* in the EDB, the insertion of the corresponding repair *Forum(a)* guarantees that *ic8* is fulfilled. Similarly, when the first to be violated is *ic8* because of the presence of a fact *Forum(a)*, the violation is repaired by the insertion of *Moderates(a,b)*, which guarantees the satisfaction of condition *ic1*.

Formally, an arc r_i from the constraint ic_i to ic_j is superfluous when $V(ic_j) = r_i\theta$ and there is some repair $R_k(ic_j)$ such that $R_k(ic_j) = V(ic_i)\theta$, where θ is a unifier of the sets $V(ic_i) \cup r_i$ and $V(ic_j) \cup R_k(ic_j)$ that assigns a different term to each distinct variable. This guarantees that ic_j is never violated after the repair of ic_i, since although the facts added by r_i potentially violate ic_j, this condition is always satisfied because its repair also belongs to the EDB. Thus, once these superfluous arcs are identified, they can be left aside since they indicate the ending of any sequence of repairs.

Let $C = (ic_1\ r_1\ ...\ ic_n\ r_n\ ic_{n+1} = ic_1)$ be an alternating sequence of vertices and arcs that define a cycle in a dependency graph G. The existence of C implies that the repair r_i of a condition ic_i may violate other conditions whose repairs could violate ic_i again.

As can be seen in Fig. 2, there are two cycles in our dependency graph, defined by the conditions (*ic3 ic14*) and (*ic3 ic14 ic5/6*). Since each constraint has a single repair, an enumeration of vertices suffices to identify each cycle. Note that the existence of superfluous arcs significantly reduces the number of cycles in the dependency graph.

3.2 Decidability of Reasoning on a Schema

Our approach to reasoning is aimed at constructing a database state which shows that a certain property holds. That is, a sample EDB where both the particular condition that defines the reasoning task and all the integrity constraints in the schema are satisfied. Therefore, our approach requires to perform integrity maintenance when trying to build such a sample EDB.

It can be seen that the constraints that form a cycle in a dependency graph are the only reason for the existence of infinite models. Clearly, a condition that does not belong to a cycle will not cause an infinite sequence of repairs, since it will not be violated again when it has been maintained once for a certain set of facts. On the contrary, constraints that belong to cycles can be violated a potentially infinite number of times, since once they have been maintained, the same facts inserted by the repairs may cause new violations and new repairs, which can result in an infinite model. Then, if we can identify which are the cycles that do not cause an infinite sequence of violations, we can determine whether a schema is suitable to perform any reasoning task in finite time.

In this section we study the cycles of the dependency graph to ensure that the process of integrity maintenance does not loop forever. To do this, we provide a set of theorems that allow to discard the presence of infinite models in the constraints that define each cycle. When all the models of a cycle of constraints are finite we call it a *finite cycle.*

A first condition that guarantees that a cycle is finite is that it includes a constraint whose violation requires facts that are not inserted in the EDB by some repair in the same cycle. This implies that the cycle will not lead to an infinite sequence of repairs, since there is necessarily a condition in the cycle that will not be violated at some time. This is formalized in theorem 4.

Theorem 4. *A cycle* $C = (ic_1\ r_1\ ...\ ic_n\ r_n\ ic_{n+1} = ic_1)$ *is a* finite cycle *if*

$$\bigcup_{i=1}^{n}\left(\bigcup_{p(\bar{X})\in r_i} p\right) \subset \bigcup_{i=1}^{n}\left(\bigcup_{q(\bar{Y})\in V(ic_i)} q\right)$$

Intuitively, since the union of repairs of the conditions in the cycle is a proper subset of the union of potential violations, at least one potential violation of a constraint ic_j in the cycle is an EDB predicate which is not updated during maintenance of the rest of the constraints. Therefore, since the set of facts in the sample EDB at the beginning of the process is finite, ic_j may always be violated only a finite number of times.

An example is the following set of constraints, which define a cycle since the repair of the first one is a potential violation of the second, and viceversa:

$$\begin{array}{l} \leftarrow \texttt{p(X)} \wedge \texttt{q(X)} \wedge \neg\texttt{r(X)} \\ \leftarrow \texttt{r(X)} \wedge \neg\texttt{aux(X)} \\ \texttt{aux(X)} \leftarrow \texttt{p(Y)} \wedge \texttt{Y} \neq \texttt{X} \end{array}$$

In this example, the potential violation *q(X)* in the first condition is not added by the repair of the second one. Thus, even when the first constraint is violated because $p(X) \wedge q(X)$ holds in the EDB for some X, the repairs of the second condition may only lead to new violations of the first one a finite number of times (one for each fact *q(a)* contained in the initial EDB when the process of maintaining the previous constraints started).

A cycle may be finite although it does not satisfy the previous condition. Examples can be found such that all the facts that are potential violations are created inside the cycle and, however, the cycle is not potentially infinite. An example is the cycle (*ic3 ic14*), which does not satisfy the previous condition but is finitely satisfiable. For instance, when a *Participant(a,b,c)* is added to the EDB, *ic3* requires the insertion of *Forum(b)* which, in turn, violates *ic14*. In order to repair this violation, the facts *Participant(a2,b,c2)*, *HasInvited(a2,a3)* and *Participant(a3,b,c3)* may be inserted, but they will never violate *ic3* again.

Definition 5. *A variable* x *is free in a repair* $\mathrm{R}_i(\mathrm{ic})$ *if* $\mathrm{x} \in \mathrm{variables}(\mathrm{R}_i(\mathrm{ic}))$ *and* $\mathrm{x} \notin \mathrm{variables}(\mathrm{V}(\mathrm{ic}))$.

Theorem 6. *A cycle* $C = (ic_1\ r_1\ ...\ ic_n\ r_n\ ic_{n+1} = ic_1)$ *is a* finite cycle *if* $\forall i,\ 1 \leq i \leq n, \forall p$ *such that* $\mathrm{p}(\mathrm{X}_1,...,\mathrm{X}_m) \in r_i$ *and* $\mathrm{p}(\mathrm{Y}_1,...,\mathrm{Y}_m) \in V(ic_{i+1})$, $\forall k,\ 1 \leq k \leq m$, X_k *is free in* $\mathrm{r}_i \Rightarrow \mathrm{Y}_k \notin variables(r_{i+1})$.

Intuitively, free variables in a repair are the source of infinity since they propagate the violations to new objects other than the ones that initially violated the constraints in the cycle. The previous condition guarantees that the free variables in the repair of the first constraint are not propagated by the repair of the second constraint. Since such a condition is required for each two consecutive constraints, it is guaranteed that the cycle will not loop forever since no new objects will be infinitely introduced by the repairs.

Applying this condition to the cycle consisting of *ic3* and *ic14* we can conclude that it is a finite cycle, since the objects added by *ic14* (the free variables in its repair) do not appear in the repair of *ic3*, which means that the new objects are not propagated in the cycle.

There is another cycle in our example, defined by the constraints (*ic14 ic5/ic6 ic3*), that does not satisfy the previous condition. However, this cycle is not infinite, since the free variables in *ic14* are propagated by *ic5/ic6* but not by *ic3*, which means that the new objects do not cause a new violation of *ic14*.

We propose another theorem to identify this kind of cycles, which determines whether all the constraints of a cycle are violated at most once.

Definition 7. *Let* $\mathrm{C} = (\mathrm{ic}_1\ \mathrm{r}_1\ ...\ \mathrm{ic}_n\ \mathrm{r}_n\ \mathrm{ic}_{n+1}{=}\mathrm{ic}_1)$ *be a cycle in* G, *where each* r_i *corresponds to some repair* $\mathrm{R}_j(\mathrm{ic}_i)$. *Let* $\mathrm{V}(\mathrm{ic}_i){=}\{\mathrm{p}_1(\bar{X}_1),...,\mathrm{p}_m(\bar{X}_m)\}$. *Then,*

$$Facts(ic_1) = (V(ic_1)\ \cup\ r_1)\theta_1$$

$$Facts(ic_i) = \bigcup_{k=1}^{t} r_i\theta_k\ \cup\ Facts(ic_{i-1}),\ i > 1$$

where θ_1 *is a substitution that bounds a distinct constant to each variable, each* $\theta_k = \theta_j\ \cup\ \theta'_j$ *is one of the* t *possible substitutions such that* $Facts(ic_{i-1}) \models (p_1(\bar{X}_1) \wedge ... \wedge p_m(\bar{X}_m))\theta_j$ *and* θ'_j *assigns a new constant to each variable* X *such that* $X \in \mathrm{variables}(\mathrm{r}_i)$ *and* $X \notin \mathrm{variables}(\mathrm{V}(\mathrm{ic}_i))$ *if* $\theta_j \neq \oslash$, *otherwise* $\theta_k = \oslash$.

Theorem 8. C *is a* finite cycle *if, for each possible starting* ic_i, $\exists k,\ 1 \leq k \leq n$, *such that* $\mathrm{Facts}(\mathrm{ic}_k) = \mathrm{Facts}(\mathrm{ic}_{k+1})$.

Intuitively, for each i>1, the set *Facts*(ic_i) extends the set *Facts*(ic_{i-1}) by taking into account the repairs required to satisfy ic_i. Therefore, the condition *Facts*(ic_k) = *Facts*(ic_{k+1}) guarantees that the constraint ic_{k+1} is not violated and, hence, the maintenance of the constraints in the cycle will not loop forever.

The previous results allow us to determine whether reasoning on a given conceptual schema will always terminate. Note, however, that this set of theorems is not complete due to the undecidability of this problem.

4 Reasoning on a Schema

Once we have determined that all the models of a conceptual schema are finite (as it happens in our example), we can take advantage of the characterization of the logic formulas obtained from our UML and OCL schemas to define a reasoning procedure that works more efficiently than in the general case. Reasoning is concerned with determining the correctness of a schema. Several reasoning tasks have been considered in the literature, such as *satisfiability* (i.e. checking whether the schema admits a non-empty state that satisfies all the constraints), *liveliness of a predicate* (i.e. determining whether a certain class or association can have at least one instance) or *reachability of partially specified states* (i.e. assessing whether certain goals conceived by the designer may be satisfied). In general, each reasoning task can be formulated in terms of a particular goal to attain.

A well-known approach to deal with this problem is to define methods whose purpose is to construct a database state (i.e. an EDB) for which the tested property holds. That is, a sample EDB where both the particular goal to attain and all the integrity constraints in the schema are satisfied. In this way, these methods can uniformly deal with all reasoning tasks.

In this section we propose a new reasoning procedure based on the previous approach. We divide it in two different steps: *goal satisfaction* and *integrity maintenance*. These steps are defined in sections 4.1 and 4.2, respectively.

4.1 Goal Satisfaction

Our method is aimed at building a sample EDB which proves that the schema fulfills a specific property defined in terms of a certain goal G to attain. We assume that G is a conjunction of (positive and negative) literals corresponding to EDB predicates and built-in literals; which suffices to handle schema satisfiability, predicate liveliness and reachability of partially specified states [8].

The first step of our method determines the EDB facts that are required to satisfy G without taking into account whether they violate any integrity constraint. Positive literals in G define facts that are necessarily required to satisfy G while negative literals in G identify facts that the sample EDB under construction must not contain. Built-in literals state conditions over the values that the variables of positive and negative literals in G may take.

One of the most difficult tasks is the assignment of concrete values to the variables appearing in G in order to construct the sample EDB. Each possible choice defines a different alternative that satisfies G, i.e. a different sample EDB. We use *Variable Instantiation Patterns (VIPs)* [9] for this purpose. These VIPs guarantee that the number of sample EDBs to be considered is kept finite, by taking into account only those variable instantiations that are relevant for the schema, without losing completeness. I.e. the VIPs guarantee that if a solution is not found by instantiating the variables in the goal using only the constants they provide, then no solution exists. VIPs are selected according to the syntactic properties of the schema considered in each test:

1. The Simple VIP: for schemas without negation and integrity constraints.
2. The Negation VIP: for schemas with negation and/or integrity constraints.
3. The Dense Order VIP: for schemas with order comparisons over a dense domain (e.g. real numbers).
4. The Discrete Order VIP: for schemas with order comparisons over a discrete domain (e.g. integer numbers).

In our example, the appropriate VIP is the Negation VIP, since the schema has negation and integrity constraints, but not order comparisons. With this VIP, each variable of the fact to be included in the EDB is instantiated either with a previously used constant or with a new one. For instance, assume that *p(X)* must be instantiated and that the only constant used up this moment is 0. Then, according to this VIP, the only relevant instantiations are *p(0)* and *p(1)*.

Step 1: Goal Satisfaction. Formally, the set *EDB* of facts required to satisfy G and the set Unw_{EDB} of facts that *EDB* may never include to fulfill the tested property are obtained as stated in definition 9. There is a different alternative *EDB* for each possible substitution θ provided by the corresponding VIP.

Definition 9. *Let* $G = \leftarrow P_1(\bar{X_1}) \wedge ... \wedge P_n(\bar{X_n}) \wedge \neg Q_1(\bar{Y_1}) \wedge ... \wedge \neg Q_m(\bar{X_m}) \wedge B_1 \wedge ... \wedge B_s$, *where* P_i, Q_j *are base predicates and* B_k *are built-in literals.*

Let θ *be one of the possible ground substitutions obtained via an instantiation of* variables(G) *and such that* $\forall i, 1 \leq i \leq s, B_i\theta$ *evaluates to true. Then,*

- *The set of* facts required *to satisfy G is* EDB $=\{P_1\theta,...,P_n\theta\}$
- *The set of* facts unwanted *to satisfy G is* $\text{Unw}_{EDB} = \{Q_1\theta,...,Q_m\theta\}$

As an example, assume that the designer wants to check the liveliness of the association *Moderates* in the conceptual schema of Fig. 1. *Moderates* will be lively if the goal $G = \leftarrow Moderates(F, U)$ succeeds for some instantiation. Applying the step 1 of our method we will obtain two different EDBs that satisfy G according to the Negation VIP: $EDB_1 = Moderates(0,0)$ and $EDB_2 = Moderates(0,1)$.

4.2 Integrity Maintenance

Once we have determined the set of EDB facts that satisfies the goal G to attain, the problem of reasoning on the schema may be reduced to that of integrity maintenance [10]. Note that, in fact, we already know that the property checked will be satisfied if the EDB resulting from Step 1 does not violate any constraint of the schema. If this is not the case, we must look for additional base facts (i.e. repairs) that make the sample EDB being constructed fulfill all constraints.

Unfortunately, we may not rely on existing integrity maintenance methods to perform this activity. On the one hand, some methods like [11,12] can only handle restricted types of integrity constraints which do not cover the kind of constraints we obtain as a result of the translation of the conceptual schema into logic. On the other hand, most methods do not provide an appropriate treatment to the existential variables that appear in the integrity constraint definition [13,14,15,16]. The general approach of these methods when instantiating an existential variable is either asking for a value from the user at run-time or assigning an arbitrarily chosen value of the corresponding data type. This is not suitable when using integrity maintenance for reasoning since only a few of the possible alternatives (just one in most cases) would be taken into account to repair a violated constraint. Therefore, this approach does not guarantee the correctness of the result since the impossibility to find a sample EDB would not necessarily imply that the tested property does not hold.

To our knowledge, the most appropriate method to perform the kind of integrity maintenance we require is the CQC-Method [9]. However, and in addition to the decidability drawbacks stated in section 1, the CQC-Method has important efficiency limitations that make questionable its use in practical situations.

Thus, we need to build a new reasoning procedure, which can take advantage both of the dependency graph and the characterization of the logic formulas obtained from our schemas to work efficiently. Since the graph shows the interactions between the constraints, it provides the order in which they should be maintained. In principle, all constraints in the graph must be considered for maintenance since all of them may be violated by the EDB obtained as a result of Step 1. Vertices with no incoming arcs or whose incoming arcs have already been maintained are selected with priority so that a constraint is not considered until all the constraints that may violate it have already been maintained.

An integrity constraint *ic* must be repaired if its potential violations hold in the sample EDB. Maintenance of *ic* results in the inclusion of its repairs in the sample EDB being constructed. Note that *ic* may be violated by several different instantiations of its potential violations. Each of them gives raise to different repairs to be added in the EDB. If a constraint with an empty set of repairs is violated, the sample EDB being constructed must be discarded since it is impossible to make it satisfy such a constraint.

The process of integrity maintenance is formalized as follows. Note that we also use the VIPs to assign concrete values to the existential variables that appear in the repairs of a constraint. Backtracking must be performed each time that the sample EDB under construction reaches a situation where the selected *ic* can not be repaired. Such backtracking involves considering a different repair of one of the constraints that has been maintained before *ic*.

	Selected constraint(s)	*Additions to the EDB*
Step 1:		{*Moderates*(0,0)}
Step 2:	ic1	{*Forum*(0)}
	ic10	{*Participant*(0,0,1)}
	ic14	{*HasInvited*(0,1), *Participant*(1,0,2)}
	ic13, ic11, ic5-6, ic3, ic8	
	ic2	{*User*(0)}
	ic9, ic7	
	ic4	{*User*(1), *User*(2) }
	ic12	
Sample EDB:		{*Forum*(0), *User*(0), *User*(1), *User*(2), *Participant*(0,0,1), *Participant*(1,0,2), *HasInvited*(0,1), *Moderates*(0,0)}

Fig. 3. A sample EDB that proves that *Moderates* is lively

Step 2: Integrity Maintenance. Let $ic = \leftarrow P_1(\bar{X_1}) \wedge ... \wedge P_n(\bar{X_n}) \wedge \neg Q_1(\bar{Y_1}) \wedge ... \wedge \neg Q_m(\bar{X_m}) \wedge B_1 \wedge ... \wedge B_s$ be the condition selected for maintenance from the *dependency graph*, where P_i, Q_j are base predicates and B_k are built-in literals. Let EDB_i be the set of required facts at that moment. Let *EvalV(ic)* and *EvalR(R_i(ic))* be the set of built-in literals that appear in the body of ic and in the body of the rule from which $R_i(ic)$ is obtained, respectively. Then EDB_{i+1} is computed as follows:

if $R_i(ic) = \oslash$ and $\text{EDB}_i \models (P_1(\bar{X_1}) \wedge ... \wedge P_n(\bar{X_n}) \wedge B_1 \wedge ... \wedge B_s)\theta_j$
  then error(ic cannot be repaired)

else $$\text{EDB}_{i+1} = \bigcup_{k=1}^{t} R_i(ic)\theta_k \cup \text{EDB}_i$$
  if $\exists Q_i \in Unw_{EDB}$ such that $Q_i \in \text{EDB}_{i+1}$
    then error(ic cannot be repaired)

where each $\theta_k = \theta_j \cup \theta'_j$ is a substitution such that $\text{EDB}_i \models (V(ic) \wedge EvalV(ic))\theta_j$ and θ'_j is one of the possible substitutions obtained from an instantiation of all the variables in *variables(R_i(ic))* $\setminus$ *variables(V(ic))* such that $EvalR(R_i(ic))\theta'_j$ evaluates to true, if $\theta_j \neq \oslash$, otherwise $\theta_k = \oslash$.

Figure 3 shows an execution of the integrity maintenance step of our method for one of the EDBs obtained in Step 1. Each row in the figure shows the integrity constraint being maintained (as selected through the order defined by the dependency graph) and the additions to the EDB required to repair the constraint, if any. A row contains several constraints when none of them is violated by the EDB under construction. As a result of the execution, our method obtains a sample EDB which confirms that the association *Moderates* is lively.

The constraint *ic1* is selected first since it is the only vertex with no incoming arcs in the graph. It is violated since V(*ic1*)=*Moderates*(F,U) holds in the EDB with substitution θ_j ={F/0, U/0}. Then, since R(*ic1*)=*Forum*(F), the repair *Forum*(0) is added to the sample EDB to ensure that it does not violate *ic1*.

The next constraint to be selected is *ic10* since all its predecessors have already been maintained. Similarly, it is violated since V(*ic10*)=*Forum*(F) holds in the EDB with substitution θ_j ={F/0}. Since *R(ic10)* = *Participant*(P,F,U), *Participant*(0,0,1) is added to the sample EDB since we assume that the substitution obtained is θ'_j ={P/0, U/1}.

The method proceeds then with *ic14* which requires considering two additional repairs whose concrete values have also been obtained via the application of a VIP. The rest of the constraints are either not violated or require repairs which are obtained in the same way than the repairs of *ic1*.

At the end, the method succeeds and it obtains the sample EDB = {*Forum*(0), *User*(0), *User*(1), *User*(2), *Participant*(0,0,1), *Participant*(1,0,2), *HasInvited*(0,1), *Moderates*(0,0)}. Note that seven additional facts have been added to the EDB to ensure that it does not violate any integrity constraint.

We do not show in this figure the unsuccessful repairs that may have happened during the execution. For instance, when determining repairs for *ic10*, *Participant(0,0,0)* could have been considered. However, this alternative does not lead to a valid solution since it violates *ic12*, which can not be repaired.

5 Related Work

In this section we review how reasoning on conceptual schemas has been addressed in the literature. As will be seen, the main contribution of our approach is to deal with more expressive conceptual schemas than previous methods.

The problem of determining the satisfiability of a schema has been widely studied in ER schemas, mostly regarding strong satisfiability of cardinality constraints. This notion was introduced in [5], where the problem was reduced to solving a linear inequality system. In [17] the problem is solved by means of a graph theoretic approach, which was extended in [3] to deal with a generalization of the concept of cardinality. In [4], satisfiability is checked by building a minimal sample database satisfying a set of global cardinality constraints.

More expressive schemas are considered in [6,7], where the finite satisfiability of object-oriented database schemas is determined. The schemas can be annotated only with specific textual constraints to restrict the value of an attribute by comparing it to another value.

The Alloy language and analyzer [18] provide interesting reasoning capabilities for more expressive schemas by searching for examples of the tested properties. However, since the search space must be limited by the user, failure to find an example does not necessarily mean that one does not exist.

The problem of checking satisfiability has also been addressed for UML conceptual schemas. Restricted UML schemas are analyzed in [19], detecting conflicts regarding disjointness and covering constraints in hierarchies, and inconsistencies in the redefinition of inherited cardinality constraints.

A different approach to reason on UML schemas is to translate them into Description Logics (DL) and perform several reasoning tasks using a DL-based system. This allows not only checking the satisfiability of the complete schema but also determining other properties such as class consistency, class equivalence or class subsumption [1], or checking whether a class is forced to have either zero or infinite objects [2]. However, OCL constraints, as well as other UML constructs such as association classes or n-ary associations, are disallowed in order to guarantee decidability.

An alternative approach is not to restrict the expressiveness of the schema and consider general constraints, but then reasoning becomes semidecidable. This direction has been followed in [8], where several reasoning tasks on a UML schema with textual OCL constraints are performed using the CQC-Method as a reasoning engine. In addition to the decidability drawback of this approach, and as far as efficiency is concerned, the reasoning procedure proposed in this paper represents an important improvement regarding the number of integrity constraints that are considered for maintenance.

6 Conclusions

We have proposed an approach to reason on UML schemas with OCL constraints. Our approach can deal with almost all the operators that can be used to define an OCL expression (all the boolean operators defined in the OCL standard, as well as `select` and `size`, that return a collection and an integer). Exceptions are those expressions resulting from recursively combining `includes`, `includesAll`, `notEmpty`, `exists` and `one`. Then, given a conceptual schema of this kind, our method allows determining whether it satisfies certain desirable properties such as schema satisfiability, predicate liveliness or reachability of partially specified states.

Our approach consists of two different tasks, which are the main contributions of our work. First, we analyze whether the schema is such that any reasoning task performed on it will terminate. This is achieved by means of the construction of the dependency graph of constraints and the definition of a set of conditions over this graph that ensure that the schema does not have any infinite model, which are the reason for undecidability.

Second, we define a procedure that allows to efficiently check whether a certain property holds by constructing a sample EDB in which the property is satisfied. Moreover, the impossibility of finding any solution implies that the property does not hold. This procedure is decomposed in two different steps: satisfying the goal that defines the tested property and maintaining all the integrity constraints of the schema to ensure that the sample EDB built is consistent.

As further work, we plan to implement the approach defined in this paper and apply it to practical situations. We would also like to extend our results to minimize the restrictions on the OCL expressions we can deal with.

Acknowledgements. This work has been partly supported by the Ministerio de Ciencia y Tecnología under projects TIN2005-06053 and TIN2005-05406.

References

1. Berardi, D., Calvanese, D., de Giacomo, G.: Reasoning on uml class diagrams. Artificial Intelligence 168(1-2), 70–118 (2005)
2. Cadoli, M., Calvanese, D., Giacomo, G.D., Mancini, T.: Finite model reasoning on uml class diagrams via constraint programming. In: AI*IA 2007: Artificial Intelligence and Human-Oriented Computing, pp. 36–47 (2007)
3. Hartmann, S.: On the Consistency of Int-cardinality Constraints. In: Ling, T.-W., Ram, S., Li Lee, M. (eds.) ER 1998. LNCS, vol. 1507, pp. 150–163. Springer, Heidelberg (1998)
4. Engel, K., Hartmann, S.: Minimal Sample Databases for Global Cardinality Constraints. In: Eiter, T., Schewe, K.-D. (eds.) FoIKS 2002. LNCS, vol. 2284, pp. 268–288. Springer, Heidelberg (2002)
5. Lenzerini, M., Nobili, P.: On the satisfiability of dependency constraints in entity-relationship schemata. Inf. Syst. 15(4), 453–461 (1990)
6. Formica, A.: Finite satisfiability of integrity constraints in object-oriented database schemas. IEEE Trans. on Knowledge and Data Eng. 14(1), 123–139 (2002)
7. Formica, A.: Satisfiability of object-oriented database constraints with set and bag attributes. Information Systems 28(3), 213–224 (2003)
8. Queralt, A., Teniente, E.: Reasoning on UML Class Diagrams with OCL Constraints. In: Embley, D.W., Olivé, A., Ram, S. (eds.) ER 2006. LNCS, vol. 4215, pp. 497–512. Springer, Heidelberg (2006)
9. Farre, C., Teniente, E., Urpí, T.: Checking query containment with the cqc method. Data and Knowledge Engineering 53(2), 163–223 (2005)
10. Moerkotte, G., Lockemann, P.C.: Reactive consistency control in deductive databases. ACM Trans. Database Syst. 16(4), 670–702 (1991)
11. Console, L., Sapino, M.L., Dupré, D.T.: The role of abduction in database view updating. J. Intell. Inf. Syst. 4(3), 261–280 (1995)
12. Lobo, J., Trajcevski, G.: Minimal and consistent evolution in knowledge bases. J. Applied Non-Classical Logics 7(1-2), 117–146 (1997)
13. Ceri, S., Fraternali, P., Paraboschi, S., Tanca, L.: Automatic generation of production rules for integrity maintenance. ACM Trans. DB Syst. 19(3), 367–422 (1994)
14. Decker, H.: An extension of sld by abduction and integrity maintenance for view updating in deductive databases. In: JICSLP, pp. 157–169 (1996)
15. Schewe, K.D., Thalheim, B.: Towards a theory of consistency enforcement. Acta Inf. 36(2), 97–141 (1999)
16. Mayol, E., Teniente, E.: Consistency preserving updates in deductive databases. Data Knowl. Eng. 47(1), 61–103 (2003)
17. Thalheim, B.: Entity-Relationship Modeling: Foundations of Database Technology. Springer, New York (2000)
18. MIT Software Design Group: The Alloy Analyzer, `http://alloy.mit.edu`
19. Kaneiwa, K., Satoh, K.: Consistency Checking Algorithms for Restricted UML Class Diagrams. In: Dix, J., Hegner, S.J. (eds.) FoIKS 2006. LNCS, vol. 3861, pp. 219–239. Springer, Heidelberg (2006)

Round-Trip Engineering for Maintaining Conceptual-Relational Mappings

Yuan An, Xiaohua Hu, and Il-Yeol Song

College of Information Science and Technology, Drexel University, USA
{yan,thu,isong}@ischool.drexel.edu

Abstract. Conceptual-relational mappings between conceptual models and relational schemas have been used increasingly to achieve interoperability or overcome impedance mismatch in modern data-centric applications. However, both schemas and conceptual models evolve over time to accommodate new information needs. When the conceptual model (CM) or the schema associated with a mapping evolved, the mapping needs to be updated to reflect the new semantics in the CM/schema. In this paper, we propose a round-trip engineering solution which essentially synchronizes models by keeping them consistent for maintaining conceptual-relational mappings. First, we define the consistency of a conceptual-relational mapping through "semantically compatible" instances. Next, we carefully analyze the knowledge encoded in the standard database design process and develop round-trip algorithms for maintaining the consistency of conceptual-relational mappings under evolution. Finally, we conduct a set of comprehensive experiments. The results show that our solution is efficient and provides significant benefits in comparison to the mapping reconstructing approach.

Keywords: Round-trip Engineering, Mapping Maintenance.

1 Introduction

Modern data-centric applications increasingly rely on mappings between conceptual models and relational schemas, i.e., *conceptual-relational mappings (a.k.a., object-relational mappings)*, to achieve interoperability [4] or to overcome the well-known *impedance mismatch* problem [13]: the differences between the data model exposed by databases and the modeling capabilities and programmability needed by the application. Essentially, a conceptual-relational mapping specifies a *semantically consistent* relationship between a conceptual model (hereafter, CM) and a relational schema. For example, a many-to-one relationship from an entity E_1 to an entity E_2 in an Entity-Relationship (ER) diagram can be mapped using some mapping formalism to a relational table that uses the identifier of E_1 as the key and referring to the identifier of E_2 as a foreign key [13]. The key and foreign key constraints reflect the semantics encoded in the relationship.

However, conceptual models and schemas evolve over time to accommodate the changes in the information they represent. Such evolution causes the existing conceptual-relational mappings to become inconsistent. For example, if

Z. Bellahsène and M. Léonard (Eds.): CAiSE 2008, LNCS 5074, pp. 296–311, 2008.

the database administrator (DBA) in charge of the aforementioned relational table has changed the key of the table from the identifier of E_1 to the combination of the identifiers of E_1 and E_2 due to new requirements, then the many-to-one relationship from E_1 to E_2 in the ER diagram is *semantically inconsistent* with the new table because some instances of the table may violate the many-to-one relationship. When conceptual models and schemas change, the conceptual-relational mappings between the conceptual models and schemas must be updated to reflect the evolution. This process is called *conceptual-relational mapping maintenance under evolution*, or *mapping maintenance* for short.

A typical solution to the mapping maintenance problem is to regenerate the conceptual-relational mapping. However, there are two major problems: first, regenerating the mapping alone sometimes cannot solve the inconsistency problem because the semantics of the conceptual model and the schema are out of synchronization, as shown by the previous example; second, the mapping generation process, even with the help of mapping generation tools [6,5], can be costly in terms of human effort and expertise, especially for complex CMs and schemas that were developed independently. A better solution would be to design algorithms that synchronize the CMs and schemas and reuse the original mappings to (semi-)automatically update them into a set of new mappings that are consistent with respect to the new CMs and schemas.

The process for synchronizing models by keeping them consistent is called *Round-Trip Engineering* (RTE) [22,17]. RTE offers a bi-directional exchange between two models. Changes to one model must at some point be reconciled with the other model. In this paper, we propose a round-trip engineering approach for maintaining the consistency of conceptual-relational mappings. Notice that round-trip engineering is **not** forward engineering, e.g., generating a relational schema from a CM, plus reverse engineering [16], e.g., generating a **new** CM from an existing schema. RTE focuses on synchronization.

1.1 Motivation

To motivate our work, we first consider a number of applications and environments in which conceptual-relational mappings are used extensively and a solution to the mapping maintenance problem will greatly benefit to the applications.

Database Design. A typical database design process begins with the development of a conceptual model such as an ER diagram and ends up with a logical database schema manipulated by a commercial database management system. Although the process of generating a logical schema from a CM is mostly automated, the translation mappings between CMs and logical schemas are not kept in automated tools, and the CMs and logical schemas may evolve independently causing the "legacy data" problem. Saving the mappings between CMs and logical schemas implied by the database design process and maintaining the mappings when CMs and schemas evolve will help reduce the "legacy data".

Data-Centric Applications. To increase the productivity of the developers of these applications, there are a number of middleware mapping technologies such as Hibernate [9], DB Visual Architect [1], Oracle TopLink [2], and Microsoft ADO.NET [3]. They provide an ease-to-use environment for generating conceptual-relational mappings. In these middleware mapping tools, when the object/conceptual models and the database schemas change, a solution is needed for maintaining the conceptual-relational mappings.

Data Integration. In data integration, a set of heterogeneous data sources are queried and accessed through a unified global and virtual view [19]. There are many ontology-based data integration applications which use ontologies as their global views. For these applications, the mappings between ontologies and local data sources are the main vehicle for data integration. Early studies have been focused on integration architectures, query answering capabilities , and global view integration. What has been missing is a solution to maintaining the mappings between ontologies and local data sources when ontologies and database schemas evolve.

The Semantic Web. On the Semantic Web, data is annotated with ontologies having precise semantics. For the "deep web" where data is stored in backend databases, the semantic annotation of the data is achieved through the mappings between web ontologies and schemas of backend databases. However, maintaining mappings on the semantic web has not yet been considered.

Although mapping maintenance is important and necessary for many applications, solutions to the problem are rare. This is due to many challenges involved, including: how to define consistency of mapping and detect inconsistency of a mapping; what is a right mapping language; how to capture changes to CMs and database schemas; how to devise a plan for reconciling the CMs and schemas according to the intent and expectation of the user; and what are the principles for systematic reconciliation. In this paper, we address these challenges and offer a systematic study and comprehensive evaluation of how round-trip engineering can be applied to solve the mapping maintenance problem.

The rest of the paper presents our principled approach. In summary, we explore the approach of using correspondences for capturing changes and develop a novel round-trip engineering approach for mapping maintenance. We demonstrate the effectiveness and efficiency of our algorithm by conducting a set of comprehensive experiments.

The remaining content is organized as follows. Section 2 summarizes studies on schema mapping adaptation, schema evolution for object-oriented databases, and other related work. Section 3 presents the formal notation used in later sections. Section 4 introduces our formalism for conceptual-relational mappings. Section 5 characterizes schema and CM evolution. Section 6 describes a solution to the problem of mapping maintenance. Section 7 presents our evaluation results. Finally, Section 8 concludes this paper.

2 Related Work

The directly related work is the study on schema mapping adaptation [23,24]. The goal of schema mapping adaptation is to automatically update a schema mapping by reusing the semantics of the original mapping when the associated schemas change. Yu & Popa [24] explore the schema mapping composition approach. Schema evolutions are captured by formal and accurate schema mappings, and schema adaptation is achieved by composing the evolution mapping with the original mapping. On the other hand, the schema change approach in [23] proposed by Velegrakis et al. incrementally changes mappings each time a primitive change occurs in the source or target schemas. Both solutions focus on reusing the semantics encoded in existing mappings for merely adapting the mappings without considering the synchronization between schemas. This is due to the nature of their problems where schema mappings are primarily used for *data exchange*, i.e., translating a data instance under a source schema to a data instance under a target schema. If a schema mapping connecting two schemas which are semantically inconsistent, then the data exchange process simply does not always produce a target instance. Our approach is different from these solutions in that we aim to maintain the *semantic consistency* of conceptual-relational mappings through model synchronzation.

Other related work is schema evolution [21]. In object-oriented databases (OODB), the problem of schema evolution is to maintain the consistency of an OODB when its schema is modified. The challenges are to update the database efficiently and minimize information loss. A variety of solutions, e.g., [8,11,15] have been proposed in the literature. Our problem is different from the schema evolution problem in OODB in that we are concerned with the semantic consistency between a schema and a CM. In AutoMed [10,14], schema evolution and integration are combined in one unified framework. Source schemas are integrated into a global schema by applying a sequence of primitive transformations to them. The same set of primitive transformations can be used to specify the evolution of a source schema into a new schema. In our approach, we do not ask users to specify a sequence of transformations. The EVE [18] investigates the view synchronization problem, which supports a limited set of changes. The work in [12] describes techniques for maintaining mapping in XML p2p databases which is different from our problem.

Another mapping maintenance problem studied in [20] mainly focuses on detecting inconsistency of simple correspondences between schema elements when schemas evolve. This problem is complementary to the problem we consider here.

3 Formal Preliminaries

A table or relation in a relational database consists of a set of tuples. The schema for a table specifies the name of the table, the name of each column (or attribute or field), and the type of each column. Furthermore, we can specify *integrity constraints*, which are conditions that the tuples in tables must satisfy. Here, we consider the *key* and *foreign key* (abbreviated as *f.k.* henceforth) constraints. A key in a table is a subset of the columns of the table that uniquely identifies a tuple. A f.k. in a table T is a set of columns F that *references* the key of another table T' and imposes a constraint that the projection of T on F is a subset of the projection of T' on the key of T'. A relational schema thus consists of a set of relational schemes (or tables for short). Formally, we use $\mathcal{R}=(R, \Sigma_R)$ to denote a relational schema $\mathcal{R}$ with a set of tables R and a set Σ_R of key and f.k. constraints.

A conceptual model (CM) describes a subject matter in terms of concepts, relationships, and attributes. In this paper, we do not restrict ourselves to any particular language for describing CMs. Instead, we use a generic conceptual modeling language (CML), which has the following specifications. The language allows the representation of *classes/concepts/entities* (unary predicates over individuals), *object properties/ relationships* (binary predicates relating individuals), and *datatype properties/ attributes* (binary predicates relating individuals with values such as integers and strings); attributes are single valued in this paper. Concepts are organized in the familiar ISA hierarchy. Relationships and their inverses (which are always present) are subject to cardinality constraints, which allow 1 as lower bounds (called *total* relationships) and 1 as upper bounds (called *functional* relationships). In addition, a subset of attributes of a concept is specified as the identifier of the concept. As in the Entity-Relationship model, a strong entity has a global identifier, while a weak entity is identified by an identifying relationship plus a local identifier. We use $\mathcal{C}=(C, \Sigma_C)$ to denote a CM $\mathcal{C}$ with a set C of concepts, attributes, and relationships and a set Σ_C of identification and cardinality constraints.

We represent a given CM as a graph called a *CM graph*. We construct the CM graph from a CM by considering concepts and attributes as nodes and relationships as edges. There are also edges between a concept node and the attribute nodes belonging to the concept. A many-to-many relationship p between concepts C_1 and C_2 will be written in text as [C_1] ---p--- [C_2]. For a functional relationship q – ones with upper bound cardinality of 1, from C_1 to C_2, we write [C_1] ---q->-- [C_2]. In a CM graph, we will represent an ISA relationship as a 1:1 functional edge.

4 Conceptual-Relational Mappings

A conceptual-relational mapping specifies a relationship between a CM and a relational schema. More specifically, a mapping consists of a set of statements each of which relates a query expression $\Phi(X, Y)$ in a language $\mathcal{L}_1$ over the CM with a query expression $\Psi(X, Z)$ in a language $\mathcal{L}_2$ over the relational schema, where the shared variables X give rise to the query results. In this paper, we consider conjunctive formulas over concepts, attributes, and relationships in a CM and conjunctive formulas over relational tables which can be translated into equivalent select, join, and project (SJP) query expressions over a relational schema. Queries are evaluated as the usual way.

In the sequel, we will use the terms "mapping" and "mapping statement" interchangeably when the context is clear. Generally, we represent a conceptual-relational mapping (or mapping statement) between a CM and a relational schema as an expression $\Phi(X,Y) = \Psi(X,Z)$, where $\Phi(X,Y)$ and $\Psi(X,Z)$ are conjunctive formulas. The following example illustrates the mapping formalism using a gene expression database and a conceptual model.

Example 1. A gene expression database contains a biosample table to record information about a biological sample which can be a tissue, cell, or RNA material that originates from a donor of a given species:

biosample(sample_ID, species, organ, pathology,..., donor_ID),

where the underlined column sample_ID is the key of the table and donor_ID is a foreign key to a table called donor.

Figure 1 shows a mapping between the biosample table and a CM containing two concepts Biosample and Person, and a relationship donation. The CM is described in the UML notation. The dashed arrows indicate the correspondences between columns of the relational table and attributes of concepts in the CM. We represent the conceptual-relational mapping between the relational table and the CM as the following expression:

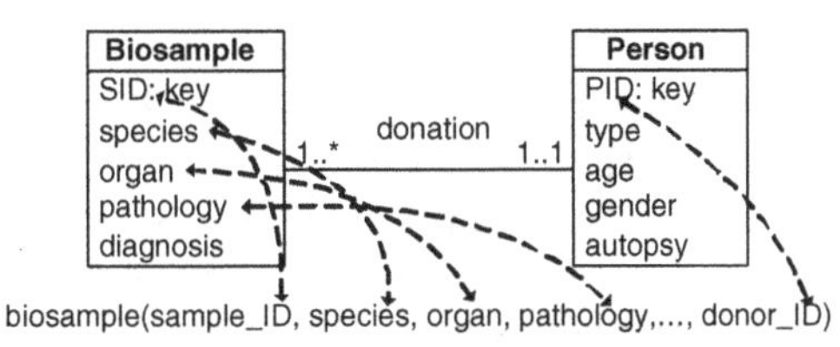

Fig. 1. A Conceptual-Relational Mapping

Biosample(x_1)∧SID(x_1, $sample\_ID$)∧ species(x_1, $species$)∧ ...∧ Person(x_2)∧ donation(x_1, x_2)∧ PID(x_2, $donor\_ID$)

= biosample($sample\_ID$, $species$,..., $donor\_ID$),

where the predicates Biosample and Person represent the concepts in the CM, the predicates SID, species,..., represent the attributes of the concepts and the relationship, and the shared variables $sample\_ID$, $species$,..., give rise to query results on both sides. ■

Consistent Conceptual-Relational Mappings. We define a consistent conceptual -relational mapping between a CM and a relational schema in terms of *legal instances* of the CM and the relational schema. For a CM $\mathcal{C}=(C,\Sigma_C)$, a legal instance I is an instance of C which satisfies the constraints Σ_C. We use $\mathcal{I}$ to denote the set of all legal instances of $\mathcal{C}$, i.e., $\mathcal{I}=\{I \mid I$ is an instance of C and $I \models \Sigma_C\}$. Likewise, for a relational schema $\mathcal{R}=(R,\Sigma_R)$, we use $\mathcal{J}$ to denote the set of all legal instances of $\mathcal{R}$, i.e., $\mathcal{J}=\{J \mid J$ is an instance of R and $J \models \Sigma_R\}$.

For a query expression $\Phi(X,Y)$ over $\mathcal{C}$, we use I^{Φ} to denote the query results over the instance I. We use J^{Ψ} to denote the query results of the query expression $\Psi(X,Z)$ over the instance J of $\mathcal{R}$. We say that a pair of legal instances $\langle I,J\rangle$ satisfies a mapping statement M:$\Phi(X,Y) = \Psi(X,Z)$ between $\mathcal{C}$ and $\mathcal{R}$, if and only if $I^{\Phi}=J^{\Psi}$, denoted as $\langle I,J\rangle \models M$.

Definition 1 (Consistent Conceptual-Relational Mapping). *For a CM $\mathcal{C}=(C,\Sigma_C)$ and a relational schema $\mathcal{R}=(R,\Sigma_R)$, a mapping M:$\Phi(X,Y) = \Psi(X,Z)$*

between $\mathcal{C}$ and $\mathcal{R}$ is consistent if and only if for every legal instance $I \in \mathcal{I}$, there is a legal instance $J \in \mathcal{J}$ such that $\langle I, J \rangle \models M$, and for every legal instance $J' \in \mathcal{J}$, there is a legal instance $I' \in \mathcal{I}$ such that $\langle I', J' \rangle \models M$.

Essentially, the consistency of a mapping dictates the "compatibility" of the constraints in the CM and the schema.

5 Changes to Schemas and CMs

A user can change a schema (or CM) in different ways: either through modifying the original schema (or CM) or by generating a new schema (or CM) directly. It is difficult to ask the user to provide a sequence of primitive actions for capturing the changes. It is probably easier to ask the user to draw a set of simple correspondences between the elements in the new schema (or CM) and the elements in the original schema (or CM). In this paper, we use a set of correspondences between columns in schemas (or attributes in CMs) to capture the commonality/differences between the new schema (or CM) and the original schema (or CM).

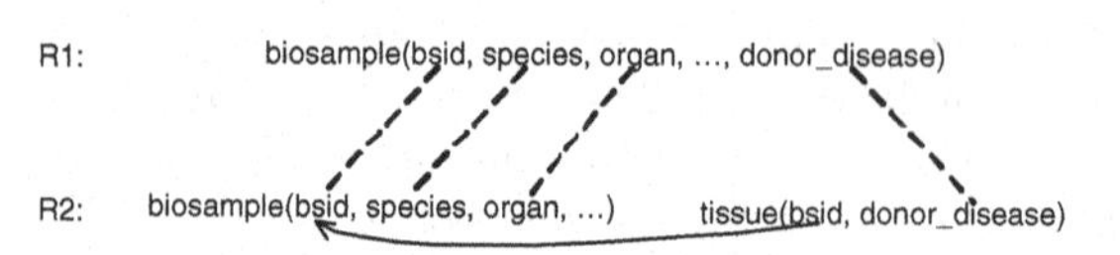

Fig. 2. Capturing Changes to a Schema

Example 2. Figure 2 shows on the top an original schema $\mathcal{R}_1$ consisting of a single table biosample. On the bottom is a new schema $\mathcal{R}_2$ containing two tables biosample and tissue. $\mathcal{R}_2$ evolved from $\mathcal{R}_1$. The dashed lines between columns in $\mathcal{R}_1$ and the columns in $\mathcal{R}_2$ capture the commonality/differences between the original schema and the new schema. The open arrow indicates that the column tissue.bsid is a foreign key referring the key biosample.bsid. ■

6 Round-Trip Engineering for Conceptual-Relational Mappings

We now develop a round-trip engineering solution for maintaining conceptual-relational mappings under evolution. The primary goal of the maintenance is to keep the mapping consistent by synchronizing the schema and the CM. To fulfill the goal, the algorithm must *understand* the existing semantics in the original mapping and carry out necessary updates based on sound principles. We begin with the exploration on the knowledge encoded in the forward engineering process.

Knowledge about the Conceptual-Relational Mappings in Standard Database Design Process. In relational database design, a standard technique (we refer to this as er2rel schema design) which is widely covered in undergraduate database courses [13] derives a relational schema from an Entity-Relationship diagram. The er2rel design implies a set of conceptual-relational mappings in the form $\Phi(X,Y)=T(X)$, where $\Phi(X,Y)$ is a conjunctive formula encoding a tree structure called *semantic tree (or s-tree)* [7] in a CM, and $T(X)$ is a relational table with columns X. Such a conceptual-relational mapping is also used in the middleware mapping technologies.

We choose to design our solution for mapping maintenance in a systematic manner by considering the behavior of our algorithm on the conceptual-relational mappings implied by the er2rel design. In our previous work [7], we have carefully analyzed the knowledge encoded in the er2rel design. We summarize the knowledge related to our study in this paper as follows.

1. The er2rel design associates a relational table with a tree structure called semantic tree (s-tree) in a CM.
2. An s-tree can be decomposed into several subtrees called *skeleton trees*: a skeleton tree corresponding to the key of the table, skeleton trees corresponding to f.k.s of the table, and skeleton trees corresponding to the rest of the columns of the table.
3. Each skeleton tree has an anchor which is the root of the skeleton tree. An anchor also corresponds to the central object for deriving a table.
4. To satisfy the semantics of the key in a table, the s-tree is connected by functional paths from the anchor of the key skeleton tree to the anchors of f.k. skeleton trees and other skeleton trees.

Example 3. In Figure 1, the mapping associates the biosample table with the s-tree `Biosample` `---donation-->-` `Person`. The s-tree is decomposed into two skeleton trees: `Biosample` with anchor Biosample for the key <u>sample_ID</u> of the table and `Person` with anchor Person for the foreign key donor_ID. (Skeleton trees for weak entities are more complex; see Example 5). The two anchors are connected by a functional edge `---donation->--`. ∎

Sketch of the Maintenance Algorithm. We first outline the algorithm for maintaining mappings which are in the form of $\Phi(X,Y)=T(X)$. We develop the complete algorithm later. Given a relational schema $\mathcal{R}$, a CM $\mathcal{C}$, a set of existing consistent conceptual-relational mappings $M=\{\Phi(X,Y)=T(X)\}$ between $\mathcal{R}$ and $\mathcal{C}$, a new schema $\mathcal{R}'$ (or CM $\mathcal{C}'$), and a set of correspondences M' between $\mathcal{R}$ and $\mathcal{R}'$ (or between $\mathcal{C}$ and $\mathcal{C}'$), the algorithm works in several steps for fulfilling the goals of mapping maintenance:

1. Analyze the existing semantics in the original mapping in terms of skeleton trees and connections between anchors of skeleton trees.
2. Discover changes through the correspondences between the new schema/CM and the original schema/CM.
3. Synchronize the associated CM/schema and adapt the mapping accordingly.

Illustrative Examples. Before fleshing out the above steps, we illustrate the algorithms using several examples on *schema evolution*. Through these examples, we lay out our principles for mapping maintenance.

Example 4 [Adding a Column]. Figure 3 (a) shows a mapping which is specified as following statement:
Sample(x_1) $\wedge$ sid(x_1, sid) $\wedge$ Person(x_2) $\wedge$ originates(x_1, x_2) $\wedge$ pid(x_2, $donor$) = sample($sid, donor$).

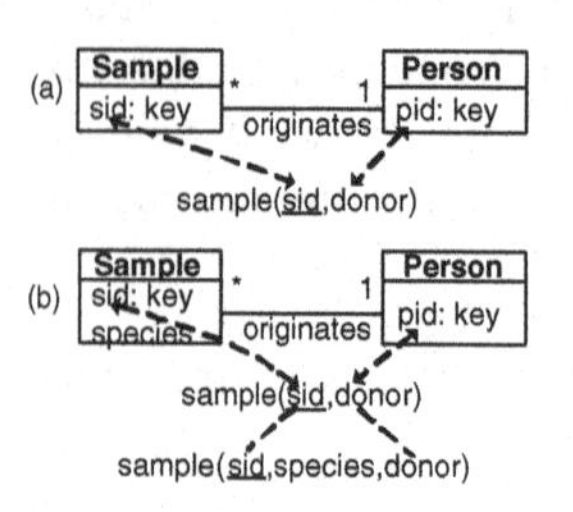

Fig. 3. Adding a Column to Schema

Figure 3 (b) shows that a column species was added to the table sample(sid, donor). *For adding an element in the schema, our goal of mapping maintenance is to add a corresponding element in the CM to maximize the coverage of the schema elements.* Since the key column sid corresponds to the identifier attribute of the Sample class and the column donor is a foreign key referring to the key of a table donor(did) (not shown in the figure) for the Person class, we synchronize the CM through adding an attribute species to the Sample class which is the anchor of the skeleton tree corresponding to the key sid. ■

The *first principle* for the mapping maintenance for schema evolution is to use the key and foreign key information in the original and new schemas through the correspondences to locate the appropriate elements in the CM for adding new attributes.

Example 5. Let us consider the case for adding a foreign key column. Figure 4 shows an original mapping enclosed whithin the rectangle:

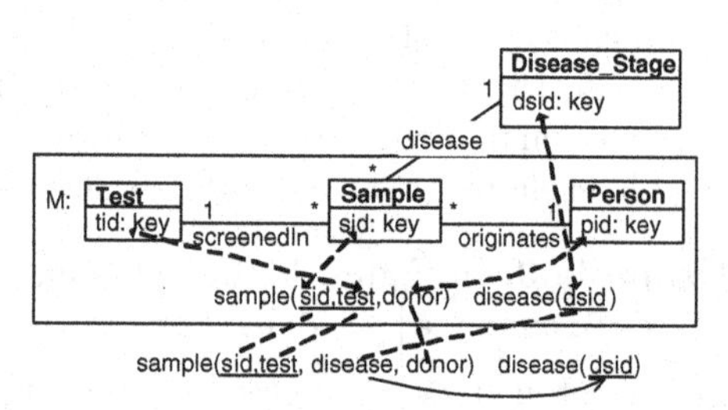

Fig. 4. Adding a Foreign Key Column to Schema

M:Test(x_1) $\wedge$ tid($x_1, test$) $\wedge$ Sample(x_2) $\wedge$ sid(x_2, sid) $\wedge$ screenedIn(x_1, x_2) $\wedge$ Person(x_3) $\wedge$ originates(x_2, x_3) $\wedge$ pid(x_3, $donor$) = sample($sid, test, donor$).

In the CM, Sample is modeled as a weak entity with an identifying functional relationship screenedIn connecting to the owner entity Test. Accordingly, the key of the table sample(sid,test,donor) is the combination of columns sid and test with test being a foreign key referring to a table test(tid) for the Test class (not shown in the figure.) On the bottom of Figure 4, the table sample(sid, test, donor) was changed to sample(sid,test,disease,donor) with the column disease being a foreign key referring to the key of the table disease(dsid) (shown as the open arrow.) To update the mapping between the new sample table and the CM, we analyze the key and foreign key structure of the table and recognize that Sample class is the *anchor* of the skeleton tree `Test` `-<--screenIn---` `Sample` for the key. The newly added foreign key disease should indicate that there is a functional relationship from the Sample class to the Disease_Stage class rather than a functional relationship from the Test class to the Disease_Stage class. Therefore, we add/discover a functional relationship disease in the original CM and update the mapping between the sample(sid, test, disease, donor) and the new CM. ■

Our *second principle* is to use key and foreign key structure in the schemas through the correspondences to locate the anchors of the appropriate skeleton trees for discovering/adding relationships.

Example 6 [Changing Constraints]. The following existing mapping associates a relational table treat(tid, sgid) with a CM `Treatment` ---appliesTo--- `Sample_Group`:

Treatment(x_1) ∧ tid(x_1, *tid*) ∧ Sample_Group(x_2) ∧ appliesTo(x_1, x_2) ∧ sgid(x_3, *sgid*) = treat(*tid*, *sgid*), where the relationship appliesTo is many-to-many.

Later, the database administrator obtained a better understanding of the application by realizing that each treatment only applies to one sample group. Consequently, the DBA changed the key of the treat table from the combination of columns tid and sgid to the single column tid. Having the change on the schema, we update the appliesTo from a many-to-many relationship to a functional relationship `Treatment` ---appliesTo->-- `Sample_Group` to keep the mapping consistent. ■

The *third principle* is to align the key and foreign key constraints in the (new) schema with the cardinality constraints in the (new) CM.

Maintenance Algorithm. In this paper, the maintenance algorithm requires that each original conceptual-relational mapping statement $\Phi(X,Y)$=$T(X)$ is consistent and associates a relational table $T(X)$ with a semantic tree $\Phi(X,Y)$ in a CM. For a general consistent conceptual-relational mapping associating a graph with a conjunctive formulas over a schema, we can first convert the graph into a tree by replicating nodes (see [4]). Then we either decompose the mapping into mappings between semantic trees and single tables or treat the entire conjunctive formula over the schema as a big table. The details for converting general mappings into mappings between semantic trees and tables are beyond the scope of this paper and will be realized in the future work.

The maintenance algorithm has two components. The first component deals with changes to schemas, and the second component deal with changes to CMs. We first focus on schema changes. The following Procedure 1 maintains the consistency of conceptual-relational mappings when schemas evolve.

Procedure 1. Maintain Mappings When Schemas Evolve
Input: A set of consistent conceptual-relational mappings M=$\{\Phi(X,Y)$=$T(X)\}$ between a CM $\mathcal{C}$ and a relational schema $\mathcal{R}$; a set of correspondences M' between columns in $\mathcal{R}$ and columns in a new schema $\mathcal{R}'$
Ouput: Synchronized CM $\mathcal{C}''$ and a set of updated mappings M'' between $\mathcal{C}''$ and $\mathcal{R}'$.

Steps:

1. Mark skeleton trees: for each mapping statement in M, decompose the semantic tree in the CM into several skeleton trees based on the key and foreign key structures of the table; mark the associations between keys/f.k.s and skeleton trees.
2. Apply the principles we have laid out above to each of the following cases for synchronizing the CM and updating the mapping (we ignore the renaming change in our algorithm):
   - **Case 1:** A new table evolved from a single table by adding columns, deleting columns, or changing constraints.
   - **Case 2:** A new table evolved from several tables by adding columns, deleting columns, or changing constraints.
   - **Case 3:** Several tables evolved from a single table by adding columns, deleting columns, or changing constraints.

We now elaborate on each case.

Case 1: If a new table evolved from a single table, then columns which are not foreign key have been changed or a foreign key has been deleted. If a new column is added, then add a new attribute to the anchor of the key skeleton tree (see Example 4). If the column becomes part of the key, then the new attribute becomes part of the identifier of the anchor. If a column is deleted, we only update the mapping by removing the reference to the deleted column in the mapping. If the key constraint has been changed, then synchronize the identifier of the anchor of the key skeleton tree accordingly.
Case 2: If a new table T evolved from several tables $\{T_1, T_2, ..., T_n\}$, then we connect the semantic trees corresponding to the original tables $\{T_1, T_2, ..., T_n\}$ into a larger semantic tree as follows. Suppose the key of the table T come from the key of table T_1. Let the skeleton trees $\{S_1, S_2, ..., S_n\}$ correspond to the keys of $\{T_1, T_2, ..., T_n\}$. Connect the anchor of S_1 to the anchors of $\{S_2, ..., S_n\}$ by functional edges. The new table is mapped to the larger tree. Example 5 illustrates the case where a new table sample(sid, test, disease, donor) evolved from two original tables sample(sid, test,donor) and disease(dsid,diagnosis). The new table is mapped to a larger semantic tree by connecting the two anchors Sample and Disease_Stage using a functional edge `---disease-->-`.
Case 3: Several tables $\{T_1, T_2, ..., T_n\}$ evolved from a single table T. Without losing generality, suppose T_1 inherit the key of T. We create new concepts $\{C_2, ..., C_n\}$ in the CM for the new tables $\{T_2, ..., T_n\}$, respectively. Let C_i be the anchor of the skeleton tree corresponding to the key of T_i. For two tables T_i and T_j, if there is a foreign key constraint from the column T_i.f to the key of T_j, then we connect C_i to C_j by a functional edge in the CM. If the column T_i.f is also the key of the table T_i, then we connect C_i to C_j by an ISA relationship (note that nodes could be merged if there are two-way f.k.s between the keys of two tables.)

Example 7 [Adding New Tables]. In Figure 5, a new schema $\mathcal{R}_2$ containing two tables biosample and tissue evolved from the original schema $\mathcal{R}_1$ with a single table biosample. The original mapping associates $\mathcal{R}_1$ with the concept Biosample. On the top of the figure is a new CM, where a new concept Tissue is added and connected to Biosample by an ISA relationship according to the f.k. constraint between the keys of tissue and biosample tables in the new schema $\mathcal{R}_2$. ∎

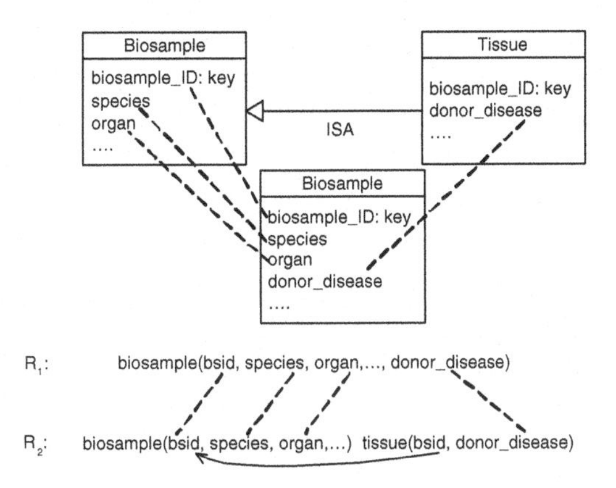

Fig. 5. New Tissue Concept for New tissue Table

We now turn to the procedure dealing with changes to CMs. Intuitively, synchronizing schemas when associated CMs change is more costly than synchronizing CMs when schemas change because synchronizing schema often results in data translation. Two strategies can be considered for maintaining mappings when CMs change. The first strategy is to design a procedure in the similar fashion as for the Procedure 1. The second is to adapt mappings to maintain consistency without automatic synchronization. We take the second approach in this paper and leave the first approach in the future work. The following Procedure 2 updates conceptual-relational mappings when the CMs evolve.

Procedure 2. Maintain Mappings When CM Evolve
Input: A set of consistent conceptual-relational mappings $M{=}\{\Phi(X,Y){=}T(X)\}$ between a CM $\mathcal{C}$ and a relational schema $\mathcal{R}$; a set of correspondences M' between attributes in $\mathcal{C}$ and attributes in a new CM $\mathcal{C}'$
Ouput: Update M to a new set of mappings M'' between $\mathcal{R}$ and $\mathcal{C}'$.

Steps:

1. Mark skeleton trees: the same as in the first step of Procedure 1.
2. For a mapping statement in M associating a semantic tree S with a table T
   (a) If the skeleton tree corresponding to the key of T has changed such that identifier attributes of the anchor were added/deleted or a cardinality constraint in the skeleton tree has changed from one to many, then drop the mapping. */*changes to the identifier information of either a strong or a weak entity will result in inconsistent mapping to the original table.*/*
   (b) Else if a cardinality constraint imposed on a relationship p in S has changed from many to one or from one to many, then remove from S the relationship edge p and the rest part connecting to the anchor through p. Update the mapping so that T is mapped to the new smaller tree.
   (c) Else compose the correspondences M' with the original mapping M to generate a new mapping M'' between $\mathcal{R}$ and $\mathcal{C}'$. */*see [24] for composition algorithm.*/*

The following states the desired property of the maintenance algorithm consisting of the steps in Procedure 1 and Procedure 2.

Proposition 1. *Let $M = \{\Phi(X,Y) = T(X)\}$ be a set of consistent conceptual-relational mappings between a CM $\mathcal{C}$ and a relational schema $\mathcal{R}$. Let $\mathcal{R}'$ (or $\mathcal{C}'$) be a new schema (or a new CM) that evolved from $\mathcal{R}$ (or $\mathcal{C}$). Let M' be a set of identity mappings between columns in $\mathcal{R}$ and columns in $\mathcal{R}'$ (or attributes in $\mathcal{C}$ and attributes in $\mathcal{C}'$.) Each mapping in the set of conceptual-relational mappings returned by the* Procedure 1 *(or* Procedure 2*) is consistent.*

7 Experience

To evaluate the performance of our round-trip solution for maintaining conceptual-relational mappings, we applied the algorithm to a set of conceptual-relational mappings drawn from a variety of domains. The purpose of our evaluation is two-fold: (1) to test the efficiency of the algorithm and (2) to measure the benefits of mapping maintenance over reconstructing consistent mappings using mapping discovery tools.

Data Sets. We selected our test data from a variety of domains. Our previous work [7] on the development of the MAONTO mapping tools generated conceptual-relational mappings for many of the test data. Subsequently, our other previous work [4] used the conceptual-relational mappings for improving traditional tools on constructing direct mappings between database schemas. It follows naturally to continue on this set of data for measuring the benefits of mapping maintenance. Table 1 summarizes the characteristics of the test data. The size of a mapping is measured by the size of the semantic tree - the number of nodes including attribute nodes.

Table 1. Characteristics of Test Data

Schema	#Tables	Avg. # Cols Per Table	CM	#Nodes in CM	Avg. Mapping Size
DBLP	22	9	Bibliographic	75	9
Mondial	28	6	Factbook	52	7
Amalgam	15	12	Amalgam ER	26	10
3Sdb	9	14	3Sdb ER	9	6
CS Dept.	8	6	KA onto.	105	7
Hotel	6	5	Hotel Onto.	7	7
Network	18	4	Network onto.	28	6

Methodology. Our experiments focused on maintaing the consistency of tested mappings under *schema evolution*. For each mapping, we applied different types of changes to the relational table. For each type of change, we ran the maintenance algorithm for measuring (1) execution time and (2) benefits in comparison to the mapping reconstructing approach. The types of changes to a table include: (*a*) adding/deleting ordinary columns; (*b*)adding/deleting key columns; (*c*) splitting a table; (*d*) merging two tables; (*e*) add/deleting f.k. columns; (*f*) moving columns from one table to another table; and (*g*) changing existing key and f.k. constraints.

To measure the benefits of mapping maintenance, we adopt the approach for measuring how much user effort can be saved when schemas evolved and a new *consistent* mapping has to be established. Both Velegrakis et al. in [23] and Yu & Popa in [24] applied the similar approach for measuring the benefits of mapping adaptation. Essentially, the user effort for obtaining a consistent mapping through mapping maintenance after the schema evolved is compared to the same type of user effort spent for reconstructing the mapping. In our study, we compared the mapping maintenance approach with the MAPONTO [7] tool for discovering mappings.

For a mapping $\Phi(X,Y)$=$T(X)$ associating a semantic tree with a relational table, let T' be the new table that evolved from T. For mapping maintenance, the user specifies a set of simple correspondence bewteen T' and T. Then the maintenance algorithm generates a new mapping between T' and, probably, an updated semantic tree. On the other hand, to reconstruct a mapping using the MAPONTO tool, the user also needs to specify a set of correspondences between T' and the CM. However, MAPONTO tool may be unable to generate the expected mappings because the CM is out of synchronization. If the expected mapping is generated by the maintenance algorithm while it is missing from the results of MAPONTO, then we assign 100% to the benefit of maintenance. Otherwise, we use the following quantity to measure the benefit:

$$1 - \frac{\#mapping_{maintenace}}{\#mappping_{MAPONTO} + \#correspondences}$$

Because specifying correspondences between a schema and CM is much more costly than specifying correspondences between an evolved schema and the original schama, we omit the effort for specifying evolution correspondences from the above quantity.

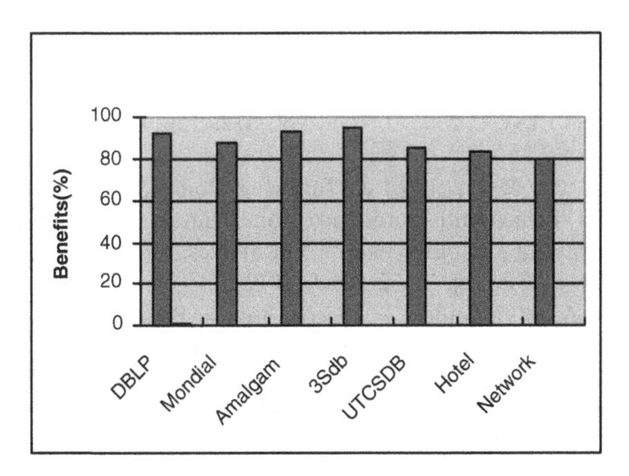

Fig. 6. Benefits of Mapping Maintenance

Results. First of all, the times used by the maintenance algorithm for synchronizing CMs and updating mappings are insignificant. For all the tested mappings, the maintenance algorithm took less than one second to generate expected results. This is comparable with the MAPONTO tool for discovering mappings between schemas and CMs. Next, in terms of benefits, Figure 6 presents the average benefits for the tested cases. The results show that the round-trip engineering solution provides significant benefits in terms of maintaining the consistency of conceptual-relational mappings under evolution.

8 Conclusions

In this paper, we studied the problem of maintaining the consistency of conceptual-relational mappings with evolving schemas and CMs. We motivated the need for synchronizing the CM and relational schema associated by a conceptual-relational mapping. We presented a novel round-trip engineering framework and developed algorithms that automatically maintain conceptual-relational mappings as schemas/CMs evolve. Our solution is unique in that we carefully compile the knowledge encoded in the widely covered methodology for database design into our approach. Experimental analysis showed that the solution is efficient and provides significant benefits for maintaining conceptual-relational mappings in dynamic environments.

References

1. Database Visual Architect, `http://www.visual-paradigm.com/product/dbva`
2. Oracle TopLink, `http://www.oracle.com/technology/product/ias/toplink`
3. Adya, A., Blakeley, J., Melnik, S., Muralidhar, S.: Anatomy of the ado.net entity framework. In: SIGMOD 2007 (2007)
4. An, Y., Borgida, A., Miller, R.J., Mylopoulos, J.: A Semantic Approach to Discovering Schema Mapping Expressions. In: Proceedings of International Conference on Data Engineering (ICDE) (2007)
5. An, Y., Borgida, A., Mylopoulos, J.: Constructing Complex Semantic Mappings Between XML Data and Ontologies. In: Gil, Y., Motta, E., Benjamins, V.R., Musen, M.A. (eds.) ISWC 2005. LNCS, vol. 3729, pp. 6–20. Springer, Heidelberg (2005)
6. An, Y., Borgida, A., Mylopoulos, J.: Inferring Complex Semantic Mappings between Relational Tables and Ontologies from Simple Correspondences. In: Proceedings of International Conference on Ontologies, Databases, and Applications of Semantics (ODBASE), pp. 1152–1169 (2005)
7. An, Y., Borgida, A., Mylopoulos, J.: Discovering the Semantics of Relational Tables through Mappings. Journal on Data Semantics VII, 1–32 (2006)
8. Banerjee, J., et al.: Semantics and Implementation of Schema Evolution in Object-Oriented Databases. In: SIGMOD 1987 (1987)
9. Bauer, C., King, G.: Java Persistence with Hibernate. Manning Publications (November 2006)

10. McBrien, P., Poulovassilis, A.: Schema Evolution in Heterogeneous Database Architectures, A Schema Transformation Approach. In: Pidduck, A.B., Mylopoulos, J., Woo, C.C., Ozsu, M.T. (eds.) CAiSE 2002. LNCS, vol. 2348, Springer, Heidelberg (2002)
11. Claypool, K.T., Jin, J., Rundensteiner, E.: SERF: Schema Evolution through an Extensible, Re-usable, and Flexible Framework. In: CIKM 1998 (1998)
12. Sartiani, C., Colazzo, D.: Mapping Maintenance in XML P2P Databases. In: Bierman, G., Koch, C. (eds.) DBPL 2005. LNCS, vol. 3774, pp. 74–89. Springer, Heidelberg (2005)
13. Elmasri, R., Navathe, S.B.: Fundamentals of Database Systems, 5th edn. Addison-Wesley, Reading (2006)
14. Poulovassilis, A., Fan, H.: Schema Evolution in Data Warehousing Environments – A Schema Transformation-Based Approach. In: Atzeni, P., Chu, W., Lu, H., Zhou, S., Ling, T.-W. (eds.) ER 2004. LNCS, vol. 3288, pp. 639–653. Springer, Heidelberg (2004)
15. Ferrandina, F., Ferran, G., Meyer, T., Madec, J., Zicari, R.: Schema and Database Evolution in the O2 Object Database System. In: VLDB 1995 (1995)
16. Hainaut, J.-L.: Database reverse engineering (1998), `http://citeseer.ist.psu.edu/article/hainaut98database.html`
17. Knublauch, H., Rose, T.: Round-trip engineering of ontologies for knowledge-based systems. In: SEKE 2000 (2000)
18. Lee, A., Nica, A., Rundensteiner, E.: The eve approach: View synchronization in dynamic distributed environment. TKDE 14(5), 931–954 (2002)
19. Lenzerini, M.: Data Integration: A Theoretical Perspective. In: Proceedings of the ACM Symposium on Principles of Database Systems (PODS), pp. 233–246 (2002)
20. McCann, R., et al.: Maveric: Mapping Maintenance for Data Integration Systems. In: VLDB 2005 (2005)
21. Rahm, E., Bernstein, P.: An on-line bibliography on schema evolution. SIGMOD Record 35(4), 30–31 (2006)
22. Sendall, S., Kuster, J.: Taming model round-trip engineering. In: Proceedings of Workshop on Best Practices for Model-Driven Software Development (2004)
23. Velegrakis, Y., Miller, R.J., Popa, L.: Mapping Adaptation under Evolving Schemas. In: Proceedings of the International Conference on Very Large Data bases (VLDB), pp. 584–595 (2003)
24. Yu, C., Popa, L.: Semantic Adaptation of Schema Mappings when Schema Evolve. In: Proceedings of the International Conference on Very Large Data bases (VLDB) (2005)

Capturing and Using QoS Relationships to Improve Service Selection

Caroline Herssens[1], Ivan J. Jureta[2], and Stéphane Faulkner[2]

[1] ISYS, LSM, Université catholique de Louvain, Belgium
[2] PReCISE, LSM, University of Namur, Belgium
caroline.herssens@uclouvain.be, iju@info.fundp.ac.be,
stephane.faulkner@fundp.ac.be

Abstract. In a Service-Oriented System (SOS), service requesters specify tasks that need to be executed and the quality levels to meet, whereas service providers advertise their services' capabilities and the quality levels they can reach. Service selectors then match to the relevant tasks, the candidate services that can perform these tasks to the most desirable quality levels. One of the key problems in QoS-aware service selection lies in managing tradeoffs among QoS expectations at runtime, that is, situations in which service requesters specify quality levels that cannot be simultaneously met. We propose a service selection approach that can deal with tradeoffs. The approach consists of: (i) rich QoS models to be used by service requesters when expressing QoS expectations and service providers when describing services' QoS, and for representing preference and priority relationships between QoS dimensions; and (ii) a multi-criteria decision making technique that uses the models for service selection.

Keywords: QoS model, service selection, QoS relationships.

1 Introduction

Engineering and managing the operation of increasingly complex information systems is a key challenge in computing. It is now widely acknowledged that degrees of automation needed in response cannot be achieved without open, distributed, interoperable, and modular systems capable of dynamic adaptation to changing operating conditions. Among the various approaches to building such systems, service-orientation stands out in terms of its reliance on the World Wide Web infrastructure, availability of standards for describing and enabling interaction between services, attention to interoperability, and uptake in industry. In a Service-Oriented System (SOS), service providers advertise the tasks that their services can perform and the Quality of Service (QoS) levels they can meet. Service requesters indicate tasks to execute and QoS levels to achieve. Service selectors (i.e., allocation mechanisms in SOS) then proceed to compare available services and select those that can execute the required tasks while achieving the most desirable feasible QoS levels.

Z. Bellahsène and M. Léonard (Eds.): CAiSE 2008, LNCS 5074, pp. 312–327, 2008.

Service selection is a fundamental issue in SOS because it determines how well the requests are satisfied [4,12,17,30]. Comparing competing services (i.e., services that can execute the same tasks) over the levels of QoS dimensions they can meet is an appropriate approach to ensuring that quality expectations are met to the "best" feasible extent.

Contributions. Using QoS dimensions in service selection requires their definition by way of a QoS model [6]. The QoS model must cover all QoS constructs needed in service selection. Two models are needed in practice. Although both the service requester and service provider models must share the primitives for the representation of QoS dimensions and characteristics, and the definition of their values, the two models must also differ given the difference in purpose: the provider model should include relationships for stating dependencies between QoS dimensions while the requester model must involve relationships for defining requesters' priorities over QoS dimensions and preferences over the values of QoS dimensions. Given such models, the service selector will be able to appropriately compare services and take better selection decisions. Decision-making in presence of potentially many QoS dimensions and requests thereon, can be performed through multi-criteria decision analysis (MCDA). The aim is to rank competing services according to the values of their QoS characteristics. We propose an approach to service selection that responds to these considerations. The approach consists of:

- Rich QoS models used by service requesters when expressing QoS expectations and service providers when describing services' QoS, and for representing preference and priority relationships between QoS dimensions. The models are defined as extensions of the UML QoS framework metamodel [21].
- A multi-criteria decision making technique that uses the models for service selection. QoS models, and more precisely, the QoS relationships are used in a fuzzy MCDA approach to build a fuzzy reference set on which interaction weights are set up, subsequently used for ranking competing services. To establish a ranking with fuzzy MCDA, the selector proceeds as follows:
  1. A reference set of service alternatives is built to compute weights of interacting criteria. The reference set is set up with help of requester's preferences and priorities and with providers' observed dependencies over QoS dimensions.
  2. Once weights are fixed, the selector applies them to the available services that correspond to the requested QoS dimensions and levels.
  3. Finally, values obtained with the application of fuzzy MCDA allow the selector to rank possible services and determine the optimal one to each service request.

Organization. Section 2 introduces the case study used in the remaining of the paper. Section 3 assesses the UML QoS Profile with our proposed subdivision into distinct models and our relationships extensions while illustrating their utilizations through the case study. Relationships identified in QoS models are used in Section 4 for an efficient selection based on fuzzy MCDA. This section also provides a selection example based on the case study. The Section 5 presents the related work of existing QoS models and exposes some existing selection approaches. Finally, Section 6 outlines conclusions and future work.

2 Case Study

In this section, we propose a case study subsequently used throughout the paper. The European Space Agency's (ESA) program on Earth observation allows researchers to access and use infrastructure operated and data collected by the agency.[1] Our case study focuses on the information provided by the MERIS instrument on the Envisat ESA satellite. MERIS is a programmable, medium-spectral resolution imaging spectrometer operating in the solar reflective range. MERIS is used in observing ocean color and biology, vegetation and atmosphere and in particular clouds and precipitation. In relation to MERIS, web services are made available by the ESA for access to the data the instrument sends and access and use of the associated computing resources.

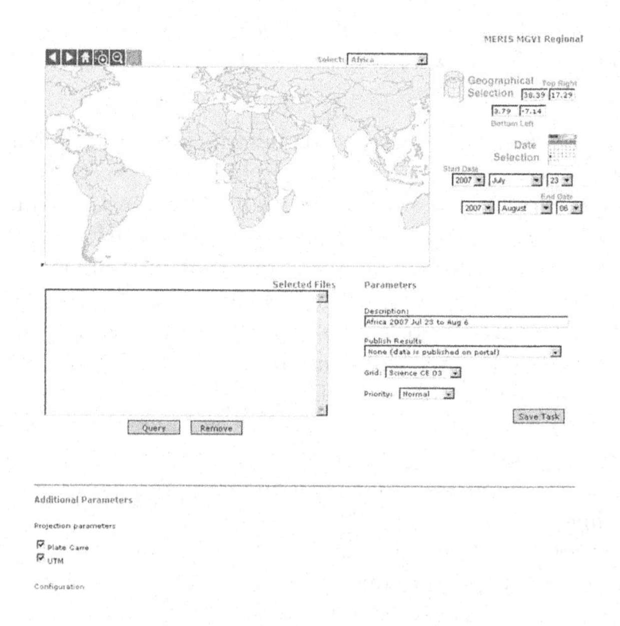

Fig. 1. Graphical user interface of the ENVISAT/MERIS MGVI web service

We are interested in the remainder about services able to provide vegetation indexes for a given region of the globe. A vegetation index measures the amount of vegetation on the Earth's surface. Below, we briefly review the functional requirements that the service satisfies. We then consider quality requirements.

[1] http://gpod.eo.esa.int

Fig. 2. An illustration of the result provided by the ENVISAT/MERIS MGVI Web Service (Guinea - Cameroon region)

Functional requirements. Services considered here process MERIS data and are able to extract the vegetation index. This processing can be selected for any time range (with the start of the satellite mission as the earliest time point); an option is available to delimit the region of the world of interest. The graphical user interface used to access the service is shown in Figure 1. Figure 2 illustrates the visualization of the output obtained for the Guinea - Cameroon region. The following are the required inputs of the service: *Time range*, *Bounding box* (to select a region of the globe), *Dataset*, *Publish site*, and *Projection type.*

Quality requirements. Due to the calculations executed by the service and its parallel use, expected delays and availability are relevant quality considerations from the user's perspective. To make an appropriate selection, quality considerations need to be expressed by users and measured and advertised by providers. We focus on three such considerations, namely availability, reliability, and latency. For now we define them as follows. We then return to each throughout the paper and illustrate how our proposed extensions to the UML QoS profile work with this case study.

- Availability indicates the duration when a component is available for queries. Its value in percent is obtained as follows [24]:

$$A = \frac{upTime}{upTime + downTime}$$

- Reliability is a measure of confidence that the service is free from errors. Its value is given in percent and calculated as follows:

$$R = \frac{succeededAttempts}{succeededAttempts + failedAttempts}$$

- Latency measures the mean time taken by the platform to return the expected result. The value is given in minutes.

$$L = \frac{\sum_1^n (networkTime + selectionTime + executionTime)}{n}$$

  where n is the total number of past executions. Latency of a such service is situated between 4 and 6 hours by day of the selected period due to the quantity of data to process. (This range is certified for a service requestor having network bandwidth of a least 15 mbits/s.)

This basic specification is incomplete. Further explanations are needed. A quality model will provide a checklist of relevant information and in this respect assist the requester and the provider in evaluating and managing the quality of the service.

3 Conceptual Foundations

This section overviews main concepts of the OMG UML QoS Framework and present our distinct models with our relationships extensions. Once instantiated, our QoS models are useful to lead service selection in Section 4.

The UML QoS Framework metamodel introduced by the OMG in [21] includes different modeling constructs describing QoS concepts. It covers submodels aiming at defining different facets of QoS. The **QoS Characteristics submodel**

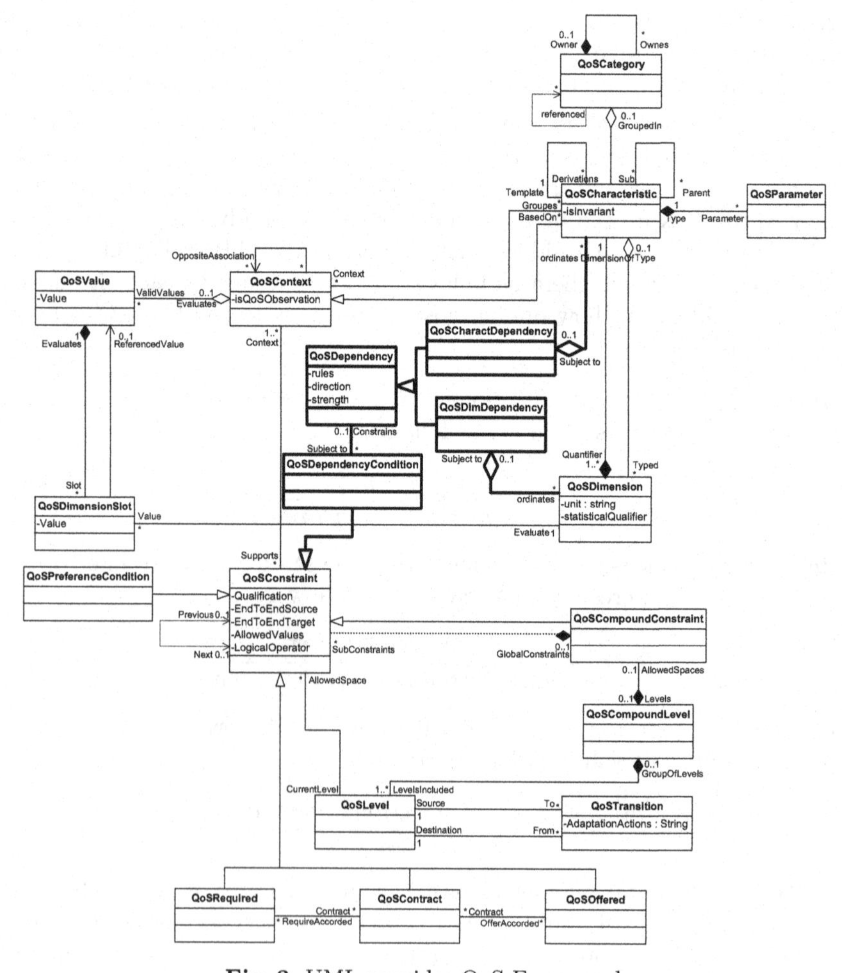

Fig. 3. UML provider QoS Framework

outlines *QoS Characteristics* that are a description for some quality considerations and *QoS Dimensions* that are measures quantifying QoS Characteristics. *QoS Categories* are used to group together QoS Characteristics related to the same abstract quality topic. **QoS Constraint submodel** main constructs are *QoS Constraints* that restrict values of QoS Characteristics while stating limitations on modeling elements identified by application requirements and architectural decisions. The **QoS Level submodel** provides *QoS Levels* that specify the working mode under which the service is executed. Complete description of constructs of these submodels is given in [21].

To make an explicit distinction between users requirements and providers advertisements, we split the OMG metamodel into two distinct metamodels. The service selector task will consist of match instantiation of the user metamodel with the one from the provider. The provider metamodel is illustrated in Figure 3 while the user metamodel is available in Figure 4. In addition to original modeling constructs, we have added some submodels aiming at express particular relationships existing over QoS Characteristics and QoS Dimensions:

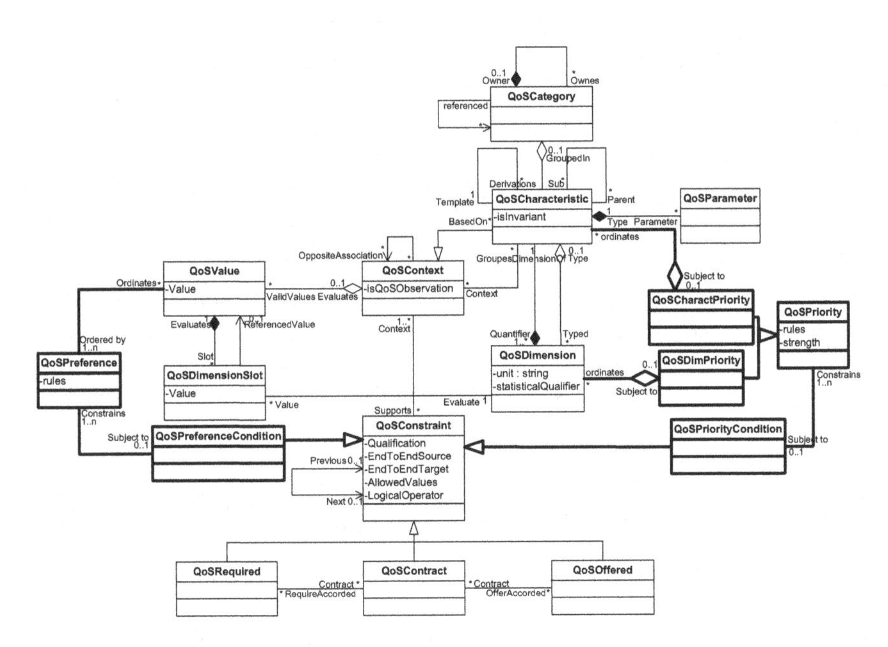

Fig. 4. UML user QoS Framework

QoS Priorities submodel. *QoS Priorities* are used to explicitly represent service requester priorities over QoS Characteristics and QoS Dimensions. Rules determine the order at which characteristics or dimensions are considered for optimization when services are being selected. The relative importance of the priority difference between elements is specified with the strength attribute . *QoS DimPriority* and *QoS CharactPriority* are specializations of QoSPriority defining specific elements for priorities over, respectively, dimensions and characteristics.

QoS PriorityCondition are constraints specifying when priorities hold. The integration of these modeling constructs in the requester QoS model is illustrated in bold in Figure 4. An utilization of the QoS Priorities submodel is proposed in Example 1.

Example 1. In our case study, due to important delays, the requester of the service awards more importance to the *availability* characteristic than to the *latency* characteristic. This particular priority is constraint to a specific condition, stating that the priority is applied only if the *reliability* is inferior to 85%. This priority and its condition are expressed with the UML user QoS Framework in Figure 5. This example will be used to build the reference set in Subsection 4.3.

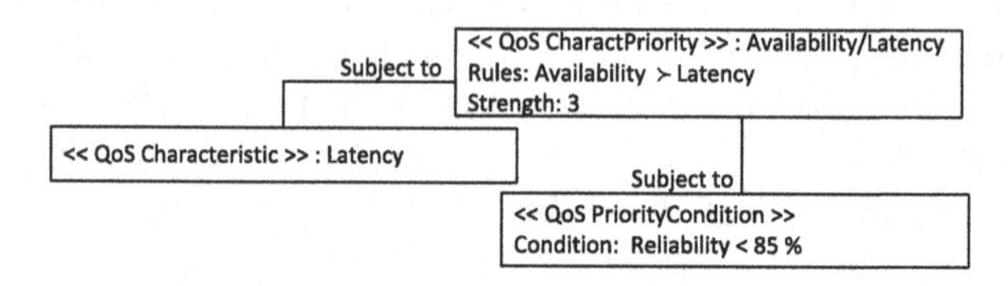

Fig. 5. An illustration of QoS Priorities submodel utilization

QoS Preferences submodel. *QoS Preferences* enable the service requester to sort values of dimensions. Rules are used to determine a precedence order over values. The *QoS PreferenceCondition* indicates conditions for the preference on values to hold. This submodel is illustrated in bold on the user QoS model in Figure 4 and an example of utilization is given in Example 2.

Example 2. About the service providing regional vegetation indexes, the user has some preferences concerning the *networkTime*, he wishes that its value belongs to a specific range as illustrated in Figure 6. The instantiation of the preferences submodel also introduces a specific condition under which the preference is available.

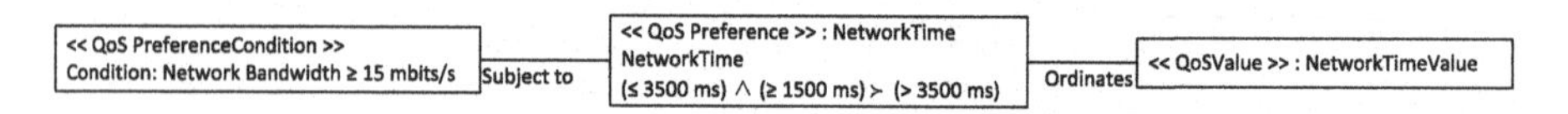

Fig. 6. An illustration of QoS Preferences submodel utilization

QoS Dependencies submodel. *QoS Dependencies* allow to express explicitly dependency relationships existing over different QoS Characteristics or QoS Dimensions while specifying the strength and the direction of the link. The direction indicates that QoS Characteristics (respectively, QoS Dimensions) involved in the dependency are parallel or opposite, meaning that their direction is correlated or anti-correlated. The strength is represented with a level value between 1 and 10, corresponding to the importance of the correlation. The *QoS DependencyCondition* is used to define specific constraints under which a QoS Dependency is applicable. These extensions to the provider QoS model are available in bold in Figure 3 while their utilization is illustrated in Example 3.

Example 3. The provider of the MGVI/Regional service will use the dependencies submodel to underline interactions between several QoS Characteristics. In Figure 7, one of these dependencies is illustrated. *FailedAttempts* is one of the dimension used to quantify *reliability* while the *downTime* is a metric used to measure the value of the *availability*. The dependency appears on dimension level with the downTime inducing the number of failedAttempts. Moreover, as exposed in the instantiation of the submodel, this dependency is subject to a particular condition on the availability value.

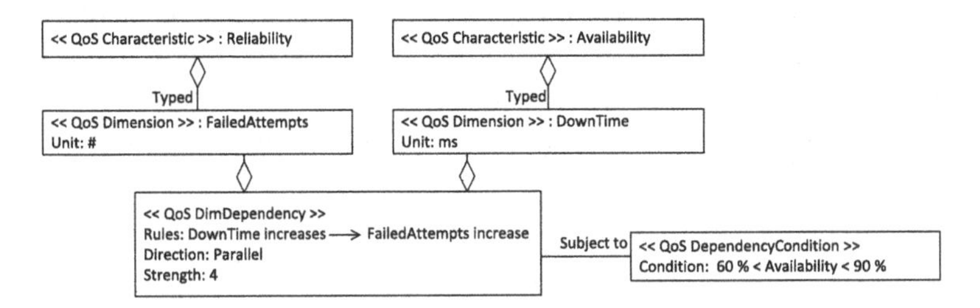

Fig. 7. An illustration of QoS Dependencies submodel utilization

4 QoS Driven Selection

Our aim is to provide a service selection approach that takes existing relationships identified over QoS Characteristics into consideration. In order to sort alternative services, we will use a particular class of methods of Multi-Criteria Decision Analysis (MCDA) [7]: the fuzzy MCDA. Fuzzy MCDA allows to establish a ranking over alternatives while accounting for multiple criteria, represented here by QoS Characteristics. Moreover, this technique introduces interaction indexes, able to express relationships over QoS Characteristics. The first step for the service selector is to build a reference set of alternatives that makes appear existing relationships over QoS Characteristics. Next, an algorithm is executed on the reference set in order to fix interaction indexes. Finally, the service selector calculates the score of existing alternatives with these interaction indexes and their ranking is established. Below, we outline how to obtain these indexes and how the ranking of available services is established with them.

In subsection 4.1, we introduce fuzzy MCDA concepts and explain how the utility value of each alternative can be calculated. In subsection 4.2, we describe how weights of criteria and interacting criteria are determined from a reference set established by the service selector and how provided services are then ranked. Finally, we present an example of utilization of our service selection approach in subsection 4.3.

4.1 Fuzzy Multiple Criteria Decision Analysis: Concepts

Fuzzy MCDA methods [7] are of particular relevance in services selection because these allow to determine weights related to criteria but also weights related to interactions between criteria. Indeed, in case of interacting criteria, the usual weighted arithmetic mean ($U(x^a) = \sum_{i=1}^n w_i x_i^a$) is extended with a Choquet integral such that the utility of alternative a is calculated by $U(x^a) = \sum_{i=1}^n x_{(i)}^a \left[\mu(A_{(i)}) - \mu(A_{(i+1)})\right]$. With $a \in A$ which is the set of all possible alternatives and n as the total number of criteria considered. We can observe that the weights w_i which are considered as independent in the usual weighted arithmetic mean have been substituted by the weights $\mu(i_1, ..., i_k)$ in the extended mean. These weights, related to all possible combinations of criteria, make possible to express dependencies between criteria. For complete description of fuzzy measures, Shapley's values and Choquet integrals, see Marichal [15,16] and cited references.

The overall importance of a criterion $i \in N$ is not solely determined by the weight $\mu(i)$ but also by all $\mu(S)$ such that $i \in S$, S being a subset of criteria related to the same subject. The importance index (Shapley value) of criterion i w.r.t. μ is defined by its Shapley's value, as in Equation 1.

$$\phi_{Sh}(i) = \sum_{T \subseteq N \backslash i} \frac{(n-t-1)!t!}{n!} \left[\mu(T \cup i) - \mu(T)\right] \tag{1}$$

To focus on interaction among subsets of criteria, the difference $a(ij) = \mu(ij) - \mu(i) - \mu(j)$ is used. The difference is 0 when the individual importances $\mu(i)$ and $\mu(j)$ add up without interfering. In this case, there is no interaction between criterion i and criterion j. If the criteria interfere in a positive way, the difference is positive and the difference is negative in case of overlap effect between i and j. The interaction indexes of criteria i and j are defined by Equation 2.

$$I(ij) = \sum_{T \subseteq N \backslash ij} \frac{(n-t-2)!t!}{(n-1)!} \left[\mu(T \cup ij) - \mu(T \cup i) - \mu(T \cup j) + \mu(T)\right] \tag{2}$$

With interaction indexes, a problem involving n criteria will require 2^n coefficients. As the user is not able to specify a such amount of information, we can confine ourself to the 2-order case that permits to model interactions between criteria while remaining simple. Only $n(n+1)/2$ coefficients are then required to define the fuzzy measure. Moreover, in a QoS based selection approach, interactions among more than two quality properties are difficulty interpretable. The coefficients are given by $\mu(i) = a(i)$, the interacting coefficients by $\mu(ij) = a(i) + a(j) + a(ij)$, $i, j \subseteq N$ and the Choquet integral of the utility of alternative x becomes $C_\mu(x) = \sum_{i \in N} a(i)x_i + \sum_{i,j \subseteq N} a(ij)(x_i \wedge x_j), x \in \mathbb{R}^n$.

4.2 Building the Reference Set and Ranking of Alternative Services

The main step in our selection approach is to derive weights of interacting criteria to apply them to existing service alternatives. These are computed on the basis of a reference set. The reference set is build by the service selector which refers on relationships information provided by the user and the provider QoS models. It consists of fictitious service alternatives and their respective QoS performances ranked with a partial order. Service QoS performances must be expressed on the same scale [7]. Indeed, QoS values are usually stated with different units, according to their respective type or modality. Some QoS Characteristics are defined in percent, others in levels and some in time unit. Moreover, some tend to be minimized while other should be maximized, in accordance with their requester QoS Preferences specification. In the aim to consider all properties on the same scale [9], a preparatory conversion must be made. This conversion consists of:

- **Unifying the unit.** The first step is to choose a common unit to all QoS considered. E.g.: the marks will be attributed on a 20 mark or in percent.
- **Setting the modality.** All quality attributes must be optimized on the same modality, i.e.: increasing or decreasing. If we choose to maximize all properties, attributes that are usually minimized (e.g.: latency, cost) will inverse their marks. E.g.: a 100 mark for the latency is the quickest latency possible.
- **Scaling of quality attributes.** The last element to consider is the scale, all attributes need to be expressed on the same basis. This basis is specified by the unit chosen, properties not directly expressible on this unit must be transformed. E.g.: if the quality property has a level unit (i.e.: as the security), the transformation function is $actual\ value \times \frac{best\ mark}{max\ value}$. If the property is expressed with a time value like the latency, the transformation function is $1 - \frac{actual\ value - min\ value}{max\ value - min\ value}$ expressed with the chosen unit.

In addition to a partial ranking of service alternatives, the reference set contains specific informations. These informations are detailed here:

- **Importance of criteria.** The relative importance of criteria in the service selection approach are compared to the priorities fixed over QoS Characteristics of each service. As these priorities have been established by the user with help of its QoS model, the strength of priorities can be integrated in the reference set with values used. It is also possible to bind these priorities to particular conditions by making them appear in the ranking of service alternatives provided by the service selector in the reference set.
- **Interaction between criteria.** This information refers in our selection approach to the dependencies specified by the provider with QoS Dependencies that appear between QoS Characteristics. These appear in the reference set established by the service selector. Theirs strengths and theirs directions can easily be expressed in the initial data of the reference set. Likewise, binded conditions can be included added to the reference set.
- **Symmetric criteria.** Symmetric criteria refer to criteria that can be exchanged without changing the aggregation mode. Characteristics belonging to the same QoS Category may sometimes appear as being substitutable, a poor performance in a parameter being compensated by good results in another. Such information needs to be explicitly attached to the parameters of the reference set.

Once all identified relationships among QoS Characteristics appear in the reference set, its corresponding Choquet integrals may be computed thanks to algorithm specified in [15,16]. Next, to establish Shapley's Value and interaction indexes, linear programming is made on Choquet integrals. Once these weights are fixed, these can be used to determine the performance of services made available by providers. The service selector restricts available services to those that satisfy user functional expectations and constraints on non-functional requirements. Their QoS score need to have been previously scaled as those of reference set alternatives. The ranking of available services is given by the sorting of their respective performance that provides the best available service satisfying user requirements.

4.3 Motivating Example

To illustrate clearly contributions and advantages of our approach, we refer to the Example 1 illustrating the utilization of the Priority submodel. Values provided by the service selector to compose the reference set are available on Table 1.

Table 1. Example: reference set

Alternative	Availability	Latency	Reliability
a	85	90	90
b	90	85	90
c	90	85	80
d	85	90	80

The ranking of alternatives constituent the reference set is given by: $a \succ b \succ c \succ d$. $b \succ c$ and $a \succ d$ are evident preferences. $a \succ b$ and $c \succ d$ are consequences of the priority condition expressed in Example 1 specifying that the availability is more important than the latency if the reliability is under 85%. Similarly, the latency will be favored to availability while reliability is above 85%.

Shapley's values and interaction indexes obtained with help of linear programming on Choquet Integrals associated to the reference set are proposed in Table 2. δ fixes the indifference threshold and has been set to 0.2.

Table 2. Example: Shapley's values and interaction indexes

Quality property	Shapley's value
Availability	0.25
Latency	0.25
Reliability	0.5

	Latency	Reliability
Availability	0	-0.5
Latency	-	0.5

The respective performance of alternatives of the reference set are given in Table 3.

Once Shapley's values and interaction indexes are known, these can be used to calculate quality score of providers' alternatives. Table 3 provides quality properties of available services and their respective scores. The best alternative is the service e which proposes a score of 89.50.

Table 3. Example: Scores of reference set's alternatives and available services and their respective performances

Alternative	Score
a	90.0
b	87.5
c	85.0
d	82.5

Alternative	Availability	Latency	Reliability	Score
e	77	87	92	89.50
f	94	78	85	76.00
g	75	87	91	89.00
h	78	94	87	87.00
i	97	86	78	87.50

The user of the Meris MGVI service will now consider four characteristics rather than three. To this aim, he adds to its specification the QoS Characteristic *integrity* and he specializes this characteristic as belonging to the same QoS Category than the reliability as illustrated in Figure 8. When characteristics are in the same category, these may be substitutable, compensating one characteristic with a poor performance by another with good results.

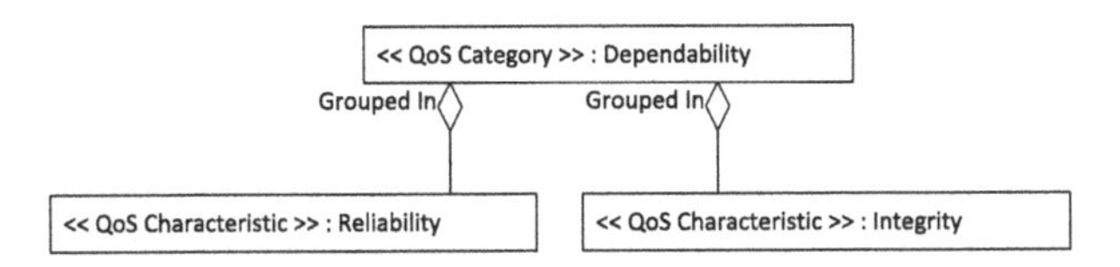

Fig. 8. Reliability and Integrity belonging to the same QoS Category

Table 4. Example: modified reference set

Alternative	Availability	Latency	Reliability	Integrity
a	85	90	90	80
b	90	85	90	80
c	90	85	80	80
d	85	90	80	80

The reference set is modified in Table 4 to account for the integrity characteristic. The indifference threshold has been fixed to 0.01 and the interaction threshold to 0.05.

The Shapley's value and interaction indexes obtained with the modified reference set are given in Table 5.

Table 5. Example: Modified Shapley's values and interaction indexes

Quality property	Shapley's value
Availability	0.2815
Latency	0.2395
Reliability	0.2395
Integrity	0.2395

	Latency	Reliability	Integrity
Availability	0.05	0.05	0.378998
Latency	-	0.378998	0.05
Reliability		-	-0.05

The performance of alternatives of the modified reference set and those of alternatives advertised by providers are given in Table 6. The best available service in response to user's specification is now the service f with a score of 84.60.

Table 6. Example: Scores of reference set's alternatives and available services and their respective performances

Alternative	Score
a	85.00
b	83.56
c	80.67
d	80.46

Alternative	Availability	Latency	Reliability	Integrity	Score
e	77	87	92	75	80.69
f	94	78	85	91	84.60
g	75	87	91	86	80.90
h	78	94	87	79	81.91
i	97	86	78	84	82.07

This example illustrates the possibilities given by fuzzy MCDA to service selection. The reference set is built on user non-functional requirements and fixed weights reflect its expectations. The obtained ranking of available services is totally lead by relationships identified over QoS Characteristics.

5 Related Work

In this section, we overview some existing models in subsection 5.1 and we place our selection approach in existing work in subsection 5.2.

5.1 QoS Models

In attempts to define formally non-functional properties of services, different QoS models have been proposed. Some introduce a description of some concepts [14,30] while others provide a complete definition of modeling constructs with a linked XML specification [6,31]. To reach a compromise between a natural conceptualization and an exhaustive formal definition of non-functional characteristics, some authors have used the Unified Modeling Language (UML) to enable QoS modeling [1,2,11,21,25]. Among them, the Object Management Group (OMG) outlines in [21] a standard, made of UML Profile extensions, to model quality of services. Our definition of two distinct models to, respectively, user and provider and their respective relationships are based on the standard proposed by the OMG.

This standard and most QoS model propositions [14,30] highlight a clear distinction of QoS characteristics and their quantification. We allow to define quality relationships at characteristics level as at their quantitative calculation level to benefit from this separation.

Our identified relationships over QoS characteristics are identified in some models:

- The preference relationship is introduced in most models with help of a direction attribute [10,14,20,27,30] indicating if a QoS characteristic has to be maximized or minimized;
- The priority relationship is defined with means of a weight attribute associated to QoS characteristics [14,30];
- The dependency relationship has only been summarily addressed in some QoS models [17,21] without specific attribute.

However, none of the cited models provides QoS constructs needed to account for all of the considerations defined in our QoS models. We have shown the relevance of these considerations, as they are needed in comparing competing services and subsequently selecting the most appropriate ones. Although some of the cited models define constructs, some of which are intended for service providers, others for service requesters, we separate the two perspectives and thereby make a clear and explicit distinction between the models of the two parties.

5.2 QoS Driven Selection

The aim of service selection is to affect the most suitable service to each service request. If the selection is based on non-functional considerations [5], in the services instance, it will result on a matching of providers QoS capabilities and requesters QoS expectations. Different techniques have been proposed to process this match, among them: multi-criteria decision making [20,26,30]; fuzzy MCDA [27]; heuristics [10]; Euclidean distances [14] and; reputation models [17]. Some propositions combine several techniques as Vu et al. [28] whose use data-mining with a reputation model and multi-criteria decision making. Some of these propositions are integrated in larger approaches that compose Web services [8,10,30] while others are confined to the definition of the most adapted service to user requirements.

Existing service selection approaches do not account for relationships over QoS Characteristics. However, the priority relationship is threated by means of weighting in some models [26,28,30]. These weights allow to indicate the relative importance of each QoS Characteristic considered during the selection step [8]. Fuzzy values are used in [19,29] to fix weights a posteriori, responding to the decision maker uncertainty. Our proposed selection approach relying on fuzzy MCDA allow for the integration of all identified relationships over QoS Characteristics. This way, the ranking established with our technique benefits from advanced concepts specified by providers and requesters.

6 Conclusions and Future Work

In a Service-Oriented System (SOS), service requesters specify tasks that need to be executed and the quality levels to meet, whereas service providers advertise their services' capabilities and the quality levels they can reach. Selecting appropriate services among competing ones requires rich QoS models for describing services' QoS dimensions and characteristics, as this information is subsequently needed to inform comparison and decision-making during selection. We introduce an approach that features complete QoS models and a service selection method that uses the QoS information available in the models. The approach consists of: (i) rich QoS models to be used by service requesters when expressing QoS expectations and service providers when describing services' QoS, and for representing preference and priority relationships between QoS dimensions; and (ii) a multi-criteria decision making technique that uses the models for service selection. The approach therefore allows us to deal with tradeoffs through priorities, and to account for stakeholders' preferences over values of QoS dimensions and characteristics.

Future effort will be focused on the automation of the approach. Namely, automated matching of provider and requester QoS model instances will be proposed in addition to an algorithm for building the reference set while accounting for existing QoS relationships.

Acknowledgments

We are grateful to Emmanuel Mathot of the European Space Agency, who provided precise information about the ESA Earth observation program and assisted our efforts in describing quality information and requirements of services related to this program.

References

1. Aagedal, J.O., Ecklund Jr., E.F.: Modelling QoS: Towards a UML Profile. In: Jézéquel, J.-M., Hussmann, H., Cook, S. (eds.) UML 2002. LNCS, vol. 2460, pp. 275–289. Springer, Heidelberg (2002)
2. Asensio, J.I., Villagra, V.A., Lopez de Vergana, J.E., Berrocal, J.J.: UML Profiles for the Specification and Instrumentation of QoS Management Information In Distributed Object-based Applications. In: Proceedings of the fifth world multi-conference on systemics, cybernetics and informatics, pp. 22–25 (2002)
3. Benetallah, B., Casati, F.: Special Issue on Web Services. Distributed and Parallel Databases 12, 115–116 (2002)
4. Casati, F., Castellanos, M., Dayal, U., Shan, M.C.: Probabilistic, context-sensitive, and goal-oriented services selection. In: ICSOC 2004: Proceedings of the 2nd international conference on Service oriented computing (2004)
5. Chung, L., Nixon, B.A., Yu, E., Mylopoulos, J.: Non-Functional Requirements in Software Engineering. The Kluwer International Series in Software Engineering, vol. 5. Springer, Heidelberg (1999)
6. D'Ambrogio, A.: A model-driven WSDL Extension for Describing the QoS of Web Services. In: Proceedings of the International Conference on Web Services (ICWS 2006) (2006)
7. Figueira, J., Greco, S., Ehrgott, M.: Multiple Criteria Decision Analysis: State of the Art Surveys. Springer, Heidelberg (2005)
8. Gu, X., Nahrstedt, K.: A Scalable QoS-Aware Service Aggregation Model for Peer-to-Peer Computing Grids. In: HPDC 2002: Proceedings of the 11th IEEE International Symposium on High Performance Distributed Computing (2002)
9. Hwang, C.L., Yoon, K.: Multi-Attribute Decision Making: Methods and Applications. Springer, Heidelberg (1981)
10. Jaeger, M.C., Rojec-Goldmann, G., Mühl, G.: QoS Aggregation for Web Service Composition using Workflow Patterns. In: EDOC 2004: Proceedings of the Enterprise Distributed Object Computing Conference, Eighth IEEE International (2004)
11. Jureta, I.J., Herssens, C., Faulkner, S.: A Comprehensive Quality Model for Service-Oriented Systems. Software Quality Journal (accepted for publication), `http://www.jureta.net/papers/QVDPdraft.pdf`
12. Kalepu, S., Krishnaswamy, S., Loke, S.W.: Verity: A QoS Metric for Selecting Web Services and Providers. In: WISEW 2003: Proceedings of the fourth International Conference on Web Information Systems Engineering Workshops (2003)

13. Keller, A., Ludwig, H.: The WSLA Framework: Specifying and Monitoring Service Level Agreements for Web Services. Journal of Network Systems Management 11(1) (2003)
14. Liu, Y., Ngu, A.H., Zeng, L.Z.: QoS computation and policing in dynamic web services selection. In: WWW Alt. 2004: Proceedings of the 13th international World Wide Web conference on Alternate track papers & posters (2004)
15. Marichal, J.-L., Roubens, M.: Determination of weights of interacting criteria from a reference set. European Journal of Operational Research 124, 641–650 (2000)
16. Marichal, J.-L.: Aggregation of interacting criteria by means of the discrete Choquet integral. Studies in Fuzziness and Soft. Computing 97, 224–244 (2002)
17. Maximilien, E.M., Singh, M.P.: Toward autonomic services trust and selection. In: ICSOC 2004: Proceedings of the International Conference on Service-Oriented Computing (2004)
18. Menascé, D.A.: QoS Issues in Web Services. IEEE Internet Computing 6(6), 72–75 (2002)
19. Mikhailov, L., Tsvetinov, P.: Fuzzy Approach to Outsourcing of Information Technology Services. In: SAC 2005: Proceedings of ACM Symposium on Applied Computing (2005)
20. Naumann, F., Freytag, J.C., Leser, U.: Quality-driven Integration of Heterogeneous Information Systems. In: Proceedings of th 25th VLDB Conference (1999)
21. The Object Management Group. UML Profile for Modeling Quality of Service and Fault Tolerance Characteristics and Mechanisms. Adopted Specification (2006)
22. O'Sullivan, J., Edmond, D., Ter Hofstede, A.: What's in a Service? Towards accurate description of non-functional service properties. Distrib. Parallel Databases 12(2-3), 117–133 (2002)
23. Papazoglou, M.P., Georgakopoulos, D.: Service-Oriented Computing. Communications of the ACM 46(10), 25–28 (2003)
24. Ran, S.: A Model for Web Services Discovery with QoS. ACM Sigecom exchanges (2003)
25. Salazar-Zarate, G., Botella, P.: Use of UML for modeling non-functional aspects. In: Proceedings of the International Conference on Software and Systems Engineering and their Applications (ICSSEA 2000) (2000)
26. Shaikh, S.E., Mehandjiev, N.: Multi-Attribute Negotiation in E-Business Process Composition. In: WETICE 2004: Proceedings of the 13th IEEE International Workshops on Enabling Technologies: Infrastructure for Collaborative Enterprises (2004)
27. Tong, H., Zhang, S.: A Fuzzy Multi-attribute Decision Making Algorithm for Web Services Selection Based on QoS. In: APSCC 2006: Proceedings of the IEEE Asia-Pacific Conference on Services Computing (2006)
28. Vu, L.-H., Hauswirth, M., Aberer, K.: QoS-Based Service Selection and Ranking with Trust and Reputation Management. In: Proceedings of the 13th International Conference On Cooperative Information Systems (CoopIS 2005) (2005)
29. Xiong, P., Fan, Y.: QoS-aware Web Services Selection by a Synthetic Weight. In: Proceedings of the International Conference on Fuzzy Systems and Knowledge Discovery (2007)
30. Zeng, L., Benatallah, B., Ngu, A.H.H., Dumas, M., Kalagnanam, J., Chang, H.: QoS-Aware Middleware for Web services composition. IEEE Trans. Softw. Eng. 30(5), 311–327 (2004)
31. Zhou, C., Chia, L.-T., Lee, B.-S.: Daml-qos ontology for web services. In: ICWS 2004: Proceedings of the International Conference on Web Services (2004)

KAF: Kalman Filter Based Adaptive Maintenance for Dependability of Composite Services

Huipeng Guo, Jinpeng Huai, Yang Li, and Ting Deng

School of Computer Science and Engineering, Bei Hang University, Beijing
guohp@act.buaa.edu.cn, huaijp@buaa.edu.cn,
{liyang,dengting}@act.buaa.edu.cn

Abstract. Service composition is fundamental in development of Web service oriented applications. Dependability of composite services is of significant importance since it directly impacts users' experience. However, dependability of a composite service may change over time as a result of inevitable changes in component services. In addition, users may also pose varying dependability requirements to meet different needs. It has become a big challenge to dynamically maintain the dependability of composite services. This paper proposes an innovative system called KAF that constructs a closed-loop control for adaptive maintenance of composite services. Modeling the control process as a Markov decision process (MDP), we further design an efficient Kalman-Filter based algorithm for service state prediction. With the availability of the precise prediction, optimal control decisions can be made. We evaluate the performance of KAF against other alternative approaches through comprehensive experiments and results demonstrate that KAF is capable for adaptive dependability maintenance.

Keywords: service composition, dependability, adaptive maintenance, Kalman filter.

1 Introduction

With the maturity of key standards such as SOAP, WSDL, and UDDI, Web service has been widely recognized as a promising technology for distributed application development. Following the service-oriented architecture, Web service supports better interoperability, higher usability, and increased reusability compared to traditional middleware technologies such as RMI and CORBA. Service composition is the fundamental process for Web service based application development, which constructs a composite service by composing basic Web services that may be distributed over the world. As this trend continues, more and more composite services will be deployed in the next few years, serving various sectors of our society.

The dependability property of composite services is of significant importance to users, which includes many critical factors, such as availability, reliability and so on. A service with higher availability promises more functional time and a more reliable service reduces the probability to fail when it is invoked. The trust property of a service is also important since a user may risk invoking a malicious service.

Z. Bellahsène and M. Léonard (Eds.): CAiSE 2008, LNCS 5074, pp. 328–342, 2008.

A key issue to service composition is that the dependability of a composite service may change dynamically over time. Users however, desire that the composite service delivers stable dependability. A stable dependability makes it possible for the users to expect the predictable results. This is particularly important when the application is mission-critical, e.g., a disaster management. When the dependability of the composite service degrades, the users may severely suffer from the decreased performance.

The intrinsic dynamics of composite services stems from the fact that a composite service is composed of many component services that potentially belong to different providers distributed in the open Internet. Each component service is subject to different environments and changes, such as varying system load and available bandwidth. The properties of such a service may therefore change from time to time. In addition, existing component services may become unavailable or new services may appear to serve. All these factors lead to the inevitable dynamic changes of the composite service. On the other hand, the dependability posed by the users or the applications may evolve to meet different needs in reality.

To obtain stable composite services with desirable dependability, the system must provide adaptive control to offset the unpredictable changes in either environments or application requirements. The Web service and service composition has been studied and several attempts have been made to construct high dependable composite services [1-4]. However, they fail to support the adaptive maintenance of dependability. To the best of our knowledge, there is no successful work for this problem.

In this paper we propose an innovative system called KAF to dynamically maintain the dependability of a composite service. KAF constructs a closed-loop control for the maintenance. By continuously monitoring the system status, KAF makes progressive control over the composition of the component service. We model this as a Markov decision process (MDP) that facilitates control decision making. Since it is highly desirable to maintain the dependability of the composite service over the one required by the users, it becomes imperative to estimate the next status of the service. Exploiting the MDP modeling, we propose a Kalman-Filter Based adaptive maintenance algorithm that precisely predicts and adjusts the service state.

We have made the following original contributions in this paper.

- We make the attempt to explore the important issue of adaptive maintenance of composite services. The innovative system KAF is proposed which implements a closed-loop control over service dependability.
- By modeling the control process as a MDP, we design a Kalman-Filter based algorithm that accurately predicts the state of the composite service such that optimal control decisions are thereafter made.
- We have preliminarily implemented KAF and conducted comprehensive trace-driven experiments. The performance of KAF is studied comparatively against other alternative approaches.

The rest of this paper is organized as follows. Section 2 introduces the background and problem. Section 3 states the KAF architecture. Section 4 describes the adaptive control method of service composition dependability maintenance. Adaptive control mechanism implement of service composition dependability is discussed in Section 5. In Section 6 we evaluate the proposed approach. And we review the related work in section 7. At last, conclusions and future work are presented in Sections 8.

2 Background and Problem Statement

In this paper, we model a composite service as a tuple (*ST*, *SC*), where *ST* is a collection of tasks combined according to the four basic service composition modes including *sequence* (•), *parallel* (||), *choice* (+) and *iteration* (∘), and *SC* is the set of component services for each task in *ST*.

There are many attributes associated with a Web service. In this paper we particularly focus on four key attributes for component services: *availability*, *reliability*, *trust* and *price*. The approach can support more flexible definition of QoS other than these attributes. As discussed before, dependability can be considered as an aggregate property of availability, reliability and trust [4-10]. These attributes of services reflect from different perspectives the ability of a service providing a specific functionality. Price directly indicates the cost that a user has to pay and thus is indispensable to be included for study in a practical setting. Therefore, each component service can be represented by a tuple ($A(t)$, $R(t)$, $T(t)$, $P(t)$), where $A(t)$, $R(t)$, $T(t)$ and $P(t)$ represent the numerical values of availability, reliability, trust and price at time t, respectively.

Service availability is defined as the probability that a service is properly functioning in a specified period of time under certain conditions. According to [10], service availability can be measured as:

$$A\left(s_{ij},t\right)=\frac{T_s\left(t\right)}{T_a\left(t\right)}\ , \tag{1}$$

where $A(s_{ij}, t)$ denotes the availability of service s_{ij} in time t; $T_s(t)$ is the total available time and $T_a(t)$ is the total measurement time.

Service reliability is the ability of executing the specific functionality under certain conditions and within a particular period. And it can be measured by the ratio of the unsuccessful execution number and the total number of executing:

$$R\left(s_{ij},t\right)=\frac{N_s\left(t\right)}{N_a\left(t\right)}\ , \tag{2}$$

where $R(s_{ij}, t)$ denotes service reliability; $N_s(t)$ is the number of the successful execution instances, $N_a(t)$ is the total number of execution.

How to calculate the trust value is a hot research topic, and there are many works [6-8]. Generally, there are two types of trust value: direct trust and recommendation trust. Based on Beth's works [6], taking into account practicability and measurability, we describe trust value of the service mainly with recommendation trust. In this paper we assume that the trust of a Web service is based on recommendations of the service information manager (SIM), which collect the feedback results of some selected honest sampling clients. In addition, other trust value calculation methods [7, 8] can also be supported.

According to [4], the availability of a composite service with redundant services can be computed given the availability information of its component services. Similarly, we can also compute the reliability and the trust of the composite services. Given the three attributes computed as above, we are able to compute the dependability d as a function of them.

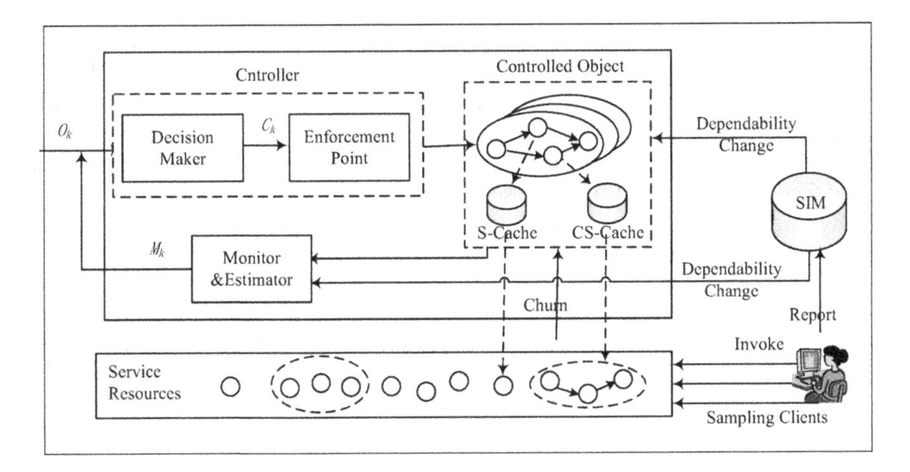

Fig. 1. KAF architecture

$$d = \alpha a + \beta r + (1 - \alpha - \beta)\tau, \ \alpha, \beta, a, r, \tau, d \in [0,1]. \tag{3}$$

Hereinto, a, r and τ are the availability, the reliability and the trust of a composite service in a certain period, respectively. Note that α, β and $(1-\alpha-\beta)$ are the weights of availability, reliability and trust. The assignment of weights can be determined by the users to reflect different emphasis on these attributes.

On the other hand, we take into account maintenance costs of a composite service. In this paper, based on our previous work [4], the attribute can be dynamically adjusted based on the instrument of redundancy provision. It is intuitive that more services are used, higher cost are introduced.

The objective of the maintenance for composite services is as follows. For a given composite service (*ST*, *SC*), and the given desired dependability (denoted as *O*), we need to keep the instant dependability of the composite service $d(t)$ as close to *O* as possible for any time *t*. Meanwhile, we should minimize the cost of the composite as much as possible over time.

3 KAF Architecture

In this section we first introduce the system architecture of KAF and then describe the modeling of the decision making process for dynamic maintenance in KAF.

3.1 System Architecture

The KAF Architecture is shown in Figure 1, which consists of three main components. The Monitor and Estimator component monitors the current attributes of all services and also estimate their further states based on the current and historical information. The information provides the basis to the Decision Maker that produces maintenance strategy. The decision output is feed into Enforcement Point that implements the decision and dynamically adjusts the composition of the composite service.

In the KAF, composite service based on redundancy, including component services cache named S-Cache and composite services cache named CS-Cache, is the controlled object. Guarantee strategy of composite services is the corresponding controller and is performed by Decision Maker and Enforcement Point. Monitor& Estimator, SIM and sampling clients form a feedback loop. Composite services, guarantee strategy and feedback loop constitute a closed-loop feedback control system.

In figure 1, O_k ($0 \leq k \leq N$) are the desired value of the dependability in a composite service system, represents dependable attributes of a composite service. M_k ($0 \leq k \leq N$) are feedback values of a composite service, denote actual values of dependability attribute. M_k are estimated by Estimator based on summarization of SIM from the report of sampling clients. C_k ($0 \leq k \leq N$) are control values and are determined by the difference between expectations and estimating value. C_k denote the dependability value needed to be adjusted. Based on adjustment strategies, the Enforcement Point choose candidate services and complete the reconstruction of services composition.

In KAF architecture, the feedback loop is composed of two mechanisms: execution monitoring and reporting. Execution monitoring gains the services available state information from the execute engine by the monitor. Reporting mechanism summarizes the using result of sampling clients and sampling clients are some honest client nodes selected by SIM. Estimator estimates dependability of component service base on the results of SIM and then computes dependability of composite service.

SIM is an extended service registry supporting the description and update of QoS information. SIM is responsible for the reporting mechanism, collecting and managing the information of service. Some function of SIM includes service dependability value monitoring, service available state detection under different strategies (such as different frequencies, different incentive incentives, and so forth), evaluation and statistical functions of sampling client monitoring reports.

To adjust the dependability effectively, the Decision Maker determines service options and update strategy, such as increasing number of backup service, replacing service decreased most and so on. In this paper, guarantee strategy is a kind of meta-strategies. According to a meta-strategy, Enforcement Point runs the APB algorithm [4] selecting new service and replacing declining or failed services to implement the composite service's re-structure. In addition to the dependability of services, we must also consider the cost of replacement and rewards of expectation, so that to make the long-term revenue maximize. For the realization of this goal, in this paper we use Markov decision process theory to support strategy choices.

3.2 MDP Modeling

Due to the dynamic nature of component services, such as verity of load, network conditions and churn, the dependability of composite service will be affected. In these conditions, we need selecting some redundant service to ensure the composite service can still satisfying the desire even dependability of some component services drop.

We design dependability factor $f = lnO/lnD$, where O is the desired value for dependable properties need to satisfied and D means the design value of dependability. Dependability factor reflecting the important degree of dependability is used to guide the choice of component service. Generally f is larger than 1, which means D usually is larger than O. We can see that, when the dependability of component services declines rarely, low dependability factor can reduce the cost of constructing composite services. When dependability dropped significantly, increasing of dependability factor can improve the quality of composite service, and the cost may increase.

To adjust the dependability factor of composite service reasonably, we select MDP (Markov Decision Process)[11] to modeling the composite service maintenance process: RSC = $\langle S, A, \{A(x) \mid x \in S\}, Q, RV \rangle$, where S is the set of all possible states and A is the set of decision strategies, $\{A(x) \mid x \in S\}$ is the strategy when the state is x, $Q(B|x, a) = Q\{X_{t+1} \in B \mid X_t = x, A_t = a\}$ specifies the probability to the next state X_{t+1} when the state is $X_t = x$ and after strategy $A_t = a$ is executed, and RV is the reward of executing the strategy. If we assume that the state of system in moment t was $X_t = x \in S$, while the decision is $A_t = a \in A(x)$, under the decision of transfer function Q, system transfers to the state X_{t+1} with the reward of $RV(a,x)$. After the transfer, the system has entered a new state, and then choose a new decision to continue decision-making process. Therefore we establish Markov decision process model of component service dependability's maintenance as follows.

Definition 1. In a period of time, dependability of a composite service CS is $d = \alpha a + \beta r + (1 - \alpha - \beta)t$, and [0,1] is divided into k parts. When $d \in [\frac{i-1}{k}, \frac{i}{k})$, we define that the state of the composite service is s_i, marked as $s_i = [\frac{i-1}{k}, \frac{i}{k})$. And the state set of composite service dependability is denoted by $S = \{s_1, \ldots, s_m\}$.

By dividing [0, 1], we can reduce the state space and reduce the complexity of decision-making in service composition.

Definition 2. Let Δf be the variety amount of dependability factor, We define the adjustment strategy set as $A = \{-\Delta f, 0, +\Delta f\}$, where the strategy $-\Delta f$ and $+\Delta f$ mean the amount of reduction or addition values of dependability factor respectively, and the strategy 0 means no changing to the dependability factor.

Definition 3. For a composite service CS, suppose that d, ψ, w and η represent the dependability, revenue, service adjustment cost and the importance degree of the CS, where η is a positive real number, we denote by $rv = \eta(-\ln(1-d)) + \psi - w$ the reward of composite service CS.

The definition 3 represents if composite service properties such as availability, reliability and trust are higher or the maintenance cost of composite service is smaller, we can get more reward. Then composite service developers will be more satisfied with the composite service, that is, the value of v will be larger. Among them, the income is related to the number of service execution and the prices of services.

Definition 4. For a composite service CS, at the moment t suppose that CS is in the state s_i, we denote by $rv = \eta(-\ln(1-d)) + \psi - w(A_t, s_i)$ the immediate reward of composite service CS after adopting strategy A_t, where $w(A_t, s_i)$ represents the cost of dependability adjustment.

While using Markov decision process, if the state transferred probability function and reward function are known, by dynamic programming methods, we can construct value function and achieve optimal decision strategy. However, in the service-oriented software development process, it is difficult to observe all the historical actions of component services. That means that transition probability function and reward function are unknown, we can not use dynamic programming techniques to determine optimal decision strategy. Hence we design Kalman-Filter based approach to estimate systemic state and choose optimal decision strategy.

4 KAF Adaptive Control

In the distributed and dynamic environment, we can not get the parameters of service state instantly and need estimate them online. So to adjust controller parameters and to keep the performance satisfying the design demand, we apply Kalman-Filter based approach to estimation service's state, then adjust controller following the estimated value.

4.1 Kalman Filter-Based Estimation

Kalman filter [12] uses the recurrence of the state equation to achieve the optimal estimate of state variables in the linear dynamic system. The Kalman filter is unbiased and has the smallest variance. It is easy to realize by computer and be suitable for on-line analysis. Furthermore, the extended Kalman filter (EKF) can be used in nonlinear systems. Therefore, this paper introduces the extended Kalman filter to estimating the dependability of component services in order to implement the adaptive control maintenance of composite service's dependability.

At moment *t*, the state of a service is a $x(t)$ and represent the dependability of the service. The state equation of the component service is:

$$x(t+1)=\Phi(t, x(t))+w(t) . \tag{4}$$

In this equation, Integer $t \geq 0$ is discrete-time variable, $x(t) \in R^{n\times 1}$ is the dependability of the component service. $\Phi(t, x(t))$is the nonlinear transition matrix of service state. We suppose that $w(t)$ is the disturbance to the dependability of the component service.

The measurement equation for the dependability of the component service is:

$$y(t)=C(t,x(t))+v(t) . \tag{5}$$

Here, $y(t) \in R^{m\times 1}(m \leq n)$ is the measurement value of the dependability. $C(t,x(t))$is measurement matrix. $v(t) \in R^{m\times 1}$ is the measurement noise.

And assume that $w(t)$, $v(t)$ and the starting states $x(0)$ have characteristic: $E\{w(t)\}=0$, $E\{v(t)\}=0$, $E\{x(0)\}=x_0$, $E\{w(t)w^T(j)\}=Q_i(t)\delta_{t,j}$, $E\{v(t)v^T(j)\}=P_i(t)\delta_{t,j}$, $E\{[x(0)-x_0][x(0)-x_0]^T\}=P_0$ and $E\{x(0)w^T(t)\}=0$, $E\{x(0)v^T(t)\}=0$, $E\{w(t)v^T(j)\}=0$.

$$\delta_{t,j}=\begin{cases}1 & t=j\\ 0 & t\neq j\end{cases}, \quad t,j \geq 0.$$

We get State transition matrix:

$$\Phi(t+1, t)=\frac{\partial\Phi(t,x)}{\partial x}\Big|_{x=\hat{x}(t|y_t)} . \tag{6}$$

Measurement matrix:

$$C(t) = \frac{\partial C(t,x)}{\partial x}\Big|_{x=\hat{x}(t|y_{t-1})} . \tag{7}$$

The observation values are *y(1),y(2),…,y(t)*, so optimal predict value of *x(t+1)* is:

$$\hat{x}(t+1 | y_t) = \Phi(t+1,\ t)\hat{x}(t | y_t) . \tag{8}$$

Suppose that we have gotten the optimal predictive value $\hat{x}(t|t\text{-}1)$ and the corresponding prediction error covariance *K(t|t-1)* before getting the dependability measure value *y(t)* of the component service. Then we can get the predict value $\hat{x}(t+1|t)$ based on the overall information and the corresponding prediction error covariance matrix *K(t+1|t)* at moment *t* by using Kalman filter and *y(t)*.

The Kalman gain is:

$$G(t) = K(t | t\text{-}1)C^T(t)[C(t)K(t | t\text{-}1)C^T(t) + Q_2(t)]^{-1} . \tag{9}$$

Assume innovation $\rho(t) = y(t) - C(t)\hat{x}(t | y_{t-1})$ and the optimal predictive value based on overall information of state *x(t)* , we can apply *ρ(t)* to modify the predictive value $\hat{x}(t+1 | t-1)$:

$$\hat{x}(t | y_t) = \hat{x}(t | y_{t-1}) + G(t)\rho(t) . \tag{10}$$

And the corresponding predict error for the covariance is:

$$K(t+1 | t) = \Phi(t+1,\ t)K(t)\Phi^T(t+1,\ t) + Q_1(t) . \tag{11}$$

$$K(t) = [I - G(t)C(t)]K(t,\ t-1) . \tag{12}$$

4.2 Adaptive Control Algorithm

According to the aforementioned analysis, we design the adaptive control algorithm for the dependability maintenance of the composite service. The basic idea is as follows. First we collect the sampling client's using report data about service execution, summarize the state and value of component services and calculate the dependability of composite services and compare with the expectation value. Then estimate the value of every selected component service by the Kalman filter formula, compute the immediate reward of every action and choose corresponding action following the MDP framework. At last we execute the APB algorithm [4] and the strategy execute module to select new service and replace some degenerate service.

The input of KAF adaptive control algorithm includes constructed composite service, available component services' properties measure value, expected dependable value D_k of composite service. The output includes dependability factor and selected services.

The first part of KAF algorithm (lines 1-4) is initialization, including read the expectation value O_k of composite services, collect parameter of component services

KAF adaptive control algorithm

input: constructed composite service, available component service properties measurement value, expected dependable value D_k of composite service.

output: dependability factor, selected services

1. Read the O_k, determine the weight of every attributes;
2. Read the initial dependable attribute values from SIM;
3. Calculate the dependable attribute values for each component service respectively;
4. Calculate d^*_k of composite service;
5. Collect sampling measure values M_k of component services;
6. Predict the value of the next period by Kalman filter following the measure values;
7. Compute the immediate reward according to MDP strategy;
8. Determine the action and modify the dependable factor α;
9. Calculate updated d_k ;
10. Calculate $\Delta d_k = d_k - d^*_k$; // Δd_k is the C_k
11. Select new service following the APB algorithm;
12. $d^*_k = d_k$;
13. **goto** step 5.

from SIM, calculate the dependable attribute values, and calculate real dependability of composite service.

Then obtain sampling measure values of component services, and according to Kalman filter calculation formulas estimate the values in the next period. Afterwards in the MDP, compute the immediate reward and determine the action of modifying the dependable factor α (line 5-8).

In the third part (lines 9-13), compute the dependable attribute verity values Δd_k and select new service following the APB algorithm [4]. At last, to measure the component services and start another prediction and control process.

5 KAF Implementation

The KAF platform provides environments where (i) composite service developers may create a new composite service and adaptive control the choice of candidate services, (ii) Monitor and SIM may get and validate the state and value verity of component services, and (iii) Estimator may predict the value of component services with the Kalman filter and then constructor implement the adaptive strategy of the controller. KAF is the extension of the WebSASE [24]. WebSASE supports Web service development, deployment, composition and is based on established standards such as SOAP, WSDL and BPEL. KAF is implemented in Java using IBM eclipse, and implement of middleware is based on the AXIS 2.0.

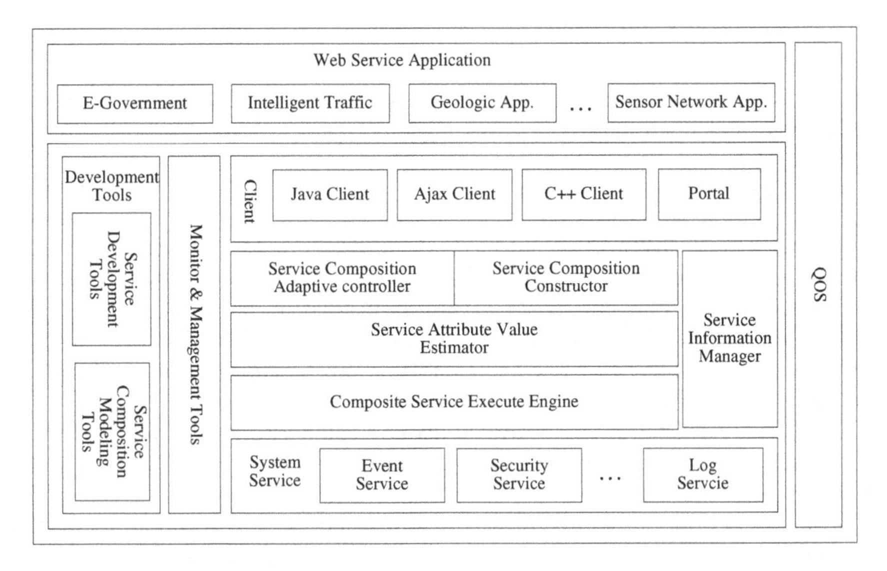

Fig. 2. The KAF implementation architecture

The KAF system introduces the notion of service composition adaptive control. The adaptive control is implemented by four main components: the *estimator*, the *controller*, the *constructor* and the *monitor*.

The *estimator* is responsible for estimating the value of services in the next period with the measure value of the currently period by the Kalman filter formula. The SIM is implemented as an extension to the UDDI to support the measure and evaluation of service quality. Through validation of the monitor, SIM chooses some honest clients as sampling nodes, collects the report of them and summarized as measure value.

According to dependability requirements and estimated value of the next period, the *controller* computes the immediate reward of every action and choosing corresponding action following the MDP framework. The amount of action to adjust the dependable factor can be edited through a visual interface by the composite service developer. The *controller* also determines actions to be taken in response to events such as service fail.

The *constructor* (Enforcement Point) constructs the composites according to component information and quality requirements from the model created by the modeling tools. Redundant services are selected and component service cache is constructed to store information of redundant components. Composite service cache is built and the BPEL documents of backup composites are created. Most important, the *constructor* modify the composite service through adding some more services or replace some services according to the adjustment of dependable factor using APB algorithms presented in [4].

The *monitor* performs the task of receiving the sampling measure value of services from SIM, receiving the event report from the execute engine and validating honesty of sampling clients. After measure value or other information is received, the *monitor* transmits these data to the *estimator* for the new estimation and control.

6 Performance Evaluation

In this section, we introduce experiments we conducted for performance evaluation and discuss evaluation results.

6.1 Simulation Setting and Metrics

To evaluate the effectiveness of our method, we design a set of simulation experiments to analysis the verity of dependability. First we generate a certain number of services and queries randomly. Each service is marked by its name, version, type, availability, reliability, trust and price etc, and these data are updated dynamically.

We adopt the following metrics and strategies: *dependability*, *cumulative reward*, *Random strategy* and *Kalman-Filter based adaptive control strategy*.

Dependability denotes an integrative ability of a composite service to provider certain functionality in a certain period. In this paper we select detailed properties including availability, reliability and trust.

Cumulative reward defines the reward of composite service execution in a certain period of time (Compute following definition 3. And in this paper the time is a day.).

Random strategy: If some component service in the composite service fails, then we select random another service to replace it.

Kalman-Filter based adaptive control strategy: In execution process of composite service, we use Kalman filter to estimate the dependability of the component services and adjust the dependable factor and then reconstruct a composite service.

Base on the above metrics and strategies, we design the adaptive control mechanism of composite service dependability, the execute process is shown in figure 3. The main function of SIM include service information registering and publishing, service available state monitoring, service dependability evaluation, service dependability properties analysis adjustment. The main function of Service Provider includes service information registering, service execution record, service information update adjustment. The main function of Composite Service Developer includes service discovery, service modeling, service construction, service execution, service execution evaluation and redundant service management.

6.2 Results Analysis

In our simulation, 3000 services are generated and distributed into selected nodes. 2.3 percent of the services fail every day, which is based on the observation of [13].

The first simulation is implemented to analyze the effect of adaptive control maintenance mechanism. 50 composite services is constructed and the dependability of execute for 100 days is recorded. The results in figure 4 show that dependability of composite services without backup services drops to a very low level after some days.

With the help of backup services, the dependability drops slowly. Results also show that relatively higher dependability can be achieved with the random replacement of failed service. With the help of adaptive maintenance mechanism based on KAF, the dependability stays steadily in a relatively high level.

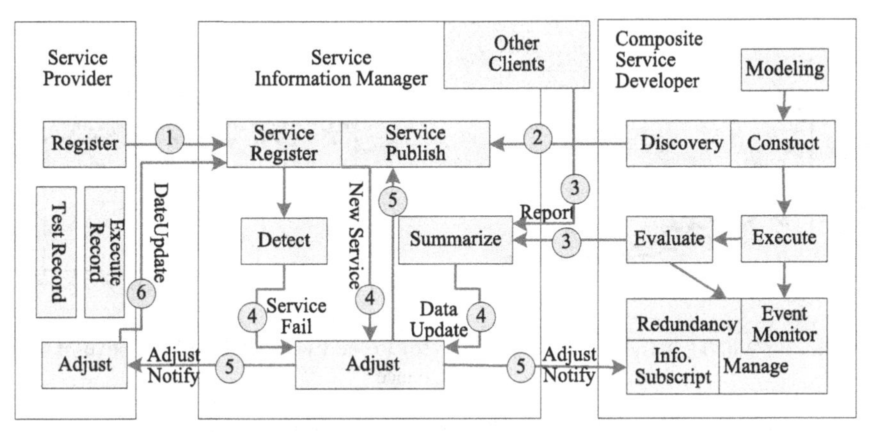

1. Service Registry, 2. Service Discovery, 3. User Report,
4. Service Data Update, New Service Register, Service Fail Message Exchange,
5. Service Information Adjust Notify, 6. Service Information Update

Fig. 3. Process of services' dependability information feedback

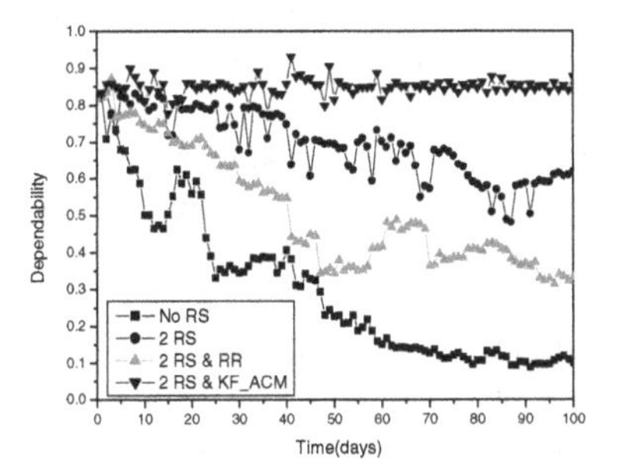

Fig. 4. Dependability maintain dynamically

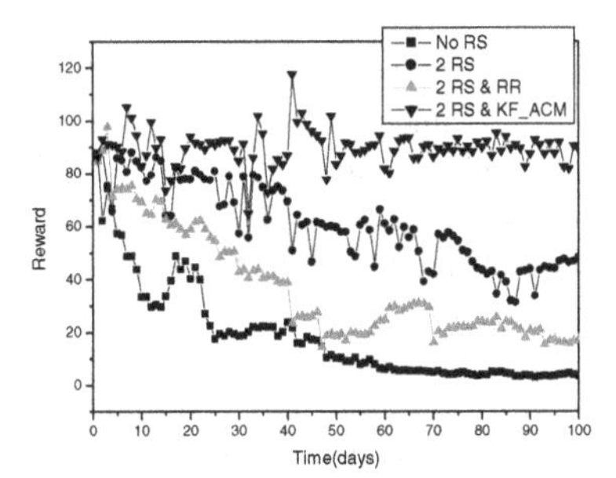

Fig. 5. Reward of maintain dynamically

The second experiment is to analyze the effect of adaptive maintenance mechanism on the Reward of composite service. The results in Figure 5 show that reward of composites without backup services drops to a very low level after some days. With the help of backup services, the reward drops slowly. Results also show that relatively higher dependability can be achieved with the random replacement of failed service. With the help of adaptive maintenance mechanism based on Kalman filter, the reward stays steadily in a relatively high level.

The third experiment is to analyze effect of dependability's important degree on the Reward of composite service. The results in Figure 6(a) show that cumulative reward of composites without backup services drops to a very low level after some days in different important degree. With the help of backup services and adaptive maintenance mechanism based on Kalman filter, the Cumulative reward stays steadily in a relatively high level. And the higher important degree is, the reward will be much.

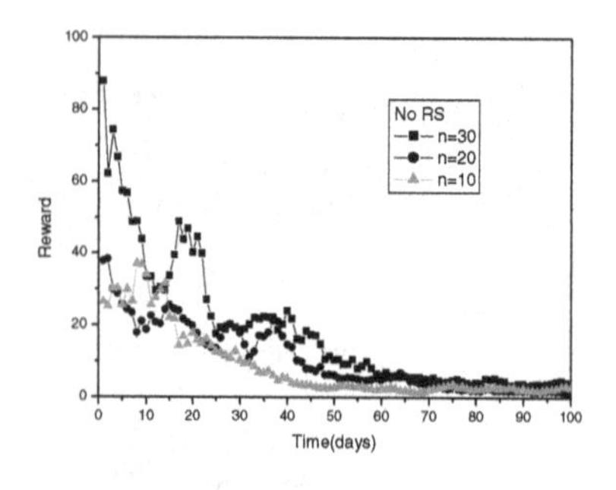

(a) Reward Without Adaptive Maintenance

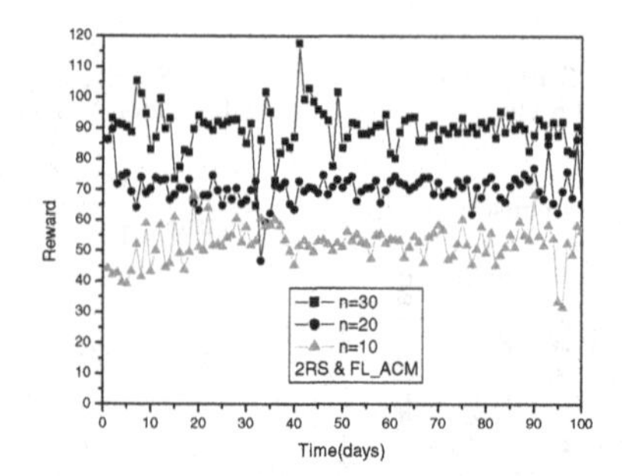

(b) Reward with KF-based Adaptive Maintenance

Fig. 6. Dependability's important degree vs. reward

7 Related Work

To achieve dependable composite service, many efforts have been made. Related work mainly involves dynamic service composition, dynamic adaptation technology, monitoring and recovery technology, evaluation, replication technology and so forth.

F. Casati and H. Sun et al. [14, 15, 16] proposed a dynamic service composition method, only required functions are defined in the definition period and component services bind and instantiate at runtime. Composite service need communicating with service registry dynamically, find necessary component services according to the pre-established strategies. Possible strategies include choice based on the QoS [9, 17], or based on the semantic [18] and so on. This method can improve the flexibility of composition and dynamic adaptability. But dynamically searching and several remote interactions with the service registry, the efficiency will be low. And the quality of services cannot be guaranteed because the existing service registries cannot guarantee the authenticity of data and the state of registered services.

J. Harney and P. Doshi et al. [1][19] proposed adaptive methods -VOC and VOC $^{\varepsilon}$. VOC method can avoid unnecessary inquiries and reduce overhead by calculating the potential value of changed information. VOC$^{\varepsilon}$ is the improvement of VOC. VOC$^{\varepsilon}$ only monitors the service out of expiration time so as to gain better efficiency. They mainly concerned achieving verify of attributes efficiently and economically. While in KAF we consider the adaptive adjustment of services selection to meeting desired dependability steadily and economically in a volatile environment.

S. Guinea et al. [20, 21] proposed a self-healing service composition method through service state monitoring and recovery of fails. F. Tartanoglu et al. [22] based on the concept of cooperative atomic action and web service composition action, proposed a forward error recovery method to achieve fault tolerance and dependable composite service. K. Birman, et al. [23] proposed an extended WS architecture to achieve the high dependability of services, to support the error monitoring, reliable message transferring through components such as health control, reliable message, event notification and so on. However, these monitoring and recovery technologies do not consider monitoring and recovery efficiency, costs and rewards, and also they do not assess the effect of the monitoring and recovery in quantitative forms.

P. Godfrey et al. [25] presented the definition and calculating method of Churn, compared and analyzed different failure node replacing strategy based on some real traces. X. Li and K. Nahrstedt [26] presents a service composition approach with QoS guarantee to minimize the interference. R. Jurca et al. [2] proposed a reliable service quality monitoring mechanism based on customers' feedback, reduce lying and anti-conspiracy.

In addition, J. Salas et al. [27] proposed a framework to increase service availability. X. Ye and Y. Shen [28] proposed a middleware supporting reliable web services based on active replication technology. These studies concern availability and reliability of service individuals and their methods are qualitative. While KAF considers the dependability of composite service and supports quantitative assessment.

8 Conclusions and Future Work

In this paper, we have studied the important problem of dynamic maintenance of desirable dependability of composite services. We have proposed KAF, a system that provides closed-loop feedback based control over the dependability of composite services. The adaptive dependability control over a composite service is modeled as a Markov Decision Process. To maintain the dependability constantly over a given desirable value, it is particularly imperative to predict the state of the composite service. We have designed a Kalman-Filter based algorithm for accurate estimation. Comprehensive experiments have been conducted and the results have demonstrated that KAF is able to achieve the dynamic maintenance of dependability. At the same time, the cost of maintenance is reduced significantly on the long run. As a result, KAF is viable and valuable for Web service based applications.

Although significant achievements have been made so far, there are several interesting directions in our future work. On the one hand, we will investigate more comprehensive and finer grained QoS models for composite services, e.g., to define parameters more specific to Web services in more flexible way. On the other hand, we will extend to richer composition model other than supporting the general mode of sequence, parallel, choice and iteration, e.g., more complex relationships, such as compensation and transaction, will be considered.

Acknowledgments. We would like to thank Yanmin Zhu of Imperial College London for his fruitful suggestions and kind help in the writing of this paper. We would also like to thank the anonymous reviewers for their constructive comments and suggestions. This research was supported in part by China NSF grants (No.90412011 and No.60525209), China 863 High-tech Programme (No.2006AA01Z19A) and China 973 Fundamental R&D Program (No.2005CB321803).

References

1. Harney, H., Doshi, P.: Speeding up Adaptation of Web Service Compositions Using Expiration Times. In: Proc. of WWW (2007)
2. Jurca, R., Binder, W., Faltings, B.: Reliable QoS Monitoring Based on Client Feedback. In: Proc. of WWW (2007)

3. Wu, K., David, J., et al.: The Applicability of Adaptive Control Theory to QoS Design: Limitations and Solutions. In: Proc. of IEEE IPDPS (2005)
4. Guo, H., Huai, J., et al.: ANGEL: Optimal Configuration for High Available Service Composition. In: Proc. of ICWS (2007)
5. Avizienis, A., Laprie, J., et al.: Basic Concepts and Taxonomy of Dependable and Secure Computing. IEEE Transaction on Dependable and Secure Computing 1(1), 11–33 (2004)
6. Beth, T., Borcherding, M., Klein, B.: Valuation of Trust in Open Networks. In: Proc. of Conference on Computer Security (1994)
7. Golbeck, J.: Computing and Applying Trust in Web-based Social Networks. PhD thesis, University of Maryland (2005)
8. Abdul-Rahman, A., Hailes, S.: A distributed trust model. In: Proc. of NSPW (1997)
9. Liu, Y., Ngu, A.H.H., Zeng, L.: QoS Computation and Policing in Dynamic Web Service Selection. In: Proc. of WWW (2004)
10. Ran, S.: A model for web services discovery with QoS. ACM SIGecom Exchanges (2003)
11. Puterman, M.: Markov Decision Processes: Discrete Stochastic Dynamic Programming. Wiley-Interscience, Chichester (1994)
12. Haykin, S.: Adaptive Filter Theory, 4th edn. Pearson Education, London (2002)
13. Kim, S., Rosu, M.: A survey of public web services. In: Proc. of WWW (2004)
14. Baresi, L., Guinea, S.: Towards Dynamic Web Services. In: Proc. of ICSE (2006)
15. Sun, H., et al.: Research and Implementation of Dynamic Web Services Composition. In: Zhou, X., Xu, M., Jähnichen, S., Cao, J. (eds.) APPT 2003. LNCS, vol. 2834, pp. 457–466. Springer, Heidelberg (2003)
16. Mennie, D., Pagurek, B.: An Architecture to Support Dynamic Composition of Service Components. In: Proc. of WCOP (2000)
17. Casati, F., et al.: eFlow: A Platform for Developing and Managing Composite e-Services. In: Proc. of AIWORC (2000)
18. Verma, K., et al.: METEOR–S WSDI: A Scalable P2P Infrastructure of Registries for Semantic Publication and Discovery of Web Services. Journal of Information Technology and management (2005)
19. Harney, J., Doshi, P.: Adaptive Web Processes Using Value of Changed Information. In: Dan, A., Lamersdorf, W. (eds.) ICSOC 2006. LNCS, vol. 4294, pp. 179–190. Springer, Heidelberg (2006)
20. Guinea, S.: Self-healing web service compositions. In: Proc. of ICSE (2005)
21. Baresi, L., et al.: Towards Self-healing Service Compositions. In: Proc. of PRISE (2004)
22. Tartanoglu, F., Issarny, V., et al.: Coordinated forward error recovery for composite Web services. In: Proc. of SRDS (2003)
23. Birman, K., Renesse, R., Vogels, W.: Adding High Availability and Autonomic Behavior to Web Services. In: Proc. of ICSE (2004)
24. Ge, S., et al.: WebSASE: A Web Service-based Application Supporting Environment. In: Proc. of the 5th Northeast Asia Symposium (2002)
25. Godfrey, P., et al.: Minimizing Churn in Distributed Systems. In: Proc. of SIGCOMM (2006)
26. Li, X., Nahrstedt, K.: Minimum User-perceived Interference Routing in Service Composition. In: Proc. of INFOCOM (2006)
27. Salas, J., et al.: WS-Replication: A Framework for Highly Available Web Services. In: Proc. of WWW (2006)
28. Ye, X., Shen, Y.: A Middleware for Replicated Web Services. In: Proc. of ICWS (2005)

SpreadMash: A Spreadsheet-Based Interactive Browsing and Analysis Tool for Data Services

Woralak Kongdenfha[1], Boualem Benatallah[1], Régis Saint-Paul[1], and Fabio Casati[2]

[1] CSE, University of New South Wales, Sydney, NSW, 2052, Australia
{woralakk,boualem,regiss}@cse.unsw.edu.au
[2] DIT, University of Trento, Via Sommarive 14, I-38050 POVO (TN), Italy
casati@dit.unitn.it

Abstract. Spreadsheets are one of the most popular end-users programming environment. Although spreadsheets provide an interactive interface for data manipulation and analysis, they are mostly used today in data entry mode and not as interactive browsing tool for data stored in underlying data sources. In this paper, we present SpreadMash, a high-level language and tool for interactive data browsing and analysis for data services. The key innovation of SpreadMash is a repository of application building blocks called *data widgets* that characterize various data importation and presentation patterns in spreadsheets. Data widgets enable the separation of end-users tasks (composing data widgets) from the tasks of data architects (creating data abstractions and data widgets). Through a series of examples we illustrate how tasks that would be challenging in existing environments are facilitated by SpreadMash.

1 Introduction

Interactive data presentation and analysis applications are applications that allow users to enter or present information stored in the underlying data sources, and possibly to perform manipulation operations on these data, such as calculations or aggregations. Examples of such applications include relational reporting applications and on-line analytical processing (OLAP) systems. Relational reporting tools, such as Crystal Reports [2], enable reporting against relational databases. They deliver information, but provide very little support for interactive analysis [3,6]. On the other hand, OLAP systems are specialized technologies that present numerical data in a multidimensional format and provide more powerful analytical capabilities than relational reporting tools. While OLAP systems deliver advanced analytics, these features exceed the skills and capabilities of the average PC users. One of the most successful paradigms for data analysis is that of spreadsheets [3,6,8,16]. The success of spreadsheets comes from:

- an interactive interface that make it easy to view, and interact with the data [14]. Spreadsheets also provide analysis and manipulation alternatives that span various application domains such as financial, statistics, etc.;

Z. Bellahsène and M. Léonard (Eds.): CAiSE 2008, LNCS 5074, pp. 343–358, 2008.

- a flexible data model that does not impose much constraints regarding the data layout [14]: data can be organized based on criteria such as subjective importance by placing important data in the top-left corner, or by placing related data elements next to each other, etc.;
- a simple programming environment that allows users to accomplish their tasks easily and intuitively. Spreadsheets eliminate many of the stumbling blocks in traditional programming environments such as data dependencies or memory management [13];
- an incremental approach for building fairly complex computations while getting immediate feedback (i.e., continuous evaluation) [13];
- a tolerant approach regarding typing and formula consistency (i.e., a spreadsheet program executes even if it contains some invalid statements) [8].

All of the above elements concur spreadsheets to be a good candidate for developing interactive data presentation and analysis applications. However, today spreadsheets are mostly used in data entry mode and not as interactive presentation and browsing tool for data stored in underlying data sources, thereby using only a fraction of the potential of the spreadsheet model. Recent work has made progress in this direction. Example of these efforts include OracleBI [3], SAPBWP [6], XL Report Builder [7] that integrate spreadsheets with both relational databases and OLAP systems.

While exposing underlying data in spreadsheets is a promising idea, the above tools still fall short in terms of data browsing and presentation. Specifically, their presentations are essentially limited to relational tables (for textual data from relational databases) or pivot tables (for numerical data from OLAP systems). They thus hinder the flexibility in data presentation as provided by spreadsheets and required by end-users. As mentioned earlier, a user might want to present data based on criteria such as subjective importance by placing important data in the top-left corner. However, this is not possible in such tools.

A further limitation of the aforementioned tools is the difference between the data model exposed by relational databases and the one required by applications. This problem, known as impedance mismatch, makes it hard to develop and maintain data analysis applications. For instance, information about customers may be represented in several database tables for normalization and performance purposes. In this case, if a user is interested to import information about a specific customer, she may have to specify a join query over these tables. However, end-users may not necessarily have the skills or the will to understand the underlying logical database schema and specify such queries. It is desirable to present underlying data to end-users at the right abstraction level and provide easy and intuitive data manipulation support.

To address the above issues, we propose SpreadMash, a high-level language and a tool for developing interactive data presentation and analysis applications. SpreadMash is interactive, provides high-level data access and enables multiple data presentation schemes. The key innovation of SpreadMash is a repository of application building blocks called *data widgets* that characterize various data importation and presentation patterns in spreadsheets. We take the view that, although concrete data importation and presentation is application-specific, in many cases it is possible to capture various types of data widgets in a generic way. Essentially each data widget is associated to a template for importing data from underlying data sources and a template for presenting query results in a spreadsheet. The design of SpreadMash embodies several key features:

- SpreadMash is based on spreadsheets to provide an interactive presentation of data stored in relational databases and web applications. Users can also use spreadsheets' built-in functions to explore their analysis.
- SpreadMash decouples end-users from logical models of underlying data sources. It allows query formulations over high-level concepts, and thus solves the impedance mismatch. This is facilitated by the integration of data services [9,10] into the framework.
- SpreadMash bridges the data services and spreadsheets with a three-layer framework: data services, data widgets and spreadsheets. The three-layer framework enables the separation of end-users tasks (composing data widgets) from the tasks of data architects (creating data services and data widgets).

The rest of this paper is organized as follows. Section 2 introduces the data and query model used throughout this paper, as well as gives an overview of SpreadMash. Section 3 presents the SpreadMash language and discusses the instantiation of data widgets. Section 4 presents how multiple widgets can be composed to generate complex spreadsheets. Section 5 reviews related work. Finally, in Section 6, we conclude the paper.

2 SpreadMash: Background and Overview

In this section, we first review the underlying models and systems used by SpreadMash: the entity-based data access model of data services (Section 2.1) and an extended spreadsheet data model of SpreadATOR [15] (Section 2.2). Then, in Section 2.3, we provide an overview of the SpreadMash proposed in this paper to bridge data services and SpreadATOR.

2.1 Entity-Based Data Access

End-users might want to access data from heterogeneous data sources to enable their analysis. Over the years, there has been major progress in providing APIs to simplify access to various types of data sources (e.g., REST, RSS and Atom for Web data). Data services [9,10] is a recent advent in this direction. They simplify developers' tasks by wrapping heterogeneous data sources (e.g., relational databases, XML documents), and provide uniform data access to them.

In addition, data services also provide a conceptual model that describes the structure of underlying data sources with higher-level constructs, i.e., entities and their relationships. In our work, we leverage the data services provided by the ADO.Net Framework, and its conceptual model known as *Entity Data Model*

(EDM) [9]. An example EDM model for DBLP data source is shown in Figure 1 and will be used throughout this paper. This EDM consists of three entity types: Publication, Author and Conference. There is a relationship between Publication and Author entity types, and another between Publication and Conference.

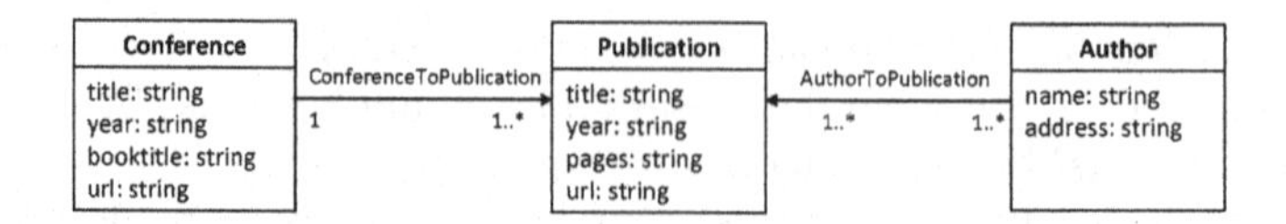

Fig. 1. An example EDM model of DBLP data source

The conceptual model offers high-level access to underlying data sources. For instance, data related to an author is represented as an entity, rather than being normalized over multiple tables in the logical model of a database. Developers can therefore access directly this pre-joined entity. Moreover, the pre-joined relationships also simplify navigation between related entities. For instance, once entity Author is accessed, the pre-joined relationship AuthorToPublication can be used directly to access publications that are written by a given author.

2.2 An Extended Spreadsheet Data Model

In Section 2.1, we have seen how professional developers can benefit from the uniform and high-level data access provided by data services. SpreadMash aims to bring such benefits to end-users by integrating data services with spreadsheets. This integration has an important challenge in the mismatch between the data model exposed by data services and the one provided by spreadsheets.

The traditional spreadsheet model consists of a set of worksheets. A worksheet is a collection of cells organized in a tabular grid. Each cell can contain either an atomic value (i.e., string, integer, datetime) or a formula. A formula can be built from atomic values, functions (e.g., ``+'', ``/''), and cell references (e.g., A5). On the other hand, data services returns composite entities (e.g., an entity of type publication with four attributes as shown in Figure 1) as query results. To support the browsing and manipulating of composite entities in spreadsheets, SpreadATOR [15] has been proposed to make composite entities as first-class cell values.

SpreadATOR provides spreadsheet-like formula language to support queries over entities of data services. A SpreadATOR formula, also called a *mapping formula*, is therefore built from entity references similar to the way traditional spreadsheet formulas are built from cell references. The syntax of mapping formulas is based on path expressions. For example, a formula `A1=Publication[001]` specifies that cell A1 references an entity of type publication, and `A2=Publications` specifies that cell A2 references a collection of publication entities. Cells containing references to entities will be evaluated by SpreadATOR, which returns a string representation of the referenced entity (obtained by the default transtyping given

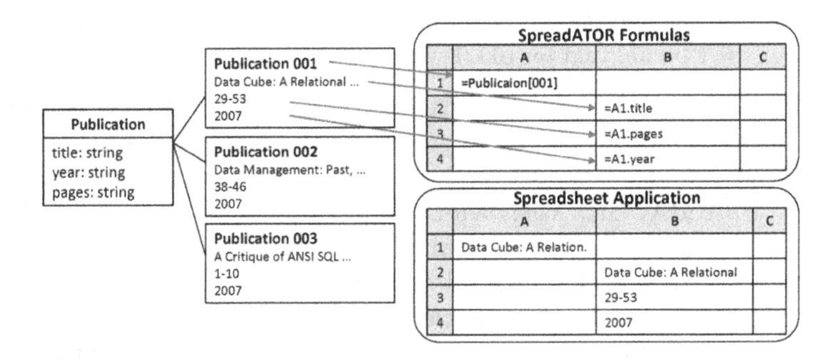

Fig. 2. An example of building data presentation with SpreadATOR formulas

by toString()). For example, the mapping formula `A1=Publication[001]`, shown in cell A1 of Figure 2, returns a string `''Data Cube: A Relational...''`.

By referencing composite entities in a cell, SpreadATOR allows users to build a presentation of composite entities on a spreadsheet by using cell references. For example, in Figure 2, a presentation of entity `Publication[001]` can be built in cells B2, B3 and B4, which respectively contain references to attributes title, pages and year of such an entity. When evaluated, the formula in cell B2 results in a string `''Data Cube: A Relational...''`, which is the value of attribute title of entity `Publication[001]`. Similarly, the formulas in cell B3 and B4 return strings 29-53 and 2007, which are values of attributes pages and year respectively.

Although SpreadATOR provides flexibility for users to build presentations of composite entities, the efficiency of the data presentation is however problematic. The straightforward way of building data presentation implies specifying a number of formulas for each referenced entity. While this approach is acceptable for small datasets, it does not scale for larger ones.

SpreadATOR proposes the notion of *templates* to address the above limitation. Specifically, in SpreadATOR, an entity type is associated with one or several templates, each of which is a complete worksheet that defines a generic data presentation pattern for instances of such an entity type. Formulas for templates are different from typical worksheets as the keyword *obj* is used instead of entity references. Consider, for example, a template T_1 associated to the entity type publication. This template consists of the following two formulas: `B3=obj.title` and `B4=obj.year`. It defines that any instances of type publication may have a presentation in a worksheet, in which cell B3 presents the value of attribute title, and cell B4 contains the value of attribute year.

To illustrate how the template mechanism works, consider an example of the worksheet W_1 with a formula `A1=Publication[001]`. From SpreadATOR point of view, cell A1 references an entity of type publication. SpreadATOR associates cell A1 with a set of templates defined for type publication. When A1 is selected, a list of publication's templates is shown in a combo-box. Assuming that the user selects a template T_1, a new worksheet W_2 is shown. The worksheet W_2 is an instantiation of the template T_1 by associating the *obj* with a reference to entity `=Publication[001]`. The instantiation of our example template is shown below:

```
B3=obj.title  ⇒ B3=Publication[001].title
B4=obj.year   ⇒ B4=Publication[001].year
```

The implementation of SpreadATOR intends to alter as few as possible existing spreadsheet applications. Hence, it does not modify spreadsheet formula language, rather acts as a middleware. Thus mapping formulas are maintained separately by SpreadATOR. As shown in Figure 2, mapping formulas are in one spreadsheet (SpreadATOR) and its evaluation in another (spreadsheet applications). SpreadATOR provides a visual assistant for constructing formulas and relies on Jscript.Net (.Net implementation of javascript) for formula evaluation.

While the notion of template partially addresses the issue of effective presentation of entities returned by querying data services, this mechanism is still limited. Specifically, the instantiation of a template results in a separated new worksheet. When users want to create a presentation of multiple entities within a single worksheet, they need to step back to the straightforward way of specifying presentation for each individual entity. To overcome such a limitation, we therefore propose in this paper SpreadMash, a model-driven approach for developing data presentation and analysis applications.

2.3 An Overview of SpreadMash

SpreadMash proposes a repository of data widgets. Elements of this repository capture various types of data presentation and importation patterns. End-users can develop data presentation and analysis applications by instantiating and composing data widgets. SpreadMash therefore enables the separation of end-users tasks (composing data widgets) from the tasks of data architects (creating data abstractions and data widgets).

A data widget is a parameterized specification of how to import data from a data service and present it in a spreadsheet. To instantiate widgets, users specify queries over the entities and relationships of a data service, as well as mappings that specify the presentation of the query results in the spatial layout of spreadsheets. SpreadMash provides four types of widgets to enable the importation and presentation of a single entity (*content widget*), a collection of entities (*repeater widget*), collections of entities associated by relationships (*hierarchical widget*), an index over entities (*index widget*).

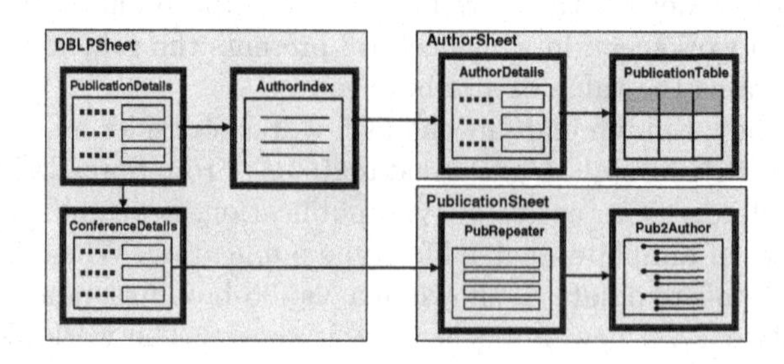

Fig. 3. An example of SpreadMash specification

Figure 3 shows an example of an application specified using SpreadMash. It consists of seven widgets (shown as solid rectangles with different icons). The content widgets (e.g., PublicationDetails, AuthorDetails, ConferenceDetails), repeater widgets (e.g., PubRepeater, PublicationTable) and hierarchical widget (e.g., Pub2Author) present the actual content of the entities. On the other hand, an index widget (e.g., AuthorIndex) presents a list of entities and allows browsing and accessing to detailed information of a selected entity.

Multiple widgets can be composed by means of *links* to build composite *worksheets*. For example, the DBLPSheet in Figure 3 is composed of three widgets: PublicationDetails, ConferenceDetails and AuthorIndex. Links connecting these widgets are based on the relationships that associate entities from which the widgets import data from. For example, the link between PublicationDetails and AuthorIndex is based on the AuthorToPublication relationship that connects the entity type Publication and Author in the underlying data service (see Figure 1). On the other hand, the link between PublicationDetails and ConferenceDetails is based on the ConferenceToPublication relationship. A set of worksheets, whose widgets are connected, compose an application. The application in Figure 3 consists of three *worksheets*: DBLPSheet, AuthorSheet, and PublicationSheet.

SpreadMash specifications are given as input to an automatic code generator, which generates code necessary for data importation and presentation. The generated code is in the form of SpreadATOR's formulas (see Section 2.2). These formulas are used by SpreadATOR to present data in a spreadsheet.

3 Widgets for Data Importation and Presentation

In this section, we discuss each type of data widgets provided by SpreadMash and its instantiation, by using the example of EDM in Figure 1.

3.1 Content Widget

Content widgets are used to specify the importation and presentation of a single entity. More than one widget can be defined for the same entity, to offer alternative points of view. The instantiation of a content widget requires *parametric query* and a *mapping definition* to generate *the corresponding spreadsheet presentation view*.

The parametric query allows the specification of a query that retrieves a single entity from a data service. A parametric query consists of two elements: source and selector. The *source* identifies the data service as well as the entity type from which data is retrieved. The *selector* is a query used to select a single entity. To offer a spreadsheet-like programming experience, SpreadMash leverages SpreadATOR [15] formula expressions to formulate queries over entities. An example of a content widget is shown in Figure 4. The user-defined values of the source and the selector properties are used to create a query `=Publication[title=''Data Cube: A Relational...'']` to retrieve data from the DBLP data service. This query returns a reference to an entity `Publication[001]`, whose value of the title attribute equals "`Data Cube: A Relational...`".

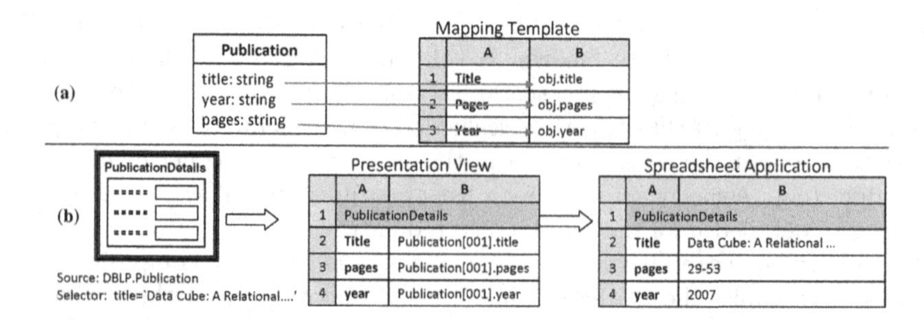

Fig. 4. Content Widgets

The role of mapping definition is to specify spatial locations in a spreadsheet where the query result will be presented. In traditional spreadsheets, spatial locations can be specified through cell or range expressions. SpreadMash leverages the notion of template, which defines a generic presentation of instances of a particular type, for mapping specification. Figure 4(b) shows an example of a content widget that uses the mapping template defined in Figure 4(a) to present entity `Publication[001]` obtained as a query result from a data service. The mapping template specifies that any instance of type publication will have a presentation in a worksheet, in which cells A1, A2, A3 contain constants `Title`, `Pages`, `Year`, and cells B1, B2, B3 contain values of attributes title, pages, year of such an entity type.

The query result and mapping template are used together to generate a presentation view of `Publication[001]` as shown in Figure 4(b). The content widget also inserts a row at the top of the presentation view, and the widget's name is rendered in the leftmost cell of this row.

3.2 Repeater Widget

A repeater widget is used to import and present a collection of entities by repeating a mapping template over such a collection. Therefore, the selector property

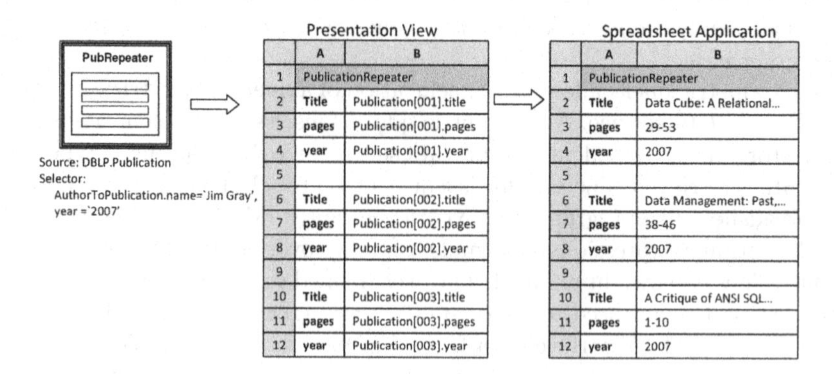

Fig. 5. Repeater Widgets

of a repeater widget is a query used to select a collection of entities. An example of user-provided source and selector properties for a repeater widget is shown in Figure 5. These two parameters are used by the repeater widget to create a query `Publication[AuthorToPublication.name = ''Jim Gray'', year=''2007'']` to retrieve its content from the DBLP data service. This query returns three entities of type publication whose author is "Jim Gray", and publication year equals to '2007'. These entities correspond to `Publication[001]`, `Publication[002]`, and `Publication[003]`.

A repeater widget presents a collection of entities all together in a single worksheet by reapplying the definition of a mapping template. It requires two levels of mappings: (i) entity-level mapping, and (ii) attribute-level mapping. The entity-level mapping specifies a range in which the details of an entity will be presented. The attribute-level mapping specifies a cell in which an attribute value will be presented. The entity-level mapping can be defined by a range expression such as `A2:A10 = Publications` to indicate that a presentation of the publication list is in a range of one column (A), and spanning from row 2 to row 10. This mapping definition is, however, not desirable since the user needs to know the number of cells to specify the presentation view. One possible solution for this issue is to use a built-in spreadsheet functions that allows the indication of a dynamic range of cells without a need to know the exact length. For example, a user can use Excel's formula like `OFFSET(A2, 0, 0, COUNT(A:A), 1) = Publications`. This formula returns a range starting at cell A2 and spanning 1 column. This range is dynamic since the number of rows is computed using function `COUNT(A:A)` which returns the number of nonempty cells in column A.

While the above built-in functions are useful, they only allow to specify a range for the entire collection. A repeater widget however requires a mapping that specifies also the location in which an individual entity will be presented. In this case, the user can specify directly a set of ranges, each for presenting one entity in the collection. In traditional spreadsheets, this can be specified by a list of noncontiguous ranges e.g., A2:A5, A7:A10, A12:A15. While this may be acceptable for small datasets, it does not scale well for larger ones.

We therefore leverage extensions proposed by SpreadATOR to allow users to specify a mapping for a repeater widget. SpreadATOR proposes four functions for indicating the boundary of a range presenting an entity: `top(l)`, `left(l)`, `right(l)`, and `bottom(l)`. Assuming that an entity `Publication[001]` has a presentation on a range B2:B4, the coordinates of this range can be obtained as follows: top(`Publication[001]`) = 2, left(`Publication[001]`) = 2, right(`Publication [001]`)= 2, and bottom(`Publication[001]`) = 4. To illustrate the advantages of using these extensions, consider the example of mappings below:

```
A2:A<bottom(Publication[001])> = Publication[001]
A<bottom(Publication[001])+2>:A<bottom(Publication[002])> = Publication[002]
A<bottom(Publication[002])+2>:A<bottom(Publication[003])> = Publication[003]
```

The first mapping states that the presentation of entity `Publication[001]` is a range of one column (A) and spanning from row 2 to the lower-most row of the range. The user therefore does not need to know the exact boundaries of the range. The second mapping states that a range presenting entity Publication[002] has one column (A). This range spans from two rows below the lower-most row of the range presenting entity Publication[001] until the lower-most row of the range presenting entity Publication[002]. The user therefore does not need to know exactly the coordinates in which the presentation of each entity starts. Rather the user can specify the location of each entity by referencing the others.

In the above example, the user still needs to specify a mapping for each individual entity. We therefore leverage another extension of SpreadATOR's formula language, i.e., the keyword *next*, for specifying the coordinates of the first cell of subsequent ranges. For example, users can specify a mapping for a collection of entities as follow: A2:A<next=bottom(Publication)+2> = Publication. This mapping specifies that each entity is presented by a range of one column (A), and spanning from row 2 to the lower-most row of the range associated to current entity of publication. The locations of ranges presenting subsequent entities are two rows after the last row (bottom) of the range presenting the current entity. In this case, the user needs to specify only one mapping for a collection of entities and has flexibility to organize the presentation.

Once an entity-level mapping has been defined, it is used by a repeater widget to iterate over a collection of entities. For each iteration, a repeater widget also requires a mapping definition of how attributes of such an entity should be presented in the range. Figure 5 shows an example of a repeater widget that uses the mapping template defined in Figure 4(b) to present a collection of publication entities. The mapping template defines a generic presentation for any instances of type publication. The repeater widget therefore instantiates such a mapping template to present the entity of the current iteration.

Repeater widgets have a specialization, namely *table widgets*, that present a collection of entities in a table format. A table widget assumes a default template in which each row presents an entity and each column contains an attribute value. The only parameter required for the mapping definition of table widgets is a single cell that will be the upper-leftmost of the table presenting the referenced entities. The number of cells used by the table depends on the number of entities returned from the query and the number of attributes comprising that entity.

Figure 6 shows an example of a table widget, with a mapping specifying that a table presenting the referenced publication entities starts at cell A1. A default template used by the table widget defines that attributes title, pages and year of publication entities are presented in columns A, B and C respectively. Each row of the generated table presents an entity of type publication, whose value

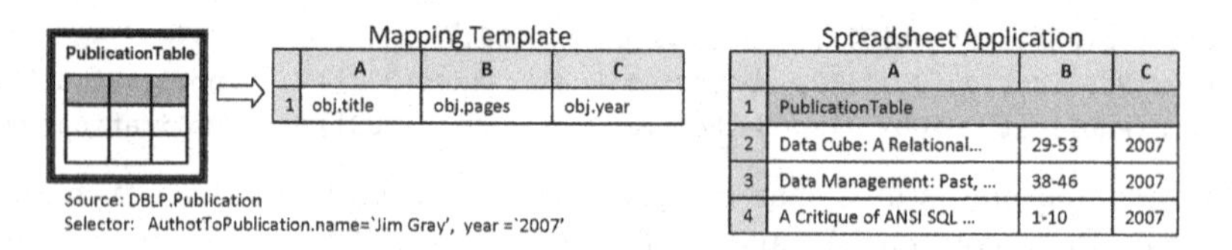

Fig. 6. Table Widgets

of attribute author equals ''Jim Gray'', and of attribute year equals '2007'. The default template of table widgets can be modified by setting the orientation property to *transpose*. The transpose keyword indicates that a table is presented with attributes as row and entities as columns.

3.3 Hierarchical Widget

A hierarchical widget specifies the importation of a collection of related entities. For each entity, the widget navigates through a relationship to import another collection of associated entities. Consider the example in Figure 7, this hierarchical widget consists of a query: =Author[address=''microsoft'']. This query returns two entities of type author, whose values of the attribute address equals ''microsoft'', i.e., Author[111] and Author[112].

For each of these author entities, the hierarchical widget uses another query to import a collection of publications associated to such an author. For instance, the hierarchical widget in Figure 7 uses a query =Author[111].AuthorToPublication to retrieve a collection of publication entities that are associated to Author[111] through a relationship AuthorToPublication. This query returns three publication entities: Publication[001], Publication[002], Publication[220].

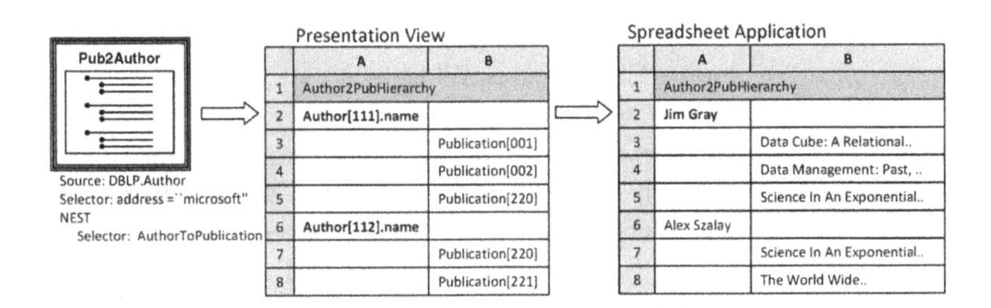

Fig. 7. Hierarchical Widgets

In Section 3.2, we have seen an example of using SpreadATOR functions for specifying a mapping for a collection of entities of the same type. However, the entities of type author in our example in Figure 7 cannot be presented one after the other since there are presentations of entities of different types in between. In particular, there are publication entities in between presentations of author entities. A hierarchical widget therefore requires a specific mapping for each individual collection of entities. In our example in Figure 7, the mapping for entities of type author can be defined as: A1:A<next=bottom(Publications)+1> = Name. This mapping specifies that the value of the first author name is presented in cell A1, and then the subsequent author names are presented in one row below the end of its related list of publication entities. On the other hand, the mapping for entities of type publication can be defined as follows: B<Name+1>:B<bottom(Publications)> = Publications. This mapping specifies that entities of type publication are presented in column B and ranging from the next row after its corresponding author name until the end of the collection.

3.4 Index Widget

An index widget specifies the importation of a collection of entities that are presented as a list without detailed information. By selecting a cell referencing an entity of this list, the user can navigate to another worksheet that contains detailed information of that given entity. We adopt the template mechanism proposed by SpreadATOR to enable such a navigation.

The query of an index widget selects a collection of entities. For example, the query in Figure 8 returns three entities of type publication, whose author is `''Jim Gray''`, and are published in `''2007''`. The mapping of an index widget is specified by a range of cells used to present a list of entities. For example, a mapping `A2:A<bottom(Publications)> = Publications` specifies that a list of publications will be generated by this index widget in column A spanning from row 2 to the lower-most row of a range presenting the publication list.

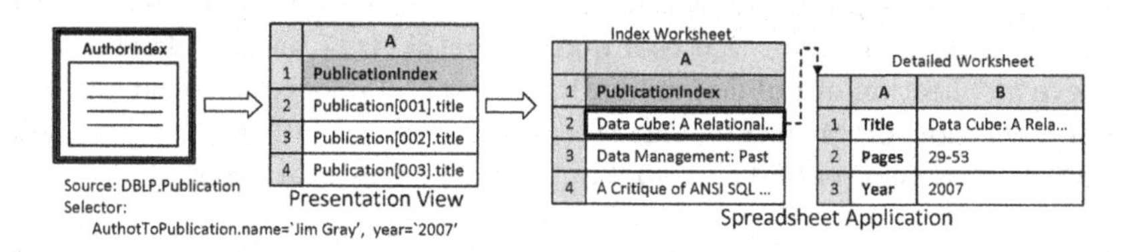

Fig. 8. Index Widgets

An index widget allows users to see detailed presentation of an entity in the list by associating each entity with a template defined for its entity type (see Section 2.2). For instance, the index widget in Figure 8 associates to each entity in the list a publication's template (as defined in Figure 4(a)). When a user selects cell A2 on the index worksheet, she can navigate to the detailed worksheet that presents details of entity `Publication[001]`.

4 Widget Composition

In the previous section, we focused on how to generate a worksheet from a single widget. In this section, we discuss how multiple widgets can be composed to generate a composite worksheet.

When composing widgets to generate a composite worksheet, there are two aspects that need to be considered: *data dependency*, and *spatial dependency*. The data dependency specifies how contents of mashup widgets depend on each others. On the other hand, the spatial dependency specifies the relative locations of contents generated from different widgets on the same worksheet.

4.1 Data Dependency

In traditional spreadsheets, the data dependency between cells is specified through a formula built from cell references. For instance, a data dependency between cells A2 and A1 can be built by a formula in cell A2 referencing cell A1 (see Section 2.2). Then the content of cell A2 will be evaluated by taking into account the content of cell A1. SpreadMash follows this cell referencing mechanism. To build data dependencies, we propose the notion of *links*. When two widgets are connected by a link, some contextual information is conveyed from the source widget to the destination widget. This contextual information is used to determine the entity or set of entities to be presented by the destination widget.

Consider as an example the DBLPSheet shown in Figure 9. This worksheet is composed of two widgets: PublicationDetails and AuthorIndex. The link (I) connecting the two widgets conveys, as the contextual information, the identifier of an entity referenced by the PublicationDetails widget (i.e., `Publication[001]`). The AuthorIndex widget then uses this contextual information to generate its query. For instance, the index widget takes the contextual information (`Publication[001]`) together with a user-defined selector property (AuthorToPublication), and generates a query `Publication[001].AuthorToPublication`. This query results in three entities of type author that are related to `Publication[001]`.

Figure 9 also shows an example of a worksheet generated from the DBLPSheet. This worksheet contains the contents of both the PublicationDetails and AuthorIndex widgets. The specification of mappings that specify how the contents of these two widgets can be presented in the spatial layout of spreadsheets is discussed in Section 4.2.

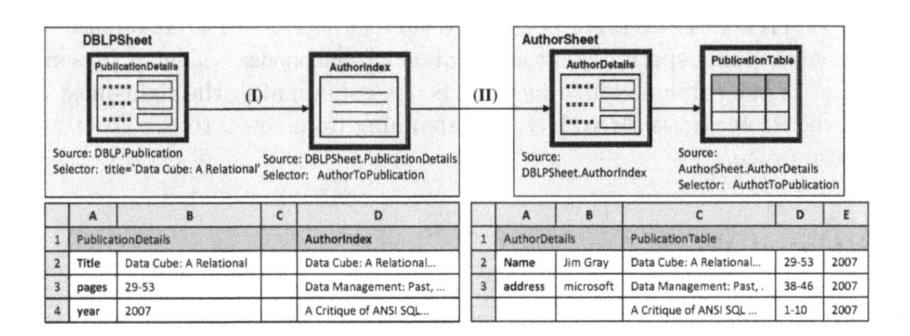

Fig. 9. Widget Composition

A link can also connect two widgets on different worksheets. For example, consider the link (II) connecting AuthorIndex on DBLPSheet and AuthorDetails on AuthorSheet. This link specifies that the content of the AuthorDetails widget depends on the content of the selected cell of the index widget. To illustrate the semantics of this link, suppose that the AuthorIndex widget generates a presentation of two entities of type author: `Author[111]`, `Author[112]`. When a user selects a cell referencing `Author[111]`, the link (II) passes some contextual information (e.g., identifier of Author[111]) to the AuthorDetails widget. The AuthorDetails widget uses this contextual information to generate a query that retrives entity `=Author[111]`. The AuthorDetails widget then build a presentation of entity

Author[111] according to a user-defined mapping definition of the AuthorDetails widget. Note that composing an index widget with a content widget is different from the template mechanism described in Section 2.2. In particular, when a cell in the index is selected, the user can navigate to a new worksheet which does not necessarly contain only detailed information of an entity. The presentation of the new worksheet depends on the specification of the widgets composing that worksheet. For example, in Figure 9, when cell A1 is selected, the user can navigate to a separated worksheet which consists of the contents of two widgets: AuthorDetails and PublicationTable.

4.2 Spatial Dependency

In Section 3.4, we have seen how a mapping of an index widget can be defined (A2:A<bottom(Publications)> = Publications). This mapping specifies that a range presenting a collection of entities referenced by the index widget corresponds to a fix number of columns (1 column) spanning from row 2 to the lower-most row of the range. However, when composing multiple widgets in a single worksheet, the user needs to define the locations in which the contents of these widgets are placed relatively to each other. We reuse the functions discussed in Section 3.2 for identifying the spatial locations of composing widgets by referencing each other. For example, the mapping of the AuthorIndex widget in Figure 9 can be defined as follows:

<right(PublicationDetails)+2>2:
<right(PublicationDetails)+2><bottom(Publications)> = Publications.

The above mapping specifies that the content of AuthorIndex widget is presented in a range of one column. This range starts one column after the right-most column of the PublicationDetails widget, and spanning from row 2 to the lower-most row of the range.

5 Related Work

We have already reviewed reporting tools and OLAP systems in the introduction. Related work in the spreadsheet research community include prototype system like Gencel [8,12]. Gencel proposes a high-level language that allows the definition of spreadsheet templates. These templates are then translated into spreadsheet applications as well as a set of update operations, which ensure the correctness of spreadsheet evolutions. Similar to this proposal, our work introduces high-level language for specifying the data presentation in spreadsheets based on widgets. Unlike this approach, our work provides a three-layer framework that bridges the data services and spreadsheets. Moreover, in Gencel, a generated spreadsheet is a single worksheet consisting of a single table. Rather we generate a spreadsheet with a collection of worksheets, and each worksheet consisting of one or more instantiated widgets.

Efforts in the area of web applications have promoted the use of the web as the fundamental data interface. Tools such as ASP.Net [1], PHP [5] and JSF [4] simplify the generation and deployment of data-intensive web applications by means of page generators. They typically extract content from data sources and present it in user-programmed page templates.

Although the above tools improve developers productivity, they require users to have web development skills. WebML [11] proposed to use data abstraction and conceptual modeling methods for web application development. However, WebML targets the delivery of information on a website, and thus provides very little support for interactive data manipulation. WebML also falls short in terms of data presentation as the provided units present data using a predefined layout, and do not support user-defined templates. Our approach overcomes this inflexibility by allowing users to define mapping definition to specify how data should be presented.

6 Conclusion and Future Work

In this paper, we have presented SpreadMash, a high-level language for developing interactive data presentation and analysis applications. SpreadMash is based on spreadsheets and provides an interactive presentation of data services. It also enables end-users to focus on their analysis as they are decoupled from the knowledge of the underlying data sources. They can develop interactive data presentation and analysis applications by reusing and composing widgets. Then a spreadsheet is generated as a collection of instantiated widgets. The SpreadMash has been described through a series of examples, each motivated by challenges of typical user tasks.

There are several directions for future work. First of all, we plan to support the transformation of imported data. We are now investigating the operators for both transforming data values and restructuring. Second, in this paper, we only provided a basic sets of widgets for importing and presenting data obtained from a single data service. Often users would like to integrate data from multiple data services, in which relationships are not pre-built. We plan to expand the widget repository to capture the data integration aspect. Finally, we plan to carry out user studies in using SpreadMash. The feedback from such a study would help us in refining the framework.

References

1. ASP.Net. http://asp.net/
2. Crystal Reports, http://www.businessobjects.com/products/reporting/crystalreports
3. Exploiting The Power of Oracle Using Microsoft Excel, www.oracle.com/technology/products/bi/pdf/BI_Spreadsheet_Addin_WP.pdf
4. Java Server Faces, http://java.sun.com/javaee/javaserverfaces/
5. PHP, http://www.php.net/
6. SAP BI Excel Add-in, www.sap.com/solutions/netweaver/businessintelligence/pdf/BWP_BI_Overview.pdf

7. XL Report Builder, `http://www.afalinasoft.com/xl-report-builder`
8. Abraham, R., Cooperstein, I., Kollmansberger, S., Erwig, M.: Automatic generation and maintenance of correct spreadsheets. In: Proc. ICSE 2005 (2005)
9. Adya, A., Blakeley, J., Melnik, S., Muralidhar, S.: Anatomy of the ADO.Net entity framework. In: SIGMOD 2007, pp. 877–888. ACM Press, China (2007)
10. Carey, M.: Data delivery in a service-oriented world: the bea aqualogic data services platform. In: SIGMOD 2006, Chicago, IL, USA, pp. 695–705. ACM Press, New York (2006)
11. Ceri, S., Fraternali, P., Bongio, A.: Web modeling language (webml): a modeling language for designing web sites. In: Proc. WWW 2000, pp. 137–157 (2000)
12. Engels, G., Erwig, M.: Classsheets: automatic generation of spreadsheet applications from object-oriented specifications. In: Proc. ASE 2005 (2005)
13. Jones, S., Blackwell, A., Burnett, M.: A user-centred approach to functions in excel. SIGPLAN J 38(9), 165–176 (2003)
14. Pemberton, J., Robson, A.: Spreadsheets in business. IMDS J 100(8), 379–388 (2000)
15. Saint-Paul, R., Benatallah, B., Vayssiére, J.: Data services in your spreadsheet? In: Proc. EDBT 2008 (2008)
16. Scaffidi, C., Shaw, M., Myers, B.: Estimating the numbers of end users programmers. In: Proc. VLHCC 2005, pp. 207–214 (2005)

Managing the Evolution of Service Specifications*

Vasilios Andrikopoulos[1], Salima Benbernou[2], and Mike P. Papazoglou[1]

[1] INFOLAB, Dept. of Information Systems and Management, Tilburg University, The Netherlands
[2] LIRIS, Université de Lyon 1, France
{v.andrikopoulos,mikep}@uvt.nl, sbenbern@liris.univ-lyon1.fr

Abstract. The ability to cope with multiple competing stakeholders, fluid requirements, emergent behavior, and susceptibility to external pressures that can cause changes across an entire organization, coupled with the ability to support service diversification, is a key to an enterprise's competitiveness. Web services equip enterprises with the potential to react to change by addressing two interrelated sets of requirements: the ability to accommodate service changes that demand rapid response and to support service variation according to customers' needs and requirements. In this paper we introduce the concept of service evolution management, which provides an understanding of change impact, service changes control, tracking and auditing of service versions, and status accounting. To achieve this, we develop a formal model and theory for service evolution that allows multiple active service versions to be created consistently and co-exist, while executing schema changes effectively.

Keywords: Web services, service versioning, service differentiation, service contracts.

1 Introduction

XML-(or Web)-based services are key technologies providing a foundation for a net-centric services environment, which reacts to change by addressing two interrelated sets of requirements: the ability to accommodate service changes that demand rapid response, and the ability to support *service variation* according to the needs and requirements of customers. These two inter-related sets of requirements place emphasis on the ability of services to co-exist in *multiple active versions* and to execute changes effectively and efficiently. They therefore epitomize the common need for constant change that challenges service applications development. Service changes may, for instance, originate from the introduction of new functionality, the modification of existing functionality to improve performance, or the inclusion of new regulatory constraints that require that the

* The research leading to these results has received funding from the European Community's Seventh Framework Programme under the Network of Excellence S-Cube - Grant Agreement n° 215483.

Z. Bellahsène and M. Léonard (Eds.): CAiSE 2008, LNCS 5074, pp. 359–374, 2008.

behavior services is altered. Such changes should not be disruptive by requiring radical modifications in the very fabric of services or the way that business is conducted.

Service evolution is a precursor to successful service adaptation. Service adaptation refers to the *a posteriori* ability of a service to modify itself in order to interact with other services by detecting potential functional or non-functional mismatches with its peer services by semi-automated means ([1], [2]). Current service adaptation approaches assume that services can evolve independently and do not constrain their mutual inter-dependencies. In contrast to this, service evolution attempts to *a priori* validate and constrain service changes and ensuing service versions, so that they are consistent and well-behaved.

Routine change increases the propensity for error. To control service development one needs to know why a change was made, what are its implications and whether the change is complete. In a Web services environment, changes only affect the Web service provider's system. Typically Web service consumers do not immediately perceive the upgraded process, particularly the detailed changes of Web services. Hence, Web service based applications may fail on the Web service client side due to changes carried out during the provider service upgrade. In order to manage changes as a whole, the Web service consumers have to be taken into consideration as well, otherwise changes that are introduced at the service producer side can create severe disruption. Eliminating spurious results and inconsistencies that may occur due to uncontrolled changes is therefore a necessary condition for the ability of services to evolve gracefully, ensure service stability, and handle variability in their behavior. Thus, any service evolution management system has to be able to handle consistently and unambiguously the *propagation*, *validation*, and *conformance* of any kind of modifications applicable to a service.

Service evolution management requires an understanding of all the points of change impact, controlling service changes, tracking and auditing all service versions, and providing status accounting. In summary, service evolution management exhibits the following characteristics:

- **identification** of all kinds of permissible changes to services and **classification** of these changes,
- **propagation analysis mechanisms** that record the status of services, analyze changes, and gather information about their effects on clients of a service version,
- **validation and conformance mechanisms** that maintain the consistency of a service by ensuring that the service is a well-behaved collection of service changes and versions, and ensure conformance with respect to service updates and version contracts,
- **version control mechanisms** that control the release of a service and the changes applied to it throughout its lifecycle, and
- **instance migration mechanisms** for associating instances of running services with new service versions.

In this paper we shall consider all above items except for the topic of instance migration. This issue is examined by the workflow community (see section 2).

The paper is organized as follows: section 2 discusses related work from a number of different fields. In section 3 we present a service specification model that acts as a reference point in the discussion about service evolution. Service evolution management characteristics are covered in section 4. Finally, we wrap up the paper with some conclusions and future work (section 5).

2 Related Work

As services grow more complex to compensate for increasing business needs, valuable lessons and techniques can be drawn from Software Configuration Management (SCM), the discipline of software engineering that deals with controlling the evolution of complex software systems [3]. More specifically, the usage of *versions* as a representation of incrementally changed software objects (in that case, services) can be especially useful. The graph models that support the various versioning schemes [4] provide an intuitive way to manage the history of different versions.

Current Web services technologies do not directly address the versioning issue, usually requiring developers to solve the problem through the application of patterns and best practices [5]. Nevertheless, elements of these techniques can be used for service evolution management, see, for example, [6], [7], [8], and [9]. The common denominator of all these approaches is that they discuss *how* to put a versioning mechanism in place using an existing set of technologies, without concerning themselves with *what* constitutes the version of a service. An application of versioning in XML Schema technology is investigated in [10], where a number of use cases are presented that describe the desirable behaviors for XML Schema versioning. This approach provides guidelines for the behavior of schema processors in face of different versions, but it deliberately avoids discussing implementation (which is the critical component of that approach) and does not guarantee change consistency.

Lessons can also be drawn from the work on heterogeneous databases in general, and in specific from *schema mappings* between disparate data sources. Changes to the schemas of the data sources have to be reflected to the mappings between them. In that case the mappings have to be adapted to compensate for the evolution of the original source material [11], [12]. The evolution of requirements in information systems, as examined in the requirements engineering domain, is also a source of useful ideas and techniques. In [13] for example, a similar to ours approach is presented that combines abstraction from specific models with a generic typology for gaps (and similarities) in order to express evolution requirements.

Finally, we can also draw on the work of evolution in the field of workflows for methodologies and ideas. The problem of workflow evolution has two facets: *static*, referring to the issue of modifying the workflow description, and *dynamic*, referring to the problem of managing running instances of a workflow whose description has been modified (instance migration). The work in this field, at least in its conception, draws heavily from the literature on o-o databases evolution for its static aspects e.g. [14], but focuses mainly on the dynamic aspect [15], [16], [17].

3 Service Specification Reference Model

To be able to identify and study the changes happening to a service during its lifetime we can either choose a specific set of technologies for service description, like for example WSDL and BPEL, or use a technology-agnostic model that could easily be translated to the current standards. To abstract away from the idiosyncracies and syntactic nuances of these standards we chose the second approach. In particular, we introduce a *Service Specification Reference Model* that exhibits the main characteristics of different service description models and technologies.

More specifically, we define the following three levels for services specification:

Abstract Service Definition Model (ASD): an abstract model containing generic concepts and their relationships that are common to all service schemas.

Service Schema Definition (SSD): the schema of a specific service, or in other words, the service specification. This consists of the elements of the service and their relationships that are *generated* by corresponding concepts and relationships found in the ASD.

Instance of Service Schema Definition (ISD): that is produced from the instantiation of the SSD during the execution of the service.

Example 1 (Running Example). Consider the case of an (aggregate) service implementing a composite Order-to-Cash (OtC) process. OtC takes care of the revenue collection after a successful sale and shipping of a product. It involves a number of sub-processes/steps like purchase order processing, advanced shipping and delivery notification, invoicing, etc. These sub-processes are themselves exposed as services, offering their functionality to a number of services apart from the OtC. In that case, the specifications of all of the above services, expressed for example in WSDL, BPEL, etc., are considered the SSDs of each service. Irrespective of which set of standards is used for their description, all these services share a common reference framework that allows them to interoperate. All of them for example, allow clients to invoke them - or they are able to invoke back clients. These fundamental assumptions about how the services work constitute the ASD. Furthermore, when these services are implemented, deployed and invoked by their clients, then a number of instances of them are created based on their specification (their ISDs).

The following sections discuss only the ASD and SSD, starting with the general relationships that connect the building blocks of each level.

3.1 Universal Relationships

In order to show relationships between concepts in the ASD and therefore also between corresponding elements in the SSD we need to define the formal semantics of conventional relationships such as *composition*, *aggregation*, and *association* found in object-oriented languages and in the AI semantic nets. These will be described using the UML class diagram notation [18]:

Composition (fig. 1(a)) $\forall y, \exists! x : \overset{c}{\overrightarrow{xy}}$: y can belong in exactly one composition relationship with x. Additionally, deleting x deletes also y (*cascading delete*).

Aggregation (fig. 1(b)) $\forall y, \exists x : \overset{a}{\overrightarrow{xy}}$: y may participate in more than one aggregation relationships with x. Deletion of x deletes also y, but only if there are no other relationships of this type in which y participates.

Association (fig. 1(c)) $\exists y, \exists x : \overset{s}{\overrightarrow{xy}}$: No further restrictions on the participation and the existence of y. We use the notation $\overset{r}{\overrightarrow{xy}}, r \in \{c, a, s\}$ to show that elements x and y have a relationship of type r.

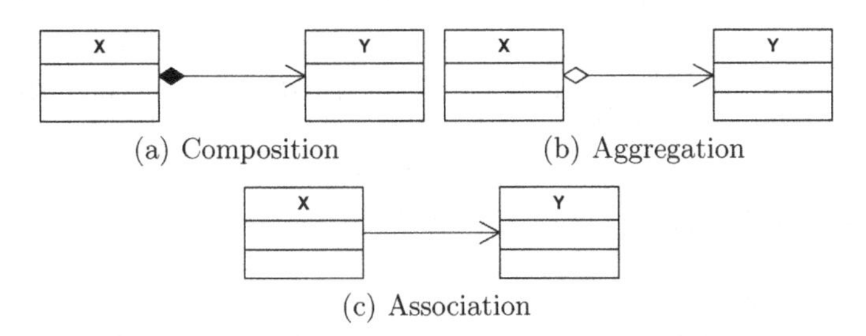

(a) Composition (b) Aggregation (c) Association

Fig. 1. Types of Relationships

Furthermore, relationships in UML have multiplicities, denoting the possible *cardinalities* of the instances of each class, that is, how many instances of it may exist at the same time, in each relationship. Putting a *1..** on the y side of the arrow for the relationship $\overset{s}{\overrightarrow{xy}}$ for example means that instance x must always have one or more associations with instances of y. The notation we introduced above to show the semantics of the relationships has therefore to be expanded to accommodate cardinalities with: $|cardinality_x||cardinality_y|^{-1}$, $cardinality \in \{0| \geq 0| \leq 1| \geq 1\}$ (≥ 0 for * in UML, ≤ 1 for *0..1*, and ≥ 1 for *1..** respectively). For example, $\overset{a}{\overrightarrow{xy}}\ |1|| \geq 1|^{-1}$ means there may be more than one instances of y that participate in this aggregation with x. If not specified explicitly, cardinality values are considered to be equal to 1.

3.2 Abstract Service Definition Model

The Abstract Service Definition Model (ASD) is a collection of *concepts* common across service schemas that have a set of *parameters* and *relationships*. Parameters can either be *property-domains* or *attributes*. An attribute, e.g. a string denoting the currency that will be used in the scope of a specific message, will be assigned a value during the *instantiation* of the service schema, i.e. the creation of a running instance of the service from its schema. Property-domain pd_i has a set of values called properties (denoted by p_i) that, as we will see in section 3.3, may restrict some of the relationships of the element generated by the concept. A specific value p_i will be selected from the domain when the SSD is generated by the ASD.

The ASD spans three layers: *Structural*, *Behavioral*, and *Regulatory*, which lean on each other in ascending order to fulfill their functionality, and two sections: *Public* and *Private* (denoted by a *visibility* attribute of the concept definition). Figure 2 illustrates the ASD using UML class diagram notation.

ASD Notation. The concepts depicted in the figure and their relationships are partly based on the models discussed in [19] and [20], which describe metamodels for services. Attributes are not contained in the figure for reasons of brevity.

Based on the above, the concepts in Figure 2 can be described using the following notation: $Concept(property-domain^*, attribute^*, relation^*, visibility)$.

Example 2. The `Message` concept as shown in this figure, has two property-domains: `messageRole`, and `type` (that is inherited from concept `Role`[1]), and an one-to-many aggregation relationship with the `Information Type` concept.

Consequently, it can be written in this notation as: $Message(messageRole, type, attributes^*, \overrightarrow{Message\ InformationType}^{a} |1|| \geq 1|^{-1}, public)$.

Definition 1. ***ASD Layers*** *The ASD concepts can be perceived horizontally as in the three distinct layers, viz. structural, behavioral and regulatory (see Figure 2). Each layer* $L_i, i = 1, 2, 3$ *is defined as* $L_i = \{\exists! \mathcal{P}_x \in ASD / \forall x \in \mathcal{P}_x \rightarrow x \in L_i\}$, *where* $\mathcal{P}_x$ *is a* partitioning *of all concepts into exactly one of the three layers. Now we can define the notion of* horizontal *and* vertical *relationships based on this partitioning:*

Horizontal: $[[\overrightarrow{xy}^{r^h}]]_k^h : x, y \in L_k$, *i.e., all concepts belong to the same layer.*

Vertical: $[[\overrightarrow{xy}^{r^v}]]_k^v : x \in L_k, y \in L_m, m \neq k$, *i.e. concepts are in different layers.*

For example, the `Policy Profile` concept has a horizontal relationship with `Policy Alternative` and `Service Policy` since they belong to the same layer (the regulatory), and a vertical relationship with the `Operation Sequence` and `Operation` concepts in the behavioral and structural layers respectively (see Figure 2).

Definition 2. ***The ASD.*** *The* $ASD = \{(c_i^j, L_i), i \leq 3, j \geq 1$ *and* $\forall c_l^n \, |\overrightarrow{c_i^j c_l^n}^{r}, (i = l \wedge r = r^h) \vee (i \neq l \wedge r = r^v)\}$ *where* c_i^j *is a concept* c^j *in the layer* L_i, *i.e. ASD is the set of all concepts and the layers they belong to.*

For this paper we have mainly concentrated on the structural and behavioral layers of the ASD.

ASD concepts. In the following we briefly present the layers and the concepts of the ASD:

---

[1] We assume that the inheritance relationship maintains its UML semantics, i.e., all attributes and parameters (but not relationships) are copied to the inherited concept.

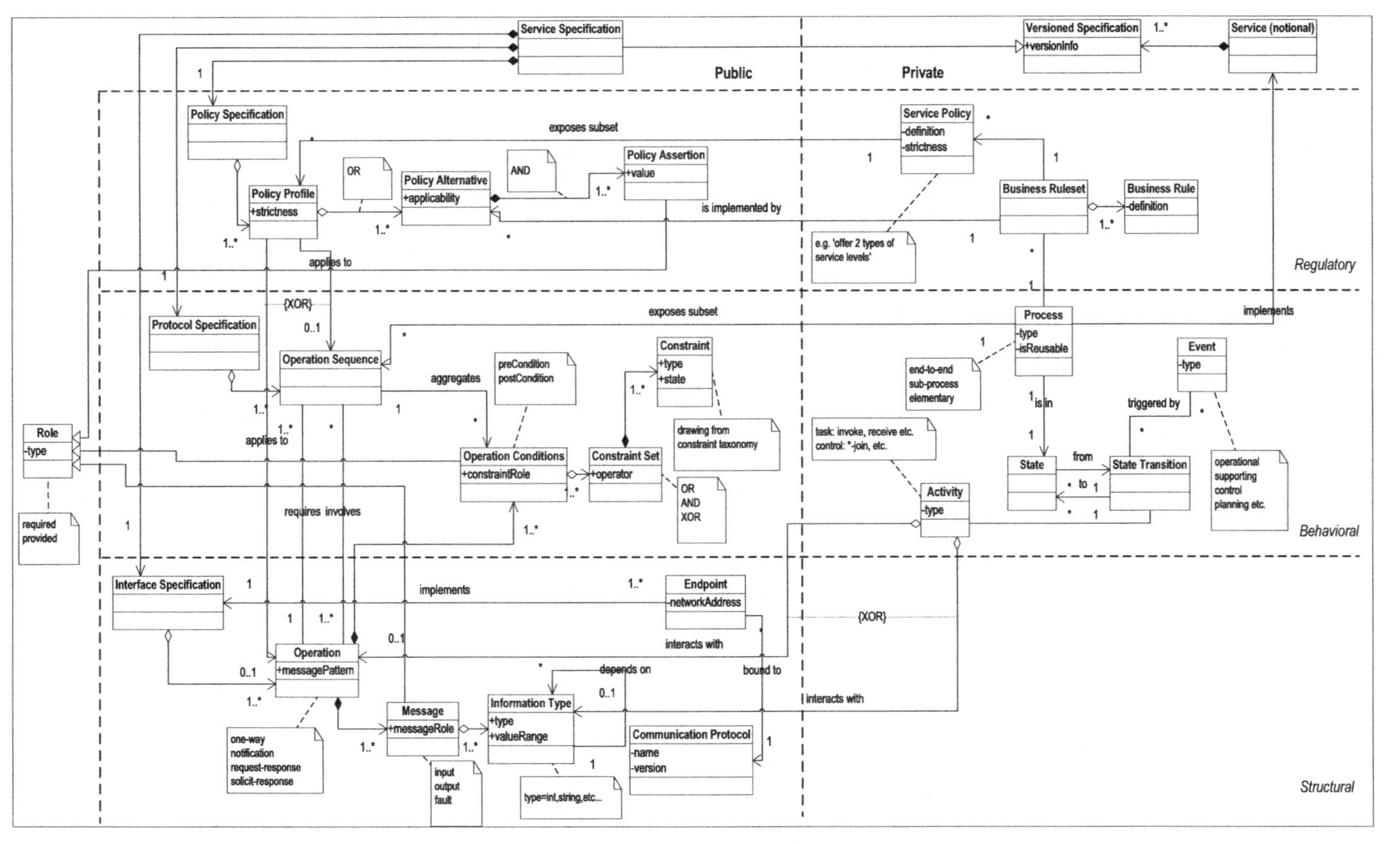

Fig. 2. Layers of the ASD

Structural layer is responsible for the structural description of the concepts that constitute a service. In that sense it contains: an `Interface Specification`, that acts as the access point for the structural signature of the service, aggregating a number of `Operations`, i.e. specific functions that the service is performing, defined in an abstract format, and `Messages` that are exchanged as part of the operations. A `Message` is defining the informational content consumed and/or produced by the service in the form of hierarchically organized `Information Types`. Furthermore, `Endpoints`, define specific URIs that the service can be reached from. Each endpoint is bound to a number of `Communication Protocols` that can be used to access it.

Behavioral layer manages the control and execution aspects of the service business logic. It contains for this purpose: a `Protocol Specification` that aggregates the necessary constructs for specifying business protocol information, i.e. `Operation Sequences` (specific sequences that have to be performed in order to achieve a part of the business logic). They aggregate sets of `Operation Conditions` that either have to be fulfilled in order for the particular sequence to be executed, or are produced by it during its execution, in terms of pre- and post-executional `Constraints` (e.g., temporal, spatial, financial, etc.). `Constraints` are organized around `Constraint Sets` that may contain overlapping constraints. `Constraint Sets` and `Constraints` are for this layer what `Messages` and `Information Types` are for the Structural: a method to express the signature of services - but this time on protocol level. In order to model the workflow aspect of the service description, the concept of `States` of the `Process` for which the service is the implementation is used. `State Transitions` govern the switch from one state to the other. The switch can either occur as the outcome of the successful execution of an `Activity` and/or triggered by an `Event`.

Regulatory layer contains the necessary elements for describing and managing both the business decision rules and policies that govern the business logic of the process, and key performance indicators, e.g., QoS factors, and requirements set by the service clients and providers. It contains: a `Policy Specification` that collects the various policies defined as part of the service operation. `Policy Profiles` define the business rules and/or the functional and non-functional requirements of complete `Operation Sequences` or atomic `Operations` through `Policy Alternatives` that are expressed as a number of `Policy Assertions`. They implement a `Business Ruleset`, which is defined as a set of `Business Rules` specified by the owner of the service. These rules constitute an abstract `Service Policy` that all versions of the service have to respect.

Auxiliary concepts Figure 2 also contains a number of auxiliary concepts that do not belong to a specific layer but are used throughout the ASD. The `Role` concept inherits a very important as we will see in later sections property-domain to a number of concepts. `Service (notional)` acts as the main reference point for all *versions* of the service, expressed as `Versioned Specifications` - `Service Specifications` that aggregate the specifications from each layer, enriched with versioning information.

Public and Private sections. A vertical distinction of concepts is necessary in order to distinguish between service elements exposed to the clients of the service (public specifications) and the private specifications that are solely used for internal service purposes. The former are the access points for service clients that want to interact with the specified service. The latter are used for example to define the workflow of the process implemented by the service, and the service policies that are common among all versions of the service. Each service uses a combination of public and private concepts to specify itself; it exposes to its context only the public ones, but at the same time it incorporates the public concepts of the services it uses in turn. This implies that a uniform representation model is used across the network of interrelated services.

3.3 Service Schema Definition

The Service Schema Definition (SSD) consists of a number of elements and their relationships. In the following we discuss the structure and the properties of the SSD.

Definition of service schema. The SSD (for a particular service) is generated from the ASD by evaluating the concepts in it, i.e., by assigning a uniquely identified name to them, deciding on a specific property p_i for each properties-domain pd_i, and instantiating their relationships by replacing concepts with elements and assigning a value to their cardinality. The SSD is defined as a set of *elements* $\mathcal{E} = \{e_i\}, i = 1, \ldots, n$ and each element e is described by a tuple as follows in BNF-format:

$$e := < name, attribute^*, (property^?, relation^?)^*, visibility >$$
$$name := string \quad property := string \quad attribute := string$$
$$relation := RT_e|c||c|^{-1}$$
$$RT_e := \overset{c}{\overrightarrow{ee_i}} \mid \overset{a}{\overrightarrow{ee_i}} \mid \overset{s}{\overrightarrow{ee_i}} \;\; where \; i \geq 1$$
$$c := integer \quad visibility := public|private$$

'*' means 0-n occurrences and '?' means 0-1 occurrences

Example 3. Consider on ASD level the `Operation` concept defined as[2]:

$Operation(messagePattern, Att^*, \overset{c}{\overrightarrow{Operation\ Message}}\ |1|| \geq 1|^{-1}, public)$.

On SSD level, the evaluation of the concept for the creation of the SSD for the purchase order processing process from example 1 would e.g. generate:

$$e_1 = < val(Operation), val(messagePattern), Att^*,$$
$$val(\overset{c}{\overrightarrow{Operation\ Message}}\ |1|| \geq 1|^{-1}), public >$$
$$= < processOrder, request - response, Att^*, \overset{c}{\overrightarrow{e_1e_2}}\ |1||1|^{-1}, \overset{c}{\overrightarrow{e_1e_3}}\ |1||1|^{-1}, public >$$
$$e_2 = < purchaseOrder, Att^*, input, \ldots, public >$$
$$e_3 = < orderAcknowledge, Att^*, output, \ldots, public >$$

[2] *Note:* we are omitting some of the relationships of the concept for brevity.

where $val(messagePattern) = request - response$ means that $request - response$ is a property in the property domain `messagePattern`.

Note: since e_1 was generated by `Operation`, it is said to be an `Operation` element. In the same manner e_2 and e_3 are `Message` elements. For the remainder of this paper we shall use this convention to identify elements by the concepts they are generated from. In the same spirit, the distinction of concepts into layers (definition 1) also applies to elements. Therefore e_1, e_2, e_3 are said to belong to the *Structural* layer.

Constraining the service schema elements. The properties defined previously may restrict *(a)* the cardinality of the relationship(s) of the element, and *(b)* some of the properties of the related element(s).

Some properties of an element therefore define a number of constraints. Let's assume $\mathcal{P}_x$ the set of the properties assigned to an element x, then $\exists p \in \mathcal{P}_x$, the relation $\overrightarrow{xy}^{r}$ must satisfy the following: $\begin{cases} |\ |^{-1} = k \\ \text{and zero or more of:} \\ pd_y = p_{y_x} \end{cases}$
where the values of the property domain pd_y are restricted by the properties p_{y_x} (a subset of the original properties domain p_y defined by the property x).

Example 4. Consider the relation between `Operation` and `Message` elements in the structural layer. `Operation` has a property domain pd_1 = `messagePattern` that takes the following values/properties $\{one - way,$
$notification, request-response, solicit-response\}$, and `Message` has two property domains $pd_2 = message-role$ and $pd_3 = type$; the latter takes the property $\{required, provided\}$ and the former takes the properties $\{input, output, fault\}$. The property `one-way` defines the constraints:

$|\ |^{-1} = 1$, and $messageRole = input, type = required$, i.e. there can only be one `Message` element related to the `Operation` element, and it has to have the properties *input* and *required.* In a similar way, property `request-response` defines the constraints:

$$|\ |^{-1} = 2, \text{ and} \begin{cases} Msg_1.messageRole = input, Msg_1.type = required \\ Msg_2.messageRole = output, Msg_2.type = provided \end{cases}$$

or

$$|\ |^{-1} = 3, \text{ and} \begin{cases} \text{(as above)}, \\ Msg_3.messageRole = fault, Msg_3.type = provided \end{cases}$$

where Msg_1, Msg_2, and Msg_3 are occurrences of the `Message` element.

4 Service Schema Evolution

The evolution of the service is taking place through a series of discrete modifications to elements in its schema that constitute *evolutionary acts* in the sense that they are the carriers of change. These acts are expressed as sets of *operations* that may have consequences both inside and across the service.

4.1 Operations on the Service Schema

The following list is a minimal classification of operations and their semantics that we have identified, in the form of primitive changes on the elements of the SSD and/or their relationships.

1. *Insertion of Relationship between Elements* $[[ADD(e_i, e_j, r)]] : \mathcal{R}_{e_i} \rightarrow \mathcal{R}_{e_i} \cup \{\overrightarrow{e_i e_j}^{r}\}$, where $\mathcal{R}_{e_i}$ is the set of relationships of element e_i.
2. *Removal of Relationship between Elements* $[[DEL(e_i, e_j, r)]] : \mathcal{R}_{e_i} \rightarrow \mathcal{R}_{e_i} - \{\overrightarrow{e_i e_j}^{r}\}$. Bilateral relationships are deleted by performing this operation once for each direction.
3. *Insertion of Element* [[ADD(e)]]: $\mathcal{E} \rightarrow \mathcal{E} \cup \{e\}$ and/or not, $ADD(e_j, e, r)$ Insertion of an element e may be accompanied by the insertion of a relationship between another (preexisting) element e_j and e.
4. *Removal of Element* [[DEL(e)]]: $\mathcal{E} \rightarrow \mathcal{E} - \{e\}$ and $if\ \exists\ \overrightarrow{e_j e}^{r_j}\ \forall r_j \in \mathcal{R}, \forall j\ then$ $DEL(e_j, e, r_j)$,
   where $\mathcal{R} = \{\overrightarrow{e_i e_j}^{r}\ \forall i, \forall j, \forall r\}$the set of all relationships for all elements in $\mathcal{E}$. Removal of element e must always be preceded by the removal of all the relationships that e is participating in.
5. *Replacement of Property* [[REP(e,p_i, p_j)]]:$\mathcal{P}_e \rightarrow \mathcal{P}_e \cup \{p_j\} - \{p_i\}$ A property can either be replaced ($p_i, p_j \in pd_k$), added (for $p_i = \oslash$), or deleted by replacing it with an empty property ($p_j = \oslash$). In case that the properties constrain the cardinality of (some) relationships, then the appropriate operations on the relationships of e must be performed too.

This set of operations is *complete*; it can be easily shown that every SSD can be constructed from another SSD by a finite sequential application of additions, deletions, and replacements to it. In that sense, richer typologies of operations like the ones in [13], [17], and elsewhere, would be more convenient but not necessary for our approach.

4.2 Service Schema Versioning

The definition of SSD as a set of elements uniquely identified by their name which is subjected to a number of modifications allows us to draw on versioning techniques from Software Configuration Management. Based on the terminology used in [4], we define the spaces of a service:

Definition 3. ***Service spaces***

1. *The elements belonging to the SSD are the* product space *of the service, denoted by* ps, *which contains the specifications of the various versions of the service.*
2. *The temporal relationships between all the versions of the elements that constitute a service is called* version space *and is denoted by* vs.

3. *A version v^e represents the state of the evolving element* e *and it is characterized by the pair $v^e = (ps^e, vs^e)$, where ps^e denotes its state in the product space (i.e. the specification of the element), and vs^e its position in the version space (denoted by a version identifier). A version v^s of a service* s *is therefore defined as the set of the versions of its elements $v^s = \{v^e, \forall e \in \mathcal{E}\}$ at a given time (-point in the version space).*

Service versioning comprises service specifications as observed at discrete points in time. These are identifiable by a version identification number; each version is agnostic of the others and managed individually. Each of the service versions is created by applying a number of changes to a previous service version, which can be thought of as the *baseline* for that version. Information regarding the baseline of each version, and how a service version differs from its baseline constitutes the *version history* of a given service. There might be, for instance, three versions of the SSD (signified by the version ids '2.1', '2.1.1', and '2.1.2'). Each version is a full-fledged service schema specification and corresponds to different (possible) active versions of the service. By examining the version space of the service, the designer is able to infer that versions '2.1.1' and '2.1.2' use '2.1' as a baseline and additionally what changes where applied to '2.1' in order to produce these two versions. In that respect, the *extensional versioning* scheme is used to record version history and is defined as follows:

Definition 4. ***Extensional versioning***

1. *Let's assume a sequence of elementary change operations $op_1 \dots op_m$ which, when applied to one version of an element v^e, v^e_i, yields another version v^e_j, denoted as $v^e_i \rhd v^e_j$.*
2. *The extensional versioning of an element is the set V^e of all versions of* e *and is defined by $V^e = \{v^e_1, v^e_1 \rhd v^e_2, v^e_2 \rhd v^e_3 \dots, v^e_m \rhd v^e_n\}, m < n$. We therefore keep track of all versions with the corresponding operation changes. By extension then, the set of all versions of a service, is defined as $V^s = \{v^s_1, \dots, v^s_n\}$.*
3. *An* evolutionary act *can therefore be defined as the set of operations that transform version v^s_i of the service into version v^s_j and is denoted by $v^s_i \rhd v^s_j, i < j$.*

4.3 Consistency of Service Schema Evolution

The operations and the versioning approach presented in the previous section are generic enough to cover all possible modifications to service elements; but they do not make any effort to ensure that these modifications are meaningful - that is, they are not destructive for the SSD. For example, the *DEL(e)* operation removes an element completely from the current version of a service; but is this a valid operation for all elements of the SSD? In order to preserve the consistency of an SSD it is necessary to define a set of *Invariants*, such as those defined in [21]. These invariants (must) hold at every state of the SSD and ensure that the SSD is never left in an *inconsistent* state (i.e. a state that violates any invariant):

$\mathcal{INV}_1$: **Validity of the SSD.** $\mathcal{ASD} \models \mathcal{SSD}$: the SSD must always be *valid* with respect to the ASD, i.e., every element and relationship in the SSD must be able to be generated from the respective ASD concepts and relationships. This also includes the preservation of the semantics of the relationships, as defined in section 3.1.

$\mathcal{INV}_2$: **Reachability of Elements.** $\forall e \in \mathcal{E}$, *then* $\exists e_j \in \mathcal{E}$, $\exists r \in \mathcal{R}$, $\overrightarrow{ee_j}^{r}$: All elements must participate in at least one (directed) relationship with another element. If there are elements without any relationships in the schema then they are automatically deleted.

$\mathcal{INV}_3$: **Cardinality Constraint Preservation.** $\exists p \in \mathcal{P}_e, \exists \overrightarrow{ee_j}^{r}$ *with* $|j|^{-1} = k$ *then* $\overrightarrow{ee_j}^{r} \rhd \overrightarrow{ee_{j\prime}}^{r}, |j\prime|^{-1} = k$: If there is a property of the element that constraints the cardinality of some relationship of the element, then this constraint must be respected by all versions of the element.

$\mathcal{INV}_4$: **Existence Constraint on Composition.** $\forall e \in \mathcal{E}$ *if* $\overrightarrow{ee_j}^{c}$ *and* $DEL(e)$ *then* $DEL(e_j), \forall j$: If an element with composition relationships is deleted, then all its related elements through composition must be deleted too. (*Note:* The case of aggregation is covered by $\mathcal{INV}_1$.)

Now we can define the notions of *consistency* and *consistent evolutionary acts*:

Definition 5. *A version of the SSD is called Consistent iff it respects the set of obligatory invariants* $\mathcal{INV}^o = \{\mathcal{INV}_i, 1 \leq i \leq 4\}$. *Consistent evolutionary acts are therefore the series of operations that preserve the consistency of the SSD.*

For example, reducing the payload of a `Message` element by deleting one of the `Information Type` elements that it is related to is considered consistent. Deleting *all* of the `Information Type`s though is inconsistent, since it violates $\mathcal{INV}_1$; as shown in Figure 2, this relationship must have cardinality *at least* 1. The former then is a consistent evolutionary act, the latter isn't.

4.4 Conformance of Service Schema Versions

In summary, consistency ensures that the evolutionary acts are valid transformations of one version of the service SSD into another version. Taking into account the fact that services work in a network environment, using each other to achieve their stated objectives, creates the added necessity for the *preservation* of the service execution result. This ensures the seamless substitution of an SSD by a new version of it, without requiring any modifications by the clients of the service (its context); in other words, the *conformance* of the two versions in terms of expected results of service execution and not (only) in terms of specification:

Definition 6. *Given two consistent versions* v_i^s *and* v_j^s *of a service, they are called Conformant iff* v_i^s, v_j^s *can be interchangeable without requiring changes in their context.*

Example 5. Consider the case of the owner of the invoicing service wanting to expand its operations to international marketplaces. For that purpose, a new version of the service is created. Among other changes, information is added to the invoice data schema about the currency that the payments are to be made in, the tax regulations that apply to the specific invoice, etc. As long as the existing clients of the service can still use the same service specification by simply ignoring this additional information, the two service versions are considered to be conformant with respect to this change.

We have identified the following invariants that could ensure conformance:

$\mathcal{INV}_5$: **Co-Variance of Required Elements.** All elements that have the property `required` can only be *restricted*. This implies a restriction in the data type and the number of arguments (represented as relationships between elements). It can be stated as follows:

- in the number of relationships: $\exists p \in \mathcal{P}_e = required\ and\ if \exists\ \overset{r}{\overrightarrow{ee_j}}\ then\ \overset{r}{\overrightarrow{ee_j}} \triangleright \overset{r}{\overrightarrow{ee_{j'}}}, |j'|^{-1} \leq |j|^{-1}$, i.e. the element must have the same number or less relationships after any change to it,
- in the value domain of its properties: if the property is defined as a range of values, then this range can only be *restricted*.

This also holds for all elements in $\mathcal{R}_e$, the set of all relationships of e.

$\mathcal{INV}_6$: **Contra-Variance of Provided Elements.** All elements that have the property `provided` can only be *extended*. This implies an extension in the data type and the number of arguments. It can be stated as follows:

- in the number of relationships: $\exists p \in \mathcal{P}_e = provided\ and\ if \exists\ \overset{r}{\overrightarrow{ee_j}}\ then\ \overset{r}{\overrightarrow{ee_j}} \triangleright \overset{r}{\overrightarrow{ee_{j'}}}, |j'|^{-1} \geq |j|^{-1}$, i.e. the element must have the same number or more relationships after any change to it,
- in the value domain of its properties: if the property is defined as a range of values, then this range can only be *expanded*.

This also holds for all elements in $\mathcal{R}_e$.

$\mathcal{INV}_7$: **Finality of Cardinality-Constraining Properties.** The properties that constraint the cardinality of (some) relationship of a given service element are *final*, i.e. no such property of any element is allowed to be modified. Properties with no constraints on relationships can be subjects of change as long they respect the previous invariants.

For example, increasing the number of `Constraints` in a `Constraint Set` (see Figure 2) is only allowed if it is related to an `Operation Conditions` element that has the property `provided`, but not if it has the property `required`. The same applies also to the property `valueRange` in `Information Type`: the property can only be replaced by a 'smaller' range in the former case, and by a 'wider' one in the latter. Modifying the `MessagePattern` property in any way is forbidden since it constraints the cardinality of the `Operation` element with element(s) `Message`. Therefore:

Definition 7. *Conformance preservation is the property of an evolutionary act to respect the set* $\mathcal{INV}^c = \{\mathcal{INV}_i, 5 \leq i \leq 7\}$. *Conformance-preserving evolutionary acts therefore create conformant versions of the service.*

Example 6. Assume that the OtC service described in example 1 is used by a Purchase-to-Pay (PtP) process of another enterprise. PtP takes care of the procurement of goods process and at a certain step uses the invoice produced by the OtC to arrange for payments. In that case, the same invoice document (more accurately, the `Information Type` element that corresponds to the invoice) is a provided element for OtC and according to $\mathcal{INV}_6$ it can be extended in the manner described above. However, since PtP uses the same element as an input (and therefore it is required for it), then $\mathcal{INV}_5$ forbids this modification. In that case, the PtP service can not use the new version of OtC and has to rely on the previous one to do business (ensuring in that way that there are no misunderstandings in the currency that the transactions take place).

What is illustrated by the previous example is the fact that $\mathcal{INV}^c$ is a set of *necessary* but not *sufficient* conditions for conformance. That is a by-product of the loosely-coupled nature of the service-oriented architecture: it is not desirable to be able to reason explicitly about the effect of a service change at provider side to its client services. This is due to the fact that client services should be oblivious to the changes that happen to a provider service. This enforces a modus operandi based on the separation of concerns: each party will decide from their own perspective whether a new version and the evolutionary act that created it is conformance-preserving with respect to their own services. In that sense, two services using each other have an implicit *contract* between them, enforced partially by each side using their interpretation of $\mathcal{INV}^c$. Every new version that is issued by a service proposes the alteration of this contract between them; it is up to the other party to decide whether the proposed changes are acceptable or not.

5 Conclusions and Future Work

In this paper we have introduced service evolution management facilities that identify and classify all kinds of permissible changes to services, analyze the propagation effects of changes, introduce version control mechanisms, validate the completeness of a change, and maintain consistency by ensuring that a service is a well-behaved collection of service changes and versions. The service evolution management facilities rely on a service specification reference model that abstracts away from the idiosyncracies and syntactic nuances of current standards and provides a theoretical approach to service evolution. The service specification reference model contains an abstract service definition model (ASD) that comprises generic concepts and inter-relationships that are common to all service schemas in three layers. Thus far, we have concentrated on representing and analyzing the behavior of multiple active service versions that are mutually conformant with respect to a contract from both the perspective of the service provider and the service client. In the future, we expect to concentrate on developing formalisms and proofs for the service regulatory layer and connect them with current work, so as to be able to prove the completeness and soundness of the overall approach. Another extension of this work is to focus on relaxed co- and contra-variance mechanisms for more flexible service evolution purposes, e.g., exception handling.

References

1. Ponnekanti, S., Fox, A.: Interoperability among independently evolving web services. In: Middleware, pp. 331–351 (2004)
2. Benatallah, B., Casati, F., Grigori, D., Nezhad, H.R.M., Toumani, F.: Developing adapters for web services integration. In: CAiSE, pp. 415–429 (2005)
3. Tichy, W.F.: Tools for software configuration management. In: SCM, pp. 1–20 (1988)
4. Conradi, R., Westfechtel, B.: Version models for software configuration management. ACM Comput. Surv. 30(2), 232–282 (1998)
5. Brown, K., Ellis, M.: Best practices for Web services versioning. IBM developerWorks White Paper (2005)
6. Russell, M.: Manage message contract changes with versioning. IBM developerWorks White Paper (2005)
7. Butek, R.: Make minor backward-compatible changes to your Web services. IBM developerWorks White Paper (2004)
8. Poulin, M.: Service Versioning For SOA. SOAWorld Magazine 6(7) (2006)
9. Kaminski, P., Litoiu, M., Müller, H.A.: A design technique for evolving web services. In: CASCON, pp. 303–317 (2006)
10. Hoylen, S.(ed.): XML Schema Versioning Use Cases. W3C XML Schema Working Group Draft (2006)
11. Velegrakis, Y., Miller, R.J., Popa, L.: Mapping adaptation under evolving schemas. In: VLDB 2003: Proceedings of the 29th international conference on Very large data bases, VLDB Endowment, pp. 584–595 (2003)
12. Yu, C., Popa, L.: Semantic adaptation of schema mappings when schemas evolve. In: VLDB 2005: Proceedings of the 31st international conference on Very large data bases, VLDB Endowment, pp. 1006–1017 (2005)
13. Salinesi, C., Etien, A., Zoukar, I.: A Systematic Approach to Express IS Evolution Requirements Using Gap Modelling and Similarity Modelling Techniques. In: Persson, A., Stirna, J. (eds.) CAiSE 2004. LNCS, vol. 3084, pp. 338–352. Springer, Heidelberg (2004)
14. Casati, F., Ceri, S., Pernici, B., Pozzi, G.: Workflow evolution. In: Thalheim, B. (ed.) ER 1996. LNCS, vol. 1157, pp. 438–455. Springer, London (1996)
15. Reichert, M., Dadam, P.: ADEPTflex - supporting dynamic changes of workflows without losing control. J. Intell. Inf. Syst. 10(2), 93–129 (1998)
16. Joeris, G., Herzog, O.: Managing evolving workflow specifications with schema versioning and migration rules (1999)
17. Weber, B., Rinderle, S., Reichert, M.: Change Patterns and Change Support Features in Process-Aware Information Systems. In: Krogstie, J., Opdahl, A., Sindre, G. (eds.) CAiSE 2007 and WES 2007. LNCS, vol. 4495, pp. 574–588. Springer, Heidelberg (2007)
18. Rumbaugh, J., Jacobson, I., Booch, G.: Unified Modeling Language Reference Manual, 2nd edn. Addison-Wesley Object Technology Series. Addison-Wesley Professional, Reading (2004)
19. Everware-CBDI Inc.: CBDI-SAE^{TM} Meta Model for SOA Version 2.0. (2007), `http://www.cbdiforum.com/public/meta_model_v2.php`
20. Dubray, J.J.: WSPER An abstract SOA framework (2007), `http://www.wsper.org/primer.html`
21. Banerjee, J., Kim, W., Kim, H.J., Korth, H.F.: Semantics and implementation of schema evolution in object-oriented databases. In: SIGMOD 1987: Proceedings of the 1987 ACM SIGMOD international conference on Management of data, pp. 311–322. ACM Press, New York (1987)

On the Definition of Service Granularity and Its Architectural Impact

Raf Haesen[1,2], Monique Snoeck[1], Wilfried Lemahieu[1], and Stephan Poelmans[2]

[1] Department of Decision Sciences & Information Management, Katholieke Universiteit Leuven, Belgium
firstName.lastName@econ.kuleuven.be
[2] Hogeschool-Universiteit Brussel, Belgium
firstName.lastName@hubrussel.be

Abstract. Service granularity generally refers to the size of a service. The fact that services should be large-sized or coarse-grained is often postulated as a fundamental design principle of service oriented architecture (SOA). However, multiple meanings are put on the term granularity and the impact of granularity on architectural qualities is not always clear. In order to structure the discussion, we propose a classification of service granularity types that reflects three different interpretations. Firstly, *functionality granularity* refers to how much functionality is offered by a service. Secondly, *data granularity* reflects the amount of data that is exchanged with a service. Finally, the *business value granularity* of a service indicates to which extent the service provides added business value. For each of these types, we discuss the impact of granularity on a set of architectural concerns, such as performance, reusability and flexibility. We illustrate each granularity type with small examples and we present some preliminary ideas of how controlling granularity may assist in alleviating some architectural issues as we encounter them in a large-sized bank-insurance company that is currently migrating to SOA.

Keywords: granularity, service oriented architecture, component based development, architectural qualities, impact analysis.

1 Introduction

Service granularity generally refers to the size of a service. The fact that services should be large-sized or coarse-grained is often postulated as a fundamental design principle of service oriented architecture (SOA). This advice is a rather obvious consequence of the quest for design artefacts that are defined at a high level of abstraction. Indeed, business people are generally not interested in fine-grained, implementation-level concepts for the construction of automated support for their work. Instead, they prefer to use and reuse automated chunks of functionality (or services) that correspond to units of work as they are used to handle them. These units are typically broader in scope than units that are processed in a software program. For example, services that provide support

Z. Bellahsène and M. Léonard (Eds.): CAiSE 2008, LNCS 5074, pp. 375–389, 2008.

for (parts of) business processes offer a high amount of functionality and are therefore labelled as coarse-grained.

It is interesting to compare services to other units of software construction that were proposed earlier, such as objects and components. The transition from objects to components and then to services is generally associated with an increase in granularity, i.e. from fine-grained objects, to coarser-grained components and even more coarse-grained services [1,2]. In what follows we briefly elaborate on these transitions.

The object oriented paradigm introduced, among others, the idea to create units of abstraction that are close to real-world concepts. However, the resulting objects turned out to be too fine-grained and biased towards implementation to be useful for the development of business applications. These issues were partly solved with the introduction of component based development (CBD), which promotes the creation of coarser-grained components. To further stress the importance of making abstractions that are recognisable for the business, the difference between generic software components and business components was made. A *business component* is generally defined as a software component that implements functionality from a particular business domain [3,4]. In general, a business component encapsulates a business-level entity or process. Therefore business components tend to be defined at higher (and hence improved) levels of abstraction.

The step towards service oriented computing (SOC) caused a further increase in granularity. While components are building blocks for applications, services are access points to an implementation that potentially covers multiple applications. As already stated, these services encapsulate business-level functionality that may even cover (parts of) enterprise-wide processes. As a consequence, their granularity is coarser than that of objects and components.

Instead of merely advocating for coarse-grained services, it is more appropriate to firstly acknowledge that *the spectrum of possible service granularity levels has become wider*. Indeed, we will show that both coarse-grained and fine-grained services can have positive impact on the architecture. As a consequence more refined judgments to control granularity are required. A few unanswered questions concerning service granularity are:

- What is the impact of service granularity on architectural qualities, such as performance, reusability and flexibility?
- How can service granularity be measured?
- Is there an upper limit for service granularity? In other words, are there any criteria that rather favour finer-grained services?

Defining granularity is quite complex since it cannot draw on theoretical groundings. Indeed, granularity can hardly be measured in terms of absolute numbers, because of the subjectivity of the related concepts that may determine the granularity in question. For example, a service may be defined in terms of an activity that is executed by that service. However, the concept ‘activity’ itself has a vague, hierarchical nature: it can represent a simple state change, the work performed by one actor in one unit of time, or even a complete business process (see e.g. [5]). This makes it far from straightforward to define granularity in terms of executed activities.

In what follows we attempt to provide initial answers to the above questions about service granularity. The paper is organised as follows. Section 2 gives an overview of related work. Section 3 classifies multiple interpretations of service granularity from an interface point of view. For each granularity type, we present some small examples and we discuss the impact of granularity on architectural qualities. In Section 4, we discuss the difference between the interface and realisation viewpoint on granularity. Section 5 briefly discusses some evaluation tracks and outlines areas of future research. Finally Section 6 concludes the paper.

2 Related Work

A multitude of scientific papers, industrial papers and web entries touch upon the topic of service and component granularity. Until now, most attention was paid to measuring and assessing the impact of granularity of *components*. As argued by Herzum and Sims [4, pg. 38], component granularity is defined recursively, since a component can be defined as the composition of finer-grained components. This recursion can be discrete or continuous, respectively depending on whether the granularity levels are predefined or not. Herzum and Sims prefer the discrete form since it caters for reduced levels of design complexity. They distinguish between system-level components, business components and distributed components in descending order of granularity. System-level components are composed of business components, while a business component is composed of distributed components.

Since component based development mainly focuses on reuse, the relationship between granularity and reusability is widely discussed. Despite the general tendency towards design artefacts of increasing granularity levels, some refined observations were made [6,7]. Firstly, coarse-grained components have high reuse efficiency (because of a high contribution to the system) but low reusability (because of highly specific problem solving capabilities). Furthermore, the coarser the granularity is, the lower the composition cost is because of the fewer number of components and interactions that are required. Finally, Wang et al. [8] argue that, if a component cannot absorb requirement changes through configuration (e.g. business rules, parameterisation, etc.), then its granularity should be decreased. Besides the impact on reusability, Vitharana et al. [9] concluded a negative correlation between granularity and other managerial goals such as cost effectiveness, customization and maintainability. On the other hand, increasing levels of granularity tend to ease component assembly.

Sims [10] gives some clues of how service granularity may be measured, i.e. (1) by counting the number of components invoked through an operation on a service interface, (2) by counting the number of function points for a component, or (3) by counting the number of database tables updated. As an alternative to the latter, the number of update operations invoked on a component can be counted or the number of types in the information model if both read and update access are relevant.

Besides these quantitative results, many authors provide an overview of general design principles to optimize service and component granularity. In what follows we give an overview of some of these principles:

- The 'right' granularity of a service or component generally varies over time [4]. A service or component that seems appropriate nowadays was maybe unsuited a few years ago because both markets and technology constantly evolve. For example, since SOA enables searching for services at runtime (e.g. facilitated by the UDDI standard), registry management and brokering are typical services that were less important before the introduction of SOA. Moreover, when particular vertical service or component standards mature, the corresponding industries can be relieved from searching for appropriate granularity levels.
- Good candidates for business components or services represent real and independent concepts to business domain people [4,11,12]. In other words, they should not be based on implementation concepts and the scope should be understandable without further context information.
- Herzum and Sims [4] give additional heuristics to identify right-sized business components: they should be easily marketable, highly usable and reusable; they should support autonomous development and should correspond to units of stability. Furthermore they should adhere to several cohesion principles, i.e. temporal (provide for development and evolution stability), functional (combine logically related functions), run-time (run e.g. computing-intensive tasks in the same address space) and actor (users of a given component should be similar) cohesion.
- If service granularity is defined in terms of the number of operations delivered [12], a service should not be too coarse as it will increase the number of consumers. Hence, a possible service change may impact many consumers. Furthermore, a huge list of operations does not provide a clear overview of which functionality is offered.
- A service should contain support for transaction integrity and compensation [13,14]. Put otherwise, all activities executed by the service should be in the scope of one transaction. If a service fails during a transaction, it should provide a compensation mechanism to undo possible changes.
- Finding the right granularity is a matter of balancing between multiple criteria [15]. For example, coarse-grained services require less network roundtrips as the execution state is contained in the message. On the other hand, small services generally require uncomplicated input data and are more easily composed.

This literature overview shows that the existing knowledge about service granularity is quite fragmented: each author takes a particular view on the subject to devise criteria for granularity optimisation, without making the considered context explicit. In the following section we attempt to consolidate and extend the insights on service granularity.

3 Service Granularity Types

In order to structure the discussion of service granularity, we propose a classification of service granularity types that reflects three different interpretations: firstly, *functionality granularity* refers to how much functionality is offered by a service. Secondly, *data granularity* reflects the amount of data that is exchanged with a service. Finally, the *business value granularity* of a service indicates to which extent the service provides added business value. For each of these types, we describe the impact of granularity on a set of architectural concerns, such as performance, reusability and flexibility.

It should be noticed that we define different granularity types only by looking at the *interface* of the service. In other words, we describe granularity from the point of view of a consumer, although we assess the impact for both the consumer and the provider. In section 4 we briefly describe service granularity from the *realisation* viewpoint, which inspects the implementation of the service. Furthermore we indicate the differences between the interface and realisation perspectives.

The classification of service granularity is schematically represented in Figure 1. Concerning data granularity, a distinction is made between data that is sent to the service (*input data granularity*) and data that is returned by the service (*output data granularity*). For functionality granularity, we distinguish between the amount of functionality that is always offered when calling the service (*default functionality granularity*) and the functionality that can optionally be offered (*parameterised functionality granularity*).

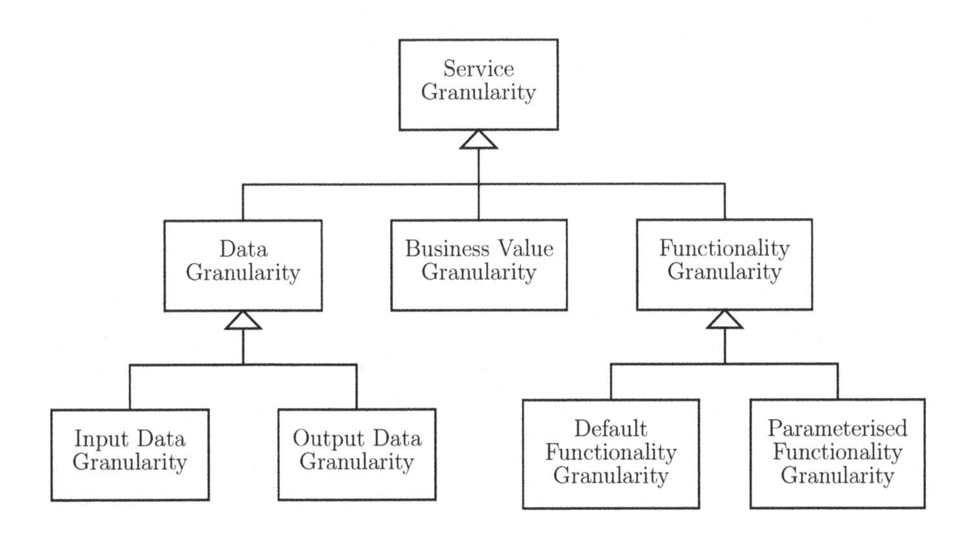

Fig. 1. Classification of service granularity types

Table 1 gives an overview of the architectural impact of coarse-grained services for each of the five granularity types. In the last column we indicate whether the impact is advantageous or disadvantageous for the consumer and the provider. In the following sections we go into more detail.

Table 1. Architectural impact of *coarse-grained* services

granularity type	**architectural impact of coarseness**	**party involved**
Input Data Granularity	less communication overhead better transactional support data possibly outdated no state lost better scalability no coordination required	p+,c+ p+ p- p+ p+ c+
Output Data Granularity	less communication overhead higher reusability	p+,c+ c+
Default Functionality Granularity	higher reuse efficiency lower reusability stability problems	c+ c- c-
Parameterised Functionality Granularity	higher reuse efficiency higher reusability no stability problems difficult implementation	c+ c+ c+ p-
Business Value Granularity	clear architecture control points consumer needs satisfied	p+ c+

Legend: p = provider, c = consumer,
+ = possitive impact, - = negative impact

3.1 Input Data Granularity

The input data granularity of a service reflects how much data is passed on to that service by a service consumer. A coarse-grained service requires one or more business objects as parameters while a fine-grained service has few or even no input parameters. Not only the number of parameters influences granularity but also their type. For example, the (data) granularity of insurance contract is bigger than that of zip code, hence when used as input parameter, they influence the input data granularity of the service accordingly. In general, a data element is coarser-grained if it is composed of other data elements and if the datatypes of its attributes are other data elements instead of primitive datatypes.

Example. With respect to input data granularity, the service ValidateContract (Contract c) is coarser-grained than the service ValidateAddress (Address a).[1]

Discussion. It is generally recommended to create coarse-grained services of this type for several reasons: Firstly, if the business objects are transferred by value, the communication overhead is reduced since the number of network transfers is decreased. Especially in the case of Web services, this overhead is high since asynchronous messaging requires multiple queuing operations and numerous XML transformations [16]. Moreover, if a service has to update multiple data elements in one transaction, it is best to pass all data at the same time, since this approach makes compensation mechanisms unnecessary. On the other hand, the input data of a coarse-grained service may be outdated if it was collected during previous service calls (i.e. not in the same transaction). Therefore the input data should be validated by the service.

It is common practice to make a service document-based, i.e. to include the entire execution context in the input message of a service, which makes the service coarse-grained. Since the provider service itself does not maintain state in this case, it is called stateless [17]. Statelessness is generally considered as a desired property for many reasons: firstly, the call of a service (operation) does not depend on previous calls, which eliminates the risk of losing state between different calls. Secondly, statelessness ensures higher scalability since more provider instances can be added if demand is high. Finally, the consumer is relieved from coordinating several fine-grained services if all data can be sent at once.

3.2 Output Data Granularity

The output data granularity of a service indicates how much data is returned to the service consumer. A coarse-grained service returns one or more (references to) business objects while a fine-grained service rather returns nothing or a few attributes. The above-mentioned remark about granularity of data elements also applies to output data granularity.

Example. With respect to output data granularity, the service Client SearchCustomer() is coarser-grained than the service Date SearchBirthDate().

Discussion. Generally it is beneficial to create services that are coarse-grained with respect to output data: similarly as for input data granularity, the number of consequent calls can be kept small if much data is returned by value. Secondly a coarse-grained service of this type doesn't hamper reuse since the superfluous part can simply be discarded by the service consumer. Although in this case

[1] All examples follow the format OutputParameterType ServiceName (InputParameterType name), whereby both the input and output parameter are only specified if they influence the corresponding level of granularity. Although all service examples are represented as a single conceptual operation, their interface might consist of multiple operations that can be invoked.

some network bandwidth might be wasted, this generally doesn't pose any severe problems, certainly not for intra-enterprise service interactions.

It is possible to make the output data granularity more dynamic by specifying a list of data elements that should be returned. However, this increases the amount of input data and may decrease the comprehensibility of the service. Alternatively it is possible to develop multiple services with different output data granularities, whereby a coarser-grained service is composed of the finer-grained services. These services are called *multi-grained* in [1, chap. 2].

3.3 Default Functionality Granularity

The default functionality granularity of a service indicates how much functionality is offered in any case, i.e. the amount of functionality that cannot be adjusted by setting some parameters. A service that performs CRUDS (create, read, update, delete, search) functionality is finer-grained than a service that also executes logic. Moreover, services that aggregate (e.g. orchestrate) other services are typically coarser-grained than their constituents. For example a service that supports a business process is coarser-grained that a service that executes a single activity of that process.

Example. With respect to default functionality granularity, the service HandleClaimProcess() is coarser-grained than the service IdentifyCustomer().

Discussion. This definition of service granularity is usually implied since it directly reflects the amount of work that is performed by the service. As we already discussed earlier, business people prefer to use and reuse services that correspond to units of work as they are used to handle them. These units of work are typically coarser-grained than the units that are processed in a software program.

The architectural consequences of coarse-grained services are similar to those of coarse-grained components, which were discussed in section 2. Firstly, the reuse efficiency is high because of the large contribution that is made by the service. Secondly, the reusability of coarse-grained services is low since the service can only be used to solve specific problems. For example the service HandleClaimProcess() will only be used in the claims domain, whereas IdentifyCustomer() may be used in multiple domains. Finally, chances are high that a change to some of the many functionalities in a coarse-grained service will cause changes to its interface. In other words, the service is unstable since it has limited capabilities to adapt to changes. The latter two arguments may be valid reasons to limit the granularity of a service.

3.4 Parameterised Functionality Granularity

The parameterised functionality granularity of a service defines the amount of functionality that optionally *can* be offered by a service. A coarse-grained (fine-grained) service offers many (a few) facilities to let the consumer configure the desired functionality, e.g. by means of input parameters. Not only the number of parameters, but also their type defines the coarseness of the service. For example the parameter may be a boolean which represents a binary choice, or it may as well be a structured file that is being interpreted by the service. With other things being the same, the former case will yield a service with a smaller parameterised functionality granularity than the latter.

Example. With respect to parameterised functionality granularity, the service HandleProcess (Process aProcess) is coarser-grained than the service WriteCredit (boolean alsoValidate).

Discussion. Since a coarse-grained service of this type makes the service rather generic, it can easily be used in different contexts. Indeed, each different combination of input parameters yields a different behaviour of the service and therefore the service is highly reusable. Schmelzer [18] argues that, if we push this line of reasoning to the extreme, we would create a service DoSomething() that fulfils every possible need. He continues that, despite the apparent advantages of this service construction method, it has a major drawback in that it shifts the problem to the implementation of the service. Additionally, the usage tends to become more complex to the consumers as well, as they need to understand how the – often complicated – parameterisation mechanism works.

Whereas a small-grained service obviously is not reuse efficient, the consumer can control the reuse efficiency of coarse-grained services through parameter setting. For example, the service HandleProcess (Process aProcess) is reuse efficient if a complex process description is provided as input, while a straightforward process with only a few activities as input will limit the contribution of the service. Finally, a coarse-grained service is typically protective to changes (or stable) since these changes can be absorbed through configuration.

3.5 Business Value Granularity

Business value granularity measures the appropriateness of a service for the business. In other words, this type of granularity indicates the value being attached to a service. The analysis of value creation is an essential part of business modelling techniques, such as the *e^3-value* approach [19] or the i^* framework [20]. In most general terms, those approaches capture value exchanges or the extent to which the creation of value (i.e. the execution of services in our case) contributes to the goals and visions of an organisation. The extent to which a service *directly* contributes to a high-level business goal can therefore be seen as a metric for business value granularity. As an example, consider the goal-oriented derivation of services as proposed by Rolland et al. [21]. More specifically, each service realises the fulfilment of an *intention* or *goal* by following a particular *strategy*. A goal can be seen as a state to be reached while a strategy represents an approach to reach a particular state. Because the resulting services have close ties to business goals, they have high levels of business value granularity by construction.

Example. With respect to business value granularity, the service ConcludeInsuranceAgreement() is coarser-grained than the service AddClient(), which is coarser-grained than the service ValidateAccountNumber().

Discussion. The business value granularity obviously is an important indicator for business people since it gives an overview of which services should receive most attention. Dreyfus and Iyer argue that, given the complexity of architecture and limited organisational resources to implement and modify the architecture, it is indispensable to choose a subset of systems that are deemed important because of their influence on the emergence of the architecture [22]. These systems support the business goals of the enterprise and are denoted as *architecture control points (ACP)*. With respect to business value granularity, coarse-grained services and their implementing systems are the ACPs of an organisation. Services with high business value are beneficial to their consumers as well since they are more likely to satisfy the needs of those consumers. On the contrary, the composition of multiple fine-grained services with respect to business value generally causes more overhead for the consumer. Therefore companies tend to bundle multiple services into one package with increased business value granularity. We refer the reader to the work of Baida for more information about service bundling [23].

One could argue that high levels of functionality granularity automatically imply high levels of business value granularity. For example, a service that supports insurance claim handling consists of many process steps (i.e. it has high functionality granularity) and that service is highly valued in the insurance domain (i.e. it has high business value granularity). However, other examples indicate a negative relationship between these two types of granularity. Firstly, consider a service that consolidates accounting data from different information systems once a month, in batch mode. As this service executes multiple steps (data retrieval, comparison, cleansing, etc.) it has a high functionality granularity. On the other hand, the business value granularity is low since it merely corrects (or even just reports on) inconsistencies between data sources. As a second example, consider an accurate and zero-latency currency conversion service that is being used inside the company as well by external clients. Although the service has a low functionality granularity, its business value granularity is high because of its high Quality of Service (QoS) and level of reuse.

4 Interface Versus Realisation View on Granularity

In the discussions of the different service granularity types we only took the interface viewpoint into account. In other words, only the externally visible properties of a service were considered during the evaluation of the influence of service granularity on both the consumer and the provider. However, this viewpoint does not reveal all architectural consequences. Indeed, granularity can also be discussed by looking at how the service is realised in the information system(s). This viewpoint is therefore of particular interest to the service provider. In what follows, we briefly discuss the differences between the interface and realisation view on the three types of granularity. By means of a few examples, we will show that both views on granularity are not always in accordance with each other.

- **Data granularity:** Many industrial consortia have proposed sets of standardised messages that can be exchanged between different parties. For example, the ACORD (Association for Cooperative Operations Research and Development) standards define messages for the insurance and related financial services industries; likewise SWIFT (Society for Worldwide Interbank Financial Telecommunication) defines messages that are exchanged between banks and other financial institutions. These messages are typically very extended since they ought to cover all data that may be relevant during a particular transaction. Since services in these particular domains may (and should) rely on standards for their data exchange, these services are coarse-grained with respect to (input and output) data granularity. Although a lot of data is exchanged, this does not imply that all data is effectively being used during the service execution. Hence from the interface viewpoint the service is coarse-grained while from the realisation point of view it may be fine-grained.
- **Functionality granularity:** We argued that an orchestration service is coarser-grained that its constituents with respect to default functionality granularity. In fact, the granularity of the former is the sum of the granularities of the orchestrated services plus the granularity of the coordination logic. From the realisation point of view however, the orchestration service only implements the coordination logic. Therefore the service can be implemented without much effort, although it is coarse-grained from the interface viewpoint.
- **Business value granularity:** The difference between the interface and realisation viewpoint is particularly relevant to business value granularity. Suppose that a provider wants to determine how much business value is attached to the services that are delivered by ICT infrastructure components. For example, consider a database management system (DBMS) that delivers data storage, data retrieval and transaction processing services. From an interface point of view, these services are fine-grained with respect to business value granularity, since they do not directly contribute to high-level business goals. Suppose that from a realisation viewpoint, not much business value would be attached to these services either. This would imply that ICT could just as well reimplement the data services for each business case that would require these services. Obviously this inefficient approach would repeatedly generate pointless ICT costs. Therefore the business should appreciate the use of a DBMS that is proven to be reliable, reusable and high-performing. In other words, from a realisation viewpoint, the business value granularity of ICT infrastructure components is high.

  Note that the distinction between the two viewpoints on business value granularity has far-reaching consequences for the interrelation between business and ICT. From the interface viewpoint, business would only be interested in the fulfilment of their requirements towards ICT without considering the approach adopted by ICT. From the implementation viewpoint though, business would appreciate the optimisation strategies that are chosen by ICT, such as the construction of reusable and flexible infrastructures. In this case costs should be distributed among all consumers that (will) use these infrastructures. This may not be a straightforward task if not all consumers are known in advance.

5 Evaluation and Future Work

The results of this work are currently being validated at KBC Bank & Insurance Group, one of the top three bankinsurers in Belgium with a key position in Central-Europe. To have control over granularity is one of the major concerns in their migration to SOA. The validation of this work consists of two parts: firstly, the presented classification is in general adopted by KBC. This means that the impact of each service under development is verified with respect to each type of granularity. Moreover, to the best of our knowledge, our classification covers all aspects of granularity that are discussed in the existing literature. The second part of validation considers each type of the granularity in more detail. Whereas the validation of functionality and business value granularity are left for future work, we already focused on data granularity.

In general it can be observed that current services research mainly focuses on the issue of flexibility because services generally represent "units of functionality" that need to be coordinated. Therefore, too little attention is paid to the data perspective on services. To alleviate this problem, we elaborated guidelines to optimise the input data granularity of services. This resulted in the active-passive hybrid data collection pattern [24], which distributes the responsibility of collecting data across the service consumer and provider. The decisions are mainly based on the properties of the data to be collected, such as their availability, visibility and accessibility.

As part of future research, we will propose concrete metrics for all granularity types in two different contexts. Firstly, we will define metrics in an event-driven SOA that is based on the MERODE methodology [25]. Although the architecture is object based (as a possible service implementation) and therefore of limited use on enterprise level, all models and concepts are formally defined, which allows inferring formal metrics as well. Secondly, we will extend our approach in the context of BECO [26]. BECO itself is an extension to MERODE that defines the enactment of business processes by means of an event-based coordination of components. This approach allows incorporating enterprise-class concerns, such as the integration of legacy and the treatment of business processes as first-class citizens.

Finally, we perform research on rules to determine which granularity levels are appropriate in a particular context. At the ICT department of KBC, all projects are firstly analysed in the 'work preparation' stage before they are effectively being implemented in the 'work execution' stage. It is obvious that the concerns of the people in the two stages are different, and yet, the same service concept is used by both. For example during work preparation, the problem is firstly assigned to a particular service domain, such as the claims domain. Subsequently, the architects have to delineate the relevant services that will be implemented in the scope of the project. Now the services should be defined at such a level of granularity that changes to the existing service portfolio can be assessed. For example, the introduction of a service for claim handling will affect other domains such as accounting, payments, etc. Finally, to enable work execution, the services must be decomposed into even more fine-grained services. For example, the service for claim handling will rely on some backend services that contain business logic, some services that maintain process state, some services that generate user interfaces, etc. We will verify how the proposed granularity types can be used to derive appropriate granularity levels in a given context.

6 Conclusion

In this paper we attempted to structure the discussion of service granularity. Although the importance of coarse-grained services is often stated, we argued that enterprise architects nowadays have to deal with a broad spectrum of possible service granularity levels for different granularity types. From an interface perspective, we distinguished between data granularity, functionality granularity and business value granularity. By means of some extreme values for each granularity type we discussed the impact on architectural concerns such as reusability, reuse efficiency, stability, performance, etc. Although the interface perspective reveals several consequences of granularity for both consumer and provider, the provider will also be interested in the realisation view on granularity. By means of some examples, we showed that both views are not always in accordance with each other. Finally we presented some preliminary ideas of how granularity may assist in alleviating some architectural issues as we currently encounter them at KBC, such as the data issues around services and a granularity-driven delineation of services.

Acknowledgements

This work was funded by the KBC-Vlekho-K.U.Leuven research chair on 'Service and Component Based Development' sponsored by KBC Bank & Insurance Group.

References

1. McGovern, J., Tyagi, S., Stevens, M., Mathew, S.: Java Web Services Architecture. Morgan Kaufmann, San Diego (2003)
2. Hanson, J.: Coarse-grained interfaces enable service composition in soa (August 2003), `http://articles.techrepublic.com.com/5100-22-5064520.html`
3. Fellner, K.J., Turowski, K.: Classification framework for business components. In: Proceedings of the 33rd Annual Hawaii International Conference on System Sciences (HICSS-33). IEEE Computer Society, Maui (2000)
4. Herzum, P., Sims, O.: Business Components Factory: A Comprehensive Overview of Component-Based Development for the Enterprise. John Wiley & Sons, Inc., New York (2000)

5. Goedertier, S., Haesen, R., Vanthienen, J.: EM-BrA^{2}CE v0.1: A vocabulary and execution model for declarative business process modeling. FETEW Research Report KBI_0728, K.U.Leuven (2007)
6. Mili, H., Mili, A., Yacoub, S., Addy, E.: Reuse-Based Software Engineering: Techniques, Organizations, and Controls. John Wiley & Sons, Chichester (2002)
7. Wang, Z., Xu, X., Zhan, D.: A survey of business component identification methods and related techniques. International Journal of Information Technology 2, 229–238 (2005)
8. Wang, Z., Zhan, D.C., Xu, X.F.: STCIM: a dynamic granularity oriented and stability based component identification method. ACM SIGSOFT Software Engineering Notes 31(3), 1–14 (2006)
9. Vitharana, P., Jain, H., Zahedi, F.: Strategy-based design of reusable business components. IEEE Transactions on Systems, Man and Cybernetics, Part C: Applications and Reviews 34(4), 460–474 (2004)
10. Sims, O.: Developing the architectural framework for SOA - part 2-service granularity and dependency management. CBDI Forum Journal (June 2005)
11. Erradi, A., Anand, S., Kulkarni, N.: SOAF: An architectural framework for service definition and realization. In: Proceedings of the IEEE International Conference on Services Computing (SCC 2006), pp. 151–158. IEEE Computer Society, Washington, DC (2006)
12. Artus, D.J.: SOA realization: Service design principles. IBM Developer Works (February 2006), http://www-128.ibm.com/developerworks/webservices/library/ws-soa-design/
13. Wang, Z., Xu, X., Zhan, D.: Normal forms and normalized design method for business service. In: ICEBE 2005: Proceedings of the IEEE International Conference on e-Business Engineering, pp. 79–86. IEEE Computer Society, Washington, DC (2005)
14. Foody, D.: Getting web service granularity right (August 2005), http://www.soa-zone.com/index.php?/archives/11-Getting-web-service-granularity-right.html
15. Wilkes, L., Veryard, R.: Service-oriented architecture: Considerations for agile systems (April 2004), http://msdn2.microsoft.com/en-us/library/aa480028.aspx
16. Bussler, C.: The fractal nature of web services. IEEE Computer 40(3), 93–95 (2007)
17. Foster, I., Frey, J., Graham, S., Tuecke, S., Czajkowski, K., Ferguson, D., Leymann, F., Nally, M., Sedukhin, I., Snelling, D., Storey, T., Vambenepe, W., Weerawarana, S.: Modeling stateful resources with web services (March 2004)
18. Schmelzer, R.: Solving the service granularity challenge (March 2006), http://www.zapthink.com/report.html?id=ZAPFLASH-200639
19. Gordijn, J., Akkermans, H.: Value based requirements engineering: exploring innovative e-commerce ideas. Requirements Engineering Journal 8(2), 114–134 (2003)
20. Yu, E.S.K.: Towards modeling and reasoning support for early-phase requirements engineering. In: Proceedings of the 3rd IEEE International Symposium on Requirements Engineering (RE 1997), pp. 226–235. IEEE Computer Society, Annapolis (1997)
21. Rolland, C., Kaabi, R.S., Kraïem, N.: On ISOA: Intentional Services Oriented Architecture. In: Krogstie, J., Opdahl, A., Sindre, G. (eds.) CAiSE 2007 and WES 2007. LNCS, vol. 4495, pp. 158–172. Springer, Heidelberg (2007)
22. Dreyfus, D., Iyer, B.: Enterprise architecture: A social network perspective. In: Proceedings of the 39th Hawaii International International Conference on Systems Science (HICSS-39), January 2006. IEEE Computer Society Press, Kauai (2006)

23. Baida, Z.: Software-aided Service Bundling - Intelligent Methods & Tools for Graphical Service Modeling. PhD thesis, Vrije Universiteit, Amsterdam, The Netherlands (2006)
24. Haesen, R., De Rore, L., Snoeck, M., Lemahieu, W., Poelmans, S.: Active-passive hybrid data collection. In: Proceedings of the 11th European Conference on Pattern Languages of Programs (EuroPLoP 2006), Irsee, Germany, Universitaetsverlag Konstanz, pp. 565–577 (2006)
25. Snoeck, M.: Object-Oriented Enterprise Modelling with Merode. Leuven University Press (1999)
26. Lemahieu, W., Snoeck, M., Goethals, F., De Backer, M., Haesen, R., Vandenbulcke, J., Dedene, G.: Coordinating cots applications via a business event layer. IEEE Software 22(4), 28–35 (2005)

Reasoning about Substitute Choices and Preference Ordering in e-Services

Sybren de Kinderen and Jaap Gordijn

VU University Amsterdam
De Boelelaan 1081
1081 HV, Amsterdam, The Netherlands
{sdkinde,gordijn}@few.vu.nl

Abstract. e-Services are just like normal services, but can be ordered and provisioned via the Internet completely. Increasingly, these e-services are offered as a multi-supplier bundle of elementary services. How to automatically compose and prioritize these multi-supplier e-service bundles is considered as a key problem. In this paper, we present the e^3*service* ontology to represent a multi-supplier e-service catalogue from a consumer need perspective. Then, we use this ontology to reason about alternative e-service bundles satisfying a particular need, and to prioritize the found bundles using the consumer benefits they provide. The ontology and the reasoning process are illustrated by a case study in the Dutch telecommunication industry.

Keywords: e-services, ontology, service bundling, consumer needs.

1 Introduction

In recent years, *customizable* e-service bundles, satisfying complex consumer needs have gained interest. We understand e-services as *commercial* services: economic activities, deeds and performances of a mostly intangible nature. Web services and web service languages, such as WSDL [6] and others, are a useful *technical* implementation platform for e-services but do not really recognize the commercial perspective on services. Consider e.g. the daily-life example of a *specific* consumer need for Internet access and email. Often, the proposition of an ISP is then a *general purpose* e-service bundle, consisting of more elementary e-services such as IP-based access, an email box, space to host a website, telephony, and access to newsgroups. However, the original, individual, consumer need *only* requires the provisioning of an IP-based access/email e-service. The latter bundle more closely matches the consumer need compared to the -fits for all- full-service bundle. Additionally, these e-services are increasingly offered by a *networked value constellation*, rather than just a single enterprise [18]. By doing so, suppliers can utilize their core competencies, while still satisfying a consumer need. In the ISP-example, the offered bundle can be a *multi-supplier* bundle: IP-access is then provided by a telecommunication operator, an email box is offered by a commercial enterprise utilizing economies of scale, as can hold for website hosting, which may be offered by yet another enterprise.

Z. Bellahsène and M. Léonard (Eds.): CAiSE 2008, LNCS 5074, pp. 390–404, 2008.

We perceive the automatic composition and provisioning of such a customized, needs-driven, multi-supplier e-service bundle as a key information system engineering problem as the e-services are provisioned by IT itself. In a future scenario we foresee that a consumer would ideally state to the web his preferences using a question-answer dialog, and the web (or some intermediate party) responds with a list of candidate multi-supplier e-service bundles, which are sorted according to how well they fit to the stated consumer preferences. After selection of a specific bundle by the consumer, the e-services in the bundle should be provisioned automatically. Guidelines on creating customized service bundles have already been studied in business literature, most notably by [12] and [16]. However, these guidelines are fairly generic (the focus is on services in general and not specifically on e-services). More importantly, they lack conceptualization and formalization so it is difficult to systematically and (semi-) automatically reason about service bundles. Such reasoning is important, because e-services, as illustrated by the ISP example, are bought and provisioned on line, *enabled by* information systems. To adequately facilitate this buying and provisioning process, the elicitation of needs, as well as the selection of *commercial* e-services that can be provisioned to satisfy such needs, should be supported by information systems as much as possible.

In earlier work [7], we have presented the e^3*service* ontology that allows for the structured creation of service bundles based upon consumer preferences. However, since often multiple, and alternative, service bundles are possible, the next question is then how to rank the bundles according to the consumer need. Therefore, in this paper we show how to (1) reason about substitute services, and (2) assign a *preference ordering* to found service-bundles, based on a consumer-given prioritization of the benefits (s)he wants to obtain. Additionally, we involve pricing of the service bundle in the reasoning process about preference ordering. The contribution of this paper therefore is that we provide a framework allowing *semi-automated* reasoning about multi-supplier, commercial-service bundles. Our approach relates to goal modeling, such as i* [19]. Rather we consider goals as consumer needs, which are problem statements of the consumer. By using various reasoning mechanisms, we search for e-services that satisfy the stated need.

This paper is structured as follows. In section 2, we provide a comprehensive overview of the e^3*service* ontology. In section 3 we apply this ontology to a case study, to create a consumer-oriented catalogue of e-services. Based upon this catalogue, we then show *how* we reason about preference ordering when creating e-service bundles. Finally, in section 4 we provide a discussion, and in section 5 we present our conclusions.

2 The e^3*service* Ontology

To make this paper self-contained, we summarize the e^3*service* ontology (see figure 1, and [7] for more details). This summary is organized by clustering the concepts in the ontology as follows: (1) the need/demand/want hierarchy,

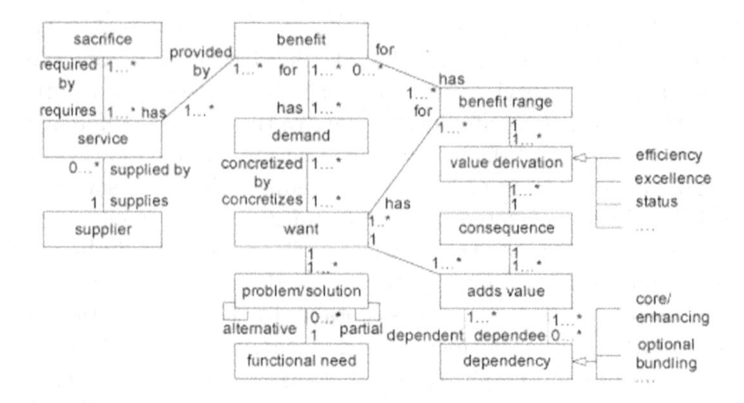

Fig. 1. The e^3*service* ontology

(2) benefits, consequences, and value derivations, (3) dependencies between want/consequence pairs, and (4) services. The ontology is based on established service marketing literature (e.g. [12] and [16]).

2.1 The Need/Want/Demand-Hierarchy

The need/want/demand-hierarchy emphasizes a gradual transition from a *need* - a problem statement - to a set of services that together provide a solution for that need, called a *demand* (see [2] and [15]). For the e^3*service* ontology, this results in the following concepts.

Functional need. A functional need represents a problem statement or goal, *independently* from a solution direction [2].

Example. A consumer may have a need to 'communicate with family abroad'. This need does not include a notion of a solution yet, as nothing is stated about how the communication will be done.

Want. A need (problem) can usually be covered by multiple *alternative* wants (solutions) [2]. Also, a need may require *multiple* wants that each satisfy a need partly, but together do so completely. The separation of problem and solutions is important to avoid that we overlook alternative wants (solutions) for needs (problems) during the elicitation process. In doing so, a want does not indicate a specific (named) supplier satisfying the stated need yet. Thereby, we explicitly separate the finding of a *general* solution for a need (a want), from finding a *specific* supplier doing so (a demand). However, as a want indicates a solution available in the market, at least *one* supplier should be willing to provide the solution.

Relations. A want *has* one or more benefit ranges (to be detailed later), which, in short, are properties of a provided service, which are economically valued by a consumer. For a want, these properties are defined independently from a specific supplier.

Example. A want satisfying the need 'communicating with family abroad' is for example 'e-mail hosting'. A benefit for 'e-mail hosting' is a certain mail-box size range (eg. 500 MB to 1 GB). An alternative want is 'instant messaging'.

Problem/solution pair. A want is a (partial) *solution* for a functional need, which is stated by a problem/solution pair. Problem/solution pairs can be related to each other to state that (1) they are *alternatives* for each other, or (2) together they form a *complete* solution for a problem.

Example. 'e-mail hosting' and 'instant messaging' are alternative wants for satisfying the need 'communicating with family abroad'. 'E-mail hosting' plus 'IP-connectivity' exemplify partial problem/solution pairs, which together satisfy the need 'communicating with family abroad'.

Demand. A 'want' is provisioned by a specific supplier as a demand. A demand differs from a want, as a demand provides supplier-specific values to the properties for benefits of a want. We use a strict distinction between wants and demands, because they refer to two different steps in the automated reasoning process about substitutes and preference ordering. In the first step, we reason about the required *generic* benefits, as contained by a want, to satisfy a need, independently from a supplier. In the second step, we reason about the *specific* suppliers who can satisfy a need in terms of a demand, with specific benefits. This simplifies the reasoning process as the customer first only focuses on choosing the required benefits (e.g. a mailbox with a small/big size) *without* a supplier in mind, and thereafter chooses specific properties benefits as offered by a specific supplier (eg. a 1MB sized box).

Relations. A demand *concretizes* a want if: (1) each benefit of the want matches with a benefit of the demand, meaning that (2) for each benefit, the supplier-specific benefit of the demand is in the specified range of the corresponding benefit for the want. Usually, a want has one or more demands, meaning that one or more suppliers can satisfy a want.

Example. 'Gmail' (from Google) is a demand that specifies the want 'e-mail hosting'. For example, 'Gmail' may have a distinguishing benefit 'mail-box size=1 GB' that would be different from the 'mail-box size=0.5 GB' as offered by 'Hotmail'. However, both benefits fall into the benefit range '500 MB to 1 GB' as specified for the want.

2.2 Benefits, Consequences and Value Derivations

Benefit range and benefits. Benefits describe properties that are of economic value to the customer in the sense of value-in-use [17]. In other words, benefits provide an increase of economic utility to the customer, through something functional, social (e.g. status) or otherwise. A benefit is also used to connect *demands*, as needed by the customer, to *services*, as provided by the supplier. Often, there is a mismatch between the *set* of benefits as contained by a customer demand, and the *set* of benefits as contained by a supplied service. In our work, we assume that the customer and the supplier use the *same* terminology to represent the customer/supplier benefit itself, so ontologically, these benefits are the same (although specific values may differ obviously). Reasoning about a match between a found consumer demand and available supplier e-services, is then about finding a multi-supplier e-service bundle with a set of benefits, that comes closest the required set of benefits as contained by the customer demand. A benefit *range* is a more general construct, which specifies a range of values a benefit may have.

Relations. First, a want *has* one or more benefit *ranges*. Since a want exists independently of a specific supplier, benefits on the want-level *do not* possess supplier-specific values. Instead, benefits on the want level have a *range* of possible values, within which a supplier-specific benefit could fall. For instance, in the case of the size of a mailbox, a range could be 500 MB-3 GB. Second, a single demand *has* one or more benefits. Since a demand is specific for a supplier, benefits of a demand have supplier-specific properties. In the case of the size of a mailbox, the size could for instance be 2.6 GB for a specific supplier. Third, a specific benefit is *for* one benefit range, and a range can *have* multiple benefits that fall-in the benefit range.

Consequence. A *consequence* represents the *subjective added value* for the end-customer if he consumes a benefit (falling into a certain benefit range). In the reasoning process as presented in section 3, deriving consequences from benefits is based upon the laddering-technique from means-end chaining [13]. In brief, this is done by asking the question 'what happens if we consume service X in which benefit Y is contained? '.

Relations. A benefit range *has* one or more consequences. Multiple benefit ranges can point to the same consequence. A consequence indirectly contributes to satisfying a need, via the benefit range, demand, and want of that need.

Example. The benefit 'web-based e-mail access' has the consequence 'cost-effective communication'. 'Cost-effective communication' contributes to satisfying the need 'communicating with family abroad'. Considering an example of a benefit with a range of values, we can define the consequence 'have a large mail box' based upon the range 1GB-3GB for the benefit 'mail box size'.

Value derivation. We reify the relation between 'benefit range' and 'consequence' by introducing the concept of *value derivation.* While *eliciting* a service catalogue, we reason about value derivation as a result of consuming a certain benefit, by using a consumer value framework presented by Holbrook et al [14]. This framework, which originates from the field of axiology, is used to explain *how* end-consumers derive value while consuming a product/service. Note that a framework as proposed by Holbrook serves as a 'plug-in'. In case of business-to-business services, value derivation will be done entirely differently, and so other frameworks should be used. Since the focus of this paper is on finding appropriate e-service bundles given a certain need, and not on eliciting the e-service catalogue itself, we do not elaborate further on value derivation.

Example. The benefit 'customized domain' from an e-mail service, can be annotated with the value derivation 'status', resulting in the consequence of 'enhancing status through personalized e-mail address'.

2.3 Dependencies between Want/Consequence Pairs

The notion of service-dependencies (see [3]) indicates that services may depend on each other. For instance, a service can serve as an option for another service, or a service may exclude meaningful consumption of another service. In [3], this relation has only been investigated from a *supplier* perspective; e.g. a paid e-mail service cannot be delivered without a billing service. We have found that such dependencies can also exist from a *consumer* perspective; e.g. a spam filter adds value for the customer if it is bundled with an e-mail hosting service.

Adds value and dependency. As benefits of wants have economic value consequences for the customer, the wants themselves also have consequences. In e^3*service* , this is represented as a reified 'adds-value' relationship between one want and one consequence. Obviously, each want and consequence may be in many of these relationships.

We have found two specific kinds of *dependencies*, which may exist between two or more 'adds value' relations (so between want/consequence pairs). In a 'Core/Enhancing' (C/E) dependency, a want/consequence pair B provides added value if bundled with a want/consequence pair A. Pair B cannot be acquired independently from A. In a 'Optional Bundling' (OB) dependency, a want/consequence pair B adds value to a want / consequence A. Yet, in case of an OB relation, A and B can also be acquired separately.

These dependencies may exist between multiple want/consequence pairs ('adds value'), as shown by the concept *dependency* in the e^3*service* ontology.

Relations. First, an 'adds value' relationship contains a single want and a single consequence. This pair represents a commercially feasible offering, plus part of the subjective value gained from consuming a benefit contained within this offering. Second, 'adds value' has a relationship with one or more other adds value relationships, via the 'dependency' concept.

Example. The pair 'e-mail' (want)/'local access to mail' (consequence) is in a Core/Enhancing dependency with pair 'spam-filter' (want)/'reduction in number of unwanted e-mails' (consequence). So, the want 'e-mail' is related to the consequence 'reduction in number of unwanted e-mails' from the want 'spam filter', where the consequence from latter want indicates *why* this relationship exists. Note that a Core/Enhancing relationship is present, because an acquisition of a spam-filter only makes sense in combination with an e-mail service.

2.4 Service

Service. A service is of economic value to the end customer, and is provisioned by a supplier. It is the smallest unit that, from a commercial point of view, can be obtained from a supplier. Services are listed in a service catalogue of a supplier. The notion of service allows for connecting the customer-oriented e^3*service* ontology to supplier-oriented ontologies (see e.g. [1]).

Relations. First, a service is *supplied by* precisely one supplier, since a service is supplier specific. Obviously, a supplier can *supply* multiple services. Second, a service *has* one or more benefits. These benefits are the source for matching supplier-services with benefits that belong to wants.

Example. An 'e-mail hosting' service is an is example of a service.

Sacrifice. A sacrifice represents something valuable to the consumer and supplier that has to be given in return, in order to acquire a service.

Relations. A service *requires* one or more sacrifices. This models that a consumer is not willing to obtain a service against any price, but rather is confronted with a budget-constraint, and therefore is limited in demands (s)he can have.

Example. Based upon a monthly fee (e.g. 40 €) (sacrifice) that has to be paid for the service '4 Mb/s Internet access', as well a contract-duration of minimally one year (also a sacrifice), a consumer may decide to revise his/her demand, such that a '1 Mb/s Internet access' service for a monthly fee of 10 €will be selected.

3 A Case Study on e-Service Substitution and Preference Ordering in the TelCo Industry

3.1 An e-Service Catalogue

We now show, for the need 'communicating with family abroad', how e^3*service* can be used to (1) make a choice between *substitute* services (different services that satisfy a similar need (e.g. 'instant messaging' and 'VoIP')), and (2) make a choice between *similar* services (e.g. two similar 'VoIP' services but of different suppliers). To this end, we first need an e-service catalogue, which is shown in figure 2. In brief, we create such a catalogue by considering the e-services as available in the market, and then by deriving the needs these services could satisfy. For a detailed description of how to create such a catalogue, see [7]. The catalogue for this case study, which due of lack of space we can only show partly (see figure 2), has been created by studying service documentation as provided by our industry partner KPN (the largest Dutch TelCo operator), and by interviewing KPN representatives. Effectively, the catalogue is an instantiation of the e^3*service* ontology. We have evaluated the catalogue afterwards with a domain expert from KPN, who is actively involved in realizing service bundles for 'VoIP', for descriptive validity. The catalogue itself is further explained as part of the following description of the substitution and preference ordering reasoning process.

3.2 Reasoning about Substitution and Preference Ordering of e-Services

We now illustrate how to derive telecom e-services from consumer needs, by considering an average 2.4 household consumer, who wants to communicate with family abroad but finds that using a traditional phone is too expensive, as a prototypical example.

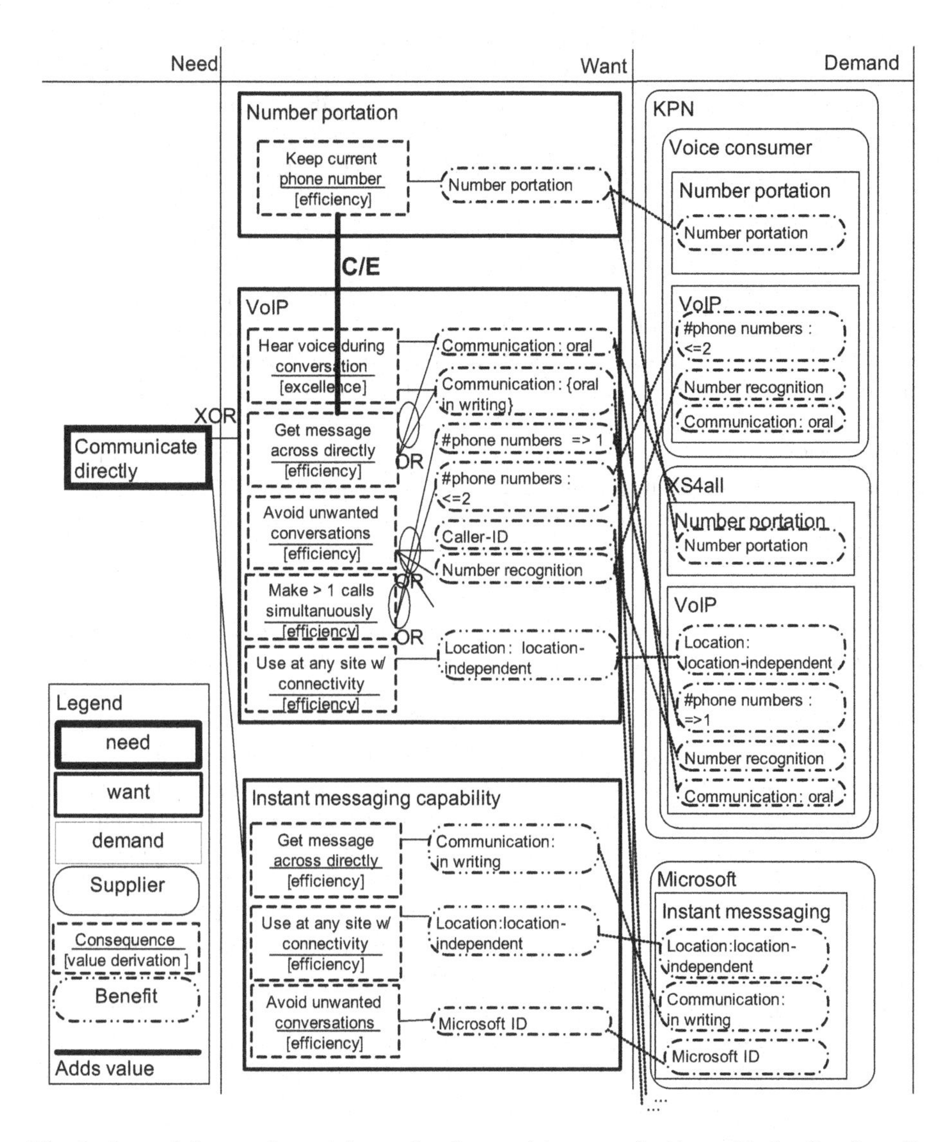

Fig. 2. A partial e-service catalogue for the need 'communicating with family abroad'

Step 1: Select a consumer need, and derive an initial set of wants satisfying this need.

In the e-service catalogue, we first traverse from a need to (alternative) wants that can (partly) satisfy this need.

Case: We assume that the consumer starts at the need 'communicate directly' since this need-definition comes closest to the need 'communicating with family abroad'. By traversing from this need to the want-level, we thus find the initial, alternative, set of wants [VoIP, instant messaging], because both can satisfy the need 'communicate directly'.

Step 2: Consider dependencies between want/consequence pairs, to elicit additional wants. The next step is to consider for each 'want' as found in step 1, any additional wants, by checking the reified 'adds-value' relationships between want/consequence pairs. If, for a considered want A (with its consequences), we can find related wants B, C, ... (with their consequences), we ask the consumer to decide upon inclusion of wants B, C, ..., by presenting him/her the consequence for wants B, C, ... As such, the consequences provide the consumer with a rationale to choose.

Case: In our catalogue (figure 2), there is a Core/Enhancing(C/E) reified 'adds-value' relationship between 'VoIP' and 'number portation'. Therefore, the consumer is presented the consequence 'keep current phone number' for the 'number portation' service. Due to all the hassle involved when changing a phone number, we assume that the consumer indicates that (s)he would want to keep her current number. Therefore the want 'number portation', which includes the consequence 'keep current phone number', is included as an additional want. So, the set of wants that is now used for our further reasoning is: [{VoIP, number portation}, instant messaging] (note that 'number portation' is only relevant in combination with 'VoIP', and not with 'instant messaging').

Step 3: Ask the consumer to assign a preference ordering to the consequences. In the third step, we ask the consumer to prioritize the consequences for the benefits of wants found so far. We do so, because consumers are not really interested in a service itself, but in the *specific benefits* that a service possesses and in how these benefits help them to fulfill their needs. Yet, the consumer might not always be able to assign a preference ordering to a benefit because the benefit is *by definition* always stated in objective (and often technical) terms (recall that a benefit is e.g. the size of mailbox, the bandwidth of a data-connection, etc.). For this reason, we ask the consumer to assign a preference ordering to a *consequence*, which is then propagated to the underlying benefit. In other words; we present the consumers with the value-in-use or *goal* that can be achieved through a benefit, rather than the benefit itself.

The consumer should have a concrete way to express his/her preference. For this purpose, we use a four-point importance scale which we based on the MoSCoW-list. The latter is often used in Rapid Application Development (RAD) software engineering projects [5] to prioritize software requirements. Our scale consists of the following categories:

- **Must have.** A consumer can assign a 'must-have' priority to one or more consequences. For a service bundle to be relevant for the consumer, all 'must-have' consequences must be satisfied by the bundle.
- **Should have.** A 'should-have' consequence should be realized by a benefit from a supplier-specific service, but as opposed to a 'must-have' consequence', realization is not a necessity.
- **Could have.** A 'could-have' consequence is something that the consumer perceives as a nice-to-have feature. A 'could-have' consequence is perceived to be less important than a 'should-have' consequence.

– **Does not matter.** As implied by name, this last category can be used by the consumer to indicate that a consequence does not have to be taken into consideration in the bundling process.

Case: The consumer is presented with the consequences for the wants {VoIP, number portation}, and {instant messaging}. Examples include 'hear (natural) voice during conversation' and 'keep current phone number', each of which can be assigned an importance ranking. We assume that our consumer assigns the importance 'must-have' to 'hear (natural) voice during conversation' because it gives more character to the conversation. Considering 'should-have' preferences, our consumer assigns such a preference to 'avoid unwanted conversations' and 'keep current phone number'. Lastly, we assume that our consumer assigns a 'could-have' preference to the consequence 'make multiple calls simultaneously', because there are multiple members in the household.

Step 4: Present consumer with the set of underlying benefits. Next, the consumer is presented with the set of benefits that belong to the consequences for which an importance ranking was done, and with an opportunity to change these benefits. Of course, these benefits are automatically assigned the same importance ranking as their consequences have. This step is in particular relevant if a consumer wants to consider the *objective* benefits of a service. This often the case for the more technically oriented consumers who want to specify exactly the benefits they expect from a service, such as the exact download speed they desire from an internet connection.

Case: Our consumer has selected the consequence 'make multiple calls simultaneously' and would therefore be presented with the benefits: '#phone numbers: <= 2' and '#phone numbers: >= 1'. As such, the consumer from our average 2.4 household has the option of specifying how many phone numbers are required. In our scenario, we assume that the consumer chooses '#phone numbers: >= 1' since there are more than 2 members in the household who each, at some point, might require their own phone number.

Step 5: Generate alternative service bundles and assign preference ordering. After all benefits have been prioritized and possibly changed by the consumer, the next step is to generate the actual e-service bundle that provides the benefits. To do so, a table is built with the prioritized benefits and the corresponding consequences on the one hand, and the e-services, taken from the catalogue, which can provide the requested benefits (see table 1) on the other hand. The 'consequences' are important in the process, since multiple benefits may have a same consequence. This allows for realizing a particular consequence in alternative ways, more specifically by providing alternative benefits. As a result, the service bundle composition problem becomes less constrained, thus increasing the chance of finding a valid solution. Alternative benefits for a same consequence are in the catalogue (figure 2) shown by the 'OR' label between a consequence and its benefits.

So, the found table is used to generate the relevant e-service bundles. The bundles are evaluated against the benefits and their preference ordering, as assigned by the consumer, in the following order.

Table 1. A preference ordering as done by the average 2.4 household

Consequence	Benefit	VoIP (KPN)	VoIP (XS4all)	VoIP (skype)	Instant messaging (Microsoft)	Number portation (KPN)	Number portation (XS4all)
Must have							
Voice during conversation	communication: oral	X	X	X			
Should have							
Keep current phone number	number portation					X	X
Avoid unwanted conversations	number recognition	X	X				
	caller-ID			X			
	Microsoft-ID				X		
Could have							
Multiple calls simult.	#Phone numbers: >= 1		X				

1. Service bundles that do not satisfy all 'must-have' benefits, are rejected.
2. For the remaining service bundles, the total number of satisfied 'should-have' benefits per bundle are used to rank the bundles. If a bundle A satisfies more 'should-have' benefits than bundle B, bundle A will be preferred over B, *independently* of the amount of 'could have' benefits. This corresponds to how MoSCoW is used in RAD software engineering projects, where first all 'should-have'requirements are implemented before, *and independently of*, the 'could have' requirements.
3. In case two or more service are ranked equally, the number of 'could-have' benefits per bundle are used for ranking.

Case: For our average 2.4 household, consider the generated table 1. Here, the 'must-have' benefits are clustered at the top rows, followed by the 'should-haves' and thereafter the 'could-haves'.

'Must-have' benefits. The 'must-have' benefits lead to the following set of *alternative* services [VoIP (KPN), VoIP(Skype), VoIP(Xs4all)], because all these services each include the benefit 'communication: oral'. Since all these services result in the same number benefits (namely one), we can not indicate a preference ordering yet.

'Should-have' benefits. We now consider the 'should-have' benefits. In our table, we have four of such benefits:

- (1) 'number portation', which inherited its 'should-have' preference from the consequence 'keep current phone number', and
- (2) 'number recognition', (3) 'caller ID', and (4) 'Microsoft ID'.

The last three benefits inherited their 'should-have' preferences from the 'avoid unwanted conversations' consequence; therefore these benefits are alternatives for the 'avoid unwanted conversations' consequence. In figure 2, this is represented by an OR-dependency between 'avoid unwanted conversations' (consequence) and the benefits realizing this consequence.

These benefits result in the services [Instant messaging (Microsoft), Number Portation (KPN), Number portation (XS4All))], to be added to the already found services as a result of the 'must-haves'. Note that the other 'should-have' benefits (Number recognition (KPN) (XS4All), Caller-ID (Skype)) are already included in the services found while considering the 'must-haves'.

'Could-have' benefits. We finally consider the 'could-have' benefit '#phone numbers: >= 1'. In our consumer specific catalogue, we find that the 'VoIP' service of XS4all is the only supplier that can provide this benefit. This service was already included as a candidate in the service bundle.

Alternative services. We now consider *alternative* services satisfying the same need. In other words: we must avoid awkward bundles, such as the bundle [VoIP (KPN), VoIP (XS4all)] because both services in this bundle act as an alternative in satisfying the same need: 'communicate directly'. However, we may find [VoIP (KPN)] and [VoIP (XS4all)] as *alternative* bundles.

Preference ordering. After checking for alternative services, we arrive at the following possible bundles, sorted according to their correspondence to the preference ordering of benefits.

1. [VoIP (XS4all), number portation(XS4all)], [VoIP(XS4all), number portation(KPN)].
2. [VoIP (skype), number portation(XS4all)], [VoIP(Skype), number portation(KPN)], [VoIP(KPN), number portation(XS4all)], [VoIP(KPN), number portation(KPN)].
3. [XS4all].
4. [VoIP(skype)], [VoIP(KPN)].

Preference ordering is done by comparing each possible bundle using the actual benefits desired by the consumer. So, for instance, the combination [VoIP (XS4all), number portation(XS4all)] is ranked higher than [VoIP(KPN), number portation(KPN)] because, supposing that the satisfaction of all other benefits is equal, [VoIP (XS4all), number portation(XS4all)] satisfies an extra 'should-have' benefit:'#Phone numbers: >= 1' . Additionally, *any* 'VoIP' service combined with 'number portation' is ranked higher than an individual service. This is because the number of 'should-have' benefits satisfied by such a combination is higher than the number of 'should-have' benefits satisfied by any individual 'VoIP' service, since it includes the benefit 'number portation'. Also, no bundles are generated that contain an 'instant messaging' service. This is because 'instant messaging' does not contain the benefit 'communication:oral'. Moreover, an 'instant messaging' service will not be offered in combination with a 'VoIP' service, because it acts as an alternative in satisfying the same need: 'communicating with family abroad'. Finally, note that we only discuss *consumer-oriented* reasoning about service bundling in this paper. Therefore, bundles such as [VoIP(Skype), number portation(XS4all)] are taken into consideration, even though they are likely not possible from a *supplier-oriented* view. Such a bundle would be rejected by supplier-oriented bundling analysis. For a more elaborate discussion on the supplier-view on service bundling, we refer to [3].

Step 6: Present the consumer with the sacrifices of services and an opportunity to change the importance rankings. Before the consumer actually chooses one of the possible bundles, (s)he considers the sacrifices of each alternative bundle. Based on the sacrifices, the consumer may change the preference orderings assigned to the benefits. We do so because the price of a service plays a significant role in a consumers' decision to actually acquire a service [9]. In our ontology, this issue is represented with the concept 'sacrifice' (To represent the pricing of a service itself, we use the pricing models of [8]). If the consumer is not satisfied with the price for a certain service bundle, (s)he has an opportunity of changing the preference ordering of consequences and therefore benefits. Changing the preference ordering of a benefit entails going back to step 3. If the consumer eventually finds that there is a balance between benefits received from a bundle and the price that has to be paid for it, (s)he selects the bundle for provisioning.

Case: Our consumer zooms in on both bundles on top of the preference ordering, and finds that acquiring the benefit 'number portation' costs 10 € in both cases. Now, for the sake of argument, say our consumer finds this too expensive and wants to change the preference of the benefit 'number portation'. For this, (s)he is presented with an opportunity to go back to step 3, where all the values already filled in are still present (such as '#Phone numbers: >= 1'.). Now the consumer changes the preference ordering on 'number portation' from 'should-have' to 'does-not-matter' and generates the alternative bundles again. A new set of service bundles is generated, this time with [XS4all] at the top of the list.

4 Discussion

Practical usefulness. The domain expert of KPN considers e^3*service* to be a useful tool for facilitating communication between marketeers and IT-personnel. This precedes our own goal, namely automated consumer-oriented e-service bundling. Marketeers, responsible for designing these bundles, do not always know whether an e-service bundle is *technically* feasible. Since e^3*service* relates benefits of e-services as experienced by consumers to a supplier-oriented catalogue of services (see also [1]), e^3*service* contributes to closing this gap.

The domain expert from KPN also pointed out that on-the-fly e-service bundling as envisioned, brings about problems that need to be considered before this idea can be *realistically* implemented. Below, we provide a selection of a few mentioned problems.

Planning of e-service provisioning. Some e-services take days of preparation before they can be provisioned to the consumer, often due to contractual and technical arrangments to be made. Therefore, the provisioning of a bundle has to be carefully planned. This calls for inclusion of (skeleton) planning techniques in the reasoning process (see e.g. [10]).

Single-point-of-contact. Often consumers want to have a single-point-of-contact in case there are problems with the provisioned service bundle (e.g. a helpdesk). A dynamic, and on-the-fly generated multi-supplier e-service bundle should have mechanisms to mitigate these single-point-of-contact services. One solution is to consider such services as e-services *themselves*, which therefore should be part of the e-service composition process. Moreover, the need for single-point-of-contact services (e.g. to repair service-failures) can be reduced by allowing for automated reconfiguration of provisioned service-bundles, e.g. facilitated by platforms for adaptable compositions of web-services [4].

e-Service pricing. In the telecommunication industry, discounts are a frequently used mechanism to attract consumers. With *single*-enterprise bundles, pricing these bundles and deciding on discounts is relatively straightforward. If however consumers create their own *multi*-supplier bundles, deciding on discounts is more difficult, due to supplier-specific pricing schemes and discount-policies.

5 Conclusions

In this paper, we showed how to reason about bundling of e-services, based upon consumer needs, in a structured and semi-automatic way. Additionally, we showed how to derive a preference ordering for the found e-service bundles, and how the pricing of a supplier-specific service - through an influence upon consumer preferences -can influence this preference ordering of bundles. We have also illustrated how e^3*service* works in practice, for a case study in the telecommunication industry. Currently, we are working on software support for the e^3*service* methodology to validate that service bundles can be semi-automatically generated from a stated consumer need.

For future research directions, we will integrate a supplier perspective on e-service bundling (specifically *e-serviguation*, see [1]), to generate e-services bundles that are not only valid from a consumer perspective, but also from a supplier perspective. Additionally, we will address the quality (or: non-functional) attributes of a service more in-depth. This is because quality also plays an important role in acquiring e-services, especially when considering B-to-B-environments where such quality aspects usually have to be strictly agreed upon by means of a SLA (see e.g. [11]).

Acknowledgements. We want to thank Leo Stout and Ron van der Kwaak from KPN for useful comments on the case presented in this paper. This research has been partly funded by NWO/STW/Jacquard as the project VITAL.

References

1. Akkermans, H., Baida, Z., Gordijn, J.: Value webs: Ontology-based bundling of real-world services. IEEE Intelligent Systems 19(44), 2332 (2004)
2. Arndt, J.: How broad should the marketing concept be? Journal of Marketing 42(1), 101–103 (1978)

3. Baida, Z.S.: Software-aided service bundling. PhD thesis, Free University Amsterdam (May 2006)
4. Baresi, L., Nitto, E.D., Ghezzi, C., Guinea, S.: A framework for the deployment of adaptable web service compositions. Service Oriented Computing and Applications 1(1), 75–91 (2007)
5. Beynon-Davies, Carne, Mackay, Tudhope: Rapid application development (rad): an empirical review. European Journal of Information Systems 8(3), 211–223 (1999)
6. Booth, D., Liu, C.K.: Web services description language (wsdl) version 2.0 (2007), http://www.w3.org/TR/2007/PR-wsdl20-primer-20070523/
7. de Kinderen, S., Gordijn, J.: e^{3}service - an ontological approach for deriving multi-supplier it-service bundles from consumer needs. In: Proceedings of the Forty-first Hawai'i International Conference on System Sciences (HICSS-41) (CD-ROM), January 7-10, Computer Society Press (2007)
8. de Miranda, B., Baida, Z., Gordijn, J.: Modeling pricing for configuring e-service bundles. In: Proceedings of The 19th Bled eCommerce Conference, June 5-7 (2006)
9. Fishbein, M.: Belief, attitude, intention and behavior: an introduction to theory and research. Addison-Wesley, Reading (1978) (third print)
10. Friedland, P.E., Iwasaki, Y.: The concept and implementation of skeletal plans. Journal of Automated Reasoning (1), 161–208 (1985)
11. Greiner, U.: Quality-Oriented execution and optimization of cooperative processes: Model and algorithm. PhD thesis, Univ. Leipzig (2006)
12. Grönroos, C.: Service Management and Marketing. Lexington Books (1990)
13. Gutman, J., Reynolds, T.J.: Laddering theory-analysis and interpretation. Journal of Advertising Research 28(1), 11 (1988)
14. Holbrook, M.B.: Consumer value; a framework for analysis and research, 1st edn. Routledge (1999)
15. Kotler, P.: Marketing Management. Prentice-Hall, Englewood Cliffs (2000)
16. Lovelock, C.: Service Marketing - People, Technology, Strategy, 4th edn. Prentice-Hall, Englewood Cliffs (2001)
17. Ramsay, J.: The real meaning of value in trading relationships. International Journal of Operations and Production Management 25(6), 549–565 (2005)
18. Tapscott, D., Ticoll, D., Lowy, A.: Digital Capital - Harnessing the Power of Business Webs. Nicholas Brealy Publishing (2000)
19. Yu, E.: Towards modelling and reasoning support for early-phase requirements engineering. In: Proceedings of the 3rd IEEE Int. Symp. on Requirements Engineering (RE 1997), pp. 226–235. IEEE Computer Science Press, Los Alamitos (1997)

Message Correlation and Business Protocol Discovery in Service Interaction Logs

Belkacem Serrour[1], Daniel P. Gasparotto[1], Hamamache Kheddouci[1], and Boualem Benatallah[2]

[1] Université de Lyon, Laboratoire LIESP, Bât. Nautibus (ex 710), 43, Bd. du 11 novembre 1918, 69622 Villeurbanne Cedex, France
{bserrour,hkheddou}bat710.univ-lyon1.fr, daniel.gasparotto@gmail.com
[2] CSE, UNSW, Sedney NSW 2052, Australia
boualem@cse.unsw.edu.au

Abstract. The problem of discovering protocols and business processes based on the analysis of log files is a real challenge. The behavior of a Web service can be specified using a Business Protocol, hence the importance of this discovery. The construction of the Business Protocol begins by correlating the logged messages into their conversations (i.e. instances of the business protocol). The accomplishment of this task is easy if we assume that the logs contain the right identifiers, which would allow us to associate every message to a conversation. But in real-world situations, this kind of information rarely exists inside the log files.

Our work consists in correlating the messages present in Web service logs into the conversations they belong to, and then generating automatically the Business Protocol that reflects the messaging behavior perceived in the log. Contrary to other approaches, we do not assume the existence of a conversation identifier. We first model logged message relations using graphs and then we use graph theory techniques to extract the conversations and finally the Business Protocol. Logs are often incomplete and contain errors. This induces some uncertainty on the results. To address this problem, we apply the Dempster-Shafer theory of evidence. Our approach is implemented and tested using synthetic logs.

Keywords: Web services, business protocols, message correlation, log analysis, graph theory.

1 Introduction

Every Web service has an interface (specified by the language WSDL[1], for example). This interface contains the realizable operations, ingoing and outgoing message types, etc. The interface specified by WSDL is only a functional interface, i.e. it describes only the various invocable methods (the ordered sequences of these are not described). To assure the dynamic and behavioral aspect of the Web services, a new interface is proposed in [1][2]: the Business Protocol. A

[1] Web Services Description Language.

Z. Bellahsène and M. Léonard (Eds.): CAiSE 2008, LNCS 5074, pp. 405–419, 2008.

Business Protocol is a specification of all possible conversations that a service can have with its partners[10]. Recent work shows the importance of specifying business protocols and also propose models to represent them. The authors used Final States Machine (FSM) to model the Business Protocol. The motivations to use FSM are that: (i) FSM is a well-known paradigm with established formal foundations, (ii) it is simple and suitable for modeling reactive behaviors, (iii) it embeds the notion of state, which is useful for applications such as monitoring. Business Protocols provide several advantages: (i) They provide developers with information on how to program the client to interact correctly with the service. (ii) They serve as a verification model of behavioral constraints (make sure that the real service corresponds well to the conception constraints). (iii) A Business Protocol is modeled in the form of a Final States Machine (FSM), which is an easily exploitable visual model by the user (addition of new functionalities, constraints, etc.)[4][10].

Business Protocol provides several advantages and is important for Web services paradigm. In spite of this, a key point is how to reconstitute a protocol with no prior specification?

The idea proposed in this paper is to analyze log files of Web services to extract the Business Protocol. The discovery models depend on the size of log file to analyze. More the log file is long more the probability to have all the possible exchanges between the service and his clients is great. Otherwise, if the log to analyze does not contain all the instances of the service, it will validate at least a part of the model.

Analyze of log files is a delicate task because they are often incomplete, uncertain and contain errors. The first difficulty of this task is the detection of errors in logs. There are two types of errors:

Log Incompleteness. In practice, conversation logs are very often incomplete in the sense that they do not contain all the possible conversations allowed by the service protocol. Incompleteness makes it difficult for a model discovery algorithm to discover even simple models.

Noises/Errors/Interruptions. Various approaches try to solve the problem of handling noise in real-world log files, which are often imperfect. In a real-world web service log file, it is normal to find problems such as omitted/lost entries, swapped timestamps, random interruption, etc. In a protocol execution instance, this will add a lot of problematic sequences that should be considered[9]. Below are some examples that illustrate the issues:

Missing Messages: The logging infrastructure may fail to record one or more messages of a conversation. For example, for a conversation *abcde*, we may have *acde* captured in the log, in which *b* is missing. This type of error happens for various reasons, including bugs in the logging infrastructure or performance degradation.

Swapping Messages: The order of messages as recorded in the log may differ from the real ordering of messages as exchanged between services. For example, for conversation *abcde*, we may find *acbde* recorded in the log. The order of *b* and

c is swapped. This type of error may be due to the granularity of time stamping, delays in the network infrastructures, etc.

Partial Conversations: We call partial a conversation that is interrupted before its completion. For instance, this can be due to network failure, client abortion or service execution exceptions. For example, if *abcd* is the message sequence of a complete conversation, only *abc* may have been exchanged and recorded in the log.

The second difficulty of log analysis task is the message correlation, i.e. grouping log messages in a set of conversations. This step is the first in the protocol discovery process (see Figure 1). The message correlation task can be easily accomplished if we have an information in logs indicating to which conversation belongs every message (this information is called conversation identifier or *cid*). But in real-world implementation, this information rarely exists in the logs.

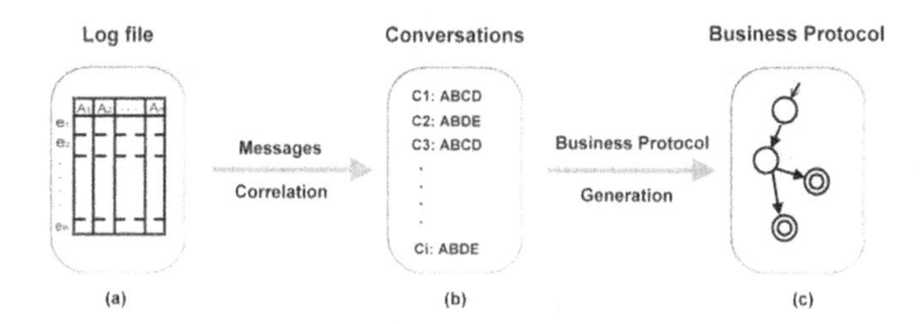

Fig. 1. Business Protocol Discovery Process

This paper is structured as follows. In the next section we give some definitions and we discuss some approaches. Then, in Section 3 we present the details of our approach. Section 4 discusses implementation aspects. We conclude in Section 5.

2 State of the Art

2.1 Definitions

WS Message Log (ML)

Definition 1. *A message log ML is a collection of entries (or events) $e = (cid, s, r, t, m)$, where cid is the conversation identifier, s and r denote the sender and the receiver of message m, and t is the timestamp.*

At this point, our definition diverges from [10]. The log files that we consider are slightly different:

A message log ML is a collection of entries $e = (s, r, t, m)$.

Note that a *cid* is not present as it was in the previous definition. The omission of the conversation identifier will not let us correlate each message to the conversation that it made part. The log is a sequence of events ordered by timestamp and with the identifier of the sender and receiver.

Conversation Log (CL)

Definition 2. *A conversation log* CL *is a collection of conversations* $CL = \{c_1, c_2, ..., c_n\}$. *Each conversation* $c_i \in CL$ *is a sequence of messages* $c_i = \langle m_1^i, m_2^i, ...m_k^i \rangle$.*(see Figure 1(b)).*

Each client exchanges a set of messages (conversation) to interact with the service.

Business Protocol

Definition 3. *A Business Protocol, as already defined, specifies the potential sequencing of messages exchanged by a particular partner with its other partners to achieve a business goal. It can be modeled as a tuple:*

$$P = (S, s_0, F, M, T)$$

where S *is the set of states of the protocol,* M *is the set of messages supported by the service,* $T \subseteq S^2 \times M$ *is the set of transitions,* s_0 *is the initial state, and* F *represents the finite set of final states. A transition from state* s_i *to state* s_j *triggered by the message* m *is denoted by the triplet* (s_i, s_j, m)*[10] (see Figure 1(c)).*

Note that in the model, conversations may contain markings indicating whether the message is sent or received by the service. This is easily detected by looking at the sender attribute of the message (If the sender is the service we say that the message is outgoing else the message will be marked as ingoing). In our work, we have not heeded that.

2.2 Approaches

The problem of logs analysis for models discovery relates to several fields. For instance, the main idea of process mining is the analysis of log files to extract knowledge that enhance and improve information systems[5][6][13]. To discover these models, it uses different techniques for the analysis of logs, especially data mining techniques. Data mining is the extraction of interesting (non-trivial, implicit, previously unknown) information or patterns from data in large databases[7].

In the Web services field, the majority of works focus on the process discovery (workflows) and protocol discovery[9][10][11].

In this paper, we are interested in the protocol discovery problem.

In our research domain, there are two kinds of problems: protocol discovery using conversation logs (messages log with conversations identifiers) and using message logs (messages log without conversations identifiers). A lot of efforts have been put in the resolution of the first problem. On the other hand, the problem of having message logs without conversation identifiers is a research field that has been gaining attention recently[11].

Protocol Discovery. The solution in [9] uses conversation logs as input (messages log with conversations identifiers). The conversations are subdivided in smaller sequences and an automatic bottom threshold is applied taking advantage of the information about the conversations. These sequences are building blocks of an oriented graph, called Message Graph. It is initially overgeneralized, because of the way it is built, meaning that refinement is needed to reach expected results. After the removal of erroneous paths (using an enhanced version of the splitting method originally), the Message Graph is then converted into an FSM, minimized using Hopcroft's algorithm and consumed by final user-refinement in order to generate an acceptable Discovered Protocol.

Message Correlation. B. Benatallah *et al.* in [11], bring up this specific problem of protocol discovery *without* message correlation information. The authors define rules (celled Correlation Rules) between messages log to construct conversations. The objective of this work consists in the identification attributes to use for the correlation. For this objective, they use some heuristics. After this, they try to specify the rules and composite rules to apply on these attributes. The authors gave some other directions for solving the problem, and making it the sole article found by us that deals with the entire process of business protocol discovery starting from uncorrelated message log files.

Due to difficulty of this task (message correlation in logs), all papers which treat this problem suppose the existence of *cid*, which facilitates the conversations extraction. Our main contribution consists in the correlation of log messages without assuming the existence of conversation identifiers and then to generate the Business Protocol.

3 Protocol Discovery Framework

The general architecture of our approach is described in Figure 2. It has as objective the Finite State Machine (FSM) that describes the protocol that would have generated the log file.

The log file to be analyzed is generated by the logging infrastructure of the service. In this log, one can find all messages exchanged by the service and its clients (sent and received messages). It is assumed that every client begins a conversation with the service by a single starting message (e.g. login message).

In following, we describe each step in the process of discovery framework.

3.1 Pre-partitioning the Log File

Each message of the log is generated by one partner (client/server) and received by another (client/server). This step consists of grouping the messages exchanged between a same pair client/server. New sub-log files containing these groups are created.

The log partitioning aims to eliminate the conversation overlaps in logs. When a service begins several simultaneous conversations with different clients causes

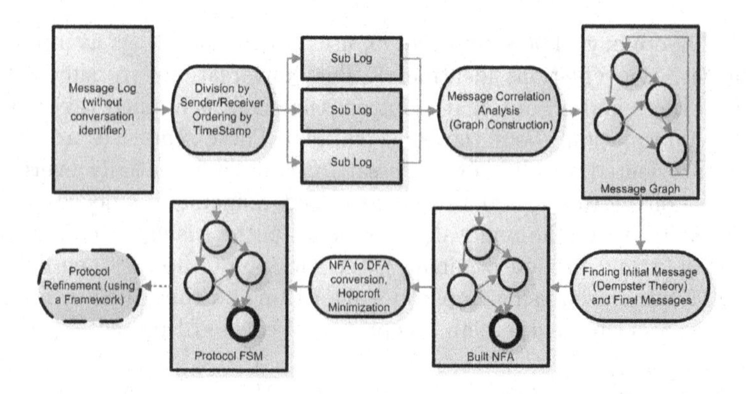

Fig. 2. Business Protocol Discovery from Message Log Files

overlaps of conversations (e.g. two messages which follow each other in this log can belong to different conversations).

Using **S** and **R**, it is possible to divide the log into pairs as $\{\mathbf{S_i},\mathbf{R_j}\}$ and with entries ordered by **T** (see example in Tables 1 and 2).

Table 1. In a first view of a log file, each line is an entry and we observe the Sender (S), Receiver (R), Timestamp (T) and MessageType (MT) already identified and n other unknown attributes

Logfile					
#	**S**	**R**	**T**	**MT**	Other Attrib. (unknown meaning yet)
1	$\mathbf{C_1}$	$\mathbf{S_1}$	**12**	**INIT**	$\{A_1, A_2, ..., A_n\}$
2	$\mathbf{C_2}$	$\mathbf{S_1}$	**15**	**INIT**	$\{A_1, A_2, ..., A_n\}$
3	$\mathbf{S_1}$	$\mathbf{C_1}$	**17**	**WLCM**	$\{A_1, A_2, ..., A_n\}$
4	$\mathbf{S_1}$	$\mathbf{C_2}$	**16**	**WLCM**	$\{A_1, A_2, ..., A_n\}$
5	$\mathbf{C_1}$	$\mathbf{S_1}$	**20**	**REQA**	$\{A_1, A_2, ..., A_n\}$
...	...	...	...	...	...

Table 2. Partitioned Logs

Log File Divided by $\{\mathbf{S_i},\mathbf{R_j}\}$ and Sorted by **T**					
#	**S**	**R**	**T**	**MT**	Other Attrib. (unknown meaning yet)
Partition $\{\mathbf{S_1},\mathbf{C_1}\}$					
1	$\mathbf{C_1}$	$\mathbf{S_1}$	**12**	**INIT**	$\{A_1, A_2, ..., A_n\}$
3	$\mathbf{S_1}$	$\mathbf{C_1}$	**17**	**WLCM**	$\{A_1, A_2, ..., A_n\}$
5	$\mathbf{C_1}$	$\mathbf{S_1}$	**20**	**REQA**	$\{A_1, A_2, ..., A_n\}$
...	...	...	...	...	...
Partition $\{\mathbf{S_1},\mathbf{C_2}\}$					
2	$\mathbf{C_2}$	$\mathbf{S_1}$	**15**	**INIT**	$\{A_1, A_2, ..., A_n\}$
4	$\mathbf{S_1}$	$\mathbf{C_2}$	**16**	**WLCM**	$\{A_1, A_2, ..., A_n\}$
...	...	...	...	...	...

After the grouping, we can order the log by timestamp and observe the message flows of each specific client with a specific server.

From now on, a simplification in the notation of the sub-log files will be used, i.e. $\{INIT, WLCM, REQA, \ldots\}$ and $\{INIT, WLCM, REQB, \ldots\}$, two large sequences of messages, each related to particular client and server pair. To simplify, we use in our examples symbolic sequences as {A, B, C, D, A, B, D, E, A, ...}, where each letter is a reference to a message type.

3.2 Message Graph

Once sub-logs are identified, we can now build a graph that represents the transitions between events and their frequency. The oriented weighted graph is denoted as $G(V, E)$, where:

- $V = \{v_1, ..., v_n\}$ is the set of vertices (messages, in this case, with n different types),
- $E = \{e_1, ..., e_k\}$ is the set of edges (correlations) ($e = (u, v)$ marks a correlation between two message types u and v, means that the message v succeed the message u in the sub-log (not in log file source)).

Both the frequency of vertices $f(v_i)$ and edges $g(e_i)$ are counted.

- $f(v_i)$: number of times that message v_i appears in the log file (for example, in figure 3, $f(A)$= 49, means that the message A appears 49 times in the log).
- $g(e_i)$: $e_i = (u, v)$, number of times that the message v succeed u in the log file (for example, in figure 3, $g(A, B) = 46$, means that the message B succeed A 46 times in log).

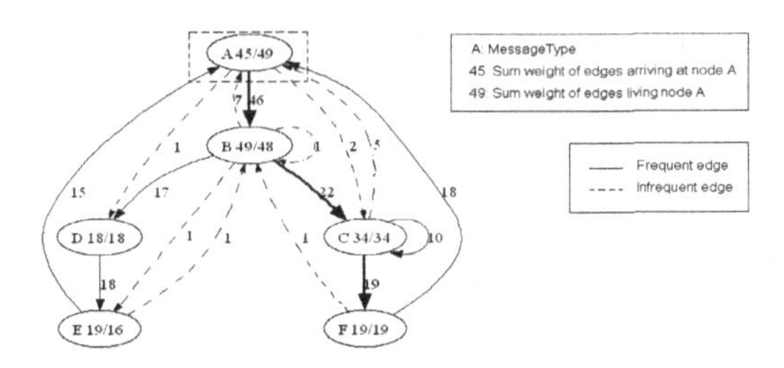

Fig. 3. Message Graph

In this graph, there is only one node for each message type. In the Figure 3, the message graph contains all the messages of the log. The creation of the graph G is given by Algorithm 1:

Algorithm 1. Message Graph

Require: $subLog$
Ensure: G
  $v \leftarrow subLog[0]$
  $f(v) \leftarrow 1$
  for $i = 0, i < (len(Log) - 1)$ /*for every event logged*/ **do**
    $v \leftarrow subLog[i]$
    $v' \leftarrow subLog[i+1]$
    if $v \notin V$ **then**
      add v to V
    end if
    if $v' \notin V$ **then**
      add v' to V
    end if
    $f(v') \leftarrow f(v') + 1$
    if $(v, v') \notin E$ **then**
      add (v, v') to E
    end if
    $g(v, v') \leftarrow g(v, v') + 1$
  end for
  return G

The union of the various generated message graphs (see Figure 3) represents the complete information contained in the Log file.

Initial Message Type. The first information that we should extract in message graph is the Initial Message Type. We must be able to infer which message is used in the beginning of the protocol execution and which messages end an execution (always considering noise and interruptions). Log files can always contain errors. Thus, nothing guarantees that the first message read in log is the initial message (it can be due to an error at the beginning of the log file or the log to analyze is only a not complete fragment; Consequently, the first message in this file has no certainty to be the initial message).

To find the initial message, we should rely on some evidences.

From the graph built in the previous step, we can extract evidences witch help to discover the initial message of the protocol:

1. P_1 : Frequency of the node: $Score(C_v) = f(v)$. Surely, all conversations must make use of the initial message. It will show a high frequency of observations when compared to other normal messages. Although, we must consider the cases of messages loops, that could increase the occurrence of regular messages when compared to the frequency of the initial message.
2. P_2 : Highest sum of frequencies of infrequent edges that arrive to the node. $Score(C_v) = \sum g(k, v), k \in V, g(k, v) < \phi(threshold)$. Another good evidence is that every time there is an interruption in the protocol, the next message seen from the client will be an initial message (a new instance of

protocol execution). We use the sum of the frequency values of all edges arriving to a message node that are below a minimum frequency threshold. In other words, we take all the weak edges arriving to a node and count their appearances. The best candidate will be the one with the highest value.

3. P_3 : Node with *weight degree out* higher than *weight degree in*. The partitioning of the log into subsets avoids returned edges of last messages of various clients towards the initial node. $Score(C_v) = weight\_degree\_out(v) - weight\_degree\_in(v)$.
4. P_4 : No Self Loops: We assume that there is no self-loop in the initial message of a protocol, so every time we find a frequent one, the message will not be a candidate.

These evidences do not always point to the correct candidate(s) of initial message. The criteria have their own situations where the chosen node is not the real initial message, so we combine the results of the evidences to achieve a better outcome. To do so, we use Dempster-Shafer's mathematical theory of evidence, based on belief functions and plausible reasoning, which is used to combine separate pieces of information (evidence) and to calculate the probability of an event. Detailed explanations this the theory can be found in [12] and a practical use of it in [3].

Each of the three criteria (P_1, P_2, P_3) to be combined gives a score to the node in the candidates set (a score function is associated to every criterion).

The set of candidate nodes, $U = \{C_1, C_2, ..., C_k\}$, to be initial node, are all nodes of the graph except those having self loops.

In order to apply DS' theory, we need to normalize the scores. The normalization equation is shown in eq. 1:

$$m(C_i) = \frac{Score(C_i)}{\sum_{i=1}^{n} Score(C_i)} \times 100 \qquad (1)$$

By the theory, the uncommitted belief (UB) is defined as an amount of belief that is not assigned to the focal elements by the evidence.

$$m'(C_i) = \left(\frac{m(C_i)}{\sum_{i=1}^{n} m(C_i)} \right) \times (100 - UB) \qquad (2)$$

This equation allows revaluing the amount of belief associated with each cluster after the introduction of the uncertainty. $m(C_i)$ is an original mass assigned to the candidate C_i without considering the uncertainty. $m'(C_i)$ is the new mass that considers for the uncertainty. UB is the uncertainty related to the criterion. $\sum m(C_i)$ is the total amount of belief affected to all the focal elements before considering the uncertainty. Note that, by the equation 2, after the revaluation $\sum m(C_i) + UB = 100$ [3].

To calculate the UB for each criterion, it was carried out as follows. We take a log file of which we know the starting message. For each criterion, we test if it manages to find the true starting message. We take each criterion alone and we do 100 simulations on this criterion. For each execution, we look if the criterion finds the right starting node. The UB associated with this criterion will the number of time that it fails to find the right node (failure rate to find the right node). The UB founded by our simulations are these:

- P_1: UB = 55,
- P_2: UB = 10,
- P_3: UB = 15.

The Dempster combination rule is provided by the theory to let us pool evidences from a variety of sources. This rule aggregates bodies of evidence two by two and at each run, and gives a new combined one. The combination rule is commutative and associative.

$$m(A) = m_1 \otimes m_2 = \frac{\sum_{B \cap C = A} m_1(B) \cdot m_2(C)}{\sum_{B \cap C = \phi} m_1(B) \cdot m_2(C)} \tag{3}$$

Note that the criterion weighting can be considered when combining evidences in order to prioritize the most significant ones. After combining the criteria and associating a score to each candidate message, our algorithm selects the highest value as the initial message of the business protocol.

Final Message Types. Now with the initial message, finding the final messages that terminate the protocol execution is trivial: we search for all the messages that have a well supported (above the threshold ϕ) edge going toward the initial message, meaning that in the log file, after this specific message, the next one is most probably the initial message.

3.3 Deriving the Business Protocol from the Message Graph

The Message Graph, after the removal of arcs below a user-defined threshold ϕ and those going toward the initial node, is *overgeneralized*, meaning that it accepts incorrect paths (see Figure 4).

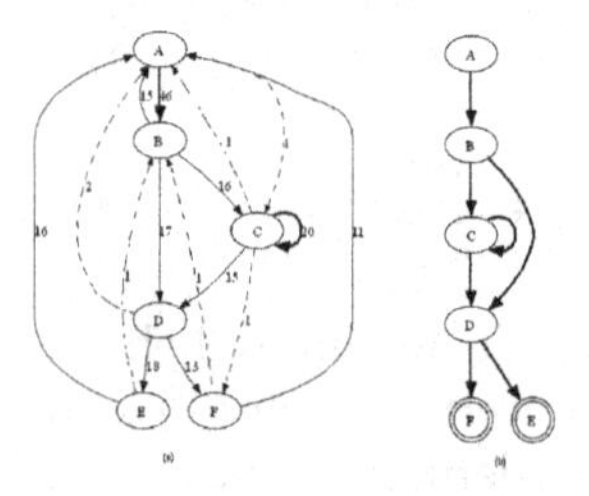

Fig. 4. (a): Message Graph, (b): Reduced Message Graph

In [10], the chosen solution is to perform a graph splitting. They check all the possible paths in the Message Graph and see if it appears in the CL. If not, the path was mistakenly accepted and should be removed from the graph. We did the inverse by building a *overfittet* FSM to our "conversations" and doing a state merging on it.

Candidates to be Conversations. With the information on the initial and final messages of the protocol, we can now extract from the Message Log what we can call "candidates" to conversations.

1. At every occurrence of an initial message, we start extracting the subsequence until finding a next initial message or reaching the end of the log. After this step we have a "candidate" conversation log file, see Figure 5(a).
2. At this point, we have to solve the problem of loop generalization: sequences as $\{A, B, C, B, C, D\}$ and $\{A, B, C, B, C, B, C, D\}$ cannot be seen as different possible executions of the business protocol. To address these situations, we transform each candidate conversation into a small Message Graph. In a graph like this, the loops are described as a way that it can be infinitely executed (see Figure 5(b)).
3. Now the conversations, as message graphs, are somehow generalized, so we group the now-identical ones and sum their frequencies.
4. It can also be supposed that the conversations $\{A, B, C, D, E, F, E, F, G\}$ and $\{A, B, C, B, C, D, E, F, E, F, G\}$ are identical executions, as they share the same set and the edges set E of the first is fully contained in the second and in the same order. So, as an arc will be missing in the graph of the first conversation ($C \rightarrow B$), but the properties are satisfied, we can join them, as described in Figure 5(c) and is a part of Algorithm 2.

Algorithm 2. Building the NFA

```
Require: ini,ends,cCL //candidates to be conversations
Ensure: FSM
  for all c in cCL do
    if last_event(c) in ends then
      mg(c)=Message_Graph(c)
    else
      del c
    end if
  end for
  for all c in cCL do
    for all c' in cCL do
      if c'! = c and (mg(c) == mg(c') or mg(c') ⊂ mg(c)) then
        f(c) = f(c) + f(c')
        del mg(c'),c'
      end if
    end for
  end for
  for all c in cCL do
    NFA.append_unique(bind_to_NFA(convert_to_FSM(mg(c))))
  end for
  return NFA
```

Finally, after reducing the number of different conversations in the conversation log, we are going to have a refined conversation log file with frequency annotation.

Create and convert a NFA into a DFA (Determinization). Every small Message Graph of every conversation candidate is transformed into a DFA by doing a simple translation, messages becoming arcs and transitions becoming states on the DFA. All the DFAs are binded together using an epsilon transition from an initial state, as depicted in Figure 5(d). With this binding, it becomes an NFA and will suffer a *determinization* process to become a DFA with its similar states merged, while still accepting the same language.

In the theory of computation, the subset construction is a standard method for converting a nondeterministic finite automaton (NFA) into a deterministic finite automaton (DFA) which recognizes the same formal language. The algorithm used to do the conversion is called subset algorithm. Its description and theory analysis is not in the scope of our work and can be seen in 1979's work of E. Hopcroft in [8].

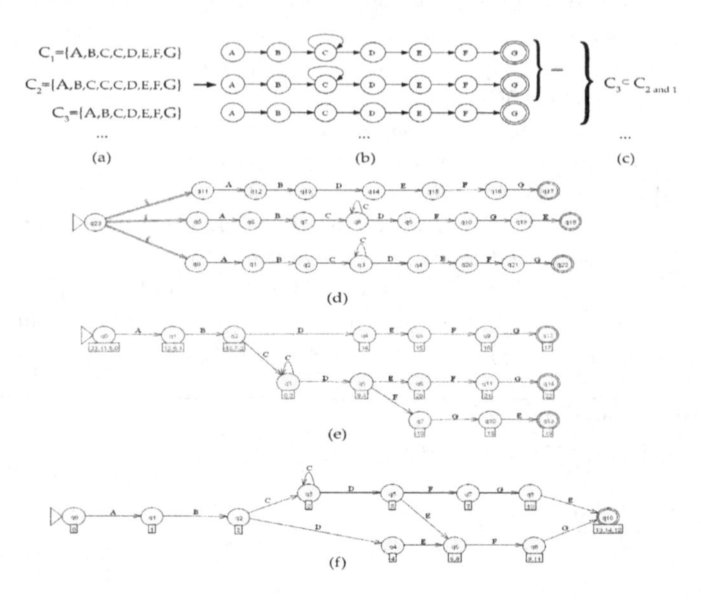

Fig. 5. in (a) we have the list of candidate conversations, in (b) the message graphs of each conversation, in (c) the process of joining the similar sequences, in (d) the NFA binding the messages through the state of the initial message and with the ϵ transitions highlighted, in (e) the NFA after the determinization (subset algorithm), now a DFA and in (f) the final DFA (Hopcroft's minimization)

Ulmann-Hopcroft's minimization algorithm and cycle detection. Ullman-Hopcroft's DFA state minimization algorithm reduces the equivalent states without modifying the accepted language. In Figure 5(d), we have the DFA minimized after the execution of Hopcroft's algorithm.

The idea of the last two algorithms is to minimize the Discovered Protocol without modifying the accepted executions. The generalization is assured when the Message Graph for each conversation candidate was created.

3.4 Refining the Discovered Protocol

Mainly, there are two reasons why user-assisted refinements will be needed: In the presence of noise and, as we use frequency thresholds to separate between noisy and correct conversations, the latter could be rejected and first ones could be accepted. Also, depending on the desired complexity of the modeled protocol, user may want to simplify its description.

4 Implementation

The best way to contribute to the already existing effort to formalize Business Protocols would be to implement our conversation and protocol discovery algorithm in a framework for the management and analysis of Web service interactions. However, in the first step, we coded an isolated solution using the Python language to execute and simulate our ideas.

For the creation of synthetic logs, simulated Business Protocols were used as inputs, message logs were generated containing multiple clients, the timestamps of the messages and the message names. Of course, the conversation identifier attribute was omitted from the log file. We use the following variables to create the synthetic log:

- Protocol model with probabilities on choices of messages,
- Number of desired conversations,
- Probability of random interruption in an execution of the protocol,
- Probability of random error (swap or omission) in the log file,
- Number of clients and servers of services making part of the log.

The errors were inserted by executing all the desired conversations (with random interruptions) and then making another run through the log, swapping and erasing events in the desired proportion.

(a) P1

P1: Frequency of Nodes			
Node	Score	Norm.	+UB(55%)
A	990	25.05	11.72
B	992	26.12	11.75
D	922	24.26	10.92
E	417	10.97	04.93
F	449	12.60	05.68

(b) P2

P2: Freq. of weak inc. arcs			
Node	Score	Norm.	+UB(10%)
A	3	42.88	38.60
B	1	14.28	12.85
D	1	14.28	12.85
E	1	14.28	12.85
F	1	14.28	12.85

(c) P3

P3: Degree_out > Degree_in			
Node	Score	Norm.	+UB(15%)
A	3	30	25
B	2	20	17
D	2	20	17
E	1	10	8.5
F	2	20	17

(d) Final Score

Final Results	
Node	Final Score
A	**45.43**
B	14.98
D	10.06
E	13.14
F	13.34
UB	03.05

Fig. 6. Application of the Dempster-Shafer Theory

Protocol models were taken from the literature so they could be as realistic as possible.

Once with the synthetic business protocols and their log files generated, we were able to run our protocol discovery algorithm to analyze it.

4.1 Simulation and Results

As an example of simulation, we take a Business Protocol. 1000 conversations were generated with 30% of probability of swaps or deletions in a message, and 10% of the conversations were randomly interrupted. An empirical, user-defined, general bottom threshold of 25% of the highest arc frequency was used for deciding if an arc was considerable or not.

Figure 4(a) show the corresponding Message Graph. The results of the Dempster-Shafter analysis, for finding the initial message, can be seen in Figure 6 (d). Note that the node C does not appear in the set of candidates nodes because it has self loops. After finding initial message, the message graph is reduced by removing the infrequent edges (see Figure 4 (b)).

The UB (uncommitted belief) of each Criterion was calculated based on the observation of the number of times that the criterion had a correct answer, taken different configurations.

The overall results of the Discovered Protocols were satisfactory: even when inserting a few interleaving, the results were as expected (Discovered Protocols very similar to the originals). The *overgeneralization* versus *overfitting* issue was well pondered by the form we did the generalization, and the accepted conversations were at most of the time the same as in the original tested protocols.

5 Conclusion and Perspectives

Message correlation in logs to discover models is a very important step for protocol discovery. That is why it is necessary to find an automatic way to correlate messages, without having an information in logs (conversations identifiers) which groups them in conversations.

Our contribution consists of two parts: Message Correlation in logs into conversations, without conversations identifiers, and generating the Business Protocol in the form of FSM (Final States Machines).

We modeled logs using graphs and we made use of graph theory techniques to extract the conversations and build the Business Protocol. We have extracted four evidences to convert our initial message graph into a reduced one. For this, we apply Dempster-Shafers mathematical theory of evidence.

As perspectives, we intend to consider other attributes for the correlation. There are yet new problems in this domain that need to be explored: as now the composition of web services are becoming a popular procedure, it should be thought about ways to solve the problem when multiple logs with communications from multiple cooperative services are taken.

References

1. Benatallah, B., Casati, F., Toumani, F.: Analysis and Management of Web Service Protocols. In: Atzeni, P., Chu, W., Lu, H., Zhou, S., Ling, T.-W. (eds.) ER 2004. LNCS, vol. 3288, pp. 524–541. Springer, Heidelberg (2004)
2. Benatallah, B., Motahari, H.: Servicemosaic project: Modeling, analysis and management of web services interactions. In: Third Asia-Pacific Conference on Conceptual Modelling (APCCM 2006), vol. 53, pp. 7–9 (2006)
3. Dekar, L., Kheddouci, H.: A cluster based mobility prediction scheme for ad hoc networks. Ad Hoc Networks 6(2), 168–194 (2008)
4. Devaurs, D., De marchi, F., Hacid, M.S.: Caractérisation des transitions temporisées dans les logs de conversation de services web. In: Extraction et Gestion des Connaissances (EGC 2007), vol. RNTI-E-9, pp. 45–56 (January 2007)

5. Dustdar, S., Gombotz, R.: Discovering web service workflows using web services interaction mining. International Journal of Business Process Integration and Management 1(4), 256–266 (2006)
6. Greco, G., Guzzo, A., Pontieri, L.: Discovering expressive process models by clustering log traces. IEEE Transactions on Knowledge and Data Engineering 18(8), 1010–1027 (2006)
7. Han, J., Kamber, M.: Data Mining: Concepts and Techniques. The Morgan Kaufmann Series in Data Management Systems (2000)
8. Hopcroft, J.E., Motwani, R., Ullman, J.D.: Introduction to automata theory, languages, and computation. In: SIGACT News, 2nd edn., vol. 32(1), pp. 60–65 (March 2001)
9. Motahari, H., Benatallah, B., Saint-Paul, R.: Protocol discovery from imperfect service interaction data. In: Proceedings of the VLDB 2006 Ph.D. Workshop (September 2006)
10. Motahari, H., Saint-Paul, R., Benatallah, B., Casati, F.: Protocol discovery from web service interaction logs. In: ICDE 2007: Proceedings of the IEEE International Conference on Data Engineering (April 2007)
11. Motahari, H., Saint-Paul, R., Benatallah, B., Casati, F., Andritsos, P.: Message correlation for conversation reconstruction in service interaction logs. Technical Report, University of Trento and University of New South Wales (2007)
12. Sentz, K., Ferson, S.: Combination of evidence in dempster-shafer theory. Technical report, Sandia National Laboratories (2002)
13. van der Aalst, W., Weijters, A., Maruster, L.: Workflow mining: Discovering process models from event logs. IEEE Transactions on Knowledge and Data Engineering 16(9), 1128–1142 (2004)

Concern-Sensitive Navigation: Improving Navigation in Web Software through Separation of Concerns

Jocelyne Nanard[1], Gustavo Rossi[2], Marc Nanard[1], Silvia Gordillo[2], and Leandro Perez[2]

[1] LIRMM, CNRS/Univ. Montpellier, 161 rue Ada, F34392 Montpellier cedex 5, France
{jnanard,mnanard}@lirmm.fr
[2] Facultad de Informática, Universidad Nacional de La Plata and Conicet Argentina
gustavo@sol.info.unlp.edu.ar

Abstract. Traditionally, the use of good techniques to improve software modularity, such as advanced separation of concerns, has no impact in the user experience, for example while navigating Web software. While the intent of these techniques is to simplify evolution and maintenance, navigation design quality is often seen as an unrelated concern. In this paper we present a novel approach for improving navigation in Web applications by using some of the core application's concerns (called navigational concerns) to derive their navigational structure. Using some realistic examples we show that, by carefully using these concerns, we can improve the user experience. Some implementation issues are discussed and a thorough comparison with related ideas in the Web Engineering field is presented.

Keywords: Separation of concerns, Concern-sensitive navigation, User experience.

1 Introduction and Motivation

Web applications have evolved from being simple information repositories to complex and ubiquitous platforms for performing complex business processes or for publishing and sharing multimedia information. Huge e-commerce sites such as Amazon.com, blogs like Youtube or Flickr or cooperative encyclopedia like Wikipedia are clear examples of this evolution. As these applications are being constantly modified, maintenance implies an additional challenge to software development methodologies. Fortunately the Web engineering community has already discussed and proposed advanced software techniques to simplify design and evolution (See for example [19], [14]); most of them are based on variants of the separation of concerns principle [12].

However, these design techniques are usually considered orthogonal to the problem of application usability. In this way for example, the quality of navigation structures is considered a completely disparate problem with respect to, for example, achieving design modularity. In other word, a Web software which has been conceived with high standards regarding evolution and maintenance does not necessary provide a good navigation experience to the final user.

Z. Bellahsène and M. Léonard (Eds.): CAiSE 2008, LNCS 5074, pp. 420–434, 2008.

In this paper we show how a wise separation of *application concerns* during modeling and design, and the information recorded during those stages can be cleverly used also towards providing a more *flexible* navigational structure and thus improving the user navigation experience. Our work aims at improving the cognitive and rhetoric access to information, which means providing the user with the needed information in each concern, and such that it is organized and presented in a more opportunistic way [13]. Suppose for example an application such as Amazon.com in which users navigate through thousands of products with different concerns (tasks or interests) in mind. In Fig. 1 we show a typical screen of a book with the corresponding information and available functionality.

Fig. 1. A Book in Amazon.com

The page for a book (like in Fig. 1) looks exactly the same independently of the reason why the user reached it. For example it could have been accessed as a book on Italy, it could result from an Amazon recommendation according to the user's buying history, or it could be also accessed from the shopping cart because the user wants to be sure about the book's contents before proceeding to pay for it. This "flat" structure, in which every object looks equally regarding the context in which it is accessed, diminishes usability and might also cause errors [3].

The contents of the book page should be improved by taking into account the "dominant" concern in which it is being accessed. For example in Fig. 2.a we show part of a possible Web page for the book when accessed as a recommended item and in Fig. 2.b, the same book when accessed from the shopping cart. In Fig. 2.a there is a link to get an explanation of why the book was recommended, and links to the previous and next recommendations. Notice that these links do not make sense when the book is accessed with other different concern in mind. In Fig. 2.b meanwhile, there is an indication that the book is already in the cart and that if added again, it will imply adding a new unit. In both cases knowing the actual user's concern helps to enrich the information on the target page with new contents and links to simplify or clarify the user's task.

Some Web Engineering approaches have solved sub-sets of this problem using specific ad-hoc techniques and/or notations. For example, OOHDM [23] uses the concept of navigational contexts to enrich hypermedia nodes when accessed in a particular set (for example the set of recommended products). In [24] a similar idea is used to restrict operations in the context of a business process, i.e. to avoid that the same product is added once more in the shopping cart while checking-out. Our approach aims at providing a more systematic context-sensitive navigation.

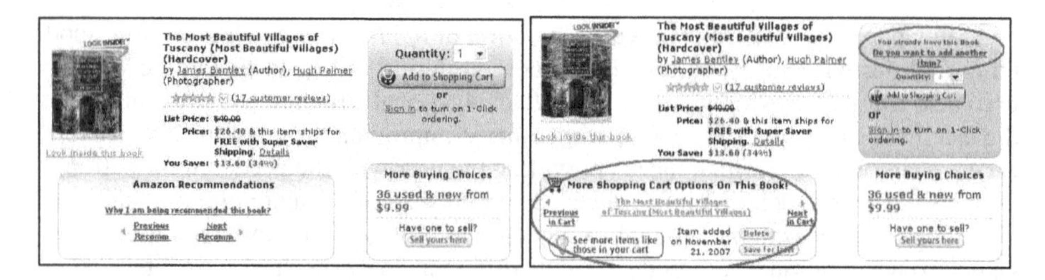

Fig. 2. (a) Book accessed in the Recommendation Concern

Fig. 2. (b) Book accessed in the Shopping Cart concern

This paper has three main contributions:

- We introduce the concept of concern-sensitive navigation (CSN) as a conceptual and practical tool towards improving the navigational structure of Web applications as perceived by the user.
- We show how to introduce CSN in Web development methodologies, emphasizing how the use of techniques for advanced separation of application concerns can be used not only to improve modularity, but also the user's navigation experience.
- We show the feasibility of CSN by briefly analyzing implementation alternatives.

The rest of the paper is organized as follows. We first introduce concern-driven navigation and illustrate the concept with simple examples. We next discuss how to engineer applications which support concern-driven navigation discussing modeling, design and implementation issues. Finally we compare our work with related approaches and present some further research we are pursuing. We show the design and implementation feasibility of these ideas by providing illustrative examples.

2 Concern-Sensitive Navigation

The main motivation of our research is to show that by separating concerns we can not only improve evolution and maintenance but also produce better navigation structures. In this section we explain deeply which are the issues one has to consider to realize CSN.

2.1 Background

According to [17] an *application concern* is defined by any coherent set of requirements, e.g. all requirements referring to a particular theme or behavioral application feature. More generally [25] defines a concern as a "matter of consideration in a software system". Concerns may reflect functional or non-functional aspects of an application such as recommendations and checkout in E-commerce, topic areas such as history or geography in an Encyclopedia. Concerns may be generic, when they appear in a broad number of applications (e.g. adaptivity, usability), domain specific when they only apply to a set of applications (payment in e-commerce), or even application specific when they only show up in a particular kind of software (e.g. Marketplace in Amazon.com).

A *navigational concern* is an application concern that affects navigation, i.e. it manifests itself in the navigational structure of the application (the exhibited contents and links), and which therefore impacts in the way users navigate the application. In this paper we focus on navigational concerns and ignore others which are nevertheless important but do not affect navigation (such as persistence or security).

Most Web applications deal with a myriad of navigational concerns and usually (for improving usability though many times only for marketing reasons) they exhibit information pertaining to more than one concern in the same page, i.e. in the same page we might find contents, links and functions which belong to different concerns. Modern software engineering techniques such as aspect-oriented development [9] promote a clear separation of concerns during specification, design and programming and their late weaving either during compilation or even execution. In this way one can diminish the impact of crosscutting concerns during software evolution. However, and as we showed before, the page in Fig. 1 looks the same independently of the concern the user has in mind when accessing it (searching a product, being recommended a product, as part of an offer, etc).

Users navigate in Web applications to perform a specific task; mature Web design methods have already prescriptive approaches and notations to map task descriptions (e.g. specified as use cases) into conceptual and navigational models [26]. Our approach complements these ideas with a strategy to enrich the navigation objects with information specific to the concern that the user is traversing. The mechanics of CSN, as well as its scope should be defined by the designer according to the user's need and convenience. In what follows, we show how to use the information on navigational concerns to improve the navigation structure of the application. For the sake of conciseness we only focus on navigation aspects.

2.2 Definition

To formalize our notion of CSN we refer to a navigation object type Nj (the realization of an atomic or composite hypermedia node type) as comprising a set of properties; these properties may be further classified in media contents, anchors for links or operations exhibited by the node and can be divided in two groups:

- properties intrinsic to the object [15] (i.e. which are present regardless the concern in which an object is accessed). We call them core properties;
- properties which, given a concern Ci, correspond to the set of perceivable properties of Nj when accessed in the Concern Ci and which is the result of applying a function P (Ci, Nj).

For each meaningful pair (Ci, **N**j) the set of properties should be a superset of the core properties of Nj. In Fig. 3 we illustrate the definition for the examples in Fig. 1 and 2 using a UML-like notation. Notice that the same node instance exhibits different properties according to the concern in which it is accessed. By adjusting the node's properties to the concern in which it is being accessed we improve the navigational structure, by making contents more focused to the actual concern the user is navigating (i.e. the intended task).

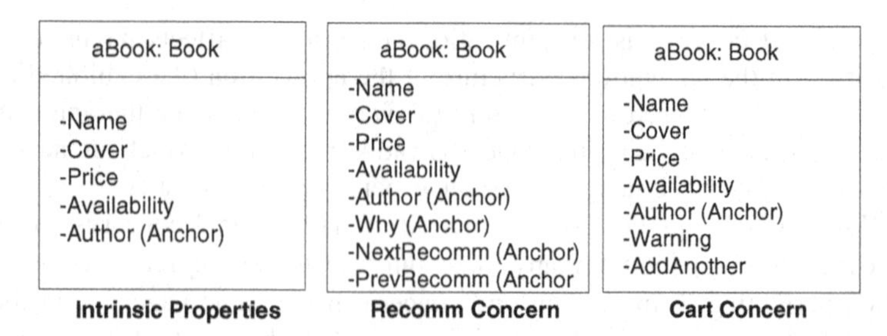

Fig. 3. Intrinsic Properties vs. Properties in Recommendation and Cart Concerns

Regarding this initial definition there are some additional issues to consider:

- Notice that the perceivable properties do not depend on the user profile or identity, which means that CSN is slightly different from adaptive navigation (See the related work section).
- Besides the so called intrinsic properties, there might be properties which pertain to different concerns and which we want to exhibit permanently (e.g. the Add To Wish List operation), i.e. regardless the navigation path. Though this is a design issue not fully related with concern-driven navigation but with concern composition, we only give an overview of it (See Section 3.2).
- Defining the concerns which affect a node type requires a clear understanding of the application concerns, their relationships and the way they reflect in navigation (See next section).

2.3 Which Categories of Concerns Affect Navigation?

As explained in [25] there are many different kinds of concerns which may arise in the process of Web software development. For the sake of conciseness we enumerate here the most important types of concerns which are exposed to the user and therefore affect navigation:

- *Task concerns* are the broadest category of navigation concerns; they abstract those concerns which relate to the different high level actions that the user can perform in a Web application, for example exploring products, managing the shopping cart, adding reviews, checking out, managing lists, etc. Some of these tasks are related with finer grained application features such as *services* offered by the application; in the case of Amazon, recommendations, marketplace, lists, etc. Notice that while involved in a service the user is always performing a task.
- *Topics:* Pure informational sites might introduce even finer-grained concerns; for example topics or themes such as in an Encyclopedia. Topic-based concerns are also present in the context of tasks; for example while searching books in Amazon.com, the genre of the book (thriller, travel, technical) or its theme area (Software Engineering, Programming, etc) might itself become a concern.
- *"Pure" navigational concerns*, like Guided Tours or sets. These are usual abstractions in navigational design and therefore can be considered also as specific concerns.

Deciding which is the appropriate level of granularity for choosing concerns during the modeling stage is, as well as choosing the "right" concerns, part of the designer job, and it is outside the scope of this paper. However, the reader can find good guidelines in the literature on Early Aspects [8], particularly in [2].

2.4 Which Kinds of Concern "Enrichment" Improve the User's Experience?

Though the answer to this question strongly depends on the specific concern, there are two broad categories of enrichments:

Basic enrichments. We found three kinds of enrichments, namely:

- *New or modified contents*: As shown in the cart concern of Fig. 3, we can enrich the node instance with new attributes
- *Anchors and Links*: Also in Fig. 3, in the Recommendation concern we added a new link and the corresponding anchor to improve navigation
- *Operations*: A node instance might exhibit additional operations when accessed in a concern; for example we could have added a (deleteFromCart) operation in the Cart concern in Fig. 3.

Enrichment Patterns. For each of the previously mentioned concern types, the following patterns are the most recurrent:

- *For Task-Based Concern*: When the concern is defined by a business process (like in [24]), and operating on the target node might conflict with the process, it is advisable either to eliminate operations which collide with the concern or to add specific warnings (e.g. the shopping cart or checkout concerns in Amazon).
- *For Thematic concerns*: when a node is accessed in that concern, add information and links specific to the topic which is related with the node. For example the book in Fig. 1 could be enriched with links to other books on Italy (or related to the higher level concern, Travel)
- *For Pure Navigational Concerns*: When the concern can be represented as a set as in OOHDM navigational contexts [23] (e.g. the set of recommendations, etc.), it is wise to enrich the node with links to the index of the current set, and to the previous and next elements of the set. Another example of this kind of enrichment can be found in tag-based navigation like in Flickr (e.g. by providing links to other photos with the same tag).

3 Engineering Web Applications Supporting CSN

Web Engineering approaches, like those in [22], support separation of the most outstanding concerns in this kind of software: requirements' capture, content or application modeling, navigation and presentation design, business process modeling, etc. (they correspond to methodology-related concerns). Some of them have also introduced elements of advanced separation of concerns (such as aspect-orientation) to deal with cross-cutting concerns [4]. Even though the kind of application concerns which might be reflected in CSN structures does not necessarily correspond to "aspects" (as they may not crosscut in the standard way), we claim that the most relevant identified concerns (e.g. following the classification in 2.3) should be designed separately. We next explain how to map concerns into navigational structures. Though we use OOHDM as the exemplary method, the ideas can be applied to other well-known approaches like UWE [14], OOWS [19] or WebML [5]. In the following, we discuss mainly the Requirement, Modeling and Navigational design issues; some Implementation aspects are then outlined. Presentation issues can be read at [10].

3.1 Requirement and Modeling Issues

In [11] we presented an approach to model navigational concerns in Web applications; the approach which derives from well-known ideas in the Early Aspect community [8] helps to elicit, identify and specify the interactions which emerge in each navigational concern. In our work, each concern is explicitly represented using a XML-template and for each use case in the concern a User Interaction Diagram (UID) is built. UIDs show how the interaction proceeds in a high level way. In Fig. 5, we show part of the definition of the Recommendation concern (on the left) and the corresponding UID (on the right). The UID shows in a simple state diagrams which items are presented to the user, either as simple structures such as Book and its attributes or as sets of structures (those which begin with "...") and the transitions corresponding to user's actions such as selecting the "Why" option in the right part, or the "Next" and "Previous" recommendation below state "C". When comparing the UID in Fig. 4 with one in the core application concern (not shown for conciseness), we will find that the information exhibited by books is slightly different (See state "C"); this information will be used to define CSN as described in Section 3.2.

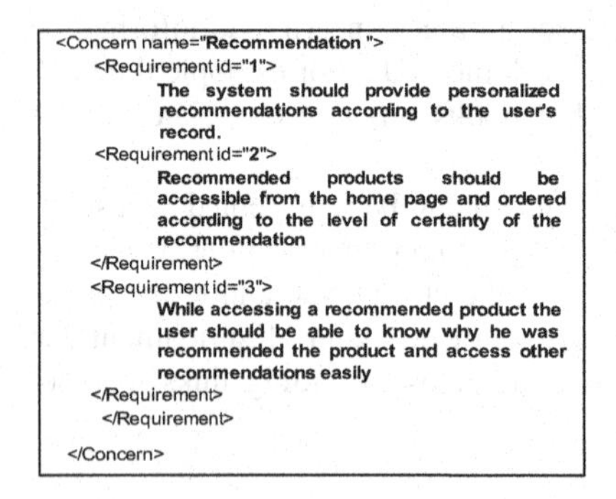

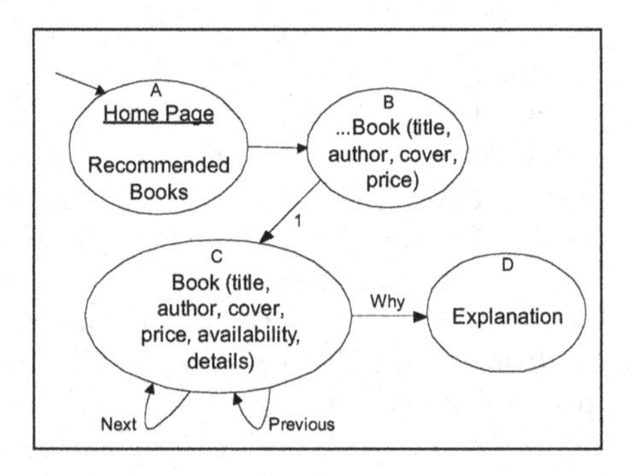

Fig. 4. Requirements corresponding to the Recommendation concern

Once the whole set of requirements have been elicited and modeled, a conceptual OOHDM model is built using the information collected from the UIDs, which as explained in [23], allow to define the attributes and methods of conceptual classes. Following the Theme approach [7] we propose to partition the conceptual model in sub-models, one for each of the relevant concerns (See [11]). However, other approaches (aspects, object decorators, etc) can be also used according to the kind of concern crosscutting; this discussion is outside the scope of this paper (See for example [16]).

For the sake of conciseness, we will concentrate on navigational concerns and use as an example a fragment of the e-store shown in previous Figures. In Fig. 5 we present a simplified OOHDM conceptual model corresponding to this example. We only present the conceptual sub-models corresponding to the Core and Recommendation concerns emphasizing class structure over relationships, as they are sufficient for illustrating the rationale and mechanics for building the concern-sensitive navigational model. Following [7], each model presents the view of the application classes according to the concern. Notice that there are two classes in the Recommendation concern which do not appear in the Core concern. When these models are weaved some classes may be transformed in a class containing the union of the definitions of each model (e.g. Product), others might evolve into aspects, etc.

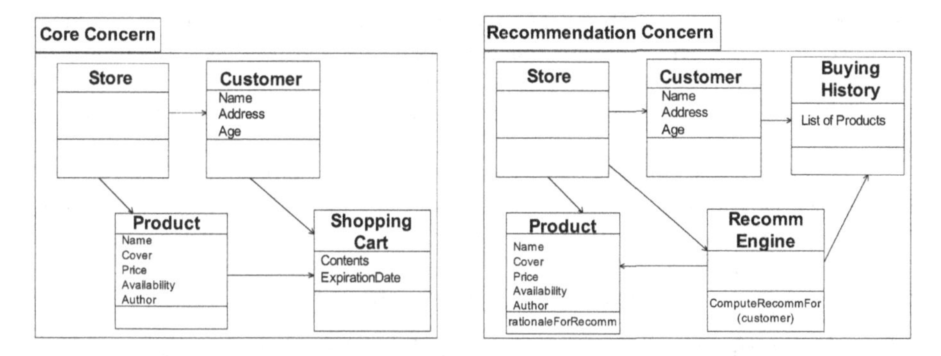

Fig. 5. Conceptual Model for the e-store site

3.2 Navigational Design

Web engineering approaches provide primitives for describing the navigational structure of the application, i.e. for defining nodes, links, indexes and higher-level structures such as landmarks, guided-tours, etc. Nodes contents are usually defined opportunistically to improve usability; for this reason, many times the same node exhibits information pertaining to different concerns (See Fig. 1). In OOHDM we do this by defining the node's attributes as views on the corresponding conceptual classes (eventually "viewing" their definitions in the different involved concerns) [23].

We are interested however, in the information and links which are only meaningful in some specific concerns. As defined in 2.3, CSN allows enriching the contents and links of a hypermedia node according to the current user's concern. A navigational model expressing concern-sensitive navigation should take into account what follows:

Identify which node types are "affected" by existing concerns. The first output of a navigational diagram (in OOHDM and other methods) is a navigational schema formed out of node and link types, representing the type of objects the user will perceive with their attributes and the navigable relationships. As said above, these types are obtained from the conceptual model using a view mechanism [23] which allows gathering information according to users' needs. A simplified "flat" navigational schema for the e-store site is presented in Fig. 6. Again, we emphasize node's structures and do not indicate link types' names.

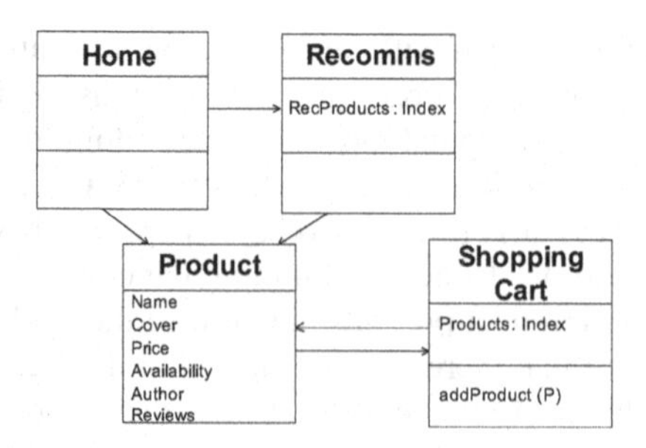

Fig. 6. Navigational Schema for the e-store site

The navigational schema shows the specification of "natural" node types (i.e. those which stand independently of any concern, See [15]), and the links which provide navigation paths between node instances. For the sake of simplicity we don't present the OOHDM navigational contexts model in which indexes are further specified (See [23]).

By analyzing the requirement model, particularly the specification of concerns and their realization with UIDs, we can identify which of the node types are affected by each concern. We do this by building a table (See Table 1) which makes explicit the function P, described in Section 2.2. Each line corresponds to a concern. For each concern, we add a new column for each node type affected in this concern.

Define the information, links and services to be added when accessing a node in each particular concern. For each pair (concern, node type) we indicate the corresponding enrichment when a node instance is accessed in that concern. This decision takes into account the nature of the concern (i.e. the current user's task); for example we might decide to add more specific information to improve user's understanding, links to related information objects (corresponding to the same concern) to improve the completion of the task, etc. As mentioned in 3.1, UIDs are the first source for this information as they collect most of the data and possible interactions corresponding to each concern. Table 1 shows a sketch of the enrichment corresponding to the Node type Product when accessed in the Recommendation concern (omitting the anchor's specification for conciseness).

We represent concern-sensitive navigational diagrams with the notation of role-enrichment. A role type (indicated as a rounded rectangle) shows, when attached to a node, the additional information and links that will be shown in the corresponding

Table 1. Table showing the enrichment for each concern and node type

Nodes / Concerns	Node Type 1	...	**Product**	...
Concern 1				
...				
Recomm			**Why (Anchor)** **Next (Anchor)** **Prev (Anchor)**	
...				

concern. Roles of a node type act as decorations adding the concern-specific information. In Fig. 7 we show how we enriched the navigational diagram of Fig. 6 with concern information, represented with roles. There are two roles, one for products accessed from the recommendation list (i.e. in the recommendation concern) and one for products accessed from the cart (i.e. in the Cart concern); in both cases the role contains the additional features as part of its specification. Notice for example the two links defined from the *RecommProduct* role into itself and the additional Explanation node type, which reflect the specification in the UID of Fig. 5 and the corresponding table entry of Table 1. Also the role *ProductInCart* adds a behavior *AddToCart* which possesses a slightly different semantics with respect to the "normal" addToCart behavior in Product, asking the user if he really wants to increase the number of units of the product. The use of roles in Web Engineering has been discussed in our previous work in [21].

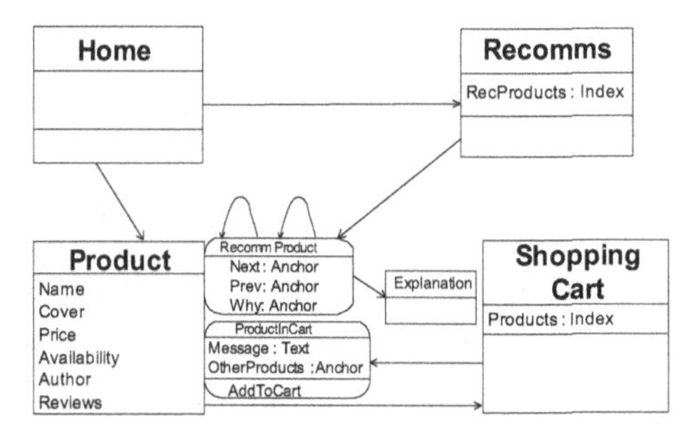

Fig. 7. A concern-sensitive navigational schema

3.3 Further Issues

An interesting modeling (and also implementation) issue arises when dealing with families of navigational concerns. There are some concerns which are "atomic" because there is only one instance of the concern, and as a consequence all nodes affected by the concern either have the same enrichment or the enrichment only depends on the affected node. The best example of this kind of concerns is the Cart concern: there is only one Shopping Cart and therefore all products navigated from the Shopping Cart (i.e. in the Cart Concern) will have the same enrichment. Meanwhile, in the recommendation concern, the enrichment depends on the node instance (the product) as links to other recommendations are a function of the product. In this case, in our modeling approach, the corresponding role type can be considered a singleton (it has only one instance) which adds information somewhat parameterized by the target node instance. This added information (See Fig. 2 and 7) is obtained by collaboration with the node instance.

Meanwhile, certain concerns, particularly Topic or Thematic concerns have usually many instances, one for each possible topic. In our example of Fig. 1 and 2, we could have been exploring books on Italy, and therefore when accessing the book, the actual concern is Italy (which is a sub-category of the concern Travel). Even though the designer should decide which the suitable level of granularity is, and eventually choose if Italy is a possible concern (e.g. Travel might be preferred), it is obvious that we expect different additional information while exploring books on Web Engineering. In any cases, the most elegant design solution is to consider that the concern role type (Topic) has many instances, one for each topic; these instances are created dynamically, when accessing the target node instance. If necessary, there might be a hierarchy of role types to cope with variants among Travel, Technical Books, Software, etc. with respect to the specific enrichment for each concern. Fig. 8 shows the e-store screen with this enrichment and the corresponding role and type specification to cope with this situation; the parameter in the role specification is used at instantiation time, i.e. when the corresponding role instance is created.

Fig. 8. Role Type parameterized with the concern instance

3.4 Implementation

We show the feasibility of mapping the previously described navigational structures onto a running application, by briefly describing two implementation alternatives.

The first alternative consists (as most Web design methods do) in translating the specification of nodes into XML files which, when being populated, are themselves translated into final interfaces by using XSL specifications. Similarly, we map the role types into XML files which are aware of the specifications they enrich. In the simplest case, producing the intended node instance requires the injection of the role file into the corresponding node instance. In [10], we described how to use XML transformations (described using XSLT) to weave two concerns together. Transformations are easy to specify and only use standard technologies. There might be cases however, in which the enrichment requires more subtle processing, for example to compute links which depend on the target node instances (as in the recommendation concern), or on the combination between the concern and the node instance (as in the case of topic-based concerns). This processing can be also specified using rules in the transformation files. Though the details are outside the scope of the paper, we show in Fig. 9 the simplified XML files for the core and recommendation concern and in Fig. 10 the XSLT transformation to weave them together; finally in Fig. 11 we present a UML activity diagram illustrating the process from the user's request to the generation of the necessary structure to realize concern-sensitive navigation.

```
<book>
<name>The Most Beautiful Villages of Tuscany </name>
<cover>TheMostBeautifulVillagesofTuscany .jpg</cover>
<price>40.00</price>
<availability >24 hs</availability>
<author>/search .do?author=JamesBentley </author>
</book>
```

```
<recommendation >
<why>/recommendation .do?id=2345</why>
<nextRecomm >/product?id=254</nextRecomm >
<prevRecomm >/product ?id=168</prevRecomm >
</recommendation >
```

Fig. 9. XML files corresponding to core and recommendation concerns

```
<xsl:template match ="/book">
<book>
<xsl:copy-of select ="*"/>
<xsl:copy-of select ="document ('recomm .xml')/recommendation /*"/>
</book>
</xsl:template>
<xsl:template match ="@*|node()" >
<xsl:copy>
<xsl:apply-templates select ="@*|node()"/>
</xsl:copy>
</xsl:template>
</xsl:stylesheet>
```

```
<book>
<name>The Most Beautiful Villages of Tuscany </name>
<cover>TheMostBeautifulVillagesofTuscany .jpg</cover>
<price>40.00</price>
<availability >24 hs</availability >
<author>/search .do?author=JamesBentley </author>
<why>/recommendation .do?id=2345</why>
<nextRecomm >/product ?id=254</nextRecomm >
<prevRecomm >/product?id=168</prevRecomm >
</book>
```

Fig. 10. XSLT transformation and the result of its application

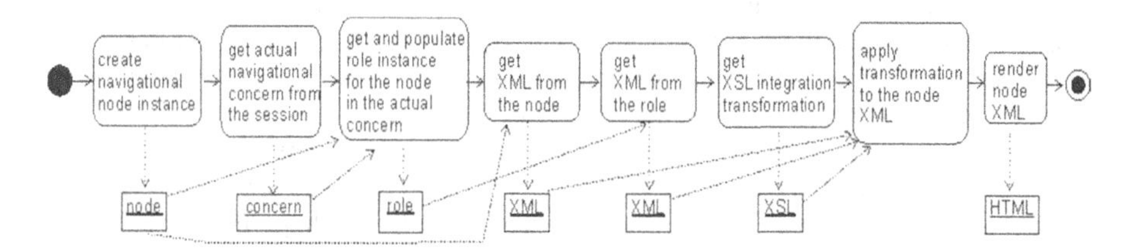

Fig. 11. Activity diagram for weaving a role into the core concern

A more systematic alternative is to use a model-driven approach [26], in which the semantics of role diagrams, such as the one in Fig. 7, feeds the transformation engine to produce the intended behavior. In [20] we presented a framework, CAZON, which injects irregular functionality into node instances. We have used CAZON to incorporate volatile functionality in node instances; such services are included for short periods of time (such as draws, promotions, holiday's offers, etc.). Insofar as CAZON uses XML files to describe nodes (both core and volatile), its underlying engine uses the same basic ideas as the previously described transformations, relieving the designer from the job of specifying them. We are currently extending CAZON to provide role enrichment, by expressing the relationships between role and node types using CAZON's built-in mechanisms, to show relationships between base and injected structures and behaviors.

As a summary of the previous explanation, we stress the importance of specifying in a declarative way the relationships between role types (as realization of concern enrichment), and node types. Even though this specification can be done using XML we think that model-driven approaches are preferable because they relieve us of implementation details.

4 Related Work

Separation of concerns has been a driving force in Software (and Web) Engineering for years. As mentioned in Section 1, the main rationale for improving separation of concerns has been simplifying software evolution and maintenance by achieving modularity. Treating concerns as "first class" artifacts in the software development process helps to better understand the underlying domain and produce composable modules which reflect the requirements in each concern and the relationships between them. Approaches like Aspect-Oriented Software Development [7], [9] guarantee modularity and seamless composition of core and aspectual functionality.

All Web Engineering approaches recognize the need to separate the design in layers which deal with clearly different concerns such as navigation, presentation and also business processes (We call this *horizontal separation of concern*) and even adaptation [4]. An interesting example of the use of advanced horizontal separation of concerns in Web Engineering is [6] which proposes a formal specification of the connections between different models during the development cycle to produce seamless weaving; the weaving model itself is described as a meta-model. The main difference between our work and existing approaches in Web engineering, is that we use also a *vertical separation of concerns* related to the essence of the application, to systematically produce better navigational structures and not just to ease evolution. While we rely on well known separation of concern techniques to specify and design each concern, we use the information collected from requirements to navigational design, to realize an improvement in the information and links perceived by the user while navigating in the context of a concern. Our work also generalizes some existing approaches to enhance navigation in specific contexts such as sets of related objects (called Navigational Contexts in OOHDM [23]) or business processes [24].

CSN has some points in common with the work on adaptive hypermedia [1]. Adaptive hypermedia approaches seek to improve user's navigation by taking into account the user's profile and needs. In an adaptive hypermedia application, nodes and links vary according to the characteristics of the user, his navigation history, etc. Adaptive hypermedia systems rely on a user model which represents the meaningful user's features and an adaptation model in which the adaptation rules and algorithms are specified. Our work meanwhile, while also producing an improvement with respect to "flat" navigation structures, does not pose additional requirements to Web software, such as recording the user's features or elaborated rules or algorithms. We rely on the use of well-known and mainstream software practices (such as separation of concerns) to generate a better navigational experience. Additionally, as the underlying designs are modular (e.g. by the use of aspects and/or roles), adding new concerns and their corresponding navigational adaptations is straightforward, as the core functionality is oblivious with respect to the new (and of course the "old") concerns.

5 Concluding Remarks and Further Work

We have presented an approach to use separation and composition of concerns, not only to enhance Web software modularity, but also to improve its cognitive and rhetoric access. While we recognize the importance of using advanced separation of concerns techniques to ease application evolution, we claim that these techniques can be further applied in the context of navigational design to obtain better navigational models; for this aim we introduced the concept of concern-sensitive navigation. Applications supporting concern-sensitive navigation can offer the user more focused information, links and services according to his actual concern. While this idea shares some of the objectives of adaptive and context-aware hypermedia approaches, it introduces an orthogonal concept: the navigational concern. We have shown, with simple examples, that producing a concern-sensitive navigational model is rather simple, and we have provided a proof of concept of its implementation feasibility. For this, we have relied on standard tools and representation techniques to demonstrate that the idea can be easily put to work without complex engines or frameworks.

We are now studying several aspects, which belong to finer grained details of concern-sensitive navigation. One of them is dealing with the extent of a concern during navigation; while the "entrance" to a concern is clearly defined at design time as shown in the diagram of Fig. 7, many times it is not obvious when the user concern changes, for example when navigating links which are defined in the core concern. Related with this issue and with the enrichment patterns (See Section 3), we are also researching on patterns for concern selection to enrich existing catalogues of Web patterns such as [27]; in some kinds of concerns, particularly topic-based concerns, the user can select the actual concern for example when choosing a menu option (e.g. categories of books, subjects in an encyclopedia, etc). Knowing these patterns can help the designer to further improve the navigational structure. We are also researching on the concern-based improvement of applications in which most contents are provided by user, such as Wikipedia. Further usage analysis is necessary to have a clear understanding of the impact of concern driven navigation in users.

We are also finally researching on different ways of implementing concern-sensitive navigation; particularly we are using our ideas of XML transformations [10] and tree transformation through grammar networks [18] to ease the process of concern enrichment both at the navigation and interface levels. Finally we are working in the process of measuring the improvement provided by CSN; this can be done by analyzing how user's tasks are simplified but will also require further experiments with real users.

References

1. Adaptive Hypermedia Reference Library, http://wwwis.win.tue.nl/ah/publications.html
2. Baniasaad, E., Clarke, S.: Finding Aspects in Requirements with Theme/Doc. In: Proc. of Workshop Early Aspects 2004, associated to the ACM Conf. AOSD (2004)
3. Baresi, L., Denaro, G., Mainetti, L., Paolini, P.: Assertions to Better Specify the Amazon Bug. In: Proc. of the 14th Int. Conf. on Software Engineering and Knowledge Engineering. ACM Int. Conference Proceeding Series, vol. 27 (2002)
4. Baumeister, H., Knapp, A., Koch, N., Zhang, G.: Modelling Adaptivity with Aspects. In: Lowe, D.G., Gaedke, M. (eds.) ICWE 2005. LNCS, vol. 3579, pp. 406–416. Springer, Heidelberg (2005)
5. Ceri, P., Fraternali, P., Bongio, A.: Web Modeling Language (WebML), A Modeling Language for Designing Web Sites. Computer Networks and ISDN Systems 33(1-6), 137–157 (2000)

6. Cicchetti, A., Di Ruscio, D., Pierantonio, A.: Weaving Concerns in Model Based Development of Data-Intensive Web Applications. In: Proceedings of the ACM Symposium on Applied Computing (SAC 2006), pp. 1256–1261. ACM Press, New York (2006)
7. Clarke, S., Baniassad, E.: Aspect-Oriented Analysis and Design. The Theme Approach. Object Technology Series. Addison-Wesley, Reading (2005)
8. Early Aspects Home: http://www.earlyaspects.net
9. Filman, R., Elrad, T., Clarke, S., Aksit, M.: Aspect Oriented Software Development. Addison-Wesley, Reading (2004)
10. Ginzburg, J., Rossi, G., Urbieta, M., Distante, D.: Transparent Interface Composition in Web Applications. In: Baresi, L., Fraternali, P., Houben, G.-J. (eds.) ICWE 2007. LNCS, vol. 4607, pp. 152–166. Springer, Heidelberg (2007)
11. Gordillo, S., Rossi, G., Moreira, A., Araujo, J., Vairetti, C., Urbeita, M.: Modeling and Composing Navigational Concerns in Web Applications. Requirements and Design Issues. In: Proc. of Latino American Conf. on the WWW (LA-Web 2006). IEEE Computer Society Press, Los Alamitos (2006)
12. Harrison, W., Ossher, H., Tarr, P.: General Composition of Software Artifacts. In: Löwe, W., Südholt, M. (eds.) SC 2006. LNCS, vol. 4089, pp. 194–210. Springer, Heidelberg (2006)
13. Horchani, M., Nanard, J., Nanard, M.: Les Hypermédias comme Paradigme d'Interfaces Adaptatives. In: Saleh, I. (ed.) Les hypermédias. Hermès, pp. 119–146 (2004)
14. Koch, N., Knapp, A., Zhang, G., Baumeister, H.: UML-Based Web Engineering. In: (22)
15. Kristensen, B.B., Osterbye, K.: Roles, Conceptual Abstraction Theory and practical Language Issues. Theory and Practice of Object Systems 2(3), 143–160 (1996)
16. Marin, M., Moonen, L., van Deursen, A.: A classification of Crosscutting Concerns. In: Proc. IEEE Conf. on Software Maintenance (ICSM 2006). IEEE Computer Society Press, Los Alamitos (2006)
17. Moreira, A., Araujo, J., Rashid, A.: A Concern-Oriented Requirements Engineering Model. In: Pastor, Ó., Falcão e Cunha, J. (eds.) CAiSE 2005. LNCS, vol. 3520, pp. 293–308. Springer, Heidelberg (2005)
18. Nanard, M., Nanard, J., King, P.R.: A structural computing approach to the production of multimedia document series. NRHM 12(2), 165–190 (2006)
19. Pastor, O., Abrahão, S., Fons, J.: An Object-Oriented Approach to Automate Web Applications Development. In: Bauknecht, K., Madria, S.K., Pernul, G. (eds.) EC-Web 2001. LNCS, vol. 2115, pp. 16–28. Springer, Heidelberg (2001)
20. Rossi, G., Nieto, A., Mengoni, L., Lofeudo, N., Nuño Silva, L., Distante, D.: Model-Based Design of Volatile Functionality in Web Applications. LA-WEB, pp. 179–188 (2006)
21. Rossi, G., Nanard, J., Nanard, M., Koch, N.: Engineering Web Applications with Roles. Journal of Web Engineering 6(1), 19–48 (2007)
22. Rossi, G., Pastor, O., Schwabe, D., Olsina, L. (eds.): Web Engineering: Modelling and Implementing Web Applications. Springer, Heidelberg (2008)
23. Rossi, G., Schwabe, D.: Modeling and Implementing Web Applications with OOHDM. In: (22)
24. Schmid, H., Rossi, G.: Modeling and Designing Processes in E-Commerce Applications. IEEE Internet Computing 8(1), 19–27 (2004)
25. Sutton, S., Rouvellou, I.: Modeling of Software Concerns in Cosmos. In: Proc. of ACM Conf. AOSD 2002. ACM Press, New York (2002)
26. Valderas, P., Fons, J., Pelechano, V.: Transforming Web Requirements into Navigational Models: AN MDA Based Approach. In: Delcambre, L.M.L., Kop, C., Mayr, H.C., Mylopoulos, J., Pastor, Ó. (eds.) ER 2005. LNCS, vol. 3716, pp. 320–336. Springer, Heidelberg (2005)
27. Van Duyne, D.K., Landay, J.A., Hong, J.I.: The Design of Sites: Patterns for Creating Winning Websites. Prentice-Hall, Englewood Cliffs (2006)

A Flexible and Semantic-Aware Publication Infrastructure for Web Services

Luciano Baresi, Matteo Miraz, and Pierluigi Plebani

Dipartimento di Elettronica e Informazione – Politecnico di Milano
Piazza Leonardo da Vinci 32, 20133 Milano (Italy)
{baresi,miraz,plebani}@elet.polimi.it

Abstract. This paper presents an innovative approach for the publication and discovery of Web services. The proposal is based on two previous works: DIRE (DIstributed REgistry), for the user-centered distributed replication of service-related information, and URBE (UDDI Registry By Example), for the semantic-aware match making between requests and available services. The integrated view also exploits USQL (Unified Service Query Language) to provide users with a higher level and homogeneous means to interact with the different registries. The proposal improves background technology in different ways: we integrate USQL as high-level language to state service requests, widen user notifications based on URBE semantic matching, and apply URBE match making to all the facets with which services can be described in DIRE. All these new concepts are demonstrated on a simple scenario.

1 Introduction

The publication and discovery of Web services [1] have been tackled in several different ways so far. While the community agrees on WSDL and BPEL, as description and composition languages, respectively, the efficient and effective *exposition* and *retrieval* of Web services are still open problems. For example, UDDI [2] and ebXML [3] are probably the two most "famous" registry solutions, proposals based description logics and ontologies [4,5,6] try to improve service discovery, while METEOR-S [7] and Pyramid-S [8] integrate registries and ontologies to offer semantically-enriched service publication and discovery. The lack of a winning solution pushed us to further analyze the problem and concentrate on the distributed publication of services as a way to improve both exposition and retrieval.

Even if all the main registry standards have moved towards distributed approaches, we think that this distribution cannot be defined a priori. We think that the information about available services must be moved closer to their possible users, and this must be done in a user-centric way. Our proposal lets users fully control their registries (i) by defining what they want to share with the others and (ii) by specifying the services potentially available on external registries they are interested in. Therefore, the paper concentrates on the distributed *user-centered* propagation of service information and on the discovery features that

Z. Bellahsène and M. Léonard (Eds.): CAiSE 2008, LNCS 5074, pp. 435–449, 2008.

such a distribution enables. The discovery, in particular, is enriched with the adoption of semantic-aware analysis to improve the responsiveness of the system and help users with solutions (services) that are close enough to what they would have liked to get (even if they do not fully match their expectations).

The proposed interaction among registries exploits a *publish and subscribe* [9] (P/S, hereafter) communication infrastructure to allow for flexible and dynamic interactions. This means that each registry can decide the services it wants to publish, that is, the services it wants to share with the others. Similarly, it can declare its interests by means of special-purpose subscriptions. The infrastructure ensures that as soon as a registry publishes the information about one of its services, this same information is propagated to (and replicated on) all the registries that had declared their interest. Subscriptions (and unsubscriptions) can be issued dynamically and thus each registry can accommodate and tailor its interests (i.e., those of its users) while in operation.

The second key message of the paper is that oftentimes users are not only interested in services that fully and exactly match their requests, but they would like to know if there are "similar" solutions, that is, services that suitably adapted can be used instead of the ones part of the original request. This requirement is tacked in the paper in two different and orthogonal ways. User requests are formulated in a technology neutral and high-level query language, called USQL (Unified Service Query Language, [10]), and are then automatically translated into subscriptions suitably distributed through the communication infrastructure. On the other hand, the dispatching is powered with matchmaking capabilities to provide the different registries with semantically-enriched notifications, that is, information about services whose match with the original request (subscription) is within a given threshold.

The work presented in this paper builds on top of two existing proposals: DIRE (DIstributed REgistry, [11]), as for the communication framework among registries and the *facet*-based [12] description of services, and URBE (UDDI Registry By Example, [13]), for the matchmaking and semantic awareness. The integration of the two proposals allows us to consider a semantically-enabled replication infrastructure that supports different registry technologies (UDDI, ebXML, and the SeCSE registry[1]) by means of JAXR (Java API for XML Registries, [14]).

Besides the obvious integration of the two proposals, the novel contributions of this paper lie in: (i) the use of USQL as high-level language to state service requests, along with its automatic translation in terms of subscriptions for the communication infrastructure, (ii) the widening of notifications based on URBE semantic matching, and (iii) the extension of the URBE matching to all the facets with which services can be described.

The rest of the paper is organized as follows. Section 2 introduces an example scenario to motivate the proposal presented in the paper, while Section 3 summarizes background technologies. Section 4 describes the proposed infrastructure, along with the new features. Section 5 surveys some related proposals and Section 6 concludes the paper.

---

[1] `http://secse.eng.it`

2 Example Scenario

Even if the UDDI Business Registries by IBM, Microsoft, and SAP are not operated anymore, alternative "global" Web service registries are still available. Among the others, XMethods[2] and Wsoogle[3] are currently used worldwide, host Web services of any kind, and provide facilities to ease their discovery. Since the number of available services is always increasing, this section introduces the approach presented in this paper as a means to better exploit these "global" registries and increase the effectiveness of service discovery.

The example scenario[4] of Fig. 1 assumes the presence of three different (classes of) users interested in the Web services advertised by these global registries. The first is a company specialized in software development for healthcare solutions, which is interested in Web services able to support as many activities as possible in this application domain. The second is a tour operator willing to improve its Web site with mash-up services, while the third is a community of chess players who want to be aware of new opportunities (Web services) to play chess over the Internet.

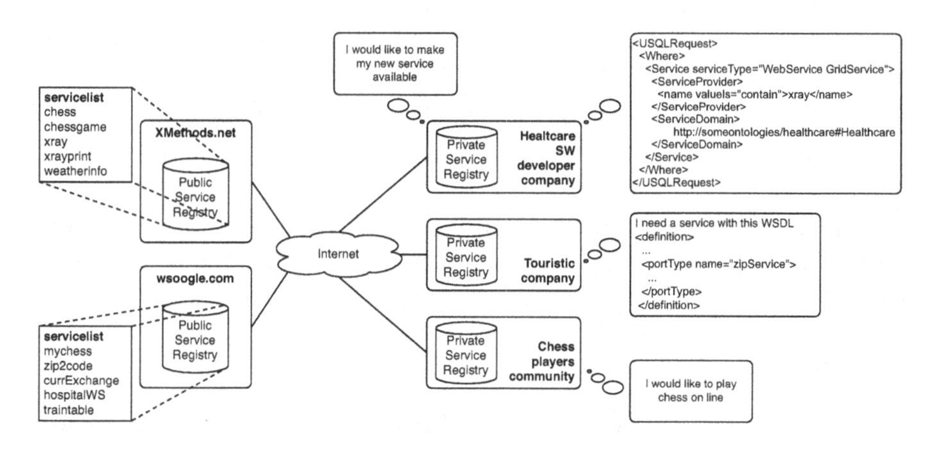

Fig. 1. Example scenario

All these three groups of users decide to run their own local registries and periodically browse XMethods and Wsoogle to find the services of interest and update their local copies. Each requester is interested in services of different categories, but also their requirements are stated in different ways. For instance, on the average, chess players do not know WSDL, and can only express their requirements using chess- and QoS-related keywords (e.g., *chess* or *free chess*

[2] `http://www.xmethods.net`

[3] `http://www.wsoogle.com`

[4] This example only aims at exemplifying how our approach works; further technical considerations behind it are outside the scope of the paper.

server). In contrast, the software company wants Web services with particular WSDL interfaces and is also interested in becoming a *quality* service provider for its clients.

All these activities are time consuming and the actual results heavily depend on the ability of who works on service discovery. Automatic ways to feed the local registries with no need for period updates would definitively ease their management, and would also help obtain better results (in terms of discovered services).

Our solution works in this direction. Each registry, be it global or local, must be connected to the communication infrastructure described in Section 4, and only has to declare its interests. The infrastructure grabs relevant services as soon as they become available directly from where they are published (mainly the two big repositories, in our example). Similarly, when one of the users (e.g., the software company) also holds the role of service provider, the infrastructure automatically publishes the new services onto the infrastructure and they (immediately) become available for the other interested registries.

Since Web services can be published, updated, and unpublished, the infrastructure is also in charge of updating the proprietary replicas as soon as new information (services) becomes available. In this scenario, all the services published in the general purpose registries are public by definition, and these registries are interested in all the public services in the local registries.

3 Background

This section briefly recalls DIRE, URBE, and USQL to provide the reader with a self-contained paper, and also highlight those elements that will be used in the next sections.

3.1 DIRE

DIRE[5] (DIstributed REgistry, [11]) provides a common service model for heterogenous registries and makes them communicate through a P/S middleware.

DIRE is in line with those approaches that tend to unify the service model (e.g., JAXR and USQL). Business data are rendered by pre-defined elements called Organizations, Services, and ServiceBindings, with the meaning that these elements usually assume in existing registries. Technical data are described by typed Facets, where each facet addresses a particular feature of the service by using an XML language. StandardFacets characterize recurring features (for example the compliance with an abstract interface), and we assume that they are shared among services. SpecificFacets describe the peculiarities of the different services (for example, particular SLAs or additional technical information). Users can attach new facets to services, even if they are not their provider, to customize the way services are perceived by the different registries (users), and

[5] `http://code.google.com/p/delivery-manager/`

to let them share this information with the other components attached to the communication bus.

The communication bus, which is based on a distributed P/S middleware called *ReDS* [15], decouples the interactions among components by means of a *dispatcher*. Each component can *publish* its messages on the dispatcher, and decide the messages it wants to listen to (*subscribe/unsubscribe*). The dispatcher forwards (*notifies*) received messages to all registered components. ReDS filters, which can both refer to shared standard facets and embed XPath expressions on the content of specific facets, let the different registries declare their interests for particular services. The goal is to disseminate the information about services based on interests and requests, instead of according to predefined rules.

A *delivery manager* is attached to each registry and acts as *facade*, that is, it is the intermediary between the registry and the bus and manages the information flow in the two directions. The adopted service model is generic enough to let different vendors create adapters for their registries. The adoption of the delivery manager does not require modifications to the publication and discovery processes used by the different users. They keep interacting with the (local) registry they were used to, but published services are distributed through the P/S infrastructure, which in turn provides information about the services published by the others (if they are of interest). In the end, each single registry is able to notify its users about the new services published in the other registries.

Notice that a registry can connect to the bus and declare its interests at any time. The infrastructure guarantees that a registry can always retrieve the information it is interested in by means of *lease* contracts. The lease period, which is configurable at run-time, guarantees that the information about services is re-transmitted periodically. This is also the maximum delay with which a registry is notified about a service. Moreover, if the description of a service changes, the lease guarantees that the new data are distributed to all subscribed registries within the period.

3.2 URBE

URBE[6] (UDDI Registry By Example) is an extension of typical UDDI Registries to support content-based queries, that is, the retrieval of services whose operations have a given input or output. Users submit the WSDL description of the requested Web service, and the system returns a ranked list of services whose signature is similar to the submitted one.

URBE supports service substitutability at both design- and run-time, and also the top-down design of BPEL processes. Traditional design approaches push the designer to identify the potential partner services and then design the BPEL process by exploiting the previously selected WSDL interfaces (bottom-up approach). URBE allows the designer to start focusing on the definition of the process before selecting the Web services that fit it.

[6] `http://black.elet.polimi.it/urbe`

URBE's *similarity engine* compares WSDL descriptions of Web services. Assuming that users express their queries using WSDL, this component compares the submitted WSDL with the WSDL of all the Web services in the registry. Each comparison relies on function $WSDLSim : (wsdl_q, wsdl_p) \rightarrow [0..1]$, where the higher the result is, the higher the similarity between the two Web services is [16]. This value is obtained recursively by analyzing the overall signature, the operations, and their parameters. For each operation in $wsdl_q$, the similarity engine finds the operation in $wsdl_p$ with maximum similarity. This similarity depends on the similarity between the operations' names (calculated by *opSim*) and the similarity of their input and output parameters (calculated by *parSim*). Finally, the similarity of parameters depends on the similarity of the parameters' names and their data types. Figure 2 shows a high level overview of the similarity evaluation process.

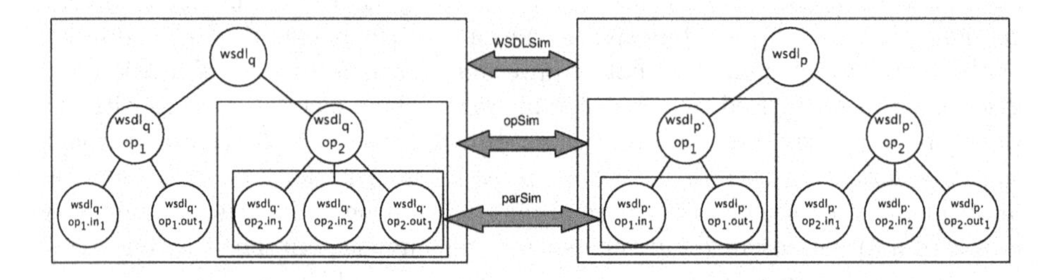

Fig. 2. Example similarity evaluation

As a consequence, the similarity between two signatures heavily depends on the names assigned to the whole services, available operations, and exchanged parameters. The comparison between terms relies on a term similarity function $termSim : (t_i, t_j) \rightarrow [0..1]$. This function returns a value that reflects how the two terms t_i and t_j are semantically close: 1 if t_i and t_j are synonym, 0 if they are antonym.

To achieve this goal, *termSim* relies on two kind of ontologies: a *domain specific ontology* and a *general purpose ontology*. The first includes terms related to a given application domain. We assume that this ontology can be built by domain experts who analyze the terms included in the Web services published in the registry. The latter includes all the possible terms (at this stage we adopt Wordnet[7]).

The domain specific ontology offers more accuracy in the relationships among terms, while the general purpose one offers wider coverage. This happens because in a general-purpose ontology, a word may have more that one *synset*, each corresponding to a different meaning. In contrast, we assume that in a domain-specific ontology each word has a unique meaning with respect to the domain itself. URBE can be configured to only use Wordnet to obtain a better coverage,

[7] `http://wordnet.princeton.edu/`

to only use the domain specific ontology to obtain better precision, and to use both of them to gain the two advantages contemporarily.

Name similarity depends on the way two names are connected inside the ontology [17]. If we assume that WSDL descriptions are generated automatically, for example from Java classes, it is possible that the names of operations and parameters reflect the naming convention usually adopted by programmers: `getData` or `currencyExhange` are more frequent than the simple names directly included in the ontology. For this reason, if the terms are composite words, *termSim* tokenizes the word and returns the average similarity among the terms.

URBE is built on top of a UDDI implementation only for historical reasons, but the similarity engine has wider applicability —as we will see in Section 4. Such a module can also be used as a stand-alone component or be embedded in complex frameworks.

3.3 USQL

USQL (Universal Service Query Language, [10]) is an XML language to express service requirements in a technology agnostic way. The language allows users to abstract the particular protocol and details used by the registry, and focus on *what* services are supposed to offer. USQL, like SQL in the database world, is thus a language for searching services understood by different registry vendors. Its simplicity, expressiveness, and extensibility make USQL a good solution for both experts and unskilled users. For example, users without technical skills can search for services provided by certain organizations, while more skilled users can search for services that offer particular operations.

A dedicated engine translates both the queries from users into the format imposed by the particular registry, and the responses from registry-dependent descriptions into a *generic service model* (GeSMO). This model adopts a layered structure: the lowest level contains the concepts common to different services, while higher layers describe properties specific to particular services. This way, we have an extensible model able to capture different service types (e.g., Web services, Grid services, and P2P services) using orthogonal metrics, like semantics, QoS, trust and security, and management.

USQL queries can exploit *syntactic* information about Web services, for example, their names or the names of the organizations behind them. They can also embed *semantic* data that belong to users' domain knowledge, and QoS elements to predicate on the non-functional requirements that the service is supposed to comply with. Obviously, we can easily mix these data to conceive complex and sophisticated queries to retrieve the services of interest.

The language is based on a simple XML dialect to describe both required services and their QoS properties. In particular, there are elements to select services with a particular name, with a particular service description, or provided by a particular service provider. As for semantics, USQL supports different taxonomy schemes such as the *North American Industry Classification System* or the *United Nations Standard Products and Services Code System*. The user is able to specify requirements on the operations the service should provide. USQL also accepts constraints on the desired quality of service in terms of price, availability, reliability, processing time, and security. These orthogonal aspects fully support the user to retrieve services with the required functional and non-functional properties.

For example, if we want a service to send SMS messages, we might think of different properties. We can specify that interesting services must contain `SMS` in their name. We could also exploit their semantic characterization to discover only services provided by phone companies, or require that the WSDL interface of the service we want must have a `send` method that accepts a phone number and a short message as inputs. Finally we can also say that we are only interested in cheap services by setting a maximum price.

4 Proposed Solution

A set of isolated registries would require interested providers to publish their services on each registry separately to proficiently advertise them and foster user awareness. This is exactly why we propose a flexible infrastructure that takes advantage of DIRE, URBE, and USQL to simplify the way services are published over a set of registries and ease their retrieval.

Once a new Web service becomes available, this information is not only stored in the registry used by the provider to publish the new service, but it is also forwarded to all the other registries interested in the same kind of services. This way, the provider can reduce the set of target registries to ideally a single one. In turn, even service retrieval becomes more effective: we move from a scenario where requesters have to browse different registries to find what they want, to a scenario where requesters only express their needs once, their requirements are spread around, and the information about interesting Web services is automatically moved onto their registries.

The proposed infrastrcture is shown in Figure 3. Its core is similar to the one adopted in DIRE, where a *communication bus*[8] connects all the companies that own a registry. Generally speaking, every registry can be used to both publish new Web services and retrieve interesting ones. Each registry is connected to the bus by means of a *delivery manager*, which is in charge of the different registry technologies and also manages the information flow in the two directions.

The first significant addition of this paper is that the *communication bus* also relies on an extended version of URBE's *similarity engine* for the comparisons between requests (subscriptions) and available services.

The figure also shows how the proposed solution works with our running example. For the sake of simplicity, we assume that XMethods and Wsoogle are connected to the communication bus via a *delivery manager*. Since we are considering general purpose registries, with a high number of services, the P/S infrastructure could become the bottleneck of the entire system. If this were the problem, "thematic" buses (e.g., about games, health, and so on) would help

[8] We can easily assume secure and reliable interactions since the P/S communication infrastructure is in charge of it.

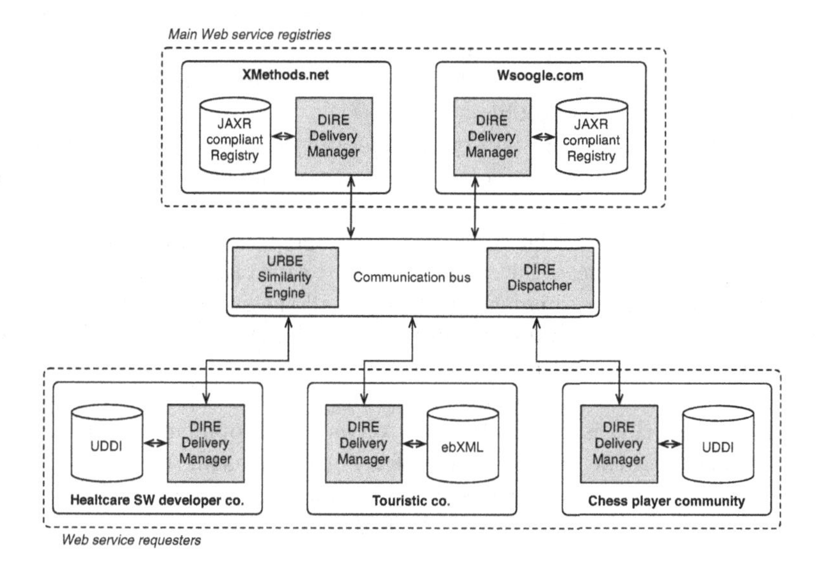

Fig. 3. Proposed infrastructure exemplified on the example scenario

split the traffic, and therefore manage the performance of the communication infrastructure. A thematic bus may also be organized around a domain ontology and, in this case, such an ontology could be used by URBE to compute the similarity among service interfaces. Said this, the explanation of the approach can easily consider a single bus without losing any significant detail. In addition, we also assume that all the three actors (already introduced in Section 2), interested in new Web services, have a proprietary service registry, along with a delivery manager properly connected to the communication bus.

The introduction of USQL and URBE aims at (i) affecting the way subscriptions can be expressed and (ii) improving the effectiveness of the filtering performed by the *dispatcher* when it has to decide whether to forward the information about a new service or not. The next sections illustrate these two aspects in detail.

4.1 USQL-Based Subscriptions

This section explains how users can interact with the *delivery manager* using USQL. As stated before, the main benefits are the independence of any particular technology and the openness towards non technical users. Our additional goal is to leverage these features inside our distributed environment, which means translating USQL queries into appropriate subscriptions to support service lookup. Since USQL queries are nothing but verbose XML documents, the presentation in this section is organized around increasingly complex examples.

Moving back to the scenario of Section 2, let us suppose that the chess players had discovered an interesting set of services provided by an organization called AcmeChess. Now, the community is looking for other services, and given the good reputation, they would like to know whether there are new services provided by AcmeChess. For example, if we think of a simple facet with tag `serviceprovider`, the filter (i.e., the XPath expression) could be: `//serviceprovider = "AcmeChess"`.

Another example considers the tourist operator that uses USQL in a smarter way and exploits the semantic facets. If we assume the existence of a standard facet that represents a shared taxonomy about travels, we can easily select all the services related to it. The subscription behind this query would predicate on the relationship between the standard facet `traveling` and the various services to allow the delivery manager to retrieve all the services in the domain of interest. Otherwise, if the ontology were more dynamic and lightweight, it could be embedded into a specific facet used to describe a particular service, and the filter would be: `contains(//service/ontology, "traveling")`.

The third actor, that is, the health-care software development company, selects services by analyzing the interface they offer. For this reason, the first query they create analyzes the operations exposed by the different services to select the ones relevant for their goals. For example, the following USQL query:

```
<USQL version="1.0" xmlns="urn:sodium:USQL">
 <USQLRequest>
  <Where>
   <Service serviceType="WebService P2PService GridService">
    <Operation minDegreeOfMatch="0.75">
     <Inputs>
      <input>
       <type valueIs="contain">integer</type>
       <semantics ontologyURI="http://sodium/ontologies/healthcare">
        http://sodium/ontologies/healthcare#ssn
       </semantics>
      </input>
     </Inputs>
     <Outputs>
      <output>
       <type valueIs="contain">string</type>
       <semantics ontologyURI="http://sodium/ontologies/healthcare">
        http://sodium/ontologies/healthcare#surname</semantics>
      </output>
     </Outputs>
    </Operation>
   </Service>
  </Where>
 </USQLRequest>
</USQL>
```

requires a service that gives the patient's name knowing his/her social security number. It also checks that the service only requires an integer, tagged with taxonomy's node `ssn`, and returns a string, tagged with node `surname`. This query is transformed into a filter with three parts. The first part is an XPath that analyses the WSDL facet to check whether there is an operation with an integer input and a string result. The second part checks whether the input parameter of the operation refers to taxonomy's node `ssn`, while the last part checks the result of the operation, and controls that it represents a `surname`.

When the query is issued at run-time, probably generated by an application component in charge of replacing a faulty service, the infrastructure must guarantee exact results (i.e., only retrieve services that can replace a previous service without human intervention). This is what the standard XPath matching technique provides. When we move the problem at design-time, users create queries to understand what Web services they can exploit. The results of these queries are usually not directly plugged in the system and approximate results better help understand the different alternatives: USQL allows us to specify the degree of matching, and URBE helps the communication bus retrieve services that match the approximation.

The health-care development company can also decide to include QoS requirements in its queries. For example, they can decide to only bind to services with high availability and low processing time. All these elements can easily be translated into both XPath queries directly, for full matches, and complete required facets, then passed to URBE for evaluation, for partial matches.

4.2 Similarity-Based Subscriptions

The extended version of URBE's *semantic matching* provides two main functions: *termSim*, to evaluate the similarity between two terms, and *facetSim*, to evaluate the similarity between two facets, which is an extension and generalization of the original match making that was limited to WSDL or SAWSDL descriptions. The infrastructure we propose exploits these two functions whenever users want to move from *exact* matches to *relaxed* ones, that is, users are satisfied even if their requirements are not totally fulfilled.

To notify the publication of a new Web service, the dispatcher was used to verify that the new description and the subscription had a perfect match. For example, if the chess player community submits a subscription with an XPath expression as `//service/type="chessgame"`, their registry will never receive Web services with a facet whose field `type` is `chess`. Since we cannot force all the actors to use the same terms, we can take advantage of function *termSim*. We introduce the clause `relaxed[sim]`, and append it to the XPath expression included in the subscription, where $sim \in [0..1]$ is a threshold that specifies the minimum admissible similarity. In the example, the subscription could be `//service/type="chessgame"relaxed[0.5]` to make the dispatcher notify the publication of new Web services with $termSim(\text{'}chessgame\text{'}, \text{'}chess\text{'}) \geq 0.5$. Notice that the definition of this similarity threshold is not easy for unskilled users as the average chess player. For this reason, we assume that the `relaxed` clauses can actually be set by transforming a qualitative scale (e.g., high, medium, and low) into the corresponding threshold values.

Function *facetSim* can be exploited in case the requester is skilled enough to know what a facet is (that is, the structure of an XML document used to describe a service). Thus, the requester wants a Web service that is not only related to a given type, but it is described in a very particular and technical way. To make the substitution possible, the substitute Web service has to expose a facet that is equal to or at least similar to the facet of the failed service.

Since a WSDL description is nothing but a particular facet, a subscription of the healthcare software company could be `//service/wsdl='http://www.hcc.org/x-rayPrinter' relaxed[0.8]` where the WSDL corresponds to a Web service able to print and deliver X-rays to patients. When a company develops a new service of this kind and publishes it onto one of the general purpose registries, the dispatcher compares its WSDL with the WSDL at `http://www.hcc.org/x-rayPrinter`. If *facetSim* returns a value greater than 0.8, the Web service is also published onto the private registry owned by the software company.

5 Related Work

Our proposal can easily be compared with two wide classes of approaches: those that concentrate on service publication and discovery and those that deal with term similarity.

As for the first group, Garofalakis et al. in [1] introduce an overview of current Web service publication and discovery mechanisms and also propose a categorization. Registry technologies support the cooperation among registries, but they imply that all registries comply with a single standard and the cooperation needs a set up phase to manually define the information contributed by each registry. For example, UDDI v.3 [18] extends the replication and distribution mechanisms offered by the previous versions to support complex and hierarchical topologies of registries. It also identifies services by means of a unique key over different registries. The standard only says that different registries can interoperate, but the actual interaction policies must be defined by the developers. In our approach, the role of the registries and the way in which they cooperate are clearly defined.

Similarly, ebXML [3] is a family of standards based on XML to provide an infrastructure to ease the online exchange of commercial information. ebXML fosters the cooperation among them by means of the idea that groups of registries share the same commercial interests or are located in the same domain, as the thematic buses do in our approach. One of such groups can then be seen as a single logical entity where all the elements are replicated on the different registries. With respect to our approach, service retrieval with ebXML registries results ineffective since users must browse pre-defined taxonomies or submit keywords to find the desired services.

METEOR-S [7] and PYRAMID-S [8] fall in the family of semantic-aware approaches for the creation of scalable peer-to-peer infrastructures for the publication and discovery of services. These works create a federation of registries using several concrete nodes. Conversely to our approach, the single node become simply a gateway to the logical registry, ensuring higher availability or better response time, but loosing its identity. In particular, the usage of a semantic infrastructure allows for the implementation of different algorithms for the publication and discovery of services, but it also forbids the complete control over the registries. The semantic layer imposes too heavy constraints on publication policies and also on the way federations can evolve dynamically. METEOR-S only supports UDDI registers, while PYRAMID-S supports both UDDI and ebXML registries. They adopt ontology-based meta-information to allow a set of registries to be federated with each registry "specialized" according to one or more categories it is associated with. This means that the publication of a new service requires the meta-information needed to categorize the service within the ontology. Services are discovered by means of semantic templates that give an abstract characterization of the service and are used to query the ontology and identify the registries that contain significant information.

Term similarity has been tackled in several different ways [17]. These algorithms usually calculate such a similarity by relying on the relationships between terms defined in a reference ontology (e.g., *is-a*, *part-of*, *attribute-of*). In contrast, we compute similarity between terms according to the approach proposed by Seco et al. [19], where the authors adapt existing approaches with the assumption that concepts with many hyponyms convey less information than concepts that have less hyponyms or any at all (i.e, they are leaves in the ontology).

About the similarity between whole signatures, our approach is closely related to the approaches studied for the retrieval of reusable components [20]. In this field, as stated by Zaremski and Wing, there are two types of methods to address this problem: signature matching [21] and specification matching [22]. In particular, signature matching considers two levels of similarity and introduces the *exact* and *relaxed* matching of signatures. As for services, Stroulia and Wang [23] propose an approach that also considers the description field usually included in WSDL specifications.

6 Conclusions and Future Work

The paper presents an innovative infrastructure for the distributed publication of Web services and for their easy retrieval. The proposal leverages previous experiences of the authors, namely DIRE and URBE, and also other initiatives (USQL) to provide a holistic solution able to govern the replication of service information by means of user requests and preferences, and also able to provide users with partial, but acceptable, solutions whose fitness is defined through semantic match making techniques. The overall framework provides users with a wide set of options.

The integrated infrastructure exists as a very first prototype, but more stable solutions are needed for its deployment in realistic settings and also for a thorough empirical evaluation of the approach. Both these directions will govern our future work. The plan is to keep working on a fully functional prototype implementation and design a complete empirical evaluation of the proposal by exploiting a distributed set of registries and the usual collections of public Web services as benchmarks. Such a new prototype will also deal with query optimization and filter maintenance.

Acknowledgments

This work has been supported by the following projects: Tekne (Italian FIRB), Discorso (Italian FAR), SeCSE (EC IP), and ArtDecò (Italian FIRB).

References

1. Garofalakis, J., Panagis, Y., Sakkopoulos, E., Tsakalidis, A.: Contemporary Web service discovery mechanisms. Journal of Web Engineering 5(3), 265–290 (2006)
2. UDDU: Universal Description, Discovery, and Integration, `http://uddi.xml.org`
3. ebXML: Electronic Business using eXtensible Markup Language, `http://www.ebxml.org/`
4. Martin, D. et al. (ed.): OWL-S: Semantic Markup for Web Services. W3C Submission (2004), `http://www.w3.org/Submission/2004/SUBM-OWL-S-20041122/`
5. WSMO Working Group: Web Service Modeling Ontology, `http://www.wsmo.org`
6. Farrel, J., Lausen, H.: Semantic annotations for WSDL and XML schema (2007), `http://www.w3.org/TR/sawsdl/`
7. Verma, K., Sivashanmugam, K., Sheth, A., Patil, A., Oundhakar, S., Miller, J.: METEOR-S WSDI: A scalable p2p infrastructure of registries for semantic publication and discovery of web services. Information Technology and Management, 6, 17–39 (2005)
8. Pilioura, T., Kapos, G., Tsalgatidou, A.: PYRAMID-S: A scalable infrastructure for semantic web services publication and discovery. In: RIDE-DGS 2004 14th Int'l Workshop on Research Issues on Data Engineering, in conjunction with the IEEE Conf. on Data Engineering (ICDE 2004) (March 2004)
9. Eugster, P.T., Felber, P.A., Guerraoui, R., Kermarrec, A.M.: The many faces of publish / subscribe. ACM Comput. Surveys 35(2), 114–131 (2003)
10. Tsalgatidou, A., Pantazoglou, M., Athanasopoulos, G.: Specification of the Unified Service Query Language (USQL). Technical report (June 2006)
11. Baresi, L., Miraz, M.: A distributed approach for the federation of heterogeneous registries. In: Dan, A., Lamersdorf, W. (eds.) ICSOC 2006. LNCS, vol. 4294, pp. 240–251. Springer, Heidelberg (2006)
12. Sawyer, P.: Specification language definition. Technical Report A1.D2.3, EC SeCSE Project (2006)
13. Plebani, P., Pernici, B.: Web service retrieval based on signatures and annotations. Technical Report 2007.47, Dipartimento di Elettronica ed Informazione - Politecnico di Milano (2007)
14. Najmi, F. (ed.): Java API for XML Registries (JAXR) (2002), `http://java.sun.com/webservices/jaxr/`
15. Cugola, G., Picco, G.P.: REDS: a reconfigurable dispatching system. In: Proc. of the 6th international workshop on Software engineering and middleware, pp. 9–16 (2006)
16. Bianchini, D., De Antonellis, V., Pernici, B., Plebani, P.: Ontology-based methodology for e-service discovery. Information Systems 31(4-5), 361–380 (2006)

17. Pedersen, T., Patwardhan, S., Michelizzi, J.: WordNet:Similarity - measuring the relatedness of concepts. In: Proc. National Conf. on Artificial Intelligence, San Jose, California, USA, July 25-29, pp. 1024–1025 (2004)
18. Clement, L., Hately, A., von Riegen, C. (eds.): T.R.: Universal Description, Discovery and Integration version 3.0.2 (2004), http://uddi.org/pubs/uddi_v3.htm
19. Seco, N., Veale, T., Hayes, J.: An intrinsic information content metric for semantic similarity in Wordnet. In: Proc. Eureopean Conf. on Artificial Intelligence (ECAI 2004), Valencia, Spain, August 22-27, pp. 1089–1090. IOS Press, Amsterdam (2004)
20. Damiani, E., Fugini, M.G., Bellettini, C.: A hierarchy-aware approach to faceted classification of objected-oriented components. ACM Trans. Softw. Eng. Methodol. 8(3), 215–262 (1999)
21. Zaremski, A., Wing, J.: Signature matching: a tool for using software libraries. ACM Trans. Softw. Eng. Methodol. 4(2), 146–170 (1995)
22. Zaremski, A., Wing, J.: Specification matching of software components. ACM Trans. Softw. Eng. Methodol. 6(4), 333–369 (1997)
23. Stroulia, E., Wang, Y.: Structural and semantic matching for assessing Web-service similarity. Int'l J. Cooperative Inf. Syst. 14(4), 407–438 (2005)

Measuring Similarity between Business Process Models

Boudewijn van Dongen[1], Remco Dijkman[1], and Jan Mendling[2]

[1] Eindhoven University of Technology, The Netherlands
{b.f.v.dongen,r.m.dijkman}@tue.nl
[2] Queensland University of Technology, Brisbane, Australia
j.mendling@qut.edu.au

Abstract. Quality aspects become increasingly important when business process modeling is used in a large-scale enterprise setting. In order to facilitate a storage without redundancy and an efficient retrieval of relevant process models in model databases it is required to develop a theoretical understanding of how a degree of behavioral similarity can be defined. In this paper we address this challenge in a novel way. We use *causal footprints* as an abstract representation of the behavior captured by a process model, since they allow us to compare models defined in both formal modeling languages like Petri nets and informal ones like EPCs. Based on the causal footprint derived from two models we calculate their similarity based on the established vector space model from information retrieval. We validate this concept with an experiment using the SAP Reference Model and an implementation in the ProM framework.

Keywords: Business Process Modeling, Event-driven Process Chains, Similarity, Equivalence.

1 Introduction

Many multi-national companies use tools such as ARIS Toolset for documenting their business processes. Due to the operational diversity of such large enterprises, there are often several thousands of processes modeled and stored in the database of the modeling tool [26]. The sheer number causes serious problems for the management and maintenance of these model: It is difficult to see the forest because there are too many trees, as a German proverb puts it. While quality aspects of process models (e.g. [15]) and process modeling languages (e.g. [10]) are quite well understood, there is a notable research gap on quality issues across models.

The similarity between business process models can be related to several of these cross-model quality issues. Consider a large organization that wants to identify redundancies in the operations of different divisions. Models are indeed helpful to discuss the overlap of two processes and the potential for integration, yet it is difficult and time-consuming to identify similarities in a process database with several thousands of models. Clearly, there is a need for automatic detection of similarities between process models to facilitate certain model management activities. There are several model management activities that would benefit from good tool support. Firstly, similar models as well as the corresponding business operations can be integrated into one process. This is interesting not only for refactoring the model database, but also to facilitate the

Z. Bellahsène and M. Léonard (Eds.): CAiSE 2008, LNCS 5074, pp. 450–464, 2008.

integration of business operations in a merger scenario. Secondly, the reference models of an ERP system vendor could be automatically compared to company processes. This way, organizations could more easily decide which packages match their current operations best. Thirdly, multi-national enterprises can identify specialized processes of some national branch which no longer comply with the procedures defined in the company-wide reference model using a similarity measurement.

In this paper, we discuss the foundations of detecting and measuring similarity between business process models. In particular, our contribution is an approach considering linguistic and behavioral aspects of process models to calculate a degree of similarity. We validate the approach using the SAP reference model. The results highlight which benefits organizations can have from tool support for similarity detection.

The remainder of the paper is organized as follows. Section 2 introduces Event-driven Process Chains (EPCs), a popular process modeling language that we use to illustrate our approach. Furthermore, we discuss one particular redundancy problem that was identified in the SAP reference model in prior research. Section 3 then presents our approach to calculate the degree of similarity between two processes based on their causal footprint. A causal footprint covers extensive behavioral information about a process without calculating its state space, but requires the identification of matching functions in the EPCs being compared. Section 4 addresses the problem of matching functions across different processes, with an emphasis on EPCs. We discuss an approach to identify matches between functions automatically. In Section 5, the presented techniques are combined, applied to a large portion of the SAP reference model, and empirically validated against human interpretations of similarity. Then, Section 6 discusses related work to our approach before Section 7 concludes the paper.

2 Background on EPCs

In this paper, we will illustrate our argument using Event-driven Process Chains (EPCs). The EPC is a popular business process modeling language that was introduced in [13]. EPCs are used by most companies that manage their process models with ARIS Toolset. This way, our results are directly applicable for these organizations.

EPCs capture the control flow of a process in terms of the temporal and logical dependencies of activities [13]. EPCs offer *function type* elements to represent these activities, *event type* elements describing pre- and post-conditions of functions, and three kinds of *connector types* including AND, OR, and XOR. Control flow arcs are used to link these elements. Connectors have either multiple incoming and one outgoing arc (join connectors) or one incoming and multiple outgoing arcs (split connectors). As a syntax rule, functions and events have to alternate on each path through the EPC, either directly or indirectly when they are linked via one or more connectors.

The informal (or intended) semantics of an EPC can be described as follows. The AND-split activates all subsequent branches in a concurrent manner. The XOR-split represents a choice between one of several alternative branches. The OR-split triggers one, two or up to all of multiple branches based on conditions. For both XOR-splits and OR-splits, the activation conditions are given in events subsequent to the connector. The AND-join waits for all incoming branches to complete, then it propagates control to the subsequent EPC element. The XOR-join merges alternative branches. The OR-join synchronizes all active incoming branches. This feature is called non-locality since the state of all transitive predecessor nodes has to be considered. For a recent discussion of formal semantics of EPCs refer to [18].

The following definition formalizes EPC. We need this definition in the section on behavioral similarity. Furthermore, we define a notion of syntactical correctness that we check before applying our approach to the SAP reference model.

Definition 2.1. (EPC)
An $EPC = (E, F, C, l, A)$ consists of three pairwise disjoint and finite sets E, F, C, a mapping $l : C \rightarrow \{and, or, xor\}$, and a binary relation $A \subseteq (E \cup F \cup C) \times (E \cup F \cup C)$ such that

- An element of E is called *event*. $E \neq \emptyset$.
- An element of F is called *function*. $F \neq \emptyset$.
- An element of C is called *connector*.
- The mapping l specifies the type of a connector $c \in C$ as and, or, or xor.
- The relation A defines the control flow as a coherent, directed graph. An element of A is called an *arc*. An element of the union $N = E \cup F \cup C$ is called a *node*.

In order to be able to discuss the events surrounding a function, or the functions surrounding an event, notations are introduced for paths and connector chains.

Definition 2.2. (Paths and Connector Chains)
Let N be a set of $nodes$ and $A \subseteq N \times N$ a binary relation over N defining the arcs. For each $node\ n \in N$, we define *path* $a \hookrightarrow b$ refers to the existence of a sequence of EPC nodes $n_1, \ldots, n_k \in N$ with $a = n_1$ and $b = n_k$ such that for all $i \in 1, \ldots, k$ holds: $(n_1, n_2), (n_2, n_3), \ldots, (n_{k-1}, n_k) \in A$. This includes the empty path of length zero, i.e., for any node $a : a \hookrightarrow a$. If $a \neq b \in N$ and $n_2, \ldots, n_{k-1} \in C$, the path $a \overset{c}{\hookrightarrow} b$ is called *connector chain*. This includes the empty connector chain, i.e., $a \overset{c}{\hookrightarrow} b$ if $(a, b) \in A$.

In this paper, we focus on syntactically correct EPCs, i.e. EPCs with at least one initial and final events, at least one function and strict alternation of functions and events on all paths. According to this definition, both example EPCs of Figure 1 are syntactically correct. Therefore, we can apply the techniques for matching functions that are discussed later in Section 4. Out of the 604 EPCs in the SAP reference model mentioned before, 556 are syntactically correct. Please note that we demand a strict alternation of functions and events, which is not included in all EPC syntax definitions.

Figure 1 gives an example of two EPCs that captures similar processes (cf. [19]). Both are taken from the aforementioned SAP Reference Model. The EPC on the left-hand side of Figure 1 stems from the Sales and Distribution branch and its name is *Customer Inquiry*. In essence, when a customer inquires about a product (denoted by the event "Customer inquires about products"), this inquiry is processed and a quotation is created which results in the fact that a customer project is needed. As an alternative, the need for a customer project can arise based on plan data which triggers a resource related quotation. The EPC on the right-hand side of Figure 1 is taken from the Project Management branch and it is called *Customer Inquiry and Quotation Processing*. It

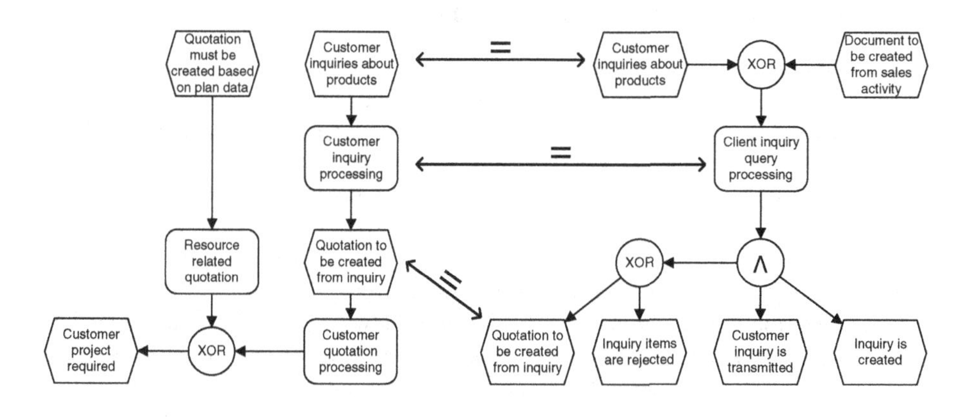

Fig. 1. *Customer Inquiry* and *Customer Inquiry and Quotation Processing* EPCs

identifies a sales activity as alternative reason to process a customer inquiry. As a result the inquiry is created and transmitted. Furthermore, either a quotation is created or the inquiry is rejected. The processes share two equivalent events and one equivalent function as depicted in Figure 1 . Since the overlapping part of the models, i.e. the sequence "customer inquiry", "inquiry processing", and "quotation to be created", can be handled by both processes, they could easily be integrated into one model, for instance using the approach defined in [19].

In Section 3, we provide a metric for determining how similar two business processes are, given that it is known which functions (or activities in the more general sense) in one model correspond to functions in the other model. In Section 4, we show how to automatically find the relations between functions of different models.

3 Similarity of Behavior

Comparing the behavior of processes using traditional notions such as bisimulation is problematic for different reasons. Firstly, most of these notions are defined as a verification property which yield as yes or no, but no degree of similarity. Secondly, process models with concurrency suffer from a state explosion problem. For some process modeling languages a formalization of the reachability graph as a transition system is even missing. Thirdly, if there are deadlocks or dead transitions in the process model, these parts are not captured in the behavioral comparison. Motivated by these problems, we defined the concept of a causal footprint [7] which is a collection of the essential behavioral constraints imposed by a process model.[1] We will use the causal footprints of two processes as a basis to calculate their similarity. Section 3.1 describes the derivation of a causal footprint, then Section 3.2 defines the degree of similarity for causal footprints.

[1] Note that this paper adopts the concept of a causal footprint from [7] where we use it for verification purposes. In contrast to [7] we use this concept for measuring similarity.

3.1 Deriving the Causal Footprint of an EPC

Before defining a causal footprint of an EPC, we first need to introduce the notion of a case as well as the semantics of look-back and look-ahead links.

A case basically captures the behavior of one particular execution sequence of functions according to the rules of a process model. Consider N as the set of nodes of an EPC. The behavior of the process Φ_{EPC} is defined as the set $W \subseteq N^*$, where N^* is the set of all sequences that are composed of zero of more nodes from N. A $\sigma \in W$ is called a *case*, i.e. a possible execution of the EPC. To denote a function at a specific index in σ, we use $\sigma[i]$, where i is the index ranging from 1 to $|\sigma|$.

The causal footprint identifies two relationships between nodes in N that are called look-back and look-ahead links. For each *look-ahead link*, we say that the execution of the source of that link leads to the execution of at least one of the targets of that link, i.e., if $(a, B) \in L_{la}$, then any execution of a is followed by the execution of some $b \in B$. A look-ahead link is denoted as a bullet with one or more outgoing arrows. Furthermore, for each *look-back link*, the execution of the target is preceded by at least one of the sources of that link, i.e., if $(A, b) \in L_{lb}$, then any execution of b is preceded by the execution of some $a \in A$. The notation of a look-back link is a bullet with one or more incoming arrows. Note that we do not give any information about when in the future or past executions took place, but only that they are there. This way of describing a process is related to work on dominance and control dependence in program analysis (see e.g. [12]), and similar to the work presented in [8]. However, by splitting up the semantics in the two different directions (i.e. forward and backward), causal footprints are more expressive. With footprints you can for example express the fact that task `A` is always succeeded by `B`, but that `B` can also occur before `A`, which is typically hard to express in other languages.

Definition 3.1. (Causal Footprint)
We define a causal footprint $G = (N, L_{lb}, L_{la})$ as a graph where, where:

- N is a finite set of *nodes* (activities),
- $L_{lb} \subseteq (\mathcal{P}(N) \times N)$ is a set of *look-back links*[2]
- $L_{la} \subseteq (N \times \mathcal{P}(N))$ is a set of *look-ahead links*.

For relating the definition of a causal footprint to the behavior of an EPC we define a notion of consistency based on the cases implied by the EPC process model.

Definition 3.2. (Consistency of Causal Footprint with EPC)
Let N be a set of nodes and $EPC = (E, F, C, l, A)$ be an EPC with behavior W. Furthermore, let $G = (N, L_{lb}, L_{la})$ be a causal footprint. We say that $G = (N, L_{lb}, L_{la})$ is consistent with the behavior of EPC, denoted by $G \in \mathcal{F}_{EPC}$, if and only if:

1. $N = F$, i.e. the nodes of the footprint represent the functions of the EPC,
2. For all $(a, B) \in L_{la}$ holds that for each $\sigma \in W$ with $n = |\sigma|$, such that there is a $0 \leq i \leq n-1$ with $\sigma[i] = a$, there is a $j : i < j \leq -1$, such that $\sigma[j] \in B$,
3. For all $(A, b) \in L_{lb}$ holds that for each $\sigma \in W$ with $n = |\sigma|$, such that there is a $0 \leq i \leq n-1$ with $\sigma[i] = b$, there is a $j : 0 \leq j < i$, such that $\sigma[j] \in A$,

[2] With $\mathcal{P}(N)$, we denote the powerset of N, where $\emptyset \notin \mathcal{P}(N)$.

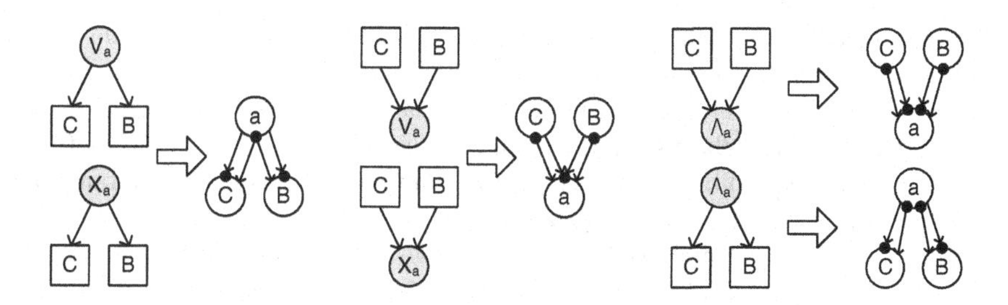

Fig. 2. Mapping of EPCs to causal footprints

While the different cases of an EPC can explicitly be generated using the semantics formalization defined in [18], there is a more efficient way. The mapping defined in [7] and depicted in Figure 2 yields a consistent causal footprint for an EPC under the assumption that no AND-join or OR-join deadlocks. Furthermore, it is clear from Definition 3.2 that a causal footprint is not unique, i.e., different processes can have common footprints. For example, $G = (N, \emptyset, \emptyset)$ is the causal footprint of any process having activities F. Therefore, we aim at footprints that are more informative without trying to capture detailed semantics. In [7] a set of rules for calculating the transitive closure of a causal footprint are introduced such that the closure is still a causal footprint that is consistent with the EPC. In Section 5, where we present the application to the SAP reference model, we used the rules of Figure 2 in combination with the transitive closure rules of [7] to obtain a causal footprint for all EPCs.

3.2 Similarity of Causal Footprints

In information retrieval the degree of similarity between a document and a query plays a very important role for ranking the returned documents according to their relevance. For calculating similarity, we use the well-known *vector model* [2, 28] which is one of the basic techniques used for information filtering, information retrieval, and the indexing of web pages. Its classical application is to determine the similarity between a query and a document. The original vector space model proposed by Salton, Wong, and Yang in [28] attaches weights based on term frequency to the so-called "document vector". We use a more liberal interpretation, where other weights are possible. However, to explain the basic mechanism we use terms originating from the domain of information retrieval, i.e., terms like "document collection", a set of "terms", and a set of "weights" relating to the terms. Later we will provide a mapping of these terms to causal footprints.

The *document collection* contains a set of documents. Each of these documents is considered to be a list of *terms* which are basically the words of the document. The union of all terms of all documents is then used to describe each document as a vector. For one specific document an entry in the vector represents that the term associated with the vector position of this entry is included in the document. In a simple case the occurrence of a term can be indicated by a one and the non-occurrence with a zero, however there is also the option to assign *weights* to terms in order to address the fact that they differ in relevance. A common choice is to use one divided by the number of occurrences of a term throughout all documents of the document collection as a weight which has the effect that scarcely used terms get a higher weight. A query can also be considered as a document, i.e., a list of terms.

The similarity between a query and a document is then calculated based on their vector representation as the cosine of the angle between the two vectors [2, 28]. Calculating this degree of similarity for each document provides a mechanism to rank them according to their relevance for the query.

Our proposal for determining the similarity of two business process models builds on the vector model and causal footprints. We consider causal footprints of two processes $G_1 = (N_1, L_{1,lb}, L_{1,la})$ and $G_2 = (N_2, L_{2,lb}, L_{2,la})$ as input for the calculation. In order to apply the vector model, we have to define (1) the document collection, (2) the set of terms, and (3) the set of weights.

The document collection includes two entries, namely the two causal footprints that need to be compared. We will refer to the first and the second causal footprint as $G_1 = (N_1, L_{1,lb}, L_{1,la})$ and $G_2 = (N_2, L_{2,lb}, L_{2,la})$).

The set of terms is build from the union over nodes, look back, and look ahead links of the two causal footprints. We define $\Theta = N_1 \cup L_{1,lb} \cup L_{1,la} \cup N_2 \cup L_{2,lb} \cup L_{2,la}$ as the set of terms and $\lambda : \Theta \to \{1, 2, \ldots |\Theta|\}$ as an indexing function that assigns a running number to each term, i.e., the set of all elements appearing in the two footprints are enumerated. (Note that we implicitly assume all sets of nodes and links to be disjoint in a single model.)

The relevance of each term is closely related to the number of tasks from which it is built. Consider for example two look ahead links $x_{la} = (a, \{g\}) \in L_{la}$ and $y_{la} = (a, \{b, c, d, e, f\}) \in L_{la}$. x_{la} refers to only two tasks: a and g. y_{la} refers to six tasks (a through f). It seems obvious that the look ahead links with fewer tasks are more informative and therefore more important. To address this we use weights depending on the number of tasks involved in a look-ahead/back link.

The weights are determined using the size of the relations. If $\theta \in \Theta$ is a single node (i.e. $\theta \in N_1 \cup N_2$), then we define the weight of θ as $w_\theta = 1$. Furthermore, since the number of potential look ahead and look back links depends upon the powerset of nodes, is seems natural to use exponentially decreasing weights. Therefore, for all links $\theta \in \Theta$, we define the weight of a link $w_\theta = 1/(2^{|\theta|-1})$, where $|\theta|$ denotes the number of tasks in the link.

For the two look ahead links $x_{la} = (a, \{g\})$ and $y_{la} = (a, \{b, c, d, e, f\})$, we get $w_{x_{la}} = 1/(2^{2-1}) = 0.5$ and $w_{y_{la}} = 1/(2^{6-1}) = 0.03125$ as their weights.

Using the document collection, the set of terms and the weights presented above, we define the document vectors, which we call *footprint vectors*.

Definition 3.3. (Footprint vectors)
Let $G_1 = (N_1, L_{1,lb}, L_{1,la})$ and $G_2 = (N_2, L_{2,lb}, L_{2,la})$ be two causal footprints, with Θ the set of terms and $\lambda : \Theta \to \mathbb{N}$ an indexing function. We define two footprint vectors, $\overrightarrow{g_1} = (g_{1,1}, g_{1,2}, \ldots g_{1,|\Theta|})$ and $\overrightarrow{g_2} = (g_{2,1}, g_{2,2}, \ldots g_{2,|\Theta|})$ for the two models as follows. For each element $\theta \in \Theta$, we say that for each $i \in \{1, 2\}$ holds that

$$g_{i,\lambda(\theta)} = \begin{cases} 0 & \text{if } \theta \notin (N_i \cup L_{i,lb} \cup L_{i,la}) \\ w_\theta = \frac{1}{2^{|\theta|-1}} & \text{if } \theta \in (L_{i,lb} \cup L_{i,la}) \\ w_\theta = 1 & \text{if } \theta \in N_i \end{cases}$$

Using the two footprint vectors, we can define the similarity between two footprints as the cosine of the angle between these two vectors.

Definition 3.4. (Footprint similarity)
Let $G_1 = (N_1, L_{1,lb}, L_{1,la})$ and $G_2 = (N_2, L_{2,lb}, L_{2,la})$ be two causal footprints, with Θ the set of terms and $\lambda : \Theta \to \mathbb{N}$ an indexing function. Furthermore, let $\overrightarrow{g_1}$ and $\overrightarrow{g_2}$ be the corresponding footprint vectors. We say that the similarity between G_1 and G_2, denoted by $sim(G_1, G_2)$ is the cosine of the angle between those vectors, i.e.

$$sim(G_1, G_2) = \frac{\overrightarrow{g_1} \times \overrightarrow{g_2}}{|\overrightarrow{g_1}| \cdot |\overrightarrow{g_2}|} = \frac{\sum_{j=1}^{|\Theta|} g_{1,j} \cdot g_{2,j}}{\sqrt{\sum_{j=1}^{|\Theta|} g_{1,j}^2} \cdot \sqrt{\sum_{j=1}^{|\Theta|} g_{2,j}^2}}$$

The value of $sim(G_1, G_2)$ ranges from 0 (no similarity) to 1 (equivalence). In this paper, we do not elaborate on this formula. If one accepts the weights that we associate to the "terms" in a causal footprint, then the cosine of the angle between these two vectors provides a generally accepted way to quantify similarity [2, 28].

The similarity $sim(G_1, G_2)$ between footprints can be calculated for any two footprints G_1 and G_2. However, for the similarity to exceed 0, there should be at least one node $n \in N_1 \cap N_2$.

Property 3.5. (Disjoint footprints have similarity 0)
Let $G_1 = (N_1, L_{1,lb}, L_{1,la})$ and $G_2 = (N_2, L_{2,lb}, L_{2,la})$ be two causal footprints, with Θ the set of terms and $\lambda : \Theta \to \mathbb{N}$ an indexing function. Furthermore, let $\overrightarrow{g_1}$ and $\overrightarrow{g_2}$ be the corresponding footprint vectors. If $N_1 \cap N_2 = \emptyset$ then $sim(G_1, G_2) = 0$.

Proof. It is sufficient to show that $\overrightarrow{g_1} \times \overrightarrow{g_2} = 0$, i.e. that $\sum_{j=1}^{|\Theta|} g_{1,j} \cdot g_{2,j} = 0$. Assume that for some $1 \leq j \leq |\Theta|$ holds that $g_{1,j} > 0$. Then, from Definition 3.3, we know that $\lambda(\theta) = j$ with either $\theta \in N_1$, or $\theta \in (L_{1,lb} \cup L_{1,la})$. Assume $\theta \in N_1$. Then we know that $g_{2,j} = 0$, since $\theta \notin N_i$. Hence $g_{1,j} \cdot g_{2,j} = 0$. Assume $\theta \in (L_{1,lb} \cup L_{1,la})$. Since Definition 3.1 shows that $L_{1,lb} \subseteq (\mathcal{P}(N_1) \times N_1)$ and $L_{1,la} \subseteq (N_1 \times \mathcal{P}(N_1))$, we know that $\theta \notin (L_{2,lb} \cup L_{2,la})$ and hence that $g_{1,j} \cdot g_{2,j} = 0$. Therefore, $\sum_{j=1}^{|\Theta|} g_{1,j} \cdot g_{2,j} = 0$ and hence $sim(G_1, G_2) = 0$. □

Property 3.5 shows that for two footprints to be considered similar, we need to identify nodes that appear in both footprints. For this, we use the notion of an equivalence mapping defined in Section 4.

4 Matching Functions

When comparing EPCs it is not realistic to assume that equivalent functions and events have labels that are the same to the letter. Figure 1 illustrates this: the functions "Customer inquiry processing" and "Client inquiry query processing" are similar from a human perspective, but they have different labels.

To determine the match between functions from different EPCs, we:

1. determine how similar pairs of functions are on a 0 to 1 scale, based on the equivalence of words in their labels (we call this the semantic similarity score);
2. determine whether a function matches another function on a true/false scale, based on the semantic similarity score;
3. determine what the best mapping is between all functions from one EPC and all functions from another, based on the semantic similarity score; and
4. extend this technique by determining the best match by not only looking at the semantic similarity score of the functions themselves, but also at the semantic similarity scores of the events that surround these functions (we call this the contextual similarity score).

These techniques are explained successively in the following subsections.

We experimented with other techniques for determining function mappings, inspired by the work of Ehrig, Koschmider and Oberweis [9]. We also experimented with different parameters for these techniques. However, we obtained the best results for the techniques and parameters explained below. A comparison is presented in the technical report that accompanies this paper [6].

4.1 Determine the Semantic Similarity Score between Two Functions

Given two functions, their semantic similarity score is the degree of similarity, based on equivalence between words in their labels. Words that are identical are given an equivalence score of 1, while words that are synonymous are given an equivalence score of 0.75, a value that was determined experimentally. We assume an exact match is preferred over a match on synonyms. Hence, the semantic similarity score is defined as follows.

Definition 4.1. (Semantic similarity)
Let $(E_1, F_1, C_1, l_1, A_1)$ and $(E_2, F_2, C_2, l_2, A_2)$ be two disjoint EPCs. Let $f_1 \in F_1$ and $f_2 \in F_2$ be two functions (and assume that f_1 and f_2 are sets of words, i.e. we denote the number of words by $|f_1|$). We define the *semantic similarity* as follows:

$$\text{sem}(f_1, f_2) = \frac{1.0 \cdot |f1 \cap f2| + 0.75 \cdot \sum_{(s,l) \in f_1 \setminus f_2 \times f_2 \setminus f_1} \text{synonym}(s, l)}{\max(|f_1|, |f_2|)}$$

Where $synonym$ is a function that returns 1 if the given words are synonyms and 0 if they aren't.

For example, consider the functions "Customer inquiry processing" and "Client inquiry query processing" from figure 1, which consist of the collections of words $f1 =$ ["Customer","inquiry","processing"] and $f2 =$["Client", "inquiry", "query", "processing"], respectively. We only need to consider a synonym mapping between $f_1 \setminus f_2$ and $f_2 \setminus f_1$, i.e. between ["Customer"] and ["Client","query"]. Therefore, the semantic similarity between f_1 and f_2 equals
$sem(f_1, f_2) = \frac{1.0 \cdot 2 + 0.75 \cdot (1+0)}{4} \approx 0.69$.

When determining equivalence between words, we disregard special symbols, and we change all characters to lower-case. Furthermore, we skip frequently occurring words, such as "a", "an" and "for". Also we *stem* words using Porter's stemming algorithm [23]. Stemming reduces words to their stem form. For example, "stemming", "stemmed" and "stemmer" are stemmed into "stem".

4.2 Determine a Semantic Match between Two Functions

The semantic similarity score of two functions is a value between 0 and 1. However, when determining equivalence, we require a boolean result stating whether or not two functions are equivalent, i.e. we need cut-off values that state when the similarity score exceeds this value then the functions are equivalent. The optimal cut-off value is the cut-off value for which the syntactic similarity degree most accurately reflects the equivalence judgements of a human.

We conducted experiments to optimize these cut-off values for use in the context of the SAP Reference Models. In particular, we compared the semantic similarity scores with human judgement for 210 function pairs from the SAP Reference Model. Their similarity degrees were evenly distributed over the 0 to 1 range and they were compared against human judgement as to whether these function pairs are equivalent of not. Based on this experiment, we determined an optimal cut-off for the similarity scores to decide whether functions match or not. We expect that these cut-off values and correctness score are typical for the SAP reference model, since other data-sets yield different values [9].

Our experiments determined that for semantic similarity, a cut-off value of 0.89 while giving synonyms a similarity score higher than 0.75 is optimal. It leads to a prediction of whether functions are a match according to humans, with a 90% accuracy.

4.3 Determine a Semantic Mapping between All Functions

So far, we only considered the similarity between two functions. However, the behavioral comparison presented in Section 3 requires a symmetric mapping between functions of two process models, i.e. we have to select pairs of functions that we consider a match, where each pair consists of a function from one model and a function from the other model.

Definition 4.2. (Equivalence mapping)
Let F_1, F_2 be two disjoint sets. Furthermore, let $s : F_1 \times F_2 \rightarrow \{0..1\}$ be a symmetric similarity function and let $c \in \{0..1\}$ be a cut-off value. A function $m : F_1 \rightarrow F_2$ is an *equivalence mapping*, if and only if:

- m is invertible ($m(f_1) = f_2$ implies that $m(f_2) = f_1$), and
- $m(f_1) = f_2$ implies that $s(f1, f2) \geq c$.

In the following section, we evaluate the degree of similarity calculation for the SAP Reference Model with different approaches to matching functions.
An *optimal equivalence mapping* $m^{opt} : F_1 \rightarrow F_2$ is an equivalence mapping, such that for all other equivalence mappings m holds that

$\sum_{(f1,f2)\in m^{opt}} s(f1, f2) \geq \sum_{(f1,f2)\in m} s(f1, f2)$.

When determining an equivalence mapping between the functions of two EPCs, each mapping satisfying Definition 4.2 is a good mapping, i.e. each element of the mapping satisfies the criterium that the similarity between the two functions exceeds the cut-off value. However, many equivalence mappings are possible. Therefore, we define the concept of an optimal equivalence mapping m^{opt}, i.e. the sum of the similarities expressed by m^{opt} is greater than the sum of the similarities of all other possible equivalence mappings[3]. An optimal equivalence mapping can be calculated in a straightforward way using integer linear programming techniques with binary variables.

4.4 Contextual Similarity

The techniques that we provided so far can be applied when comparing any two business process models. However, we are specifically considering EPCs, where each function has a preset and a postset of events. We define a second similarity metric based on this pre- and postset, which we call the contextual similarity metric. This metric produces better results than the semantic similarity metric.

Given two functions the contextual similarity technique returns the degree of similarity, based on the similarity of the events that precede and succeed them. We call these input and output events the input and output context of a function, respectively.

Definition 4.3. (Input and output context)
Let (E, F, C, l, A) be an EPC. For a function $f \in F$, we define the input context $f^{in} = \{e \in E \mid e \overset{c}{\hookrightarrow} f\}$ and the output context $f^{out} = \{e \in E \mid f \overset{c}{\hookrightarrow} e\}$

Now, we use the concept of equivalence mappings to determine the contextual similarity between functions.

Definition 4.4. (Contextual similarity)
Let $(E_1, F_1, C_1, l_1, A_1)$ and $(E_2, F_2, C_2, l_2, A_2)$ be two disjoint EPCs. Let $f_1 \in F_1$ and $f_2 \in F_2$ be two functions. Furthermore, let $m_{in}^{opt} : f_1^{in} \to f_2^{in}$ and $m_{out}^{opt} : f_1^{out} \to f_2^{out}$ be equivalence mappings between the input and output contexts of f_1 and f_2 respectively. We define the contextual similarity as follows:

$$con(f_1, f_2) = \frac{|\{m_{in}^{opt}\}|}{2 \cdot \sqrt{|f_1^{in}|} \cdot \sqrt{|f_2^{in}|}} + \frac{|\{m_{out}^{opt}\}|}{2 \cdot \sqrt{|f_1^{out}|} \cdot \sqrt{|f_2^{out}|}}$$

A full implementation of the function matching and the similarity degree calculation is available in the Process Mining framework ProM, which can freely be downloaded from www.processmining.org. In the following section we evaluate our approach using the data generated by this tool.

[3] Note that there might be more optimal equivalence mappings, however they all express a good mapping and we have no way of distinguishing between them, so any optimal equivalence mapping will suffice.

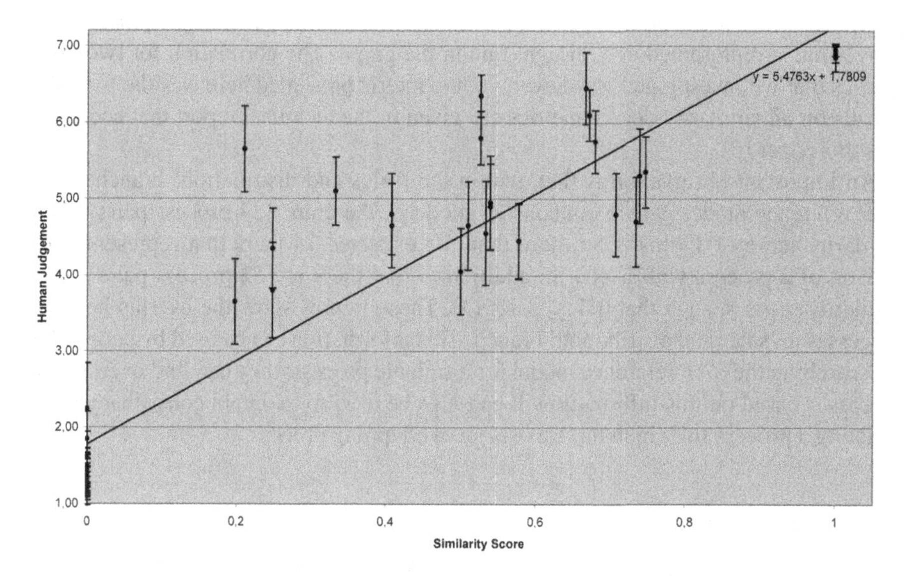

Fig. 3. Correlation between Similarity Score and Human Judgement.

5 Empirical Validation

We validated our approach to calculate the degree of similarity by computing its correlation with a similarity assessment of process modelers.

We obtained the similarity assessment using an online questionnaire that was distributed among academic process modelers. This questionnaire consisted of 48 pairs of process models from the SAP reference model database. For each pair of models, we asked the participants whether they agreed or disagreed (on a 1 to 7 Likert scale) with the proposition: 'These processes are similar.' To obtain a representative collection of model pairs, we selected the model pairs to be evenly distributed over the 0 to 1 similarity degree range. More details on how a representative collection of processes was obtained is described in the technical report that accompanies this paper [6].

We computed the correlation of the human assessment with various similarity degree metrics, which we obtained by varying cut-off values and relative importance of the syntactic, semantic and contextual similarity. We observed the best correlation for a similarity score metric that:

- does not consider syntactic similarity,
- uses a cut-off value of 0.89 for semantic similarity of events,
- uses a relative importance of semantic:contextual similarity of 1:2 and a cut-off value of 0.90 for similarity of functions.

Figure 3 shows the correlation between the similarity degree (computed using the settings described above) and the similarity assessment as obtained from the questionnaire. Each point in the graph represent a pair of processes, with a similarity degree as indicated by its x-value and a human similarity assessment as indicated by its y-value. The confidence intervals are also plotted (with a 90% confidence). For this metric we got a high (Pearson) correlation coefficient of 0.84 with the human judgement. The correlation is represented as a straight line in the graph. The correlation for two other metrics that we investigated was lower, i.e. the metric presented here was the best one. Details on all similarity degree metrics are given in the technical report that accompanies this paper [6].

An important observation is that, within the 'sales and distribution' branch of the SAP reference model (which contains 74 models), there are 124 process pairs with a similarity score of 1 (this is 50 more than the expected 74 pairs that represent comparison of a process with itself). In addition to that there are 52 process pairs with a similarity score s, such that $0.5 \leq s < 1.0$. These figures show the overlap between processes in 'sales and distribution' branch. This information can be used by people that are searching the SAP reference model for a suitable process; they can find overlapping processes based on this information. It can also be used to maintain consistency when updating a process for which there exists an overlapping process.

6 Related Work

This paper mainly relates to two streams of research, namely (1) similarity of business process models and (2) quality of business process models.

Existing work in the context of determining *similarity* between process models can be assigned to three categories: verification, behavioral similarity, and textual similarity. There are different notions of equivalence of process models that are subject to *verification* such as trace equivalence and bisimulation. While trace equivalence is based on a comparison of the sets of completed execution traces, bisimulation also considers at which point of time which decisions are taken, i.e., bisimulation is a stricter notion of equivalence. Details on different equivalence notions are given e.g. in [1]. A general problem of such verification approaches is that it provides a true-false answer to the question whether two models are similar. While some work has been done on determining a degree of behavioral similarity that measures the fitness of a set of event logs relative to a process model [1], we compare causal footprints [7] of two process models. Since causal footprints capture constraints instead of the state space, this approach relates to declarative approaches to process modeling and verification [8,17,22]. Beyond that, there are some works on textual or metadata similarity of process models (e.g. [9,14,20]). In this paper we adapt some concepts from this area for matching function labels, and we combine this approach with the calculation of behavioral similarity.

While there has been intensive research into quality aspects of process models and process modeling languages [3, 10, 15], there is little work on quality issues across models. The guidelines of modeling [3] touch this area by stressing the importance of a systematic design. The novelty of our approach is that systematic design in terms of non-overlapping models can now be checked automatically. This might prove valuable for providing tool support for process model normalization as defined in [21]. Beyond that, the quantification of a degree of behavioral similarity between process models could be a useful contribution for the area of process model *integration*. While there are several approaches reported on integration issues [5] and regarding *how* two models are integrated (e.g. [11,19,24]) the similarity degree gives an answer to the question *which* two process models might be good candidates for integration, e.g. in a merger situation. The redundancies that we identified in the SAP reference model underline the need for techniques and tools to manage process model variants such as defined in [25, 27]. Furthermore, there is clearly a need for a view concept on business process models in order to avoid anomalies [4] as they were identified in database research before.

7 Conclusion

In this paper, we presented a novel approach for measuring the degree of similarity of business process models. This approach builds on the vector model from information retrieval, an abstract representation of process behavior as causal footprints, and an automatic matching of functions across process models. While quality aspects of single process models and process modeling languages are well understood, this work contributes to a better foundation of those quality aspects across models that relate to similarity. Our approach has been validated using the SAP Reference Model, and a respective implementation is available as part of the ProM framework.

The results that we obtained for the SAP Reference Model clearly highlight the need for an automatic detection of similarity for supporting refactoring activities of a process model database. In future research we will investigate the benefits of our approach in various case studies. In particular, we aim to use the degree of similarity to detect operational overlap between companies that engage in a merger. While the application for the SAP Reference Model could build on a presumably homogeneous vocabulary of function labels, we assume that synonyms in function labels might play a more important role in a merger. Furthermore, there are some practical issues with reading the similarity matrix for a large set of models that need to be addressed. Once there is commercial tool support available, companies will find it easier to maintain large databases of process models.

References

1. van der Aalst, W.M.P., Alves de Medeiros, A.K., Weijters, A.J.M.M.: Process Equivalence: Comparing two process models based on observed behavior. In: Dustdar, S., Fiadeiro, J.L., Sheth, A. (eds.) BPM 2006. LNCS, vol. 4102, pp. 129–144. Springer, Heidelberg (2006)
2. Baeza-Yates, R.A., Ribeiro-Neto, B.A.: Modern Information Retrieval. ACM Press, New York (1999)
3. Becker, J., Rosemann, M., von Uthmann, C.: Guidelines of Business Process Modeling. In: van der Aalst, W.M.P., Desel, J., Oberweis, A. (eds.) Business Process Management. Models, Techniques, and Empirical Studies, pp. 30–49. Springer, Berlin (2000)
4. Biskup, J.: Achievements of relational database schema design theory revisited. In: Libkin, L., Thalheim, B. (eds.) Semantics in Databases 1995. LNCS, vol. 1358, pp. 29–54. Springer, Heidelberg (1998)
5. Dijkman, R.: A Classification of Differences between Similar Business Processes. In: Proceedings of the 11th IEEE EDOC Conference (EDOC 2007), pp. 37–50 (2007)
6. van Dongen, B.F., Dijkman, R.M., Mendling, J.: Detection of similarity between business process models. BETA Working Paper 233, Eindhoven University of Technology (2007)
7. van Dongen, B.F., Mendling, J., van der Aalst, W.M.P.: Structural Patterns for Soundness of Business Process Models. In: Proceedings of the 10th IEEE International EDOC Conference (EDOC 2006), pp. 116–128. IEEE, Los Alamitos (2006)

8. Eertink, H., Janssen, W., Oude Luttighuis, P., Teeuw, W.B., Vissers, C.A.: A business process design language. In: Wing, J.M., Woodcock, J.C.P., Davies, J. (eds.) FM 1999. LNCS, vol. 1708, pp. 76–95. Springer, Heidelberg (1999)
9. Ehrig, M., Koschmider, A., Oberweis, A.: Measuring similarity between semantic business process models. In: Roddick, J.F., Hinze, A. (eds.) Proceedings of the Fourth Asia-Pacific Conference on Conceptual Modelling (APCCM 2007), pp. 71–80 (2007)
10. Green, P., Rosemann, M.: Integrated Process Modeling. An Ontological Evaluation. Information Systems 25(2), 73–87 (2000)
11. Grossmann, G., Ren, Y., Schrefl, M., Stumptner, M.: Behavior based integration of composite business processes. In: van der Aalst, W.M.P., Benatallah, B., Casati, F., Curbera, F. (eds.) BPM 2005. LNCS, vol. 3649, pp. 186–204. Springer, Heidelberg (2005)
12. Johnson, R., Pearson, D., Pingali, K.: The program structure tree: Computing control regions in linear time. In: Proceedings of the ACM SIGPLAN'94 Conference on Programming Language Design and Implementation. SIGPLAN Notices, vol. 29(6), pp. 171–185 (1994)
13. Keller, G., Nüttgens, M., Scheer, A.-W.: Semantische Prozessmodellierung auf der Grundlage Ereignisgesteuerter Prozessketten (EPK). Heft 89, Institut für Wirtschaftsinformatik, Saarbrücken, Germany (1992)
14. Klein, M., Bernstein, A.: Toward high-precision service retrieval. IEEE Internet Computing 8(1), 30–36 (2004)
15. Krogstie, J., Sindre, G., Jørgensen, H.D.: Process models representing knowledge for action: a revised quality framework. Europ. J. of Information Systems 15(1), 91–102 (2006)
16. Levenshtein, I.: Binary code capable of correcting deletions, insertions and reversals. Cybernetics and Control Theory 10(8), 707–710 (1966)
17. Manna, Z., Pnueli, A.: The Temporal Logic of Reactive and Concurrent Systems: Specification. Springer, New York (1991)
18. Mendling, J., van der Aalst, W.M.P.: Formalization and Verification of EPCs with OR-Joins Based on State and Context. In: Krogstie, J., Opdahl, A., Sindre, G. (eds.) CAiSE 2007 and WES 2007. LNCS, vol. 4495, pp. 439–453. Springer, Heidelberg (2007)
19. Mendling, J., Simon, C.: Business Process Design by View Integration. In: Eder, J., Dustdar, S. (eds.) BPM Workshops 2006. LNCS, vol. 4103, pp. 55–64. Springer, Heidelberg (2006)
20. Momotko, M., Subieta, K.: Process query language: A way to make workflow processes more flexible. In: Benczúr, A.A., Demetrovics, J., Gottlob, G. (eds.) ADBIS 2004. LNCS, vol. 3255, pp. 306–321. Springer, Heidelberg (2004)
21. Pankratius, V., Stucky, W.: A formal foundation for workflow composition, workflow view definition, and workflow normalization based on petri nets (2005)
22. Pesic, M., Schonenberg, M.H., Sidorova, N., van der Aalst, W.M.P.: Constraint-based workflow models: Change made easy, pp. 77–94 (2007)
23. Porter, M.F.: An algorithm for suffix stripping. Program 14(3), 130–137 (1980)
24. Preuner, G., Conrad, S., Schrefl, M.: View integration of behavior in object-oriented databases. Data & Knowledge Engineering 36(2), 153–183 (2001)
25. Recker, J., Mendling, J., Rosemann, M., van der Aalst, W.M.P.: Model-driven Enterprise Systems Configuration. In: Dubois, E., Pohl, K. (eds.) CAiSE 2006. LNCS, vol. 4001, pp. 369–383. Springer, Heidelberg (2006)
26. Rosemann, M.: Potential pitfalls of process modeling: part b. Business Process Management Journal 12(3), 377–384 (2006)
27. Rosemann, M., van der Aalst, W.: A Configurable Reference Modelling Language. Information Systems 32, 1–23 (2007)
28. Salton, G., Wong, A., Yang, C.S.: A Vector Space Model for Automatic Indexing. Communications of the ACM 18(11), 613–620 (1975)

How Much Language Is Enough? Theoretical and Practical Use of the Business Process Modeling Notation

Michael zur Muehlen[1] and Jan Recker[2]

[1] Stevens Institute of Technology, Howe School of Technology Management, Castle Point on Hudson, Hoboken, NJ 07030 USA
Michael.zurMuehlen@stevens.edu
[2] Queensland University of Technology, Faculty of Information Technology, 126 Margaret Street, Brisbane QLD 4000, Australia
j.recker@qut.edu.au

Abstract. The Business Process Modeling Notation (BPMN) is an increasingly important industry standard for the graphical representation of business processes. BPMN offers a wide range of modeling constructs, significantly more than other popular languages. However, not all of these constructs are equally important in practice as business analysts frequently use arbitrary subsets of BPMN. In this paper we investigate what these subsets are, and how they differ between academic, consulting, and general use of the language. We analyzed 120 BPMN diagrams using mathematical and statistical techniques. Our findings indicate that BPMN is used in groups of several, well-defined construct clusters, but less than 20% of its vocabulary is regularly used and some constructs did not occur in any of the models we analyzed. While the average model contains just 9 different BPMN constructs, models of this complexity have typically just 4-5 constructs in common, which means that only a small agreed subset of BPMN has emerged. Our findings have implications for the entire ecosystems of analysts and modelers in that they provide guidance on how to reduce language complexity, which should increase the ease and speed of process modeling.

Keywords: BPMN, Language Analysis, Process Modeling.

1 Introduction

The Business Process Modeling Notation (BPMN) [1] is emerging as a standard language for capturing business processes, especially at the level of domain analysis and high-level systems design. A growing number of process design, enterprise architecture, and workflow automation tools provide modeling environments for BPMN. The development of BPMN was influenced by the demand for a graphical notation that complements the BPEL standard for executable business processes. Although this development gives BPMN a technical focus, the intention of the BPMN designers was to develop a modeling language that can equally well be applied to typical business modeling activities. This is clearly visible in the specification document, which

Z. Bellahsène and M. Léonard (Eds.): CAiSE 2008, LNCS 5074, pp. 465–479, 2008.

separates the BPMN constructs into a set of core graphical elements and an extended, more specialized set. BPMN's developers envisaged the core set to be used by business analysts for the essential, intuitive articulation of business processes in very easy terms. The full set of constructs would then enable users to specify even complex process scenarios with a level of detail that facilitates process simulation, evaluation or even execution. This separation mirrors an emerging tendency in industry to separate business-focused process modeling from implementation-oriented workflow implementation.

The evolution of BPMN closely mirrors the emergence of another modeling standard, UML [2]. Both have been ratified by the standardization body OMG. Both contain a larger set of constructs in contrast to competing languages, and offer a multitude of options for conceptual modeling. Both have been found in analytical studies to be not only semantically richer but also theoretically more complex than other modeling languages, [e.g., 3, 4]. And, in UML's case, this complexity motivated users to deliberately reduce the set of constructs for system analysis and design tasks. Related studies found that frequently not even 20% of the constructs are used in practice [5, 6].

The apparent complexity of the BPMN standard seems to be similar to the UML standard, which raises a number of questions: Are BPMN users able – and willing – to cope with the complexity of the language? Does the separation into core and extended constructs provided by the specification hold in modeling practice? And – really – how exactly is BPMN used in practice?

While BPMN has been receiving significant attention not only in practice but also in academia, virtually all contributions have been made on an analytical or conceptual level, [7, 8]. There are only few empirical insights into how BPMN is used in practice – exceptions are reported in [9] and [10].

Accordingly, our research imperative has been to provide empirical evidence on the usage of BPMN in real-life process modeling practice. The *aim of this paper* is to examine, using statistical techniques, which elements of BPMN are used in practice. We collected a large set of BPMN diagrams from three different application areas (i.e., consulting, education, process re-engineering) and analyzed the models regarding their construct usage. This study is a first step to determine the most commonly used set of BPMN constructs and to provide the ecosystem of process modelers with specific advice which elements of BPMN to use when. BPMN training programs could benefit from a structure that introduces students to the most commonly used subset first before moving on to advanced modeling concepts.

We proceed as follows: The next section briefly introduces the background of our research, viz., BPMN and our data sources, and presents our research design. Section 3 presents the analysis results and discusses them. Section 4 concludes this paper with a discussion of contributions, implications and limitations, and provides an outlook to future research.

2 Background

2.1 Introduction to BPMN

The Business Process Modeling Notation [1] is a recently published notation standard for business processes. Its development has been based on the revision of other notations including UML, IDEF, ebXML, RosettaNet, LOVeM and Event-driven Process Chains.

BPMN was developed by an industry consortium (BPMI.org), whose constituents represented a wide range of BPM tool vendors but no end users. The standardization process took six years and more than 140 meetings, both physical and virtual. The BPMN working group developed a specification document that differentiates the BPMN constructs into a set of core graphical elements and an extended specialized set. The complete BPMN specification defines 50 constructs plus attributes, grouped into four basic categories of elements, viz., Flow Objects, Connecting Objects, Swimlanes and Artefacts. *Flow Objects*, such as events, activities and gateways, are the most basic elements used to create BPMN models. *Connecting Objects* are used to inter-connect Flow Objects through different types of arrows. *Swimlanes* are used to group activities into separate categories for different functional capabilities or responsibilities (e.g., different roles or organizational departments). *Artefacts* may be added to a model where deemed appropriate in order to display further related information such as processed data or other comments. For further information on BPMN refer to [1].

Existing research related to BPMN includes, *inter alia*, analyses and evaluations, [e.g., 9, 11], use in combination with other grammars, especially BPEL [7], or its support for workflow concepts and technologies [8]. This and other research is mostly analytical in nature. Few insights exist into the practical use of BPMN, which has motivated our study.

2.2 Data Sources

In order to arrive at an informed opinion about the use of BPMN in practice we collected BPMN models from three types of sources: A search using Internet search engines for "BPMN model" resulted in 57 BPMN diagrams, obtained from organizations' web sites, from practitioner forums and similar sites. These diagrams were labeled in a variety of languages, but since our study focuses on the modeling constructs and not their content this was no hindrance. We collected an additional 37 BPMN diagrams from consulting projects to which we had access. These diagrams depicted as-is and to-be processes from business improvement projects or software deployment projects. An additional 26 diagrams were collected through BPMN education seminars taught by the authors. These diagrams were created by seminar participants and depicted business processes from the participants' organization. Overall, our data set consists of 126 BPMN models approximating the use of BPMN for a variety of purposes including process (re-) design, education, consulting, and software and workflow engineering. 6 models were excluded from the analysis because they explicitly illustrated nonsensical diagrams or were duplicates.

While by no means do we claim our data set to be statistically representative of the overall use of BPMN in practice, it nevertheless gives us an informed opinion about the *real* use of BPMN beyond the examples typically given by developers or tool vendors.

2.3 Research Design

Having obtained a large set of BPMN models, our next step was to prepare these models for analysis. We created an Excel spread sheet counting the type of BPMN constructs in use per model. Each occurrence of a BPMN construct was marked as 1, otherwise 0. This coding allowed us to treat the individual models as binary strings for further analysis. In our coding effort, we kept track of the data sources for each model, which, for analysis purposes, we labeled 'web' (those models that we obtained from Internet search engines), 'consulting' (those that we obtained from consulting engagements) and 'seminar' (those obtained from educational seminars).

The resulting tables provided the basis for the application of statistical techniques such as cluster analysis, frequency analysis, covariance analysis and distribution analysis. We employed analysis techniques available in Excel (frequency counts), Mathematica (covariance matrices, Hamming distances) and R (cluster analysis). The following sections provide further details about the exact application of the various techniques used, and discuss the results we obtained.

3 Analysis and Discussion

3.1 Overall Use of BPMN Constructs

BPMN offers 50 modeling constructs, ranging from Task and Sequence Flow to Compensation Associations and Transaction Boundaries. Our first question was: Which of these symbols are used in practice and how frequently?

Fig. 1 shows the frequency distribution of the individual BPMN constructs, separated by the three sample sets and ranked by overall frequency. Generally speaking, the distribution of constructs follows a power-law distribution, with only four constructs being common to more than 50% of the diagrams: Sequence Flow, Task, End Event, and Start Event. Notably, these constructs all belong to the originally specified BPMN core set [1].

Fig. 1 shows that every model contained the Sequence Flow construct, and nearly every model contained the basic Task construct (the diagrams that did not contain the Task construct used the Subprocess construct). The majority of Web and Seminar models contained Start and End Events, while the Consulting models replaced these with more specific event types (e.g., Message or Timer Events for Start Events, Terminate, Message, or Link, for End Events). The other BPMN constructs were unevenly distributed. A visual inspection of Fig. 1 leads to a number of interesting observations:

While the majority of consulting models contained Data-based XOR Gateways (77%), Pools (81%) and Lanes (69%), these constructs were much less frequent in the other two sample sets (57%, 30%, 21% and 23%, 56%, 16% respectively for web and seminar models). This indicates that the consulting models depict organizational structure in more detail than the random web sample. The majority of consulting models contained detailed Gateway constructs, whereas only ¼ of the seminar models did not used them. This implies that beginning modelers tend to create diagrams with few alternative or parallel flows.

The Web diagrams use (non-specific) Gateways frequently (observed in 55% of the models), whereas the consulting and seminar sets make much less use of this symbol (5% and 12%, respectively). Models in the web sample express the control flow logic of the diagrams in plain text (which can be inserted into the basic Gateways), rather than the more formal XOR, AND, and Inclusive OR constructs.

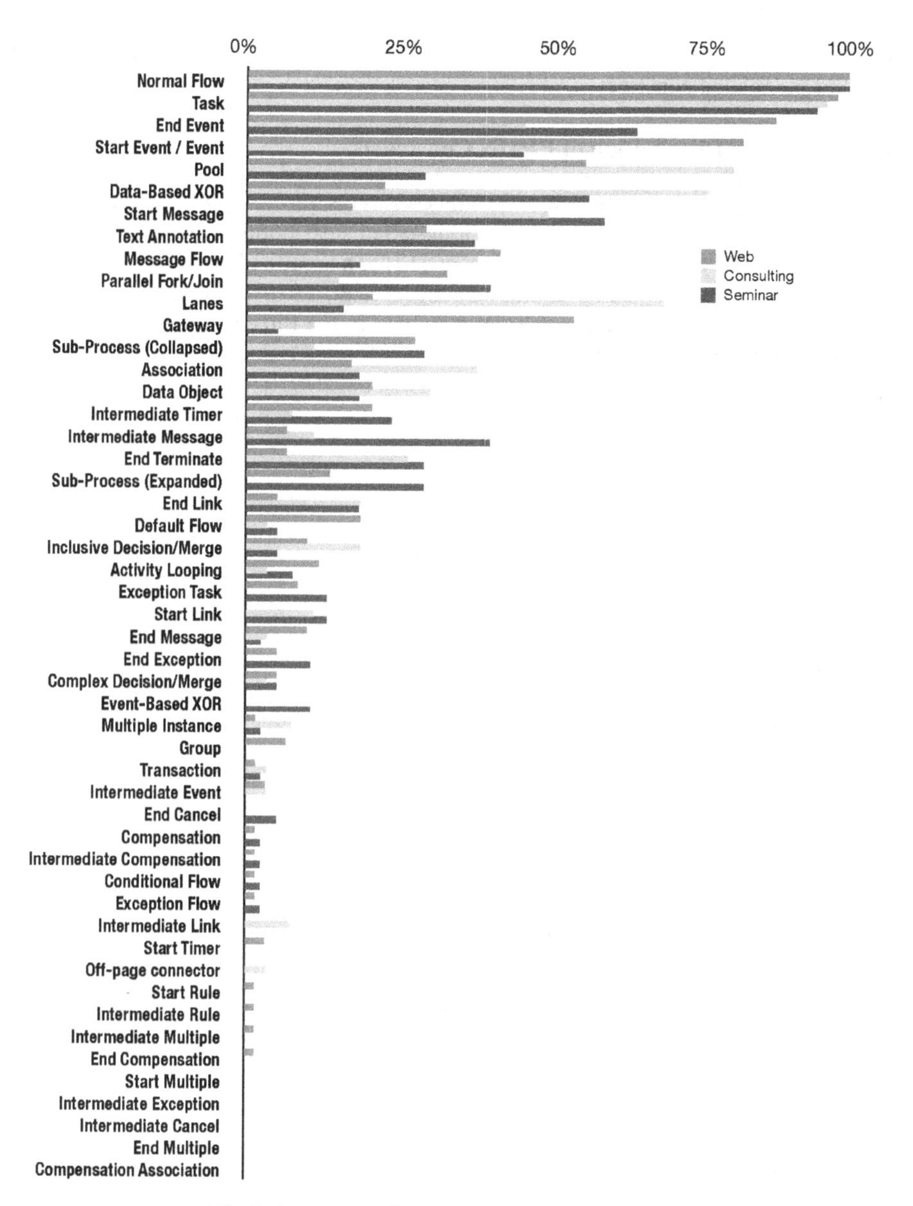

Fig. 1. Occurrence Frequency of BPMN Constructs

A sizable fraction of seminar models contain Intermediate Message constructs (41%) whereas only 7% of web models and 12% of consulting models contain this construct. This indicates that this construct is emphasized in BPMN classes but not very common in practice. A potential explanation may stem from the underlying design paradigm for process choreography in BPMN, which typically requires a lot of time to explain in classrooms. Practitioners in general may not be fully confident in the use of these choreography concepts, which could be explain the less frequent usage of the related constructs.

3.2 Frequency Distribution of BPMN Constructs

The ranked frequency distribution of BPMN constructs generally follows an exponential (power-law) distribution, similar to long-tailed distributions that have been observed as a result of preferential attachment [12]. This particular shape has been observed previously in studies of natural languages, [e.g., 13, 14]. Fig. 2 shows a plot of the frequency distribution of the BPMN elements in the three sample sets compared with the Zipfian distribution [14].

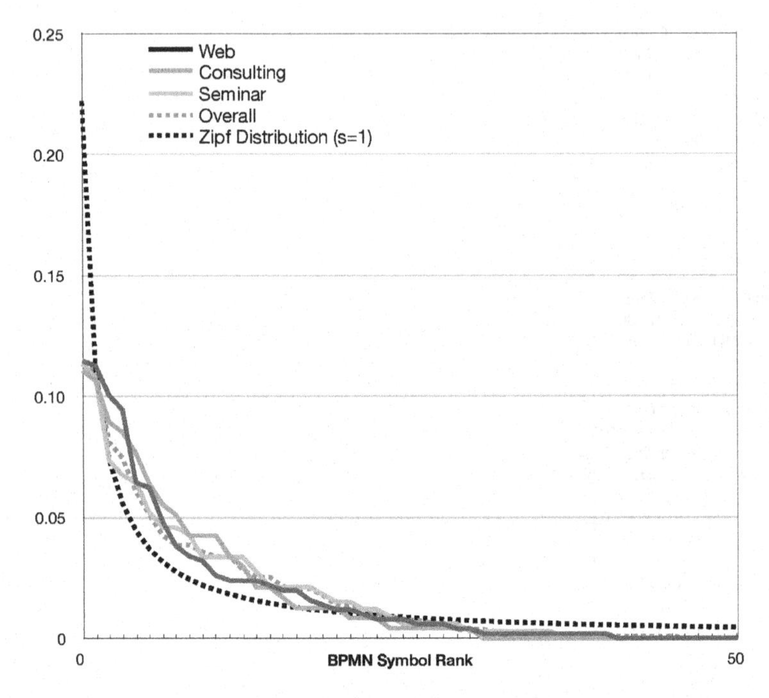

Fig. 2. Frequency Plot of BPMN Constructs by Rank

Zipf's Law states that the frequency of words in natural languages is inverse to their rank (in other words, the second most frequent word is used 1/2 the time of the first, the third most frequent word 1/3 of the time, and so on) and has been observed in numerous contexts [see, for instance, 13]. While not a perfect fit, the BPMN subsets exhibit a distribution that is very close to the distribution of word usage in natural languages. This suggests that the use of BPMN constructs to design (graphical) statements about organizational or system processes mirrors the use of natural languages.

This finding is of importance for future research on the way users learn, retain, and use BPMN constructs, and – really – any other graphical modeling language. For instance, linguistics research could be used to formulate conjectures about appropriate modeling training programs – a still under-researched aspect of modeling research in

IS. In general terms, the distribution of BPMN constructs shows that BPMN – as many natural languages – has a few essential constructs, a wide range of constructs commonly used, and an abundance of constructs virtually unused. Based on this observation, training and usage guidelines can be designed to reduce the complexity of the language to inexperienced analysts and to deliberately build such models that can safely be assumed to depict the core essence of a process without adding too much complexity.

3.3 BPMN Construct Correlations

Having determined the most frequent set of BPMN constructs in use, we turn to some related questions: Which of the BPMN constructs are typically used in combination? Which are used in alternation? In order to answer these questions, we used Mathematica to generate covariance matrices, which allowed us to examine pairs of BPMN constructs with regard to their combined or alternative use. Those pairs of constructs with negative covariance ($p < -0.05$) indicate alternatively used constructs while those with positive covariance ($p > 0.05$) indicate constructs used in combination. Table 1 summarizes the results.

Table 1. Combined and alternative use of BPMN constructs

Constructs with p > 0.05	Constructs with p < -0.05
Data Object → Association	Start Event → Start Message
Pool → Message Flow	Gateway → Data-based XOR
Start Event → End Event	Text Annotation → Message Flow
Start Message → Data-based XOR	Start Message→ End Event
Start Message → Intermediate Message	Start Message → Gateway
Start Message → End Terminate	Start Event → Data-based XOR
Pool → Lane	End Event → Data-based XOR
Lane → Message Flow	

Our findings present some interesting implications regarding BPMN modeling practice. Looking at the combined use of BPMN constructs (left column in Table 1), most correlations confirm that BPMN modeling practice obeys the grammatical rules of BPMN. For instance, Data Objects need to be linked to flow objects via the Association constructs, Pools can only communicate with other Pools via message flow, Lanes require Pools, and BPMN models require both Start and End Event. However, at least two interesting observations emerge. First, the positive correlation of Start Message events with End Terminate events indicates a more sophisticated level of BPMN modeling, suggesting that when users start using the differentiated event constructs, they tend to use a variety of these. Similarly, the combined use of Start Message events with the Data-based XOR constructs indicates an advanced use of the language for models in which different types of messages lead to different variants of a process, depending on the actual content of the arriving message.

Looking at the alternative use of BPMN constructs (right column in Table 1), we can identify additional interesting patterns of BPMN use. For instance, the negative correlation between Gateway and Data-based XOR suggests that when modelers refine the semantics of their models they choose the data-based XOR over the unspecific

Gateway in order to clarify the control flow semantics of their models. The negative correlation between Text Annotation and Message Flow suggests that at initial stages, modelers avoid choreography concepts and instead use free-form text to indicate message exchange. More advanced modeling relies on the provided semantic constructs instead of simple textual additions. Similarly, the negative correlations between Start Message event and the Gateway construct, and the Start/End Event and the Data-based XOR imply that modelers who refine the event constructs have achieved a level of sophistication of language use at which they avoid the use of the non-descriptive gateways altogether and instead rely on the more differentiated gateway and event subtypes.

3.4 BPMN Construct Clusters

In addition to identifying pairs of constructs that are used alternatively or in combination, we were also interested in uncovering whether clusters of BPMN constructs can be found in practice. To that end, we performed a hierarchical cluster analysis using the Euclidian distance measure in order to classify the set of BPMN constructs into distinct subsets. Fig. 3 shows the resulting dendrogram.

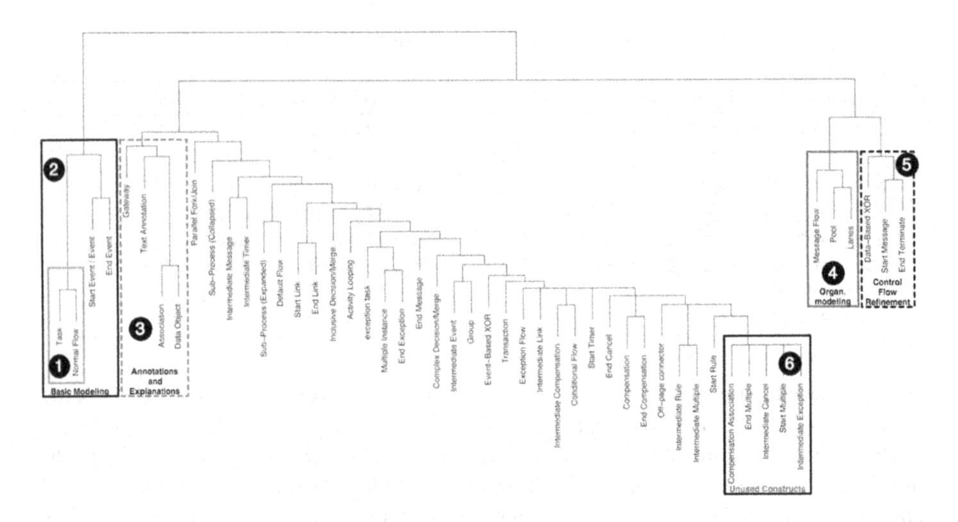

Fig. 3. Cluster Dendrogram of BPMN Constructs

In Fig. 3 six construct clusters are highlighted. First, the Task and Normal Flow cluster depicts the core of process modeling – the orchestration of activities that constitute a business process. Together with start and event conditions (through the use of events), these clusters indicate the simplest form of depicting the essence of a process in a graphical model. A third cluster is comprised of elements that are used to embellish and explain such process models through the use of text annotations, gateways (that specify control flow conditions of sequences of tasks) and data processing information. Clusters four and five essentially denote additions to these core modeling concepts by adding information about the organizational task allocation schemes, required roles and responsibilities as well as choreography information in collaborative scenarios, or refinements to the orchestration of the flow of the process through different types of event and gateway constructs. The sixth cluster we found denotes the set of constructs that are very simply not used at all (e.g., compensation association, end message, etc).

The clustering of BPMN constructs provides a promising starting point for a complete ecosystem of BPMN users – vendors, consultants, coaches and end users alike. These users can be guided in their efforts to learn and apply BPMN in an effective and efficient manner. Training programs, for instance, could focus on the 'basic modeling' clusters first before teaching advanced concepts such as organizational modeling and control flow orchestration. Coaches and consultants in charge of modeling conventions are guided by delineating the most common – and most frequently avoided – BPMN constructs.

3.5 Core or Extended Set?

According to the BPMN specification, BPMN modelers are envisaged to choose either the core set of ten BPMN constructs, or an extended set in which these core constructs are modified (i.e., revised and extended). Our questions are: Do modelers use core or extended constructs? Do they comply with the differentiation?
In order to answer these questions we split the modeling constructs into 10 sets:

- Tasks are split into Basic Tasks and an extended task set which contains the constructs for Subprocesses (collapsed and expanded) as well as Tasks with additional semantics, such as Multiple Instance Tasks, Compensations, or Transactions.
- Sequence flow constructs are split into a basic set (the Normal Flow) and an extended set (consisting of Default Flow, Conditional, and Exception flow).
- Gateways are split into the Basic (blank) XOR Gateway, and an extended Gateway set, which comprises Data- (X-labeled) and Event-based XOR, Inclusive-OR, and Parallel Gateways. We contrast these two sets with the representation of routing information through the Conditional Sequence Flow construct.
- Events are split into the Basic Events, and an extended Event set including constructs such as Messages, Rule Events, Links, etc.
- In addition we distinguished from these constructs Layout elements such as off-page connectors and the Grouping construct.

For these sets, we performed three separate frequency counts, for each of the three data sets. The results are shown in Fig. 4.

The usage patterns exhibited in Fig. 4 shed some light on when users turn to elements from the extended set of BPMN constructs. First, while users tend to employ basic task and sequence flow constructs, they mostly employ an extended set of gateway constructs. Especially the sequence flow extensions are rarely used in practice. In terms of event constructs, basic and extended sets appear to be equally utilized. The following additional observations can be made from the frequency analysis:

- Consultants especially avoid extended task constructs and use mainly basic tasks. On the other hand, they largely utilize the set of specialized gateway constructs.

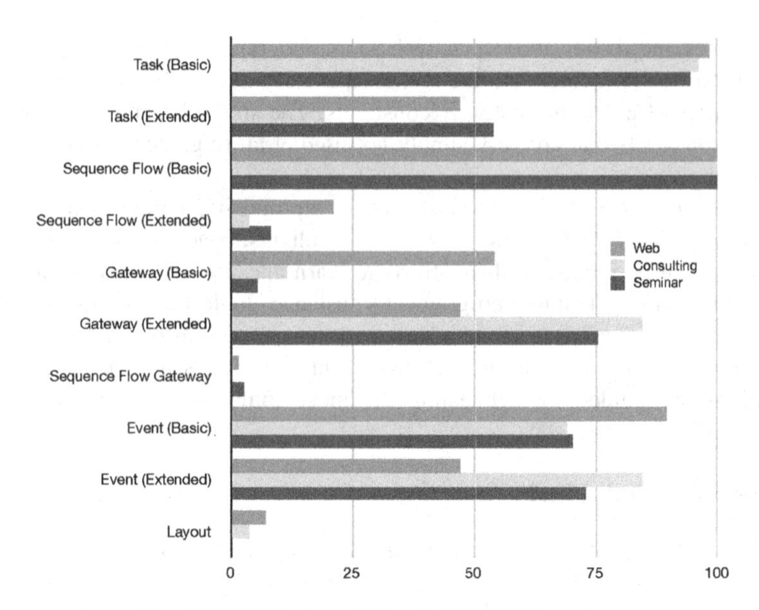

Fig. 4. Use of Core and Extended BPMN Constructs

- Decision Sequence Flow constructs are very rarely used. This would suggest that BPMN users prefer the explicit decision routing representation capacity of Gateways over the alternative, rather implicit way of annotating sequence flows.
- Basic Gateways are dominant on the web. However, neither consulting nor seminar models use them in large numbers. This suggests that formal training (as exercised through seminar courses or trained consultants) leads to the use of precise semantics for articulating process orchestration.
- Layout constructs are very rarely used. This suggests two things. First, language users often use tool functionality to annotate diagrams (e.g., meta-tags, free form tags, navigation capacity). Second, it may be worthwhile externalizing such constructs from a modeling language in order to reduce their complexity.

3.6 Complexity of BPMN Models

Previous studies on the usage of UML [5, 6] uncovered that the theoretical complexity of a language (as measured by the number of constructs originally specified) often considerably differs from the practical complexity (the number of constructs actually used in a model). We are interested in whether a similar situation exists in the case of BPMN. In other words, while the theoretical complexity of BPMN is standardized by its specification [1], we wanted to measure the practical complexity of BPMN (i.e., the vocabulary used in practice). To that end, we contrasted the semantic complexity of the BPMN models we obtained (i.e., the size of the models) with their syntactic complexity (i.e., the number of semantically different BPMN constructs used in these models). Fig. 5 illustrates the results of this analysis.

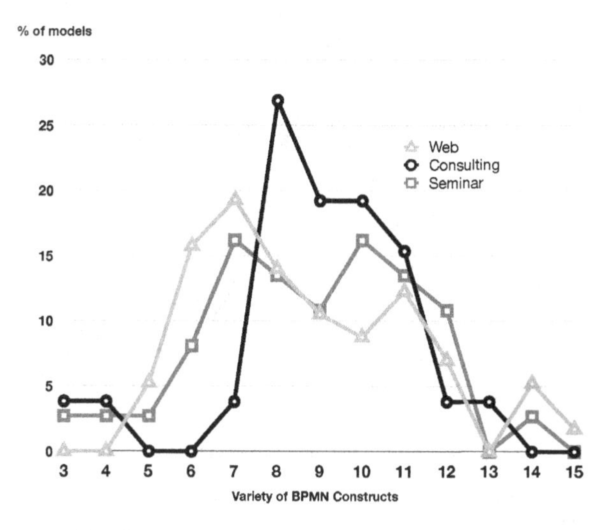

Fig. 5. Syntactic Complexity of BPMN Models

While the 50 BPMN constructs theoretically allow for 2^{50} permutations, the actual number of usable subsets is much smaller. All BPMN models obviously require the use of Tasks and Sequence Flow. Since the majority of models we observed used a BPMN vocabulary of between 6 and 12 constructs, the number of possible BPMN vocabulary subsets in practice is between $\binom{48}{4}$ =194,580 and $\binom{48}{10}$ = 6,540,715,896. Given that 9 constructs in our sample were used by fewer than two models we can exclude these from the search space and arrive at a theoretical range from $\binom{39}{10}$ = 82,251 to $\binom{39}{4}$ = 635,745,396. On average, we found the average number of semantically different BPMN constructs to be 9 (consulting), 8.78 (web), and 8.7 (seminar), respectively. However, while this finding indicates the size of the average BPMN vocabulary used in practice, it does not mean that every model with 9 BPMN constructs uses the exact same BPMN subset. In fact, a pair wise comparison of the 120 models revealed only 6 pairs of models that shared the same BPMN subset between each pair (i.e., there were 6 identical pairs of construct sets).

3.7 Variety of BPMN Subsets

In order to determine the variety of BPMN subsets, we computed the Hamming Distance [15] for each model vocabulary. Originally, the Hamming distance between two strings of equal length is the number of positions for which the corresponding symbols are different. In other words, it measures the minimum number of substitutions required to change one into the other. In the case of BPMN, we treated each model vocabulary as a 50-bit binary string, where a positive bit at position i signals the usage of BPMN construct [i]. The Hamming Distance between two model vocabularies then indicates the number of bits that differ between the two vocabularies, in other words the discrepancy between the BPMN constructs used in the creation of two models. The results are visualized in Fig. 6.

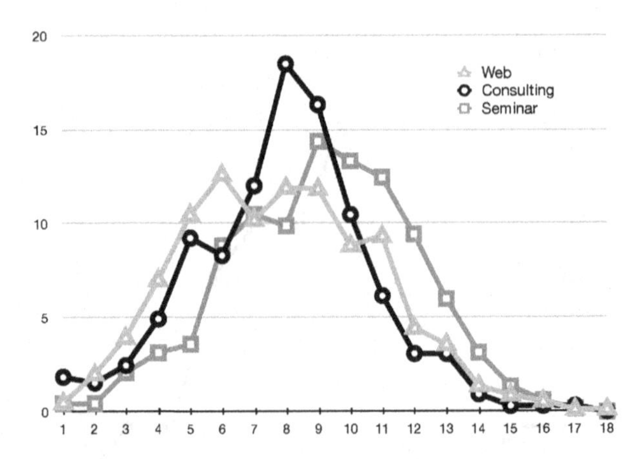

Fig. 6. Hamming Distance of BPMN Vocabularies

The average Hamming distance for the three subsets was 7.6 (web), 7.5 (consulting), and 8.8 (seminar), indicating a slightly more diverse use of BPMN constructs by novice modelers, whereas the web and consulting sets were slightly more homogeneous (but not by much). These metrics indicate that the average dissimilarity between two BPMN subsets is 7-8 constructs. A common scenario would be that one model uses 4 BPMN constructs that the other model does not exhibit and vice versa. As BPMN becomes more prevalent we plan on observing this metric over time, to see whether the commonly used vocabularies become more homogeneous over time. Annotating these BPMN subsets with context information (e.g., the process modeling purpose), in turn, could provide a starting point for deriving the most suitable BPMN subsets for a variety of application areas.

3.8 The Common Core of BPMN

Our evaluation thus far has focused on the individual elements and their grouping into core and extended constructs. However, one of our questions relates to the subset of BPMN constructs that are shared by different models. While we found six pairs of models that each share a complete set of constructs, there are subsets that are shared by more than two models. Figure Fig. 7 shows a Venn diagram of different BPMN construct combinations. The number in the corner of each grouping indicates the number of models that contained this specific subset of the language. We included combinations of constructs that were shared by more than 10 models.

The most apparent subset is the combination of Tasks and Sequence Flow – 97% of the models we analyzed shared this subset, and those that did not used a representation for tasks from the extended BPMN set (e.g., Subprocess). The addition of Start and End Events is the next most common subset – used by more than half of the models we analyzed. The following subsets show an interesting pattern: Either modelers focus on *process orchestration* through by adding gateways and their refinement to their models, or they focus on *process choreography* and add related organizational constructs, such as Pools and Lanes. While the addition of Pools leads to a subset that is common in nearly 30% of all models, the addition of Lanes halves this fraction.

Adding Basic Gateways or Parallel Gateways to the core set leads to a subset that is shared by 20% of all models. The popularity of the Data-based XOR Gateway and the Parallel Gateway construct indicate that they are a core element in many modeler vocabularies, even though the BPMN specification places them in the extended set of the language. The same situation holds for Message and Timer Events (both Start Events and Intermediate Events). While other event types were used very infrequently, these two event types were the most popular addition to the core modeling set in lieu of unspecified events.

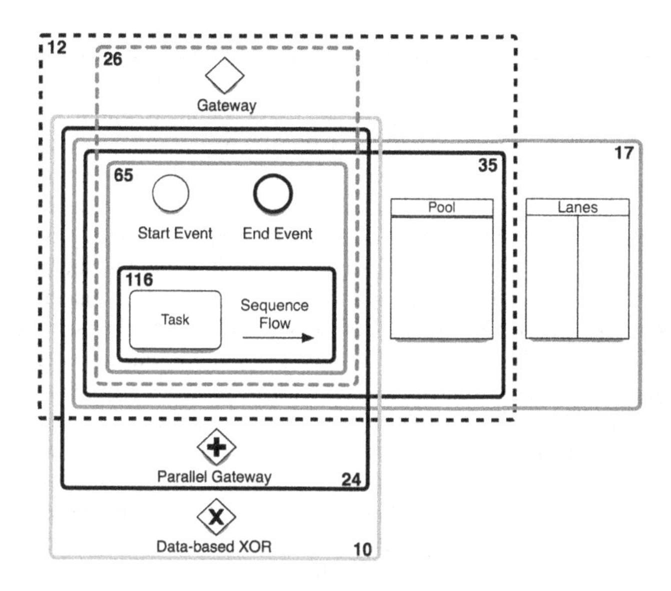

Fig. 7. Most popular BPMN Vocabulary Subsets

Overall, BPMN models appear to fall into two main sets (indicated in Fig. 7 by horizontal versus vertical grouping). The horizontal groups contain tasks, basic events plus constructs for separating organizational duties and responsibilities (Pools and Lanes). Consultants will use these types of models will most likely for organizational (re-)engineering and process improvement. The vertical groups add to this set of constructs refined constructs for specifying the exact control flow of processes (through various gateway types) as well as the exact event conditions pertaining to a process (i.e., various event construct types). This is not shown in Fig. 7 in the interest of clarity. Overall, this set of BPMN construct combinations can be expected to be favored by designers and analysts seeking to articulate the precise flow conditions, for instance, in the context of workflow engineering or process simulation rather than the organizational responsibilities (depicted by Lanes or Pools).

An interesting property of the BPMN subsets is their frequency distribution. The ranked frequency distribution again follows an exponential distribution, mirroring the behavior of individual BPMN constructs. This suggests that modelers use blocks or subsets of BPMN constructs in a similar fashion as they use individual constructs.

Combinations of BPMN constructs can thus be treated as metawords and be analyzed as such.

4 Contributions, Limitations, and Outlook

In this paper we studied the use of BPMN in actual process modeling practice. We obtained 126 (120 considered) BPMN models and used a wide range of statistical techniques to shed light onto the practical complexity afforded by the use of BPMN. Our paper makes a key contribution to the growing area of process modeling by reflecting on empirical data about the use of a rising industry standard. The most important finding is that the complexity of BPMN in practice differs considerably from its theoretical complexity. This, in turn, suggests that future research should take this distinction into account when considering BPMN's expressive power, complexity or other features or characteristics. Our study shows that the frequency of BPMN constructs follows an exponential distribution, both at the elementary level and the subset level. This means that the practical use of a formal modeling language shows similarities to the use of natural language, and suggests that linguistic techniques can be applied to better understand the formation and use of languages in conceptual modeling overall. We see an opportunity for replicating our study with other standardized modeling approaches (e.g., UML) to obtain further evidence for this conjecture.

Our findings have major implications, both for language developers and the organizational ecosystems in which modeling languages are used. Our findings point to some areas of concern in current language standardization practices, which appear to prefer language extensions (more expressive languages) to language revision (more lean languages). Our findings indicate that this may be to some extent contradictory to practical usage. Also, our findings motivate organizations to invest resources into *conventions management* in order to be able to manage and limit the complexity brought to bear by the languages employed for process modeling.

The presented research findings have to be contextualized in light of some *limitations*. First, the source of empirical evidence is limited to three sets of data sources and 126 BPMN models overall. We also did not consider any longitudinal data (e.g., the evolution of BPMN models through various iterations). However, we made an effort to collect data from multiple application areas and to consider these in our analysis. While we grouped the models by origin, we did not have sufficient information about the model content to analyze the models based on their intended use. We performed a hierarchical cluster analysis on the models themselves, but did not identify significant clusters. While this supports the random nature of our sample, it contradicts one of our expectations – that there is a clear differentiation between BPMN models depending on their intended use.

In future research, we will continue our data collection and extend it with more context-related information, e.g., for what purpose were the models created, what types of modelers created the models etc. This will allow us to triangulate our findings with contextual variables so as to arrive at informed opinions about BPMN usage across a wide range of application areas. In a related stream of research, we will apply a number of complexity metrics [e.g., 16] to the identified BPMN clusters to make a statement about how complex the frequently used BPMN constructs subsets are.

References

1. BPMI.org, OMG: Business Process Modeling Notation Specification. Final Adopted Specification. Object Management Group (2006), http://www.bpmn.org
2. Fowler, M.: UML Distilled: A Brief Guide To The Standard Object Modelling Language, 3rd edn. Addison-Wesley Longman, Boston, Massachusetts (2004)
3. Siau, K., Cao, Q.: Unified Modeling Language: A Complexity Analysis. Journal of Database Management 12, 26–34 (2001)
4. Rosemann, M., Recker, J., Indulska, M., Green, P.: A Study of the Evolution of the Representational Capabilities of Process Modeling Grammars. In: Dubois, E., Pohl, K. (eds.) CAiSE 2006. LNCS, vol. 4001, pp. 447–461. Springer, Heidelberg (2006)
5. Siau, K., Erickson, J., Lee, L.Y.: Theoretical vs. Practical Complexity: The Case of UML. Journal of Database Management 16, 40–57 (2005)
6. Kobryn, C.: UML 2001: A Standardization Odyssey. Communications of the ACM 42, 29–37 (1999)
7. Ouyang, C., Dumas, M., ter Hofstede, A.H.M., van der Aalst, W.M.P.: Pattern-based Translation of BPMN Process Models to BPEL Web Services. International Journal of Web Services Research 5, 42–61 (2008)
8. Recker, J., Rosemann, M., Krogstie, J.: Ontology- versus Pattern-based Evaluation of Process Modeling Languages: A Comparison. Communications of the Association for Information Systems 20, 774–799 (2007)
9. Recker, J., Indulska, M., Rosemann, M., Green, P.: How Good is BPMN Really? Insights from Theory and Practice. In: Ljungberg, J., Andersson, M. (eds.) Proceedings of the 14th European Conference on Information Systems. Association for Information Systems, Goeteborg, Sweden, pp. 1582–1593 (2006)
10. zur Muehlen, M., Ho, D.T.-Y.: Service Process Innovation: A Case Study of BPMN in Practice. In: Sprague Jr., R.H. (ed.) Proceedings of the 41th Annual Hawaii International Conference on System Sciences, Waikoloa, Hawaii (2008)
11. Wahl, T., Sindre, G.: An Analytical Evaluation of BPMN Using a Semiotic Quality Framework. In: Siau, K. (ed.) Advanced Topics in Database Research, vol. 5, pp. 102–113. Idea Group, Hershey, Pennsylvania (2006)
12. Barabási, A.-L., Bonabeau, E.: Scale-Free Networks. Scientific American 288, 50–59 (2003)
13. Li, W.: Random Texts Exhibit Zipf's-Law-Like Word Frequency Distribution. IEEE Transactions on Information Theory 38, 1842–1845 (1992)
14. Zipf, G.K.: On the Dynamic Structure of Concert Programs. Journal of Abnormal and Social Psychology 41, 25–36 (1946)
15. Hamming, R.W.: Error Detecting and Error Correcting Codes. Bell System Technical Journal 26, 147–160 (1950)
16. Rossi, M., Brinkkemper, S.: Complexity Metrics for Systems Development Methods and Techniques. Information Systems 21, 209–227 (1996)

On a Quest for Good Process Models: The Cross-Connectivity Metric

Irene Vanderfeesten[1], Hajo A. Reijers[1], Jan Mendling[2], Wil M.P. van der Aalst[1,2], and Jorge Cardoso[3]

[1] Technische Universiteit Eindhoven,
Department of Technology Management,
PO Box 513, 5600 MB Eindhoven, The Netherlands
{i.t.p.vanderfeesten,h.a.reijers,w.m.p.v.d.aalst}@tue.nl
[2] Queensland University of Technology,
Faculty of Information Technology,
Level 5, 126 Margaret Street, Brisbane, Australia
j.mendling@qut.edu.au
[3] SAP Research CEC, SAP AG
Chemnitzer Strasse 48, 01187 Dresden, Germany
jorge.cardoso@sap.com

Abstract. Business process modeling is an important corporate activity, but the understanding of what constitutes good process models is rather limited. In this paper, we turn to the cognitive dimensions framework and identify the understanding of the structural relationship between any pair of model elements as a hard mental operation. Based on the weakest-link metaphor, we introduce the cross-connectivity metric that measures the strength of the links between process model elements. The definition of this new metric builds on the hypothesis that process models are easier understood and contain less errors if they have a high cross-connectivity. We undertake a thorough empirical evaluation to test this hypothesis and present our findings. The good performance of this novel metric underlines the importance of cognitive research for advancing the field of process model measurement.

Keywords: business process modeling, quality metrics, connectivity, EPCs.

1 Introduction

Business process models are widely used for a variety of purposes, such as system development, training, process enactment, costing and budgeting. In many business applications their primary purpose is to act as a *means of communication* such that a process model facilitates the understanding of complex business processes among various stakeholders [16,19,26]. A process model may be used towards this end much as an architect will use a model to ascertain the views of users, to communicate new ideas, and to develop a shared understanding amongst participants. Beyond that, process models are also used as a formal

Z. Bellahsène and M. Léonard (Eds.): CAiSE 2008, LNCS 5074, pp. 480–494, 2008.

specification for the development of information systems. Altogether, it is highly desirable that process models do not contain execution errors such as deadlocks and that they are easy to understand for the involved stakeholders.

Even though theoretical quality frameworks [18] and practical modeling guidelines [2] are available for quite some time, it is only a very recent development that *empirical* insights emerge into the factors that influence the quality of process models. For instance, recent studies suggest that larger, real-world process models tend to contain more formal flaws (such as e.g. deadlocks) than smaller models [22,24]. The other study worth mentioning supports the notion that when model size is kept constant (i) a higher density of arcs between the nodes in a model and (ii) a larger number of paths through a model's logical connectors negatively affect its understandability [23].

These results are important stepping stones to what we think is a highly desirable asset for process modelers: Concrete guidelines on how to create process models in such a way that they are easy to understand for people while reducing the risk on errors. It is important to realize that a reengineering project within a multinational company may already involve the creation of thousands of process models [30]. This implies that effective modeling guidelines may lead to substantial economic benefits. This is of particular importance since most modelers are non-experts and hardly familiar with sophisticated design issues [29]. It is a considerable problem for these application areas of process modeling in practice that the current situation in understanding measurable factors of process model quality is still immature. While the mentioned experiments have progressed process model measurement, existing metrics tend to explain not more than half of the variability in a subject's understanding of process models [23]. Clearly, there is a need for a more theoretical stance to advance the design of process model metrics. In this paper, we build on insights from cognitive research into visual programming languages for the development of a new metric, the *Cross-Connectivity (CC) metric*, that aims to capture the cognitive effort to understand the relationship between any pair of process model elements.

The structure of the paper is as follows. In the next section, we will provide the motivation for the CC metric and its formalization. In Section 3 we will describe the empirical evaluation of this metric. Then, we will give an overview of related work, before giving reflections and conclusions in the final section.

2 The Cross-Connectivity Metric

Up to now, little work exists on measuring business process models that considers the cognitive effort of a model user for understanding it. One of the few examples is the research on the Control Flow Complexity (CFC) metric. In its motivation Cardoso refers to the *mental states* that may be generated by a process model and the different types of routing elements [7]. Beyond that, a recent survey into complexity metrics identifies the cognitive motivation as a potential backbone [9]. Most other existing model metrics, however, are adaptations of software artifact quality metrics that do not dig too deep into cognitive foundations. In such cases, the theoretical basis for their application on process models is indirect at best.

To break away from this tendency, we draw inspiration from the Cognitive Dimensions Framework, as first introduced in [11]. The motivation behind this framework is to use research findings from applied psychology for assisting designers of notational systems. Designers can use the framework to evaluate their designs with respect to the impact that they will have on the users of these designs. Since its introduction, it has gained widespread adoption in the evaluation and design of information artifacts; for an overview of results, see [4].

For the purpose of this paper, the most important dimension of this framework consists of *the hard mental operations* that may be incurred through a particular notation, i.e. the high demand on a user's cognitive resources. Reading a process model implies some hard mental operations in this regard that behavioral relationships between model elements have to be constructed in the mind of the reader. In particular, it is quite difficult – even for experts – to understand whether pairs of activities in a model with lots of parallelism and choices are exclusive or not. Furthermore, even if activities are on a directed path, it is not directly clear on which other elements they depend if there are lots of routing elements in between them. The Cross-Connectivity (CC) metric that we define below aims to quantify the ease of understanding this interplay of any pair of model elements. It builds on the weakest-link metaphor assuming that the understanding of a relationship between an element pair can only be as easy, in the best case, as the most difficult part. Therefore, we identify suitable weights for nodes and arcs along a path between two model elements. Our assertion then is *that a lower (higher) CC value is assigned to those models that are more (less) likely to include errors, because they are more (less) difficult to understand for both stakeholders and model designers.*

Below a set of definitions is given, which together form the basis of the Cross-Connectivity metric. The term 'Cross-Connectivity' is chosen because the strength of the *connections* between nodes is considered *across* all nodes in the model. To appreciate the formalization below, it is important to note firstly that the CC metric expresses the sum of the connectivity between all pairs of nodes in a process model, relative to the theoretical maximum number of paths between all nodes (see Definition 5). Secondly, we assume that the path with the highest connectivity between two nodes determines the strength of the overall connectivity between those nodes (see Definition 4). Thirdly, the tightness of a path (i.e., degree of connectivity) is determined by the product of the valuations of the links connecting the nodes on the path (see Definition 3). So, a single weak link has its effect on the *entire* connection. Finally, differences in the types of nodes that a path consists of determine the tightness of the arcs connecting nodes (see Definitions 1 and 2). For example, an AND connector on a path gives a stronger relation than an XOR connector. At the end of the formalization, illustrative example are given of the application of the CC metric for small process models, showing how the metric can be used to select from alternatives.

Definition 1 (Weight of a Node)
Let a process model be given as a graph consisting of a set of nodes ($n_1, n_2, ... \in N$) and a set of directed arcs ($a_1, a_2, ... \in A$). A node can be of one of two types: (i) task, e.g. $t_1, t_2 \in T$, and (ii) connector, e.g. $c_1, c_2 \in C$. Thus, $N = T \cup C$. The weight of a node n, $w(n)$, is defined as follows:

$$w(n) = \begin{cases} 1 & , \text{ if } n \in C \wedge n \text{ is of type AND} \\ \frac{1}{d} & , \text{ if } n \in C \wedge n \text{ is of type XOR} \\ \frac{1}{2^d-1} + \frac{2^d-2}{2^d-1} \cdot \frac{1}{d} & , \text{ if } n \in C \wedge n \text{ is of type OR} \\ 1 & , \text{ if } n \in T \end{cases}$$

with d the degree of the node (i.e. the total number of ingoing and outgoing arcs of the node).

There are three remarks we would like to make. In the first place, note that the definition above assumes that the process model consists of tasks and connectors. Tasks have at most one input and output arc while connectors can have multiple input and output arcs. A connector of type AND with multiple input arcs is a so-called AND-join, i.e., it synchronizes the various flows leading to the join. The OR-split connector has a behavior in-between an XOR-split (one output arc is selected) and AND-split (all output arcs are chosen). A connector can be both a join and a split (i.e. having multiple input and multiple output arcs), provided that both are of the same type. Secondly, note that we treat all model nodes as unique elements, even though their (business) semantics may be the same. In this way, for example, we support the inclusion of duplicate tasks. Finally, Definition 1 does not correspond to a concrete process modeling language with well-defined semantics. It captures those routing elements that can be expressed with standard process modeling languages such as EPCs, UML Activity Diagrams, Petri nets, BPMN, or YAWL [1].

Most of the values for $w(n)$ in Definition 1 are straightforward given the intent of this metric, e.g., arcs connected to an AND connector will have a higher weight than arcs connected to an XOR connector because the latter involves considering optionality. The only value that requires some explanation is the value for the OR connector. For the OR connector it is not clear upfront how many of the arcs will be traversed during an execution of the process, e.g., in case of an OR split with two outgoing arcs either one of the arcs can be traversed, or both of the arcs might be used. This behavior is reflected in the definition of the weight for an OR connector. The number of all possible combinations of d arcs is: $2^d - 1$. Only one of those combinations (i.e. 1 out of $2^d - 1$) is similar to the situation in which the node would have been an AND, namely the situation in which all arcs are traversed. This particular combination gets a weight of 1 (since that is the weight for an AND connector from Definition 1). Therefore, the first part of the formula for the OR connector is: $\frac{1}{2^d-1} \cdot 1 = \frac{1}{2^d-1}$. All other combinations of arcs can be seen as separate XOR nodes with weight $\frac{1}{d}$. Thus, in $2^d - 2$ out of $2^d - 1$ combinations a weight of $\frac{1}{d}$ is added, which leads to the second part of the formula.

The following definition shows that the weight of an arc is based on the weight of the corresponding nodes.

Definition 2 (Weight of an Arc)
Let a process model be given by a set of nodes (N) and a set of directed arcs (A). Each directed arc (a) has a source node (denoted by src(a)) and a destination node (denoted by dest(a)).
The weight of arc a, W(a), is defined as follows:

$$W(a) = w(src(a)) \cdot w(dest(a))$$

Definition 3 (Value of a Path)
Let a process model be given by a set of nodes (N) and a set of directed arcs (A). A path p from node n_1 to node n_2 is given by the sequence of directed arcs that should be followed from n_1 to n_2: $p =< a_1, a_2, ..., a_x >$. The value for a path p, v(p), is the product of the weights of all arcs in the path:

$$v(p) = W(a_1) \cdot W(a_2) \cdot ... \cdot W(a_x)$$

Definition 4 (Value of a Connection)
Let a process model be given by a set of nodes (N) and a set of directed arcs (A) and let P_{n_1,n_2} be the set of paths from node n_1 to n_2. The value of the connection from n_1 to n_2, $V(n_1, n_2)$, is the maximum value of all paths connecting n_1 and n_2:

$$V(n_1, n_2) = \max_{p \in P_{n_1,n_2}} v(p)$$

If no path exists between node n_1 and n_2, then $V(n_1, n_2) = 0$. Also note that loops in a path should not be considered more than once, since the value of the connection will not be higher if the loop is followed more than once in the particular path.

Based on the above valuation of connectivity (i.e., tightness of the connection between two nodes), we define the *Cross-Connectivity metric*.

Definition 5 (Cross-Connectivity (CC))
Let a process model be given by a set of nodes (N) and a set of directed arcs (A). The Cross-Connectivity metric is then defined as follows:

$$CC = \frac{\sum_{n_1,n_2 \in N} V(n_1, n_2)}{|N| \cdot (|N| - 1)}$$

Example 1. To illustrate the use of the CC metric an example is elaborated. Figure 1 contains a process model with five tasks (i.e. $T = \{A, B, C, D, E\}$), three connectors (i.e. $C = \{XOR, AND, OR\}$) and seven directed arcs (i.e. $A = \{a_1, a_2, a_3, a_4, a_5, a_6, a_7\}$). To calculate the value for Cross-Connectivity the weight for each node is calculated first (see Table 1).

Table 1. The degrees and weights for the nodes in the process model of Figure 1

Node (n)	Degree (d)	Weight ($w(n)$)
A	1	1
B	1	1
C	1	1
D	1	1
E	1	1
XOR	3	$\frac{1}{3}$
AND	3	1
OR	3	$\frac{1}{2^3-1} + \frac{2^3-2}{2^3-1} \cdot \frac{1}{3} = \frac{3}{7}$

Then, the weight for each arc is calculated:

$$W(a_1) = w(A) \cdot w(XOR) = 1 \cdot \frac{1}{3} = \frac{1}{3}$$
$$W(a_2) = w(B) \cdot w(XOR) = 1 \cdot \frac{1}{3} = \frac{1}{3}$$
$$W(a_3) = w(XOR) \cdot w(AND) = \frac{1}{3} \cdot 1 = \frac{1}{3}$$
$$W(a_4) = w(C) \cdot w(AND) = 1 \cdot 1 = 1$$
$$W(a_5) = w(AND) \cdot w(OR) = 1 \cdot \frac{3}{7} = \frac{3}{7}$$
$$W(a_6) = w(OR) \cdot w(D) = \frac{3}{7} \cdot 1 = \frac{3}{7}$$
$$W(a_7) = w(OR) \cdot w(E) = \frac{3}{7} \cdot 1 = \frac{3}{7}$$

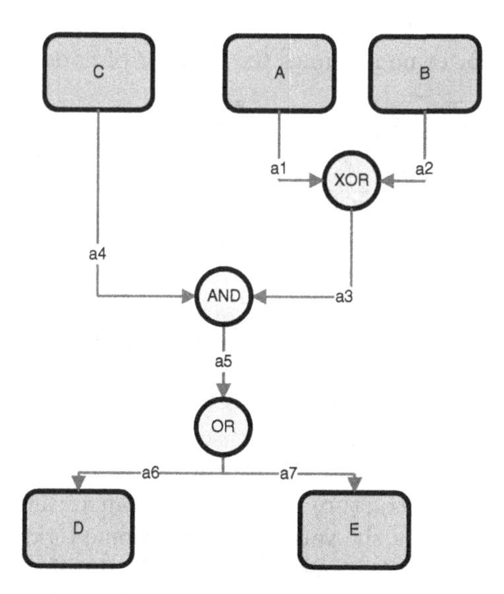

Fig. 1. A simple example with five tasks and three connectors. $T = \{A, B, C, D, E\}$, $C = \{XOR, AND, OR\}$, $A = \{a_1, a_2, a_3, a_4, a_5, a_6, a_7\}$.

The paths between each pair of nodes are determined and the value for the connection between the pair of nodes is computed. For example node A and node D are connected through the path $< a_1, a_3, a_5, a_6 >$. In this case, this is the only path from A to D. Thus, the value of this path is the maximum value over all paths from A to D:

$$V(A,D) = v(< a_1, a_3, a_5, a_6 >) = W(a_1) \cdot W(a_3) \cdot W(a_5) \cdot W(a_6) = \frac{1}{3} \cdot \frac{1}{3} \cdot \frac{3}{7} \cdot \frac{3}{7} = \frac{1}{49}.$$

Similarly, the value for the connection from the XOR-node to the OR-node is computed:

$$V(XOR, OR) = v(< a_3, a_5 >) = W(a_3) \cdot W(a_5) = \frac{1}{3} \cdot \frac{3}{7} = \frac{1}{7}.$$

For all values, see Table 2.

Table 2. Table showing the values for the connections between all pairs of nodes

	A	B	C	D	E	XOR	AND	OR	Total
A	0	0	0	$\frac{1}{49}$	$\frac{1}{49}$	$\frac{1}{3}$	$\frac{1}{9}$	$\frac{1}{21}$	$\frac{235}{441}$
B	0	0	0	$\frac{1}{49}$	$\frac{1}{49}$	$\frac{1}{3}$	$\frac{1}{9}$	$\frac{1}{21}$	$\frac{235}{441}$
C	0	0	0	$\frac{9}{49}$	$\frac{9}{49}$	0	1	$\frac{3}{7}$	$\frac{88}{49}$
D	0	0	0	0	0	0	0	0	0
E	0	0	0	0	0	0	0	0	0
XOR	0	0	0	$\frac{3}{49}$	$\frac{3}{49}$	0	$\frac{1}{3}$	$\frac{1}{7}$	$\frac{88}{147}$
AND	0	0	0	$\frac{9}{49}$	$\frac{9}{49}$	0	0	$\frac{3}{7}$	$\frac{39}{49}$
OR	0	0	0	$\frac{3}{7}$	$\frac{3}{7}$	0	0	0	$\frac{6}{7}$

Finally, the CC value is determined as the sum of the values for all connections, divided by the number of nodes times the number of nodes minus one:

$$CC = \frac{\frac{235}{441} + \frac{235}{441} + \frac{88}{49} + 0 + 0 + \frac{88}{147} + \frac{39}{49} + \frac{6}{7}}{8 \cdot 7} = \frac{\frac{2255}{441}}{56} \approx 0.09131$$

Now the mechanics behind the CC metric have been dealt with, it is worthwhile to explore how it can help to distinguish models that are preferable. This will clarify that our interest at this point is rather with a model's CC value *relative* to that of another model.

Consider the two models that are shown in Figure 2. Both models express the same business logic as the initial model in Figure 1, but their CC values are different. The model at the left-hand side in Figure 2 is block-structured, i.e., it differs from the initial model in the sense that the AND-join and XOR-join at the top of the model are matched by corresponding splits. Intuitively, one may expect that a block-structure will positively affect model comprehension. Indeed, as the links between the various nodes become tighter, this is expressed by a higher CC value of 0.12486 versus a value of 0.09131 of the initial model.

The model at the right-hand side in Figure 2 is different from the initial model of Figure 1 in the sense that it reorders the top connectors: It expresses

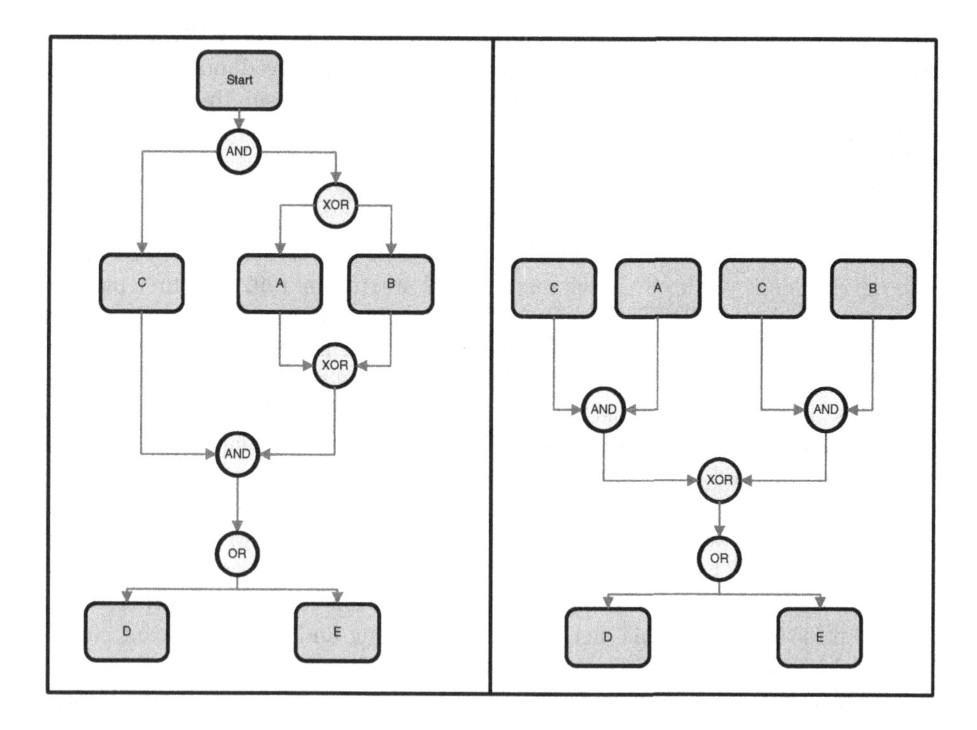

Fig. 2. Two alternatives to the example of Fig. 1

(C AND A) XOR (C AND B) instead of C AND (A XOR B), the former being the more elaborate expression with a duplicate for task C.[1] So, the model at the right-hand side can be expected to be slightly more difficult to understand than the initial model, which is supported by a lower CC value of 0.08503 versus the initial model's CC value of 0.09131.

3 Evaluation

In this section, we report on how the CC metric has been subjected to a thorough empirical evaluation. First, we will describe the evaluation with respect to the metric's capability to predict error probabilities in process models. Next, we will present on its suitability to explain which process models are easier to understand than others.

3.1 Validation for Error Prediction

An indication for a metric's predictive power is that it can accurately distinguish between models with errors and without errors. Because this evaluation uses a

[1] Recall that in the computation of the CC metric all model elements are treated as unique elements.

large set of Event-driven Process Chains (EPCs), we use the EPC soundness criterion as defined in [21] for determining whether an involved model has errors or not and assume that a decrease in CC is likely to result in more errors. Therefore, our hypothesis is:

H1: A decrease in *CC* implies an increase in error probability.

To evaluate this hypothesis, the EPCs of the *SAP Reference Model* are used. The development of the SAP reference model started in 1992 and first models were presented at CEBIT'93 [17, p.VII]. Since then, it was developed further until version 4.6 of SAP R/3, which was released in 2000. The SAP reference model includes 604 non-trivial EPCs. The advantage of considering this set of models is that there is extensive literature available that explains its creation, e.g., [17]. Furthermore, it is frequently referenced in research papers as a typical reference model and used in previous quantitative analyses, as e.g. reported in [20,22,24]. This way, our results can be compared to these related works.

As a first step, we use *correlation analysis*. In particular, we investigated to what extent the CC metric is capable to rank non-error and error models. This capability can be estimated using the rank correlation coefficient by Spearman. For CC it is -0.434. For this metric there is a strong and 99% significant correlation, which matches the expectation of the hypothesis, i.e. H1 holds.

In a second step, we use *multivariate logistic regression*. This approach estimates the coefficients B of a linear combination of input parameters for predicting event versus non-event based on a logistic function. In our case, we predict error versus non-error for the EPCs in the SAP reference model based on the CC metric and a constant. The accuracy of the estimated model is assessed based on the significance level of the estimated coefficients, the percentage of cases that are classified correctly, and the share of the variation that is explained by the regression. This share is typically measured using the Nagelkerke R^2 ranging from 0 to 1 (1 being the best possible value). The estimated coefficient should have a Wald statistic that is below 5% signalling that it is significantly different from zero. For technical details of logistic regression we refer to [13]. For applications in predicting errors in process models see [20,22,24].

We calculated a univariate logistic regression for CC first. Table 3 shows that CC alone already yields a high Nagelkerke R^2 of 0.586. The negative coefficient matches the expectation of hypothesis H1. Furthermore, we stepwise introduced other metrics to the model. We used those metrics that were found in [22] as the best combination to predict errors in EPCs. In this context, it is interesting to note that adding these metrics yields quite similar coefficients for them as in the predicting function of [22]. This suggests that the CC metric indeed measures a process model aspect that is orthogonal to metrics that have been defined before.

3.2 Validation for Understandability

To evaluate the capability of the CC metric to explain which process models are easier to understand than others, we used the empirical data described in

Table 3. Multivariate Logistic Regression Models with CC

	Parameter	Coefficient	Std.Error	Wald	Sig.	Nagelkerke	Classification
Step 1	CC	-13.813	1.229	126.386	0.000	0.586	0.791
...	...	...	...	...	...	...	...
Step 5	CC	-10.478	2.931	12.783	0.000	0.847	0.916
	Structuredness	-9.500	1.028	85.328	0.000		
	Diameter	0.139	0.032	18.829	0.000		
	Cyclicity	6.237	1.857	11.281	0.001		
	CNC	5.541	0.935	35.145	0.000		

[23]. This data was obtained in a project that aims at the analysis of the impact of both model and personal characteristics on the understandability of process models. In particular, a set of 20 model characteristics were investigated, which have been proposed and formally defined in [20].

In total, 73 students filled out a questionnaire in the fall of 2006. A set of 12 process models from practice, each having the same number of tasks (25), formed the basis of the questionnaire. As part of the models' evaluation, students were asked to answer questions like "If task K is executed for a case, can task L be executed for the same case?" The evaluation of the 12 models by the 73 students led to a total of 847 complete model evaluations. On this basis, a SCORE variable could be calculated per model as the mean sum of correct answers it received. This SCORE variable served as a way to make understandability operational.

From the earlier analysis of these results [23], the following main conclusions were drawn with respect to model characteristics:

1. From the 20 factors considered, five model factors exhibited the hypothesized relation with SCORE, i.e. (1) #OR-JOINS, (2) DENSITY, (3) AVERAGE CONNECTOR DEGREE, (4) MISMATCH, and (5) CONNECTOR HETEROGENEITY.
2. From these five model characteristics, only the correlations between DENSITY (the ratio between the actual number of arcs and the theoretical maximal number of arcs) and SCORE (-0.618) and between AVERAGE CONNECTOR DEGREE (the average number of input and output arcs of the routing elements in a model) and SCORE (-0.674) correlated significantly, with respective P-values of 0.032 and 0.016.
3. From all linear regression models on the basis of a combination of these five model factors, the regression model that only used AVERAGE CONNECTOR DEGREE displayed the best explanatory power for the variability in SCORE, with an adjusted R^2=45% (Nagelkerke's coefficient of determination).

To evaluate the CC metric, it was incorporated in the above analysis. We arrive at the following conclusions:

- Just like the five model factors that emerged from the original analysis, the CC metric displays the expected relation with SCORE.

- Unlike the density and average connector degree factors, the correlation between SCORE and the CC metric (0.549) is *not* significant at a 95% confidence interval as the P-value of 0.065 slightly exceeds the 0.05 confidence interval.
- A regression model with a *much better* explanatory power for the variation in SCORE could be developed by *including* the CC metric: the adjusted R^2 from the original model increased from 45% to 76% in the new regression model. In particular, by combining the #OR-JOINS, DENSITY, AVERAGE CONNECTOR DEGREE, MISMATCH factors and the CC metric this result could be achieved. A visualization of this regression model can be seen in Figure 3.

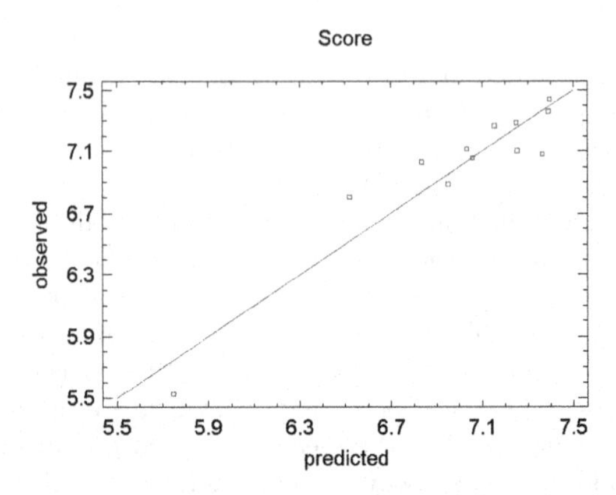

Fig. 3. Linear regression model explaining the mean SCORE for the 12 process models

What this analysis suggests is that CC on its own is slightly less powerful as an indicator for process model understandability than the two best candidate metrics available, but that it can deliver a superior explanation of the variation in understandability across models when combined with existing metrics.

4 Related Work

This section briefly describes the related work for business process metrics. In essence, related work can be organized in two categories: process model metrics inspired by software measurement and experimental work on process model metrics. In this section, we focus in particular on metrics that consider overall structural aspects of the process model beyond simple count metrics. For an overview of process model metrics in general refer to [9,12,20,33].

The early development of process model metrics is greatly inspired by and based on software quality metrics. These metrics aim at obtaining program designs that are less error-prone, easier to comprehend and easier to maintain. A survey of existing software metrics can be found in e.g. [14,35]. A number of studies demonstrate the significant correlation of software quality metrics with errors in the software design (e.g. [3,5,15,31,32]). In the tradition of this work, there are some works in the 1990s that are mainly rooted in software quality measurement.

Daneva et al. [10] introduce a set of complexity indicators for EPCs based on the visual attributes of the model: *function cohesion*, *event cohesion* and *cohesion of a logical connector*. From their validation with 11 EPCs they conclude that their metrics help to identify error-prone model fragments. Morasca proposes a set of simple metrics for software specifications designed with Petri-nets [25]. He identifies size, length, structural complexity, and coupling as interesting attributes of a design without striving for an empirical validation. The works by Reijers and Vanderfeesten extend this research stream by introducing a coupling-cohesion metric for guiding the design of a workflow process [27,28]. This approach is based on the data flow in a business process and uses the network structure of the product as a starting point rather than the process model. In [34], Vanderfeesten, Cardoso, and Reijers propose a weighted coupling metric, which puts a weight to the different types of connections between two activities in the process model. While this metric lacks a thorough cognitive motivation, it was used as a blueprint for the CC metric. Cardoso has developed a Control Flow Complexity (CFC) metric [7] which was validated against *Weyuker's* complexity axioms [6] and tested with respect to their correlation with perceived complexity [8]. In contrast to the CC metric, it does not consider the connections between different model elements, but focuses on routing elements in isolation.

Mendling et al. take an experimental approach towards process model metrics that is driven by the explanatory power of a metric in an empirical setting. In [20,22] Mendling et al. have tested 28 business process metrics (including size, density, structuredness, coefficient of connectivity, average connector degree, control flow complexity, and others) as error predictors on a set of over 2000 process models from different samples. All metrics, except for density and the maximum degree of a connector, are confirmed to be correlated to errorproneness as expected. Another result of this study is a logistic regression model is able to classify 90% of the process models correctly. Finally, a survey on understandability of process models is reported by Mendling, Reijers and Cardoso in relation to the set of metrics mentioned in the previous study [23]. The main results of this survey are already described in Section 3.2.

While the metrics used in these experiments are motivated theoretically, most of them are not explicitly rooted in cognitive research. The CC metric considers hard mental operations as defined in the cognitive dimensions framework [11] as the main factor that drives understanding a process model.

5 Conclusion

In this paper, we motivated, formalized, and validated the Cross-Connectivity metric for process models. The metric expresses how tightly the nodes in a process model are connected building on a weakest-link metaphor. The definition of the metric builds on the assumption that a higher value is associated with an easier understanding of the model, which implies as a consequence a lower errorprobability. As follows from our evaluation of this metric for both these aspects, it performs similarly well as the best available alternative model metrics. On top of that, our results suggest that the CC metric adds a new cognitive perspective on process model quality, which helps to deliver a better explanatory power when it is combined with the existing ones.

In reflecting on the development of business process model metrics, it is fair to say that it is a research area in development. Initially, proposals for process model metrics were highly conceptual, on the basis of the perhaps tempting idea that if metrics are useful to analyze software programs it should be equally applicable for process models. By now, we have progressed to the stage where model metrics are put to the test for determining their effectiveness in reality. The good performance of the CC metric clearly shows that a more cognitive theoretical stance is needed to advance the field of process model measurement.

Overall, feedback from empirical validations has improved the quality of process model metrics: The metrics proposed in recent works, e.g. [20,22] and this paper, perform much better in explaining the variation of understanding and occurrence of errors in process models. In our future work, we will continue towards further improvements. In particular, we aim at evaluating model quality metrics on a wider scale, by considering larger sets of real-world models. In order to achieve that, we are collaborating with consultancy companies that practice process modeling on a day-to-day basis for their clients. Since most empirical research has been done with EPC models, we are very much interested in BPMN and Petri-net process models. Furthermore, we are investigating additional factors that contribute to a comprehensive understanding of process model quality as, for example, the visual layout a process model graph and the importance of preliminary knowledge about the domain that is captured in the model. As the ultimate goal of our research, we envision the development of a set of concrete guidelines for process modelers, substantiated by solid theoretical foundations and empirical evidence, which will help to create better process models in practice.

Acknowledgement

This research is partly supported by the Technology Foundation STW, applied science division of NWO and the technology programme of the Dutch Ministry of Economic Affairs.

References

1. van der Aalst, W.M.P., ter Hofstede, A.H.M., Kiepuszewski, B., Barros, A.P.: Workflow Patterns. Distributed and Parallel Databases 14(1), 5–51 (2003)
2. Becker, J., Rosemann, M., von Uthmann, C.: Guidelines of Business Process Modeling. In: van der Aalst, W.M.P., Desel, J., Oberweis, A. (eds.) Business Process Management. LNCS, vol. 1806, pp. 30–49. Springer, Berlin (2000)
3. Bieman, J.M., Kang, B.-K.: Measuring Design-level Cohesion. IEEE Transactions on Software Engineering 24(2), 111–124
4. Blackwell, A.F.: Ten Years of Cognitive Dimensions in Visual Languages and Computing. Journal of Visual Languages and Computing 17(4), 285–287 (2007)

5. Card, D.N., Church, V.E., Agresti, W.W.: An Empirical Study of Software Design Practices. IEEE Transactions on Software Engineering 12(2), 264–271
6. Cardoso, J.: Control-flow Complexity Measurement of Processes and Weyuker's Properties. In: Proceedings of the 6th International Enformatika Conference (IEC 2005), pp. 213–218. International Academy of Sciences (2005)
7. Cardoso, J.: How to Measure the Control-flow Complexity of Web Processes and Workflows. In: Fischer, L. (ed.) Workflow Handbook 2005, Future Strategies, Lighthouse Point (2005)
8. Cardoso, J.: Process Control-flow Complexity Metric: an Empirical Validation. In: IEEE International Conference on Services Computing (IEEE SCC 2006), pp. 167–173. IEEE Computer Society Press, Los Alamitos (2006)
9. Cardoso, J., Mendling, J., Neumann, G., Reijers, H.A.: A Discourse on Complexity of Process Models. In: Eder, J., Dustdar, S. (eds.) BPM Workshops 2006. LNCS, vol. 4103, pp. 115–126. Springer, Heidelberg (2006)
10. Daneva, M., Heib, R., Scheer, A.-W.: Benchmarking Business Process Models. IWi Research Report 136, Institute for Information Systems, University of the Saarland, Germany (1996)
11. Green, T.R.G., Petre, M.: Usability Analysis of Visual Programming Environments: A 'Cognitive Dimensions' Framework. Journal of Visual Languages and Computing 7(2), 131–174 (1996)
12. Gruhn, V., Laue, R.: Complexity Metrics for Business Process Models. In: Proceedings of the 9th international conference on business information systems (BIS 2006). Lecture Notes in Informatics, vol. 85 (2006)
13. Hosmer, D., Lemeshow, S.: Applied Logistic Regression, 2nd edn. Wiley & Sons, Chichester (2000)
14. Kafura, D.: A Survey of Software Metrics. In: ACM 1985: Proceedings of the 1985 ACM annual conference on The range of computing: mid-80's perspective, pp. 502–506. ACM Press, New York (1985)
15. Kang, B.-K., Bieman, J.M.: A Quantitative Framework for Software Restructuring. Journal of Software Maintenance 11, 245–284 (1999)
16. Kawalek, P., Kueng, P.: The Usefulness of Process Models: A Lifecycle Description of how Process Models are used in Modern Organisations. In: Siau, K., Wand, Y., Parsons, J. (eds.) Proceedings of the Second CAiSE/IFIP8.1 International Workshop on Evaluation of Modelling Methods in Systems Analysis and Design, pp. 1–12 (1997)
17. Keller, G., Teufel, T.: Sap R/3 Process Oriented Implementation: Iterative Process Prototyping. Addison-Wesley Longman Publishing Co., Inc, Boston (1998)
18. Krogstie, J., Sindre, G., Jørgensen, H.: Process Models Representing Knowledge for Action: a Revised Quality Framework. European Journal of Information Systems 15(1), 91–102 (2006)
19. Lindsay, A., Downs, D., Lunn, K.: Business processes: attempts to find a definition. Information and Software Technology 45(15), 1015–1019 (2003)
20. Mendling, J.: Detection and Prediction of Errors in EPC Business Process Models. PhD thesis, Vienna University of Economics and Business Administration, Vienna, Austria (May 2007)
21. Mendling, J., van der Aalst, W.M.P.: Formalization and Verification of EPCs with OR-Joins Based on State and Context. In: Krogstie, J., Opdahl, A., Sindre, G. (eds.) CAiSE 2007 and WES 2007. LNCS, vol. 4495, pp. 439–453. Springer, Berlin (2007)

22. Mendling, J., Neumann, G., van der Aalst, W.M.P.: Understanding the Occurrence of Errors in Process Models based on Metrics. In: Meersman, R., Tari, Z. (eds.) OTM 2007, Part I. LNCS, vol. 4803, pp. 113–130. Springer, Heidelberg (2007)
23. Mendling, J., Reijers, H.A., Cardoso, J.: What Makes Process Models Understandable? In: Alonso, G., Dadam, P., Rosemann, M. (eds.) BPM 2007. LNCS, vol. 4714, pp. 48–63. Springer, Berlin (2007)
24. Mendling, J., Verbeek, H.M.W., van Dongen, B.F., van der Aalst, W.M.P., Neumann, G.: Detection and Prediction of Errors in EPCs of the SAP Reference Model. Data and Knowledge Engineering 64(1), 312–329 (2008)
25. Morasca, S.: Measuring Attributes of Concurrent Software Specifications in Petri-nets. In: Proceedings of the 6th International Symposium on Software Metrics, pp. 100–110. IEEE Computer Society, Los Alamitos (1999)
26. Ould, M.A.: Business Processes: Modelling and Analysis for Re-engineering and Improvement. Wiley, Chichester (1995)
27. Reijers, H.A.: A Cohesion Metric for the Definition of Activities in a Workflow Process. In: Proceedings of the 8th CAiSE/IFIP8.1 International workshop on Evaluation of Modeling Methods in Systems Analysis and Design (EMMSAD 2003), pp. 116–125 (2003)
28. Reijers, H.A., Vanderfeesten, I.T.P.: Cohesion and Coupling Metrics for Workflow Process Design. In: Desel, J., Pernici, B., Weske, M. (eds.) BPM 2004. LNCS, vol. 3080, pp. 290–305. Springer, Berlin (2004)
29. Rosemann, M.: Potential Pitfalls of Process Modeling: Part A. Business Process Management Journal 12(2), 249–254 (2006)
30. Rosemann, M.: Potential Pitfalls of Process Modeling: Part B. Business Process Management Journal 12(3), 377–384 (2006)
31. Selby, R.W., Basili, V.R.: Analyzing Error-Prone System Structure. IEEE Transactions on Software Engineering 17, 141–152 (1991)
32. Shen, V.Y., Yu, T.-J., Thebaut, S.M., Paulsen, L.R.: Identifying Error-Prone Software. IEEE Transactions on Software Engineering 11, 317–324 (1985)
33. Vanderfeesten, I., Cardoso, J., Mendling, J., Reijers, H.A., van der Aalst, W.M.P.: Quality Metrics for Business Process Models. In: Fischer, L. (ed.) BPM and Workflow Handbook 2007, Future Strategies, USA, May 2007, pp. 179–190 (2007)
34. Vanderfeesten, I., Cardoso, J., Reijers, H.A.: A Weighted Coupling Metric for Business Process Models. In: Eder, J., Tomassen, S.L., Opdahl, A., Sindre, G. (eds.) Proceedings of the CAiSE 2007 Forum, CEUR Workshop Proceedings, vol. 247, pp. 41–44 (2007)
35. Xenos, M., Stavrinoudis, D., Zikouli, K., Christodoulakis, D.: Object-Oriented Metrics - A Survey. In: Proceedings of the FESMA 2000, Federation of European Software Measurement Associations, pp. 1–10 (2000)

Information Systems Engineering Supported by Cognitive Matchmaking

S.J. Overbeek[1], P. van Bommel[2], and H.A. (Erik) Proper[2,3]

[1] e-office B.V., Duwboot 20, 3991 CD Houten, The Netherlands, EU
Sietse.Overbeek@e-office.com
[2] Institute for Computing and Information Sciences, Radboud University Nijmegen, Toernooiveld 1, 6525 ED Nijmegen, The Netherlands, EU
P.vanBommel@cs.ru.nl
[3] Capgemini Nederland B.V., Papendorpseweg 100, 3528 BJ Utrecht, The Netherlands, EU
E.Proper@acm.org

Abstract. In daily practice, discrepancies may exist in the suitability match of actors and the tasks that have been allocated to them. Formal theory and the prototype of a cognitive matchmaker system are introduced as a solution to improve the fit between actors and tasks. A case study has been conducted to clarify how the proposed cognitive matchmaker system can be utilized in information systems engineering. The inductive-hypothetical research strategy has been applied when performing the case study.

Keywords: cognitive characteristics, matchmaking, task allocation.

1 Introduction

Globalization, the emergence of virtual communities and organizations, and growing product complexity has an impact on how actors (i.e. a human or a computer) fulfill tasks in organizations. Notably due to these developments, an actor working on a task may experience an increase in cognitive load while task performance decreases [1,2]. The system discussed in this paper matches cognitive characteristics supplied by actors and the cognitive characteristics required to fulfill tasks. This may achieve a better fit between actors and tasks. Cognitive characteristics can be the willpower to fulfill a task or maintaining awareness of the requirements to fulfill a task for example. These characteristics are also referred to as *volition* and *sentience*, respectively in cognitive literature [2,3]. Within the enterprise, the benefits of cognitive matchmaking can be found in at least three areas: Multi-agent systems, workflow management and business process reengineering (BPR). (1) Multi-agent systems incorporate several software agents that may work together to assist humans in performing their tasks [4]. One way of providing assistance is to match tasks with human actors to understand which tasks fit best with which human actors. (2) The primary task of a workflow management system is to enact case-driven business processes by

Z. Bellahsène and M. Léonard (Eds.): CAiSE 2008, LNCS 5074, pp. 495–509, 2008.

joining several perspectives [5]. One of these perspectives is the *task* perspective. This perspective describes the elementary operations performed by actors while executing a task for a specific case. An example of a case is a tax declaration. Integration of cognitive matchmaking in a workflow management system may prescribe which available actors fit best with the tasks that are part of a case. This may improve the allocation of tasks to actors while enacting a business process. (3) BPR consists of computer-aided design of processes and automatic generation of process models to improve customer service [6]. The design and creation of processes and process models may be improved if the business process modeler knows beforehand which available actors best fit the tasks that need to be fulfilled as part of a newly designed business process.

The focus of the research reported in this paper, however, is to analyze how cognitive matchmaking can provide support for a project during which an information system is engineered. Therefore, a case study has been carried out at e-office, a company specialized in providing computer-aided support for human actors to assist them during office work. First, the framework for cognitive matchmaking that has been developed in earlier work [7] is introduced in section 2. Section 3 discusses the prototype, that has been developed proceeding from the framework. The conducted case study is explained in section 4. Section 5 briefly compares our study with other approaches in the field and outlines the benefits of our approach. Section 6 concludes this paper and gives an overview of future work.

2 Framework for Cognitive Matchmaking

The main goals of this paper are to discuss the prototype and the case study. However, it is necessary to briefly introduce the framework for cognitive matchmaking as is elaborated in [7]. First, the framework is illustrated in figure 1.

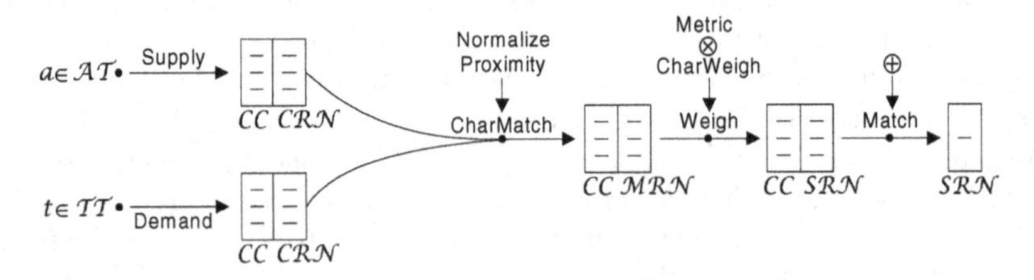

Fig. 1. Framework for cognitive matchmaking

The different concepts shown in figure 1 are functions that are necessary to calculate the eventual suitability match of an actor fulfilling a task. Even though the formal signature of these functions are not exhaustively repeated in this section, we will show some examples for clarification. First, the *supply* function shows the level on which an actor type, that is characterized by certain cognitive characteristics, offers a characteristic during task execution. The levels on which an actor type supplies a cognitive characteristic may vary over the natural numbers from 0 up to and including 10. These levels are part of the *characteristic rank domain* indicated by the set $\mathcal{CRN}$. This ranking domain includes the rank values that can be used to indicate the level on which a characteristic can be supplied by an actor or demanded by a task. The *demand* function depicted in figure 1 shows the level on which a task of a certain type requires a certain cognitive characteristic if an actor wishes to fulfill the task.

The characteristic match or *CharMatch* function shown in figure 1 matches supply and demand of a specific characteristic. There is an optimal characteristic match if an actor offers a cognitive characteristic at the same level as a certain task requires the characteristic. A characteristic match is calculated for every cognitive characteristic that is supplied by an actor type and demanded by a task type. The result is part of the *match rank domain*, which may vary over the real values from 0 up to and including 1. An optimal characteristic match is indicated by the match rank value 0.5. This is because 0 indicates complete underqualification (an actor is not able to supply a certain characteristic at all) and 1 indicates complete overqualification (the supply of a certain characteristic is not necessary at all for a task whilst an actor supplies that certain characteristic at the highest level).

The weighed characteristic match function or *Weigh* function weighes the result of the characteristic match function. The user of the system may provide a weigh value to give more importance to a characteristic match result than another. The result is part of the *suitability rank domain*, which may vary over the real values from 0 up to and including 10. The results of the weigh function are then summated by the *Match* function which shows the *suitability match*. This suitability match is also expressed by a value from the suitability rank domain. To show how we have formalized the functions of the framework, the formal signature of e.g. the match function is modeled as follows [7]: $\mathsf{Match} : \mathcal{AT} \times \mathcal{TT} \rightarrow \mathcal{SRN}$. Note that the set $\mathcal{AT}$ contains actor types, the set $\mathcal{TT}$ contains task types and the set $\mathcal{SRN}$ contains suitability rank values. This function can be defined using the aforementioned functions:

$$\mathsf{Match}(\texttt{expert}, \texttt{synthesis}) \triangleq \bigoplus_{c \in \mathcal{CC}} \mathsf{Weigh}(c, \mathsf{CharMatch}(\texttt{expert}, \texttt{synthesis}))$$

For this example the suitability match of the *expert* actor type [7] and the *synthesis* task type [8] has been calculated. The expert uses his own knowledge to solve a problem. The expert is also able to combine and modify knowledge while solving a problem and is able to learn from that. A researcher often acts as an expert. A synthesis task is related with the actual utilization of acquired knowledge. An example is a student who utilizes knowledge (acquired by reading a book) while performing an exam. The definition of the match function shows that for every characteristic the weighed characteristic match function is executed and the results are then summated. The latter is shown by the $\oplus$ operator. This operator is used instead of the large Sigma because soft (linguistic) suitability rank values can also be used instead of hard (numerical) values. The match function can be expressed as follows: $\mathsf{Match}(\texttt{expert}, \texttt{synthesis}) = \mathbf{4.25}$, which shows that the numerical suitability match of the expert fulfilling the synthesis task is 4.25. This is a fairly good result, knowing that 5 is the best suitability match that

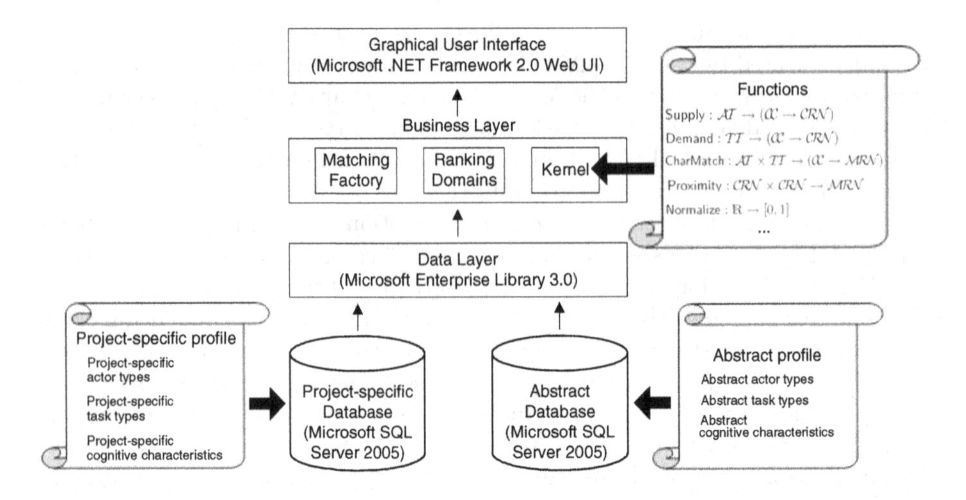

Fig. 2. Cognitive matchmaker system architecture

can be achieved. Section 3 contains a screen shot of the prototype in which this match result is shown. An implementation of the match function is shown as a code snippet in section 3. This is to illustrate that the formalisms mentioned in figure 1 are implemented in program code. The program code has been tested by means of unit tests after the implementation of all the formalisms included in the framework. This is to make sure that individual methods in the code are working properly.

Finally, a function has been introduced to determine the degree of certainty that an actor is suitable to fulfill a task [7]: $\mu : \mathbb{R} \to [0, 1]$. A linear certainty degree function can be defined as follows:

$$\mu(u) \triangleq \begin{cases} \frac{2}{\texttt{min}+\texttt{max}} \cdot u & \texttt{min} \leq u \leq \frac{\texttt{min}+\texttt{max}}{2} \\ \frac{-2}{\texttt{min}+\texttt{max}} \cdot u + 2 & \frac{\texttt{min}+\texttt{max}}{2} \leq u \leq \texttt{max} \end{cases}$$

In the implementation of the prototype the minimum and maximum values of a suitability match are equated to 0 respectively 10. Thus, $\texttt{min} = 0$ and $\texttt{max} = 10$. The degree of certainty that the expert is suitable to fulfill the synthesis task is: $\mu(4.25) = \frac{2}{0+10} \cdot 4.25 = 0.85$. This can be interpreted as being 85% sure that the expert is suitable enough to fulfill the synthesis task. It might be a good choice to let this actor fulfill the task, unless an available actor provides a better match.

3 Prototype of the Cognitive Matchmaker System

The prototype of the cognitive matchmaker system has been designed as a Web application according to the three tier software architecture depicted in figure 2. The graphical user interface is based on the Microsoft .NET Framework 2.0 Web UI namespace that provides classes and interfaces to create user interface elements. The business layer includes the main components of the application. The most important one is the kernel, which is an implementation of the formal functions shown in figure 1 and in [7]. Furthermore, the 'matching factory' instantiates all the objects involved when a suitability match should be calculated and enables the application to follow the flow of the matchmaking process as depicted in figure 1. The business layer also includes an implementation of the possible ranking domains that can include characteristic ranks, match ranks and suitability ranks. The data layer includes code to interact with connected databases. The architecture shows that it is possible to include a project-specific database as well as a database including abstract types and characteristics. This signifies that the cognitive matchmaker system can compute matches between project-specific actor types and task types as well as between the abstract actor types and task types we have defined in [7]. Project-specific actor types and task types can be defined to categorize all the actors and tasks that are part of a specific project. For instance, a person called 'John Doe' can be categorized as a project-specific actor type 'developer', meaning that he acts as a software developer in a specific project. Section 4.1 includes the project-specific actor types and task types as part of the elaborated case study. On the contrary, the abstract types categorize actors and tasks based on the supplied respectively demanded cognitive characteristics. These types are explained in section 4.2. Dependent of the choice the user of the cognitive matchmaker system makes, the system communicates with one of the available databases to calculate matches. The data layer is based on the Microsoft Enterprise Library 3.0 that contains chunks of source code for e.g. data access.

The user of the system has to walk through six steps to calculate a suitability match. In the first step, the user should select an actor type and a task type for which a suitability match should be calculated. Suppose that the user selects the *expert* actor type and the *synthesis* task type. This causes the application to generate a list of all the cognitive characteristics that have been used to characterize the expert actor type and the synthesis task type. In the following step, the application displays on which level the expert supplies the involved characteristics and on which level the synthesis task demands the characteristics for successful task fulfillment. The next part shows the characteristic match results for all cognitive characteristics. Then, the user can provide the weigh values for the cognitive characteristics by entering them for each characteristic involved. Figure 3 shows the eventual suitability match result with the corresponding graph. Due to space limitations a screen shot of the suitability match screen is shown only. The resulting graph shows that in this case the suitability match of the expert fulfilling the synthesis task is 4.25. The certainty that the expert is able to fulfill the synthesis task is 85%. The implementation of the prototype is based on the framework shown in figure 1. For example, the code implementation of the suitability match function depicted in section 2 is shown in figure 4. The code implementation obviously shows that the match function takes an actor type and a task type as input parameters and a suitability rank value as output parameter just like the formal match function in our framework prescribed. Then, the results of the weighed characteristic match function are summated for each cognitive characteristic involved in the process of computing

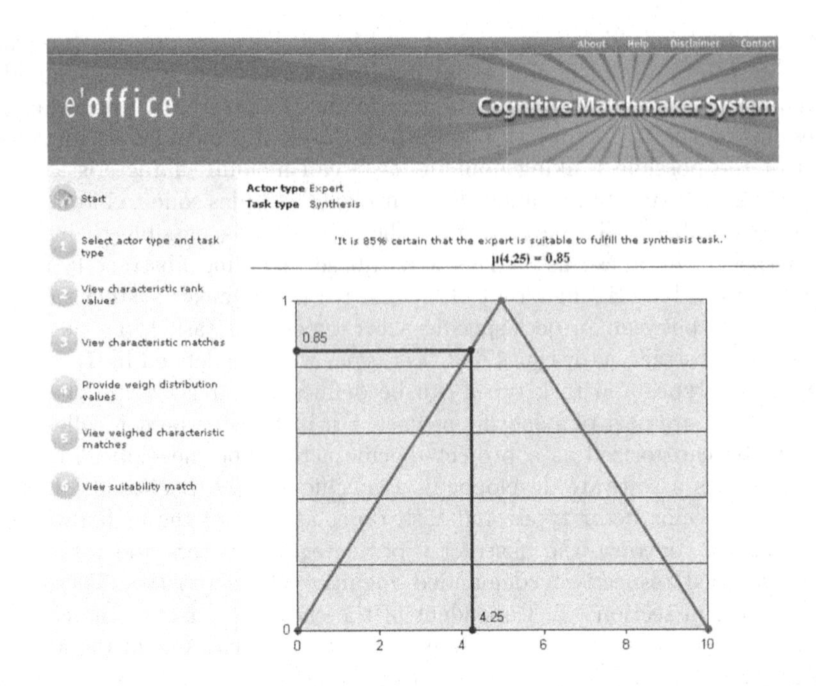

Fig. 3. Suitability match screen

```
public static SuitabilityRank Match(TaskType taskTypeObject, ActorType actorTypeObject) {

  SuitabilityRank SuitabilityRankObject = new SuitabilityRank();

  foreach (Characteristic CharacteristicObj in _matching.RetrieveCharacteristics()) {
    SuitabilityRankObject.RankValue += Weigh(CharacteristicObj,
    CharMatch(actorTypeObject, taskTypeObject,
    CharacteristicObj)).RankValue;
  }

  return SuitabilityRankObject;
}
```

Fig. 4. Source code of the suitability match function

the suitability match. This also corresponds with the definition of the suitability match function as shown in section 2.

4 Case Study and Evaluation

The case study that has been conducted is related with a recently completed information systems engineering (ISE) project at 'e-office'. This project has been concerned with the development of an 'Action Reporting Tool' (ART for short) for an international provider of banking and insurance services. ART is a software application that can generate risk reports for the user. This tool should assist risk management to better monitor and control insurance risks. The research strategy for the case study has been derived from the inductive-hypothetical research strategy [9], which consists of five phases. Empirical knowledge of the problem domain is elicited in the *initiation* phase. Elicited empirical knowledge is applied in a descriptive conceptual model in the *abstraction* phase. The *theory formulation* phase is necessary to make the descriptive conceptual model prescriptive. The prescriptive conceptual model is empirically tested in the *implementation* phase. A comparison of phase 1 with the prescriptive empirical model of phase 4 is needed to fulfill the *evaluation* phase. The derived research strategy includes the following steps. (1) Description of the project phases in which the ISE project has been divided. The description includes project-specific actor types and task types and relations between them. (2) Abstraction of the results of phase 1 of the research strategy to our abstract cognitive model of actor types and task types. (3) Formulation of how the cognitive matchmaker system can be utilized in every project phase related to the actor types and task types involved in the project. (4) Analysis to identify the benefits if the system had been applied in the studied ISE project. (5) Evaluation by comparing phase 1 with phase 4.

4.1 Initiation

The ART project is based on the Microsoft Solutions Framework (MSF) information systems engineering method. The resulting tool is a Web application running on the Microsoft Office SharePoint Server 2007 platform (MOSS 2007 for short). Applications based on this platform are aimed to facilitate organizational collaboration, content management and business process management. The following project phases are determined as part of the ART project: The definition phase, the development phase, the acceptance phase and the implementation phase. The *definition* phase incorporates requirements engineering by means of interviews with the future users of the tool and conducting interactive workshops. The requirements engineering process is proceeded by the creation of several use cases to determine the interactions between the users and the tool. These use cases can be used as input to create screen mockups. The tool is developed in an iterative way during the *development* phase. The results after every iteration are tested before proceeding to the next iteration. The tool is integrally tested during the *acceptance* phase in conformity with a test plan. The acceptance test has been carried out by the banking and insurance service provider. Eventually, the final version of the tool is implemented at the service provider during the *implementation* phase. The actors participating in the project have been categorized into several project-specific actor types based on the MSF method. First, an *integrated program management (IPM) officer* can be identified. The IPM officer is responsible for organizational scheduling, planning and resource allocation. The *project manager* develops project plans, iteration plans and status reports. The project manager also mitigates risks. The *product manager* monitors the budget and the realization of the business case. The *infrastructure architect* focuses on server deployment and the services which run on them. Furthermore, the *solution architect* is responsible for defining both the organizational and physical structure of the application. Finally, the *lead developer* and the *developer* actor types can be identified. The lead developer lends experience and skills to fellow developers. The developer is mainly responsible for building the product. The project manager of the ART project has created breakdowns of the tasks to be fulfilled. Using this documentation a project-specific task type categorization can be described together with the fulfilled tasks. Analysis of the project documentation also reveals which project-specific actor type performs a project-specific task type. The results of this analysis are shown in table 1.

Table 1. Project-specific actor types and task types

Task instance	Project phase	Project-specific task type	Project-specific actor type
Conduct interview with stakeholder	Definition Development	Elicitation task	Project manager Product manager Solution architect
Conduct workshop with stakeholders	Definition Development	Elicitation task	Project manager Product manager Solution architect
Design use case	Definition Development	Design task	Solution architect Developer
Design mockup	Definition	Design task	Developer
Design risk report	Definition	Design task	Lead developer Developer
Write technical tool description	Definition Development	Documentation task	Developer
Write project initiation document	Definition	Documentation task	Project manager Product manager
Determine hardware requirements	Definition	Documentation task	Infrastructure architect
Write security plan	Definition Development	Documentation task	Infrastructure architect
Write project plan	Definition Development	Documentation task	Project manager
Attend project meeting	All phases	Meeting task	All actor types
Attend steering committee meeting	All phases	Meeting task	IPM officer
Set up MOSS 2007 environment	Development	Code development task	Infrastructure architect
Build custom Web part	Development	Code development task	Developer
Configure Web part	Development	Code development task	Developer
Create risk report	Development	Code development task	Lead developer Developer
Implement security for tool	Development	Code development task	Infrastructure architect
Commit partial system test	Acceptance	System test task	Lead developer Developer
Commit integral system test	Acceptance	System test task	Lead developer Developer
Deploy completed tool	Implementation	Deployment task	Lead developer Developer

4.2 Abstraction

When performing the second phase of the inductive-hypothetical research strategy, it is possible to abstract the project-specific actor types and task types. First, it is shown how the project-specific task types can be abstracted to the abstract task types mentioned in [8]. Second, this section discusses how the project-specific actor types can be abstracted to the actor types mentioned in [7].

The distinguished abstract task types are the acquisition, synthesis and testing types. The project-specific task types depicted in table 1 can be abstracted to these types as follows. The mentioned elicitation tasks are typical knowledge *acquisition* tasks. The actors executing an elicitation task acquire knowledge by means of interviews or workshops. Design tasks can be abstracted as *synthesis* tasks. In a design task, the actor applies already acquired knowledge when designing a use case, mockup or risk report. Documentation tasks can also be classified as synthesis tasks. The documentation tasks mentioned in table 1 are related with the application of knowledge when writing a technical tool description, project initiation document, hardware requirements report, security plan and project plan. Next, meeting tasks are abstracted to acquisition tasks. During project meetings and steering committee meetings it is intended to acquire knowledge about e.g. project planning, project status and the remaining budget. Code development tasks can be viewed as synthesis tasks. These tasks are necessary to build the action reporting tool itself. The build process consisted of setting up the programming environment, and the creation of Web parts and risk reports. Web parts are the visual components that are part of a Microsoft SharePoint application which include functionality, such as: Listed announcements, a calendar, a discussion part, etc. *Testing* tasks are related with the project-specific system test tasks. In a testing task, earlier applied knowledge is thoroughly examined inducing an improvement of the specific knowledge applied. The partial and integral system tests are needed to identify and correct flaws in the action reporting tool. Finally, the deployment task can be abstracted to a synthesis task. Here, all relevant knowledge that is applied is related with a successful deployment of the system.

The distinguished abstract actor types are the collaborator, experiencer, expert, integrator and the transactor. Only the expert type has been mentioned up till now. The other types are explained as follows. The *collaborator* has the ability to exert an influence on state changes of knowledge involved during task fulfillment. During task fulfillment the collaborator is also able to improve its own cognitive abilities. However, a collaborator does not have complete awareness of all required knowledge to fulfill a task and requires others. An *experiencer* is aware of all the knowledge requirements to fulfill some task. An *integrator* is able to fulfill a task by working together and is able to initiate state changes of knowledge involved during task fulfillment. An integrator primarily wishes to acquire and apply knowledge of the highest possible quality. A *transactor* can fulfill a task without collaborating with others and is not required to cause modifications in the knowledge acquired and applied during task fulfillment. A project-specific actor can be classified as a certain abstract type while performing an abstract task. For instance, a project manager may be classified as a collaborator when fulfilling an acquisition task. However, if a project manager executes a synthesis task he may act differently and may be classified as a transactor instead. These abstractions materialize in the following phase of the research strategy.

4.3 Theory Formulation

The results of the previous phase of the research strategy are discussed throughout this section. The cognitive matchmaker system can be utilized in the four phases of the ART project. However, only the definition phase is elaborated in this section to understand how ISE can be supported by cognitive matchmaking. The approach can be used for the remaining phases in an identical manner. Based on sections 4.1 and 4.2 it is now possible to calculate the certainty that the actors involved in the definition phase of the ART project can successfully

Table 2. Cognitive matchmaking in the definition phase

Project-specific actor type	Task type	Actor type	Weigh values	Suitability match	Certainty
IPM officer	Acquisition	Collaborator	2, 1.5, 0.5, 3, 3	4.3	86%
Project manager	Acquisition	Collaborator	2, 1, 1, 3, 3	4.4	88%
	Synthesis	Transactor	1, 1.5, 1.5, 3, 3	4.85	97%
Product manager	Acquisition	Collaborator	2, 1, 1, 3, 3	4.4	88%
	Synthesis	Transactor	1, 1.5, 1.5, 3, 3	4.85	97%
Infrastructure architect	Acquisition	Experiencer	4, 3, 3	3.5	70%
	Synthesis	Expert	1.5, 1, 1, 1, 1.5, 2, 2	4.575	91.5%
Solution architect	Acquisition	Collaborator	1.5, 2, 1.5, 3, 2	4.4	88%
	Synthesis	Expert	1, 1.5, 1, 0.5, 1, 2, 3	4.8	96%
Lead developer	Acquisition	Experiencer	4, 3, 3	3.5	70%
	Synthesis	Expert	1, 2, 1, 1.5, 1, 1.5, 2	4.6	92%
Developer	Acquisition	Experiencer	3, 4, 3	3.3	66%
	Synthesis	Collaborator	2, 1.5, 2.5, 2, 2	4.4	88%

fulfill the tasks allocated to them. The results are depicted in table 2. The system can be utilized to calculate the matches between the actor types and task types involved in the definition phase. For this purpose, the way to abstract the project-specific actor types and task types as discussed in section 4.2 should be applied. The results of table 2 can be explained as follows. The solution architect, for instance, acts as a collaborator when working on an acquisition task and acts as an expert when working on a synthesis task respectively. In the definition phase, the solution architect conducts interviews and workshops, and attends project meetings. These tasks can be regarded as knowledge acquisition tasks. Five weigh values have to be provided by the user of the cognitive matchmaker system when calculating the suitability match of the collaborator fulfilling an acquisition task. The weigh values express the importance of the involved cognitive characteristics. The following weigh values are provided for the five characteristics involved: 1.5, 2, 1.5, 3 respectively 2. The cognitive characteristics used to characterize actor types and task types are further explained in [7,8]. At the moment, the weigh values have to be provided manually by the user. However, the next version of the prototype should include an algorithm that determines these weigh values dependent of how important a cognitive characteristic is in a certain combination of an actor type and a task type. The highest weigh value of 3 has been applied to the *satisfaction* characteristic. This is to make absolutely sure that the solution architect is pleased with the knowledge acquired and that no additional need for knowledge remains [8]. The cognitive matchmaker system then sums up the resulting weighed characteristic matches resulting in a suitability match of 4.4. The certainty that the solution architect acting as a collaborator can successfully fulfill an acquisition task is: $\mu(4.4) = \frac{2}{0+10} \cdot 4.4 = 0.88$ or $0.88 \cdot 100\% = 88\%$. The solution architect acts as an expert when working on a synthesis task during the definition phase. These synthesis tasks are related with the design of use cases. The solution architect should be able to use his own knowledge about use cases to correctly design them. The architect should also be able to combine and modify his own knowledge while designing use cases and he should also be able to learn from that process. The expert actor type matches very well with the synthesis task, because the result of the suitability match calculation is 4.8. This results in a certainty of 96%.

4.4 Implementation

The results from the theory formulation phase are now utilized to describe how ISE can be supported by cognitive matchmaking. The utilization of the system in ISE is described using three viewpoints. (1) The *design time* viewpoint embraces the situation before the project is initiated (before the definition phase starts). First, the project-specific actor types and the project-specific task types need to be conceived. If this is done, there are two options to choose from: Use the project-specific profile as a starting point or the abstract profile. The latter has been done in the case study as is elaborated in section 4.2. When using a project-specific profile as input for the system, a project-specific profile of actors and tasks should be generated. This has also been done in section 4.1. If not already entered in the project-specific database as is shown in figure 2, the actor and task data should be provided as a next step. The person who needs to allocate tasks to actors, the project manager for instance, can now calculate the suitability matches. Based on these results he can allocate tasks to actors before starting the project. (2) The *runtime* viewpoint is related to the recalculation of suitability matches if changes to task allocations are necessary during the enactment of an ISE project. This may be the case if a different actor needs to work on a task than the one specified in the project plan. The cognitive matchmaker system can then be used again to recalculate the suitability match. New tasks may also be introduced during the project that need to be allocated to actors. This may entail the need to calculate additional suitability matches during project enactment. (3) From a *post-mortem* point of view, task allocations in the project as a whole can be analyzed. The suitability matches for every actor / task combination in the ISE project may be compared to the actual results brought forward by the actors. Lessons learned should be recorded for future projects. This may help to better decide which actor types are suitable to work with which types of tasks.

4.5 Evaluation

In this section, the results of the initiation phase are compared with the results of the implementation phase. The evaluation of the initiation phase is related to the three viewpoints of the implementation phase. At *design time*, the choices leading to the project-specific actor types as shown in table 1 have not been argued in the ART project documentation. Recall that the project-specific actor types originate from the Microsoft Solutions Framework ISE method. Entering the actor types that MSF distinguishes in the project-specific database of the cognitive matchmaker system enables a better argued decision of which actor types to use in a project. For instance, the system test tasks shown in table 1 are performed by the (lead) developer actors. However, the MSF method also distinguishes the tester and test manager types. Including these actor types in the ART project may have improved the suitability matches related with the system test tasks. A difficulty is that the MSF method does not provide a clear description of the cognitive characteristics that characterize an actor type. The MSF method, however, provides a natural language description of each actor type included in the method. Proceeding from these descriptions the administrator of the cognitive matchmaker system should be able to characterize the project-specific actor types by adding or reusing cognitive characteristics to the project-specific database.

At *runtime*, the results of the theory formulation phase included suitability matches for every actor / task combination differentiated to a specific project phase. These suitability matches may be reviewed after every project phase. The lowest certainty percentages shown in table 2 deserve special attention to discover the reasons of the lowest match results. For instance, table 2 shows that the developer acting as an experiencer has a certainty of 66% to successfully fulfill an acquisition task. When viewing table 1 it can be interpreted that the acquisition task performed by the developer in the definition phase is related with the attendance of project meetings. This may be caused in case a meeting is not very relevant for a developer. For instance, when a large part of a certain meeting is about project management issues a developer may not have a satisfied feeling after the meeting. Letting developers attend the most relevant meetings may increase the suitability matches for these acquisition tasks. In the same way, the other calculated matches can be analyzed for every project phase.

For instance, the testing tasks shown in table 1 deserve attention when comparing the actual project results with the suitability matches from a *post-mortem* viewpoint. According to table 1 the partial and integral system tests are conducted by the developer and lead developer types. Assume that it is 78.5% certain that the developer can successfully fulfill these testing tasks. The MSF method includes the tester actor type that may be more suitable to fulfill testing tasks in general. According to the MSF, a key goal for the tester is to find and report the significant bugs in the product by testing the product. Obviously, more bugs could have been found and solved after testing each iteration and the overall product by the tester actor type. In the current project situation, the developer has the responsibility for code development and testing as well. Usability issues also arose during the system test tasks. What can be seen in table 1 is that the developer is also responsible for designing the mockups. The responsibility of the developer to design, develop as well as test the system may have contributed to the existence of some usability problems. The MSF advocates the addition of a user experience architect in the project to increase the usability of the tool. According to the MSF, the user experience architect is responsible for the form and function of the user interface, its aesthetics and the overall product usability. Recall that designing mockups is a synthesis task. Assume the certainty that the developer can successfully fulfill a synthesis task in the definition phase is 88%. This is not a low percentage, but may further increase when the main focus of a developer is on developing code. For future projects it may be a smart idea to introduce a tester and a user experience architect.

5 Discussion

Literature indicates that cognitive matchmaking can be found in several areas of computer science. One of these initiatives is Cognitive Match Interface Design

(COMIND) [10]. COMIND is the designing of system processes so that they proceed and interact with the user in a manner that parallels the flow of the user's own thought processes. It consists of several principles, such as: The user should be able to express his needs to the computer with constructs which mirror the user's own thought processes. Another principle is the readiness of a computer to solve problems of the user in his / her area of need. Also, the computer should sanction flexibility just like the mind. The mind is regarded as a versatile and flexible problem solver. The authors tried to apply these principles when designing a medical information system. Unfortunately, a method for interface design that incorporates COMIND is not introduced. Only the medical information system case is elaborated. Creation of a COMIND framework including the proposed cognitive principles for user interface design would have possibly enabled reuse of COMIND in different areas. The existence of our cognitive matchmaker system framework does enable its specific application in many different areas.

Another interesting study is the cognitive matchmaking of students with e-learning system functionality [11]. A way of working is presented to design e-learning systems that better adapt to the cognitive characteristics of students. First, a taxonomy of learning styles is selected to classify the user. Next, techniques should be developed to introduce the adaptation into the system that fits the learning styles. The designed adaptation is then implemented on a computer. Finally, a selection of the technologies is made that are adequate for the adaptation. Besides this described way of working, a cognitive method or a system to match students and e-learning systems is not proposed. The mentioned concept of reflection can be very useful for our own work, though. Reflection is defined as the capability of a computational system to adjust itself to changing conditions. This can be seen on e.g. `http://maps.google.com`. The process of adaptation is made stronger since it is possible to create specific code depending on the supplied characteristics of the user when using the system. Adding reflection to our system may take situational elements into account when determining a match, for instance. Concretely, the actual availability of actors during the ART project may be included when allocating tasks to actors.

Jaspers et al. [12] argue that early involvement of cognitive matchmaking in ISE may be of importance to design systems that fully support the user's work practices. From this perspective, cognitive matchmaking is used for requirements engineering to match system requirements with the user's task behavior. To understand the task behavior of future users of a clinical system, the think aloud method has been applied [12]. Think aloud is a method that requires subjects to talk aloud while performing a task. This stimulates understanding of the supplied cognitive characteristics when performing a task. Unfortunately, the method has only been utilized to design a user interface for a clinical information system. The study lacks a more abstract framework that can be reused to design interfaces in general that better match task behavior of its users. Task-analysis methods such as the think aloud method can be useful when refining our research. For instance, these methods may be very valuable to improve the way we have characterized the abstract actor types and task types based on cognitive characteristics. Jaspers et al. [12] also included a simplified model of the human cognitive system. Studying that model may further improve the way we interpret cognitive matchmaking processes.

6 Conclusions and Future Work

This paper describes how actors and tasks can be matched based on cognitive characteristics. Therefore, the framework for cognitive matchmaking developed in earlier work is mentioned. Proceeding from this framework the prototype implementation of a cognitive matchmaker system is demonstrated. An information systems engineering project provided the breeding ground for the case study in which the system has been evaluated. The ISE project has been concerned with the development of an 'Action Reporting Tool' for an international provider of banking and insurance services. This tool is a Web application that can generate risk reports. The suitability matches of the tasks allocated to the actors in the definition project phase have been determined using the cognitive matchmaker system. It can be concluded that the system can provide support for task allocation in at least three different ways: Before project initiation (at design time), during project enactment (at runtime) and after the project has finished (post-mortem). At design time, the person that needs to allocate tasks to actors can calculate the suitability matches. Tasks can be allocated to actors before starting the project based on these results. At runtime, suitability matches can be recalculated if changes to task allocations are necessary. The system can also be utilized to evaluate task allocations after every project phase. The calculated suitability matches can be compared with actual task performance. From a post-mortem point of view, the suitability matches for every actor / task combination in the project can be compared to the actual task results. Lessons learned may help to better decide which actor types are suitable to work with which task types. In this case, the cognitive matchmaker system is related with ISE. The system may also be usable in other areas, such as: Multi-agent systems, workflow management and BPR. Future work is concentrated on improving the theory as well as the prototype and further evaluation in case studies. More efforts of how the prototype could prove the usefulness of the framework can be called for. This may include testing the prototype in real settings with real users. At this moment, it is only possible to calculate a match based on one actor type and one task type. However, there are situations imaginable that multiple actors are working together to fulfill a set of tasks. It might be interesting to determine a match based on the total amount of actors and the total amount of tasks that the actors are fulfilling as a group. Besides these additions, the future system may consider situational elements. This may include the availability of actors as well as personal preferences and goals. Finally, the notion of human knowledge can be considered to determine which aspects of human knowledge, its development and its synergy in team work can be taken into consideration in the current version of the framework and the system.

References

1. Staab, S., Studer, R., Schnurr, H., Sure, Y.: Knowledge processes and ontologies. IEEE Intelligent Systems 16(1), 26–34 (2001)
2. Weir, C., Nebeker, J., Bret, L., Campo, R., Drews, F., LeBar, B.: A cognitive task analysis of information management strategies in a computerized provider order entry environment. Journal of the American Medical Informatics Association 14(1), 65–75 (2007)
3. Kako, E.: Thematic role properties of subjects and objects. Cognition 101(1), 1–42 (2006)
4. Shakshuki, E., Prabhu, O., Tomek, I.: FCVW agent framework. Information and Software Technology 48(6), 385–392 (2006)
5. van der Aalst, W., ter Hofstede, A.: Verification of workflow task structures: A Petri-net-based approach. Information Systems 25(1), 43–69 (2000)
6. R-Moreno, M., Borrajo, D., Cesta, A., Oddi, A.: Integrating planning and scheduling in workflow domains. Expert Systems with Applications 33(2), 389–406 (2007)
7. Overbeek, S., van Bommel, P., Proper, H., Rijsenbrij, D.: Matching cognitive characteristics of actors and tasks. In: Meersman, R., Tari, Z. (eds.) OTM 2007, Part I. LNCS, vol. 4803, pp. 371–380. Springer, Heidelberg (2007)
8. Overbeek, S., van Bommel, P., Proper, H., Rijsenbrij, D.: Characterizing knowledge intensive tasks indicating cognitive requirements – Scenarios in methods for specific tasks. In: Ralyté, J., Brinkkemper, S., Henderson-Sellers, B. (eds.) Proceedings of the IFIP TC8 / WG8.1 Working Conference on Situational Method Engineering: Fundamentals and Experiences., Geneva, Switzerland, vol. 244, pp. 100–114. Springer, Boston, USA (2007)
9. Sol, H.: Simulation in Information Systems. PhD thesis, University of Groningen, The Netherlands, EU (1982)
10. Coll, R., Coll, J.: Cognitive match interface design, a base concept for guiding the development of user friendly computer application packages. Journal of Medical Systems 13(4), 227–235 (1989)
11. Ruiz, M., Díaz, M., Soler, F., Pérez, J.: Adaptation in current e-learning systems. Computer Standards & Interfaces 30(1–2), 62–70 (2008)
12. Jaspers, M., Steen, T., van den Bos, C., Geenen, M.: The think aloud method: A guide to user interface design. International Journal of Medical Informatics 73(11–12), 781–795 (2004)

On Modeling and Analyzing Cost Factors in Information Systems Engineering

Bela Mutschler[1,2] and Manfred Reichert[2,3]

[1] Daimler AG, Group Research, P.O. Box 2360, 89013 Ulm, Germany
bela.mutschler@daimler.com
[2] Information Systems Group, University of Twente, The Netherlands
m.u.reichert@utwente.nl
[3] Institute of Databases and Information Systems, University of Ulm, Germany
manfred.reichert@uni-ulm.de

Abstract. Introducing *enterprise information systems* (EIS) is usually associated with high costs. It is therefore crucial to understand those factors that determine or influence these costs. Though software cost estimation has received considerable attention during the last decades, it is difficult to apply existing approaches to EIS. This difficulty particularly stems from the inability of these methods to deal with the dynamic interactions of the many technological, organizational and project-driven cost factors which specifically arise in the context of EIS. Picking up this problem, we introduce the EcoPOST framework to investigate the complex cost structures of EIS engineering projects through qualitative cost evaluation models. This paper extends previously described concepts and introduces design rules and guidelines for cost evaluation models in order to enhance the development of meaningful and useful EcoPOST cost evaluation models. A case study illustrates the benefits of our approach. Most important, our EcoPOST framework is an important tool supporting EIS engineers in gaining a better understanding of the critical factors determining the costs of EIS engineering projects.

Keywords: Information Systems Engineering, Cost Analysis, Evaluation Models, Simulation.

1 Introduction

While the benefits of *enterprise information systems* (EIS) are usually justified by improved process performance [1], there exist no approaches for systematically analyzing related cost factors and their dependencies. Though software cost estimation has received considerable attention during the last decades [2] and has become an essential task in information systems engineering, it is difficult to apply existing approaches to EIS, particularly if the considered EIS shall provide active business process support. This difficulty stems from the inability of these approaches to cope with the numerous technological, organizational and project-driven cost factors which have to be considered in the context of process-aware EIS (and which do only partly exist in data- or function-centered information systems) [3]. As an example consider the significant costs for redesigning business processes. Another challenge deals with the many dependencies existing between different cost factors. Activities for *business process*

Z. Bellahsène and M. Léonard (Eds.): CAiSE 2008, LNCS 5074, pp. 510–524, 2008.

redesign, for example, can be influenced by intangible impact factors like available *process knowledge* or *end user fears*. These dependencies, in turn, result in dynamic effects which influence the overall costs of EIS engineering projects. Existing evaluation techniques [4] are typically unable to deal with such dynamic effects as they rely on too static models based upon snapshots of the considered software system.

What is needed is an approach that enables project managers and EIS engineers to model and investigate the complex interplay between the many cost and impact factors that arise in the context of EIS. This paper presents the EcoPOST methodology, a sophisticated and practically validated, model-based methodology to better understand and systematically investigate the complex cost structures of EIS engineering projects. Specifically, this paper extends our previous work on EcoPOST [5] by introducing model design rules and modeling guidelines, which enhance the development of meaningful and useful evaluation models.

Section 2 summarizes the EcoPOST methodology. Section 3 introduces rules for designing evaluation models and Section 4 describes modeling guidelines. Section 5 summarizes results from one of our case studies in order to illustrate the benefits of the EcoPOST approach. It further discusses issues related to validation research from a more general perspective. Section 6 concludes with a summary.

2 The EcoPOST Cost Analysis Methodology

Our EcoPOST methodology was designed to ease the realization of process-aware EIS in the automotive industry (and was, consequently, also validated and piloted in several EIS engineering projects in this domain). The EcoPOST methodology comprises seven steps (cf. Fig. 1). *Step 1* concerns the comprehension of an evaluation scenario. This is crucial for developing problem-specific evaluation models. The following two steps (Steps 2 and 3) deal with the identification of two different kinds of *Cost Factors* representing costs that can be quantified in terms of money (cf. Table 1): *Static Cost Factors* (SCFs) and *Dynamic Cost Factors* (DCFs).

Step 4 deals with the identification of *Impact Factors* (ImFs), i.e., intangible factors that influence DCFs and other ImFs. We distinguish between organizational, project-specific, and technological ImFs. ImFs cause the value of DCFs (and other ImFs) to

Table 1. Cost Factors

	Description
SCF	*Static Cost Factors* (SCFs) represent costs whose values do not change during an EIS engineering project (except for their time value, which is not further considered in the following). Typical examples: software license costs, hardware costs and costs for external consultants.
DCF	*Dynamic Cost Factors* (DCFs), in turn, represent costs that are determined by activities related to an EIS engineering project. The (re)design of business processes prior to the introduction of EIS, for example, constitutes such an activity. As another example consider the performance of interview-based process analysis. These activities cause measurable efforts which, in turn, vary due to the influence of intangible *impact factors*. The DCF "Costs for Business Process Redesign" may be influenced, for instance, by an intangible factor "Willingness of Staff Members to Support Process (Re)Design Activities". Obviously, if staff members do not contribute to a (re)design project by providing needed information (e.g., about process details), any redesign effort will be ineffective and result in increasing (re)design costs. If staff willingness is additionally varying during the (re)design activity (e.g., due to a changing communication policy), the DCF will be subject to even more complex effects. In the EcoPOST framework, intangible factors like the one described are represented by *impact factors*.

change, making their evaluation a difficult task to accomplish. As examples consider factors such as "End User Fears", "Availability of Process Knowledge", or "Ability to (re)design Business Processes". Also, ImFs can be static or dynamic (cf. Table 2).

Table 2. Impact Factors

	Description
Static ImF	Static ImFs do not change, i.e., they are assumed to be constant during an EIS engineering project; e.g., when there is a fixed degree of user fears, process complexity, or work profile change.
Dynamic ImF	Dynamic ImFs may change during an EIS engineering project, e.g., due to interference with other ImFs. As examples consider process and domain knowledge which is typically varying during an EIS engineering project (or a subsidiary activity).

It is important to mention that – unlike SCFs and DCFs – the values of ImFs are not quantified in monetary terms. Instead, they are "quantified" by experts[1] using qualitative scales describing the degree of an ImF. As known from software cost estimation models, such as Boehm's COCOMO [2], the qualitative scales we use comprise different "values" (typically ranging from "very low" to "very high"). These values are used to express the strength of an ImF on a given cost factor (just like in COCOMO).

Generally, dynamic evaluation factors (i.e., DCFs and dynamic ImFs) are difficult to comprehend. In particular, intangible ImFs (i.e., their appearance and impact in EIS engineering projects) are not easy to follow. When evaluating the costs of EIS engineering projects, therefore, DCFs and dynamic ImFs constitute a major source of misinterpretation and ambiguity. To better understand and to investigate the dynamic behavior of DCFs and dynamic ImFs, we introduce the notion of *evaluation models* as basic pillar of the EcoPOST methodology (*Step 5*; cf. Section 2.2). These evaluation models can be simulated (*Step 6*) to gain insights into the dynamic behavior (i.e., evolution) of DCFs and dynamic ImFs (*Step 7*). This is important to effectively control the design and implementation of EIS as well as the costs of respective projects.

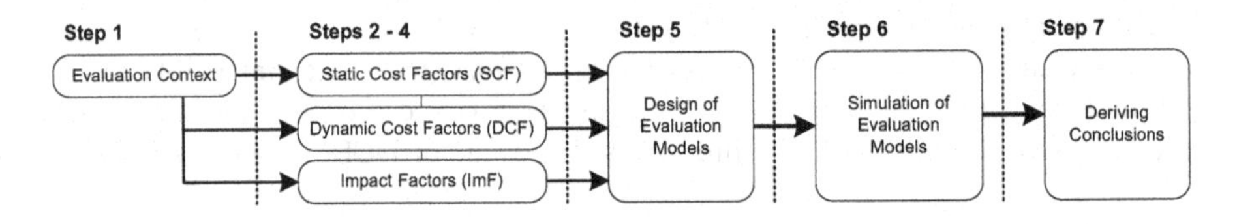

Fig. 1. Main Steps of the EcoPOST Methodology

2.1 Evaluation Models

In EcoPOST, dynamic cost/impact factors are captured and analyzed by evaluation models which are specified using the System Dynamics [6] notation (cf. Fig. 2). An evaluation model comprises SCFs, DCFs, and ImFs corresponding to model variables. Different types of variables exist. *State variables* can be used to represent dynamic factors, i.e., to capture changing values of DCFs (e.g., the "Business Process Redesign

[1] The efforts of these experts for making that quantification is not explicitly taken into account in EcoPOST, though this effort also increases information system development costs.

Costs"; cf. Fig. 2A) and dynamic ImFs (e.g., "Process Knowledge"). A state variable is graphically denoted as rectangle (cf. Fig. 2A), and its value at time t is determined by the accumulated changes of this variable from starting point t_0 to present moment t ($t > t_0$); similar to a bathtub which accumulates – at a defined moment t – the amount of water poured into it in the past. Typically, state variables are connected to at least one *source* or *sink* which are graphically denoted as cloud-like symbols (except for state variables connected to other ones) (cf. Fig. 2A). Values of state variables change through inflows and outflows. Graphically, both flow types are depicted by twin-arrows which either point to (in the case of an inflow) or out of (in the case of an outflow) the state variable (cf. Fig. 2A). Picking up again the bathtub image, an *inflow* is a pipe that adds water to the bathtub, i.e., inflows increase the value of state variables. An *outflow*, by contrast, is a pipe that purges water from the bathtub, i.e., outflows decrease the value of state variables. The DCF "Business Process Redesign Costs" shown in Fig. 2A, for example, increases through its inflow ("Cost Increase") and decreases through its outflow ("Cost Decrease"). Returning to the bathtub image, we further need "water taps" to control the amount of water flowing into the bathtub, and "drains" to specify the amount of water flowing out. For this purpose, a *rate variable* is assigned to each flow (graphically depicted by a valve; cf. Fig. 2A). In particular, a rate variable controls the inflow/outflow it is assigned to based on those SCFs, DCFs, and ImFs which influence it. It can be considered as an interface which is able to merge SCFs, DCFs, and ImFs.

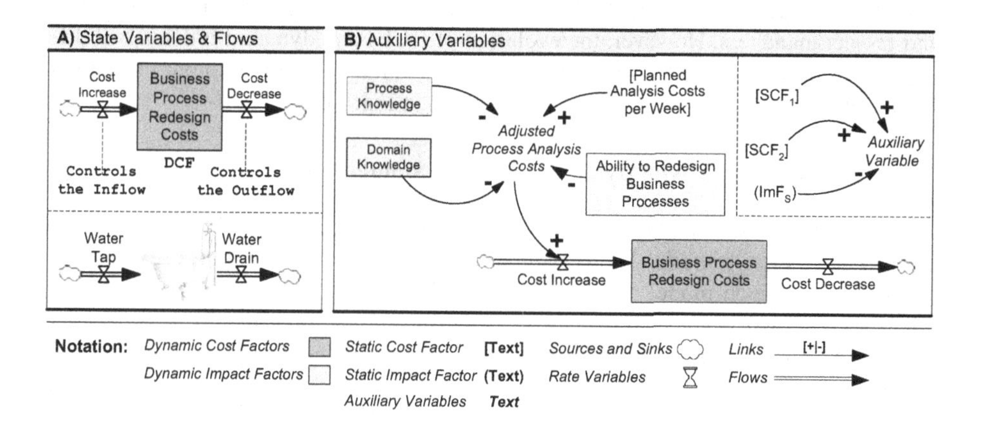

Fig. 2. Evaluation Model Notation and initial Examples

Besides state variables, evaluation models may comprise *constants* and *auxiliary variables*. Constants are used to represent static evaluation factors, i.e., SCFs and static ImFs. Auxiliary variables, in turn, represent intermediate variables and typically bring together – like rate variables – cost and impact factors, i.e., they merge SCFs, DCFs, and ImFs. As an example consider the auxiliary variable "Adjusted Process Analysis Costs" in Fig. 2B, which merges the three dynamic ImFs "Process Knowledge", "Domain Knowledge", and "Ability to Redesign Business Processes" and the SCF "Planned Analysis Costs per Week". Both constants and auxiliary variables are integrated into an evaluation model with *links* (not flows), i.e., labeled arrows. A *positive link* (labeled with "+") between x and y (with y as dependent variable) indicates that y will tend in the same direction if a change occurs in x. A *negative link* (labeled with "-") expresses that the dependent variable y will tend in the opposite direction if the value of x changes. Altogether, we define:

Definition 2.1 (Evaluation Model). *A graph EM = (V, F, L) is denotes as evaluation model, if the following holds:*

- $V := S \dot{\cup} X \dot{\cup} R \dot{\cup} C \dot{\cup} A$ *is a set of model variables with*
  - *S is a set of state variables,*
  - *X is a set of sources and sinks,*
  - *R is a set of rate variables,*
  - *C is a set of constants,*
  - *A is a set of auxiliary variables,*
- $F \subseteq ((S \times S) \cup (S \times X) \cup (X \times S))$ *is a set of edges representing flows,*
- $L \subseteq ((S \times A \times Lab) \cup (S \times R \times Lab) \cup (A \times A \times Lab) \cup (A \times R \times Lab) \cup (C \times A \times Lab) \cup (C \times R \times Lab))$ *is a set of edges representing links with* $Lab := \{+, -\}$ *being the set of link labels:*
  - $(q_i, q_j, +) \in L$ *with* $q_i \in (S \dot{\cup} A \dot{\cup} C)$ *and* $q_j \in (A \dot{\cup} R)$ *denotes a positive link,*
  - $(q_i, q_j, -) \in L$ *with* $q_i \in (S \dot{\cup} A \dot{\cup} C)$ *and* $q_j \in (A \dot{\cup} R)$ *denotes a negative link.*

The EcoPOST evaluation models presented so far are already useful for EIS engineers and project managers. However, the evolution of DCFs and dynamic ImFs is still difficult to comprehend. Thus, we have added a simulation component to our evaluation framework for analyzing this evolution (cf. Step 6 in Fig. 1).

2.2 Understanding Model Dynamics through Simulation

To enable simulation of an evaluation model we need to formally specify its behavior by means of a *simulation model*. We use *mathematical equations* for this purpose. Thereby, the behavior of each model variable is specified by one equation (cf. Fig. 3), which describes how a variable is changing over time from simulation start.

Fig. 4A shows a simple evaluation model.[2] Assume that the evolution of the DCF "Business Process Redesign Costs" (triggered by dynamic ImF "End User Fears") shall be analyzed. End user fears can lead to emotional resistance of users and, in turn, to a lack of user support when redesigning business processes (e.g., during an interview-based process analysis). For model variables, which represent an SCF or static ImF, the equation specifies a constant value for the model variable; i.e., SCFs and static ImFs are specified by single numerical values in *constant equations*. As example consider EQUATION A in Fig. 4B. For model variables representing DCFs, dynamic ImFs, or rate/auxiliary variables, the corresponding equation describes how the value of the model variable evolves over time (i.e., during simulation). Thereby, the evolution of DCFs and dynamic ImFs is characterized by *integral equations* [7]. This allows us to

[2] It is the basic goal of this toy example to illustrate simulation of evaluation models. Generally, evaluation models are much more complex. Due to lack of space we do not provide a more extensive example.

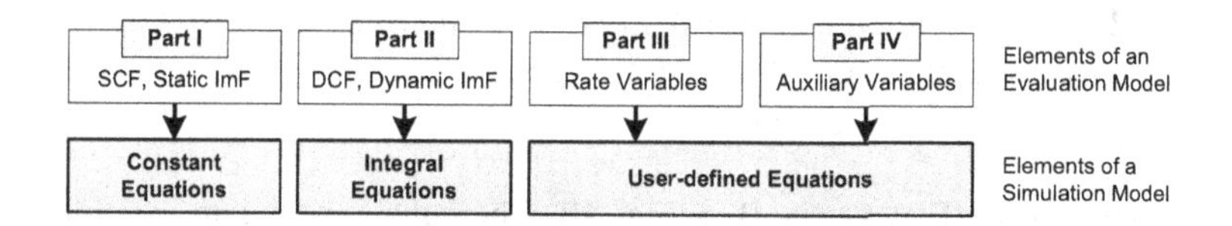

Fig. 3. Elements of a Simulation Model

capture the accumulation of DCFs and dynamic ImFs from the start of a simulation run (t_0) to its end (t):

Definition 2.2 (Integral Equation). *Let EM be an evaluation model (cf. Definition 2.1) and S be the set of all DCFs and dynamic ImFs defined by EM. An integral equation for a dynamic factor $v \in S$ is defined as follows:*

$$v(t) = \int_{t_0}^{t} [inflow(s) - outflow(s)]ds + v(t_0) \text{ where}$$

- *t_0 denotes the starting time of the simulation run,*
- *t represents the end time of the simulation run,*
- *$v(t_0)$ represents the value of v at t_0,*
- *$inflow(s)$ represents the value of the inflow at any time s between t_0 and t,*
- *$outflow(s)$ represents the value of the outflow at any time s between t_0 and t.*

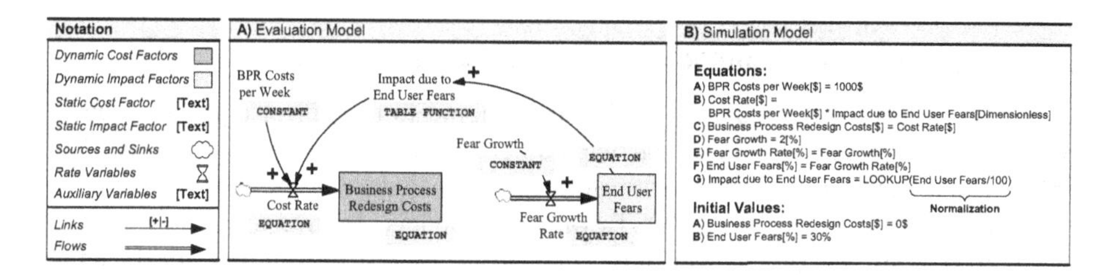

Fig. 4. Dealing with the Impact of End User Fears

As example consider EQUATION C in Fig. 4B which specifies the increase of the DCF "Business Process Redesign Costs" (based on only one inflow). Note that in Fig. 4B the equations for the DCF "Business Process Redesign Costs" and the dynamic ImF "End User Fears" are presented in the way they are specified in Vensim [8], the tool we use in EcoPOST, and not as real integral equations.

Rate and auxiliary variables are both specified in the same way, i.e., as user-defined functions defined over the variables preceding them in the evaluation model. In other words, rate as well as auxiliary variables are used to merge static and dynamic cost/impact factors. During simulation, values of rate and auxiliary variables are dynamic, i.e., they change along the course of time. Reason is that they are not only influenced by SCFs and static ImFs, but also by evolving DCFs and dynamic ImFs. The behavior of rate and auxiliary variables is specified in the same way:

Definition 2.3 (User-defined Equation). *Let EM be an evaluation model (cf. Def. 2.1) and X be the set of rate/auxiliary variables defined by EM. An equation for $v \in X$ is a user-defined function $f(v_1, ..., v_n)$ with $v_1, ..., v_n$ being the predecessors of v in EM.*

As example consider EQUATION B in Fig. 4B. The equation for rate variable "Cost Rate" merges the SCF "BPR Costs per Week" with the auxiliary variable "Impact due to End User Fears". Assuming that activities for business process redesign are scheduled for 32 weeks, Fig. 5A shows the values of all dynamic evaluation factors of the evaluation model over time when performing simulation. Fig. 5B shows the outcome of the simulation. As can be seen there is a significant negative impact of end user fears on the costs of business process redesign.

A) Computing a Simulation Run

TIME	Change ($)	BPR Costs ($)	Cost Rate ($)	Change (%)	User Fears (%)
00	-	0	1000	-	30
01	1000	1000	1010	2	32
02	1010	2010	1020	2	34
03	1020	3030	1030	2	36
04	1030	4060	1040	2	38
05	1040	5100	1050	2	40
06	1050	6150	1060	2	42
...	...	...	...	...	...
30	1840	38300	1900	2	90
31	1900	40200	2020	2	92
32	2020	42220	2140	2	94

B) Graphical Diagramm illustrating Simulation Outcome

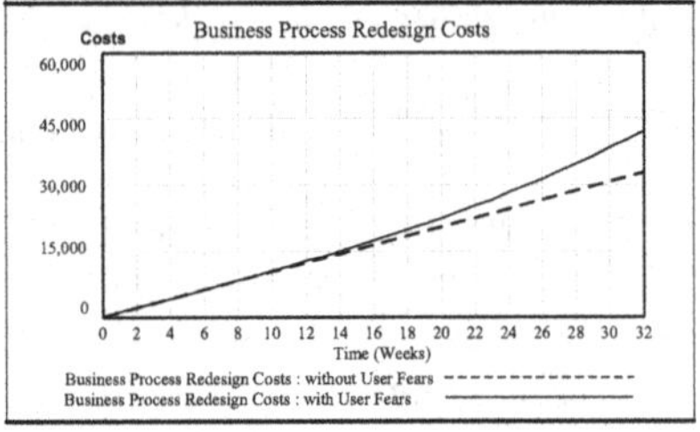

Costs
Business Process Redesign Costs
60,000
45,000
30,000
15,000
0
0 2 4 6 8 10 12 14 16 18 20 22 24 26 28 30 32
Time (Weeks)
Business Process Redesign Costs : without User Fears
Business Process Redesign Costs : with User Fears

Fig. 5. Dealing with the Impact of End User Fears

2.3 Sensitivity Analysis and Reuse of Evaluation Information

Generally, results of a simulation enable EIS engineers to gain insights into causal dependencies between organizational, technological, and project-specific factors. This helps them to better understand resulting effects and to develop a concrete "feeling" for the dynamic implications of EcoPOST evaluation models. To investigate how a given evaluation model "works" and what might change its behavior, we simulate the dynamic implications described by it – a task which is typically too complex for the human mind. In particular, we conduct "behavioral experiments" based on series of simulation runs. During these simulation runs selected simulation parameters are manipulated in a controlled manner to systematically investigate the effects of these manipulations, i.e., to investigate how the output of a simulation will vary if its initial condition is changed. This procedure is also known as *sensitivity analysis*. Simulation outcomes can be further analyzed using graphical charts.

Designing evaluation models can be a complicated and time-consuming task. Evaluation models can become complex due to the high number of potential cost and impact factors as well as the many causal dependencies that exist between them. Taking the approach described so far (cf. Section 2), each evaluation and simulation model has to be designed from scratch. Besides the additional efforts, this results in an exlusion of existing modeling experience, and prevents the reuse of both evaluation and simulation models. In response to this problem, in [5,9] we have introduced a set of reusable *evaluation patterns* (EP). EPs do not only ease the design and simulation of evaluation models, but also enable the reuse of evaluation information. This is crucial to foster the practical applicability of the EcoPOST framework.

3 Model Design Rules

Overall benefit of EcoPOST evaluation models depends on their quality. The latter, in turn, is determined by the syntactical as well as the semantical correctness of the evaluation model. Maintaining correctness of an evaluation model, however, can be a difficult task to accomplish. This section picks up this problem.

3.1 Modeling Constraints for Evaluation Models

Rules for the correct use of flows and links are shown in Fig. 6A and Fig. 6B. By contrast, Fig. 7A – Fig. 7F show examples of *incorrect* models.

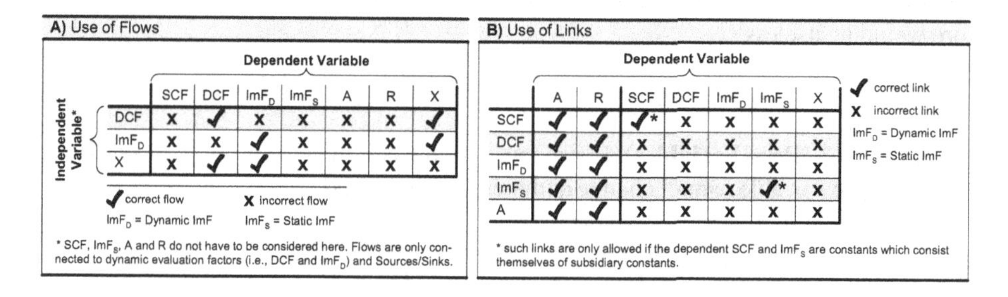

A) Use of Flows

Independent Variable* / Dependent Variable

	SCF	DCF	ImF_D	ImF_S	A	R	X
DCF	X	✓	X	X	X	X	✓
ImF_D	X	X	✓	X	X	X	✓
X	X	✓	✓	X	X	X	X

✓ correct flow X incorrect flow
ImF_D = Dynamic ImF ImF_S = Static ImF

* SCF, ImF_S, A and R do not have to be considered here. Flows are only connected to dynamic evaluation factors (i.e., DCF and ImF_D) and Sources/Sinks.

B) Use of Links

Dependent Variable

	A	R	SCF	DCF	ImF_D	ImF_S	X
SCF	✓	✓	✓*	X	X	X	X
DCF	✓	✓	X	X	X	X	X
ImF_D	✓	✓	X	X	X	X	X
ImF_S	✓	✓	X	X	X	✓*	X
A	✓	✓	X	X	X	X	X

✓ correct link
X incorrect link
ImF_D = Dynamic ImF
ImF_S = Static ImF

* such links are only allowed if the dependent SCF and ImF_S are constants which consist themselves of subsidiary constants.

Fig. 6. Using Flows and Links in our Evaluation Models

Dynamic evaluation factors, for example, may be only influenced by flows and not by links as shown in Fig. 7A. Likewise, flows must be not connected to auxiliary variables or constants (cf. Fig. 7B). Links pointing from DCFs (or auxiliary variables) to SCFs or static ImFs (cf. Fig. 7C and Fig. 7D) are also not valid as SCFs as well as static ImFs have constant values which cannot be influenced. Finally, flows and links connecting DCFs with dynamic ImFs (and vice versa) are also not considered as correct (cf. Fig. 7E and Fig. 7F).

Several other constraints have to be taken into account as well when designing evaluation models. In the following let EM = (V, F, L) be an evaluation model (cf. Definition 2.1). Then:

Design Rule 1 **(Binary Relations).** Every model variable must be used in at least one binary relation. Otherwise, it is not part of the analyzed evaluation context and can be omitted:

$$\forall v \in (S \dot{\cup} X) : \exists q \in (S \dot{\cup} X) \wedge ((v,q) \in F \vee (v,q) \in F) \quad (1)$$

$$\forall v \in (A \dot{\cup} C) : \exists q \in (A \dot{\cup} R) \wedge \exists \, (q, v, [+|-]) \in L \quad (2)$$

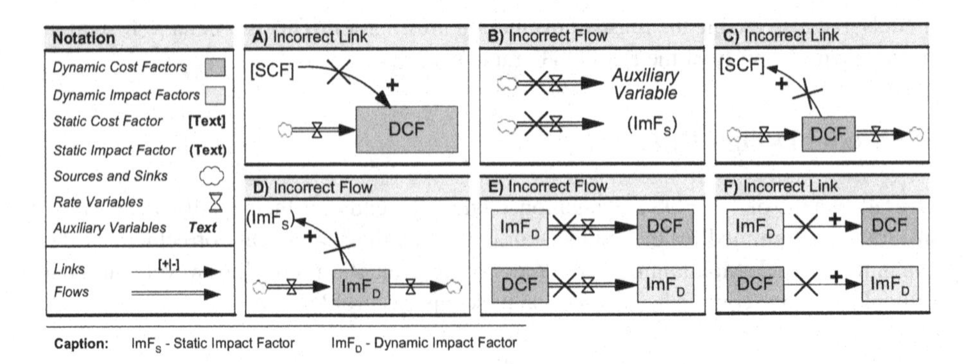

Fig. 7. Examples of Incorrect Modeling

Design Rule 2 **(Sources and Sinks).** Every state variable must be connected to at least one source, sink or other state variable. Otherwise it cannot change its value and therefore would be useless:

$$\forall v \in S : \exists q \in (S \dot{\cup} X) \wedge ((v,q) \in F \vee (q,v) \in F) \tag{3}$$

Design Rule 3 **(Rate Variables).** Every rate variable is influenced by at least one link; otherwise the variable cannot change and therefore is useless:

$$\forall v \in R : \exists q \in (S \dot{\cup} A \dot{\cup} C) \wedge \exists \, (q,v,[+|-]) \in L \tag{4}$$

Design Rule 4 **(Feedback Loops).** There are no cycles consisting only of auxiliary variables, i.e., cyclic feedback loops must at least contain one state variable (cycles of auxiliary variables cannot be evaluated if an evaluation model is simulated):

$$\neg\exists < q_0, q_1, ..., q_r > \in A^{r+1} \; with \; (q_i, q_{i+1}, [+|-]) \in L \; for$$
$$i = 0, \ldots, r-1 \wedge q_0 = q_r \wedge \; q_k \neq q_l \; for \;\; k,l = 1, \ldots, r; k \neq l \tag{5}$$

Design Rule 5 **(Auxiliary Variables).** An auxiliary variable has to be influenced by at least two other static or dynamic evaluation factors or auxiliary variables (except for auxiliary variables used to represent table functions [9]):

$$\forall v \in A : \exists p,q \in (A \dot{\cup} S \dot{\cup} C) \wedge ((q,v,[+|-]) \in L \wedge (p,v,[+|-]) \in L) \tag{6}$$

These modeling constraints provide basic rules for EcoPOST users to construct syntactically correct evaluation models.

3.2 Semantical Correctness of Evaluation Models

While syntactical model correctness can be ensured, this is not always possible for the semantical correctness of evaluation models. Yet, we can provide additional model design rules increasing the meaningfulness of our evaluation models.

Design Rule 6 **(Transitive Dependencies).** Transitive link dependencies (i.e., indirect effects described by chains of links) are restricted. As example consider Fig. 8. Fig. 8A reflects the assumption that increasing end user fears result in increasing emotional resistance. This, in turn, leads to increasing business process costs. Consequently, the modeled transitive dependency between "End User Fears" and "Business Process Redesign Costs" is not correct, as increasing end user fears do not result in decreasing business process (re)design costs. The correct transitive dependency is shown in Fig. 8B. Fig. 8C illustrates the assumption that increasing process knowledge results in an increasing ability to (re)design business processes. An increasing ability to (re)design business processes, in turn, leads to decreasing process definition costs. The modeled transitive dependency between "Process Knowledge" and "Process Definition Costs", however, is not correct, as increasing process knowledge does not result in increasing process definition costs (assuming that the first 2 links are correct). The correct transitive dependency is shown in Fig. 8D.

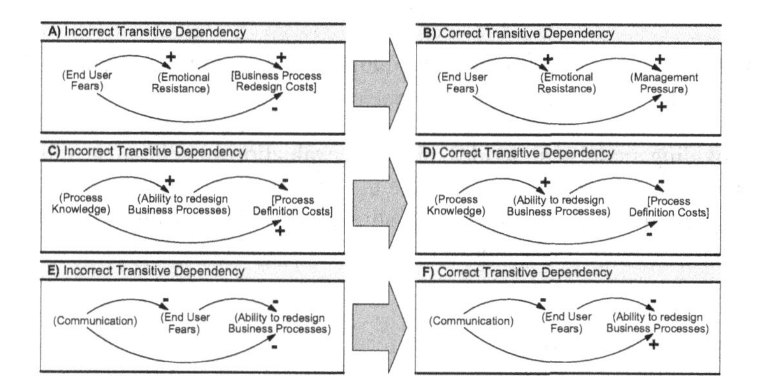

Fig. 8. Transitive Dependencies (Simplified Evaluation Models)

Finally, Fig. 8E deals with the impact of communication (e.g., the goals of an EIS project) on the ability to redesign business processes. Yet, the transitive dependency shown in Fig. 8E is not correct. The correct one is shown in Fig. 8F.

Altogether, two causal relations ("+" and "-") are used in the context of our evaluation models. Correct transitive dependencies can be described based on a *multiplication operator*. More precisely, transitive dependencies have to comply with the following three *multiplication laws* for transitive dependencies (for any $x, y \in \{+,-\}$):

$$+ * y = y \tag{7}$$

$$- * - = + \tag{8}$$

$$x * y = y * x \tag{9}$$

The evaluation models shown in Fig. 8A and Fig. 8C violate Law 1, whereas the model shown in Fig. 8E violates the second one. Law 3 states that the "*" is commutative.

Design Rule 7 **(Dual Links I).** A constant cannot be connected to the same auxiliary variable with both a positive and negative link:

$$\forall v \in C, \forall q \in A : \neg \exists\, l_1, l_2 \in L\, with l_1 = (v,q,-) \wedge l_2 = (v,q,+) \quad (10)$$

Design Rule 8 **(Dual Links II).** A state variable cannot be connected to the same auxiliary variable with both a positive and negative link:

$$\forall v \in S, \forall q \in A : \neg \exists\, l_1, l_2 \in L\, with l_1 = (v,q,-) \wedge l_2 = (v,q,+) \quad (11)$$

Design Rule 9 **(Dual Links III).** An auxiliary variable cannot be connected to another auxiliary variable with both a positive and negative link:

$$\forall v \in A, \forall q \in A : \neg \exists\, l_1, l_2 \in L\, with l_1 = (v,q,-) \wedge l_2 = (v,q,+) \quad (12)$$

Finally, there exist two additional simple constraints:

Design Rule 10 **(Representing Cost Factors).** A cost factor cannot be represented both as SCF and DCF in one evaluation model.

Design Rule 11 **(Representing Impact Factors).** An impact factor cannot be represented both as static and dynamic ImF in one evaluation model.

Without providing model design rules, incorrect evaluation models can be quickly modeled. This, in turn, does not only aggravate the derivation of plausible evaluations, but also hampers the use of the modeling and simulation tools [5] which have been developed as part of the EcoPOST framework.

4 Modeling Guidelines

To further facilitate the use of our methodology, governing guidelines and best practices are provided. This section summarizes two categories of EcoPOST governing guidelines: (1) guidelines for evaluation models and (2) guidelines for simulation models.

In general, EcoPOST evaluation models can become large, e.g., due to a potentially high number of evaluation factors to be considered or due to the large number of causal dependencies existing between them. To cope with this complexity, we introduce guidelines for designing evaluation models (cf. Table 3). Their derivation is based on experiences we gathered during the development of our approach, its initial use in practice, and our study of general System Dynamics (SD) guidelines [10]. As example consider guideline EM-1 from Table 3. The distinction between SCFs and DCFs is a fundamental principle in the EcoPOST framework. Yet, it can be difficult for the user to decide whether a cost factor shall be considered as static or dynamic. As example take an evaluation scenario which deals with the introduction of a new EIS "CreditLoan" to support the granting of loans at a bank. Based on the new EIS, the entire loan offer process shall be supported. For this purpose, the EIS has to leverage internal (i.e., within the bank) and external (e.g., a dealer) trading partners as well as other legacy applications for customer information and credit ratings. Among other things, this necessitates the integration of existing legacy applications. In case this integration is done by external suppliers, resulting costs can be represented as SCFs as they can be clearly quantified

Table 3. Guidelines for Designing Evaluation Models

GL	Description
EM-1	Carefully distinguish between SCFs and DCFs.
EM-2	If it is unclear how to represent a given cost factor represent it as SCF.
EM-3	Name feedback loops.
EM-4	Use meaningful names (in a consistent notation) for cost and impact factors.
EM-5	Ensure that all causal links in an evaluation model have unambiguous polarities.
EM-6	Choose an appropriate level of detail when designing evaluation models.
EM-7	Do not put all feedback loops into one large evaluation model.
EM-8	Focus on interaction rather than on isolated events when designing evaluation models.
EM-9	An evaluation model does not contain feedback loops comprising only auxiliary variables.
EM-10	Perform empirical and experimental research to generate needed data.

based on a contract or service agreement. If integration is done in-house, however, integration costs should be represented as DCFs as costs might be influenced by additional ImFs in this case. Other guidelines are depicted in Table 3.

To simulate EcoPOST evaluation models constitutes another complex task. The guidelines from Table 4 are useful to deal with it. Guideline SM-7, for example, claims to assess the usefulness of an evaluation model and related simulation results always in comparison with mental or descriptive models needed or used otherwise. In our experience, there often exists controversy on the question whether an evaluation model meets reality. However, such controversies miss the first purpose of a model, namely to provide insights that can be easily communicated.

Table 4. Guidelines for Developing Simulation Models

GL	Description
SM-1	Ensure that all equations of a simulation model are dimensionally consistent.
SM-2	Do not use embedded constants in equations.
SM-3	Choose appropriately small time steps for simulation.
SM-4	All dynamic evaluation factors in a simulation model must have initial values.
SM-5	Use appropriate initial values for model variables.
SM-6	Initial values for rate variables need not be given.
SM-7	The validity of evaluation models and simulation outcomes is a relative matter.

The sketched governing guidelines and best practices represent a basic set of clues and recommendations for users of the EcoPOST framework. They support the modeler in designing evaluation models, in building related simulation models, and in handling dynamic evaluation factors. Yet, it is important to mention that the consideration of these guidelines does not automatically result in better evaluation and simulation models or in the derivation of more meaningful evaluation results. Notwithstanding, taking the guidelines increases the probability of developing meaningful models.

5 Case Study

In previous work, we already showed how experimental [?] and empirical research [?,13,14,15] contributes to the derivation of good quality evaluation models. This section summarizes results from an additonal case study, this time focusing on the overall applicability of the EcoPOST framework. We also discuss model validation from a more general viewpoint. Due to space limitations we cannot decsribe the complete case study in detail (for details see [9]).

5.1 Research Design

We apply the EcoPOST framework to a complex EIS engineering project from the automotive domain in which we participated. We investigate cost overruns observed during the introduction of a large information system for supporting the development of *electrical and electronic (E/E) systems* (e.g., a multimedia unit in the car). Based on real project data, interviews with project members (e.g., requirements engineers, software architects, software developers), online surveys among end users, and practical experiences gathered in the respective EIS engineering project, we develop a set of EcoPOST evaluation models and analyze these models using simulation.

An initial business case for the considered EIS engineering project is developed prior to project start in order to convince senior management to fund the project. This business case is based on data about similar projects provided by competitors (evaluation by analogy) as well as on rough estimates on planned costs and assumed benefits of the project. The business case comprises six main cost categories: (1) *project management*, (2) *process management*, (3) *IT system realization*, (4) *specification and test*, (5) *roll-out and migration*, and (6) *implementation of interfaces*.

In a first project review (i.e., measurement of results), it turns out that originally planned project costs are not realistic, i.e., cost overruns are observed – particularly concerning cost categories (2) and (3). In our case study, we analyze cost overruns in three cost categories in detail using the EcoPOST methodology (cf. Table 5).

To be able to build evaluation and simulation models for the three analyzed cost categories, we need to collect data. This data is based on four information sources (cf. Table 6), which allow us to identify relevant cost and impact factors, i.e., evaluation factors that need to be included in the evaluation models to be developed. Likewise, the information sources also enable us to spot important causal dependencies between cost and impact factors and to derive evaluation models.

Table 5. Analyzed Cost Categories

	Description
1	***Process Management Costs***: This category deals with costs related to the (re)design of the business processes to be supported. This includes both the definition of new and the redesign of existing processes. As example of a process to be newly designed consider an E/E data provision process to obtain needed product data. As example of an existing process to be redesigned consider the basic E/E release management process. Among other things, process management costs include costs for performing interview-based process analysis and costs for developing process models.
2	***IT System Realization Costs***: This category deals with costs for implementing the new EIS on top of process management technology. In our case study, we focus on the analysis of costs related to the use of the process management system, e.g., costs for specifying and implementing the business functions and workflows to be supported as well as costs for identifying potential user roles and implementing respective access control mechanisms.
3	***III. Online Surveys among End Users***: We conduct two online surveys among two user groups of the new EIS (altogether 80 survey participants). The questionnaires are distributed via a web-based delivery platform. They slightly vary in order to cope with the different work profiles of both user groups. Goal of the survey is to confirm the significance of selected ImFs like "End User Fears" and "Emotional Resistance of End Users".
IV	***Specification and Test Costs***: This cost category sums up costs for specifying the functionality of the EIS as well as costs for testing the coverage of requirements. This includes costs for eliciting and documenting requirements as well as costs for performing tests on whether requirements are met by the EIS.

Table 6. Data Collection

	Description
I	***Project Data***: A first data source is available project data; e.g., estimates about planned costs from the initial business case. Note that we did not participate in the generation of this business case.
II	***Interviews***: We interview 10 project members (2 software architects, 4 software developers, 2 usability engineers, and 2 consultants participating in the project). Our interviews are based on a predefined, semi-structured protocol. Each interview lasts about 1 hour and is accomplished on a one-to-one basis. Goal of the interviews is to collect data about causal dependencies between cost and impact factors in each analyzed cost category.
III	***Online Surveys among End Users***: We conduct two online surveys among two user groups of the new EIS system (altogether 80 survey participants). The questionnaires are distributed via a web-based delivery platform. They slightly vary in order to cope with the different work profiles of both user groups. Goal of the survey is to confirm the significance of selected ImFs like "End User Fears" and "Emotional Resistance of End Users".
IV	***Practical Experiences***: Finally, our evaluation and simulation models also build upon practical experiences we gathered when participating in the investigated EIS engineering project. We have worked in this project as requirement engineers for more than one year and have gained deep insights during this time. Besides the conducted interviews, these experiences are the major source of information when designing our evaluation models.

5.2 Lessons Learned

Based on the derived evaluation models and simulation outcomes, we have been able to show that costs as estimated in the initial business case are not realistic. The simulated costs for each analyzed cost category exceed the originally estimated ones. Moreover, our evaluation models provide valuable insights into the reasons for the occurred cost overruns, particularly into causal dependencies and resulting effects on the costs of the analyzed EIS engineering project.

Table 7. Lessons Learned

LL	Description
LL-1	Our case study confirms that the EcoPOST framework enables EIS engineers to gain valuable insights into causal dependencies and resulting cost effects in EIS engineering projects.
LL-2	EcoPOST evaluation models are useful for domain experts and can support IT managers and policy makers in understanding an EIS engineering project and decision-making.
LL-3	EIS engineering projects are complex socio-technical feedback systems which are characterized by a strong nexus of organizational, technological, and project-specific parts. Hence, all evaluation models include feedback loops.
LL-4	Our case study confirms that evaluation models can become complex due to the large number of potential SCFs, DCFs and ImFs as well as the many causal dependencies existing between them. Governing guidelines (cf. Section 4) help to avoid too complex evaluation models.
LL-5	Though our simulation models have been build upon data derived from four different data sources, it has turned out that it is inevitable to rely on hypotheses to build simulation models.

Regarding the overall goal of the case study, i.e., the investigation of the practical applicability of the EcoPOST framework and its underlying evaluation concepts, our experiences confirm the expected benefits. More specifically, we can summarize our experiences by means of five lessons learned (cf. Table 7).

6 Summary

This paper summarizes the EcoPOST cost analysis methodology, a practically approved, model-based methodology to better understand and systematically investigate the complex cost structures of EIS engineering projects. We sketch our qualitative

EcoPOST methodology, introduce model design rules and describes modeling guidelines. We also summarize a case study illustrating the benefits of our approach.

Currently, our methodology is used in various information system engineering projects, mainly in the automotive domain. In future, we want to further validate our approach and aim at increasing the number of EcoPOST evaluation patterns [9].

References

1. Reijers, H.A., van der Aalst, W.M.P.: The Effectiveness of Workflow Management Systems - Predictions and Lessons Learned. Int'l. J. of Inf. Mgmt. 25(5), 457–471 (2005)
2. Boehm, B., Abts, C., Brown, A.W., Chulani, S., Clark, B.K., Horowitz, E., Madachy, R., Reifer, D., Steece, B.: Software Cost Estimation with Cocomo 2. Prentice-Hall, Englewood Cliffs (2000)
3. Mutschler, B., Reichert, M., Bumiller, J.: Designing an Economic-driven Evaluation Framework for Process-oriented Software Technologies. In: Proc. 28th ICSE, pp. 885–888 (2006)
4. Mutschler, B., Zarvic, N., Reichert, M.: A Survey on Economic-driven Evaluations of Information Technology. Technical Report, TR-CTIT-07, University of Twente (2007)
5. Mutschler, B., Reichert, M., Rinderle, S.: Analyzing the Dynamic Cost Factors of Process-aware Information Systems: A Model-based Approach. In: 19th CAiSE, pp. 589–603 (2007)
6. Richardson, G.P., Pugh, A.L.: System Dynamics - Modeling with DYNAMO (1981)
7. Forrester, J.W.: Industrial Dynamics, Industrial Dynamics. Productivity Press (1961)
8. Ventana Systems, Inc.: Vensim (2006), http://www.vensim.com/
9. Mutschler, B.: Analyzing Causal Dependencies on Process-aware Information Systems from a Cost Perspective. PhD Thesis, University of Twente (2008)
10. Sterman, J.D.: Business Dynamics. McGraw-Hill, New York (2000)
11. Mutschler, B., Weber, B., Reichert, M.: Workflow Management versus Case Handling: Results from a Controlled Software Experiment. In: Proc. ACM SAC, Special Track on Coordination Models, Languages and Architectures, pp. 82–89 (2008)
12. Mutschler, B., Reichert, M., Bumiller, J.: Unleashing the Effectiveness of Process-Oriented Information Systems: Problem Analysis, Critical Success Factors, and Implications. IEEE Transactions on Systems, Man, and Cybernetics—Part C: Applications and Reviews 38(3), 280–291 (2008)
13. Mutschler, B., Rijkpema, M., Reichert, M.: Investigating Implemented Process Design: A Case Study on the Impact of Process-aware Information Systems on Core Job Dimensions. In: Proc. 8th Int'l. BPMDS Workshop, pp. 379–384 (2007)
14. Mutschler, B., Reichert, M.: A Survey on Evaluation Factors for Business Process Management Technology. Technical Report, TR-CTIT-06-63, University of Twente (2006)
15. Mutschler, B., Reichert, M., Bumiller, J.: Why Process-Orientation is Scarce: An Empirical Study of Process-oriented Information Systems in the Automotive Industry. In: Proc. 10th IEEE EDOC, pp. 433–438 (2006)

Computer-Aided Method Engineering: An Analysis of Existing Environments

Ali Niknafs and Raman Ramsin

Department of Computer Engineering, Sharif University of Technology, Tehran, Iran
niknafs@ce.sharif.edu, ramsin@sharif.edu

Abstract. Analogous to Computer-Aided Software Engineering (CASE), which aims to facilitate Software Engineering through specialized tools, Computer-Aided Method Engineering (CAME) strives to support a wide range of activities carried out by method engineers. Although there is consensus on the importance of tool support in method engineering, existing CAME environments are incomplete prototypes, each covering just a few steps of the method engineering process. This paper summarizes the history and the state of the practice in CAME technology, and provides criteria-based critique on existing CAME environments, thus highlighting their strengths and weaknesses.

Keywords: Software Development Methodologies, Method Engineering, Computer-Aided Method Engineering, Criteria-Based Analysis.

1 Introduction

"If it says one size fits all, it doesn't fit anyone": Although it is safe to assume that every methodology fits at least one project situation, this variant of the Murphy's Law stresses the fact that there is no general-purpose methodology applicable to all different situations. This motivates the development of project-specific methodologies, using an approach known as Situational Method Engineering (SME) [1], a complex and error-prone process that cannot be properly performed without automated support. The automated support required is provided by Computer-Aided Method Engineering (CAME) environments [1, 2, 3, 4, 5, 6]. A CAME environment is composed of a set of correlated tools aiming to facilitate, in its ideal form, the entire SME process. CAME technology dates back to the early days of method engineering, when several academic prototypes were first introduced.

A method is composed of two parts: The *product part* which captures the product-related knowledge, and the *process part* encompassing the activity-related aspects of the method. Due to this division, two types of method fragments can be defined: *Product* fragments are artifacts such as models, diagrams, and documents, whereas activities, stages, and tasks are considered *Process* fragments. To enable computerized support for SME, method fragments need to be stored in a repository called the Method Base. They thus need to be described in a formal way. Several method representation languages have been proposed for this purpose, which are either textual or graphical, or both. Graphical languages are called *meta-modeling languages*, e.g. GOPPRR in MetaEdit+ [7]. Object Z [8] is a textual language, whereas the Method

Z. Bellahsène and M. Léonard (Eds.): CAiSE 2008, LNCS 5074, pp. 525–540, 2008.

Engineering Language (MEL) [9] is both textual and graphical. Meta-modeling languages are more popular than textual ones. This popularity is mainly because they are easier to use, learn, and implement.

As shown in Fig. 1, CAME environments are made up of two parts: The CAME part provides facilities for method engineering, whereas the CASE part offers means for the generation of CASE tools and process support environments. The set of Method Engineering Tools and the Method Base form the main elements of the CAME part. The Method Engineering Toolset offers tools for facilitating the work of method engineers, e.g. for extracting components of existing methods and storing them in the Method Base. The Method Base upon which a CAME environment is built is the kernel of the CAME environment. The method obtained from the CAME part will be fed as input to the CASE part. The CASE Generator gets the product part of methods and generates the project specific CASE tool. Process-centered Software Engineering Environments (PSEEs) are used for generating process support environments based on the process part of methods.

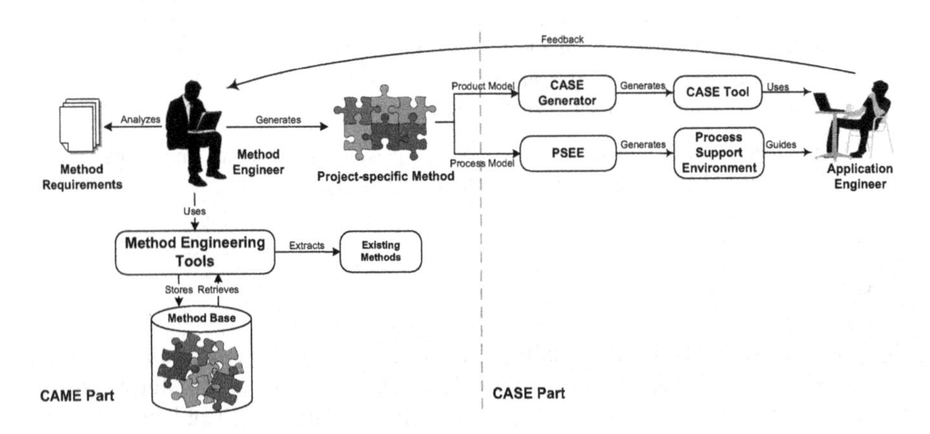

Fig. 1. General architecture of CAME environments

In [10], three distinct approaches to SME are proposed. The *assembly-based* approach is the most common and consists of three steps: specifying method requirements, selecting method fragments, and assembling them into a method. *Extension-based* SME aims at adapting and extending an existing method with new features; whereas in the *paradigm-based* approach a new method is developed by instantiating, abstracting or adapting an existing meta-model. As we will see in the following sections, almost all the existing CAME environments support the assembly-based approach, and other approaches are almost completely overlooked.

CAME environments can be classified as *product-oriented* and *process-oriented*, depending on the way they facilitate the enactment of the method engineering process. Those that focus on modeling the product-related issues of methods, and provide less support for the method's process model and its enactment, are classified as product-oriented CAME environments. Process-oriented CAME environments deal with the process-related issues of methods and support the enactment of the process model.

Most of the existing CAME environments fit into the first class, due to their emphasis on modeling the product part of methods.

Coverage of the method engineering process is one of the major shortcomings of existing CAME environments. Several method engineering processes have been proposed in the literature [10, 11, 12, 13], yet they can all be considered as consisting of the following generic phases:

1. *Method Requirements Analysis (MRA)*: focuses on the identification of important features of the method under construction. Method requirements are those features that are expected to be present in an Information Systems Development (ISD) method, such as traceability to requirements or support for umbrella activities [14]. In the MRA phase, the method requirements should be defined in a formal way, and should therefore precisely describe the features that the desired method needs to offer.
2. *Method Design (MD)*: focuses on determining a blueprint for the method, based on the requirements defined in the previous phase.
3. *Method Implementation (MI)*: focuses on selecting suitable method fragments and assembling them, instantiating an existing meta-model or process pattern, or modifying or extending an existing method. The result of this phase is a set of CASE tools providing means to support the method's product model, and process support environments to guide the application engineer during the ISD project.
4. *Method Test (MT)*: focuses on the verification and validation of the newly developed method. The results of the MRA phase are fed as input to this phase. Testing a newly developed ISD method is similar to testing any other type of system: develop test cases (in this case, sample systems), perform verification and validation, and correct the defects detected. Testing the resulting ISD method is a weak point of existing CAME environments: CAME environments do not offer adequate means for determining whether a newly developed method realizes the predefined method requirements or not.

The aim of this paper is to present a brief overview of past research conducted on CAME environments and the state of the practice as reported and documented by researchers, identifying the shortcomings and thereby offering suggestions for future research. The remainder of the paper is structured as follows: in section 2, several existing CAME environments will be briefly described; these will be analyzed in section 3, based on analysis criteria adapted from the attributes presented in the ISO/IEC 9126 quality model; section 4 contains the conclusions, and outlines our plans for furthering this research.

2 CAME Environments

Several CAME environments have been proposed, yet even though the achievements of these environments have been remarkable, none of them provide a comprehensive set of means for enacting the method engineering process. In this section, we provide concise descriptions for several existing CAME environments, limiting our review to those environments for which adequate documentation is available.

All the environments discussed come from research communities, with very little support from the industry. This is the reason why most of them do not have a long history of usage, even though they enjoy extensive documentation and many years of investigation. There exist tools and environments, such as Rumi [15], that because of little or no available documentation, are very hard to assess and have therefore not been included in this research.

2.1 MERET

The *Methodology Representation Tool (MERET)* is a forerunner of present-day CAME environments, and focuses on method engineering in a product-oriented fashion, dealing with the adaptation and customization of existing methods. MERET presents a comprehensive *methodology representation model* [16] used for the specification of methods. This representation model uses a semantic data-model called ASDM [16] for representing method knowledge. ASDM provides a powerful means for modeling objects and their interrelationships. The root concept used in the methodology representation model is the so-called *MERET object*, which provides the attributes common to all other objects (e.g. name). A MERET object can be a *methodology object* or a *guideline object*: the former consists of all the objects needed for method specification, whereas the latter describes the rules, constraints and experiences relating to the former. Rules are represented in a formal way by means of Horn clauses. Methodology objects are partitioned into *Methods* and *Techniques*, where a method consists of techniques used to develop products. The process model is described by means of *actors, milestones*, *processes* and *deliverables*. A process can be either a *phase* or an *activity*. A technique consists of several *resources*, which are non-human requirements such as any CASE tool feature, and *representation types*, which define the representation of deliverables (e.g. text or a special diagram such as a DFD). MERET provides means for automatic application of consistency checks on the method specification, and the integration of different methods, and the customization of the method produced to specific projects.

2.2 MethodBase

MethodBase is one of the first academic CAME prototypes introduced. The aim of this environment is to facilitate method customization, rather than assembly-based SME. MethodBase assists the method engineer in the selection of a method that best fits the project at hand [1].

The MethodBase system's database consists of complete methods. Therefore, methods can be selected and be customized to fit a project situation. Its data model is divided into product and process parts. The product part consists of the concepts *State*, *Event*, *Data*, *Entity*, and *Association*, whilst the concepts *Activity* and *Activity-relationship*, constitute the process part. By means of the process part, the method engineer can define guidelines to support the enactment of the method process model.

2.3 MetaEdit+

MetaEdit+ is the result of the MetaPHOR project initiated in 1990, and was originally developed as a metaCASE environment. A commercial version of MetaEdit+ has also been released. MetaEdit+ uses techniques similar to those used in assembly-based method engineering. It uses the GOPPRR [7] conceptual data model as its method specification language, which is an evolutionary extension of the OPRR and GOPRR models [17]. The basic constructs of the GOPPRR model are *Graph, Object, Port, Property, Relationship,* and *Role*. Graph is the top-level structure of the meta-model, which is an aggregate concept, composed of objects and their relationships. Object types are the design objects that typically appear as shapes in diagrams. Examples of objects are *Class* in Class Diagrams and *Entity* in Entity Relationship Diagrams. Associations between objects are regarded as their relationships. Each object has a role in the relationships in which it participates. Ports allow additional semantics or constraints on how objects can be connected. Ports can be used as parts of objects, to which roles can be attached. Properties are the characterizing attributes attached to each of the types. MetaEdit+ incorporates a specialized tool for creating and maintaining each of these basic types.

The OPRR meta-modeling language and its extensions – GOPRR and GOPPRR – only deal with the product aspects of methods. However, a process meta-modeling language called GOPRR-p [18, 19] has also been proposed as an extension to GOPRR; this language provides concepts and integration rules for defining different *Process Modeling Languages (PML).*

MetaEdit+ consists of several tool families, among which are *Method Management Tools* [17], aiming at providing CAME functionalities. This tool family consists of the following main parts:

- The *Method Base*: consists of method fragments, symbols needed for representing object types, and generic reports used by the report generator tool to deliver several reports on methods.
- The *Method Assembly System*: consists of tools needed for method assembly, such as *Meta-model Editors*, which provide various tools for specifying the GOPPRR constructs and their connections. The resulting method will be checked for incompleteness and inconsistency by means of the *Consistency Checking System*. The *Symbol Editor* is a drawing tool used for specifying symbols for each object type. A number of reports on the newly developed method can be generated using the report generator tool included in the *Metrics & Statistics System*. Metrics reports are used for analyzing the properties of methods (e.g. the number of objects therein).
- The *Environment Generation System*: is the CASE part of MetaEdit+ offering several generators for delivering a CASE tool, an online help, and a number of reports on the models.

2.4 Decamerone

Decamerone extends the Maestro II metaCASE environment with CAME capabilities, taking advantage of the features already present in Maestro II. Decamerone uses the *Object Management System (OMS)* [1, 21], which is the online object-oriented DBMS of Maestro II. The architecture of Decamerone is shown in Fig 2. As expected, the *Method Base* is the central repository containing method fragments and their relationships. The *Selected Method Fragments Repository (SMFR)* is a subset of the Method Base, containing the selected method fragments for integration into a new method.

The *Situational Method Database* is another subset of the Method Base containing the assembled method. The *CASE tool repository* stores all the products used and produced during the project.

At the core of Decamerone is the *Method Base Management System (MBMS)* [21], which provides facilities for the specification, storage and selection of method fragments and their assembly into a new method. The MBMS is an interface for accessing the OMS databases. This will help the method engineer avoid the low-level complexities of actually accessing an OMS database.

The novel feature of Decamerone is the *Method Engineering Language (MEL)* [9], which is not only used for the representation of method fragments, but also offers constructs for their manipulation operations. Thus, MEL supports the administration, selection and manipulation of fragments. As mentioned before, MEL is both textual and graphical; however, its textual form is much more powerful than the graphical form. Decamerone's user interface consists of three parts: The *MEL Command Line Interface*, which is a text editor used for the selection and manipulation of method fragments in a high-level method engineering language; the graphical editors of the *Concept Structure Diagram (CSD)* and the *Process Structure Diagram (PSD)*, which aid in the specification and assembly of product and process fragments respectively; and the *MEL Editor*, which aids in creating MEL specifications for the graphical forms of the method fragments produced, in order to construct finer grained method fragments.

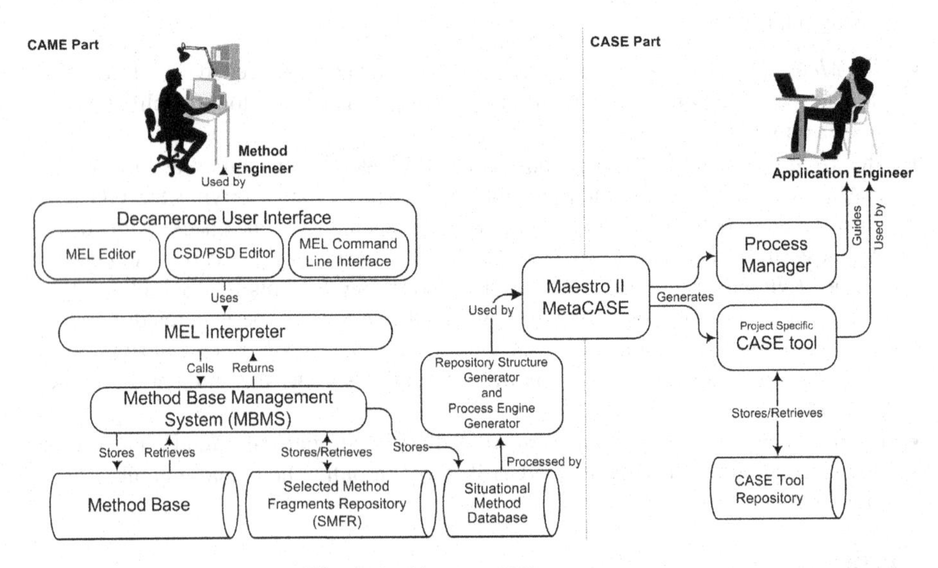

Fig. 2. Architecture of Decamerone

Consistency and completeness of method fragments are checked by the MEL Editor and the CSD/PSD Editors. The *MEL Interpreter* acts as an interface for MBMS and gets user commands and translates them into MBMS function calls. Thus, the MBMS is only called by, and returns values to the MEL interpreter. The newly developed method will be given to Maestro II for CASE tool generation. The *Project and Configuration Management System (PCMS)* is a part of Maestro II used for defining

Process Managers, which enact the method's process model. PCMS offers functionalities for configuration management, project scheduling and estimation. The *Repository Generator* takes the product part of the method as input and generates the CASE tool repository. Notational symbols of the elements manipulated through diagram editors are specified using the *Tool Customizing Interface (TCI)*.

Decamerone provides facilities for defining the semantics of method fragments. An *ontology* is defined for product fragments, as well as a *process classification system* [1] for all method fragments, thereby specifying the semantic aspects of the method fragments. The proposed ontology is called the *Methodology Data Model (MDM)* [1] which consists of the basic concepts of ISD products and the associations between them. The process classification system employs the notion of *goal*, which is represented as a tuple *(Action, Measure, Product)*. Goals are taken from a process classification, consisting of a set of basic actions in ISD, a set of measures, and a set of product types required in ISD. Basic *actions* are those actions in ISD which have the same effect; a *product* type is a class of products in ISD with the same purpose; and a *measure* is a qualifier of a product, to indicate temporal state, level of detail, or level of abstraction.

2.5 MENTOR

The core component of MENTOR is its *Guidance Engine* [22] which provides guidance to both method engineers and application engineers; MENTOR is therefore a guidance-centered environment. MENTOR uses the NATURE contextual approach to describe method fragments. In the NATURE approach, a method is viewed as a set of method fragments which can either be a *forest*, a *tree* or a *context*. A forest is a set of trees where trees are hierarchies of contexts. Contexts are pairs of the form <*situation*, *decision*>, where *decision* states the intention of the method engineer, and the state of the product that the *decision* can be taken on forms the *situation*. Method fragments, or *method chunks*, are of two kinds: *Components*, which are parts of the product/process model of methods; and generic method construction *Patterns* [23, 24], which can be instantiated to new method fragments. The main components of MENTOR are:

- The Method Engineering Environment, which consists of a set of tools, editors, and browsers for facilitating the work of method engineers [25]. The *product editor* and the *process editor* allow the graphical specification of the product model and the process model respectively. The *method generator* aids in the automatic instantiation of predefined generic patterns stored in the method base. *Browsers* are also provided to help retrieve the necessary method fragments.
- The Application Engineering Environment, which constitutes the CASE part of MENTOR, providing tools for supporting the enactment of methods.
- The Guidance Engine, which advises the method engineer in his method engineering activities and guides the application engineer by executing the resulting process model.
- The Repository, which is organized in three interrelated levels [25]: The *Application Knowledge level*, which is the lower level consisting of the process model and the products under development; the *Method Knowledge level*, which is composed

of method fragments; and the *Method Meta Knowledge level*, which deals with the semantics of the method fragments. Product and process meta-models are placed in the Method Meta Knowledge level, whereas product models and process models are placed into the Method Knowledge level. These two latter levels constitute the Method Base of MENTOR.

2.6 MERU

The main feature of *Method Engineering Using Rules (MERU)* that distinguishes it from other existing CAME environments, is a technical document describing method requirements called the *Method Requirements Specification (MRS)* [11]. MRS is implementation-independent and only expresses the nature of an ISD method. MRS is based on a meta-model called MVM. In MVM, method concepts, which are called *things*, are partitioned into *links*, *constraints* and *product elements*. A *link* is any *thing* of the product that connects two product entities together. *Constraints* are those *things* that can be used by application engineers to specify properties of *links* and *product entities*. Finally, any *thing* that is not a *link* or *constraint* is a *product entity*. The interrelationship between concepts are captured through two relationships: *is composed of*, which identifies concepts that are made up of other concepts; and *is mapped to*, which relates together concepts of two different models. The meta-model proposes to partition things into product entities, constraints and links. The procedure of ME performed by means of MERU consists of 3 main steps. In the first step, called the *Method Requirements Engineering (MRE)* phase, the method engineer expresses the preferred method requirements in the form of an MRS. A language based on the MVM meta-model, called *Method Requirements Specification Language (MRSL)*, is developed to express the MRS. In contrast to other CAME environments, in which a specification language is developed to express the method fragments, MRSL is used only for describing the MRS made by the method engineer. This step is supported by the *MRS Creator*. The MRS thus obtained will then be checked for inconsistencies such as incompleteness and non-conformity with the MVM; this is performed through the *Method Analyzer*. The analysis results are used to provide guidance for refining the MRS. After obtaining the desired MRS, *Method Design* is performed as the next step. Method Design focuses on the translation of the MRS into an instantiation of the MVM. In order to perform this instantiation, for every concept of the given MRS, decisions need to be made as to the type, relationships, and attributes of the concept. The output of this step is called a *plan of instantiation*, which can be modified by the method engineer to obtain the preferred instantiation.

The *Method Construction and Implementation (MCI)* step is then commenced, in which method fragments are created by means of the *Component Builder*. In the method assembly approach used in MERU, method fragments are generated automatically, based on the given MRS. Method fragments are described in terms of *MRS Components (MRSCs)*, which only consider the product part of the methods. The Component Builder uses several predefined rules to identify MRSCs by retrieving the appropriate method fragments from the method base (hence the tool's name). The step is concluded by giving the resulting method description to a metaCASE called RAPID, which generates the appropriate CASE tool.

2.7 Method Editor

Method Editor takes advantage of UML as its meta-modeling technique for expressing the method fragments [3]. Class diagrams are used for the specification of product fragments, while process fragments are described by means of activity diagrams. The process part of a method is attached to each corresponding product fragment, i.e. each product fragment and its development procedure will be shown as a pair of diagrams, a class diagram and an activity diagram.

Method Editor is complemented by a CASE part, so that the Method Editor's output, the resulting ISD method, is fed to the CASE part as input. The CASE part consists of a *Diagram Generator* as the CASE generator, and a *Navigator Generator* which develops a *Navigation Browser* guiding the application engineer through the process of software development. An OCL checker [26] is provided as a part of the resulting CASE tool to check method fragments against the predefined constraints. Any inconsistency seen in the development process will affect the continuation of the whole process, i.e. the process part of methods will be controlled dynamically, forcing adherence to the predefined rules of the method.

The recent version of Method Editor is extended by means of a Version Control System [26], thereby supporting version control and change management of methods or their parts.

3 Analysis of Existing CAME Environments

In this section, we examine the CAME environments introduced in the previous section. Table 1 is a summary of the major features and characteristics of existing CAME environments. In Table 2, the environments have been analyzed and compared with each other based on a few general criteria. In analyzing the environments based on their *Number of Features*, environments with numerous implemented features, partially implemented features, and very few implemented features have been marked as High, Average and Low respectively. The number and importance of the innovations

Table 1. Summary of existing CAME environments

Environment	Coverage of ME Process				SME Approach	Method Representation Language			Process Enactment Support	CASE Tool Generator	Product-Oriented	Process-Oriented
	MRA	MD	MI	MT		Textual Language	Meta-model	Semantic Data-Model				
Decamerone	☒	☑	☑	☒	Assembly-based	MEL	MDM	MDM	☑	☑	☑	☒
MENTOR	☑	☑	☑	☒	Assembly-based, Paradigm-based	-	NATURE	-	☑	☑	☒	☑
MERET	☒	☑	☒	☒	Method Customization	Methodology Representation Model	ASDM	ASDM	☒	☒	☑	☒
MERU	☑	☑	☑	☒	Assembly-based	MRSL	MVM	-	☑	☑	☑	☒
MetaEdit+	☒	☑	☑	☒	Assembly-based	-	GOPPRR	-	☒	☑	☑	☒
MethodBase	☒	☑	☑	☒	Method Customization	Object Z	-	-	☒	☑	☑	☒
Method Editor	☒	☑	☑	☒	Assembly-based	MEL	UML	-	☑	☑	☑	☒

that each CAME environment has offered is evaluated and rated as its *Contributions*. We have also strived to provide a measure of the documentation available on each environment: If more than one author have published more than one paper on a CAME environment at different levels of detail, its *Available Literature* is marked as High; if more than one author have published papers on a CAME environment but they do are not much different as to their span and/or level of detail, it has been marked as Average; and if available publications on a CAME environment are rare or do not have the needed level of detail, it has been marked as Low.

In order to provide a more detailed analysis of CAME environments, we propose the ISO/IEC 9126 quality model [27] as a useful evaluation framework. ISO/IEC 9126 is one of a large group of internationally recognized standards applicable across a wide range of applications. We have instantiated the model to fit the CAME domain, and the CAME environments described above have been evaluated based on this adapted model, with the results tabulated for enhanced legibility.

Table 2. General analysis and comparison of existing CAME environments

Environment	Use	Number of Features	Contributions	Available Literature	Year of Introduction
Decamerone	Research	High	High	Average	1995
MENTOR	Research	Average	Average	Average	1996
MERET	Research	Low	Average	Low	1992
MERU	Research	High	High	Low	2001
MetaEdit+	Research and Commercial	High	Average	High	1994
MethodBase	Research	Low	Average	Low	1992
Method Editor	Research	Average	Average	Average	2003

3.1 The ISO/IEC 9126 Quality Model

ISO/IEC 9126 was originally developed in 1991 by the International Organization of Standards to provide a framework for the evaluation of software quality. However, ISO/IEC 9126 does not provide requirements for software, but defines a quality model which is applicable to any kind of software. This model defines six product characteristics which are further subdivided into a number of sub-characteristics (See Table 3). These characteristics and sub-characteristics constitute a detailed model for evaluating any software system. To be able to take different requirements of different systems into account, the model needs to be instantiated for each concrete domain by weighing the different characteristics and sub-characteristics accordingly.

3.2 A Quality Model for CAME Environments

Our quality model for CAME environments is an adaptation of ISO/IEC 9126; i.e. we have applied the model to the domain of method engineering. Table 4 illustrates our CAME quality model. The three characteristics of Functionality, Usability and Portability of the original quality model can be assessed based on the available literature; we have therefore focused on these characteristics. We use these quality characteristics and sub-characteristics to evaluate the CAME environments discussed earlier in this paper.

Table 3. ISO/IEC 9126 characteristics and sub-characteristics [27]

Characteristic	Sub-Characteristics	Definition
Functionality	Suitability	The presence of the required functions
	Accurateness	The correctness of the results
	Interoperability	Ability of software to interact with other systems
	Security	The ability of software to prevent unauthorized access
Reliability	Maturity	The frequency of failure by faults in the software
	Fault Tolerance	The capability of software to maintain its level of performance under stated conditions for a stated period of time
	Recoverability	The capability of software to resume working and recover the data after failure
Usability	Understandability	The effort needed for use the software
	Learnability	The easiness of learning how the software works
	Operability	The effort needed for operating the software
	Attractiveness	The quality of the user interface
Efficiency	Time Behaviour	The response and processing times
	Resource Utilisation	The resource utilisation
Maintainability	Analysability	The effort needed for diagnosis of faults
	Changeability	The effort needed for modification
	Stability	The risk of modification effects
	Testability	The effort needed for testing the modified software
Portability	Adaptability	The opportunity for moving the software to other environments
	Installability	The easiness of software installation
	Co-existence	The ability of software to coexist with other software systems in a common environment
	Replaceability	The effort needed for replacing other software
All characteristics	Compliance	The compliance of software with regulations and rules

Table 4. The CAME Quality Model

Characteristic	Sub-Characteristics	Criteria Description
Functionality	Suitability	▪ Evaluates if the CAME environment offers a suitable toolkit for the development of project-specific CASE tools. ▪ Evaluates if the CAME environment supports various SME approaches. ▪ Evaluates if the CAME environment supports process enactment. ▪ Evaluates if the CAME environment offers facilities to define semantics of method fragments.
	Accurateness	▪ Evaluates if ample knowledge is available as to the results of own tests or tests published by third parties that indicate the degree of effectiveness of the CAME environment.
	Functionality compliance	▪ Evaluates if the CAME environment supports standards and techniques such as: UML, XML ...
Usability	Understandability	▪ Evaluates the level of understandability and usability of the interfaces. ▪ Evaluates the level of understandability and usability of the method representation language.
	Learnability	▪ Evaluates if the CAME environment has adequate documentation. ▪ Evaluates the level of learnability of the method representation language.
	Operability	▪ Evaluates if the CAME environment has graphical tools that facilitate the development of Method fragments.
	Attractiveness	▪ Evaluates if the CAME environment has attractive graphical design.
Portability	Installability	▪ Evaluates if the provider provides technical support and online help for the installation of the CAME environment.
	Co-existence	▪ Evaluates the capacity of the CAME environment to coexist with other independent CAME or MetaCASE environments in a common environment sharing common resources. For example, whether other MetaCASE tools can be installed to satisfy the CASE generation functionality.

3.3 Evaluation Results

The results are summarized into a matrix relating the characteristics and sub-characteristics to the features offered by the CAME environments reviewed (See Table 5) Deficiencies identified during the evaluation are indicated by a number, and an explanation is given in the legend below of how the system failed to meet the criteria in these cases.

Table 5. Evaluation of the CAME environments using the CAME Quality Model

	Quality Characteristics								
	Functionality			Usability				Portability	
Environments	Suitability	Accuracy	Functionality compliance	Simplicity	Learnability	Operability	Attractiveness	Installability	Co-existence
Decamerone	3,4	×	×	7	✓	✓	×	×	✓
MENTOR	4	×	×	✓	✓	✓	×	×	✓
MERET	1,2,3,5	×	×	✓	✓	-	×	×	-
MERU	3,4	✓	×	×	-	✓	×	×	✓
MetaEdit+	3,4,5	✓	✓	✓	✓	✓	✓	✓	✓
MethodBase	2,3,4	×	×	6	-	-	×	×	-
Method Editor	3,4	×	✓	✓	✓	✓	✓	×	-

Legend:
✓ Supported to a good extent
× Not supported
- Inadequate information to assess

1. Lack of CASE tool generation facilities
2. Partial coverage of the ME process
3. Inadequate support for SME approaches
4. Lack of semantic definition features for method fragments
5. Poor process support
6. Does not provide graphical meta-modeling language
7. Poor graphical meta-modeling language

4 Conclusion and Future Work

In this paper, we have summarized the main efforts performed in the development of CAME environments. Although CAME technology dates back to the early days of method engineering, it is not mature enough to support the whole process of situational method engineering. Each current CAME prototypes has its own advantages and shortcomings. In the following, the main shortcomings that current CAME technology suffers from are listed:

- Weak process enactment support: Even though product-related issues of ISD methods are fully considered and have been provided with computerized support, the process-related issues still need to be researched in order to find suitable ways for representing method process models. *Process Modeling Languages (PML)* [28, 29, 30] can be considered as suitable means for process representation; however, guidelines should be attached to a process described in a PML in order to support process enactment in actual ISD projects.

- Lack of support for situational method engineering approaches: The assembly-based approach is the only one adequately addressed. Paradigm-based and Extension-based approaches should also be supported by CAME environments.
- Partial coverage of method engineering process: Although method design and implementation phases are properly supported, there are still severe shortcomings as to support for method requirements analysis and method test.
- Method verification: Verifying the newly built method may be the last phase of the method engineering process, but it is never the least. Method verification requires a criterion set which a method can be checked against. But the difficult part of the task is determining how to perform the evaluation. Due to this difficulty, method verification is one of the hardest to automate. Current CAME prototypes perform method test through prompting feedback from the users of the method. Therefore, the newly developed method would not be verified until it is tested in an actual project situation.
- Weak method representation mechanisms: As mentioned in [31], there is no ultimate method representation language. Therefore, method representation languages are composed of fragments originating from several languages in a bid to obtain a purpose-fit language. This leads to a situation which is called *Method Engineering of Method Engineering Languages*. New method engineering languages need to be developed to support method verification.
- Lack of support for semantic definitions of method fragments: We believe that semantic meta-models should be an integral part of any CAME environments' Method Base, but few of the existing CAME environments address this issue. The lack of means for capturing and specifying the semantic aspects of method fragments leads to complications; examples are the selection and assembly of method fragments that may not be semantically composable into a method [32]. Describing the semantics of method fragments is one of the major problems in SME. To overcome this problem, method fragments need to be described in a complete and unambiguous way. However, as stated in [1], since methods and their semantics are interpreted differently by different human beings, there is no unique meaning for a method fragment. Nevertheless, method fragments can be anchored, i.e. described in terms of unambiguously defined concepts and relationships between those concepts, in a system for which the meaning is defined. Such systems are defined as ontologies in Decamerone and MERET.

Our future work focuses on the development of a CAME environment supporting the Hybrid Methodology Design approach [14]. This approach to methodology design uses alternative method engineering approaches for different parts of the process and at different levels of abstraction. It also provides an iterative and incremental framework allowing flexible application of four method development approaches, namely:

- *Instantiation approach:* with the focus on instantiating an already available process meta-model.
- *Artifact-oriented approach:* devising a seamless complementary chain of artifacts and building the process around it.
- *Composition approach:* using one of the already available libraries of process patterns.

- *Integration approach*: integrating features, ideas and techniques from existing methods.

Two of these approaches, *Instantiation* and *Composition*, are analogous to the *Paradigm-based* and *Assembly-based* approaches of method engineering, whereas the *Integration* and *Artifact-oriented* approaches are relatively novel in this context. The *Integration* approach is particularly nonconformist in comparison to usual method engineering practices, in that it promotes integrating ideas and techniques directly from existing methods, instead of first dissecting the methods into method fragments and then storing them in a method repository (as is common practice in the assembly-based method engineering approach); the motivation behind this stance is the observation that "breaking down the methods into fragments may result in loss of synergy and functional capacity" [14].

Acknowledgments. We wish to thank the Research Vice-Presidency of Sharif University of Technology and Iran Telecommunication Research Center (ITRC) for sponsoring this research. Also, special thanks to the anonymous reviewers of this paper for their helpful feedback.

References

1. Harmsen, A.F.: Situational Method Engineering. Moret Ernst & Young, Utrecht (1997)
2. Rolland, C.: A Primer for Method Engineering. In: Proceedings of the INFormatique des ORganisations et Systèmes d'Information et de Décision (INFORSID 1997), Toulouse (1997)
3. Saeki, M.: CAME: The First Step to Automated Method Engineering. In: Workshop on Process Engineering for Object-Oriented and Component-Based Development, Anaheim, CA (2003)
4. Arni-Bloch, N.: Towards a CAME Tools for Situational Method Engineering. In: Proceedings of the 1st International Conference on Interoperability of Enterprise Software and Applications, Geneva (2001)
5. Dahanayake, A.N.W.: Computer-Aided Method Engineering: Designing CASE Repositories for the 21st Century. Idea Group Publishing, Delft (2001)
6. Kumar, K., Welke, R.J.: Methodology engineering: a proposal for situation-specific methodology construction. In: Cotterman, W.W., Senn, J.A. (eds.) Systems Analysis and Design: A Research Agenda, pp. 257–268. John Wiley & Sons, Chichester (1992)
7. MetaCase Consulting: Method Workbench User's Guide, MetaCase Consulting, Jyväskylä, Finland (2005), `http://www.metacase.com/support/40/manuals/mwb40sr2a4.pdf`
8. Saeki, M., Wenyin, K.: Specifying software specification and design methods. In: Wijers, G., Wasserman, T., Brinkkemper, S. (eds.) CAiSE 1994. LNCS, vol. 811, pp. 353–366. Springer, Heidelberg (1994)
9. Brinkkemper, S., Saeki, M., Harmsen, F.: A Method Engineering Language for the Description of Systems Development Methods. In: Dittrich, K.R., Geppert, A., Norrie, M.C. (eds.) CAiSE 2001. LNCS, vol. 2068, pp. 473–476. Springer, Heidelberg (2001)
10. Ralyté, J., Deneckère, R., Rolland, C.: Towards a Generic Model for Situational Method Engineering. In: Eder, J., Missikoff, M. (eds.) CAiSE 2003. LNCS, vol. 2681, pp. 95–110. Springer, Heidelberg (2003)

11. Gupta, D., Prakash, N.: Engineering Methods from Method Requirements Specifications. J. Requirements Engineering 6(3), 135–160 (2001)
12. Leppanen, M.: Conceptual Analysis of Current ME Artifacts in Terms of Coverage: A Contextual Approach. In: 1st Workshop on Situational Engineering Processes, Paris (2005)
13. Prakash, N., Goyal, S.B.: Towards a Life Cycle for Method Engineering. In: 12th Workshop on Exploring Modeling Methods in Systems Analysis and Design (2007)
14. Ramsin, R.: The Engineering of an Object-Oriented Software Development Methodology. Ph.D. Thesis, University of York (2006), http://www.cs.york.ac.uk/ftpdir/reports/YCST-2006-12.pdf
15. Tekinerdoğan, B.: Synthesis-Based Software Architecture Design. Ph.D. Thesis, University of Twente (2000)
16. Heym, M., Osterle, H.: A Semantic Data Model for Methodology Engineering. In: 5th Workshop on Computer-Aided Software Engineering, pp. 142–155. IEEE Press, Los Alamitos (1992)
17. Kelly, S., Lyytinen, K., Rossi, M.: MetaEdit+ A Fully Configurable Multi-User and Multi-Tool CASE and CAME Environment. In: Constantopoulos, P., Vassiliou, Y., Mylopoulos, J. (eds.) CAiSE 1996. LNCS, vol. 1080, pp. 1–21. Springer, Heidelberg (1996)
18. Tolvanen, J.P.: Incremental Method Engineering with Modeling Tools. Ph.D. Thesis, University of Jyväskylä (1998)
19. Koskinen, M., Marttiin, P.: Process Support in MetaCASE: Implementing the Conceptual Basis for Enactment Process Models in MetaEdit+. In: Ebert, J., Lewerentz, C. (eds.) Software Engineering Environments, pp. 110–123. IEEE Computer Society Press, Los Alamitos (1997)
20. Koskinen, M.: Beyond Process Modelling Languages: A Metamodelling Approach to Customizable Concepts and Enactability in MetaCASE. In: Proceedings of the 4th Doctoral Consortium on Advanced Information Systems Engineering, Barcelona (1997)
21. Brinkkemper, S., Harmsen, F.: Design and Implementation of a Method Base Management System for a Situational CASE Environment. In: Proceedings of the 2nd Asia-Pacific Software Engineering Conference, pp. 430–438. IEEE Computer Society, Los Alamitos (1995)
22. Si-Said, S., Rolland, C., Grosz, G.: MENTOR: A Computer Aided Requirements Engineering Environment. In: Constantopoulos, P., Vassiliou, Y., Mylopoulos, J. (eds.) CAiSE 1996. LNCS, vol. 1080, pp. 22–43. Springer, Heidelberg (1996)
23. Plihon, V., Rolland, C.: Genericity in Method Construction. In: Proceedings of the 4th Asia-Pacific Software Engineering Conference, pp. 302–311. IEEE Computer Society, Washington, DC (1997)
24. Rolland, C., Plihon, V.: Using Generic Method Chunks to Generate Process Model Fragments. In: Proceedings of the 2nd International Conference on Requirements Engineering (ICRE 1996), pp. 173–181. IEEE Computer Society, Colorado (1996)
25. Plihon, V.: MENTOR: An Environment Supporting the Construction of Methods. In: Proceedings of the 3rd Asia-Pacific Software Engineering Conference, pp. 384–392. IEEE Computer Society, Washington, DC (1996)
26. Saeki, M.: Configuration Management in a Method Engineering Context. In: Dubois, E., Pohl, K. (eds.) CAiSE 2006. LNCS, vol. 4001, pp. 384–392. Springer, Heidelberg (2006)
27. International Organization for Standardization (ISO), International Electrotechnical Commission (IEC): ISO/IEC: 9126: Software engineering - Product quality; Parts 1-4. Geneva (2004)

28. Cugola, G., Ghezzi, C.: Software processes: a retrospective and a path to the future Software Process. J. Improvement and Practice 4(3), 101–123 (1998)
29. Zamli, K.Z., Lee, P.A.: Taxonomy of process modeling languages. In: ACS/IEEE International Conference on Computer Systems and Applications, pp. 435–437. IEEE Computer Society, Washington, DC (2001)
30. Zamli, K.Z.: Process Modeling Languages: A Literature Review. Malaysian Journal of Computer Science 14(2), 26–37 (2001)
31. Harmsen, A.F., Saeki, M.: Comparison of Four Method Engineering Languages. In: Proceedings of the IFIP TC8, WG8.1/8.2 working conference on method engineering: principles of method construction and tool support, pp. 209–231. Chapman & Hall, London (1996)
32. Brinkkemper, S., Saeki, M., Harmsen, F.: Meta-modeling based assembly techniques for situational method engineering. J. Information Systems 24(3), 209–228 (1999)

Adapting Secure Tropos for Security Risk Management in the Early Phases of Information Systems Development

Raimundas Matulevičius[1], Nicolas Mayer[1,2], Haralambos Mouratidis[3], Eric Dubois[2], Patrick Heymans[1], and Nicolas Genon[1]

[1] PReCISE, Computer Science Faculty, University of Namur, Belgium
{rma,phe,nge}@info.fundp.ac.be
[2] CRP Henri Tudor - CITI, Luxembourg
{nicolas.mayer,eric.dubois}@tudor.lu
[3] School of Computing and Technology, University of East London, UK
H.Mouratidis@uel.ac.uk

Abstract. Security is a major target for today's information systems (IS) designers. Security modelling languages exist to reason on security in the early phases of IS development, when the most crucial design decisions are made. Reasoning on security involves analysing risk, and effectively communicating risk-related information. However, we think that current languages can be improved in this respect. In this paper, we discuss this issue for Secure Tropos, the language supporting the eponymous agent-based IS development. We analyse it and suggest improvements in the light of an existing reference model for IS security risk management. This allows for checking Secure Tropos concepts and terminology against those of current risk management standards, thereby improving the conceptual appropriateness of the language. The paper follows a running example, called eSAP, located in the healthcare domain.

Keywords: Risk management, information system, security, Secure Tropos.

1 Introduction

Information systems (ISs) undoubtedly play an important role in today's society are more and more at the heart of critical infrastructures. ISs are also facing an increasing complexity because of their interoperability with other systems and of their operation in open, distributed and mobile environments. In such contexts, secure issues are vital and are still reinforced in many sectors with the introduction of new regulations, such as Basel II [1] or SOX [2]. Risk management is considered as central by IS professionals. The risk management does not only support security officers in the handling of security vulnerabilities but it also provides a framework that allows evaluation of the return on investment of the security solutions against the economic and business consequences of not implementing them. There are more than 200 risk management methods making it a

Z. Bellahsène and M. Léonard (Eds.): CAiSE 2008, LNCS 5074, pp. 541–555, 2008.

challenge to select the most adequate one. In a previous analysis [3] we identified some important points for possible improvements. Firstly, elements are related to the nature of the artefacts produced with such methods. These artefacts are largely informal and typically consist of natural language documents, complemented with tables and ad hoc diagrams for structuring the information. The powerful abstraction mechanisms and visualisations offered by conceptual modelling techniques are thus underexploited. Secondly, they are often designed for assessing the way existing systems handle risk in an auditing mode. This view is no longer sustainable in the context of todays ISs that need to constantly adapt to new environments and handle evolution with minimum human intervention. This is an additional argument for the use of more formal languages supporting the reasoning, evolution, monitoring and traceability of risk related information.

In this paper we report on a research related to the design of a suitable modelling language for supporting security risk management (SRM) activities. Central to this research is to first achieve a deep understanding of the SRM domain, then to design an adequate language with suitable constructs and associated semantics for that domain. A central focus of risk management methods is to consider security issues from the very early phases, a.k.a. *requirements engineering* (RE), of ISs development. The associated scientific literature features a number of modelling languages specifically dedicated to security sensitive contexts; however the risk concepts are only partially supported. This advocates for the design of 'yet another' modelling language. However, defining a new and complete notation does not appear to us as a viable option from a sustainability perspective for the modelling community. As demonstrated for example with UML in software engineering, a consensus over unified and common notations has been proven to be a big push for the adoption of modelling practices in public and private companies. At the RE level we plead for a similar approach and rather than to develop a totally new language we improve existing languages, offering an ontological basis sufficiently close to the risk management domain.

With respect to the above objective, we have identified Secure Tropos [4], which uses the concept of security constraint and methods such as security attack scenarios to analyse security requirements, as a suitable candidate language. The selection of Secure Tropos results from a detailed analysis of the adequacy of its concepts to the *information system security risk management* (ISSRM) reference model [3]. This reference model defines the fundamental concepts of ISSRM as gathered from a quantity of standards and other sources, e.g., [5] [6] [7]. The overall approach is illustrated throughout this paper reusing the example of the electronic Single Assessment Process (eSAP) [8].

The structure of the paper is as follows: in Section 2 we provide theoretical background for our research. In Section 3 we outline our research method and apply Secure Tropos in the running example. In Section 4 we describe how Secure Tropos is aligned with the concepts of the ISSRM reference model. Finally Section 5 discusses the findings and presents conclusions of the study.

2 Theory

2.1 Security Risk Domain

The ISSRM Reference model [3] presented in Fig. 1 results from a consolidation of existing security standards, e.g., [5], [6], [7]. In this section we summarise some core definitions of ISSRM concepts.

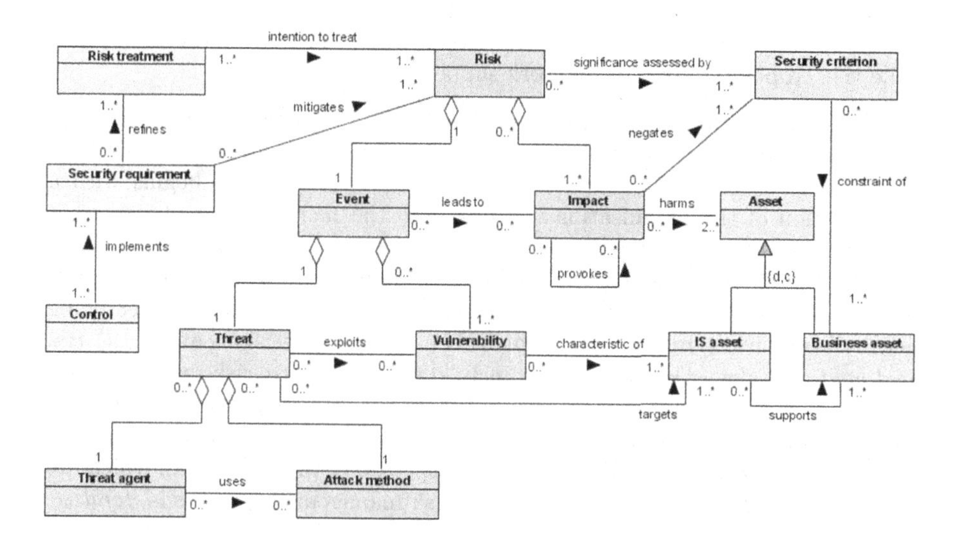

Fig. 1. The ISSRM Reference Model [3] [9]

Asset-related concepts describe assets and the criteria which guarantee asset security [3] [9]. An *asset* is anything that has value to the organisation and is necessary for achieving its objectives. A *business asset* describes information, processes, capabilities and skills inherent to the business and core mission of the organisation, having value for it. An *IS asset* is a component of the IS supporting business assets like a database where information is stored. A *security criterion* characterises a property or constraint on business assets describing their security needs, usually for confidentiality, integrity and availability.

Risk-related concepts present how the risk itself is defined [3] [9]. A *risk* is the combination of a threat with one or more vulnerabilities leading to a negative impact harming the assets. An *impact* describes the potential negative consequence of a risk that may harm assets of a system or an organisation, when a threat (or the cause of a risk) is accomplished. An *event* is the combination of a threat and one or more vulnerabilities. A *vulnerability* describes a characteristic of an IS asset or group of IS assets that can constitute a weakness or a flaw in terms of IS security. A *threat* characterises a potential attack or incident, which targets one or more IS assets and may lead to the assets being harmed. A *threat agent* is an agent that can potentially cause harm to IS assets. An *attack method* is a standard means by which a threat agent carries out a threat.

Risk treatment-related concepts describe what decisions, requirements and controls should be defined and implemented in order to mitigate possible risks [3] [9]. A *risk treatment* is an intentional decision to treat identified risks. A *security requirement* is the refinement of a treatment decision to mitigate the risk. *Controls* (countermeasures or safeguards) are designed to improve security, specified by a security requirement, and implemented to comply with it.

Like the Tropos Goal-risk framework [10], the ISSRM reference model addresses risk management at three different levels, combining asset, risk, and risk treatment views. However the ISSRM reference model focuses on the *IS* security while the Tropos Goal-risk framework supports risk in general.

Security risk management process. The ISSRM activities follow the general risk management process [3] [9]. This process originates from the risk management standards (e.g., [5], [6], [7]) and consists of six steps. It begins with a (*a*) definition of the organisation's *context* and the *identification of its assets.* Next one needs to determine the (*b*) *security objectives* (confidentiality, integrity and/or availability), based on the level of protection required for the assets. During (*c*) *risk assessment* one elicits which risks are harming assets and threatening security objectives. Once risk assessment is performed, decisions about (*d*) *risk treatment* are taken. Decisions might include risk avoidance, risk reduction, risk transfer and risk retention. *Security requirements* (*e*) on the IS can thus be determined as security solutions to mitigate the risks. Requirements are instantiated into (*f*) *security controls,* i.e. system specific countermeasures, which are implemented within the organisation. The risk management process is *iterative.* Each step can be repeated to obtain an outcome of higher quality. Furthermore, after determination of the security controls new risks, that overcome or are not addressed by these security controls, can emerge.

2.2 Security Modelling Languages

At different IS development phases, security can be addressed using various modelling languages. Abuse frames [11] suggest means to consider security during the early RE. Abuse cases [12], misuse cases [13], and mal-activity diagrams [14] address security concerns through negative scenarios executed by the attacker. SecureUML [15] and UMLsec [16] consider security during system design.

Goal modelling languages have also been adapted to security. Secure *i** [17] addresses security trade-offs. KAOS [18] was augmented with *anti-goal models* designed to elicit attackers' rationales. In [19] [20] Tropos has been extended with the notions of *ownership, permision* and *trust.* Here we investigate Secure Tropos [4] that models security using *security constraints* and *attack methods.*

All these languages are candidates for supporting largely or partially the SRM activities. In this paper we specifically target security risk management in the *early* IS development. Thus, we have chosen Secure Tropos, which incrementally introduces security concerns from the requirements phases. However, the final analysis of the security concerns takes place only during the design phases [21]. Therefore by aligning Secure Tropos with the ISSRM reference model, we suggest improvements needed for the SRM in the early (requirements) IS phases.

2.3 Secure Tropos

Secure Tropos enriches a set of Tropos [22] [23] constructs (*actor*, *goal*, *softgoal*, *plan*, *resource*, and *belief*) with security constructs such as *security constraint*, and *threat*. An *actor* (see Fig. 3) describes an entity that has strategic goals and intentions within the system or within the organisational setting [22]. A *hardgoal* or simply *goal* hereafter (see Fig. 3), represents an actor's strategic interests. A *softgoal* (see Fig. 5) unlike a *goal*, does not have clear criteria for deciding whether it is satisfied or not and therefore it is subject to interpretation (goals are said to be *satisfied* while softgoals are said to be *satisficed*). A *plan*(see Fig. 4) represents a way of doing things. A *resource* (see Fig. 3) represents an informational or physical entity. A *belief* (see Fig. 7) is the actor's knowledge of the world. All these constructs are present in both Tropos [22] [23] and Secure Tropos [8] [21] [24]. In addition Secure TROPOS introduces *security constraints* and *threats*. A *security constraint* represents a restriction related to security that the system must have and actors must respect (see Fig. 3) [4] [24]. A *threat* (see Fig. 6) "represents circumstances that have the potential to cause loss or problems that can put in danger the security features of the system" [4].

Constructs are combined using relationships: *dependency*, *decomposition*, *means-ends*, *contribution*, *restricts* and *attacks*. In the *actor model* one represents the network of relationships between actors. The relationships are captured using the *dependency* links. *Dependency* between two actors indicates that one actor (the depender) depends for some reason (dependum) on another actor (the dependee) in order to achieve a goal, to execute a plan, or to deliver a resource [22]. *Secure dependency* introduces security constraint(s) that must be respected by actors for the dependency to be satisfied [25]. This means that "the depender expects from the dependee to satisfy the security constraint(s) and also that the dependee will make effort to deliver the dependum by satisfying the security constraint(s)" [24]. The *goal model* allows a deeper understanding of how the actors reason about goals to be fulfilled, plans to be performed and available resources [23]. The goal model uses the *means-ends*, *decomposition* and *contribution* relationships. The *means-ends* relationship (see Fig. 4) permits to link a *means* (plan/goal/resource) with an *end* (goal). The *decomposition* relationship (see Fig. 4) permits to define a finer structure of a plan. A *contribution* link (see Fig. 5) describes a positive or negative impact that one element has on another. To facilitate security analysis Secure Tropos introduces *restricts* and *attacks*. The *restricts* relationship (see Fig. 3) describes how goal achievement is restricted by security constraints. The *attacks* link (see Fig. 7) shows what is the target of an attacker's plan.

3 Research Method

3.1 Method for Aligning Secure Tropos and ISSRM

In order to align Secure TROPOS with the ISSRM reference model, the method shown in Fig. 2 is applied. Our approach is based on the definition of the Secure

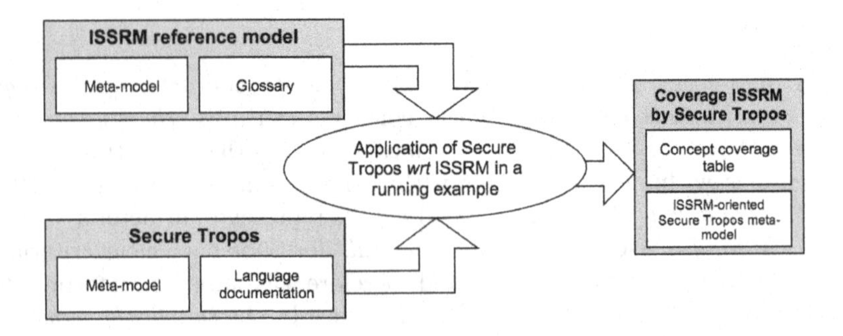

Fig. 2. Research Method

Tropos language as it is derived from the Secure Tropos meta-model and the description of the language in the literature [4] [8] [21] [24] [25].

In this paper we use a running example to explain our analysis of the alignment of Secure Tropos and the ISSRM. The running example is initially used to illustrate the use of the language. We then consider the concepts of Secure Tropos *wrt* how they were used to address ISSRM. The outcome of this comparison is the concept alignment between Secure Tropos and the ISSRM reference model. We document the final results of our alignment artefacts in Fig. 9. At the same time, an "ISSRM-oriented" Secure TROPOS meta-model is produced. By "ISSRM-oriented", we mean a meta-model [26][1] aligned on the ISSRM reference model and thus showing only concepts and relationships semantically equivalent to those of the ISSRM reference model.

3.2 Running Example

To demonstrate the applicability of our work in a practical and realistic environment we use it to analyse the electronic Single Assessment Process (eSAP) [27]. The eSAP is an IS to support integrated assessment of the health and social care needs of elderly. It is based on the Single Assessment Process, which is part of the National Service Framework for Older People Services of the English Department of Health. The eSAP is suitable to demonstrate our work for two main reasons: (*i*) security and risk are two important factors in its development and implementation; (*ii*) the security of the system have been successfully analysed using the Secure Tropos methodology [28]. Therefore, by revisiting the running example, we are able to identify the exact contributions of this paper. Due to space limitations, we focus on one of the most important aspects to make the eSAP: the Patient personal information.

(a) Context and asset identification. A Social Worker is in charge of the health care to patients. In order to fulfill her work, she needs the Patient personal information. In Fig. 3 the Social Worker depends on a goal Collected care

[1] Due to space requirements we did not include the Secure Tropos meta-model nor the ISSRM-oriented Secure Tropos meta-model.

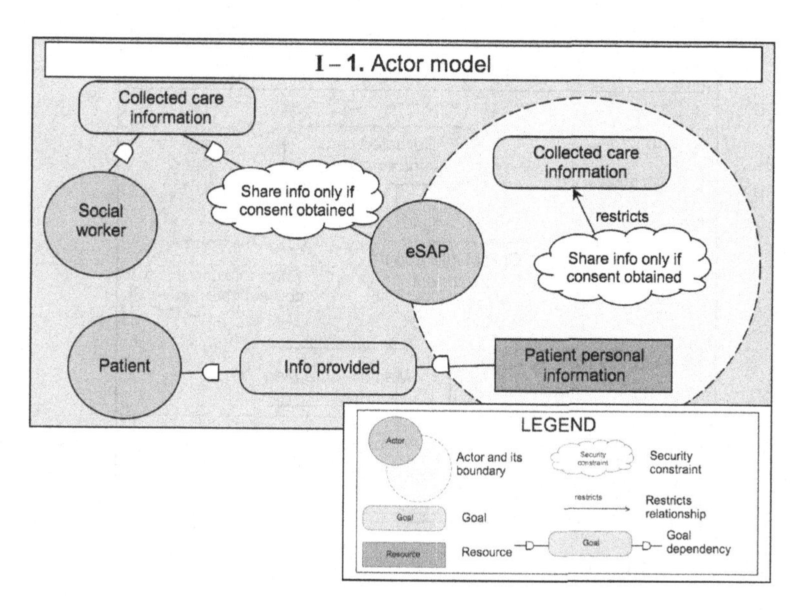

Fig. 3. Actor model

information held by the eSAP system. As the information is a valuable business asset, achievement of the goal Collected care information is restricted by a security constraint assuring that the consent has to be obtained before the personal information can be sent. The goal Collected care information can be achieved by executing the plan Collect info about treatment, which needs to gather the Patient personal information and to perform the Manage care plan, see Fig. 4.

*(b) **Security objective determination.*** The plan Check data for consent contributes positively to the security constraint Share info only if consent obtained (Fig. 5). This plan also realises the goal Consent has been obtained. In our example we strive for privacy of the Patient personal information, thus the goal Consent has been obtained takes part in the decomposition of the plan Perform authorisation checks. The latter plan is the means to fulfill the goal System privacy ensured and contributes positively to the security constraint Keep system data privacy.

*(c) **Risk analysis and assessment.*** Fig. 6 focuses on a possible risk event. We identify an Authentication attack (modelled using the *threat* construct). It describes a situation where a threat agent fakes his identity to pass himself off as a trusted actor in order to damage the business assets (e.g., Patient personal information). The Authentication attack has a negative impact on Privacy. On the other hand the constraint Keep system data privacy mitigates the possible risk difficult to realise. Note that the Authentication attack does not depend on the existence of an actor whose assets are threatened.

In Fig. 7 we present the view of an Attacker whose aim is to get the Patient personal information. The Attacker poses a threat (the goal Info about patient received and plan Collect info about breaking the system in Fig. 7). The plan is decomposed into two parts: (*i*) the attacker has to get the consent for the

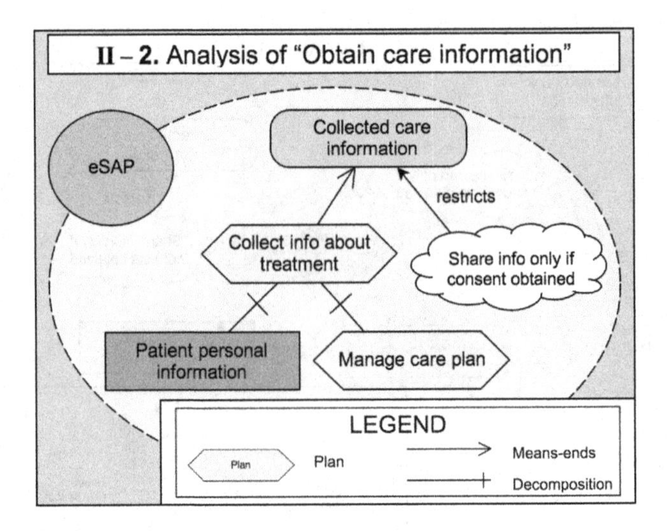

Fig. 4. Analysis of "Obtain care information"

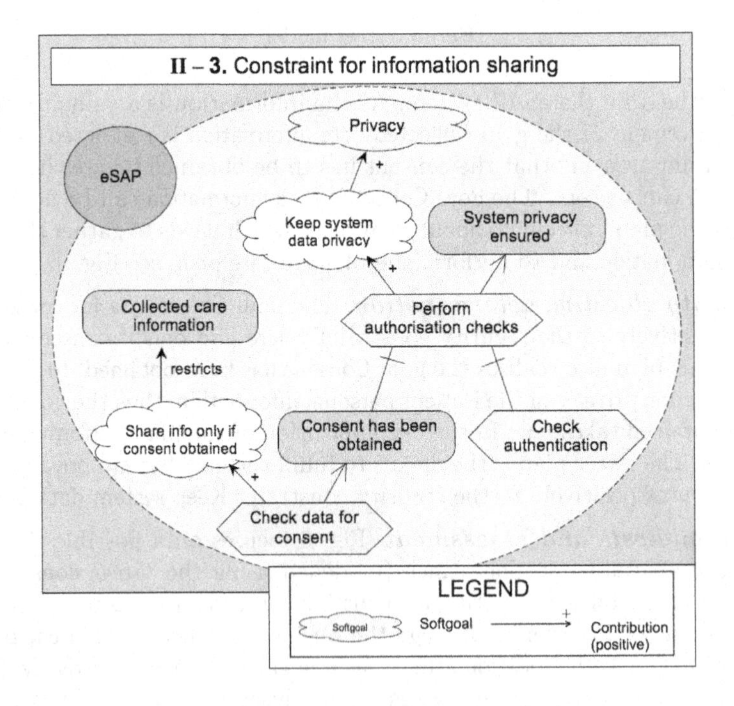

Fig. 5. Constraint for information sharing

Patient personal information; and (*ii*) he needs to find the authentication code for the system. To get the consent, the attacker can Steal data from a social worker or Buy data from the untrusted social worker. Here, belief Possible to check eSAP access repeatedly corresponds to a vulnerability, known by the attacker.

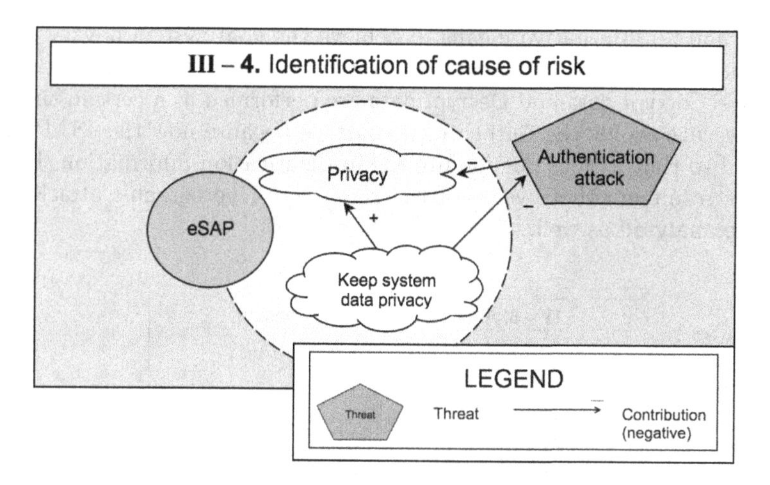

Fig. 6. Identification of an authentication risk

The vulnerability contributes positively to the decomposition between two plans Collect info about breaking the system and Check eSAP access repeatedly. Fig. 7 can be seen as the refinement of the cause of the risk identified in Fig. 6.

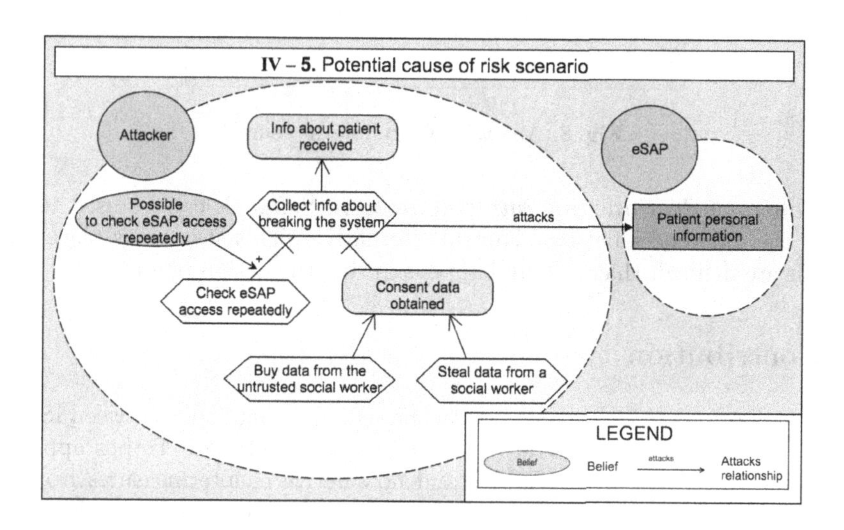

Fig. 7. Potential attack scenario

(*d*) ***Risk treatment.*** Several risk treatment decisions are suggested in [29]. In the example we apply *goal/plan substitution*, meaning that we choose different goals to be fulfilled and plans to be executed to mitigate the risk. This produces a different system design but allows avoiding the Authentication attack.

(*e*) ***Security requirements definition.*** The next step is the elicitation of the countermeasures that help to mitigate the actual risk. With respect to Fig. 5, we try to find an alternative means to achieve the goal System privacy ensured. Our solution is to Perform cryptographic procedures (Fig. 8). To realise the countermeasure, Encrypt data and Decrypt data are performed at a certain time. Our countermeasure avoids the Authentication attack because now the eSAP system is designed so that it does not require the authentication information. However this might result in other events of the risk (e.g., Cryptographic attack) which need to be analysed as well.

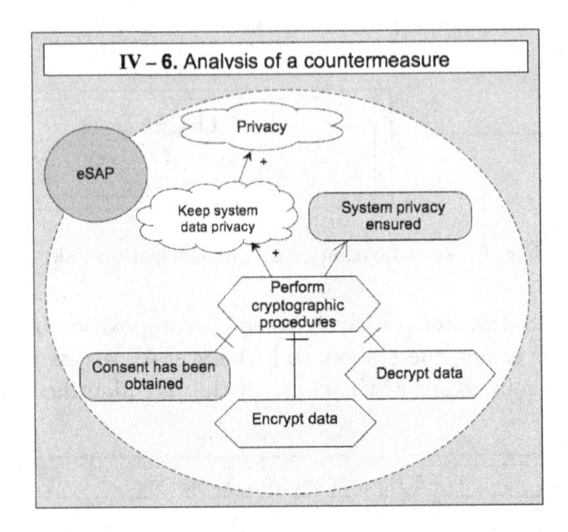

Fig. 8. Analysis of a countermeasure

(f) Control selection and implementation. Softgoals can be used to reason on the differences between control alternatives. This step takes place after controls are defined, that usually happens during the design phase.

4 Contribution

The contribution of the analysis is with the semantic alignment between ISSRM and Secure Tropos. We illustrate how we can use the Secure Tropos approach to analyse possible attack scenarios and how derive countermeasures from attack scenarios. We summarise the discussion on alignment in Fig. 9. First two columns list the concepts of the ISSRM reference model, the third column provides synonyms of the ISSRM concepts found in the Secure Tropos literature [4] [8] [21] [24] [25]. The fourth column lists the Secure TROPOS constructs used to address the ISSRM concepts. The last column illustrates the Secure TROPOS concepts used in the running example in Section 3.2.

Asset-related concepts describe what assets are important to protect, and what criteria guarantee asset security [3]. In Secure TROPOS we identify that the *actor*, *goal*, *resource* and *plan* constructs (and appropriate relationships among them) are used to model both *business* and *IS assets*. For instance, on the one

The ISSRM model		Secure TROPOS		
		Synonyms found in literature*	Language constructs	Elements from example
Asset-related concepts	Asset	–	Actor, Goal, Softgoal, Plan, Resource	–
	Business asset	–		**Actor**[Patient]; **Actor**[Social worker]; **Goal**[Obtain care information]; **Goal**[Info provided]; **Resource**[Patient personal information]; **Plan**[Collect info about treatment]; **Plan**[Manage care plan]
	IS asset	–		**Actor**[eSAP]; **Goal**[System privacy ensured]; **Plan**[Perform authorization check]; **Plan**[Check authentication]; **Goal**[Consent has been obtained]; **Plan**[Check data for consent]
	Security criteria	Security feature, Protection property	Security constraint, Softgoal	**Security constraint**[Share info only if consent obtained]; **Security constraint**[Keep system data privacy] **Softgoal**[Privacy]
Risk-related concepts	Risk	–	–	–
	Impact	–	Contribution between the threat and softgoal	**Contribution** between **Threat**[Authentication attack] and **Softgoal**[Privacy]
	Event	–	Threat	**Threat**[Authentication attack]
	Threat	–	Goal, Plan	**Plan**[Collect info about breaking the system]; **Goal**[Consent data obtained]
	Vulnerability	–	Belief **	**Belief**[Possible to check eSAP access repeatedly]
	Threat agent	Attacker	Actor	**Actor**[Attacker]
	Attack method	–	Plan, Relationship attacks	**Plan**[Collect info about breaking the system]; **Plan**[Check eSAP access repeatedly]; **Plan**[Steal data from a social worker]; **Plan**[Buy data from untrusted social worker]; **Plan**[Collect info about breaking the system] **attacks**
Risk treatment-related concepts	Risk treatment	–	–	–
	Security requirement	Secure goal, secure plan, secure resource, protection objective	Actor, Goal, Softgoal, Plan, Resource Security constraint	**Plan**[Perform cryptographic procedures]; **Plan**[Encrypt data]; **Plan**[Decrypt data]
	Control	–	New model which implements security requirements	Cryptographic module in the eSAP system

Fig. 9. Alignment between the ISSRM reference model and Secure Tropos. * – literature includes [8] [4] [21] [25]; ** – look for discussion about belief in Section 4.

hand the *actors* Patient and Social worker (see Fig. 3), the *goals* Obtain care information and Info provided and the *plans* Collect info about treatment and Manage care plan (see Fig. 4) describe the process necessary for the organisation (health care centre) to achieve its objectives. On the other hand the *resource* Patient personal information characterises the valuable information. All the mentioned examples are identified as *business assets* with respect to the ISSRM reference model [3].

The business processes and information management are supported by the IS, which in our example is the eSAP. In more details (see Fig. 5) the support for the *business assets* is described by the *goals* System privacy ensured and Consent has been obtained and the *plans* Perform authorisation check, Check authentication and Check data for consent. The concepts which describe how a component or part of the IS is necessary in supporting *business assets*, are called *IS assets*.

The ISSRM *security criteria* are properties or constraints on business assets characterising their security needs [3]. In Secure Tropos *softgoals* (e.g. Privacy) can help identify higher level security criteria, like privacy, integrity and availability. Depending on the context it might be necessary to specify other *security criteria*, like we do by using the *security constraints* Share info only if consent obtained and Keep system data privacy (see Fig. 5).

Risk-related concepts present how the risk itself is defined, and what major principles should be taken into account when defining the possible risks [3]. Risk is described by the event of the risk, corresponding to the Authentication attack in Fig. 6. The potentional negative consequence of the risk, identified by a negative contribution link between the Authentication attack and the *security constraint* Privacy is called *impact of the risk*. Here the *impact* negates the security criteria and compromises the *business asset* private.

In Fig. 7 a combination of the *goal* Info about patient received and the *plan* Collect info about breaking the system corresponds to the *threat* describing the potential attack targeting the *business asset* Patient personal information. The threat is triggered by the *threat agent* Attacker who *knows* about the possibility to check the eSAP access repeatedly as identified by the *belief* in Fig. 7. To break into the eSAP system the Attacker carries an *attack method* consisting of the plans Check eSAP access repeatedly and Steal data from a social worker.

Note that in Fig. 9 *belief* only partially corresponds to ISSRM *vulnerability*. Firstly, the fact that the *actor* (who has the role of the *attacker*) thinks he knows, might be true. In this case the *belief* will correspond to *vulnerability* in the sense of the ISSRM. However, it does not allow lining to a system design solution because this solution might not exist in the early IS development phase. Secondly, facts known by the *attacker* might be wrong: in this case there is no corresponding concept in the ISSRM. Finally, *belief* does not represent *vulnerabilities* which exist in the system but is not known by the *attacker*.

Risk treatment-related concepts describe what decisions, requirements and controls should be defined and implemented in order to mitigate possible risks [3]. According to [18] [29] in our example we use *goal/plan substitution* which leads to a different eSAP design avoiding the identified threat. New *security requirements* (see Fig. 8) that mitigate the risk are identified as *plans* Perform cryptographic procedures, Encrypt data, and Decrypt data. We illustrate the countermeasure only using the Secure Tropos *plan* construct, however we must note that, depending on the selected *risk treatment decision*, the combination of *actor*, *goal*, *resource* and *plan* might result in different security *control* systems.

5 Discussion and Conclusion

In this paper we have analysed how Secure Tropos can be applied to analyse security risks at the early IS development. Based on an illustrative example, we showed how a Secure Tropos model can be created following the security risk management process. Our purpose was not to develop the example in detail (for instance we do not detail how the plan Check data for consent in Fig. 5 has to be performed), but rather to investigate how different language constructs can be used to model security risks. We focus on the early phase (early and late requirements) of IS development. This means that the analysis of Secure Tropos is not complete *wrt* the late development, for instance we do not consider *capabilities* which are the notion used during IS design.

We know that our research method and results could hold a certain degree of subjectivity regarding the selection of the Secure TROPOS language's constructs at the modelling stage, their application and their comparison with ISSRM. To deal with the subjectivity within the team we (*i*) looked at the Secure Tropos meta-model, clarified unclear use of language constructs; (*ii*) collectively agreed on decisions made when creating the running example; (*iii*) discussed and reasoned about the Secure Tropos and ISSRM alignment.

The alignment suggests a number of improvements for Secure Tropos in the context of security risk management activities:

- Secure Tropos has to provide guidelines as to when and how to use each constructs in order to avoid misinterpretations of the ISSRM concepts. One improvement could be inclusion of tags in the label of a construct. For example, the *plan* construct can be used to model *business assets*, *IS assets*, *threats* and *security requirements*. Thus, labels such as [BS] could indicate *business assets*; [IS]- *IS assets*; [Th]- *threat*; and [SR]- *security requirements*. In our running example we deal with this limitation by decomposing the model into separate diagrams: we use the *plan* construct to represent *business assets* in Fig. 4, *IS assets* in Fig. 5, *threats* in Fig. 7, and *security requirements* in Fig. 8.
- Secure Tropos could be improved with additional constructs to better cover the concepts of ISSRM. Fig. 9 indicates that several concepts such as *risk*, *risk treatment*, and *control* are not in the Secure Tropos approach.
- The semantics of individual modelling constructs should be adapted so that they adequately represent ISSRM concepts. For example, as discussed, the *belief* construct only partially covers *vulnerability*. A possible improvement is recently suggested in [17] by introducing *vulnerable points* in the modelled IS. But some future research is needed to answer if a relationship between *vulnerable points* and *belief* is possible.

Note that the *research method* used for alignment between language constructs and the ISSRM reference model can be used to evaluate of any security modelling language. In addition to Secure Tropos we also investigated *KAOS extended to security* [26] and *misuse cases* [9]. We envision that after analysing a number of security languages it will be possible to facilitate model transformation and language interoperability. This would allow representing ISs using different perspectives, also ensuring IS sustainability.

Acknowledgment. This work is partially funded by the Interuniversity Attraction Poles Programme, Belgian State, Belgian Science Policy. We also thank A. Classen for proofread of the paper.

References

1. Basel Committee on Banking Supervision: International Convergence of Capital Measurement and Capital Standards. Bank for International Settlements (2004)
2. United States Senate and House of Representatives in Congress: Sarbanes-Oxley Act of 2002. Public Law 107-204 (116 Statute 745) (2002)
3. Mayer, N., Heymans, P., Matulevičius, R.: Design of a Modelling Language for Information System Security Risk Management. In: Proceedings of the 1st International Conference on Research Challenges in Information Science (RCIS 2007), pp. 121–131 (2007)
4. Mouratidis, H., Giorgini, P.: Secure Tropos: A Security-oriented Extension of the Tropos Methodology. International Journal of Software Engineering and Knowledge Engineering (IJSEKE) 17(2), 285–309 (2007)
5. DCSSL: EBIOS–Expression of Needs and Identification of Security Objectives (2004)
6. ENISA: Inventory of Risk Assessment and Risk Management Methods (2004)
7. ISO: Information Technology–Security Techniques–Information Security Management Systems–Requirements, International Organisation for Standardisation (2005)
8. Mouratidis, H., Giorgini, P., Manson, G.: Using Tropos Methodology to an Model Integrated Health Assessment System. In: Proceedings of the Fourth International Bi-Conference on Agent-oriented Information Systems (AOIS 2002) (2002)
9. Matulevičius, R., Mayer, N., Heymans, P.: Alignment of Misuse Cases with Security Risk Management. In: Proceedings of the ARES 2008 Symposium on Requirements Engineering for Information Security (SREIS 2008), pp. 1397–1404. IEEE Computer Society, Los Alamitos (2008)
10. Asnar, Y., Giorgini, P.: Modelling Risk and Identifying Cuntermeasure in Organizations. In: Proceedings of the 1st Interational Workshop on Critical Information Intrastructures Security, pp. 55–66. Springer, Heidelberg (2006)
11. Lin, L., Nuseibeh, B., Ince, D., Jackson, M.: Using Abuse Frames to Bound the Scope of Security Problems. In: Proceedings of the 12th IEEE international Conference on Requirements Engineering (RE 2004), pp. 354–355. IEEE Computer Society, Los Alamitos (2004)
12. McDermott, J., Fox, C.: Using Abuse Case Models for Security Requirements Analysis. In: Proceedings of the 15th Annual Computer Security Applications Conference (ACSAC 1999), p. 55 (1999)
13. Sindre, G., Opdahl, A.L.: Eliciting Security Requirements with Misuse Cases. Requirements Engineering Journal 10(1), 34–44 (2005)
14. Sindre, G.: Mal-activity Diagrams for Capturing Attacks on Business Processes. In: Sawyer, P., Paech, B., Heymans, P. (eds.) REFSQ 2007. LNCS, vol. 4542, pp. 355–366. Springer, Heidelberg (2007)
15. Lodderstedt, T., Basin, D.A., Doser, J.: SecureUML: A UML-based Modeling Language for Model-driven Security. In: Jézéquel, J.-M., Hussmann, H., Cook, S. (eds.) UML 2002. LNCS, vol. 2460, pp. 426–441. Springer, Heidelberg (2002)
16. Jurjens, J.: UMLsec: Extending UML for Secure Systems Development. In: Jézéquel, J.-M., Hussmann, H., Cook, S. (eds.) UML 2002. LNCS, vol. 2460, pp. 412–425. Springer, Heidelberg (2002)
17. Elahi, G., Yu, E.: A Goal Oriented Approach for Modeling and Analyzing Security Trade-Offs. In: Parent, C., Schewe, K.-D., Storey, V.C., Thalheim, B. (eds.) ER 2007. LNCS, vol. 4801, pp. 87–101. Springer, Heidelberg (2007)

18. van Lamsweerde, A.: Elaborating Security Requirements by Construction of Intentional Anti-models. In: Proceedings of the 26th International Conference on Software Engineering (ICSE 2004), pp. 148–157. IEEE Computer Society, Los Alamitos (2004)
19. Giorgini, P., Massacci, F., Mylopoulos, J., Zannone, N.: Modeling Security Requirements Through Ownership, Permision and Delegation. In: Proceedings of the 13th IEEE International Conference on Requirements Engineering (RE 2005). IEEE Computer Society, Los Alamitos (2005)
20. Giorgini, P., Massacci, F., Mylopoulos, J., Zannone, N.: Modelling social and individual trust in requirements engineering methodologies. In: Proceedings of the 3nd International Conference on Trust Management. LNCS, pp. 161–176. Springer, Heidelberg (2005)
21. Mouratidis, H., Jurjens, J., Fox, J.: Towards a Comprehensive Framework for Secure Systems Development. In: Dubois, E., Pohl, K. (eds.) CAiSE 2006. LNCS, vol. 4001, pp. 48–62. Springer, Heidelberg (2006)
22. Bresciani, P., Giorgini, P., Giunchiglia, F., Mylopoulos, J., Perini, A.: TROPOS: an Agent-oriented Software Development Methodology. Journal of Autonomous Agents and Multi-Agent Systems 8, 203–236 (2004)
23. Castro, J., Kolp, M., Mylopoulos, J.: Towards Requirements-Driven Information Systems Engineering: The TROPOS Project. Information Systems 27, 365–389 (2002)
24. Mouratidis, H., Giorgini, P., Manson, G.A.: When Security Meets Software Engineering: a Case of Modelling Secure Information Systems. Information Systems 30(8), 609–629 (2005)
25. Mouratidis, H., Giorgini, P., Manson, G.: Integrating Security and Systems Engineering: Towards the Modelling of Secure Information Systems. In: Eder, J., Missikoff, M. (eds.) CAiSE 2003. LNCS, vol. 2681, pp. 63–78. Springer, Heidelberg (2003)
26. Genon, N.: Modelling Security during Early Requirements: Contributions to and Usage of a Domain Model for Information System Security Risk Management. Master thesis, University of Namur (2007)
27. Mouratidis, H., Philp, I., Manson, G.: A Novel Agent-Based System to Support the Single Assessment Process of Older People. Journal of Health Informatics 9(3), 149–162 (2003)
28. Mouratidis, H.: A Security Oriented Approach in the Development of Multiagent Systems: Applied to the Management of the Health and Social Care Needs of Older People in England. PhD thesis, Department of Computer Science, University of Sheffield, UK (2004)
29. van Lamsweerde, A., Letier, E.: Handling Obstacles in Goal-oriented Requirements Engineering. Transactions on Software Engineering 26(10), 978–1005 (2000)

Probabilistic Entity Linkage for Heterogeneous Information Spaces

Ekaterini Ioannou, Claudia Niederée, and Wolfgang Nejdl

L3S Research Center/Leibniz Universität Hannover
Appelstr. 9a, 30167 Hannover, Germany
{ioannou,niederee,nejdl}@L3S.de

Abstract. Heterogeneous information spaces are typically created by merging data from a variety of different applications and information sources. These sources often use different identifiers for data that describe the same real-word entity (for example an artist, a conference, an organization). In this paper we propose a new probabilistic *Entity Linkage* algorithm for identifying and linking data that refer to the same real-world entity.

Our approach focuses on managing entity linkage information in heterogeneous information spaces using probabilistic methods. We use a Bayesian network to model evidences which support the possible object matches along with the interdependencies between them. This enables us to flexibly update the network when new information becomes available, and to cope with the different requirements imposed by applications build on top of information spaces.

Keywords: entity linkage, data integration, metadata management.

1 Introduction

Many applications rely on rich *information spaces* - collections of data from a variety of different applications and information sources. Information spaces are found in various systems of different research areas. For example Semantic Web with applications analyzing social networks such as identifying conflicts of interests [1], or researcher's influence in a community [16]. Also, Personal Information Management (PIM) with systems such as Beagle^{++} [8], and Haystack[1], as well as information integration approaches for Information Systems [19]. The success of these applications depends on a clear picture about the data and their relationships that the underling information spaces provide.

When compiling information spaces from heterogeneous sources, data referring to the same real-world entity (for example to a researcher or an artist, a conference or an organization) often use different identifiers. Different attribute sets (e.g. 'hasName', 'author'), the use of naming variants (e.g., 'Wolfgang Nejdl', 'Nejdl W.'), and most importantly, the lack of a global coordination for identifier assignment, forces each source to create and use its own identifiers.

[1] http://haystack.lcs.mit.edu/

Z. Bellahsène and M. Léonard (Eds.): CAiSE 2008, LNCS 5074, pp. 556–570, 2008.

Furthermore, each source describes entities in a way most adequate for its purpose. A publication will describe a person using name and affiliation, whereas an email will use the email address. It is the goal of *Entity Linkage* to identify data describing the same real-world entity, and link their corresponding identifiers.

In order to cope with uncertainties inherent in entity linkage, in this paper we propose a new probabilistic entity linkage algorithm, which is able to compute the probability for each possible match between data according to the evidences currently available in the information space. Our approach addresses the special characteristics and resulting **challenges** of entity linkage in information spaces for PIM:

- We can distinguish two main directions for identifying possible matches in information spaces. The first is based on observing similarities between text values of the data participating in a potential match. The second direction relies on identifying relationships between the data. $\overset{C1}{\longmapsto}$ Entity linkage should be able to follow both directions, and incorporate the observed evidences.
- Information spaces for PIM constantly change and evolve through interactions of the user with his/her desktop. This changes the information available to the entity linkage algorithm. $\overset{C2}{\longmapsto}$ An entity linkage solution should support incremental computation and adaptation of entity matching information. Also, since entity linkage results are never finalized, the original data of the information space should not be modified.
- A wide variety of applications can be executed on top of integrated information spaces. Each application might have different requirements for the entity linkage solution. For example, one application might need only certain matches, whereas other applications might accept uncertain matches based on only few evidences. $\overset{C3}{\longmapsto}$ Matches should be accompanied with a metrics, indicating the belief we have that the corresponding identifiers refer to the same real-world entity, based on the evidences in the current information space.

The strong interconnections in information spaces are a valuable source of entity matching evidence. Previous approaches operating in such information spaces, such as [10] and [3], did not restrict themselves to entity attributes but also systematically exploit the context of the entity, taking into account associations with other entities. In particular, [10] uses association properties of entities in combination with normal attributes for computing record linkage. Also, [3] exploits the link structure of Web pages about persons as an indicator for entity (person) relationships. Our approach is most similar to the approach presented in [10], which uses entity context in the form of relationships and propagation of matching evidence information. However, we go further by also addressing the other two PIM characteristics described above, while achieving comparable precision and recall performance.

The main contribution of our work is an innovative entity linkage algorithm, that addresses all three challenges listed, by: (i) clearly separating data of the information space with data representing decisions for matches, (ii) enabling incremental update of matches, when new information becomes available in the information space, and (iii) associating each match with a probability indicating the belief (confidence) we have for the existence of the specific match. Our algorithm receives entity metadata and computes matching probabilities by constructing and maintaining a Bayesian network with matching evidences. As a further contribution, we introduce and explain the problem as it appears in heterogeneous PIM information spaces, as a set of requirements to be addressed by our algorithm.

The rest of the document is organized as follows. Section 2 gives an overview over related work. Section 3 provides a formal formulation of the entity linkage problem for PIM, and Section 4 presents our algorithm. Section 5 presents our evaluation experiments, showing good precision and recall on real-life test collections. Finally, Section 6 concludes and discusses future extensions.

2 Related Work

Variants of the problem we address in this paper have been investigated in different research areas. Traditionally, the database community proposed algorithms to detect database tuples that refer to the same real-world entity [20,12]; a problem known as record linkage, data integration, and merge/purge. More recently, algorithms in the data mining community propose identifying real-world entities as a way to perform data cleaning, and clustering tasks. These algorithms try to identify the data that describe the same real-world through interconnections. Merging objects is done by calculating the *distance* between data that possibly describe the same real-world entity [5,6], or computing the interconnection strength of their alternative connection paths [15,14].

The most relevant algorithms related to our approach are the ones that identify entities through interconnections between objects found in a given dataset. Ananthakrishna et al. [2] exploit dimensional hierarchies to detect fuzzy duplicates in dimensional tables. The hierarchies are build by following the links between the data of one table to data of other tables. Entities are matched when the information along these generated hierarchies is found similar. The most recent algorithm, motivated by a Personal Information Management scenario, is the *Reference Reconciliation* algorithm by Dong et al. [10]. The authors use interconnections to identify and merge data that possibly describe the same real-world entity. Information about these merges is propagated into the rest of the dataset (reconciliation propagation), along with the exchange of information between the two merged references (reference enrichment). A modified version of this algorithm [1] is used for detecting conflict of interests in paper reviewing processes.

The DBLP system[2] faces a similar problem with author names. To solve it, they construct a co-author graph (nodes show authors, links show common publications). Merging authors is done using *edit distance* algorithms, based on

[2] http://dblp.uni-trier.de/

comparisons such as Levenshtein distance and soundex. The TAP system [11] uses a *Semantic Negotiation* process through which the common descriptions (if any) between the different resources are identified. These common descriptions are then used to create a unified view of a given data set. Swoosh [4] is another related system. Here, the authors focus on identifying the different properties that affect the efficiency of such algorithms, and introduce different approaches to address the possible combinations of the found properties.

3 Problem Formulation

Entity linkage is concerned with relations of objects described in heterogeneous information spaces, with corresponding entities existing in the real-world. We start with an information space consisting of metadata describing resources, for example emails or publications. The metadata include descriptions of *objects* such as persons or conferences. Objects that refer to the same real-world entities will often have different identifiers. Relating objects of the information space to entities of the real-world allows us to discover the objects which should have the same identifier. We will give a formal definition of this problem in the following paragraphs, and provide a summary of the used notation Table 1.

Table 1. A summary of the algorithm's notation

Symbol	Description
$\mathcal{D}$	The heterogeneous information space
r	A Resource (for example an email, a publication)
$\mathcal{M}(r)$	The metadata describing resource r
$e(t_i, t_j, \phi)$	Evidence showing the similarity of t_i with t_j, as given by function ϕ
$rep_{r,k}$	The representation of object k in resource r
$\mathrm{d}(rep_{r,k})$	The entity mapping for representation $rep_{r,k}$
$\mathrm{P}(\mathrm{d}(rep_{r_i,k})=\mathrm{d}(rep_{r_j,m}))$	A match between two entity mappings

Definition 1. *The information space $\mathcal{D}$, on which we execute our algorithm, is a set of metadata describing resources $\mathcal{R}_D$. It is defined as $\mathcal{D} = \{\mathcal{M}(r_i) \mid r_i \in R_\mathcal{D}\}$, where $\mathcal{M}(r_i)$ denotes the metadata describing an individual resource r_i.*

Definition 2. *Metadata for resource r_i is represented as a set of tuples t. These tuples describe the resource along with the objects found in the context of the resource. It is defined as $\mathcal{M}(r_i) = \{\, t_j \mid j \leq sizeof(t_{r_i})\}$.*

In this paper, we assume that $\mathcal{D}$ complies with the RDF data model, and thus a tuple t_i is a triple of the form $\langle u, p, o\rangle$. Symbol u denotes a URI, p denotes a property, and o an RDF-object which can either be a literal, or a URI. URIs are used as identities of resources. Ideally, the same identity would be used whenever describing the same object. However, since the URI assignment is not globally coordinated, multiple URIs are used for single real-world entities (see

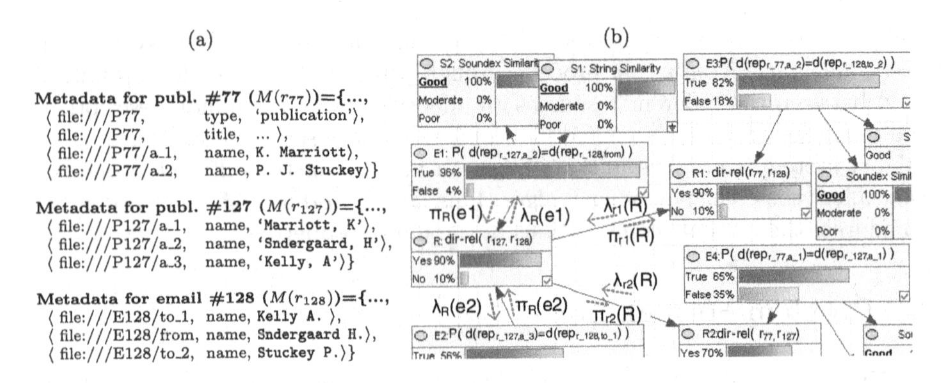

Fig. 1. (a) Metadata for desktop resources, and (b) corresponding Bayesian network

[7] for more details). Therefore, even if URIs provide an appropriate formalism for unique identifiers, the entity linkage problem is still present.

Definition 3. *Let object representation $rep_{r,k}$ denote the subset of the tuples from $\mathcal{M}(r)$ which refer to/describe the same object k in r. It is defined as* $rep_{r,k} = \{t \mid t = \langle id_k, p, o\rangle \vee t = \langle s, p, id_k\rangle\}$, *where id_k is the local identifier (URI) assigned to object k.*

Figure 1 (a) shows some example metadata describing three desktop resources together with several representations for the contained objects. Two examples of object representation are:
- rep_{r_{77},a_1} = {⟨ file:///P77/a_1, name, K. Marriott⟩}, and
- rep_{r_{127},a_1} = {⟨ file:///P127/a_1, name, 'Marriott, K'⟩}

Definition 4. *The entity mapping d(rep) semantically maps an object representation rep to the corresponding real-world entity.*

Obviously, $\mathcal{D}$ may consist of different representations $rep_{r_i,k}, \ldots, rep_{r_j,m}$ which semantically refer to the same real-world entity, i.e., $\mathrm{d}(rep_{r_i,k}) = \ldots = \mathrm{d}(rep_{r_j,m})$. The challenge is that the mapping *d* is not known to the entity linkage algorithm. To overcome this, we have to compute the *match*, $\mathrm{P}(\mathrm{d}(rep_{r_i,k}) = \mathrm{d}(rep_{r_j,m}))$, for pairs of object representations. Each match expresses the probability with which specific entity mappings refer to the same real-world entity.

Definition 5. *The probability of each match depends on the supporting information in $\mathcal{D}$. We represent each such supporting information item with an evidence $e(t_i, t_j, \phi)$, where t_i and t_j denote metadata tuples of the entity mappings in our match, and ϕ the function which reports similarity between t_i and t_j.*

Going back to our example, the previous representations correspond to the real-world entities $\mathrm{d}(rep_{r_{77},a\_1})$ and $\mathrm{d}(rep_{r_{127},a\_1})$. To decide whether these correspond to the same entity we apply our algorithm and calculate the probability of match $\mathrm{P}(\mathrm{d}(rep_{r_{77},a\_1}) = \mathrm{d}(rep_{r_{127},a\_1}))$, based on evidences for this match

from the information space. An evidence for this match can be obtained using a string similarity function. For example:
$e(\langle$...P77/a_1, name, K. Marriott$\rangle$, $\langle$...P127/a_1, name, 'Marriott, K'$\rangle$, StringSim).

4 The Entity Linkage Algorithm

4.1 A Brief Reminder of Bayesian Networks and Inference

Bayesian networks [18,13] are probabilistic graphical models for reasoning under uncertainty, using cause-effect relationships modeled as a directed acyclic graphs. Each node in the graph represents a variable with two or more possible states. Each edge from parent node X to child node Y, represents a cause-effect relationship with X being the cause and Y the effect, whenever the state of Y is directly influenced by the state of X. Each node X is accompanied with a *local probability distribution* $P(X \mid U_1, .., U_m)$, showing the conditional probability of all states in X given the states of its parents U_1,...,U_m. Nodes without parents are associated with an unconditional $P(X)$, representing prior probabilities.

Bayesian networks represent the *joint probability distribution* over all variables, defined as the product of all local probability distributions, as follows:

$$P(X_1, X_2, ..., X_n) = \prod_{i=1}^{n} P(X_i \mid parent(X_i)),$$

where $P(X_i \mid parent(X_i))$ corresponds to the local probability distribution of node X_i, and $parent(X_i)$ to the parent nodes of X_i.

Bayesian networks successfully determine the conditional probabilities of cause nodes based on the current probabilities of the effect nodes, a task called *probabilistic inference*. Given any new effects (evidences), probabilistic inference recomputes the probability of the cause nodes which are responsible for these effects. One well-known algorithm for probabilistic inference is *message-passing* by Pearl [18]. Pearl's algorithm is iterative, and in each iteration calculates the belief of a node based on messages exchanged by the node X with its parents $U_1, ..., U_m$ and its children $Y_1, ..., Y_n$. When node X is activated, and receives all messages $\pi_X(U_i)$ from its parents, and $\lambda_{Y_j}(X)$ from its children, it calculates its own belief as:

$$BEL(X) = \alpha\lambda(X)\pi(X), \quad \text{where } \alpha \text{ is a normalization constant,}$$

$$\pi(X) = \sum_{U_1,...,U_m} P(X|U_1, ..., U_m) \prod_{i=1}^{m} \pi_X(U_i), \text{ and } \lambda(X) = \prod_{j=1}^{n} \lambda_{Y_j}(X) .$$

After calculating its belief, node X computes and sends new messages $\lambda_X(U_i)$ to its parents, and $\pi_{Y_j}(X)$ to its children. These messages are:

$$\pi_{Y_j}(X) = \alpha\, \pi(X) \prod_{k \neq j} \lambda_{Y_k}(X)$$

$$\lambda_X(U_i) = \sum_{X} \lambda(X) \sum_{U_k : k \neq i} P(X|U_1, ..., U_k) \prod_{k \neq i} \pi_x(U_k)$$

4.2 Structure of Our Bayesian Network

The goal of our algorithm, is to compute the probability of each match. We perform this task using a Bayesian network constructed from information related to the matches. The information is encoded using the following node types:

Entity Nodes. These nodes represent a match, e.g., $P(d(rep_{r_{77},a\_1}) = d(rep_{r_{127},a\_1}))$. As explained in Section 3, $\mathcal{D}$ does not have the information to directly compute matches, thus we can not specify the states of entity nodes. The probability of their states is computed through probabilistic inference based on the cause-effect relationships in the network.

Evidence Nodes. These nodes represent evidences for entity nodes. It is the first type of nodes we use to represent evidence for entity nodes. Each evidence node represents an evidence $e(t_i, t_j, \phi)$, with t_i and t_j being tuples from the two representations constructing the match. Evidence nodes form the prior probabilities in our network, since they have no parent nodes upon which they depend. The unconditional probability distribution is determined by the similarity function ϕ used for creating them. In our current approach, we rely on the comparison of literals from t_i and t_j to measure compatibility between the corresponding properties. An extension of our approach for operation in more heterogeneous contexts is the inclusion of a comparison of the properties themselves and the inclusion of these similarities into the aggregation of the matching probabilities.

Direct-Relation Nodes. These nodes rely on the fact that two resources are related when their descriptions contain the same object. It is the second type of nodes that influence the states of entity nodes. For example, match $P(d(rep_{r_{77},a\_1}) = d(rep_{r_{127},a\_1}))$ implies a relation between resources r_{77} and r_{127}, which we encode in the direct-relation node dir-rel(r_{77}, r_{127}). Finding evidences for more shared objects between these resources (e.g., additional common authors) increases the belief for the relation of r_{77} with r_{127}, and consequently the belief in the corresponding matches. Direct-relation nodes are the effect we can observe for entity nodes and for deductive-relation nodes (explained bellow). The local probability distribution of their states is directly influenced by their relationships with these node types.

Deductive-Relation Nodes. These nodes represent the indirect relation between two resources, inferred by combining the information of two nodes, either direct-relation or deductive-relation. Combining direct-relation nodes dir-rel(r_{77}, r_{127}) and dir-rel(r_{77}, r_{128}) for example, implies a new relation between r_{127} with r_{128} (due to the common resource r_{127}), which we encode in deductive-relation node del-rel(r_{127}, r_{128}).

An important aspect of the Bayesian network is the cause-effect relationships between the nodes. We have already explained these together with the different types of nodes. Table 2 gives a summary.

Table 2. The possible cause-effect relationships used in our Bayesian network

Effect Nodes:		(1) Evidence	(2) Dir.-Rel.	(3) Ded.-Rel.
Cause	(1) Entity	√	√	
Nodes:	(2) Ded.-Rel.		√	√

4.3 Incremental Computation of the Network

To compute the probability of matches we collect evidence, positive and negative. Then, we calculate the probability of matches by constructing a Bayesian network, modeling matches and related evidence. Starting point for this computation is an incrementally growing metadata set, which are added to $\mathcal{D}$. We have to update the Bayesian network incrementally, after addition of new metadata.

Upon addition of a new set of metadata, the algorithm performs the following four steps: (a) Process the object representations contained in the metadata added. By comparing the new representations with previous representations, the algorithm identifies similarities, and updates the network with new evidence and entity nodes. (2) Create direct-relation nodes to represent the effects we observed which could cause these new entity nodes. (3) Analyze the updated network and generate new information about the relation of resources, represented using deductive-relation nodes. (4) Perform probabilistic inference on the Bayesian network, and generate the updated probabilities for the matches. The remaining paragraphs of this section describe these steps in more detail.

Step 1 - Adding Entity & Evidence Nodes. The new metadata contain one or more object representations. In the first step, the algorithm updates the Bayesian network with new evidences generated using these representations. We start with similarity computations to identify resemblance between tuples from the new representation $rep(r_{new}, k)$ with compatible tuples from the existing representations $rep(r_{exists}, m)$. In the current version of our algorithm, similarities are detected using two functions. The first algorithm is *String Similarity*, detecting string resemblance between literals of tuples[3]. The second algorithm is *Soundex Similarity*, which detects the resemblance in pronunciation between literals[4]. Whenever similarity is above a given threshold, we consider it as evidence for the match $\mathrm{P}(\mathrm{d}(rep(r_{new,k}))=\mathrm{d}(rep(r_{exists,m})))$.

An evidence node is created for each similarity identified. Since the current version of our algorithm includes two similarity algorithms, we create one or two evidence nodes for each match. All evidence nodes have three states, *Good*, *Moderate*, and *Poor*, which we set based on computed similarity.

An entity node is created to represent the identified match $\mathrm{P}(\mathrm{d}(rep(r_{new,k}))$ $= \mathrm{d}(rep(r_{exists,m})))$, if such node does not yet exists. The relation between the newly created evidence nodes with the entity node is represented by introducing cause-effect relationships. All entity nodes have two possible states, *Exists* to

[3] For String Similarity we use the JaroWinkler method from the SecondString API [9].

[4] For Soundex Similarity we use the Apache Codec API.

indicate that the corresponding match exists, and *Exists_Not* to indicate that the match does not exist. The probabilities of these states are computed by probabilistic inference.

Step 2 - Adding Direct-Relation Nodes. Direct-relation nodes represent the observed effect that entity nodes could cause. They are created using only information from the matches. For each match P(d($rep(r_{new,k})$)=d($rep(r_{exists,m})$)) we extract its resources, and use them to create a direct-relation node del-rel(r_{new},r_{exists}), if this node does not yet exist. If there is more than one match referring to the same two resources, we represent this through a cause-effect relationships created between the entity nodes and the corresponding direct-relation node. The direct-relation nodes have two possible states, *Yes* to indicate that the two resources are related, and *No* to indicate that the resources are not related. The probabilities of these states are again computed by probabilistic inference.

Step 3 - Adding Deductive-Relation Nodes. This step analyzes the current status of the network to extract indirect relations between the resources. The underlying idea is similar to the one represented by the direct-relation nodes. To identify possible indirect relations, our algorithm inspects the direct-relation and deductive-relation nodes. Each node is considered as a transitive, binary relation (b-relation) between the two participating resources. For example, dir-rel(r_{77},r_{127}) corresponds to b-relation (r_{77},r_{127}), and dir-rel(r_{77},r_{128}) to b-relations (r_{77},r_{128}). The algorithm extracts more relations by transitively combining b-relations. For example, b-relation (r_{127},r_{127}) is the transitive combination of our two previous b-relations. We encode the new b-relation using a ded-rel node, for example del-rel(r_{127},r_{127}).

Since computing transitive b-relations is a recursive process, we need an appropriate stopping criterium. In the current version of our algorithm, we enforce a fixed ratio between entity nodes and deductive-relation nodes. This approach allows us handle specific characteristics possibly present in $\mathcal{D}$. If $\mathcal{D}$ contains only few matches, the algorithm will be forced to search for evidence by incorporating many deductive-relation nodes. On the other hand, if $\mathcal{D}$ contains a relatively big number of matches, the algorithm will include only a small subset of them, enough to increase the belief for the specific node without overloading the network with nodes.

Step 4 - Updating the Matches. Once the network is updated with nodes representing new matches and evidences, we need to recalculate the probability for the states of each node. This task is performed through probabilistic inference which updates all nodes according to the current status of the network. To minimize the time needed for doing this, we execute probabilistic inference only on the newly added nodes and nodes related to them.

As explained in Section 4.1, computing the probability of a node requires information from its neighbor nodes. Computed results are propagated back to the neighbor nodes to allow them to recompute their probability. For example, consider node R from the Bayesian network of Figure 1 (b). Once node R is activated, and receives messages $\lambda_{r_1}(R)$, $\lambda_{r_2}(R)$, $\pi_R(e1)$, and $\pi_R(e2)$ from its parent and children nodes, it computes: (i) its own belief as $BEL(R) = \alpha\lambda(R)\pi(R)$ (marked as eq. 1 in the following equation list), and (ii) new messages to send to its parent nodes (eq. 2), and children nodes (eq. 3). These messages are as follow:

$$\lambda(R) = \lambda_{r_1}(R)\lambda_{r_2}(R), \text{ and } \pi(R) = P(R|e1, e2)\pi_R(e1)\pi_R(e2) \quad (1)$$

$$\lambda_R(e1) = P(R|e1, e2)\pi_R(e2)\lambda(R) \quad (2)$$

$$\pi_{r2}(R) = \pi(R)\lambda_{r1}(R) \quad (3)$$

The message computation in this example shows the main benefit of using cause-effect relationships between nodes. Although node $e1$ is not directly connected to node $e2$, the algorithm is able to propagate information from one node to the other, through their cause-effect relationships with node R. Consequently, a high belief of node $e2$ affects the belief of node R (eq. 1), which is reflected in the message node R sends to node $e1$ (eq. 2). Finally, node $e1$ is affected when it recomputes its belief using the message sent to it by node R.

Entity Linkage information for $M(r_{127})$:
⟨ el:///E1, object_rep, file:///P127/a_2 ⟩,
⟨ el:///E1, object_rep, file:///P128/from ⟩,
⟨ el:///E1, belief, 0.96 ⟩, ...

Fig. 2. Part of the entity linkage information generated by our algorithm, for the metadata of Figure 1

After executing probabilistic inference, we have an updated set of matches that reflect the metadata present in the information space. Different representations of the results matches are possible (i.e., include or do not include the probability of each match), and the selected representation depends on the needs of the specific system. Figure 2 shows one possible representation for part of the results of the metadata from Figure 1. In this example, additional metadata are generated to represent each match in the Bayesian network using the corresponding object representations and belief.

5 Experimental Evaluation

We evaluated our approach using a JAVA implementation of the entity linkage algorithm, including all features we described in the previous sections. For performing probabilistic inference[5]on the Bayesian network we used the *jSMILE API*[6], and for creating a database to store internal information we used *MySQL* 5.0[7]. The following paragraphs present the effectiveness of our algorithm on two datasets, the Cora and a PIM dataset.

[5] For efficiency reasons we use the 'Backward simulation' algorithm; a modified version of Pearl's algorithm that performs approximate inference.

[6] `http://genie.sis.pitt.edu/`

[7] `http://www.mysql.com/`

5.1 Cora Dataset

The *Cora dataset*[8] is a collection of publications collected from CiteSeer. Each publication contains title and author names, using different forms for the names (e.g., 'J. Antonisse', 'Antonisse , H. J.', 'Antonisse', 'Jim Antonisse'). The dataset was manually processed to accompany each publication author with an identifier that indicates the corresponding real-world entity.

We processed the Cora dataset and converted each publication into RDF triples. Our process generated 14392 triples describing title and authors for 1563 resources (publications). A total of 2882 triples described authors, with 9768 matches between these authors. Following the definitions of Section 3, we use as object representation rep_{A_i,C_k}, the triples describing author A_i as given in the triples generated for publication C_k. The task of our algorithm is then to compute the probability of entity mappings of author A_i from publication C_k with author A_j from publication C_n, represented by $\mathrm{P}(\mathrm{d}(rep_{A_i,C_k})=\mathrm{d}(rep_{A_j,C_n}))$.

The goals of our Cora dataset experiments were twofold: (i) evaluate the effectiveness of our algorithm in identifying the entities, and (ii) compare the effectiveness of our algorithm with the one given by the basic similarity functions we use for generating the evidence nodes. We measured effectiveness, as usual in information retrieval, by computing precision and recall. These measures were calculated in respect to the actual real-world entities, as specified by the unique identifier given for the authors of each publication in the Cora dataset.

We executed the experiments, by adding the triples generated from the Cora dataset incrementally into an information space, which uses our algorithm for entity linkage. After adding triples for 100 publications, we performed probabilistic inference on the Bayesian network generated by our algorithm. The following table shows the number of the matches that correspond to the different numbers of publications.

Publications	1000	1100	1200	1300	1400	1563
Matches	4129	4620	5050	6036	7337	9774

Entity Linkage Effectiveness. Figure 3 shows the plots for precision and recall under different probability thresholds, for several publications groups. The plots do not include groups that contain less than 1000 publications because the number of the corresponding matches is too small. Small values of the probability threshold (θ <0.4) are not included in the plots since the results are similar to θ=0.4.

As shown in Figure 3, our algorithm is able to maintain the same values for precision and recall for the different probability thresholds. For lower probability thresholds (i.e., θ=0.4, and θ=0.5) we see that recall is very high and precision is already quite satisfactory (around 0.9). Moving toward higher probability thresholds (i.e. θ=0.6, θ=0.7,) we see precision values increasing and, as expected, decreasing recall values. Precision does not 'automatically' increase with groups that have more publications —more data, and thus entities are available— but rather reflects our belief for the entities in the current data.

[8] We used the version from `http://www.cs.umd.edu/~indrajit/ER/index.html`

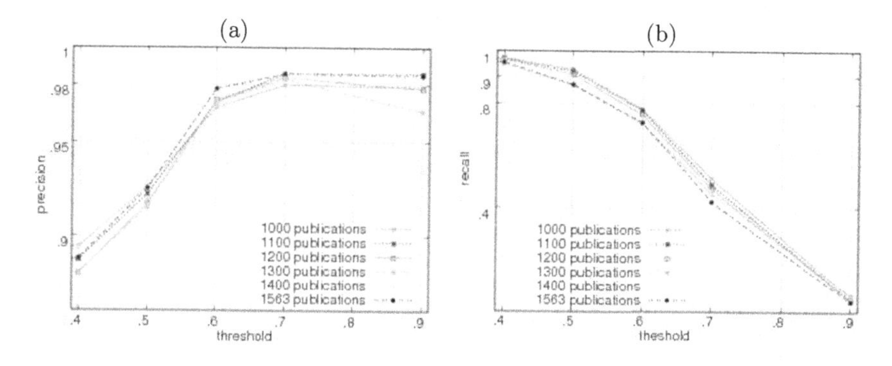

Fig. 3. (a) Precision, and (b) Recall vs. different thresholds

The results of these plots follow exactly the behavior explained in the analysis of our algorithm. It is clear that external algorithms are able to control the precision/recall of the entities by selecting an appropriating value of the probability threshold. For example, an application that needs only very certain matches will choose a high probability threshold, whereas an application that accepts uncertain matches a lower.

We also used our Cora dataset experiments to compare with previous approaches described in the literature. The authors of [10] reported precision 0.994 with recall 0.985, the authors of [17] had precision 0.842 with recall 0.909. To compare these numbers with our results, we considered only matches generated by our algorithm that exceed a preselected low probability threshold (e.g., θ=0.5). As shown in our two plots, these matches have high precision and high recall, similar to the ones given by these other algorithms. Our algorithm offers two additional advantages: (i) identified matches do not alter original metadata, (ii) our algorithm is able to further classify these matches according to the belief we have for their existence.

Comparison with basic similarity functions. In this experiment we performed a comparison of the effectiveness of our algorithm with the basic similarity functions used for generating evidence nodes. The algorithms we considered were Soundex Similarity and String Similarity, as described in Section 4.3.

Table 3 shows precision and recall values given by the two similarity functions on different publications groups. In all cases we assume that the real-world entities are the ones those probability is above threshold 0.7. Our evaluation shows our entity linkage clearly outperforms the effectiveness of the basic similarity functions.

Table 3. Precision/Recall of the entity linkage, and the basis similarity functions we used for generating the evidence nodes (θ=0.7)

Publications	Entity Linkage	String Similarity	Soundex Similarity
200	0.219/0.969	0.892/0.081	0.482/0.362
400	0.218/0.977	0.422/0.065	0.246/0.346
600	0.358/0.982	0.329/0.05	0.181/0.220

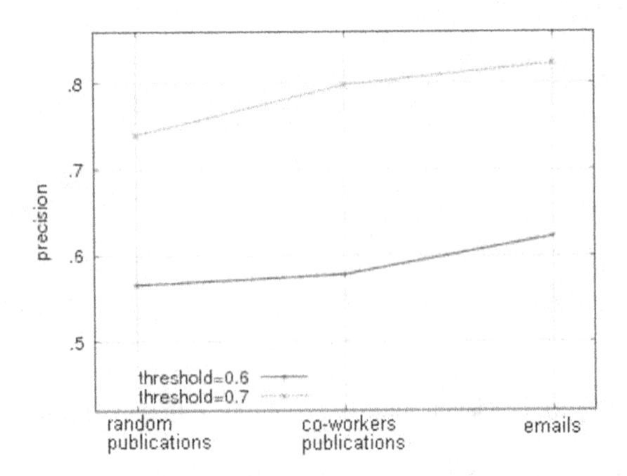

Fig. 4. (a) Precision, and (b) Recall vs. different thresholds

5.2 PIM Dataset

As a second dataset for evaluating our algorithm we use metadata generated in a personal information management environment, for desktop resources. As there is no publicly available PIM dataset, we created a suitable collection of metadata by simulating the behavior of a PIM application. Our PIM dataset included metadata describing desktop resources from three groups:

- The first group contains publications randomly selected from the DBLP system, to simulate arbitrary publications downloaded from the Web. This group resulted in metadata describing 700 imported resources, with 1326 triples corresponding to authors.
- The second group contains publications imported into the PIM environment from the DBLP system for which one of the authors is our co-worker at L3S. The results of this import were metadata for 250 resources, with 480 triples describing authors.
- The third group contains personal emails from one of the author's email client. Our goal was to identify and link authors from the publications with the corresponding person sending emails. The entity linkage problem in this case is somehow limited since persons are usually accompanied with email addresses which can act as unique identifiers for them. For this reason, we applied our entity linkage algorithm only on the existing email address, person name pairs. To capture the connections between these people, we randomly selected a small portion of available emails. The result was metadata for 200 resources, with 400 triples describing people. This group contains the most heterogeneous data, since each person has various email addresses and name variants.

We evaluated our algorithm on this PIM dataset, by adding metadata incrementally into an information space, which uses our algorithm for entity linkage.

We performed probabilistic inference on the Bayesian network generated by our algorithm, after adding the metadata of these three groups. Figure 4 shows the precision of the results generated by our algorithm. As shown, adding emails does not reduce the precision of the generated results, which is what we would have excepted since the emails contain the most heterogeneous data. As such, the results indicate that our algorithm is able to handle the heterogeneous instances of persons referenced in emails, and successfully link them with the author instances gained from the publications.

6 Conclusions

In this paper, we addressed the problem of identifying and linking heterogeneous data referring to the same real-world entity. This problem appears in a variety of situations, where we have to merge and integrate heterogeneous data from different information sources. Our algorithm uses a Bayesian network to explicitly model evidences supporting possible matches between different references, along with interconnections between these matches. The algorithm runs incrementally and does not modify existing data. Our evaluations showed that our algorithm successfully achieves our goal of efficiently and effectively linking data in heterogeneous information spaces.

We are currently investigating several directions for additional improvements and extensions of our entity linkage algorithm. First, we want to continue our experiments with other data sets and information sources. This will also include more general scenarios including unknown data schemes and data extracted from the Web. Finally, we will examine the use of other similarity functions for generating evidence nodes, and investigate the possibility of using different similarity functions for different data scenarios.

Acknowledgements

This work was supported by the NEPOMUK project, funded by the European Commission under the 6th Framework Programme (IST Contract No. 027705).

References

1. Aleman-Meza, B., Nagarajan, M., Ramakrishnan, C., Ding, L., Kolari, P., Sheth, A.P., Arpinar, I.B., Joshi, A., Finin, T.: Semantic analytics on social networks: experiences in addressing the problem of conflict of interest detection. In: WWW 2006 (2006)
2. Ananthakrishna, R., Chaudhuri, S., Ganti, V.: Eliminating fuzzy duplicates in data warehouses. In: VLDB (2002)
3. Bekkerman, R., McCallum, A.: Disambiguating web appearances of people in a social network. In: WWW 2005 (2005)
4. Benjelloun, O., Garcia-Molina, H., Menestrina, D., Su, Q., Whang, S.E., Widomr, J., Jonas, J.: Swoosh: A generic approach to entity resolution. Technical report, Stanford InfoLab (2006)

5. Bhattacharya, I., Getoor, L.: Deduplication and group detection using links. In: Workshop on Link Analysis and Group Detection, ACM SIGKDD 2004 (2004)
6. Bhattacharya, I., Getoor, L.: Iterative record linkage for cleaning and integration. In: DMKD (2004)
7. Bouquet, P., Stoermer, H., Mancioppi, M., Giacomuzzi, D.: OkkaM: Towards a Solution to the "Identity Crisis" on the Semantic Web. In: Italian Semantic Web Workshop, SWAP (2006)
8. Brunkhorst, I., Chirita, P.A., Costache, S., Julien Gaugaz, E.I., Iofciu, T., Minack, E., Nejdl, W., Paiu, R.: The beagle^{++} toolbox: Towards an extendable desktop search architecture. In: Semantic Desktop Workshop, ISWC (2006)
9. Cohen, W., Ravikumar, P., Fienberg, S.: A comparison of string distance metrics for name-matching tasks. In: Workshop on Inf. Integration on the Web (2003)
10. Dong, X., Halevy, A.Y., Madhavan, J.: Reference reconciliation in complex information spaces. In: SIGMOD Conference (2005)
11. Guha, R.V., McCool, R.: Tap: a semantic web platform. Computer Networks (2003)
12. Hernández, M.A., Stolfo, S.J.: Real-world data is dirty: Data cleansing and the merge/purge problem. Data Min. Knowl. Discov. (1998)
13. Jensen, F.V.: Bayesian Networks and Decision Graphs. Springer, New York (2001)
14. Kalashnikov, D.V., Mehrotra, S.: Domain-independent data cleaning via analysis of entity-relationship graph. ACM Trans. Database Syst. (2006)
15. Kalashnikov, D.V., Mehrotra, S., Chen, Z.: Exploiting relationships for domain-independent data cleaning. In: SDM (2005)
16. Li, J.-Z., Tang, J., Zhang, J., Luo, Q., Liu, Y., Hong, M.: Eos: expertise oriented search using social networks. In: WWW (2007)
17. Parag, Domingos, P.: Multi-relational record linkage. In: MRDM (2004)
18. Pearl, J.: Probabilistic reasoning in intelligent systems: networks of plausible inference. Morgan Kaufmann Publishers Inc., San Francisco (1988)
19. Weis, M., Manolescu, I.: Declarative xml data cleaning with xclean. In: CAiSE (2007)
20. Winkler, W.E.: The state of record linkage and current research problems. Technical report (1999)

Product Based Workflow Support: Dynamic Workflow Execution

Irene Vanderfeesten, Hajo A. Reijers, and Wil M.P. van der Aalst

Technische Universiteit Eindhoven, Department of Technology Management,
PO Box 513, 5600 MB Eindhoven, The Netherlands
{i.t.p.vanderfeesten,h.a.reijers,w.m.p.v.d.aalst}@tue.nl

Abstract. Product Based Workflow Design (PBWD) is a successful new approach to workflow process support. A description of the product, the Product Data Model (PDM), is central to this approach. While other research so far has focused on deriving a process model from the PDM, this paper presents a way to directly execute the PDM. This leads to a more dynamic and flexible support for the workflow process.

Keywords: Workflow Management, Product Data Model, Process Modeling, Process Execution Strategies.

1 Introduction

Product Based Workflow Design (PBWD) [1,4,5] is a successful new approach to workflow process design in which a description of the workflow product is central. So far, PBWD research has mainly focused on generating process models from a product structure, either manually or automatically. The manual derivation of a process model has turned out to be very time consuming [4]. Experiences with the automatic generation of process models [7] triggered a new idea to provide flexible and dynamic support for process execution *directly* on the basis of the product structure, i.e. without first deriving a process model that describes the desirable flow of work. We will refer to this concept as Product Based Workflow Support (PBWS).

2 Product Based Workflow Support

The product of a workflow process is usually an *informational* product, e.g. the decision on an insurance claim or the allocation of a subsidy. The structure of the workflow product can be described by a tree-like structure similar to a Bill-of-Material from manufacturing [2]. Such a description of a workflow product is called a Product Data Model (PDM). Figure 1(a) shows a very small example of a PDM. Because of space limitations we refer to [7] for the complete explanation of this example.

In general, a PDM consists of a number of data elements (depicted as circles) that are linked to each other through operations (depicted as arcs). Each

Z. Bellahsène and M. Léonard (Eds.): CAiSE 2008, LNCS 5074, pp. 571–574, 2008.

operation can have one or more input data elements and produces exactly one output data element. The operation on the input elements can be e.g. a calculation, an assessment by a human, or a rule-based decision to determine the output element. An operation is executable when all of its input elements are available. Moreover, several operations can have the *same* output element while having a different set of input elements. These operations represent alternative ways to produce the output product. Finally, operations can have a number of attributes such as the execution cost, processing time, failure probability and execution conditions.

The basic idea of PBWS is that dynamically, during each step of the process execution, all data elements are determined that are available for a case. At run time, it can then be decided what would be the most proper next step in the execution of the process, considering the information available for the specific case, the underlying product specification, and the desired performance.

2.1 Runtime Execution of a PDM

When the workflow process is executed for a particular case some data elements are initially provided by the client or can be retrieved from other systems. Based on these available data elements, new information is produced step-by-step by executing enabled operations. Figure 1 illustrates how the runtime execution of our example PDM works. Suppose that at the start of the process input data elements B, E, and F are available (see Figure 1(b)). The operations that are now enabled for execution are $Op1$ and $Op2$, since all of their input elements are available (Figure 1(c)). Operation $Op3$ is not executable because data element C is not available yet and $Op4$ is not executable since D is not present. Now, we have to choose which of the two executable operations ($Op1$, $Op2$) we select. Suppose we select $Op1$. Then, data element C is produced (Figure 1(d)). The executable operations are calculated again: $Op2$ and $Op3$. And one of those operations is selected. Suppose we select $Op3$. Then, the end product A is determined and the process ends.

In many situations more than one operation is executable, e.g. in the first step of the example we could have chosen for $Op2$ in stead of $Op1$, which would have led to the end product immediately, but also could have had different outcomes in terms of performance (e.g. cost, throughput time). For example, when the processing time of $Op2$ is less than the processing time of $Op1$, selecting $Op2$ as a next step would perhaps be a better decision. Now, the question arises how to select the best operation from the set of executable operations to proceed. We define 'best' with respect to the single case, i.e. the performance goal of the case in isolation (e.g. cost, total processing time) is optimized.

2.2 Execution Strategies

We have identified several selection strategies to find the best candidate from the set of enabled operations. These strategies are related to the attributes and

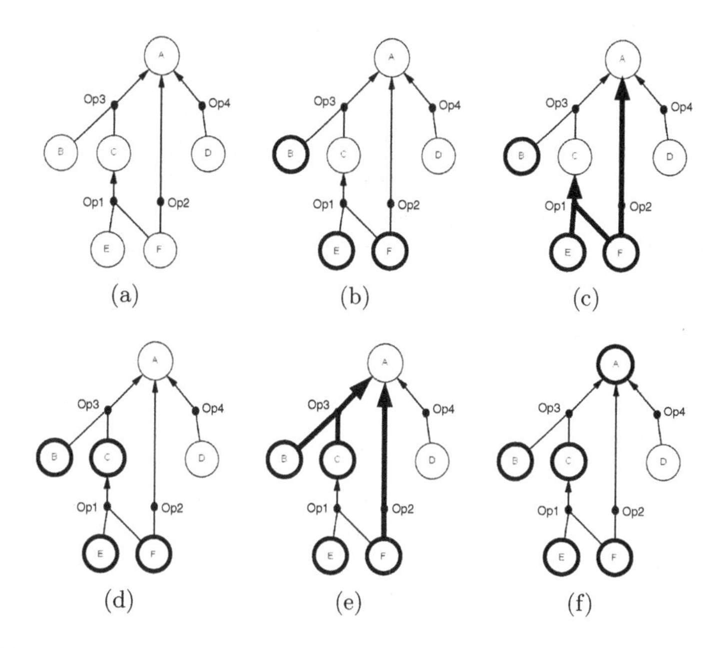

Fig. 1. The execution of a PDM: (a) The PDM, (b) Input data elements (B, E, F) are available, (c) Executable operations in the first step, (d) Data element C is produced, (e) Executable operations in step two, (f) The end product (A) is determined

properties of the operations. We drew inspiration from *sequencing and scheduling rules* in the field of logistics and production planning [3,6]:

- Random - The best candidate is randomly selected (cf. Random [3]).
- Lowest cost - The best candidate is the operation with the lowest cost.
- Shortest processing time - The operation with the shortest duration is chosen (cf. SPT [3]).
- Distance to root element - The distance of an operation to the root element is the 'shortest path' from this operation to an operation that produces the root element. The distance to the root element can be measured as the minimal number of operations to the root element (cf. FOPNR [3]).
- Shortest remaining processing time - The shortest remaining processing time is another form of the distance to the root element. In this case the processing times of the operations on the path to the root element are added up (cf. SR (shortest remaining processing time) [3]).

These execution strategies have been implemented in a prototype of a PBWS system. The system is able to load a PDM and generate case execution recommendations based on the PDM and the selected strategy. More details of this prototype are described in [7].

3 Conclusion

This paper presents Product Based Workflow Support: a dynamic approach to workflow execution on the basis of a product data model. In contrast to conventional workflow management support, there is no need for a process model that guides the execution. Therefore, a more dynamic and flexible support is possible. Based on the data elements readily available for a specific case on the one hand and a selected strategy (i.e. lowest cost, shortest processing time, etc.) on the other hand, this approach recommends the next step that should be performed for the case. *In contrast to conventional languages there is a clear separation of concerns: the product data model is based on functional requirements while the selected strategy focuses on performance* (e.g., minimize costs).

Using the strategies presented, the selection of the best candidate is only optimized locally (i.e. within the set of executable operations); the effect of the selected operation on future steps is not taken into account. Thus, such a strategy does not necessarily lead to the best overall path to the end product. To overcome this problem of local optimization, the use of the theory on Markov Decision Processes is a promising direction. With this analytical method, it is possible to completely compute the optimal strategy.

Acknowledgement

This research is supported by the Technology Foundation STW, applied science division of NWO and the technology programme of the Dutch Ministry of Economic Affairs.

References

1. van der Aalst, W.M.P.: On the Automatic Generation of Workflow Processes based on Product Sstructures. Computers in Industry 39, 97–111 (1999)
2. Orlicky, J.A.: Structuring the Bill of Materials for MRP. Production and Inventory Management, 19–42 (December 1972)
3. Panwalkar, S.S., Iskander, W.: A Survey of Scheduling Rules. Operations Research 25, 45–61 (1977)
4. Reijers, H.A.: Design and Control of Workflow Processes. LNCS, vol. 2617. Springer, Berlin (2003)
5. Reijers, H.A., Limam Mansar, S., van der Aalst, W.M.P.: Product-based Workflow Design. Journal of Management Information systems 20(1), 229–262 (2003)
6. Silver, E., Pyke, D.F., Peterson, R.: Inventory Management and Production Planning and Scheduling. John Wiley and Sons, Chichester (1998)
7. Vanderfeesten, I., Reijers, H.A., van der Aalst, W.M.P.: Product Based Workflow Support: A Recommendation Service for Dynamic Workflow Execution. BPM Center Report BPM-08-03, BPMcenter.org (2008)

Location-Based Variability for Mobile Information Systems

Raian Ali, Fabiano Dalpiaz, and Paolo Giorgini

University of Trento - DISI, 38100, Povo, Trento, Italy
{raian.ali,fabiano.dalpiaz,paolo.giorgini}@disi.unitn.it

Abstract. Advances in size, power, and ubiquity of computing, sensors, and communication technology make possible the development of mobile or nomadic information systems. Variability of location and system behavior is a central issue in mobile information systems, where software behavior has to change and re-adapt to the different location settings. In this paper, we motivate the need for integration of variable location and variable software behavior. We adapt the goal-oriented framework *i**/Tropos to model and analyze the alternative goal satisfaction strategies and the location where each alternative can be adopted. We introduce analysis techniques for the proposed location-based models.

1 Introduction

Advances in computing and communication technology have recently led to the growth of interest in Mobile Information Systems (hereafter MobIS). MobISs emphasize mobility concerns (space, time, personality, society, environment, and so on) often not considered by traditional desktop systems [1]. Technology advances do not necessarily imply the easiness of exploiting it, rather more challenges are introduced. Nomadic user expects smarter information systems, able to adapt their behavior without human intervention. MobIS has to reason about the surrounding location, including user itself, and adapt *autonomously* their behavior to location settings. Consequently, we need to model and analyze the variable location and the variable behavior and define how location influences behavior.

Behavioral and location variability are complementary. Supporting two alternative behaviors, without specifying when to adopt each of them, arises the question *"why do we support two alternatives and not just one?"*. Conversely, considering location variability without supporting alternative behaviors arises the question *"what can we do if location changes?"*. We use *i**/Tropos [2,3] goal-oriented framework to model alternative strategies for MobIS to satisfy a goal, and specify location properties that apply to each alternative. This allows us to support the decision making process when deriving a location-tailored MobIS instance and make possible different kinds of reasoning. The intended automated reasoning allows to answer questions like: *"are all MobIS objectives achievable in a given location?"*, *"what is the optimal alternative to achieve an objective in a given location?"*, and *"what is the optimal modification that is needed in one location to satisfy some MobIS objectives?"*.

Z. Bellahsène and M. Léonard (Eds.): CAiSE 2008, LNCS 5074, pp. 575–578, 2008.

2 Location-Based Goal Models

In i^*/Tropos, the system is modeled as a set of inter-dependent actors having goals, and that can commit to strategies to satisfy their goals. Autonomous selection among goal satisfaction strategies requires criteria an actor builds its decision upon. One alternative can be recommended in a certain location, while it can be even unapplicable in others. The criteria to select among alternatives

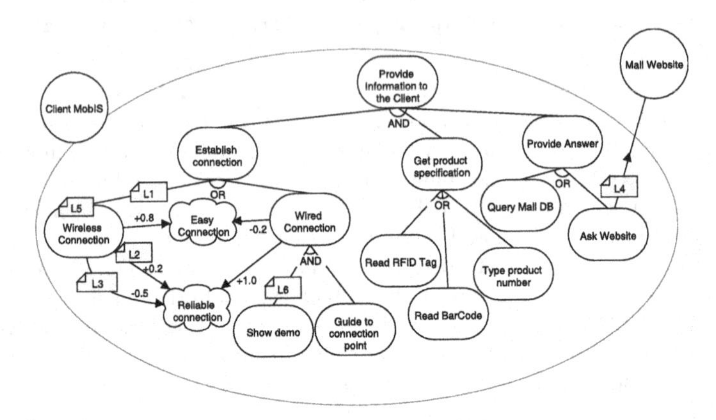

Fig. 1. A location-based goal model

is not explicitly modeled in the current i^*/Tropos goal model. Fig.1 shows a partial goal model of a PDA MobIS intended for a client in a shopping mall. As a step to support location-based variability, i^*/Tropos can attach location properties to its following variability points:

1. *Location-based Or-decomposition*: Or-decomposition is the basic variability construct; in current i^*/Tropos the choice of a specific Or-alternative is left to actor intention, without considering location properties that can inhibit some alternatives. E.g. (from Fig. 1): goal *Establish connection* can be achieved using *Wireless Connection* only if the mall has a wireless network and client is authorized to access it, and client's PDA supports WiFi (L1).
2. *Location-based contribution to soft-goals*: the value of contributions to soft-goals can vary from one location to another. E.g. the contribution from goal *Wireless Connection* to soft-goal *Reliable Connection* changes depending on the level of received signal: if user is in a location where the signal coming from the WiFi access point is high (L2), the contribution will be positive, while if the client is far from the WiFi access point and the signal level is poor (L3), the contribution will be negative.
3. *Location-based dependency*: in certain locations, an actor is unable to satisfy a goal using its own strategies. In such case, the actor might delegate this goal to another actor that is able to satisfy it. E.g. the MobIS can satisfy goal *Provide Answer* by fulfilling *Query Mall DB*; while if the database is offline and a mall website exists and has a mobile devices version (L4), the MobIS can delegate the goal to *Mall Website* browsing that website.

4. *Location-based goal activation*: an actor, when location settings change, might find necessary or possible triggering (or stopping) the desire of satisfying a goal. E.g. if the MobIS has adopted the alternative *Wired Connection* to establish a connection, and while the client is getting to one cable-based terminal, the PDA detects a wireless signal (L5), the goal *Wireless Connection* could be triggered to better satisfy the soft-goals.
5. *Location-based And-decomposition*: a sub-goal might (or might not) be needed in certain location, that is some sub-goals are not always mandatory to fulfill the top-level goal in And-decomposition. E.g. to satisfy the goal *Wired Connection*, the MobIS has first to show a demo to client only if the client is using the system for the first time (L6).

3 Defining, Eliciting and Modeling Location

We refer by "location" to an environment with high degree of commonality, like shopping malls, museums, or airports. The commonality concerns location constructs: resources (physical and informational); actors having responsibilities, objectives, and relations with resources and other actors; and rules that coordinate the interaction among actors and the use of resources. Using $i*$/Tropos concepts, we define location from the perspective of an actor as: "*the set of available actors and resources that can be employed to achieve actor goals*". Goal analysis will capture location properties that are needed at each variability point, and this in turn will enable us to construct location model. In our broad vision, location will be the input that guides MobIS derivation process: MobIS will be instantiated according to the location model instance as shown in Fig. 2.

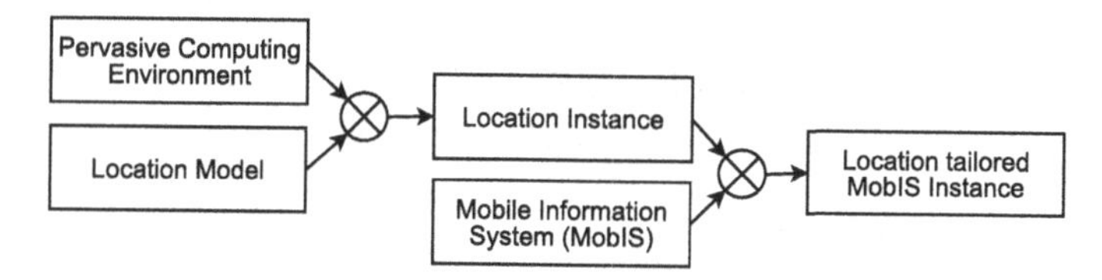

Fig. 2. The process of instantiating a location-tailored MobIS instance

4 Analysing Location-Based Models

The proposed location-based goal model has two components: (1) the goal model that describes how a goal can be satisfied and (2) the location properties that constrain each alternative. Location properties are predicates specified over a location model, whose truth values can be either true or false at a certain location. By formalizing location model and location-based goal model, we can do several analysis. We outline now three types of such automated analysis:

1. *Location-based goal satisfiability (LGS)*: it verifies whether a goal is achievable through one alternative in a specific location.

2. *Location properties satisfiability (LPS)*: this analysis checks if the current location structure is compliant with the MobIS goals. It is exploited to identify what is missing in a particular location where some top-level goals have been identified as unsatisfiable by LGS. When a goal cannot be satisfied, LPS will identify the denying conditions and find ways to solve the problem.
3. *Preferences analysis (PA)*: this type of analysis requires the specification of preferences over alternatives. Preferences can be specified using soft-goals as in [4]. We need this analysis in two cases: 1) when there are several alternatives to satisfy a goal: the selection will be based on the contributions to preferred soft-goals. 2) when there is no applicable alternative: in this case, LPS might provide several proposals about the needed location modifications. The adopted modifications are those leading to better satisfying preferences expressed over soft-goals.

5 Discussion and Future Work

We have briefly shown how to integrate goal satisfaction strategies with the concept of location, and what kind of analysis we can do over the location-based models. More details and a concrete example can be found in our technical report ([5]). For the future, we need to define a modeling language for location, and to study how to capture location model and integrate it with system behavior variability at different levels (goal satisfaction is one of them). Formalization is a basic need, since location is perceived and needed to perform reasoning. We will look for an appropriate formalism to automate the analysis techniques.

Acknowledgement

This work has been partially funded by EU Commission, through the SERENITY project, by MIUR, through the MEnSA project (PRIN 2006), and by the Provincial Authority of Trentino, through the STAMPS project.

References

1. Krogstie, J., Lyytinen, K., Opdahl, A.L., Pernici, B., Siau, K., Smolander, K.: Research areas and challenges for mobile information systems. International Journal of Mobile Communications 2(3), 220–234 (2004)
2. Yu, E.: Modelling strategic relationships for process reengineering. Ph.D. Thesis, University of Toronto (1995)
3. Bresciani, P., Perini, A., Giorgini, P., Giunchiglia, F., Mylopoulos, J.: Tropos: An agent-oriented software development methodology. Autonomous Agents and Multi-Agent Systems 8(3), 203–236 (2004)
4. Liaskos, S., McIlraith, S., Mylopoulos, J.: Representing and reasoning with preference requirements using goals. Technical report, Dept. of Computer Science, University of Toronto (2006), `ftp://ftp.cs.toronto.edu/pub/reports/csrg/542`
5. Ali, R., Dalpiaz, F., Giorgini, P.: Location-based variability for mobile information systems. Technical Report DISI-08-008, DISI, University of Trento, `http://eprints.biblio.unitn.it/archive/00001351/`

Modelling, Simulation, and Performance Analysis of Business Processes Involving Ubiquitous Systems

Patrik Spieß[1], Dinh Khoa Nguyen[1], Ingo Weber[1], Ivan Markovic[1], and Michael Beigl[2]

[1] SAP Research, CEC Karlsruhe, Germany
{patrik.spiess,dinh.khoa.nguyen,ingo.weber,ivan.markovic}@sap.com
http://sap.com/research/
[2] IBR, Braunschweig, Germany
http://www.ibr.cs.tu-bs.de/dus/

Abstract. A recent trend in Ubiquitous Computing is that embedded software (e.g. in production machines, wired or wireless networked sensors and actuators, or RFID readers) directly offers Web Services. This allows it to directly participate in IT business processes. Our work suggests an approach to assess and compare the performance of such hybrid process models. It critically evaluates BPEL, the standard instrument for creating executable process descriptions, for its support of the dynamic nature of ubiquitous services, e.g. due to device mobility and unreliable wireless communication. [1]

1 Introduction

The near future vision of industrial production (which we see merely as a special case of Ubiquitous Computing) is moving towards an Internet of things, in which smart machinery and smart semi-finished products are integrated into enterprise business processes, to enable more flexibility and adaptability by including some business intelligence already at a very low, technical level.

Following a SOA concept, we argue that the functionalities of those devices should be made available as Web Services, which could be consumed by other devices or composed into higher-level Web Services. As the shop-floor devices are now powerful enough regarding hardware resources such as computing power and memory, they can be designed to host Web Services (SOA-ready devices) and hence can easily be accessed from within business processes.

Given a completely service-oriented device landscape, the next challenges are how to describe business processes orchestrating such hybrid systems using executable process description languages like BPEL [1] and what are appropriate metrics to assess the operational performance of a given process model. In this

[1] The authors would like to thank the European Commission and the partners of the European IST projects SOCRADES (www.socrades.eu) and SUPER (www.ip-super.org), for their support.

Z. Bellahsène and M. Léonard (Eds.): CAiSE 2008, LNCS 5074, pp. 579–582, 2008.

paper, we propose the modelling and simulation approaches for standard, executable business process descriptions together with ubiquitous systems. We explain also how to use simulation results for performance analysis of such hybrid business processes.

2 Modelling and Simulation Approach

In order to evaluate the performance of an executable process description in a ubiquitous environment, we suggest explain the benefits an problem when using the BPEL standard, introduce a model of the heterogeneous device landscape and lay out our simulation approach.

2.1 Business Process Description and Execution

For the modelling of an executable business process, we suggest to use BPEL as the de-facto industry standard. A BPEL process description contains references to all services that it interacts with. This limitation is against the nature of a ubiquitous environment, where devices (and consequently, the services hosted on them) can appear and disappear unpredictably at run-time. Our solution to overcome this problem is to extend the BPEL engine, providing the ability to resolve generic partner link descriptions with concrete endpoints stored in the device landscape model at run-time.

Another deficiency of BPEL, mentioned by Wohed et. al. in [5], is the missing support for broadcasting a message to several partners through partner links. Constrained, embedded devices, especially when using a wireless link, would benefit from this functionality. We mitigate this limitation by suggesting and supporting modifications in BPEL and the engine executing it. Our work is based on the engine presented in [3] where Hackman et al. introduced a new BPEL element called *Partner Group*. During run-time, a BPEL engine supporting partner groups features addition and removal of partner links to a partner group, as well as binding and unbinding a partner link dynamically at run-time. For instance, when a message from a previously unknown service is received, its endpoint is mapped (bound) to a currently unbound partner link and this partner link can be added to a partner group. After that, the partner link can be unbound again and is ready for the next incoming messages. Reply messages can be sent to a whole partner group.

2.2 Modelling the Service Landscape

The service landscape model is a data structure, designed in XML, that describes a landscape containing hierarchical layers, groups of nodes, nodes, and services hosted on a node. Each service is specified by a service endpoint and a service type. By discovering the service type of a BPEL partner link, the simulation engine can map this partner link to an existing endpoint in the landscape.

Using simulated services generated from the service landscape definition is useful in the following situations: A process model should be tested with more devices than physically available, or a process needs to be repeated many times to get a stable average assessment which might take too long. Our simulation approach is straightforward: Replacing the ubiquitous devices with a web application server and letting the server provide the services instead of the devices. If left uncompensated, this would however lead to unrealistically high performance measurements. To compensate, we introduce technical virtual costs for service invocations.

The virtual costs of service invocations that cross the boundary between standard PC-based back-end subsystem and the embedded subsystem should be set considerably higher than communications within these subsystems. The cost for interactions within the embedded subsystem should be set higher than the cost for interactions within the back-end subsystem. For more complex systems, more than two layers may exist, each with its own cost of internal service interactions. For convenience, a default access cost can be assigned at each level of the landscape: on the whole landscape, on a layer, on a group of nodes, and on a node. Each service on a node can have a specific extra cost. The costs defined for a simulation must reflect the resources that are used for each step of the process execution (in the best case known by measurements) and their priority for the individual application. A time critical application would e.g. assign a higher cost to high-latency steps; an application that uses a volume-based 3G subscription would assign high cost to steps that send large messages over the 3G link.

Instead of measuring the performance of the simulated devices, we let the simulation engine sum up the virtual cost of each service invocation. The total cost (and other statistical meta data) of each process instance (from instantiation to termination) under varying environmental conditions (i.e. varying return values of simulated services) is summed up during the simulation. We assume the estimation to be Gaussian distributed, thus we are able to compute a reliable averaged performance indicator by averaging the results over many runs.

2.3 Simulation Approach

For simulating the process model on top of virtual devices and services (modelled in a landscape definition document) we need a business process engine that supports the following non-standard features. (1) Dynamic mapping of a partner link at runtime with a service binding stored in the device landscape. (2) Reusing a partner link at runtime, i.e., unbinding a partner link and then rebinding it to a new device. (3) Using the landscape definition to get cost values of cross-layered service invocations in order to calculate the total cost of process instance.

As a basis for our implementation, we chose the Sliver [4] open source BPEL engine, which can parse and execute a BPEL process description and covers most of the BPEL concepts. It stores all partner links in a hash map, and provides getter and setter methods to modify them at runtime. That means partner links can be mapped (or bound), unbound, and reused dynamically at runtime. We extended the engine for many other purposes, especially to link it with the device landscape in order to resolve partner links at run-time and to calculate invocation costs during process execution.

To simulate the connection and disconnection of devices, we modify the device landscape while a process instance is running. New devices can join or leave the device landscape providing new or taking away existing services. At the end of the simulation, when the process ends successfully or in an erroneous state, the calculated total cost is returned to a GUI for users to evaluate the process. The simulation engine can be instructed to repeat this process n times and return the average costs.

3 Conclusion and Future Work

In this paper we briefly described our analysis of the suitability of BPEL to support processes that interact with ubiquitous systems and sketch the suggestions for extensions to that standard. We also introduced an approach for estimating and comparing the performance of the execution of a process description. The concepts presented in this paper are based on an ongoing implementation by extending an experimental, open source BPEL engine.

This work is part of a larger project that is intended to facilitate the modelling of the behaviour of a hybrid system including services provided by both enterprise applications and ubiquitous systems. Process modelling experts should be enabled to model freely without the need to care for the resource constraints of ubiquitous systems. An automatic optimization step will partition the process and deploy the connected fragments e.g. on small execution engines running on the ubiquitous devices themselves [2]. The distributed version of the process will feature more locality during its execution and will use less system resources although it shows the same behaviour as the original one. The work presented in this paper will be used to evaluate the performance gain of the automatic optimization.

References

1. Web Services Business Process Execution Language Version 2.0 (OASIS Standard) (April 2007), `http://docs.oasis-open.org/wsbpel/2.0/wsbpel-v2.0.html`
2. Spieß, P., Karnouskos, S.: Maximizing the business value of networked embedded systems through process-level integration into enterprise software. In: Proc. Second Intl. Conf. on Pervasive Computing and Applications (July 2007)
3. Hackmann, G., Gill, C., Roman, G.-C.: Extending BPEL for interoperable pervasive computing. In: Proceedings of the 2007 IEEE International Conference on Pervasive Services, pp. 204–213 (2007)
4. Hackmann, G., Haitjema, M., Gill, C.D., Roman, G.-C.: Sliver: A BPEL Workflow Process Execution Engine for Mobile Devices. In: ICSOC, pp. 503–508 (2006)
5. Wohed, P., van der Aalst, W.M.P., Dumas, M., ter Hofstede, A.H.M.: Pattern based analysis of bpel4ws. Technical Report Technical Report FIT-TR-2002-04, Faculty of Information Technology, Queensland University of Technology (December 2002)

Open Source Workflow: A Viable Direction for BPM?

Extended Abstract

Petia Wohed[1], Nick Russell[2], Arthur H.M. ter Hofstede[3], Birger Andersson[1], and Wil M.P. van der Aalst[2,3]

[1] Stockholm University/KTH, Stockholm, Sweden
{petia,ba}@dsv.su.se
[2] Eindhoven University of Technology, Eindhoven, The Netherlands
{n.c.russell,w.m.p.v.d.aalst}@tue.nl
[3] Queensland University of Technology, Brisbane, Australia
a.terhofstede@qut.edu.au

With the growing interest in open source software in general and business process management and workflow systems in particular, it is worthwhile investigating the state of open source workflow management. The plethora of these offerings (recent surveys such as [4,6], each contain more than 30 such systems) triggers the following two obvious questions: (1) how do these systems compare to each other; and (2) how do they compare to their commercial counterparts. To answer these questions we have undertaken a detailed analysis of three of the most widely used open source workflow management systems [1]: jBPM[1], OpenWFE[2], and Enhydra Shark[3]. Another obvious candidate would have been the open-source workflow management system YAWL (www.yawlfoundation.org). However, given the authors' close involvement in the development of YAWL, we did not include it in our evaluation.

This analysis was based on the *workflow patterns* framework [2]. This framework provides a collection of generic constructs which recur in a workflow context. It is divided into control-flow, data, and resource patterns based on the process perspectives outlined in [3]. A patterns-based analysis is guided by explicit evaluation criteria which are identified for each pattern. It aims to investigate the ability of a workflow system to support each of the patterns that have been identified and is based on the premise that each pattern describes a feature that it is *desirable* to support in a business process context. Hence, the workflow patterns framework is not concerned with expressive power, but rather with *suitability* (see e.g. [5]).

We choose to use the workflow patterns as the basis for our investigation because it is a well established framework that is widely used for WFMS evaluations (as evidenced by the numerous references to it). There are already a substantial number of evaluations of contemporary offerings based on the patterns and they provide an effective means of comparing the capabilities of differing systems on a neutral basis. For the purposes of this analysis, the results from some of these earlier evaluations (i.e., Staffware, WebSphere MQ and Oracle BPEL PM) are added to the results from the analysis of open source systems summarized here.

[1] www.jboss.com/products/jbpm
[2] www.openwfe.org
[3] www.enhydra.org/workflow

Z. Bellahsène and M. Léonard (Eds.): CAiSE 2008, LNCS 5074, pp. 583–586, 2008.

Table 1. Support for the Control-flow Patterns in **A**–Staffware 10, **B**–WebSphere MQ 3.4, **C**–Oracle BPEL PM 10.1.2, **1**–JBOSS jBPM 3.1.4, **2**–OpenWFE 1.7.3, and **3**–Enhydra Shark 2.0

Basic Control–flow	A	B	C	1	2	3	Termination	A	B	C	1	2	3
1. Sequence	+	+	+	+	+	+	11. Implicit Termination	+	+	+	+	+	+
2. Parallel Split	+	+	+	+	+	+	43. Explicit Termination	–	–	–	–	–	–
3. Synchronization	+	+	+	+	+	+	Multiple Instances						
4. Exclusive Choice	+	+	+	+	+	+	12. MI without Synchronization	+	–	+	+	+	+
5. Simple Merge	+	+	+	+	+	+	13. MI with a pri. Design Time Knl	+	–	+	–	+	–
Advanced Synchronization							14. MI with a pri. Runtime Knl.	+	–	+	–	+	–
6. Multiple Choice	–	+	+	–	+/–	+	15. MI without a pri. Runtime Knl.	–	–	+/–	–	–	–
7. Str Synchronizing Merge	–	+	+	–	–	–	27. Complete MI Activity	–	–	–	–	–	–
8. Multiple Merge	–	–	–	+	–	–	34. Static Partial Join for MI	–	–	–	–	+	–
9. Structured Discriminator	–	–	–	–	+	–	35. Static Canc. Partial Join for MI	–	–	–	–	+	–
28. Blocking Discriminator	–	–	–	–	–	–	36. Dynamic Partial Join for MI	–	–	–	–	–	–
29. Cancelling Discriminator	–	–	–	–	+	–	State-Based						
30. Structured Partial Join	–	–	–	–	+	–	16. Deferred Choice	–	–	+	+	–	–
31. Blocking Partial Join	–	–	–	–	–	–	39. Critical Section	–	–	+	–	–	–
32. Cancelling Partial Join	–	–	–	–	+	–	17. Interleaved Parallel Routing	–	–	–	–	+/–	–
33. Generalized AND-Join	–	–	–	+	–	–	40. Interleaved Routing	–	–	–	–	+	–
37. Local Sync. Merge	–	+	+	–	+/–	–	18. Milestone	–	–	+/–	–	–	–
38. General Sync. Merge	–	–	–	–	–	–	Cancellation						
41. Thread Merge	–	–	+/–	+/–	–	–	19. Cancel Activity	+	–	+/–	+	–	–
42. Thread Split	–	–	+/–	+/–	–	–	20. Cancel Case	–	–	+	–	+/–	+
Iteration							25. Cancel Region	–	–	+/–	–	–	–
10. Arbitrary Cycles	+	–	–	+	+	+	26. Cancel MI Activity	+	–	+	–	–	–
21. Structured Loop	–	+	+	–	+	–	Trigger						
22. Recursion	+	+	–	–	+	+	23. Transient Trigger	+	–	–	+	+	–
							24. Persistent Trigger	–	–	+	–	–	–

The investigation was undertaken as follows. Solutions for each of the 126 patterns were sought in each of the tools evaluated. Where successfully identified, they were deployed and tested. The initial results were summarised and each of the system vendors/developers was invited to provide feedback on their accuracy. On the basis of these responses, a final set of results were agreed upon and they were comprehensively documented in the form of a technical report [7]. Tables 1- 3 summarise the main findings.

Overall, one can conclude that the range of constructs supported by the three systems is somewhat limited, although OpenWFE tends to offer a considerably broader range of features than jBPM and Enhydra Shark.

From a control-flow standpoint, jBPM and Enhydra Shark support a relatively limited set of control-flow operators (offering little support for patterns other than those related to basic control-flow). OpenWFE offers broader support for variants of the partial join and discriminator constructs and also for controlled task concurrency (i.e. multiple instance tasks).

For the data perspective, all three offerings support a limited range of data element bindings and rely heavily on case-level data elements. However, whilst simplistic, the data passing strategies employed in all three systems are reasonably effective and include consideration of important issues such as inline data manipulation when data elements are being passed. There are limited capabilities for handling external data interaction without utilising programmatic extensions. Another area of concern relates to shortcomings when dealing with parallelism of data manipulation (i.e. data is lost either because parallel updates on it are ignored, or because some of the updates are overwritten).

Table 2. Support for the Data Patterns in **A**–Staffware 9, **B**–WebSphere MQ 3.4, **C**–Oracle BPEL PM 10.1.2, **1**–JBOSS jBPM 3.1.4, **2**–OpenWFE 1.7.3, and **3**–Enhydra Shark 2.0

Data Visibility	A	B	C	1	2	3	Data Interaction-External (cont.)	A	B	C	1	2	3
1. Task Data	–	+/–	+/–	+/–	–	+/–	21. Env. to Case–Push	+/–	+/–	–	–	–	–
2. Block Data	+	+	–	–	+	+	22. Case to Env.–Pull	–	–	–	–	–	–
3. Scope Data	–	–	+	–	+/–	–	23. Workflow to Env.–Push	–	+/–	–	–	–	–
4. MI Data	+/–	+	+/–	–	+	+	24. Env. to Process–Pull	+/–	–	–	–	–	–
5. Case Data	+/–	+	+	+	+	+	25. Env. to Process–Push	–	+/–	–	–	–	–
6. Folder Data	–	–	–	–	–	–	26. Process to Env.–Pull	+	+	–	–	–	–
7. Global Data	+	+	+	–	+	–	Data Transfer						
8. Environment Data	+	+/–	+	+/–	+	+/–	27. by Value–Incoming	–	+	+	–	–	+/–
Data Interaction-Internal							28. by Value–Outgoing	–	+	+	–	–	+/–
9. Task to Task	+	+	+	+	+	+	29. Copy In/Copy Out	–	–	+	+	+	+
10. Block to Subpr. Dec.	+	+	–	–	+	+	30. by Reference–Unlocked	+	–	+	–	–	–
11. Subpr. Dec. to Block	+	+	–	–	+	+	31. by Reference–Locked	–	–	–	–	+	–
12. to MI Task	–	–	+/–	–	+	–	32. Data Transf.–Input	+/–	–	–	+	+	+
13. from MI Task	–	–	+/–	–	–	–	33. Data Transf.–Output	+/–	–	–	+	+	+
14. Case to Case	+/–	+/–	–	+/–	+/–	+/–	Data-based Routing						
Data Interaction-External							34. Task Precond.–Data Exist.	+	–	–	–	+	–
15. Task to Env.–Push	+	+/–	+	+/–	+	+	35. Task Precond.–Data Value	+	–	+	–	+	–
16. Env. to Task–Pull	+	+/–	+	+/–	+	+	36. Task Postcond.–Data Exist.	+/–	+	–	–	–	–
17. Env. to Task–Push	+/–	+/–	+	–	–	–	37. Task Postcond.–Data Val.	+/–	+	–	–	–	+/–
18. Task to Env.–Pull	+/–	+/–	+	–	–	–	38. Event-based Task Trigger	+	+/–	+	–	–	–
19. Case to Env.–Push	–	–	–	–	–	–	39. Data-based Task Trigger	–	–	–	–	–	–
20. Env. to Case–Pull	–	–	–	–	–	–	40. Data-based Routing	+/–	+	+	+/–	+/–	+

Table 3. Support for the Resource Patterns in **A**–Staffware 9, **B**–WebSphere MQ 3.4, **C**–Oracle BPEL PM 10.1.2, **1**–JBOSS jBPM 3.1.4, **2**–OpenWFE 1.7.3, and **3**–Enhydra Shark 2.0

Creation Patterns	A	B	C	1	2	3	Pull Patterns, continuation	A	B	C	1	2	3
1. Direct Allocation	+	+	+	+	–	+	24. Sys.-Determ. WL Mng.	+	–	–	–	–	–
2. Role-Based Allocation	+	+	+	–	+	+	25. Rrs.-Determ. WL Mng.	+	+	+	–	–	–
3. Deferred Allocation	+	+	+	+	+	+	26. Selection Autonomy	+	+	+	+	+	+
4. Authorization	–	–	–	–	–	–	Detour Patterns						
5. Separation of Duties	–	+	–	–	–	–	27. Delegation	+	+	+	–	–	–
6. Case Handling	–	–	+	–	–	–	28. Escalation	+	+	+	–	+	–
7. Retain Familiar	–	+	+	+	–	–	29. Deallocation	–	–	+	–	+	+
8. Capability-based Alloc.	–	–	+	–	–	–	30. Stateful Reallocation	+/–	+	+	–	+	–
9. History-based Alloc.	–	–	+/–	–	–	–	31. Stateless Reallocation	–	–	–	–	–	–
10. Organizational Alloc.	+/–	+	+/–	–	–	–	32. Suspension/Resumption	+/–	+/–	+	+	–	–
11. Automatic Execution	+	–	+	+	+	+	33. Skip	–	+	+	–	–	–
Push Patterns							34. Redo	–	–	–	–	+/–	–
12. Distr. by Offer-Single Rsr.	–	–	+	–	–	+	35. Pre-Do	–	–	–	–	–	–
13. Distr. by Offer-Multiple Rsr.	+	+	+	–	+	+	Auto-start Patterns						
14. Distr. by Alloc.-Single Rsr.	+	+	+	+	–	–	36. Comm. on Creation	–	–	–	–	–	–
15. Random Allocation	–	–	+/–	–	–	–	37. Comm. on Allocation	–	+	–	–	–	+
16. Round Robin Alloc.	–	–	+/–	–	–	–	38. Piled Execution	–	–	–	–	–	–
17. Shortest Queue	–	–	+/–	–	–	–	39. Chained Execution	–	–	–	–	–	–
18. Early Distribution	–	–	–	–	–	–	Visibility Patterns						
19. Distribution on Enablement	+	+	+	+	+	+	40. Config. Unalloc. WI Vis.	–	–	–	–	+/–	–
20. Late Distribution	–	–	–	–	–	–	41. Config. Alloc. WI Vis.	–	–	–	–	+/–	–
Pull Patterns							Multiple Resource Patterns						
21. Rsr.-Init. Allocation	–	–	–	–	–	–	42. Simultaneous Execution	+	+	+	–	–	–
22. Rrs.-Init. Exec.-Alloc. WI	+	+	+	+	–	–	43. Additional Resources	–	–	+	–	–	–
23. Rsr.-Init. Exec.-Offered WI	+	+	+	–	+	+							

For the resource perspective, only simple notions of work distribution are supported and typically only one paradigm exists for work item routing in each offering. There is no support for any form of work distribution based on organizational criteria, resource capabilities or execution history. All three offerings provide relatively simple facilities for work item management e.g., (for two of them) there is no ability to configure work lists at resource or system level, no notion of concurrent work item execution and no facilities for optimizing work item throughput (e.g. automated work item commencement, chained execution). One area where OpenWFE demonstrates noticeably better facilities is in terms of the range of detour patterns (e.g. deallocation, reallocation) that it supports.

When it comes to comparing the state-of-the-art in open source workflow systems to that in proprietary systems, the results in Tables 1- 3 show that none of the offerings stands out as being clearly superior to the others, although it can be argued that Oracle BPEL PM demonstrates a marginally wider range of features, whilst Enhydra Shark and jBPM clearly lag behind in terms of overall patterns support. Oracle BPEL PM and OpenWFE tend to demonstrate broader pattern support in their corresponding tool classes (i.e. open-source vs proprietary), especially in the control-flow perspective. Moreover, it can also be observed that the proprietary tools are generally better equipped in the resource perspective and better able to support interaction with the external environment, whereas the open-source systems essentially rely on their users having programming experience (e.g., Java) to achieve the required integration with other systems. In the data perspective jBPM clearly lags behind the other offerings.

Overall one can conclude that the open source systems are geared more towards developers than towards business analysts. If one is proficient with Java, jBPM may be a good choice, although if not, choosing jBPM is less advisable. Similarly, whilst OpenWFE has a powerful language for workflow specification in terms of its support for the workflow patterns, we postulate that it will be difficult to understand by non-programmers. Finally, Endydra Shark's minimalistic support for the workflow patterns may require complicated work arounds for capturing nontrivial business scenarios.

Acknowledgement. We would like to thank John Mettraux for prompt and helpful responses through the OpenWFE help forum and Saša Bojanic for constructive and valuable feedback on Enhydra Shark.

References

1. Harmon, P.: Exploring BPMS with Free or Open Source Products. BPTrends 5(14) (July 2007)
2. Workflow Patterns Initiative. Workflow Patterns - homepage. www.workflowpatterns.com, (last accessed September 27, 2007)
3. Jablonski, S., Bussler, C.: Workflow Management: Modeling Concepts, Architecture and Implementation. Thomson Computer Press, London, UK (1996)
4. Java-source.net. Open Source Workflow Engines in Java. java-source.net/open-source/workflow-engines, (last accessed September 27, 2007)
5. Kiepuszewski, B.: Expressiveness and Suitability of Languages for Control Flow Modelling in Workflows. PhD thesis, Queensland University of Technology, Brisbane, Australia (2003) http://www.workflowpatterns.com/documentation/documents/phd_bartek.pdf
6. Manageability. Open Source Workflow Engines Written in Java. www.manageability.org/blog/stuff/workflow_in_java, (last accessed September 27, 2007)
7. Wohed, P., Andersson, B., ter Hofstede, A.H.M., Russell, N.C., van der Aalst, W.M.P.: Patterns-based Evaluation of Open Source BPM Systems: The Cases of jBPM, OpenWFE, and Enhydra Shark. BPM Center Report BPM-07-12, BPMcenter.org (2007)

Author Index

Lecture Notes in Computer Science

Sublibrary 3: Information Systems and Application, incl. Internet/Web and HCI

For information about Vols. 1–4611
please contact your bookseller or Springer

Vol. 5075: C.C. Yang, H. Chen, M. Chau, K. Chang, S.-D. Lang, P.S. Chen, R. Hsieh, D. Zeng, F.-Y. Wang, K. Carley, W. Mao, J. Zhan (Eds.), Intelligence and Security Informatics. XXII, 522 pages. 2008.

Vol. 5074: Z. Bellahsène, M. Léonard (Eds.), Advanced Information Systems Engineering. XVII, 588 pages. 2008.

Vol. 5053: R. Meier, S. Terzis (Eds.), Distributed Applications and Interoperable Systems. XI, 303 pages. 2008.

Vol. 5034: R. Fleischer, J. Xu (Eds.), Algorithmic Aspects in Information and Management. XI, 350 pages. 2008.

Vol. 5033: H. Oinas-Kukkonen, P. Hasle, M. Harjumaa, K. Segerståhl, P. Øhrstrøm (Eds.), Persuasive Technology. XIV, 287 pages. 2008.

Vol. 5024: M. Ferre (Ed.), Haptics: Perception, Devices and Scenarios. XXIII, 950 pages. 2008.

Vol. 5021: S. Bechhofer, M. Hauswirth, J. Hoffmann, M. Koubarakis (Eds.), The Semantic Web: Research and Applications. XIX, 897 pages. 2008.

Vol. 5017: T. Nanya, F. Maruyama, A. Pataricza, M. Malek (Eds.), Service Availability. XII, 225 pages. 2008.

Vol. 5013: J. Indulska, D.J. Patterson, T. Rodden, M. Ott (Eds.), Pervasive Computing. XIV, 315 pages. 2008.

Vol. 5006: R. Kowalczyk, M. Huhns, M. Klusch, Z. Maamar, Q.B. Vo (Eds.), Service-Oriented Computing: Agents, Semantics, and Engineering. X, 154 pages. 2008.

Vol. 4997: B. Monien, U.-P. Schroeder (Eds.), Algorithmic Game Theory. XI, 363 pages. 2008.

Vol. 4993: H. Li, T. Liu, W.-Y. Ma, T. Sakai, K.-F. Wong, G. Zhou (Eds.), Information Retrieval Technology. XIII, 685 pages. 2008.

Vol. 4976: Y. Zhang, G. Yu, E. Bertino, G. Xu (Eds.), Progress in WWW Research and Development. XVIII, 699 pages. 2008.

Vol. 4956: C. Macdonald, I. Ounis, V. Plachouras, I. Ruthven, R.W. White (Eds.), Advances in Information Retrieval. XXI, 719 pages. 2008.

Vol. 4952: C. Floerkemeier, M. Langheinrich, E. Fleisch, F. Mattern, S.E. Sarma (Eds.), The Internet of Things. XIII, 378 pages. 2008.

Vol. 4947: J.R. Haritsa, R. Kotagiri, V. Pudi (Eds.), Database Systems for Advanced Applications. XXII, 713 pages. 2008.

Vol. 4936: W. Aiello, A. Broder, J. Janssen, E.. Milios (Eds.), Algorithms and Models for the Web-Graph. X, 167 pages. 2008.

Vol. 4932: S. Hartmann, G. Kern-Isberner (Eds.), Foundations of Information and Knowledge Systems. XII, 397 pages. 2008.

Vol. 4928: A.H.M. ter Hofstede, B. Benatallah, H.-Y. Paik (Eds.), Business Process Management Workshops. XIII, 518 pages. 2008.

Vol. 4903: S. Satoh, F. Nack, M. Etoh (Eds.), Advances in Multimedia Modeling. XIX, 510 pages. 2008.

Vol. 4900: S. Spaccapietra (Ed.), Journal on Data Semantics X. XIII, 265 pages. 2008.

Vol. 4892: A. Popescu-Belis, S. Renals, H. Bourlard (Eds.), Machine Learning for Multimodal Interaction. XI, 308 pages. 2008.

Vol. 4882: T. Janowski, H. Mohanty (Eds.), Distributed Computing and Internet Technology. XIII, 346 pages. 2007.

Vol. 4881: H. Yin, P. Tino, E. Corchado, W. Byrne, X. Yao (Eds.), Intelligent Data Engineering and Automated Learning - IDEAL 2007. XX, 1174 pages. 2007.

Vol. 4877: C. Thanos, F. Borri, L. Candela (Eds.), Digital Libraries: Research and Development. XII, 350 pages. 2007.

Vol. 4872: D. Mery, L. Rueda (Eds.), Advances in Image and Video Technology. XXI, 961 pages. 2007.

Vol. 4871: M. Cavazza, S. Donikian (Eds.), Virtual Storytelling. XIII, 219 pages. 2007.

Vol. 4858: X. Deng, F.C. Graham (Eds.), Internet and Network Economics. XVI, 598 pages. 2007.

Vol. 4857: J.M. Ware, G.E. Taylor (Eds.), Web and Wireless Geographical Information Systems. XI, 293 pages. 2007.

Vol. 4853: F. Fonseca, M.A. Rodríguez, S. Levashkin (Eds.), GeoSpatial Semantics. X, 289 pages. 2007.

Vol. 4836: H. Ichikawa, W.-D. Cho, I. Satoh, H.Y. Youn (Eds.), Ubiquitous Computing Systems. XIII, 307 pages. 2007.

Vol. 4832: M. Weske, M.-S. Hacid, C. Godart (Eds.), Web Information Systems Engineering – WISE 2007 Workshops. XV, 518 pages. 2007.

Vol. 4831: B. Benatallah, F. Casati, D. Georgakopoulos, C. Bartolini, W. Sadiq, C. Godart (Eds.), Web Information Systems Engineering – WISE 2007. XVI, 675 pages. 2007.

Vol. 4825: K. Aberer, K.-S. Choi, N. Noy, D. Allemang, K.-I. Lee, L. Nixon, J. Golbeck, P. Mika, D. Maynard, R. Mizoguchi, G. Schreiber, P. Cudré-Mauroux (Eds.), The Semantic Web. XXVII, 973 pages. 2007.

Vol. 4823: H. Leung, F. Li, R. Lau, Q. Li (Eds.), Advances in Web Based Learning – ICWL 2007. XIV, 654 pages. 2008.

Vol. 4822: D.H.-L. Goh, T.H. Cao, I.T. Sølvberg, E. Rasmussen (Eds.), Asian Digital Libraries. XVII, 519 pages. 2007.

Vol. 4820: T.G. Wyeld, S. Kenderdine, M. Docherty (Eds.), Virtual Systems and Multimedia. XII, 215 pages. 2008.

Vol. 4816: B. Falcidieno, M. Spagnuolo, Y. Avrithis, I. Kompatsiaris, P. Buitelaar (Eds.), Semantic Multimedia. XII, 306 pages. 2007.

Vol. 4813: I. Oakley, S.A. Brewster (Eds.), Haptic and Audio Interaction Design. XIV, 145 pages. 2007.

Vol. 4810: H.H.-S. Ip, O.C. Au, H. Leung, M.-T. Sun, W.-Y. Ma, S.-M. Hu (Eds.), Advances in Multimedia Information Processing – PCM 2007. XXI, 834 pages. 2007.

Vol. 4809: M.K. Denko, C.-s. Shih, K.-C. Li, S.-L. Tsao, Q.-A. Zeng, S.H. Park, Y.-B. Ko, S.-H. Hung, J.-H. Park (Eds.), Emerging Directions in Embedded and Ubiquitous Computing. XXXV, 823 pages. 2007.

Vol. 4808: T.-W. Kuo, E. Sha, M. Guo, L.T. Yang, Z. Shao (Eds.), Embedded and Ubiquitous Computing. XXI, 769 pages. 2007.

Vol. 4806: R. Meersman, Z. Tari, P. Herrero (Eds.), On the Move to Meaningful Internet Systems 2007: OTM 2007 Workshops, Part II. XXXIV, 611 pages. 2007.

Vol. 4805: R. Meersman, Z. Tari, P. Herrero (Eds.), On the Move to Meaningful Internet Systems 2007: OTM 2007 Workshops, Part I. XXXIV, 757 pages. 2007.

Vol. 4804: R. Meersman, Z. Tari (Eds.), On the Move to Meaningful Internet Systems 2007: CoopIS, DOA, ODBASE, GADA, and IS, Part II. XXIX, 683 pages. 2007.

Vol. 4803: R. Meersman, Z. Tari (Eds.), On the Move to Meaningful Internet Systems 2007: CoopIS, DOA, ODBASE, GADA, and IS, Part I. XXIX, 1173 pages. 2007.

Vol. 4802: J.-L. Hainaut, E.A. Rundensteiner, M. Kirchberg, M. Bertolotto, M. Brochhausen, Y.-P.P. Chen, S.S.-S. Cherfi, M. Doerr, H. Han, S. Hartmann, J. Parsons, G. Poels, C. Rolland, J. Trujillo, E. Yu, E. Zimányie (Eds.), Advances in Conceptual Modeling – Foundations and Applications. XIX, 420 pages. 2007.

Vol. 4801: C. Parent, K.-D. Schewe, V.C. Storey, B. Thalheim (Eds.), Conceptual Modeling - ER 2007. XVI, 616 pages. 2007.

Vol. 4797: M. Arenas, M.I. Schwartzbach (Eds.), Database Programming Languages. VIII, 261 pages. 2007.

Vol. 4796: M. Lew, N. Sebe, T.S. Huang, E.M. Bakker (Eds.), Human–Computer Interaction. X, 157 pages. 2007.

Vol. 4794: B. Schiele, A.K. Dey, H. Gellersen, B. de Ruyter, M. Tscheligi, R. Wichert, E. Aarts, A. Buchmann (Eds.), Ambient Intelligence. XV, 375 pages. 2007.

Vol. 4777: S. Bhalla (Ed.), Databases in Networked Information Systems. X, 329 pages. 2007.

Vol. 4761: R. Obermaisser, Y. Nah, P. Puschner, F.J. Rammig (Eds.), Software Technologies for Embedded and Ubiquitous Systems. XIV, 563 pages. 2007.

Vol. 4747: S. Džeroski, J. Struyf (Eds.), Knowledge Discovery in Inductive Databases. X, 301 pages. 2007.

Vol. 4744: Y. de Kort, W. IJsselsteijn, C. Midden, B. Eggen, B.J. Fogg (Eds.), Persuasive Technology. XIV, 316 pages. 2007.

Vol. 4740: L. Ma, M. Rauterberg, R. Nakatsu (Eds.), Entertainment Computing – ICEC 2007. XXX, 480 pages. 2007.

Vol. 4730: C. Peters, P. Clough, F.C. Gey, J. Karlgren, B. Magnini, D.W. Oard, M. de Rijke, M. Stempfhuber (Eds.), Evaluation of Multilingual and Multi-modal Information Retrieval. XXIV, 998 pages. 2007.

Vol. 4723: M. R. Berthold, J. Shawe-Taylor, N. Lavrač (Eds.), Advances in Intelligent Data Analysis VII. XIV, 380 pages. 2007.

Vol. 4721: W. Jonker, M. Petković (Eds.), Secure Data Management. X, 213 pages. 2007.

Vol. 4718: J. Hightower, B. Schiele, T. Strang (Eds.), Location- and Context-Awareness. X, 297 pages. 2007.

Vol. 4717: J. Krumm, G.D. Abowd, A. Seneviratne, T. Strang (Eds.), UbiComp 2007: Ubiquitous Computing. XIX, 520 pages. 2007.

Vol. 4715: J.M. Haake, S.F. Ochoa, A. Cechich (Eds.), Groupware: Design, Implementation, and Use. XIII, 355 pages. 2007.

Vol. 4714: G. Alonso, P. Dadam, M. Rosemann (Eds.), Business Process Management. XIII, 418 pages. 2007.

Vol. 4704: D. Barbosa, A. Bonifati, Z. Bellahsène, E. Hunt, R. Unland (Eds.), Database and XML Technologies. X, 141 pages. 2007.

Vol. 4690: Y. Ioannidis, B. Novikov, B. Rachev (Eds.), Advances in Databases and Information Systems. XIII, 377 pages. 2007.

Vol. 4675: L. Kovács, N. Fuhr, C. Meghini (Eds.), Research and Advanced Technology for Digital Libraries. XVII, 585 pages. 2007.

Vol. 4674: Y. Luo (Ed.), Cooperative Design, Visualization, and Engineering. XIII, 431 pages. 2007.

Vol. 4663: C. Baranauskas, P. Palanque, J. Abascal, S.D.J. Barbosa (Eds.), Human-Computer Interaction – INTERACT 2007, Part II. XXXIII, 735 pages. 2007.

Vol. 4662: C. Baranauskas, P. Palanque, J. Abascal, S.D.J. Barbosa (Eds.), Human-Computer Interaction – INTERACT 2007, Part I. XXXIII, 637 pages. 2007.

Vol. 4658: T. Enokido, L. Barolli, M. Takizawa (Eds.), Network-Based Information Systems. XIII, 544 pages. 2007.

Vol. 4656: M.A. Wimmer, J. Scholl, Å. Grönlund (Eds.), Electronic Government. XIV, 450 pages. 2007.

Vol. 4655: G. Psaila, R. Wagner (Eds.), E-Commerce and Web Technologies. VII, 229 pages. 2007.

Vol. 4654: I.-Y. Song, J. Eder, T.M. Nguyen (Eds.), Data Warehousing and Knowledge Discovery. XVI, 482 pages. 2007.

Vol. 4653: R. Wagner, N. Revell, G. Pernul (Eds.), Database and Expert Systems Applications. XXII, 907 pages. 2007.

Vol. 4636: G. Antoniou, U. Aßmann, C. Baroglio, S. Decker, N. Henze, P.-L. Patranjan, R. Tolksdorf (Eds.), Reasoning Web. IX, 345 pages. 2007.